GREEK ALPHABET

Alpha	A	α	Iota	I	ι	Rho	P	ρ
Beta	B	β	Kappa	K	κ	Sigma	Σ	σ
Gamma	Γ	γ	Lambda	Λ	λ	Tau	T	τ
Delta	Δ	δ	Mu	M	μ	Upsilon	Υ	υ
Epsilon	E	ε	Nu	N	ν	Phi	Φ	ϕ
Zeta	Z	ζ	Xi	Ξ	ξ	Chi	X	χ
Eta	H	η	Omicron	O	o	Psi	Ψ	ψ
Theta	Θ	θ	Pi	Π	π	Omega	Ω	ω

CONVERSION TABLE FOR UNITS

Length

meter (SI unit)	m	
centimeter	cm	$= 10^{-2}\,\text{m}$
ångström	Å	$= 10^{-10}\,\text{m}$
micron	μ	$= 10^{-6}\,\text{m}$

Volume

cubic meter (SI unit)	m^3	
liter	L	$= \text{dm}^3 = 10^{-3}\,\text{m}^3$

Mass

kilogram (SI unit)	kg	
gram	g	$= 10^{-3}\,\text{kg}$

Energy

joule (SI unit)	J	
erg	erg	$= 10^{-7}\,\text{J}$
rydberg	Ry	$= 2.179\,87 \times 10^{-18}\,\text{J}$
electron volt	eV	$= 1.602\,18 \times 10^{-19}\,\text{J}$
inverse centimeter	cm^{-1}	$= 1.986\,45 \times 10^{-23}\,\text{J}$
calorie (thermochemical)	Cal	$= 4.184\,\text{J}$
liter atmosphere	l atm	$= 101.325\,\text{J}$

Pressure

pascal (SI unit)	Pa	
atmosphere	atm	$= 101325\,\text{Pa}$
bar	bar	$= 10^5\,\text{Pa}$
torr	Torr	$= 133.322\,\text{Pa}$
pounds per square inch	psi	$= 6.894\,757 \times 10^3\,\text{Pa}$

Power

watt (SI unit)	W	
horsepower	hp	$= 745.7\,\text{W}$

Angle

radian (SI unit)	rad	
degree	°	$= \dfrac{2\pi}{360}\,\text{rad} = \left(\dfrac{1}{57.295\,78}\right)\text{rad}$

Electrical dipole moment

C m (SI unit)		
debye	D	$= 3.335\,64 \times 10^{-30}\,\text{C m}$

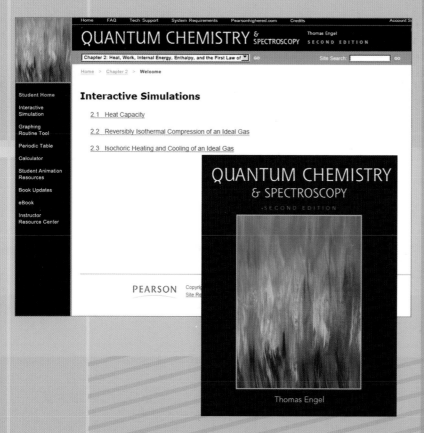

Quantum Chemistry & Spectroscopy

Quantum Chemistry & Spectroscopy

Second Edition

Thomas Engel
University of Washington

Chapter 15, "Computational Chemistry,"
was contributed by
Warren Hehre
CEO, Wavefunction, Inc.

Prentice Hall

New York   Boston   San Francisco
London   Toronto   Sydney   Tokyo   Singapore   Madrid
Mexico City   Munich   Paris   Cape Town   Hong Kong   Montreal

Library of Congress Cataloging-in-Publication Data

Engel, Thomas, 1942-
    Quantum chemistry & spectroscopy / Thomas Engel.—2nd ed.
       p. cm.
    Rev. ed. of: Quantum chemistry and spectroscopy. c2006.
    Includes bibliographical references and index.
    ISBN 0-321-61504-2
  1. Quantum chemistry—Textbooks. 2. Spectrum analysis—Textbooks.
    I. Hehre, Warren J. II. Engel, Thomas, 1942- Quantum chemistry and spectroscopy.
    III. Title. IV Title: Quantum chemistry and spectroscopy.
    QD462.E53 2010
    541'.28—dc22   2009008647

Acquisitions Editor: *Dan Kaveney*
Assistant Editor: *Carol DuPont*
Editor in Chief, Chemistry and Geosciences: *Nicole Folchetti*
Marketing Manager: *Erin Gardner*
Editorial Assistant: *Kristen Wallerius*
Managing Editor, Chemistry and Geosciences: *Gina M. Cheselka*
Associate Media Producer: *Kristin Mayo*
Senior Production Supervisor, Media: *Liz Winer*
Media Project Manager: *Natasha Wolfe*
Art Editor: *Connie Long*
Art Studio: *Precision Graphics*
Art Director: *Maureen Eide*
Interior and Cover Design: *Maureen Eide*
Senior Operations Supervisor: *Alan Fischer*
Production Supervision/Composition: *Prepare, Inc.*
Cover Credit: *Corbis/Superstock*

Printed in the United States of America.

10 9 8 7 6 5 4 3 2 1

ISBN-13: 978-0-321-61504-6
ISBN-10: 0-321-61504-2

Prentice Hall
is an imprint of

www.pearsonhighered.com

This book is dedicated to my parents,
Walter and Juliane, who were my first teachers,
and to my cherished family, Esther and Alex,
with whom I am still learning.

Thomas Engel

Brief Contents

Contents

About the Author

Thomas Engel has taught chemistry at the University of Washington for more than 20 years, where he is currently Professor Emeritus of Chemistry. Professor Engel received his bachelor's and master's degrees in chemistry from the Johns Hopkins University, and his Ph.D. in chemistry from the University of Chicago. He then spent 11 years as a researcher in Germany and Switzerland, in which time he received the Dr. rer. nat. habil. degree from the Ludwig Maximilians University in Munich. In 1980, he left the IBM research laboratory in Zurich to become a faculty member at the University of Washington.

Professor Engel's research interests are in the area of surface chemistry, and he has published more than 80 articles and book chapters in this field. He has received the Surface Chemistry or Colloids Award from the American Chemical Society and a Senior Humboldt Research Award from the Alexander von Humboldt Foundation.

Preface

The second edition of this book builds on user and reviewer comments on the first edition. Many chapters have been extensively revised and I have included many additional end-of-chapter concept questions and problems. The target audience remains undergraduate students majoring in chemistry, biochemistry, and chemical engineering as well as many students majoring in the atmospheric sciences and the biological sciences. The following objectives, illustrated with brief examples, outline my approach to teaching physical chemistry.

- **Focus on learning outcomes, not covering material:** A very good understanding of quantum mechanics can be obtained from four basic systems: the particle in the box, the harmonic oscillator, the rigid rotor and the hydrogen atom. Fewer topics explored in greater depth provide a deeper learning experience for students.

- **Show that quantum mechanics is needed to explain the world around us:** Many everyday phenomena can't be understood without quantum mechanics. The particle in the box model is used to explain why metals conduct electricity and why valence electrons rather than core electrons are important in chemical bond formation. Spectroscopy as we know it and lasers used for reading barcodes, and CDs are unimaginable without discrete energy levels.

- **Spectroscopy is the practical implementation of quantum mechanics in chemistry:** Few students of quantum mechanics will be developers of the science, but all will be users of spectroscopy. In this book, many forms of spectroscopy are explained and are presented together with the basic science rather than being relegated to separate chapters. For example, the He-Ne laser, the quantum mechanics of laser isotope separation, Auger electron spectroscopy, and X-ray photoelectron spectroscopy are discussed together with many-electron atoms. Vibrational spectroscopy using FTIR and cavity ringdown techniques are discussed together with diatomic molecules. Fluorescence spectroscopy is discussed using the sheath flow gel capillary technique used in the sequencing of the human genome. The chapter on NMR contains explanations of 2-D NMR as well as solid-state NMR and the use of NMR in imaging.

- **Present the exciting new science in the field:** The applications of quantum mechanics are growing rapidly, and many of them are easily understood with only a basic understanding of quantum mechanics. Recent applications such as scanning tunneling microscopy, band-gap engineering, quantum dots, quantum wells, teleportation, and quantum computing are discussed.

- **Relate quantum chemistry to descriptive chemistry:** Quantum chemistry becomes more relevant if it is tied to organic and inorganic chemistry. The importance of tunneling in chemical reactions is discussed. Orbital energies derived from Hartree–Fock theory are related to electronegativity. Effective nuclear charges are related to atomic diameters. The enhanced reactivity of electronically excited molecules and atoms is discussed together with term symbols to emphasize an understanding of the formalism. Localized and delocalized models of chemical bonding, hybridization, Lewis structures, and VSEPR are discussed in the language of quantum chemistry.

- **Discuss the philosophical underpinnings of quantum mechanics:** The puzzling aspects of quantum mechanics that seem contrary to experience are exciting for students. My students have been fascinated by the consequences of the Heisenberg uncertainty principle, tunneling, the "collapse" of the wavefunction, entanglement, and teleportation.

- **Use web-based simulations to avoid a math overload:**   The finite depth box is much more useful in discussing chemistry than the infinite depth box, but it is a mathematically challenging system. To deal with this issue, web-based simulations have been incorporated as end-of-chapter problems throughout the book so that the student can focus on the science and avoid a math overload. The simulations are also well suited as lecture demonstrations. For example, the student can vary the length and depth in the finite depth box and determine how many bound states the potential has through experiment. More than 30 such web-based problems are available on the course website covering topics such as tunneling, the Heisenberg uncertainty principle, and rotational-vibrational spectroscopy. An important feature is that each problem has been designed as an assignable exercise with a printable answer sheet that the student can submit to the instructor. The course website also includes a graphing routine with a curve-fitting capability, which allows graphical data to be printed and submitted.

- **Learning problem-solving skills is an essential part of physical chemistry:** Many example problems are worked through in each chapter. The end-of-chapter problems cover a range of difficulties suitable for students at all levels.

- **Introduce students to computational chemistry:**   This book also includes a chapter on computational chemistry, written by Warren Hehre, who has a distinguished record as a developer of computational chemistry software and who has brought these techniques into the educational environment. His chapter includes a large number of problems, better called computational experiments, that supplement the discussion in the chapter and illustrate the application of computational chemistry to molecules of real chemical interest. It is my experience that students welcome this material.

  Many other chapters in the second edition include computational problems for which detailed instructions for the student are available on the book website. These instructions will give students the practice needed to solve the computational experiments in Hehre's chapter.

- **Use color to make learning physical chemistry more interesting:**   Four-color images are used to display atomic and molecular orbitals both quantitatively and attractively as well as to make complex images such as the symmetry elements of ethene understandable.

   This text contains more material than can be covered in a semester, and this is entirely intentional. Effective use of the text does not require one to proceed sequentially through the chapters, or to include all sections. Some topics are discussed in supplemental sections that can be omitted if they are not viewed as essential to the course. Although the chapters on molecular symmetry, computational chemistry, and NMR spectroscopy are placed at the end of the text, they can be introduced earlier in the course. I welcome the comments of both students and instructors on how the material was used and on how the presentation can be improved.

Thomas Engel
University of Washington

ACKNOWLEDGMENTS

Many individuals have helped me to bring the text into its current form. Students have provided me with feedback directly and through the questions they have asked, which has helped me to understand how they learn. Many of my colleagues including Peter Armentrout, Doug Doren, Gary Drobny, Graeme Henkelman, Lewis Johnson, Tom Pratum, Bill Reinhardt, Peter Rosky, George Schatz, Michael Schick, Gabrielle Varani, and especially Wes Borden and Bruce Robinson have been invaluable in advising me. I am also fortunate to have access to some end-of-chapter problems that were originally presented in *Physical Chemistry,* 3rd edition, by Joseph H. Noggle and in *Physical Chemistry,* 3rd edition, by Gilbert W. Castellan. The reviewers, who are listed separately, have made many suggestions for improvement, for which I am very grateful. All those involved in the production process have helped to make this book a reality through their efforts. Special thanks are due to Jim Smith, who helped initiate this project, to my editor Dan Kaveney, and to the staff at Prentice Hall, who have guided the production process.

PRESCRIPTIVE REVIEWERS:

David L. Cedeño,
Illinois State University
Rosemarie Chinni,
Alvernia College
Allen Clabo,
Francis Marion University
Lorrie Comeford,
Salem State College
John M. Jean,
Regis University
Martina Kaledin,
Kennesaw State University

Daniel Lawson,
University of Michigan-Dearborn
Dmitrii E. Makarov,
University of Texas at Austin
Enrique Peacock-López,
Williams College
Anthony K. Rappe,
Colorado State University
Markku Räsänen,
University of Helsinki
Richard W. Schwenz,
University of Northern Colorado

Jie Song,
University of Michigan-Flint
Michael E. Starzak,
Binghamton University
Liliya Vugmeyster,
University of Alaska Anchorage
James E. Whitten,
University of Massachusetts Lowell

ART REVIEWER:

Lorrie Comeford,
Salem State College

MATH REVIEWER:

Leon Gerber,
St. John's University

MANUSCRIPT REVIEWERS:

Alexander Angerhofer,
University of Florida
Martha Bruch,
SUNY at Oswego
Stephen Cooke,
University of North Texas
Douglas English,
University of Maryland, College Park
Sophya Garashchuk,
University of South Carolina

Cynthia Hartzell,
Northern Arizona University
George Kaminski,
Central Michigan University
Herve Marand,
Virginia Technical Institute
Thomas Pentecost,
University of Colorado
Rajeev Prabhakar,
University of Miami

Sanford Safron,
Florida State University
Ali Sezer,
California University of Pennsylvania
Andrew Teplyakov,
University of Delaware
Daniel Zeroka,
Lehigh University

1

From Classical to Quantum Mechanics

As scientists became able to investigate the atomic realm, results were obtained that were inconsistent with classical physics. Classical physics predicted that all bodies at a temperature other than zero kelvin radiate an infinite amount of energy. It incorrectly predicted that the kinetic energy of electrons produced upon illuminating a metal surface in vacuum with light is proportional to the light intensity, and it could not explain the diffraction of an electron by a crystalline solid. Rutherford's laboratory showed that atoms consist of a small positively charged nucleus surrounded by a diffuse cloud of electrons. Classical physics, however, predicted that such an atom was unstable and that the electrons would spiral into the nucleus while radiating energy to the environment. These inconsistencies between classical theory and experimental observations provided the stimulus for the development of quantum mechanics.

1.1 WHY STUDY QUANTUM MECHANICS?

Imagine how difficult it would be for humans to function in a world governed by underlying principles without knowing what they were. If you couldn't calculate the trajectory of a projectile, you couldn't launch a satellite. Without understanding how energy is transformed into work, you couldn't design an automobile that gets more mileage for a given amount of fuel. Technology arises from an understanding of how matter and energy interact, which argues for a broad understanding of scientific principles.

Chemistry is a molecular science; the goal of chemists is to understand macroscopic behavior in terms of the properties of individual molecules. For example, H_2 is a good fuel because the energy released in forming H_2O is much greater than that needed

1

to break the bonds in the reactants O_2 and H_2. As you will learn in this chapter, in the first decade of the 20th century, scientists learned that an atom consisted of a small positively charged nucleus surrounded by a diffuse electron cloud. However, classical physics was unable to explain why the electrons did not follow a spiral trajectory to end in the nucleus. Classical physics was also unable to explain why graphite conducts electricity and diamond does not, why the light emitted by a hydrogen discharge lamp appears at only a small number of wavelengths, and why the bond angle in H_2O is different from that in H_2S.

These deficiencies in classical physics made it clear that another physical model was needed to describe matter at the nanometer scale of atoms and molecules. Over a period of about 20 years, quantum mechanics was developed and scientists have found that the phenomena just cited that cannot be understood within classical physics can be explained using quantum mechanics. As you will learn, the central feature that distinguishes quantum and classical mechanics is wave-particle duality. At the atomic level, electrons, protons, and light all behave as waves or particles. It is the experiment that determines whether wave or particle behavior will be observed.

Although you may not know it, you are already a user of quantum mechanics. You take the stability of the atom with its central positively charged nucleus and surrounding electron cloud, the laser in your CD player, the integrated circuit in your computer, and the chemical bond for granted. You know that infrared spectroscopy provides a useful way to identify chemical compounds and that nuclear magnetic resonance spectroscopy provides a powerful tool to image internal organs. However, spectroscopic techniques would not be possible if atoms and molecules could have *any* value of energy as predicted by classical physics. Quantum mechanics predicts that atoms and molecules can only have discrete energies and provides a basis for understanding all spectroscopies.

Technology is increasingly based on quantum mechanics. For example, quantum computing, in which a state can be described by zero *and* one rather than zero *or* one, is a very active area of research. If quantum computers can ultimately be realized, they will be much more powerful than current computers. Quantum mechanical calculations of chemical properties of biologically important molecules are now sufficiently accurate that molecules can be designed for a specific application before they are tested at the laboratory bench. As many sciences such as biology become increasingly focused on the molecular level, more scientists will need to be able to think in terms of quantum mechanical models. Therefore, quantum mechanics is an essential part of the chemist's knowledge base.

1.2 QUANTUM MECHANICS AROSE OUT OF THE INTERPLAY OF EXPERIMENTS AND THEORY

Scientific theories gain acceptance if they can make the world around us understandable. A key feature of validating theories as useful models is to compare the result of a new experiment with the prediction of current and new theories. If the experiment and the theory agree, we gain confidence in the model underlying the theory; if not, the model needs to be modified. At the end of the 19th century, Maxwell's electromagnetic theory unified existing knowledge in the areas of electricity, magnetism, and waves. This theory, combined with the well-established field of classical mechanics, ushered in a new era of maturity for the physical sciences. Many scientists of that era believed that there was little left in the natural sciences to learn. However, the growing ability of scientists to probe natural phenomena at an atomic level soon showed that this presumption was incorrect. The field of quantum mechanics arose in the early 1900s as scientists became able to investigate natural phenomena at the newly accessible atomic level. A number of key experiments showed that the predictions of classical physics were inconsistent with experimental outcomes. Several of these experiments are described in more detail in this chapter in order to show the important role that experiments have had—and continue to have—in stimulating the development of theories to describe the natural world. These experiments stimulated the leading scientists of the era to formulate quantum mechanics.

In the rest of this chapter, experimental evidence is presented for two key properties that have come to distinguish classical and quantum physics. The first of these is **quantization**. Energy at the atomic level is not a continuous variable, but comes in discrete packets called *quanta*. The second key property is **wave-particle duality**. At the atomic level, light waves have particle-like properties, and atoms as well as subatomic particles such as electrons have wave-like properties. Neither quantization nor wave-particle duality were known concepts until the experiments described in Sections 1.3 through 1.7 were conducted.

1.3 BLACKBODY RADIATION

Think of the heat that you feel from the embers of a fire. The energy that your body absorbs is radiated from the glowing coals. An idealization of this system that is more amenable to theoretical study is a red-hot block of metal with a spherical cavity in its interior that can be observed through a hole small enough that the conditions inside the block are not perturbed. An **ideal blackbody** is shown in Figure 1.1. Under the condition of equilibrium between the radiation field inside the cavity and the glowing piece of matter, classical electromagnetic theory can predict what frequencies, ν, of light are radiated and their relative magnitudes. The result is

$$\rho(\nu, T) \, d\nu = \frac{8\pi\nu^2}{c^3} \overline{E}_{osc} \, d\nu \tag{1.1}$$

In this equation, ρ is the spectral density, which has the units of energy $\times$ (volume)$^{-1}$ $\times$ (frequency)$^{-1}$. The spectral density is a function of the temperature, T, and the frequency ν. The speed of light is c, and $\overline{E}_{osc}$ is the average energy of an oscillating dipole in the solid. In words, the **spectral density** is the energy stored in the electromagnetic field of the blackbody radiator at frequency ν per unit volume and unit frequency. The basis of this model is that oscillating electric dipoles, which may be thought of as atomic nuclei and their associated electrons, radiate energy.

The factor $d\nu$ is used on both sides of this equation because we are asking for the energy density observed within the frequency interval of width $d\nu$ around the frequency ν. Classical theory further predicts that the average energy of an oscillator is simply related to the temperature by

$$\overline{E}_{osc} = kT \tag{1.2}$$

in which k is the Boltzmann constant. Combining these two equations results in an expression for $\rho(\nu, T) \, d\nu$, the amount of energy per unit volume in the frequency range between ν and $\nu + d\nu$ in equilibrium with a blackbody at temperature T:

$$\rho(\nu, T) \, d\nu = \frac{8\pi kT\nu^2}{c^3} \, d\nu \tag{1.3}$$

It is possible to measure the spectral density of the radiation emitted by a blackbody, and the results are shown in Figure 1.2 for several temperatures together with the result predicted by classical theory. The experimental curves have a common behavior. The spectral density is peaked in a broad maximum and falls off to both lower and higher frequencies. The shift of the maximum to higher frequencies with increasing temperatures is consistent with our experience that if more power is put into an electrical heater, its color will change from dull red to yellow (increasing frequency).

The comparison of the spectral density distribution predicted by classical theory with that observed experimentally for $T = 6000$ K is particularly instructive. The two curves show similar behavior at low frequencies, but the theoretical curve keeps on increasing with frequency as Equation (1.3) shows. Because the area under the $\rho(\nu, T)$ versus ν curves gives the total energy per unit volume of the field of the blackbody, classical theory predicts that a blackbody will emit an infinite amount of energy at all temperatures above absolute zero! It is clear that this prediction is incorrect, but

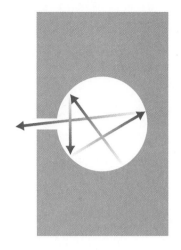

FIGURE 1.1

An idealized blackbody. The surface of a spherical cavity in a solid at a high temperature emits photons from an interior spherical surface. The photons reflect several times before emerging through a narrow channel. The reflections ensure that the radiation is in thermal equilibrium with the solid.

1.1 and **1.2** Blackbody Radiation

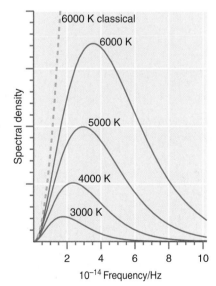

FIGURE 1.2

The red curves show the light intensity emitted from an ideal blackbody as a function of the frequency for different temperatures. The dashed curve shows the predictions of classical theory for $T = 6000$ K.

scientists were greatly puzzled about where the theory went wrong at the beginning of the 20th century.

In looking at data such as that shown in Figure 1.2, the German physicist Max Planck was able to develop some important insights that ultimately led to an understanding of **blackbody radiation**. It was understood at the time that the origin of blackbody radiation was the vibration of electric dipoles that emit radiation at the frequency at which they oscillate. Planck saw that the discrepancy between experiment and classical theory occurred at high and not at low frequencies. The absence of high-frequency radiation at low temperatures showed that the high-frequency dipole oscillators were active only at high temperatures. Because high temperatures correspond to high energies, Planck reasoned that Equation (1.2) was not correct. This led him to postulate that the energy of the radiation was proportional to the frequency. This was a radical departure from classical theory, in which the energy stored in electromagnetic radiation is proportional to the square of the amplitude, but independent of the frequency. Planck's line of reasoning was as follows: unless a large amount of energy is put into the blackbody (high temperature) it will not be possible to excite the high-energy (high-frequency) oscillators.

Planck found that he could obtain agreement between theory and experiment only if he assumed that the energy radiated by the dipoles was given by the relation

$$E = nh\nu \tag{1.4}$$

Planck's constant, h, was initially an unknown proportionality constant and n is a positive integer ($n = 0, 1, 2, \dots$). The frequency ν is continuous, but for a given ν, the energy is *quantized* according to Equation (1.4). This relationship between energy and frequency ushered in a new era of physics. Energy in classical theory is a *continuous* quantity, which means that it can take on all values. Equation (1.4) states that the energy radiated by a blackbody is not continuous in frequency, but can take on only a set of *discrete* values for each frequency. This assumption was a radical departure from classical theory, and its main justification was that agreement between theory and experiment could be obtained. Using Equation (1.4) and some classical physics, Planck obtained the following relationship:

$$\overline{E}_{osc} = \frac{h\nu}{e^{h\nu/kT} - 1} \tag{1.5}$$

It is useful to obtain an approximate value for $\overline{E}_{osc}$ from this equation in two limits, namely, at high temperatures, where $h\nu/kT \ll 1$, and at low temperatures, where $h\nu/kT \gg 1$. At high temperatures, the exponential function in Equation (1.5) can be expanded in a Taylor-Maclaurin series (see the Math Supplement, Appendix A), giving

$$\overline{E}_{osc} = \frac{h\nu}{(1 + h\nu/kT + \dots) - 1} \approx kT \tag{1.6}$$

just as classical theory had predicted. However, for low temperatures corresponding to $h\nu/kT \gg 1$, the denominator in Equation (1.5) becomes very large, and $\overline{E}_{osc}$ approaches zero. The high-frequency oscillators do not contribute to the radiated energy at low and moderate temperatures.

Using Equation (1.5), in 1901 Planck obtained the following general formula for the spectral radiation density from a blackbody:

$$\rho(\nu, T)\, d\nu = \frac{8\pi h\nu^3}{c^3} \frac{1}{e^{h\nu/kT} - 1}\, d\nu \tag{1.7}$$

The value of the constant h was not known and Planck used it as a parameter to fit the data. He was able to reproduce the experimental data at all temperatures with the single adjustable parameter, h, which through more accurate measurements currently has the value $h = 6.6260755 \times 10^{-34}$ J s. Obtaining this degree of agreement using a single adjustable parameter was a remarkable achievement. However, Planck's explanation, which relied on the assumption that the energy of the radiation came in discrete packets or quanta, was not accepted initially. Soon afterward, Einstein's explanation of the photoelectric effect gave support to Planck's hypothesis.

1.4 THE PHOTOELECTRIC EFFECT

Imagine a copper plate in a vacuum. Light incident on the plate can be absorbed, leading to the excitation of electrons to unoccupied energy levels. Sufficient energy can be transferred to the electrons such that some leave the metal and are ejected into the vacuum. The electrons that have been emitted from the copper upon illumination can be collected by another electrode in the vacuum system, called the *collector*. This process is called the **photoelectric effect**. A schematic apparatus is shown in Figure 1.3. Invoking conservation of energy, we conclude that the absorbed light energy must be balanced by the energy required to eject an electron at equilibrium and the kinetic energy of the emitted electron, because the energy of the system is constant. Classical theory makes the following predictions:

- Light is incident as a plane wave over the whole copper plate. Therefore, the light is absorbed by many electrons in the solid. Any one electron can absorb only a small fraction of the incident light.
- Electrons are emitted to the collector for all light frequencies, provided that the light is sufficiently intense.
- The kinetic energy per electron increases with the light intensity.

The results of the experiment can be summarized as follows:

- The number of emitted electrons is proportional to the light intensity, but their kinetic energy is independent of the light intensity.
- No electrons are emitted unless the frequency ν is above a threshold frequency ν_0 even for high light intensities.
- The kinetic energy of the emitted electrons depends on the frequency in the manner depicted in Figure 1.4.
- Electrons are emitted even at such low light intensities that all the light absorbed by the entire copper plate is barely enough to eject a single electron, based on energy conservation considerations.

Just as for blackbody radiation, the inability of classical theory to correctly predict experimental results stimulated a new theory. In 1905, Albert Einstein hypothesized that the energy of the light was proportional to its frequency:

$$E = \beta \nu \qquad (1.8)$$

where β is a constant to be determined. This is a marked departure from classical electrodynamics, in which there is no relation between the energy of a light wave and its frequency. Invoking energy conservation, the energy of the electron, E_e, is related to that of the light by

$$E_e = \beta \nu - \phi \qquad (1.9)$$

The binding energy of the electron in the solid, which is analogous to the ionization energy of an atom, is designated by ϕ in this equation and is called the **work function**. In words, this equation says that the kinetic energy of the photoelectron that has escaped from the solid is smaller than the photon energy by the amount with which the electron is bound to the solid. Einstein's theory gives a prediction of the dependence of the kinetic energy of the photoelectrons on the light frequency that can be compared directly with experiment. Because ϕ can be determined independently, only β is unknown. It can be obtained by fitting the data points in Figure 1.4 to Equation (1.9). The results shown by the red line in Figure 1.4 not only reproduce the data very well, but they yield

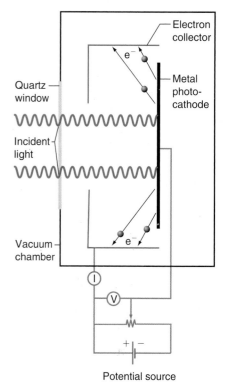

FIGURE 1.3

The electrons emitted by the surface upon illumination are incident on the collector, which is at an appropriate electrical potential to attract them. The experiment is carried out in a vacuum chamber to avoid collisions and capture of electrons by gas molecules.

⟹ Analysis is the most important !!!

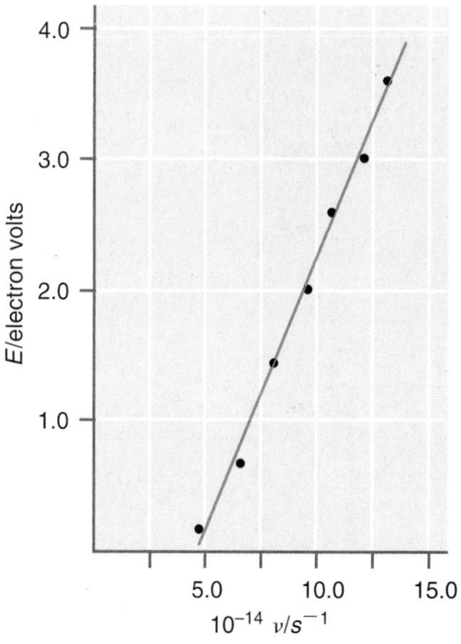

FIGURE 1.4

The energy of photo-ejected electrons is shown as a function of the light frequency. The individual data points are well fit by a straight line as shown.

the striking result that β, the slope of the line, is identical to the Planck constant, h. The equation which relates the energy of light to its frequency

$$E = h\nu \tag{1.10}$$

is one of the most widely used equations in quantum mechanics and earned Albert Einstein a Nobel Prize in physics. A calculation involving the photoelectric effect is carried out in Example Problem 1.1.

The agreement between the theoretical prediction and the experimental data validates Einstein's fundamental assumption that the energy of light is proportional to its frequency. This result also suggested that h is a "universal constant" that appears in seemingly unrelated phenomena. Its appearance in this context gained greater acceptance for the assumptions Planck used to explain blackbody radiation.

EXAMPLE PROBLEM 1.1

Light with a wavelength of 300 nm is incident on a potassium surface for which the work function, ϕ, is 2.26 eV. Calculate the kinetic energy and speed of the ejected electrons.

Solution

The wavelength and frequency are related by $\lambda\nu = c$ where c is the speed of light. Using Equation (1.9), we write $E_e = h\nu - \phi = (hc/\lambda) - \phi$ and converting the units of ϕ from electron-volts to joules: $\phi = (2.26\,\text{eV})(1.602 \times 10^{-19}\,\text{J/eV}) = 3.62 \times 10^{-19}$ J. Electrons will only be ejected if the photon energy, $h\nu$, is greater than ϕ. The photon energy is calculated to be

$$\frac{hc}{\lambda} = \frac{6.626 \times 10^{-34}\,\text{J s} \times 2.998 \times 10^{8}\,\text{m s}^{-1}}{300. \times 10^{-9}\,\text{m}} = 6.62 \times 10^{-19}\,\text{J}$$

which is sufficient to eject electrons.

Using Equation (1.9), we obtain $E_e = (hc/\lambda) - \phi = 3.00 \times 10^{-19}$ J. Using $E_e = 1/2mv^2$, we calculate that

$$v = \sqrt{\frac{2E_e}{m}} = \sqrt{\frac{2 \times 3.00 \times 10^{-19}\,\text{J}}{9.109 \times 10^{-31}\,\text{kg}}} = 8.12 \times 10^{5}\,\text{m s}^{-1}$$

Another important conclusion can be drawn from the observation that even at very low light intensities, photoelectrons are emitted from the solid. More precisely, photoelectrons are detected even at intensities so low that all the energy incident on the solid surface is only slightly more than the threshold energy required to yield a single photoelectron. This means that the light that liberates the photoelectron is not uniformly distributed over the surface. If this were true, no individual electron could receive enough energy to escape into the vacuum. The surprising conclusion of this experiment is that all of the incident light energy can be concentrated in a single electron excitation. This led to the coining of the term **photon** to describe a spatially localized packet of light. Because this spatial localization is characteristic of particles, the conclusion that light can exhibit particle-like behavior under some circumstances was unescapable.

Many experiments have shown that light exhibits wave-like behavior. We have long known that light can be diffracted by an aperture or slit. However, the photon in the photoelectric effect that exhibits particle-like properties and the photon in a diffraction experiment that exhibits wave-like properties are one and the same. This recognition forces us to conclude that light has a wave-particle duality, and depending on the experiment, can manifest as a wave or as a particle. This important recognition leads us to the third fundamental experiment to be described: the diffraction of electrons by a crystalline solid. Because diffraction is proof of wave-like behavior, if particles can be diffracted, they exhibit a particle-wave duality just as light does.

1.5 PARTICLES EXHIBIT WAVE-LIKE BEHAVIOR

In 1924, Louis de Broglie suggested that a relationship that had been derived to relate momentum and wavelength for light should also apply to particles. The **de Broglie relation** states that

$$\lambda = \frac{h}{p} \tag{1.11}$$

in which h is the by now familiar Planck constant and p is the particle momentum given by $p = mv$, in which the momentum is expressed in terms of the particle mass and velocity. This proposed relation was confirmed in 1927 by Davisson and Germer, who carried out a diffraction experiment. Diffraction is only observed under conditions for which the characteristic dimension of the diffraction grating is on the order of one wavelength. Putting numbers in Equation (1.11) should convince you that it is difficult to obtain wavelengths much longer than 1 nm even with particles as light as the electron, as shown in Example Problem 1.2. Therefore, diffraction requires a grating with atomic dimensions, and an ideal candidate is a crystalline solid. Davisson and Germer observed diffraction of electrons from crystalline NiO in their classic experiment to verify the de Broglie relation. Diffraction of He and H_2 from crystalline surfaces has also been observed in the intervening years.

 *grating. 光栅.*

This is Science !!!

EXAMPLE PROBLEM 1.2

Electrons are used to determine the structure of crystal surfaces. To have diffraction, the wavelength, λ, of the electrons should be on the order of the lattice constant, which is typically 0.30 nm. What energy do such electrons have, expressed in electron-volts and joules?

$E = \frac{p^2}{2m}$

Solution

Using Equation (1.11) and the expression $E = p^2/2m$ for the kinetic energy, we obtain

$$E = \frac{p^2}{2m} = \frac{h^2}{2m\lambda^2} = \frac{(6.626 \times 10^{-34}\ \text{J s})^2}{2 \times 9.109 \times 10^{-31}\ \text{kg} \times (3.0 \times 10^{-10}\ \text{m})^2}$$

$$= 2.7 \times 10^{-18}\ \text{J} \quad \text{or} \quad 17\ \text{eV}$$

The Davisson–Germer experiment was critical in the development of quantum mechanics, in that it showed that particles exhibit wave behavior. If this is the case, there must be a wave equation that relates the spatial and time dependencies of the wave amplitude for the (wave-like) particle. This equation could be used to describe an atomic scale system rather than Newton's 2nd law $F = ma$. It was Erwin Schrödinger who formulated this wave equation, which will be discussed in Chapter 2.

1.6 DIFFRACTION BY A DOUBLE SLIT

There is probably no single experiment that exhibits the surprising nature of quantum mechanics as well as the diffraction of particles by a double slit. An idealized version of this experiment is described, but everything in the following explanation has been confirmed by experiments carried out with particles such as neutrons. We first briefly review classical diffraction of waves.

Diffraction is a phenomenon that is widely exploited in science. For example, the atomic level structure of DNA was in large part determined by analyzing the diffraction

1.3 Diffraction of Light

FIGURE 1.5

Diffraction of light of wavelength λ from a slit whose long axis is perpendicular to the page. The arrows from the left indicate parallel rays of light incident on an opaque plate containing the slit. Instead of seeing a sharp image of the slit on the screen, a diffraction pattern will be seen. This is schematically indicated in a plot of intensity versus distance. In the absence of diffraction, the intensity versus distance indicated by the blue lines would be observed.

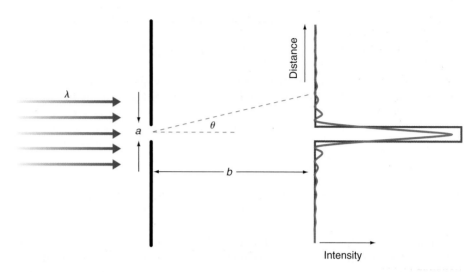

FIGURE 1.6

Each segment of a slit through which light is diffracted can be viewed as a source of waves that interfere with one another. **(a)** The waves that emerge perpendicular to the slit are all in phase and give rise to the principal maximum in the diffraction pattern. **(b)** Successive blue and red waves that emerge at the angle shown are exactly out of phase. They will interfere destructively and a minimum intensity will be observed. **(c)** Blue and red waves are out of phase and yellow waves have an intermediate phase. Destructive interference with a reduction in intensity will be observed in this case. The wavelength and slit width are not drawn to scale.

of X-rays from crystalline DNA samples. Diffraction is a general phenomenon that can occur with sound waves, water waves, and electromagnetic (light) waves. Figure 1.5 illustrates diffraction of light from a thin slit in an otherwise opaque wall.

It turns out that the analysis of this problem is much simpler if the distance b of the screen on which the image is projected is much greater than the width a of the slit. Mathematically, this requires that $b \gg a$. In ray optics, which is used to determine the focusing effect of a lens on light, the light incident on the slit from the left in Figure 1.5 would give a sharp image of the slit on the screen. In this case parallel light is assumed to be incident on the slit and, therefore, the image and slit dimensions are identical. The expected intensity pattern is that shown by the blue lines in the figure. Instead, an intensity distribution like that shown by the red curve is observed if the light wavelength is comparable in magnitude to the slit width.

The origin of this pattern of alternating maxima and minima (which lies well outside the profile expected from ray optics) is wave interference. Its origin can be understood by treating each point in the plane of the slit as a source of cylindrical waves (Huygens' construction). Maxima and minima arise as a result of a path difference between the sources of the cylindrical waves and the screen, as shown in Figure 1.6. The condition that the minima satisfy is

$$\sin \theta = \frac{n\lambda}{a}, \quad n = \pm 1, \pm 2, \pm 3, \pm \ldots \quad (1.12)$$

Diffraction :

If in the middle of slit
is off-phase with the top.
then minimum.

$\frac{a}{2} \cdot \sin\theta = \frac{n\lambda}{2}$. $\sin\theta = n\lambda$.

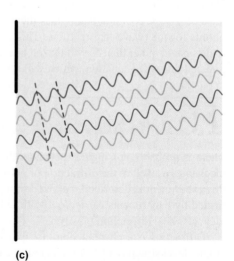

b)   **(c)**

This equation helps us to understand under what conditions we might observe diffraction. The wavelength of light in the middle of the visible spectrum is about 600. nm or 6.00×10^{-4} mm. If this light is allowed to pass through a 1.00-mm-wide slit and the angle calculated at which the first minimum will appear, the result is $\theta = \sin^{-1} 0.000600 = 0.0300°$ for $n = 1$. This minimum is not easily observable because it lies so close to the maximum, and we expect to see a sharp image of the slit on our screen, just as in ray optics. However, if the slit width is decreased to 1.00×10^{-2} mm, then $\theta = \sin^{-1} 0.0600 = 3.40°$. This minimum is easily observable and successive bands of light and darkness will be observed instead of a sharp image. Note that there is no clear demarcation between ray optics and diffraction. The crossover between the two is continuous and depends on the resolution of our experimental techniques. The exact same behavior is observed in wave-particle duality in quantum mechanics. If the slit is much larger than the wavelength, diffraction will not be observed and ray optics holds. This conclusion is drawn from experiments carried out using light waves, but it also applies to particles for which the wavelength is given by the de Broglie relation of Equation (1.11).

Consider the experimental setup designed to detect the diffraction of particles shown in Figure 1.7. The essential feature of the apparatus is a metal plate in which two rectangular slits of width a have been cut. The long axis of the rectangles is perpendicular to the plane of the page. Why two slits? Rather than detecting the diffraction from the individual slits, the apparatus is designed to detect diffraction from the *combination of the two slits*. Diffraction will only be observed for Case 2 if the particle passes through both slits simultaneously. Think about why the previous sentence is difficult to accept from the vantage point of classical physics.

First, we need a source of particles, for instance, an electron gun. By controlling the energy of the electron, the wavelength is varied. Each electron has a **random phase angle** with respect to every other electron. Consequently, two electrons can never interfere with one another to produce a diffraction pattern. One electron gives rise to the diffraction pattern, but many electrons are needed to amplify the signal so that we can see the pattern. A more exact way to say this is that the intensities of the electron waves add together rather than the amplitudes.

A phosphorescent screen that lights up when energy from an incident wave or particle is absorbed (as in a television picture tube) is mounted behind the plate with the slits. The electron energy is adjusted so that diffraction by the single slits of width a results in broad maxima with the first intensity minimum at a large diffraction angle. The distance between the two slits, b, has been chosen such that we will observe a number of intensity oscillations for small diffraction angles. The diffraction patterns in Figure 1.7 (Case 2) were calculated for the ratio $b/a = 5$.

1.4 Diffraction from Double Slit

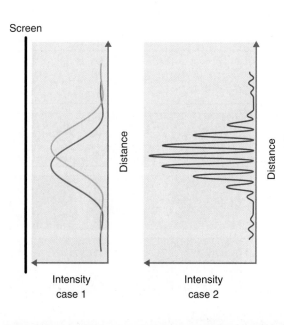

FIGURE 1.7

The double-slit diffraction experiment. Case 1 describes the outcome of the diffraction when one or the other of the slits is blocked. Case 2 describes the outcome when both slits are open.

Now let's talk about the results. If one slit is closed and an observer looks at the screen, he will see the broad intensity versus distance pattern shown as Case 1 in Figure 1.7. (i) Depending on which slit has been closed, the observer sees one of the two diffraction patterns shown, and concludes that the electron acts like a wave. (ii) If he measures the electron current, he finds that exactly 50% land on the screen and 50% land on the device that blocks one slit for a large number of electrons. (iii) When working with very sensitive phosphors, the arrival of each individual electron at the screen is detected by a flash of light localized to a small area of a screen. Viewed from classical physics, which of these three results is consistent with both wave *and* particle behavior, and which is only consistent with wave *or* particle behavior?

In the next experiment, both slits are left open. The result of this experiment is shown in Figure 1.7 as Case 2. For very small electron currents, the observer again sees individual light flashes localized to small areas of the screen in a random pattern. This looks like particle behavior. Figure 1.8 shows what the observer would see if the results of a number of these individual electron experiments are stored. However, unmistakable diffraction features are seen if we accumulate the results of many individual light flashes. This shows that a wave (a single electron) is incident on *both* of the slits simultaneously.

How can the results of this experiment be understood? The fact that diffraction is seen from a single slit as well as from the double slit shows the wave-like behavior of the electron. Yet, individual light flashes are observed on the screen, which is what we expect from particle trajectories. To add to the complexity, the spatial distribution of the individual flashes on the screen is what we expect from waves rather than from particles. The measurement of the electron current to the slit blocker seems to indicate that the electron *either* went through one slit *or* through the other. However, this conclusion is inconsistent with the appearance of a diffraction pattern because a diffraction pattern only arises if one and the same electron goes through both slits!

Regardless of how you turn these results around, you will find that all the results are inconsistent with the logic of classical physics. In particle physics, the electron must go through one slit and not the other, whereas in wave physics it must go through both slits. This classical logic can't explain the results! In a quantum mechanical description, the electron wave function is a superposition of wave functions for waves going through the top slit and the bottom slit, which is equivalent to saying that the electron can go through both slits. We will have much more to say about wave functions in the next few chapters. The act of measurement, such as blocking one slit, changes the wave function such that the electron goes through either the top slit or the bottom one. The results represent a mixture of particle and wave behavior. Individual electrons move through the slits and generate points of light on the screen. This behavior is particle-like. However,

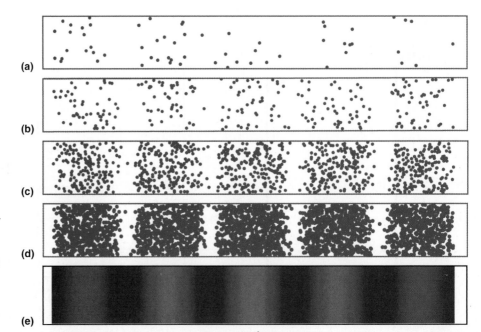

FIGURE 1.8

Simulation of the diffraction pattern observed in the double-slit experiment for (top to bottom) **(a)** 60., **(b)** 250., **(c)** 1000., and **(d)** 3000. particles. The bottom panel **(e)** shows what would be expected for a wave incident on the apparatus. Bright red corresponds to high intensity and blue corresponds to low intensity. Note that the diffraction pattern only becomes obvious after a large number of particles have passed through the apparatus, although intensity minima are evident even for 60 incident particles.

the location of the points of light on the screen is not what is expected from classical trajectories; it is governed by the diffraction pattern. This behavior is wave-like. Whereas in classical mechanics, the operative word concerning several possible modes of behavior is *or*, in quantum mechanics it is *and*. If all of this seems strange to you at first sight, welcome to the crowd! In 1997, this classic double-slit experiment was carried out using a collimated beam of He atoms. A diffraction pattern was observed exactly as described earlier, showing that each He atom also goes through both slits.

1.7 ATOMIC SPECTRA AND THE BOHR MODEL OF THE HYDROGEN ATOM

The most direct evidence of energy quantization comes from the analysis of the light emitted from highly excited atoms in a plasma. The structure of the atom was not known until fundamental studies using the scattering of alpha particles were carried out in Ernest Rutherford's laboratory beginning in 1910. These experiments showed that the positive and negative charges in an atom were separated. The positive charge is contained in a small volume called the *nucleus*, whereas the negative charge of the electrons occupies a much greater volume that is centered at the nucleus. In analogy to our solar system, the first picture that emerged of the atom was of electrons orbiting the nucleus.

However, this picture of the atom is inconsistent with electrodynamic theory. An electron orbiting the nucleus is constantly accelerating and must therefore radiate energy. In a classical picture, the electron would continually radiate away its kinetic energy and eventually fall into the nucleus. This clearly was not happening, but why? We will answer this question when we discuss the hydrogen atom in Chapter 9. Even before Rutherford's experiments, it was known that if an electrical arc was placed across a vacuum tube with a small partial pressure of a gas like hydrogen, light is emitted. Our present picture of this phenomenon is that the atom takes up energy from the electrical discharge and makes a transition to an excited state. The excited state has a limited lifetime, and when the transition to a state of lower energy occurs, light is emitted. An apparatus used to obtain atomic spectra and a typical spectrum are shown schematically in Figure 1.9.

How did scientists working in the 1890s explain these spectra? The most important experimental observation that these scientists made is that over a wide range of wavelengths,

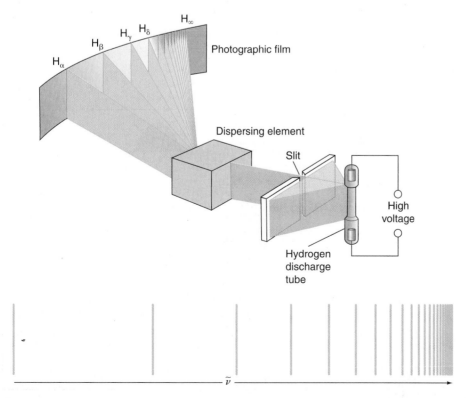

FIGURE 1.9

Light emitted from a hydrogen discharge lamp is passed through a narrow slit and separated into its component wavelengths by a dispersing element which separates light into its constituent wavelengths. As a result, multiple images of the slit, each corresponding to a different wavelength, are seen on the photographic film. One of the different series of spectral lines for H is shown. $\tilde{\nu}$ represents the inverse wavelength [see Equation (1.13)].

light is only observed at certain discrete wavelengths, that is, it is *quantized*. This result was not understandable on the basis of classical theory because in classical physics, energy is a continuous variable. Even more baffling to these first spectroscopists was that they could derive a simple relationship to explain all of the frequencies that appeared in the emission spectrum. For the emission spectra observed, the inverse of the wavelength, $1/\lambda = \tilde{\nu}$ of all lines in an atomic hydrogen spectrum is given by equations of the type

$$\tilde{\nu}/\text{cm}^{-1} = \frac{R_H}{\text{cm}^{-1}}\left(\frac{1}{n_1^2} - \frac{1}{n^2}\right), \quad n > n_1 \tag{1.13}$$

in which only a single parameter, n_1, appears. In this equation, n is an integer that takes on the values $n_1 + 1, n_1 + 2, n_1 + 3, \ldots$, and R_H is called the **Rydberg constant**. For hydrogen, R_H has the value 109,677.581 cm^{-1}. What gives rise to such a simple relationship and why does n take on only integral values?

Niels Bohr, who played a seminal role in the development of quantum mechanics, proposed a model for the hydrogen atom that explained its emission spectrum in 1911. Even though Bohr's model was superseded by the Schrödinger model described in Chapter 9, it offered the first explanation of how quantized energy levels arise in atoms as a result of wave-particle duality. Bohr assumed a simple model of the hydrogen atom in which an electron revolved around the nucleus in a circular orbit. The orbiting electron experiences two forces: a Coulombic attraction to the nucleus, and a centrifugal force which is opposite in direction. In a stable orbit, these two forces are equal.

$$\frac{e^2}{4\pi\varepsilon_0 r^2} = \frac{m_e v^2}{r} \tag{1.14}$$

In Equation (1.14), e is the charge on the electron, m_e and v are its mass and speed, respectively, and r is the orbit radius.

Bohr next introduced wave-particle duality which is equivalent to asserting that the electron had the de Broglie wavelength $\lambda = h/p$. He made a new assumption that the length of an orbit had to be an integral number of wavelengths.

$$2\pi r = n\lambda = n\frac{h}{p} \tag{1.15}$$

which leads to the condition

$$m_e v r = n\hbar, \quad \text{where } n = 1, 2, 3, \ldots \tag{1.16}$$

We have introduced the symbol $\hbar$ for $h/2\pi$. The rationale for Equation (1.15) is shown in Figure 1.10.

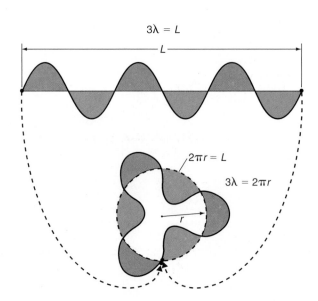

FIGURE 1.10

In analogy to a wave on a string (upper image), Bohr postulated a wave traveling on a circular orbit (lower image). Unless the circumference of the orbit is an integral number of wavelengths, the wave will cancel itself out.

Bohr reasoned that unless the orbit length is an integral number of wavelengths, the wave will destructively interfere with itself, and the amplitude will decrease to zero in a few orbits. The assertion that there is a stable orbit for the electron is not consistent with classical physics which predicts that the electron will radiate energy and spiral into the nucleus as shown schematically in Figure 1.11. Therefore, using Equation (1.16) to define a stable orbit goes beyond classical physics.

Solving Equation 1.16 for v and substituting the result in Equation (1.14) gives the following expression for the orbit radius, r.

$$r = \frac{\varepsilon_0 h^2 n^2}{\pi m_e e^2} = \frac{4\pi\varepsilon_0 \hbar^2 n^2}{m_e e^2} \tag{1.17}$$

Equation (1.17) shows that the electron can only have certain discrete values for the orbit radii, each corresponding to a different value of n. We next show that the discrete set of orbit radii gives rise to a discrete set of energy levels.

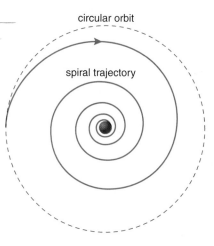

circular orbit

spiral trajectory

FIGURE 1.11

Classical particle-based physics predicts that an electron in a circular orbit will lose energy by radiation and spiral into the nucleus.

EXAMPLE PROBLEM 1.3

Calculate the radius of the electron in H in its lowest energy state, corresponding to $n = 1$.

Solution

$$r = \frac{4\pi\varepsilon_0 \hbar^2 n^2}{m_e e^2}$$

$$= \frac{4\pi \times 8.85419 \times 10^{-12}\,\mathrm{C^2\,N^{-1}\,m^{-2}} \times (1.0555 \times 10^{-34}\,\mathrm{J\,s})^2 \times 1^2}{9.109 \times 10^{-31}\,\mathrm{kg} \times (1.6022 \times 10^{-19}\,\mathrm{C})^2}$$

$$= 5.292 \times 10^{-11}\,\mathrm{m}$$

The total energy of the electron in the hydrogen atom is the sum of its kinetic and potential energies.

$$E_{total} = E_{kinetic} + E_{potential} = \frac{1}{2} m_e v^2 - \frac{e^2}{4\pi\varepsilon_0 r} \tag{1.18}$$

We transform Equation (1.18) into a more useful form by first eliminating v using Equation (1.14).

$$E_{total} = \frac{1}{2}\left(\frac{e^2}{4\pi\varepsilon_0 r}\right) - \left(\frac{e^2}{4\pi\varepsilon_0 r}\right) = -\frac{e^2}{8\pi\varepsilon_0 r} \tag{1.19}$$

We next eliminate r using Equation (1.17), obtaining Equation (1.20) which shows that the energy levels in the Bohr model are discrete.

$$E_n = -\frac{m_e e^4}{8\varepsilon_0^2 h^2 n^2} \quad n = 1, 2, 3, \ldots \tag{1.20}$$

All energy values have negative values because the zero of energy, which is arbitrary, corresponds to $n \rightarrow \infty$, for which the electron is at infinite separation from the proton. The ground state energy is the lowest energy that a hydrogen atom can have, which corresponds to $n = 1$.

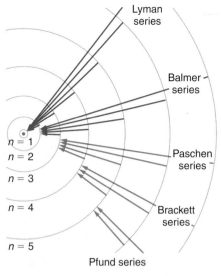

Lyman series

Balmer series

$n = 1$
$n = 2$
$n = 3$
$n = 4$
$n = 5$

Paschen series

Brackett series

Pfund series

FIGURE 1.12

Transitions in the Bohr model giving rise to light emission are shown. The series differ in the quantum number of the final state.

Because the energy of the electron can have only certain discrete values, the light emitted when an electron makes a transition from a higher to a lower energy level has a discrete set of frequencies:

$$\nu_{2\to1} = \frac{m_e e^4}{8\varepsilon_0^2 h^3}\left(\frac{1}{n_1^2} - \frac{1}{n_2^2}\right) \quad n_2 > n_1 \qquad \textbf{(1.21)}$$

We see that Equation (1.21) provides a rationale for the empirical formula given by Equation (1.13), and the calculated and measured frequencies for the hydrogen atom are in quantitative agreement. This agreement between theory and experiment appears to justify the assumptions made. A number of possible energy transitions in the Bohr model are shown in Figure 1.12.

Although the Bohr model predicts the absorption and emission frequencies observed in the hydrogen atom, it does not give quantitative agreement with spectra observed for any atom containing more than one electron for reasons that will become clear in Chapters 10 and 11. There is also a fundamental flaw in the Bohr model which was discovered by Werner Heisenberg 15 years after the model was introduced. Heisenberg showed that it is not possible to simultaneously know the electron orbit radius and its momentum as Bohr had assumed. We discuss Heisenberg's work on the uncertainty principle in Chapter 6 in the context of noncommuting operators.

Vocabulary

blackbody radiation
de Broglie relation
ideal blackbody
photoelectric effect

photon
Planck constant
quantization
random phase angle

Rydberg constant
spectral density
wave-particle duality
work function

Questions on Concepts

Q1.1   Why is there an upper limit to the photon energy that can be observed in the emission spectrum of the hydrogen atom?

Q1.2   Why were investigations at the atomic and subatomic levels required to detect the wave nature of particles?

Q1.3   Classical physics predicts that there is no stable orbit for an electron moving around a proton. What criterion did Niels Bohr use to define special orbits that he assumed were stable?

Q1.4   You observe light passing through a slit of width a and pass from the range $\lambda \gg a$ to $\lambda \ll a$. Will you observe a sharp transition between ray optics and diffraction? Explain why or why not.

Q1.5   Which of the experimental results for the photoelectric effect suggests that light can display particle-like behavior?

Q1.6   Is the intensity observed from the diffraction experiment depicted in Figure 1.6 the same for the angles shown in parts (b) and (c)?

Q1.7   What feature of the distribution depicted as Case 1 in Figure 1.7 tells you that the broad distribution arises from diffraction?

Q1.8   Why does the analysis of the photoelectric effect based on classical physics predict that the kinetic

energy of electrons will increase with increasing light intensity?

Q1.9   In the double-slit experiment, researchers found that an equal amount of energy passes through each slit. Does this result allow you to distinguish between purely particle-like and purely wave-like behavior?

Q1.10   The inability of classical theory to explain the spectral density distribution of a blackbody was called the *ultraviolet catastrophe*. Why is this name appropriate?

Q1.11   In the diffraction of electrons by crystals, the volume sampled by the diffracting electrons is on the order of 3 to 10 atomic layers. If He atoms are incident on the surface, only the topmost atomic layer is sampled. Can you explain this difference?

Q1.12   Why is a diffraction pattern generated by an electron gun formed by electrons interfering with themselves rather than with one another?

Q1.13   What did Einstein postulate to explain that the kinetic energy of the emitted electrons in the photoelectric effect depends on the frequency? How does this postulate differ from the predictions of classical physics?

Q1.14 How did Planck conclude that the discrepancy between experiments and classical theory for blackbody radiation was at high and not low frequencies?

Q1.15 Write down formulas relating the wave number with the frequency, wavelength, and energy of a photon.

Problems

Problem numbers in **red** indicate that the solution to the problem is given in the *Student's Solutions Manual*.

P1.1 When a molecule absorbs a photon, momentum is conserved. If a H_2 molecule with speed $v_{rms} = \sqrt{3kT/m}$ at 750. K absorbs an ultraviolet photon of wavelength 225 nm, what is the change in its speed, Δv? Calculate the relative change in speed $\Delta v/v_{rms}$.

P1.2 A more accurate expression for $\overline{E}_{osc}$ would be obtained by including additional terms in the Taylor-Maclaurin series. The Taylor-Maclaurin series expansion of $f(x)$ in the vicinity of x_0 is given by (see Math Supplement)

$$f(x) = f(x_0) + \left(\frac{df(x)}{dx}\right)_{x=x_0} (x - x_0)$$
$$+ \frac{1}{2!}\left(\frac{d^2f(x)}{dx^2}\right)_{x=x_0} (x - x_0)^2$$
$$+ \frac{1}{3!}\left(\frac{d^3f(x)}{dx^3}\right)_{x=x_0} (x - x_0)^3 + \dots$$

Use this formalism to better approximate $\overline{E}_{osc}$ by expanding $e^{h\nu/kT}$ in powers of $h\nu/kT$ out to $(h\nu/kT)^3$ in the vicinity of $h\nu/kT = 0$. Calculate the relative error, $(\overline{E}_{osc} - kT)/\overline{E}_{osc}$, if you had not included the additional terms for $\nu = 1.00 \times 10^{12}$ s^{-1} at temperatures of 800., 500., and 250. K. Explain the trend you see.

P1.3 The observed lines in the emission spectrum of atomic hydrogen are given by

$$\tilde{\nu}/\text{cm}^{-1} = \frac{R_H}{\text{cm}^{-1}}\left(\frac{1}{n_1^2} - \frac{1}{n^2}\right), \quad n > n_1$$

In the notation favored by spectroscopists, $\tilde{\nu} = 1/\lambda = E/hc$ and $R_H = 109{,}737$ cm^{-1}. The Lyman, Balmer, and Paschen series refers to $n_1 = 1$, 2, and 3, respectively, for emission from atomic hydrogen. What is the highest value of $\tilde{\nu}$ and E in each of these series?

P1.4 Calculate the speed that a gas-phase nitrogen molecule would have if it had the same energy as an infrared photon ($\lambda = 10^4$ nm), a visible photon ($\lambda = 500$ nm), an ultraviolet photon ($\lambda = 100$ nm), and an X-ray photon ($\lambda = 0.1$ nm). What temperature would the gas have if it had the same energy as each of these photons? Use the root mean square speed, $v_{rms} = \langle v^2 \rangle^{1/2} = \sqrt{3kT/m}$, for this calculation.

P1.5 Calculate the highest possible energy of a photon that can be observed in the emission spectrum of H.

P1.6 What is the maximum number of electrons that can be emitted if a potassium surface of work function 2.40 eV absorbs 7.75×10^{-3} J of radiation at a wavelength of 275 nm? What is the kinetic energy and speed of the electrons emitted?

P1.7 Show that the energy density radiated by a blackbody

$$\frac{E_{total}(T)}{V} = \int_0^\infty \rho(\nu, T)\, d\nu = \int_0^\infty \frac{8\pi h\nu^3}{c^3}\frac{1}{e^{h\nu/kT} - 1}\, d\nu$$

depends on the temperature as T^4. (Hint: Make the substitution of variables $x = h\nu/kT$.) The definite integral $\int_0^\infty [x^3/(e^x - 1)]\, dx = \pi^4/15$. Using your result, calculate the energy density radiated by a blackbody at 1350. and 5250. K, which is nearly the surface temperature of the sun.

P1.8 What speed does a N_2 molecule have if it has the same momentum as a photon of wavelength 180. nm?

P1.9 A newly developed substance that emits 315 W of photons with a wavelength of 275 nm is mounted in a small rocket such that all of the radiation is released in the same direction. Because momentum is conserved, the rocket will be accelerated in the opposite direction. If the total mass of the rocket is 6.75 kg, how fast will it be traveling at the end of 365 days in the absence of frictional forces?

P1.10 In our discussion of blackbody radiation, the average energy of an oscillator $\overline{E}_{osc} = h\nu/(e^{h\nu/kT} - 1)$ was approximated as $\overline{E}_{osc} = h\nu/[(1 + h\nu/kT) - 1] = kT$ for $h\nu/kT \ll 1$. Calculate the relative error $= (E - E_{approx})/E$ in making this approximation for $\nu = 4.000 \times 10^{12}$ s^{-1} at temperatures of 6000., 2000., and 500. K. Can you predict what the sign of the relative error will be without a detailed calculation?

P1.11 Using the root mean square speed, $v_{rms} = \langle v^2 \rangle^{1/2} = \sqrt{3kT/m}$, calculate the gas temperatures of H and Xe for which $\lambda = 0.35$ nm, a typical value needed to resolve diffraction from the surface of a metal crystal. On the basis of your result, explain why Xe atomic beams are not suitable for atomic diffraction experiments.

P1.12 Electrons have been used to determine molecular structure by diffraction. Calculate the speed of an electron for which the wavelength is equal to a typical bond length, namely, 0.175 nm.

P1.13 For a monatomic gas, one measure of the "average speed" of the atoms is the root mean square speed, $v_{rms} = \langle v^2 \rangle^{1/2} = \sqrt{3kT/m}$, in which m is the molecular mass and k is the Boltzmann constant. Using this formula, calculate the de Broglie wavelength for H and Ne atoms at 250. and at 750. K.

P1.14 The distribution in wavelengths of the light emitted from a radiating blackbody is a sensitive function of the temperature. This dependence is used to measure the temperature of hot objects, without making physical contact with those objects, in a technique called *optical pyrometry*. The maximum in a plot of $\rho(\lambda, T)$ versus λ is given by $\lambda_{max} \approx hc/5kT$. At what wavelength does the maximum in $\rho(\lambda, T)$ occur for $T = 850., 1300.,$ and $5500.$ K?

P1.15 A beam of electrons with a speed of 4.75×10^4 m/s is incident on a slit of width 235 nm. The distance to the detector plane is chosen such that the distance between the central maximum of the diffraction pattern and the first diffraction minimum is 0.375 cm. How far is the detector plane from the slit?

P1.16 If an electron passes through an electrical potential difference of 1 V, it has an energy of 1 electron-volt. What potential difference must it pass through in order to have a wavelength of 0.225 nm?

P1.17 Calculate the longest and the shortest wavelength observed in the Balmer series.

P1.18 X-rays can be generated by accelerating electrons in a vacuum and letting them impact with atoms on a metal surface. If the 1525-eV kinetic energy of the electrons is completely converted to the photon energy, what is the wavelength of the X-rays produced? If the electron current is 2.25×10^{-5} A, how many photons are produced per second?

P1.19 The following data were observed in an experiment on the photoelectric effect from potassium:

10^{19} Kinetic Energy (J)	4.49	3.09	1.89	1.34	0.700	0.311
Wavelength (nm)	250.	300.	350.	400.	450.	500.

Graphically evaluate these data to obtain values for the work function and the Planck constant.

P1.20 The power (energy per unit time) radiated by a blackbody per unit area of surface expressed in units of W m^{-2} is given by $P = \sigma T^4$ with $\sigma = 5.67 \times 10^{-8}$ W m^{-2} K^{-4}. The radius of the sun is 6.95×10^5 km and the surface temperature is 5750. K. Calculate the total energy radiated per second by the sun. Assume ideal blackbody behavior.

P1.21 The work function of tungsten is 4.50 eV. What is the minimum frequency of light required to observe the photoelectric effect on W? If light with a 225-nm wavelength is absorbed by the surface, what is the velocity of the emitted electrons?

P1.22 Assume that water absorbs light of wavelength 3.50×10^{-6} m with 100% efficiency. How many photons are required to heat 2.50 g of water by 1.00 K? The heat capacity of water is 75.3 J mol^{-1} K^{-1}.

P1.23 Calculate the longest and the shortest wavelength observed in the Lyman series.

P1.24 A 1000-W gas discharge lamp emits 5.25 W of ultraviolet radiation in a narrow range centered near 325 nm. How many photons of this wavelength are emitted per second?

P1.25 The power per unit area emitted by a blackbody is given by $P = \sigma T^4$ with $\sigma = 5.67 \times 10^{-8}$ W m^{-2}K^{-4}. Calculate the energy radiated per second by a spherical blackbody of radius 0.325 m at 850. K. What would the radius of a blackbody at 3250. K be if it emitted the same energy as the spherical blackbody of radius 0.325 m at 850. K?

P1.26 A ground-state H atom absorbs a photon and makes a transition to the $n = 4$ energy level. It then emits a photon of energy 4.085×10^{-19} J. What is the final energy and n value of the atom?

P1.27 Pulsed lasers are powerful sources of nearly monochromatic radiation. Lasers that emit photons in a pulse of 10.-ns duration with a total energy in the pulse of 0.10 J at 1000. nm are commercially available.

a. What is the average power (energy per unit time) in units of watts (1 W = 1 J/s) associated with such a pulse?

b. How many 1000.-nm photons are emitted in such a pulse?

Web-Based Simulations, Animations, and Problems

W1.1 The maximum in a plot of the spectral density of blackbody radiation versus T is determined for a number of values of T using numerical methods. Using these results, the validity of the approximation $\lambda_{max} = hc/5kT$ is tested graphically.

W1.2 The total radiated energy of blackbody radiation is calculated numerically for the temperatures of W1.1. Using these results, the exponent in the relation $E = CT^\alpha$ is determined graphically.

W1.3 Diffraction of visible light from a single slit is simulated. The slit width and light wavelength are varied using sliders. The student is asked to draw conclusions about how the diffraction pattern depends on these parameters.

W1.4 Diffraction of a particle from single and double slits is simulated. The intensity distribution on the detector plane is updated as each particle passes through the slits. The slit width and light wavelength are varied using sliders. The student is asked to draw conclusions about how the diffraction pattern depends on these parameters.

2

The Schrödinger Equation

The key to understanding why classical mechanics does not provide an appropriate framework for understanding phenomena at the atomic level is the recognition that wave-particle duality needs to be integrated into the physics. Rather than solving Newton's equations of motion for a particle, an appropriate wave equation needs to be solved for the wave-particle. Erwin Schrödinger was the first to formulate such an equation successfully. Operators, eigenfunctions, wave functions, and eigenvalues are key concepts that arise in a viable framework to solve quantum mechanical wave equations. The eigenvalues correspond to the possible values of measured results, or observables, in an experiment. These new concepts are introduced in this chapter, although a full explanation will be deferred to Chapters 3 and 4.

2.1 WHAT DETERMINES IF A SYSTEM NEEDS TO BE DESCRIBED USING QUANTUM MECHANICS?

Quantum mechanics was viewed as a radically different way of looking at matter at the molecular, atomic, and subatomic level in the 1920s. However, the historical distance we have from what was a revolution at the time makes the quantum view much more familiar today. It is important to realize that classical and quantum mechanics are not two competing ways to describe the world around us. Each has its usefulness in a different regime of physical properties that describe reality. Quantum mechanics merges seamlessly into classical mechanics, and one can show that classical mechanics can be derived from quantum mechanics in the limit in which allowed energy values are continuous rather than discrete. Some of these complexities will require hard thinking from you as you gain an understanding of quantum mechanics. For instance, it is not correct to say that whenever one talks about atoms, a quantum mechanical description must be used.

To illustrate this point, consider a container filled with argon gas at a low pressure. At the molecular level, the origin of pressure is the collision of rapidly and randomly moving argon atoms with the container walls. Even at the level of considering the force exerted by a single argon atom upon colliding with the wall, classical mechanics gives a perfectly good description of the origin of pressure, although we are talking about atoms. However, if we pass ultraviolet light through hydrogen gas and ask how much energy can be taken up by a H_2 molecule, we must use a quantum mechanical description. At first, this seems puzzling—why do we need quantum mechanics in one case but not the other? On further consideration, we discover that a very few important relationships govern whether a classical description suffices in a given case. We next discuss these relationships in order to develop an understanding of when to use a classical description and when to use a quantum description for a given system.

The essence of quantum mechanics is that particles and waves are not really separate and distinct entities. Waves can show particle-like behavior as illustrated by the photoelectric effect. Particles can also show wave-like properties as is shown by the diffraction of atomic beams from surfaces. How can we develop criteria that tell us when a particle (classical) description of an atomic or molecular system is sufficient and when we need to use a wave (quantum mechanical) description? Two criteria are used: the magnitude of the wavelength of the particle relative to the dimensions of the problem and the degree to which the allowed energy values form a **continuous energy spectrum**.

A good starting point is to think about diffraction of light of wavelength λ passing through a slit of width a. Ray optics is a good description as long as $\lambda \ll a$. Diffraction is only observed when the wavelength is comparable to the slit width. How big is the wavelength of a molecule? Of a macroscopic mass like a baseball? By putting numbers into Equation (1.11), you will find that the wavelength for a room temperature H_2 molecule is about 10^{-10} m and that for a baseball is about 10^{-34} m. Keep in mind that because p rather than v appears in the denominator of Equation (1.11), the wavelength of a toluene molecule with the same velocity as a H_2 molecule is about a factor of 50 smaller. As we learned in discussing the Davisson–Germer experiment in Chapter 1, crystalline solids have regular spacings that are appropriate for the diffraction of electrons as well as light atoms and molecules. Particle diffraction is a demonstration of wave-particle duality. To see the wave character of a baseball, we need to come up with a diffraction experiment. We will not see diffraction of a baseball because we cannot construct an opening whose size is $\sim 1 \times 10^{-34}$ m. This does not mean that wave-particle duality breaks down for macroscopic masses; it simply means that the wave character of a baseball can't be demonstrated. There is no sharp boundary so that above a certain value for the momentum we are dealing with a particle and below it we are dealing with a wave. The degree to which each of these properties is exhibited flows smoothly from one extreme to the other. Consider the second example cited above. Adding energy to hydrogen molecules using UV light can't be treated classically, because energy is taken up by the electrons in H_2. The localization of the electrons to a small volume around the nuclei brings out their wave-like character, and therefore the process must be described using quantum mechanics.

We next discuss the second criterion for determining when we need a quantum mechanical description of a system. It is based on the energy spectrum of the system. Because all values of the energy are allowed for a classical system, it is said to have a continuous energy spectrum. In a quantum mechanical system, only certain values of the energy are allowed, and such a system has a **discrete energy spectrum**. To make this criterion quantitative, we need to discuss the Boltzmann distribution.

You will learn more about Boltzmann's work in statistical thermodynamics. At this point, we attempt to make his most important result plausible so that we can apply it in our studies of quantum mechanics. Consider a one-liter container filled with an ideal atomic gas at the standard conditions of 1 bar of pressure (10^5 pascal) and a temperature

$\lambda = \dfrac{h}{p}$

of 298.15 K. Because the atoms have no rotational or vibrational degrees of freedom, all of their energy is in the form of translational kinetic energy. At equilibrium, not all of the atoms have the same kinetic energy. In fact, the atoms exhibit a broad range of energies. To define the distribution of atoms having a given energy, descriptors, such as the mean, the median, or the most probable energy per atom are used. For the atoms under consideration, the root mean square energy is simply related to the absolute temperature, T, by

$$E_{rms} = \frac{3}{2}kT \qquad \qquad (2.1)$$

$$k = \frac{R}{N_A}$$

The Boltzmann constant, k, is the familiar ideal gas law constant, R, divided by Avogadro's number.

We said that there is a broad distribution of kinetic energy in the gas for the individual atoms. What governs the probability of observing one value of the energy as opposed to another? This question led Ludwig Boltzmann to one of the most important equations in physics and chemistry. Looking specifically at our case, it relates the number of atoms n_i that have energy ε_i, to the number of atoms n_j that have energy ε_j by the equation

$$\frac{n_i}{n_j} = \frac{g_i}{g_j}e^{-[\varepsilon_i - \varepsilon_j]/kT} \qquad \qquad (2.2)$$

This formula is called the **Boltzmann distribution**. An important concept to keep in mind is that a formula is just a shorthand way of describing phenomena that occur in the real world. It is critical that you understand what lies behind the formula. Take a closer look at this equation. It says that the ratio of the number of atoms having the energy ε_i to the number having the energy ε_j depends on three things. It depends exponentially on the difference in the energies and the reciprocal of the temperature. This means that this ratio varies rapidly with temperature and $\varepsilon_i - \varepsilon_j$. The equation also states that it is the ratio of the energy difference to kT that is important. What is kT? It has the units of energy as it must, and it is approximately the average energy that an atom has at temperature T. We can understand this exponential term as telling us that the larger the temperature, the closer the ratio n_i/n_j will be to unity. This means that the probability of an atom having a given energy falls off exponentially with increasing energy.

The third factor that influences the ratio n_i/n_j is the ratio g_i/g_j. The quantities g_i and g_j are the degeneracies of the energy levels i and j. The **degeneracy** of an energy level counts the number of ways that an atom can have an energy ε within the interval $\varepsilon_i - \Delta\varepsilon < \varepsilon < \varepsilon_i + \Delta\varepsilon$. The degeneracy can depend on the energy. In our example, degeneracy can be illustrated as follows. The energy of an atom $\varepsilon_i = \frac{1}{2}mv_i^2$ is determined by $v_i^2 = v_{xi}^2 + v_{yi}^2 + v_{zi}^2$. We have explicitly written that the energy depends only on the speed of the atom and not on any of its individual velocity components. For a fixed value of $\Delta\varepsilon$, there are many more ways of combining different individual velocity components to give the same speed at large speeds than there are for low speeds. Therefore, the degeneracy corresponding to a particular energy ε_i increases with the speed.

The importance of these considerations will become clearer as we continue to apply quantum mechanics to atoms and molecules. As already stated, a quantum mechanical system has a discrete rather than continuous energy spectrum. If kT is small compared to the spacing between allowed energies, the distribution of states in energy will be very different from a classical system, which has a continuous energy spectrum. On the other hand, if kT is much larger than the energy spacing, classical and quantum mechanics will give the same result for the relative numbers of atoms or molecules of different energy. This can occur in either of two limits: large T or small $\varepsilon_i - \varepsilon_j$. This illustrates how it is possible to have a continuous transition between classical and quantum mechanics. A large increase in T could cause a system that exhibited quantum behavior at low temperatures to exhibit classical behavior at high temperatures. A calculation using the Boltzmann distribution for a two-level system is carried out in Example Problem 2.1.

EXAMPLE PROBLEM 2.1

Consider a system of 1000. particles that can only have two energies, ε_1 and ε_2, with $\varepsilon_2 > \varepsilon_1$. The difference in the energy between these two values is $\Delta\varepsilon = \varepsilon_2 - \varepsilon_1$. Assume that $g_1 = g_2 = 1$.

a. Graph the number of particles, n_1 and n_2, in states ε_1 and ε_2 as a function of $kT/\Delta\varepsilon$. Explain your result.

b. At what value of $kT/\Delta\varepsilon$ do 750. of the particles have the energy ε_1?

Solution

Using information from the problem and Equation (2.2), we can write down the following two equations: $n_2/n_1 = e^{-\Delta\varepsilon/kT}$ and $n_1 + n_2 = 1000$. We solve these two equations for n_2 and n_1 to obtain

$$n_2 = \frac{1000.\,e^{-\Delta\varepsilon/kT}}{1 + e^{-\Delta\varepsilon/kT}} \quad \text{and} \quad n_1 = \frac{1000.}{1 + e^{-\Delta\varepsilon/kT}}$$

If these functions are plotted as a function of $kT/\Delta\varepsilon$, the following graphs result:

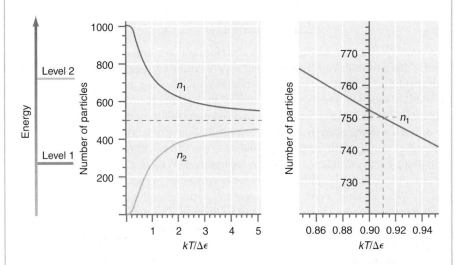

We see from the left graph that as long as $kT/\Delta\varepsilon$ is small, the vast majority of the particles have the lower energy. How can we interpret this result? As long as the thermal energy of the particle, which is about kT, is much less than the difference in energy between the two allowed values, the particles with the lower energy are unable to gain energy through collisions with other particles. However, as $kT/\Delta\varepsilon$ increases (which is equivalent to a temperature increase for a fixed energy difference between the two values), the random thermal energy available to the particles enables some of them to jump up to the higher energy value. Therefore, n_1 decreases and n_2 increases. For all finite temperatures, $n_1 > n_2$. As T approaches infinity, n_1 becomes equal to n_2.

Part (b) is solved graphically. n_1 is shown as a function of $kT/\Delta\varepsilon$ on an expanded scale on the right side of the preceding graphs, and we see that $n_1 = 750.$ for $kT/\Delta\varepsilon = 0.91$.

Example Problem 2.1 shows that the population of states associated with the energy values ε_i and ε_j are very different if

$$\frac{(\varepsilon_i - \varepsilon_j)}{kT} \gg 1 \tag{2.3}$$

and very similar if the inequality is reversed. What are the consequences of this result?

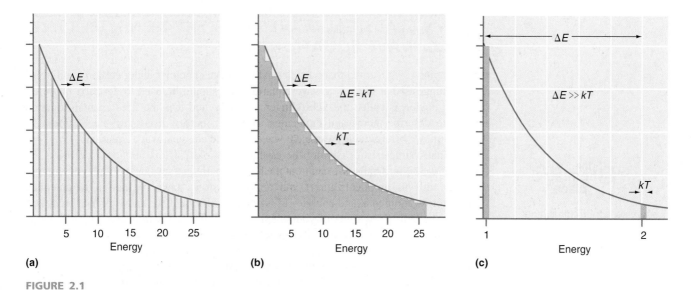

FIGURE 2.1

The relative population in the different energy levels designated by the integers 1–25 (vertical axis) is plotted at constant T as a function of energy for **(a)** sharp energy levels, **(b)** for $\Delta E \approx kT$, and **(c)** for $\Delta E \gg kT$. In parts (a), (b), and (c), the Boltzmann distribution describes the relative populations. However, the system behaves as if it has a continuous energy spectrum only if $\Delta E \approx kT$ or if $\Delta E < kT$.

Consider a quantum mechanical system, which, unlike a classical system, has a discrete energy spectrum. The allowed values of energy are called the energy levels. Anticipating a system that we will deal with in Chapter 7, we refer specifically to the vibrational energy levels of a molecule. The allowed levels are equally spaced with an interval ΔE. These discrete energy levels are numbered with integers, beginning with one. Under what conditions will this quantum mechanical system *appear* to follow classical behavior? It will do so if the discrete energy spectrum *appears* to be continuous. How can this occur? In a gas at equilibrium, the total energy of an individual molecule fluctuates within a range $\Delta E \approx kT$ through collisions of molecules with one another. Therefore, the energy of a molecule with a particular vibrational quantum number fluctuates within a range of width kT. A plot of the relative number of molecules having a vibrational energy E as a function of E is shown in Figure 2.1 for sharp energy levels and for the two indicated limits, $\Delta E \approx kT$ and $\Delta E \gg kT$. The plot is generated using the Boltzmann distribution.

For $\Delta E \approx kT$, each discrete energy level is sufficiently broadened by energy fluctuations that adjacent energy levels can no longer be distinguished. This is indicated by the overlap of the yellow bars representing individual states shown in Figure 2.1b. In this limit, any energy that we choose in the range shown lies in the yellow area. It corresponds to an allowed value and therefore the discrete energy spectrum *appears* to be continuous. Classical behavior will be observed under these conditions. However, if $\Delta E \gg kT$, an arbitrarily chosen energy in the range lies in the blue area with high probability, because the yellow bars of width kT are widely separated. The blue area corresponds to forbidden energies and, therefore, the discontinuous nature of the energy spectrum is observable. Quantum mechanical behavior is observed under these conditions.

If the allowed energies form a continuum, classical mechanics is sufficient to describe *that feature* of the system. If they are discrete, a quantum mechanical description is needed. The words "that feature" require emphasis. The pressure exerted by the H_2 molecules in the box arises from momentum transfer governed by the molecules' translational energy spectrum, which *appears* to be continuous, as we will learn in Chapter 4. Therefore, we do not need quantum mechanics to discuss the pressure in the box. However, if we discuss light absorption by the same H_2 molecules, a quantum mechanical description of light absorption is required. This is so because light absorption involves an electronic excitation of the molecule, and the spacing between electronic energy levels is much larger than kT, so that these levels remain discrete at all reasonable temperatures.

2.2 CLASSICAL WAVES AND THE NONDISPERSIVE WAVE EQUATION

In Chapter 1, we learned that particles exhibit wave character under certain conditions. This suggests that there is a wave equation that should be used to describe particles. This equation is called the Schrödinger equation, and it is the fundamental equation used to describe atoms and molecules. However, before discussing the Schrödinger equation, we briefly review classical waves and the classical wave equation.

What characteristics capture the essence of waves? Think about the collision between two billiard balls. We can treat the balls as point masses (any pool player will recognize this as an idealization) and apply Newton's laws of motion to calculate trajectories, momenta, and energies as a function of time if we know all the forces acting on the balls. Now think of yourself shouting. Often you will hear an echo. What is happening here? Your vocal cords create a local compression of the air in your larynx. This compression zone propagates away from its source as a wave with the speed of sound. The louder the sound, the larger the pressure is in the compressed zone. The pressure variation is the amplitude of the wave and the energy contained in this wave is proportional to the square of the amplitude. The sound reflects from a surface and comes back to you as a weakened local compression of the air. When the wave is incident on your eardrum, a signal is generated that you recognize as sound. Note that this energy transfer is fundamentally different from the direct transfer that occurs in the collision of the two billiard balls. The energy associated with the sound wave is only localized at its origin in your larynx. A further important characteristic of a wave is that it has a characteristic velocity and frequency with which it propagates. The velocity and frequency govern the variation of the amplitude with time for a fixed position of the observer with respect to the source and the variation of the amplitude with distance between the source and the observer at a fixed time.

A wave can be represented pictorially by a succession of **wave fronts**, corresponding to surfaces over which the amplitude of the wave has a maximum or minimum value. A point source emits **spherical waves** as shown in Figure 2.2b, and the light passing through a rectangular slit can be represented by cylindrical waves as shown in Figure 2.2c. The waves sent out from a faraway source such as the sun when viewed from the Earth are spherical waves with such little curvature that they can be represented as **plane waves** as shown in Figure 2.2a.

Mathematically, the amplitude of a wave can be described by a **wave function**. The wave function describes how the amplitude of the disturbance depends on the variables x and t. The variable x is measured along the direction of propagation. For convenience, only sinusoidal waves of wavelength λ and the single **frequency** $\nu = 1/T$, where T is the **period**, are considered. The velocity, v, frequency, ν, and **wavelength**, λ, are related by $v = \lambda\nu$. The peak-to-peak amplitude of the wave is $2A$:

$$\Psi(x, t) = A \sin 2\pi\left(\frac{x}{\lambda} - \frac{t}{T}\right) \tag{2.4}$$

In this equation we have arbitrarily chosen our zero of time and distance such that $\Psi(0,0) = 0$. This equation represents a wave that is moving in the direction of positive x. You can convince yourself of this by considering how a specific feature of this wave changes with time. The wave amplitude is zero for

$$2\pi\left(\frac{x}{\lambda} - \frac{t}{T}\right) = n\pi \tag{2.5}$$

where n is an integer. Solving for x, the location of the nodes is obtained:

$$x = \lambda\left(\frac{n}{2} + \frac{t}{T}\right) \tag{2.6}$$

Note that x increases as t increases, showing that the wave is moving in the direction of positive x. Figure 2.3 shows a graph of the wave functions. To graph this function in two dimensions, one of the variables is kept constant.

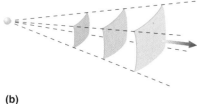

the Chemistry place™

2.1 Transverse, Longitudinal, and Surface Waves

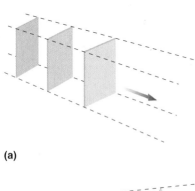

(a)

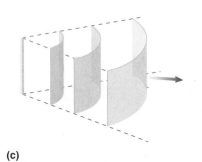

(b)

(c)

FIGURE 2.2

Waves can be represented by a succession of surfaces over which the amplitude of the wave has its maximum or minimum value. The distance between successive surfaces is the wavelength. Representative surfaces are shown for **(a)** plane waves, **(b)** spherical waves, and **(c)** cylindrical waves. The direction of propagation of the waves is perpendicular to the surfaces as indicated by the blue arrows.

The functional form in Equation (2.4) appears so often that it is convenient to combine some of the constants and variables to write the wave amplitude as

$$\Psi(x, t) = A \sin(kx - \omega t) \tag{2.7}$$

The quantity k is called the **wave vector** and is defined by $k = 2\pi/\lambda$. The quantity $\omega = 2\pi\nu$ is called the **angular frequency**. Both will appear often in the rest of this book.

Because the wave amplitude is a simple sine function in our case, it has the same value as the argument changes by 2π. The choice of a zero in position or time is arbitrary and is chosen at our convenience. To illustrate this, consider Equation (2.7) rewritten in the form

$$\Psi(x, t) = A \sin(kx - \omega t + \phi) \tag{2.8}$$

in which the quantity ϕ has been added to the argument of the sin function. This is appropriate when $\Psi(0,0) \neq 0$. The argument of the wave function is called the **phase** and a change in the initial phase ϕ shifts the wave function to the right or left relative to the horizontal axes in Figure 2.3 depending on the sign of ϕ. Figure 2.3 shows the amplitude as a function of either x or t with the other variable held constant for a plane wave.

When two or more waves are present in the same region of space, their time-dependent amplitudes add together, and the waves are said to interfere with one another. The **interference** between two waves gives rises to an enhancement in a region of space (**constructive interference**) if the wave amplitudes are both positive or both negative. It

$$\frac{1}{T} = \nu$$

$$\frac{2\pi}{T} = \omega.$$

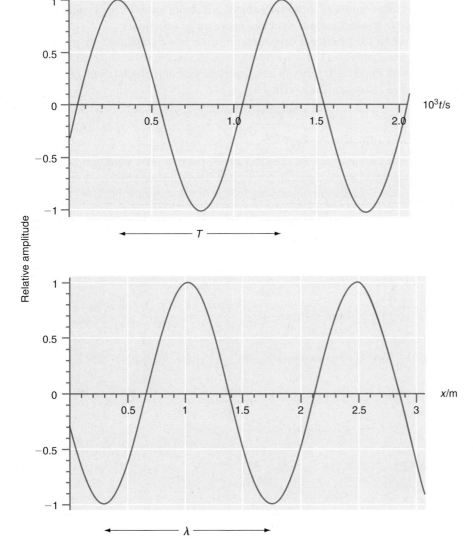

FIGURE 2.3

The upper panel shows the wave amplitude as a function of time at a fixed point. The wave is completely defined by the period, the maximum amplitude, and the amplitude at $t = 0$. The lower panel shows the analogous information when the wave amplitude is plotted as a function of distance for a given time. $\lambda = 1.46$ m and $T = 1.00 \times 10^{-3}$ s.

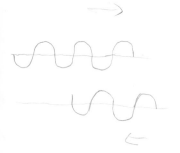

can also lead to a cancellation of the wave amplitude in a region of space (**destructive interference**) if the wave amplitudes are opposite in sign and equal in amplitude. At the constructive interference condition, maxima of the wave functions from the two sources line up (constructive interference) because the phases of the two functions are the same to within an integral multiple of 2π. They are out of phase at the destructive interference condition where the phases differ by $(2n + 1)\pi$, where n is an integer.

Interference can also result in a very different time dependence of the wave amplitude than was discussed for a traveling wave, namely, the formation of spatially fixed nodes where the amplitude is zero at all times. Consider the superposition of two waves of the same frequency and amplitude that are moving in opposite directions. The resultant wave amplitude is the sum of the individual amplitudes:

$$\Psi(x, t) = A[\sin(kx - \omega t) + \sin(kx + \omega t)] \qquad \textbf{(2.9)}$$

Using the standard trigonometric identity $\sin(\alpha \pm \beta) = \sin \alpha \cos \beta \pm \cos \alpha \sin \beta$, Equation (2.9) can be simplified to

$$\Psi(x, t) = 2A \sin kx \cos \omega t = \psi(x) \cos \omega t \qquad \textbf{(2.10)}$$

This function of x and t is a product of two functions, each of which depends only on one of the variables. Therefore, the position of the nodes, which is determined by $\sin kx = 0$, is the same at all times. This property distinguishes **standing waves** from **traveling waves** in which the whole wave, including the nodes, propagates at the same velocity.

The form that the standing wave amplitudes take is shown in Figure 2.4. Standing waves arise if the space in which the waves can propagate is bounded. For instance, plucking a guitar string gives rise to a standing wave because the string is fixed at both ends. Standing waves play an important role in quantum mechanics because, as demonstrated later, they represent **stationary states**, which are states of the system in which the measurable properties of the system do not change with time.

We return to the functional dependencies of the wave amplitude on time and distance for a traveling wave. For wave propagation in a medium for which all frequencies propagate with the same velocity (a nondispersive medium), the variation of the amplitude with time and distance are related by

$$\frac{\partial^2 \Psi(x, t)}{\partial x^2} = \frac{1}{v^2} \frac{\partial^2 \Psi(x, t)}{\partial t^2} \qquad \textbf{(2.11)}$$

Equation (2.11) is known as the **classical nondispersive wave equation** and v designates the velocity at which the wave propagates. This equation provides a starting point in justifying the Schrödinger equation, which is the fundamental quantum mechanical wave equation. (See the Math Supplement, Appendix A, for a discussion of partial differentiation.) Example Problem 2.2 demonstrates that the form of the traveling wave that we have used is a solution of the nondispersive wave equation.

FIGURE 2.4

Time evolution of a standing wave at a fixed point. The time intervals are shown as a function of the period T. The vertical lines indicate the nodal positions x_0. Note that the wave function has temporal nodes for $t = T/4$ and $3T/4$.

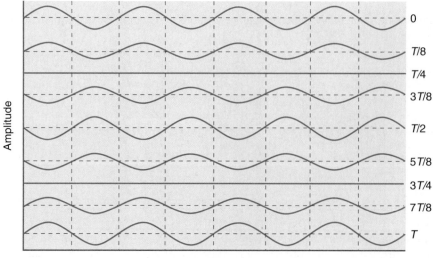

EXAMPLE PROBLEM 2.2

The nondispersive wave equation in one dimension is given by

$$\frac{\partial^2 \Psi(x,t)}{\partial x^2} = \frac{1}{v^2} \frac{\partial^2 \Psi(x,t)}{\partial t^2}$$

Show that the traveling wave $\Psi(x,t) = A \sin(kx - \omega t + \phi)$ is a solution of the nondispersive wave equation. How is the velocity of the wave related to k and ω in this case?

Solution

$$\frac{\partial^2 \Psi(x,t)}{\partial x^2} = \frac{1}{v^2} \frac{\partial^2 \Psi(x,t)}{\partial t^2}$$

$$\frac{\partial^2 A \sin(kx - \omega t + \phi)}{\partial x^2} = -k^2 A \sin(kx - \omega t + \phi)$$

$$\frac{1}{v^2} \frac{\partial^2 A \sin(kx - \omega t + \phi)}{\partial t^2} = \frac{-\omega^2}{v^2} A \sin(kx - \omega t + \phi)$$

Because $v = \omega/k$, the two sides of the equation are equal.

2.3 WAVES ARE CONVENIENTLY REPRESENTED AS COMPLEX FUNCTIONS

It turns out that the mathematics of dealing with wave functions is much simpler if they are represented in the complex number plane. As discussed, a wave traveling in the positive x direction can be described by the function

$$\Psi(x,t) = A \sin(kx - \omega t + \phi) = A \cos(kx - \omega t + \phi - \pi/2)$$
$$= A \cos(kx - \omega t + \phi') \tag{2.12}$$

where $\phi' = \phi - \pi/2$. Using Euler's formula, $e^{i\alpha} = \exp(i\alpha) = \cos \alpha + i \sin \alpha$, Equation (2.12) can be written as

$$\Psi(x,t) = \text{Re}\left(A e^{i(kx - \omega t + \phi')}\right)$$
$$= \text{Re}\left(A \exp i(kx - \omega t + \phi')\right) \tag{2.13}$$

in which the notation Re indicates that we are considering only the real part of the complex function that follows.

Whereas the wave functions considered previously (for example, sound waves) have real amplitudes, quantum mechanical wave functions can have complex amplitudes. Working with only the real part of the functions makes some of the mathematical treatment more cumbersome, so that it is easier to work with the whole complex function knowing that we can always extract the real part if we wish to do so.

The wave function of Equation (2.13) can then be written in the form

$$\Psi(x,t) = A \exp i(kx - \omega t + \phi') \tag{2.14}$$

where A is a constant. All quantities that fully characterize the wave, namely, the maximum amplitude, the wavelength, the period of oscillation, and the phase angle for $t = 0$ and $x = 0$, are contained in Equation (2.14). Operations such as adding two waves to form a superposition and operations such as differentiation and integration are much easier when working with complex exponential notation than with real trigonometric wave functions.

The following bullets list the properties of complex numbers that will be used frequently in this book. Example Problems 2.3 and 2.4 show how to work with complex numbers. See the Math Supplement (Appendix A) for a more detailed discussion of complex numbers.

- A complex number or function can be written as $a + ib$, where a and b are real numbers or real functions and $i = \sqrt{-1}$, or equivalently in the form $re^{i\theta}$, where $r = \sqrt{a^2 + b^2}$ and $\theta = \sin^{-1}(b/r)$.

- The **complex conjugate** of a complex number or function f is denoted f^*. The complex conjugate is obtained by substituting $-i$ in f wherever i occurs. The complex conjugate of the number or function $a + ib$ is $a - ib$ and the complex conjugate of $re^{i\theta}$ is $re^{-i\theta}$.

- The magnitude of a complex number or function f is a real number or function denoted $|f|$, where $|f| = \sqrt{f^*f}$. For $f = a + ib$ or $re^{-i\theta}$, $|f| = \sqrt{a^2 + b^2}$, or r, respectively.

EXAMPLE PROBLEM 2.3

a. Express the complex number $4 - 4i$ in the form $re^{i\theta}$.

b. Express the complex number $3e^{i3\pi/2}$ in the form $a + ib$.

Solution

a. The magnitude of $4 + 4i$ is $[(4 + 4i)(4 - 4i)]^{1/2} = 4\sqrt{2}$. The phase is given by

$$\sin\theta = \frac{-4}{4\sqrt{2}} = -\frac{1}{\sqrt{2}} \quad \text{or} \quad \theta = \sin^{-1}\left(-\frac{1}{\sqrt{2}}\right) = -\frac{\pi}{4} \quad \text{or} \quad \frac{7\pi}{4}$$

Therefore, $4 - 4i$ can be written $4\sqrt{2}\,e^{-i(\pi/4)}$.

b. Using the relation $e^{i\alpha} = \exp(i\alpha) = \cos\alpha + i\sin\alpha$, $3e^{i3\pi/2}$ can be written

$$3\left(\cos\frac{3\pi}{2} + i\sin\frac{3\pi}{2}\right) = 3(0 - i) = -3i$$

EXAMPLE PROBLEM 2.4

Determine the magnitude of the following complex numbers:

a. $(1 + i)(\sqrt{2} + 5i)$

c. $\dfrac{e^{\sqrt{2}i\pi}e^{-3i\pi}}{4e^{i\pi/4}}$

b. $\dfrac{1 + \sqrt{3}i}{11 - 2i}$

d. $\dfrac{1 + 6i}{i}$

Solution

The magnitude of a complex number or function f is $\sqrt{f^*f}$. Note that the magnitude of a complex number is a real number.

a. $\sqrt{(1 + i)(\sqrt{2} + 5i)(1 - i)(\sqrt{2} - 5i)} = 3\sqrt{6}$

b. $\sqrt{\dfrac{1 + \sqrt{3}i}{11 - 2i}\dfrac{1 - \sqrt{3}i}{11 + 2i}} = \dfrac{2}{5\sqrt{5}}$

c. $\sqrt{\dfrac{e^{\sqrt{2}i\pi}e^{-3i\pi}}{4e^{i\pi/4}}\dfrac{e^{-\sqrt{2}i\pi}e^{+3i\pi}}{4e^{-i\pi/4}}} = \dfrac{1}{4}$

d. $\sqrt{\dfrac{1 + 6i}{i}\dfrac{1 - 6i}{-i}} = \sqrt{37}$

2.4 QUANTUM MECHANICAL WAVES AND THE SCHRÖDINGER EQUATION

In this section, we justify the time-independent Schrödinger equation by combining the classical nondispersive wave equation and the de Broglie relation. For classical standing waves, we showed in Equation (2.10) that the wave function is a product of two functions, one of which depends only on spatial coordinates, and the other depends only on time:

$$\Psi(x, t) = \psi(x) \cos \omega t \qquad (2.15)$$

If this function is substituted in Equation (2.11), we obtain

$$\frac{d^2\psi(x)}{dx^2} + \frac{\omega^2}{v^2}\psi(x) = 0 \qquad (2.16)$$

$$\frac{\partial^2 \Psi}{\partial x^2} = \frac{1}{v^2}\frac{\partial^2 \psi}{\partial t^2}$$

The time-dependent part $\cos \omega t$ cancels because it appears on both sides of the equation after the derivative with respect to time is taken. Using the relations $\omega = 2\pi\nu$ and $\nu\lambda = v$, Equation (2.16) becomes

$$\frac{d^2\psi(x)}{dx^2} + \frac{4\pi^2}{\lambda^2}\psi(x) = 0 \qquad (2.17)$$

To this point, everything that we have written is for a classical wave. We introduce quantum mechanics by using the de Broglie relation, $\lambda = h/p$ for the wavelength. The momentum is related to the total energy, E, and the potential energy, $V(x)$, by

$$\frac{p^2}{2m} = E - V(x) \quad \text{or} \quad p^2 = 2m(E - V(x)). \qquad (2.18)$$

Introducing this expression for the momentum into the de Broglie relation, and substituting the expression obtained for λ into Equation (2.17), we obtain

$$\frac{d^2\psi(x)}{dx^2} + \frac{8\pi^2 m}{h^2}[E - V(x)]\psi(x) = 0 \qquad (2.19)$$

Using the abbreviation $\hbar = h/2\pi$ and rewriting Equation (2.19), we obtain the **time-independent Schrödinger equation** in one dimension:

$$-\frac{\hbar^2}{2m}\frac{d^2\psi(x)}{dx^2} + V(x)\psi(x) = E\psi(x) \qquad (2.20)$$

This is the fundamental equation used to study the stationary states of quantum mechanical systems. The stationary states are those states for which the values of the observables (see Section 2.5) are independent of time. The familiar $1s$ and $2p_z$ orbitals of the hydrogen atom are examples of stationary states obtained from the time-independent Schrödinger equation.

There is an analogous quantum mechanical form of the time-dependent classical nondispersive wave equation. It is called the **time-dependent Schrödinger equation** and has the following form:

$$i\hbar\frac{\partial\Psi(x, t)}{\partial t} = -\frac{\hbar^2}{2m}\frac{\partial^2\Psi(x, t)}{\partial x^2} + V(x, t)\Psi(x, t) \qquad (2.21)$$

This equation relates the temporal and spatial derivatives of $\Psi(x, t)$ with the potential energy function $V(x, t)$. It is applied in systems in which the energy changes with time. For example, the time-dependent equation is used to model infrared spectroscopy, in which the energy of a molecule changes as it absorbs a photon.

These two equations that Schrödinger formulated are the basis of all quantum mechanical calculations. Their validity has been confirmed by countless experiments carried out during the last 80 years. The equations look very different from Newton's equations of motion. The mass of the particle appears in both forms of the Schrödinger equation, but what meaning can be attached to $\Psi(x,t)$ and $\psi(x)$? This question will be discussed in some detail in Chapter 3. For now, we say that $\Psi(x,t)$ and $\psi(x)$ represent the amplitude of the wave that describes the particle or system of particles under consideration. To keep the notation simple, only one spatial coordinate is included, but note that, in general, the spatial part of $\Psi(x,t)$, denoted $\psi(x)$, depends on all spatial coordinates.

Our main focus is on the stationary states of a quantum mechanical system. For these states, both the time-dependent and time-independent Schrödinger equations are satisfied. In this case,

$$i\hbar\frac{\partial\Psi(x,t)}{\partial t} = E\Psi(x,t) \tag{2.22}$$

For stationary states, $\Psi(x,t) = \psi(x)f(t)$. Substituting this expression in Equation (2.22) gives

$$i\hbar\frac{df(t)}{dt} = Ef(t) \quad \text{or} \quad \frac{df(t)}{dt} = -i\frac{E}{\hbar}f(t) \tag{2.23}$$

Solving this equation, we obtain $f(t) = e^{-i(E/\hbar)t}$. We have shown that wave functions that describe states whose energy is independent of time have the form

$$\Psi(x,t) = \psi(x)e^{-i(E/\hbar)t} \tag{2.24}$$

Note that $\Psi(x,t)$ is the product of two functions, each of which depends on only one variable. If you go back and look at what we wrote for standing waves in Equation (2.10), you will see the same result. That is not a coincidence, because stationary states in quantum mechanics also correspond to standing waves.

2.5 SOLVING THE SCHRÖDINGER EQUATION: OPERATORS, OBSERVABLES, EIGENFUNCTIONS, AND EIGENVALUES

Now that we have introduced the quantum mechanical wave equation, we need to learn how to work with it. In this section, we develop this topic by introducing the language used in solving the Schrödinger equation. The key concepts introduced are those of *operators, observables, eigenfunctions*, and *eigenvalues*. These terms are defined later. A good understanding of these concepts is necessary to understand the quantum mechanical postulates in Chapter 3.

Both forms of the Schrödinger equation are differential equations whose solutions depend on the potential energy $V(x)$. Our emphasis in the next chapters will be on using the solutions of the time-independent equation for various problems such as the harmonic oscillator or the H atom to enhance our understanding of chemistry. We do not focus on how differential equations are solved. However, it is very useful to develop a general understanding of the formalism used to solve the time-independent Schrödinger equation. This initial introduction is brief, because our primary goal is to obtain a broad overview of the language of quantum mechanics. As we work with these new concepts in successive chapters, they will become more familiar.

We begin by illustrating the meaning of the term *operator* in the context of classical mechanics. Think about how you would describe the time evolution of a system consisting of a particle on which a force is acting. The velocity at time, t_1, is known and you wish to know the velocity at a later time, t_2. You write down Newton's second law

$$m\frac{d^2x}{dt^2} = F(x,t) \tag{2.25}$$

and integrate it to give

$$v(t_2) = v(t_1) + \frac{1}{m} \int_{t_1}^{t_2} F(x, t)\, dt \qquad (2.26)$$

In words, one could describe this process as the series of operations:

- Integrate the force acting on the particle over the interval t_1 to t_2.
- Multiply by the inverse of the mass.
- Add this quantity to the velocity at time t_1.

These actions have the names *integrate, form the inverse, multiply*, and *add*, and they are all classified as operators. Note that we started at the right-hand side of the equation and worked our way to the left.

How are operators used in quantum mechanics? To every measurable quantity (**observable**), such as energy, momentum, or position, there is a corresponding **operator** in quantum mechanics. Quantum mechanical operators usually involve differentiation with respect to a variable such as x or multiplication by x or a function of the energy such as $V(x)$. Operators are denoted by a caret: $\hat{O}$.

Just as a differential equation has a set of solutions, an operator $\hat{O}$ has a set of eigenfunctions and eigenvalues. This means that there is a set of wave functions ψ_n with the index n such that

$$\hat{O}\psi_n = a_n\psi_n \qquad (2.27)$$

The operator acting on these special wave functions returns the wave function multiplied by a number. These special functions are called the **eigenfunctions** of the operator and the a_n are called the **eigenvalues**. The eigenvalues for quantum mechanical operators are always real numbers, because they correspond to the values of observables that are measured in an experiment. There are in general an infinite number of eigenfunctions for a given operator for the specific system under consideration. For example, the eigenfunctions for the total energy operator (kinetic plus potential energy) for the hydrogen atom are the wave functions that describe the orbitals that you know as $1s, 2s, 2p_x, \ldots$ The set of these eigenfunctions is infinite in size. The corresponding eigenvalues are the $1s, 2s, 2p_x, \ldots$ orbital energies.

You should now recognize that the time-independent Schrödinger equation is an eigenvalue equation for the total energy, E

$$\left\{ \frac{-\hbar^2}{2m} \frac{\partial^2}{\partial x^2} + V(x) \right\} \psi_n(x) = E_n\psi_n(x) \qquad (2.28)$$

where the expression in the curly brackets { } is the total energy operator. This operator is given the $\hat{H}$ symbol in quantum mechanics and is called the Hamiltonian for historical reasons. With this notation, Equation (2.28) can be written in the form

$$\hat{H}\psi_n(x) = E_n\psi_n(x) \qquad (2.29)$$

The operator acting on one of its eigenfunctions returns the eigenfunction multiplied by the corresponding eigenvalue. In Example Problem 2.5, this formalism is applied for two operators. Solving the time-independent Schrödinger equation is equivalent to finding the set of eigenfunctions and eigenvalues that are the solutions to the eigenvalue problem of Equation (2.29). In this chapter, we consider only a single operator acting on a function. In Chapter 6, we will show that the outcome of two sequential operations on a wave function can depend on the order in which the operations occur. This fact has important implications for the measurement process in quantum mechanics.

EXAMPLE PROBLEM 2.5

Consider the operators d/dx and d^2/dx^2. Is $\psi(x) = Ae^{ikx} + Be^{-ikx}$ an eigenfunction of these operators? If so, what are the eigenvalues? Note that A, B, and k are real numbers.

Solution

To test if a function is an eigenfunction of an operator, we carry out the operation and see if the result is the same function multiplied by a constant:

$$\frac{d(Ae^{ikx} + Be^{-ikx})}{dx} = ik\,Ae^{ikx} - ik\,Be^{-ikx} = ik(Ae^{ikx} - Be^{-ikx})$$

In this case, the result is not $\psi(x)$ multiplied by a constant, so $\psi(x)$ is not an eigenfunction of the operator d/dx unless either A or B is zero.

$$\frac{d^2(Ae^{ikx} + Be^{-ikx})}{dx^2} = (ik)^2\,Ae^{ikx} + (-ik)^2\,Be^{-ikx}$$

$$= -k^2(Ae^{ikx} + Be^{-ikx}) = -k^2\psi(x)$$

This equation shows that $\psi(x)$ is an eigenfunction of the operator d^2/dx^2 with the eigenvalue $-k^2$.

In general, a quantum mechanical operator such as $\hat{H}$ has an infinite number of eigenfunctions. How are the eigenfunctions of a quantum mechanical operator related to one another? We discuss two of the most important properties in the next sections, namely, *orthogonality* and *completeness*.

2.6 THE EIGENFUNCTIONS OF A QUANTUM MECHANICAL OPERATOR ARE ORTHOGONAL

You are probably familiar with the concept of orthogonal vectors. For example, orthogonality in three-dimensional Cartesian coordinate space is defined by

$$\mathbf{x} \cdot \mathbf{y} = \mathbf{x} \cdot \mathbf{z} = \mathbf{y} \cdot \mathbf{z} = 0 \qquad (2.30)$$

in which the scalar product between the unit vectors along the x, y, and z axes is zero. In function space, the analogous expression that defines **orthogonality** between the eigenfunctions $\psi_i(x)$ and $\psi_j(x)$ of a quantum mechanical operator is

$$\int_{-\infty}^{\infty} \psi_i^*(x)\psi_j(x)\,dx = 0 \quad \text{unless } i = j \qquad (2.31)$$

Example Problem 2.6 shows that graphical methods can be used to determine if two functions are orthogonal.

EXAMPLE PROBLEM 2.6

Show graphically that $\sin x$ and $\cos 3x$ are orthogonal functions over the interval $[-2j\pi, 2j\pi]$ where for the purposes of our discussion, j is a very large integer. Also show graphically that $\int_{-2j\pi}^{2j\pi}(\sin mx)(\sin nx)\,dx \neq 0$ for $n = m = 1$.

Solution

The functions are shown in the following graphs. The vertical axes have been offset to avoid overlap and the horizontal line indicates the zero for each plot. Because the functions are periodic, we can draw conclusions about their behavior in an infinite interval by considering their behavior in any interval that is an integral multiple of the period.

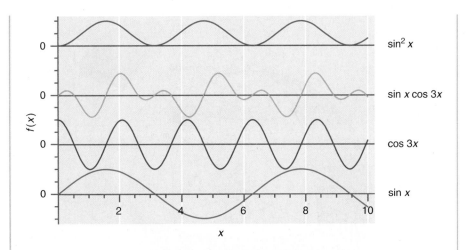

The integral of these functions equals the sum of the areas between the curves and the zero line. Areas above and below the line contribute with positive and negative signs, respectively, and indicate that $\int_{-\infty}^{\infty} \sin x \cos 3x \, dx = 0$ and $\int_{-\infty}^{\infty} \sin x \sin x \, dx > 0$. By similar means, we could show that any two functions of the type $\sin mx$ and $\sin nx$ or $\cos mx$ and $\cos nx$ are orthogonal unless $n = m$. Are the functions $\cos mx$ and $\sin mx (m = n)$ orthogonal?

Recall that the superscript * on a function indicates the complex conjugate. The product $\psi_i^*(x)\psi_j(x)$ rather than $\psi_i(x)\psi_j(x)$ occurs in Equation (2.31) because wave functions in quantum mechanics can be complex functions of x and t. If in addition to Equation (2.31), the integral has the value one for $i = j$, we say that the functions are normalized and form an **orthonormal** set. As we will see in Chapter 3, wave functions must be normalized so that they can be used to calculate probabilities. We show how to normalize wave functions in Example Problems 2.7 and 2.8.

EXAMPLE PROBLEM 2.7

Normalize the function $a(a - x)$ over the interval $0 \le x \le a$.

Solution

To normalize a function $\psi(x)$ over the given interval, we multiply it by a constant N, and then calculate N from the equation $N^2 \int_0^a \psi^*(x)\psi(x) \, dx = 1$. In this particular case,

$$N^2 \int_0^a [a(a - x)]^2 \, dx = 1$$

$$N^2 a^2 \int_0^a [a^2 - 2ax + x^2] \, dx = 1$$

$$N^2 \left(a^4 x - a^3 x^2 + a^2 \frac{x^3}{3} \right)_0^a = 1$$

$$N^2 \frac{a^5}{3} = 1 \quad \text{so that} \quad N = \sqrt{\frac{3}{a^5}}$$

The normalized wave function is $\sqrt{\dfrac{3}{a^5}} a(a - x)$

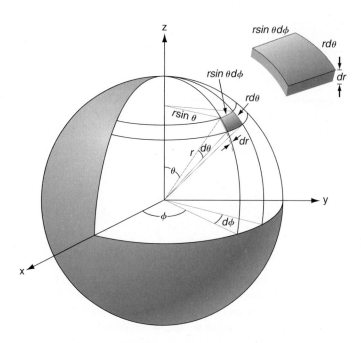

FIGURE 2.5

Defining variables and the volume element in spherical coordinates.

Up until now, we have considered functions of a single variable. This restricts us to dealing with a single spatial dimension. As we will see in Chapter 7, important problems such as the harmonic oscillator can be solved in a one-dimensional framework. The extension to three independent variables becomes important in describing three-dimensional systems. The three-dimensional system of most importance to us is the atom. Closed-shell atoms are spherically symmetric, so that we might expect atomic wave functions to be best described by spherical coordinates, as shown in Figure 2.5. Therefore, you should become familiar with integrations in these coordinates. The Math Supplement (Appendix A) provides a more detailed discussion of working with spherical coordinates. Note in particular that the volume element in spherical coordinates is $r^2 \sin \theta \, dr \, d\theta \, d\phi$ and not $dr \, d\theta \, d\phi$. A function is normalized in spherical coordinates in Example Problem 2.8.

EXAMPLE PROBLEM 2.8

Normalize the function e^{-r} over the interval $0 \le r \le \infty; 0 \le \theta \le \pi;$ $0 \le \phi \le 2\pi$.

Solution

We proceed as in Example Problem 2.7, remembering that the volume element in spherical coordinates is $r^2 \sin \theta \, dr \, d\theta \, d\phi$:

$$N^2 \int_0^{2\pi} d\phi \int_0^{\pi} \sin \theta \, d\theta \int_0^{\infty} r^2 e^{-2r} \, dr = 1$$

$$4\pi N^2 \int_0^{\infty} r^2 e^{-2r} dr = 1$$

Using the standard integral $\int_0^{\infty} x^n e^{-ax} \, dx = n!/a^{n+1}$ ($a > 0$, n is a positive integer), we obtain

$$4\pi N^2 \frac{2!}{2^3} = 1 \quad \text{so that } N = \sqrt{\frac{1}{\pi}}. \text{ The normalized wave function is } \sqrt{\frac{1}{\pi}} e^{-r}.$$

Note that the integration of any function involving r where $r = \sqrt{x^2 + y^2 + z^2}$ requires integration over all three variables, even if it does not explicitly involve θ or ϕ.

2.7 THE EIGENFUNCTIONS OF A QUANTUM MECHANICAL OPERATOR FORM A COMPLETE SET

The eigenfunctions of a quantum mechanical operator have another very important property that we will use frequently in later chapters, namely that the eigenfunctions of a quantum mechanical operator form a **complete set**. The idea of a complete set might be familiar to you from the three-dimensional Cartesian coordinate system. Because any three-dimensional vector can be expressed as a linear combination of the three mutually perpendicular unit vectors **x**, **y**, and **z**, we say that these three unit vectors form a complete set,

Completeness is also an important concept in function space. To say that the eigenfunctions of any quantum mechanical vector form a complete set means that any well-behaved wave function, $f(x)$, can be expanded in the eigenfunctions of any of the quantum mechanical operators of interest to us defined in the same space, x in this case:

$$\psi(x) = \sum_{n=1}^{\infty} b_n \phi_n(x) \tag{2.32}$$

Before we expand wave functions in a complete set of functions in later chapters, we first illustrate how to expand a simple sawtooth function in a **Fourier sine and cosine series**. See the Math Supplement (Appendix A) for a more detailed discussion of Fourier series.

We approximate the sawtooth function shown in Figure 2.6 by a linear combination of the mutually orthogonal functions $\sin(n\pi x/b)$ and $\cos(n\pi x/b)$. These functions form an infinitely large complete set for $n = 1, 2, 3, \ldots \infty$. Because these functions form a complete set only as $n \to \infty$, the approximation becomes exact as $n \to \infty$. The degree to which the approximation approaches the exact function depends only on how many terms we include in the sum. Because we are interested in knowing how good our approximation is, we start with the sawtooth function, approximate it by the finite sum in Equation (2.33), and evaluate how well we succeed for different values of m.

$$f(x) = d_0 + \sum_{n=1}^{m} \left[c_n \sin\left(\frac{n\pi x}{b}\right) + d_n \cos\left(\frac{n\pi x}{b}\right) \right] \tag{2.33}$$

We first need a way to calculate the coefficients d_0, c_m, and d_m. In our case, the function is even, $f(x) = f(-x)$, so that all the coefficients c_m are identically zero. For the Fourier series, the values for the d_m are easily obtained using the mutual orthogonality property of the sine and cosine functions demonstrated in Example Problem 2.6.

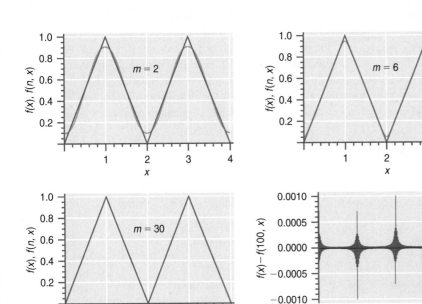

FIGURE 2.6

The sawtooth function (red curve) is compared with the finite Fourier series defined by Equation (2.34) (blue curve) containing m terms for $m = 2, 6,$ and 30. The difference between the sawtooth function and the finite Fourier series for $m = 100$ is shown in the bottom right panel as a function of x.

To obtain the d_m, we multiply both sides of Equation (2.33) by one of the expansion functions, for example $\cos(m\pi x/b)$, and integrate over the interval $-b, b$:

$$\int_{-b}^{b} f(x) \cos\left(\frac{m\pi x}{b}\right) dx = \int_{-b}^{b} \cos\left(\frac{m\pi x}{b}\right)$$

$$\times \left(d_0 + \sum_n d_n \cos\left(\frac{n\pi x}{b}\right) \right) dx$$

$$= \int_{-b}^{b} \left(\cos\left(\frac{m\pi x}{b}\right) \right) d_m \cos\left(\frac{m\pi x}{b}\right) dx = bd_m$$

Only one term in the summation within the integral gives a nonzero contribution because all cosine functions for which $m \neq n$ are orthogonal. We have used one of the standard integrals listed in the Math Supplement (Appendix A) to obtain this result. We conclude that the optimal values for the coefficients are given by

$$d_m = \frac{1}{b} \int_{-b}^{b} f(x) \cos\left(\frac{m\pi x}{b}\right) dx, \quad m \neq 0 \quad \text{and}$$

$$d_0 = \frac{1}{2b} \int_{-b}^{b} f(x)\, dx$$

Using these equations to obtain the optimal coefficients will make the finite sum of Equation (2.33) nearly exact if enough terms can be included in the sum. How good is the approximation if m is finite? This question is answered in Figure 2.6, which shows that the sawtooth function can be described reasonably well for $m < 30$. To make this statement more quantitative, we graph $f(x) - \left(d_0 + \sum_{n=1}^{100} \left[c_n \sin\left(\frac{n\pi x}{b}\right) + d_n \cos\left(\frac{n\pi x}{b}\right) \right] \right)$

versus x in Figure 2.6 for a 101 term series. We see that the difference is less than 0.1% of the maximum amplitude for the sawtooth function. Generalizing this result to other functions, the maximum error occurs at the points for which the slope of the function is discontinuous.

the Chemistry place™

2.4 Expanding Functions in Fourier Series

2.8 SUMMING UP THE NEW CONCEPTS

In this chapter we introduced a number of the key tools in quantum mechanics that are used to solve the Schrödinger equation. The time-dependent and time-independent Schrödinger equations play the role in solving quantum mechanical problems that Newton's laws play in classical mechanics. Operators, eigenfunctions, and observables form the framework for solving the time-independent Schrödinger equation. All of these concepts will be applied to problems of chemical interest in the next few chapters. However, we will first introduce and discuss the five postulates of quantum mechanics in Chapter 3.

Vocabulary

angular frequency	continuous energy spectrum	frequency
Boltzmann distribution	degeneracy	interference
classical nondispersive wave equation	destructive interference	observable
complete set	discrete energy spectrum	operator
completeness	eigenfunction	orthogonality
complex conjugate	eigenvalue	orthonormal
constructive interference	Fourier sine and cosine series	period

phase

plane wave

spherical wave

standing wave

stationary state

time-dependent Schrödinger equation

time-independent Schrödinger equation

traveling wave

wave front

wave function

wave vector

wavelength

Questions on Concepts

Q2.1 One source emits spherical waves and another emits plane waves. For which source does the intensity measured by a detector of fixed size fall off more rapidly with distance? Why?

Q2.2 What is the relationship between evaluating an integral and graphing the integrand?

Q2.3 A traveling wave with arbitrary phase ϕ can be written as $\psi(x, t) = A \sin(kx - \omega t + \phi)$. What are the units of ϕ? Show that ϕ could be used to represent a shift in the origin of time or distance.

Q2.4 Why is it true for any quantum mechanical problem that the set of wave functions is larger than the set of eigenfunctions?

Q2.5 By discussing the diffraction of a beam of particles by a single slit, justify the statement that there is no sharp boundary between particle-like and wave-like behavior.

Q2.6 Redraw Figure 2.2 to show surfaces corresponding to both minimum and maximum values of the amplitude.

Q2.7 Is it correct to say that because the de Broglie wavelength of a H_2 molecule at 300. K is on the order of atomic dimensions that all properties of H_2 are quantized?

Q2.8 Why is it necessary in normalizing the function re^{-r} in spherical coordinates to integrate over θ and ϕ even though it is not a function of θ and ϕ?

Q2.9 If $\psi(x, t) = A \sin(kx - \omega t)$ describes a wave traveling in the plus x direction, how would you describe a wave moving in the minus x direction?

Q2.10 In Figure 2.6 the extent to which the approximate and true functions agree was judged visually. How could you quantify the quality of the fit?

Q2.11 Why does a quantum mechanical system with discrete energy levels behave as if it has a continuous energy spectrum if the energy difference between energy levels ΔE satisfies the relationship $\Delta E \ll kT$?

Q2.12 Distinguish between the following terms applied to a set of functions: orthogonal and normalized.

Q2.13 Why can we conclude that the wave function $\psi(x, t) = \psi(x)e^{-i(E/\hbar)t}$ represents a standing wave?

Q2.14 What is the usefulness of a set of complete functions?

Q2.15 Can the function $\sin kx$ be normalized over the interval $-\infty < x < \infty$? Explain your answer.

Problems

Problem numbers in **red** indicate that the solution to the problem is given in the *Student's Solutions Manual*.

P2.1 A wave traveling in the z direction is described by the wave function $\Psi(z,t) = A_1 \mathbf{x} \sin(kz - \omega t + \phi_1) + A_2 \mathbf{y} \sin(kz - \omega t + \phi_2)$, where $\mathbf{x}$ and $\mathbf{y}$ are vectors of unit length along the x and y axes, respectively. Because the amplitude is perpendicular to the propagation direction, $\Psi(z, t)$ represents a transverse wave.

a. What requirements must A_1 and A_2 satisfy for a plane polarized wave in the x-z plane? The amplitude of a plane polarized wave is non-zero only in one plane.

b. What requirements must A_1 and A_2 satisfy for a plane polarized wave in the y-z plane?

c. What requirements must A_1 and A_2 and ϕ_1 and ϕ_2 satisfy for a plane polarized wave in a plane oriented at 45° to the x-z plane?

d. What requirements must A_1 and A_2 and ϕ_1 and ϕ_2 satisfy for a circularly polarized wave? The phases of the two components of a circularly polarized wave differ by $\pi/2$.

P2.2 Because $\int_0^d \cos(n\pi x/d) \cos(m\pi x/d)\, dx = 0$, $m \neq n$, the functions $\cos(n\pi x/d)$ for $n = 1, 2, 3, \ldots$ form

an orthogonal set. What constant must these functions be multiplied by to form an orthonormal set?

P2.3 Determine in each of the following cases if the function in the first column is an eigenfunction of the operator in the second column. If so, what is the eigenvalue?

a. x^3 d^3/dx^3

b. xy $x(\partial/\partial x) + y(\partial/\partial y)$

c. $\sin \theta \cos \phi$ $\partial^2/\partial\theta^2$

P2.4 If two operators act on a wave function as indicated by $\hat{A}\hat{B}f(x)$, it is important to carry out the operations in succession with the first operation being that nearest to the function. Mathematically, $\hat{A}\hat{B}f(x) = \hat{A}(\hat{B}f(x))$ and $\hat{A}^2 f(x) = \hat{A}(\hat{A}f(x))$. Evaluate the following successive operations $\hat{A}\hat{B}f(x)$. The operators $\hat{A}$ and $\hat{B}$ are listed in the first two columns and $f(x)$ is listed in the third column.

a. $\dfrac{d}{dx}$ x xe^{-ax^2}

b. x $\dfrac{d}{dx}$ xe^{-ax^2}

c. $y\dfrac{\partial}{\partial x}$ $x\dfrac{\partial}{\partial y}$ $e^{-a(x^2+y^2)}$

Note that your answers to parts (a) and (b) are not identical. As we will learn in Chapter 6 the fact that switching the order of the operators x and d/dx changes the outcome of the operation $\hat{A}\hat{B}f(x)$ is the basis for the Heisenberg uncertainty principle.

P2.5   Let $(1, 0)$ and $(0, 1)$ represent the unit vectors along the x and y directions, respectively. The operator

$$\begin{pmatrix} \cos\theta & -\sin\theta \\ \sin\theta & \cos\theta \end{pmatrix}$$

effects a rotation in the x-y plane. Show that the length of an arbitrary vector

$$\begin{pmatrix} a \\ b \end{pmatrix} = a\begin{pmatrix} 1 \\ 0 \end{pmatrix} + b\begin{pmatrix} 0 \\ 1 \end{pmatrix}$$

which is defined as $\sqrt{a^2 + b^2}$, is unchanged by this rotation. See the Math Supplement (Appendix A) for a discussion of matrices.

P2.6   Carry out the following coordinate transformations:

a. Express the point $x = 4$, $y = 2$, and $z = 3$ in spherical coordinates.

b. Express the point $r = 7$, $\theta = \dfrac{\pi}{8}$, and $\phi = \dfrac{5\pi}{8}$ in Cartesian coordinates.

P2.7   Operators can also be expressed as matrices and wave functions as column vectors. The operator matrix

$$\begin{pmatrix} \alpha & \beta \\ \delta & \varepsilon \end{pmatrix}$$

acts on the wave function $\begin{pmatrix} \alpha \\ \beta \end{pmatrix}$ according to the rule

$$\begin{pmatrix} \alpha & \beta \\ \delta & \varepsilon \end{pmatrix}\begin{pmatrix} a \\ b \end{pmatrix} = \begin{pmatrix} \alpha a + \beta b \\ \delta a + \varepsilon b \end{pmatrix}$$

In words, the 2×2 matrix operator acting on the two-element column wave function generates another two-element column wave function. If the wave function generated by the operation is the original wave function multiplied by a constant, the wave function is an eigenfunction of the operator. What is the effect of the operator

$$\begin{pmatrix} 0 & 1 \\ 1 & 0 \end{pmatrix}$$

on the column vectors $(1, 0)$, $(0, 1)$, $(1, 1)$, and $(-1, 1)$? Are these wave functions eigenfunctions of the operator? See the Math Supplement (Appendix A) for a discussion of matrices.

P2.8   Show that

$$\frac{a + ib}{c + id} = \frac{ac + bd + i(bc - ad)}{c^2 + d^2}$$

P2.9   Express the following complex numbers in the form $re^{i\theta}$.

a. $7 - 3i$

b. $-5i$

c. $\dfrac{7 - i}{5 + 3i}$

d. $\dfrac{4 + i}{1 - 2i}$

P2.10   Show that the set of functions $\phi_n(\theta) = e^{in\theta}$, $0 \le \theta \le 2\pi$, is orthogonal if n is an integer. To do so,

you need to show that the integral $\int_0^{2\pi} \phi_m^*(\theta)\phi_n(\theta)\, d\theta = 0$ for $m \ne n$ if n and m are integers.

P2.11   Operate with (a) $\dfrac{\partial}{\partial x} + \dfrac{\partial}{\partial y} + \dfrac{\partial}{\partial z}$ and (b) $\dfrac{\partial^2}{\partial x^2} + \dfrac{\partial^2}{\partial y^2} + \dfrac{\partial^2}{\partial z^2}$ on the function $Ae^{-ik_1x}\, e^{-ik_2y}\, e^{-ik_3z}$. Is the function an eigenfunction of either operator? If so, what is the eigenvalue?

P2.12   Which of the following wave functions are eigenfunctions of the operator d^2/dx^2? If they are eigenfunctions, what is the eigenvalue?

a. $ae^{-3x} + be^{-3ix}$

b. $\sin^2 x$

c. e^{-ix}

d. $\cos ax$

e. e^{-ix^2}

P2.13   Does the superposition $\psi(x, t) = A\sin(kx - \omega t) + 2A\sin(kx + \omega t)$ generate a standing wave? Answer this question by using trigonometric identities to combine the two terms.

P2.14   Determine in each of the following cases if the function in the first column is an eigenfunction of the operator in the second column. If so, what is the eigenvalue?

a. $3\cos^2\theta - 1$ ⠀⠀ $\dfrac{1}{\sin\theta}\dfrac{d}{d\theta}\left(\sin\theta\dfrac{d}{d\theta}\right)$

b. $e^{-(x^2/2)}$ ⠀⠀ $\dfrac{d^2}{dx^2} - x^2$

c. $e^{-4i\phi}$ ⠀⠀ $\dfrac{d^2}{d\phi^2}$

P2.15   Show by carrying out the integration that $\sin(m\pi x/a)$ and $\cos(m\pi x/a)$, where m is an integer, are orthogonal over the interval $0 \le x \le a$. Would you get the same result if you used the interval $0 \le x \le 3a/4$? Explain your result.

P2.16   To plot $\Psi(x, t) = A\sin(kx - vt)$ as a function of one of the variables x and t, the other variable needs to be set at a fixed value, x_0 or t_0. If $\Psi(x_0, 0)/\Psi_{max} = -0.260$, what is the constant value of x_0 in the upper panel of Figure 2.3? If $\Psi(0, t_0)/\psi_{max} = -0.325$, what is the constant value of t_0 in the lower panel of Figure 2.3? (*Hint:* The inverse sine function has two solutions within an interval of 2π. Make sure that you choose the correct one.)

P2.17   Determine in each of the following cases if the function in the first column is an eigenfunction of the operator in the second column. If so, what is the eigenvalue?

a. $e^{-i(3x+2y)}$ ⠀⠀ $\dfrac{\partial^2}{\partial x^2}$

b. $\sqrt{x^2 + y^2}$ ⠀⠀ $(1/x)(x^2 + y^2)\dfrac{\partial}{\partial x}$

c. $\sin\theta\cos\theta$ ⠀⠀ $\sin\theta\dfrac{d}{d\theta}\left(\sin\theta\dfrac{d}{d\theta}\right) + 6\sin^2\theta$

P2.18   Assume that a system has a very large number of energy levels given by the formula $\varepsilon = \varepsilon_0 l^2$ with $\varepsilon_0 = 2.34 \times 10^{-22}$ J, where l takes on the integral values 1, 2, 3, ... Assume further that the degeneracy of a level is given by $g_l = 2l$. Calculate the ratios n_3/n_1 and n_{15}/n_1 for $T = 250.$ K and $T = 900.$ K, respectively.

P2.19 Is the function $3x^2 - 1$ an eigenfunction of the operator $-(1 - x^2)(d^2/dx^2) + 2x(d/dx)$? If so, what is the eigenvalue?

P2.20 Find the result of operating with $d^2/dx^2 - 4x^2$ on the function e^{-ax^2}. What must the value of a be to make this function an eigenfunction of the operator?

P2.21 Determine in each of the following cases if the function in the first column is an eigenfunction of the operator in the second column. If so, what is the eigenvalue?

a. $\sin\theta\cos\phi$ $\partial/\partial\phi$

b. $e^{(-x^2/2)}$ $(1/x)\,d/dx$

c. $\sin\theta$ $(\sin\theta/\cos\theta)\,d/d\theta$

P2.22 Find the result of operating with $d^2/dx^2 + d^2/dy^2 + d^2/dz^2$ on the function $x^2 + y^2 + z^2$. Is this function an eigenfunction of the operator?

P2.23 Using the exponential representation of the sine and cosine functions, $\cos\theta = \frac{1}{2}(e^{i\theta} + e^{-i\theta})$ and $\sin\theta = \frac{1}{2i}(e^{i\theta} - e^{-i\theta})$, show that

a. $\cos^2\theta + \sin^2\theta = 1$

b. $d(\cos\theta)/d\theta = -\sin\theta$

c. $\sin\left(\theta + \dfrac{\pi}{2}\right) = \cos\theta$

P2.24 If two operators act on a wave function as indicated by $\hat{A}\hat{B}f(x)$, it is important to carry out the operations in succession with the first operation being that nearest to the function. Mathematically, $\hat{A}\hat{B}f(x) = \hat{A}(\hat{B}f(x))$ and $\hat{A}^2f(x) = \hat{A}(\hat{A}f(x))$. Evaluate the following successive operations $\hat{A}\hat{B}f(x)$. The operators $\hat{A}$ and $\hat{B}$ are listed in the first and second columns and $f(x)$ is listed in the third column.

a. $\dfrac{d}{dx}$ $\dfrac{d}{dx}$ $x^2 + e^{ax^2}$

b. $\dfrac{\partial^2}{\partial y^2}$ $\dfrac{\partial}{\partial x}$ $(\cos 3y)\sin^2 x$

c. $\dfrac{\partial}{\partial\theta}$ $\dfrac{\partial^2}{\partial\phi^2}$ $\dfrac{\cos\phi}{\sin\theta}$

P2.25 Make the three polynomial functions a_0, $a_1 + b_1x$, and $a_2 + b_2x + c_2x^2$ orthonormal in the interval $-1 \le x \le +1$ by determining appropriate values for the constants a_0, a_1, b_1, a_2, b_2, and c_2.

P2.26 Consider a two-level system with $\varepsilon_1 = 3.10 \times 10^{-22}$ J and $\varepsilon_2 = 6.10 \times 10^{-21}$ J. If $g_2 = g_1$, what value of T is required to obtain $n_2/n_1 = 0.225$? What value of T is required to obtain $n_2/n_1 = 0.875$?

P2.27 Find the result of operating with $(1/r^2)(d/dr)$ $(r^2\,d/dr + 2/r)$ on the function Ae^{-br}. What must the values of A and b be to make this function an eigenfunction of the operator?

P2.28 Normalize the set of functions $\phi_n(\theta) = e^{in\theta}$, $0 \le \theta \le 2\pi$. To do so, you need to multiply the functions by a normalization constant N so that the integral $NN^*\int_0^{2\pi}\phi_m^*(\theta)\phi_n(\theta)\,d\theta = 1$ for $m = n$.

P2.29 In normalizing wave functions, the integration is over all space in which the wave function is defined. The following examples allow you to practice your skills in two- and three-dimensional integration.

a. Normalize the wave function $\sin(n\pi x/a)\sin(m\pi y/a)$ over the range $0 \le x \le a$, $0 \le y \le b$. The element of area in two-dimensional Cartesian coordinates is $dx\,dy$; n and m are integers and a and b are constants.

b. Normalize the wave function $e^{-(r/a)}\cos\theta\sin\phi$ over the interval $0 \le r < \infty$, $0 \le \theta \le \pi$, $0 \le \phi \le 2\pi$. The volume element in three-dimensional spherical coordinates is $r^2\sin\theta\,dr\,d\theta\,d\phi$ and a is a constant.

P2.30 Operate with (a) $\dfrac{\partial}{\partial x} + \dfrac{\partial}{\partial y} + \dfrac{\partial}{\partial z}$ and (b) $\dfrac{\partial^2}{\partial x^2} + \dfrac{\partial^2}{\partial y^2} + \dfrac{\partial^2}{\partial z^2}$ on the function $A\cos k_1x\cos k_2y\cos k_3z$. Is the function an eigenfunction of either operator? If so, what is the eigenvalue?

P2.31 Form the operator $\hat{A}^2$ if $\hat{A} = y + d/dy$. Be sure to include an arbitrary function on which the operator acts.

P2.32 Use a Fourier series expansion to express the function $f(x) = x$, $-b \le x \le b$, in the form

$$f(x) = d_0 + \sum_{n=1}^{m} c_n\sin\left(\frac{n\pi x}{b}\right) + d_n\cos\left(\frac{n\pi x}{b}\right)$$

Obtain d_0 and the first five coefficients c_n and d_n.

P2.33 Is the function $(16y^4 - 48y^2 + 12)e^{-y^2/2}$ an eigenfunction of the operator $-d^2/dy^2 + y^2$? If so, what is the eigenvalue?

P2.34 Show that the following pairs of wave functions are orthogonal over the indicated range.

a. $e^{(-1/2)\alpha x^2}$ and $(2\alpha x^2 - 1)e^{(-1/2)\alpha x^2}$, $-\infty \le x < \infty$ where α is a constant that is greater than zero

b. $(2 - r/a_0)e^{-r/2a_0}$ and $(r/a_0)e^{-r/2a_0}\cos\theta$ over the interval $0 \le r < \infty$, $0 \le \theta \le \pi$, $0 \le \phi \le 2\pi$

P2.35 Express the following complex numbers in the form $a + ib$.

a. $2e^{i\pi/2}$

b. $2\sqrt{5}\,e^{-i\pi/2}$

c. $e^{i\pi}$

d. $\dfrac{3\sqrt{2}}{5 + \sqrt{3}}e^{i\pi/4}$

P2.36 Which of the following wave functions are eigenfunctions of the operator d/dx? If they are eigenfunctions, what is the eigenvalue?

a. $ae^{-3x} + be^{-3ix}$ d. $\cos ax$

b. $\sin^2 x$ e. e^{-ix^2}

c. e^{-ix}

P2.37 Form the operator $\hat{A}^2$ if $\hat{A} = d^2/dy^2 + 3y(d/dy) - 5$. Be sure to include an arbitrary function on which the operator acts.

Web-Based Simulations, Animations, and Problems

W2.1   The motion of transverse, longitudinal, and surface traveling waves is analyzed by varying the frequency and amplitude.

W2.2   Two waves of the same frequency traveling in opposite directions are combined. The relative amplitude is changed with sliders and the relative phase of the waves is varied. The effect of these changes on the superposition wave is investigated.

W2.3   Two waves, both of which are standing waves, are combined. The effect of varying the wavelength, period, and phase of the waves on the resulting wave using sliders is investigated.

W2.4   Several functions are approximated by a Fourier series in which the number of sine and cosine terms is varied. The degree to which the approximate function differs from the exact function is assessed.

3

The Quantum Mechanical Postulates

Quantum mechanics can be formulated in terms of six postulates. These postulates provide a convenient framework for summarizing the basic concepts of quantum mechanics. The quantum mechanical postulates have been extensively tested since they were proposed in the 1930s. No case has been found in which they predict an outcome that is in conflict with the result of an experiment.

The previous chapters focused on the classical mechanics of particles and on the mathematical description of waves. Wave-particle duality and the conditions under which the wave character of a particle (which is always present) becomes evident were discussed. We briefly discussed quantum mechanical wave functions and quantum mechanical operators and showed that values for the **observables** are obtained by operating on the wave function with the relevant operator. The rules for how information is obtained from wave functions can be summarized in a few **postulates**. The test of any set of postulates is their ability to explain the world around us.

In this chapter, five postulates are stated and explained. In the following chapters, we apply these postulates to model systems and compare the results with those obtained from classical mechanics. In Chapter 10, a sixth postulate is introduced.

3.1 THE PHYSICAL MEANING ASSOCIATED WITH THE WAVE FUNCTION

POSTULATE 1

The state of a quantum mechanical particle is completely specified by a wave function $\Psi(x, t)$. The probability that the particle will be found at time t_0 in a spatial interval of width dx centered at x_0 is given by $\Psi^*(x_0, t_0)\Psi(x_0, t_0)\, dx$.

For a sound wave, the **wave function** $\Psi(x, t)$ is associated with the pressure at a time t and position x. For a water wave, $\Psi(x, t)$ is the height of the wave. What meaning does $\Psi(x, t)$ have as a solution of the Schrödinger equation? For a particle (which also has wave character), the probability $P(x_0, t_0)$ of finding the particle at position x_0 at time t_0 within an interval dx is

$$P(x_0, t_0) = \Psi^*(x_0, t_0)\Psi(x_0, t_0)\, dx = |\Psi(x_0, t_0)|^2\, dx \tag{3.1}$$

Unlike the classical mechanics of wave motion, the wave amplitude $\Psi(x, t)$ itself has no physical meaning in quantum mechanics. Because the probability is related to the *square of the magnitude* of $\Psi(x, t)$, given by $\Psi^*(x, t)\Psi(x, t)$, the wave function can be complex or negative and still be associated with a probability that lies between zero and one. The wave amplitude $\Psi(x, t)$ can be multiplied by -1, or its phase can be changed by multiplying it by a complex number of magnitude one such as $e^{i\theta}$, without changing $\Psi^*(x, t)\Psi(x, t)$. Therefore, all wave functions with a different phase angle θ are indistinguishable in that they generate the same observables. The wave function is a complete description of the system in that any measurable property (observable) can be obtained from the wave function as will be described later.

The association of the wave function with the probability has an important consequence, called **normalization**. The probability that the particle is found in an interval of width dx centered at the position x must lie between zero and one. The sum of the probabilities over all intervals accessible to the particle is one, because the particle is somewhere in its range. Consider a particle that is confined to a one-dimensional space of infinite extent. The requirement that the particle is somewhere in the interval leads to the following normalization condition:

$$\int_{-\infty}^{\infty} \Psi^*(x, t)\Psi(x, t)\, dx = 1 \tag{3.2}$$

Such a definition is obviously meaningless if the integral does not exist. Therefore, $\Psi(x, t)$ must satisfy several mathematical conditions to ensure that it represents a possible physical state. These conditions are as follows:

- The wave function must be a **single-valued function** of the spatial coordinates. If this were not the case, a particle would have more than one probability of being found in the same interval. For example, for the ellipse depicted in Figure 3.1a, $f(x)$ has two values for each value of x except the two points at which the tangent line is vertical. If only the part of the ellipse is considered for which $f(x) < 1$, as in Figure 3.1b, $f(x)$ has only one value for each value of x.

- The second derivative must exist and be well behaved. If this were not the case, we could not set up the Schrödinger equation. This is not the case if the wave function and/or its first derivative are discontinuous. As shown in Figure 3.2, $\sin x$ is a **continuous function** of x, but $\tan x$ has discontinuities at values of x for which the function becomes infinite.

- The wave function cannot have an infinite amplitude over a finite interval. If this were the case, the wave function can not be normalized. For example, the function $\Psi(x, t) = e^{-i(E/\hbar)t}\dfrac{1}{x^2}\sin\dfrac{2\pi x}{a}$, $0 \le x \le a$ cannot be normalized.

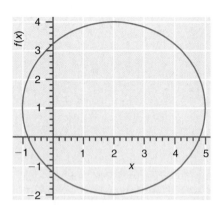

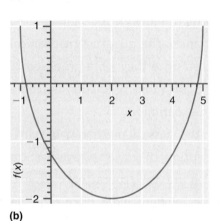

FIGURE 3.1

(a) $f(x)$ has two values for nearly all values of x and **(b)** $f(x)$ has only one value for each value of x.

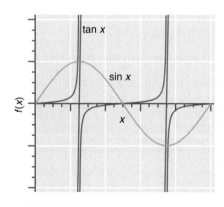

FIGURE 3.2

Examples of continuous and discontinuous functions.

3.2 EVERY OBSERVABLE HAS A CORRESPONDING OPERATOR

POSTULATE 2

For every measurable property of a classical mechanical system such as position, momentum, and energy, there exists a corresponding operator in quantum mechanics. An experiment in the laboratory to measure a value for such an observable is simulated in the theory by operating on the wave function of the system with the corresponding operator.

All quantum mechanical **operators** belong to a mathematical class called **Hermitian operators** that have real eigenvalues. For a Hermitian operator $\hat{A}$, $\int \psi^*(x)\left[\hat{A}\psi(x)\right] dx = \int \psi(x)\left[\hat{A}\psi(x)\right]^* dx$. The most important observables in classical mechanics, the corresponding quantum mechanical operators, and the symbols for these operators are listed in Table 3.1. To simplify the notation, only one spatial coordinate is considered, except for angular momentum. Partial derivatives have been retained because the wave function depends on both position and time. The total energy operator is called the Hamiltonian for historical reasons and is given the symbol $\hat{H}$. For the position and potential energy operators, the operation is "multiply on the left by the position or potential energy." Operators act on a wave function from the left, and the order of operation is important. For example, $\hat{p}_x\hat{x}$ operating on the wave function $\sin x$ gives $-i\hbar(\sin x + x \cos x)$, whereas $\hat{x}\hat{p}_x$ operating on the same wave function gives $-i\hbar x \cos x$. As discussed in Chapter 6, operators for which the order is unimportant have a particular role in quantum mechanics.

TABLE 3.1   OBSERVABLES AND THEIR QUANTUM MECHANICAL OPERATORS

Observable	Operator	Symbol for Operator
Momentum	$-i\hbar\dfrac{\partial}{\partial x}$	$\hat{p}_x$
Kinetic energy	$-\dfrac{\hbar^2}{2m}\dfrac{\partial^2}{\partial x^2}$	$\hat{E}_{kinetic} = \dfrac{1}{2m}(\hat{p}_x)\hat{p}_x$
Position	x	$\hat{x}$
Potential energy	$V(x)$	$\hat{E}_{potential}$
Total energy	$-\dfrac{\hbar^2}{2m}\dfrac{\partial^2}{\partial x^2} + V(x)$	$\hat{H}$
Angular momentum	$-i\hbar\left(y\dfrac{\partial}{\partial z} - z\dfrac{\partial}{\partial y}\right)$	$\hat{l}_x$
	$-i\hbar\left(z\dfrac{\partial}{\partial x} - x\dfrac{\partial}{\partial z}\right)$	$\hat{l}_y$
	$-i\hbar\left(x\dfrac{\partial}{\partial y} - y\dfrac{\partial}{\partial x}\right)$	$\hat{l}_z$

3.3 THE RESULT OF AN INDIVIDUAL MEASUREMENT

POSTULATE 3

In any single measurement of the observable that corresponds to the operator $\hat{A}$, the only values that will ever be measured are the eigenvalues of that operator.

This postulate states, for example, that if the energy of the hydrogen atom is measured, the only values obtained are the energies that are the **eigenvalues** of the time-independent Schrödinger equation:

$$\hat{H}\Psi_n(x,t) = E_n\Psi_n(x,t) \tag{3.3}$$

This makes sense because the energy levels of the hydrogen atom are discrete and, therefore, only those energies are allowed. What gives pause for thought is that the wave function need not be an **eigenfunction** of $\hat{H}$ because the eigenfunctions are a subset of the infinite number of functions that satisfy all the requirements to be an acceptable wave function. We address this issue in the following postulate.

3.4 THE EXPECTATION VALUE

POSTULATE 4

If the system is in a state described by the wave function $\Psi(x,t)$, and the value of the observable a is measured once each on many identically prepared systems, the average value (also called the **expectation value**) of all of these measurements is given by

$$\langle a \rangle = \frac{\displaystyle\int_{-\infty}^{\infty} \Psi^*(x,t)\hat{A}\Psi(x,t)\,dx}{\displaystyle\int_{-\infty}^{\infty} \Psi^*(x,t)\Psi(x,t)\,dx} \tag{3.4}$$

For the special case in which $\Psi(x,t)$ is normalized, the denominator in this expression has the value one. Wave functions are usually normalized and in Equations (3.5) through (3.8), this is assumed to be the case. This postulate requires some explanation. As we know, two cases apply with regard to $\Psi(x,t)$: it either is or is not an eigenfunction of the operator $\hat{A}$. These two cases need to be examined separately.

In the first case, $\Psi(x,t)$ is a normalized eigenfunction of $\hat{A}$, for example, $\phi_j(x,t)$. Because $\hat{A}\phi_j(x,t) = a_j\phi_j(x,t)$,

$$\langle a \rangle = \int_{-\infty}^{\infty} \phi_j^*(x,t)\hat{A}\phi_j(x,t)\,dx = a_j \int_{-\infty}^{\infty} \phi_j^*(x,t)\phi_j(x,t)\,dx$$
$$= a_j \tag{3.5}$$

If $\Psi(x,t)$ is $\phi_j(x,t)$, all measurements will give the same answer, namely, a_j.

Now consider the second case, in which $\Psi(x,t)$ is not an eigenfunction of the operator $\hat{A}$. Because the eigenfunctions of $\hat{A}$ form a complete set, $\Psi(x,t)$ can be expanded in terms of these eigenfunctions:

$$\Psi(x,t) = \sum_n b_n\phi_n(x,t) \tag{3.6}$$

Because $\Psi(x,t)$ is normalized, $\sum_m b_m^*b_m = \sum_m |b_m|^2 = 1$. The expression for $\Psi(x,t)$ in Equation (3.6) can be inserted in Equation (3.4), giving

$$\langle a \rangle = \int \Psi^*(x,t)\hat{A}\Psi(x,t)dx$$
$$= \int_{-\infty}^{\infty} \left[\sum_{m=1}^{\infty} b_m^*\phi_m^*(x,t)\right]\left[\sum_{n=1}^{\infty} a_n b_n\phi_n(x,t)\right]dx$$
$$= \sum_{m=1}^{\infty}\sum_{n=1}^{\infty} b_m^*b_n a_n \int_{-\infty}^{\infty} \phi_m^*(x,t)\phi_n(x,t)\,dx \tag{3.7}$$

This complicated expression can be greatly simplified by making use of the property that the eigenfunctions of a quantum mechanical operator are orthogonal. Because the eigenfunctions of $\hat{A}$ form an **orthonormal set**, the only terms in this double sum for which the integral is nonzero are those for which $m = n$. The integral has the value one for these terms, because the eigenfunctions of $\hat{A}$ are normalized. Therefore,

$$\langle a \rangle = \sum_{m=1}^{\infty} a_m b_m^* b_m = \sum_{m=1}^{\infty} |b_m|^2 a_m \qquad (3.8)$$

What are the b_m? They are the **expansion coefficients** of the wave function in the complete set of the eigenfunctions of the operator $\hat{A}$. The coefficient b_m is a measure of the extent to which the wave function "looks like" the mth eigenfunction of the operator $\hat{A}$. To illustrate this point, consider the case in which $\Psi(x, t) = \phi_n(x, t)$. In this case, all of the b_m except the one value corresponding to $m = n$ are zero and $\langle a \rangle = a_n$. So if $\phi(x, t)$ is one of the eigenfunctions of $\hat{A}$, only one of the b_m is nonzero and the average value is just the eigenvalue corresponding to that eigenfunction. If only three of the b_m are nonzero, for example b_2, b_8, and b_{11}, then $b_2^2 + b_8^2 + b_{11}^2 = 1$ and $\langle a \rangle$ is given by

$$\langle a \rangle = |b_2|^2 a_2 + |b_8|^2 a_8 + |b_{11}|^2 a_{11} \qquad (3.9)$$

Note that $\langle a \rangle$ is not simply an average of these three eigenvalues; instead, it is a **weighted average**. The weighting factor $|b_m|^2$ is directly related to the contribution of the mth eigenfunction to the wave function $\Psi(x, t)$.

The fourth postulate states what will be measured in a large number of measurements, each carried out only once on a large number of identically prepared systems. What will be measured in each of these individual measurements? The third postulate says that the only possible result of a single measurement is one of the eigenvalues a_n. However, it does not tell us which of the a_n will be measured. The answer is that there is no way of knowing the outcome of an **individual measurement**, and that the outcomes from identically prepared systems are not the same. This is a sharp break with the predictability we have come to depend on in classical mechanics.

Consider a hypothetical example. Suppose that a single hydrogen atom could be isolated in a box and the electronic wave function prepared such that it is in a superposition of the ground state, in which the electron is in the $1s$ orbital, and the excited states, in which the electron is in the $2s$, $2p_x$, and $3s$ orbitals. Assume that the wave function for this **superposition state** is

$$\Psi_{electronic} = b_1 \Psi_{1s} + b_2 \Psi_{2s} + b_3 \Psi_{2p_x} + b_4 \Psi_{3s} \qquad (3.10)$$

An example of a superposition state is the particle in the double-slit experiment going through both slits simultaneously. We now prepare a large number of these systems, each of which has the same wave function, and carry out a measurement of the total energy of the atom. The results that would be obtained are illustrated in Figure 3.3. Even though the systems are identical, the same value is not obtained for the energy of the atom in each measurement.

$$\Psi_{electronic} = b_1 \Psi_{1s} + b_2 \Psi_{2s} + b_3 \Psi_{2p_x} + b_4 \Psi_{3s}$$

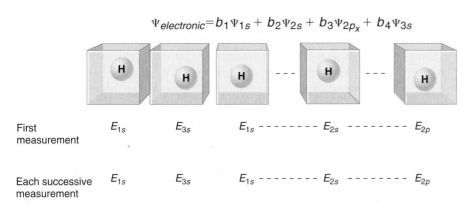

First measurement E_{1s} E_{3s} E_{1s} - - - - - - - E_{2s} - - - - - - - E_{2p}

Each successive measurement E_{1s} E_{3s} E_{1s} - - - - - - - E_{2s} - - - - - - - E_{2p}

FIGURE 3.3

A large number of identically prepared systems consist of a single hydrogen atom in a three-dimensional box. The atom is described by the superposition state $\Psi_{electronic} = b_1 \Psi_{1s} + b_2 \Psi_{2s} + b_3 \Psi_{2p_x} + b_4 \Psi_{3s}$. Consider a hypothetical experiment that measures the total energy and is completed in such a short time that transitions to the ground state can be neglected. The result of the first measurement on each system is probabilistic, whereas successive measurements are deterministic.

More generally, the particular value observed in one measurement could be any one of the eigenvalues a_n for which the corresponding b_n is nonzero. This is a **probabilistic outcome**, similar to asking what the chance is of rolling a six with one throw of a die. In this more familiar case, there is no way to predict the outcome of a single throw. However, if the die is thrown a large number of times, the six will land facing up a proportion of times that almost always approaches 1/6. The equivalent case to the die for our wave function is that all of the coefficients b_m have the same value. In the particular case under consideration, we have only four nonzero coefficients and, therefore, we will only measure one of the values E_{1s}, E_{2s}, E_{2p}, or E_{3s} in an individual measurement, but we have no way of knowing which of these values we will obtain. The certainty that we are familiar with from classical mechanics—that identically prepared systems all have the same outcomes in a measurement—is replaced in quantum mechanics by the probabilistic outcome just described.

More can be said about the outcome of a large number of measurements than about the outcome of a single measurement. Consider the more general result stated in Equation (3.8): the average number of a large number of measurements carried out once on identically prepared systems is given by a sum containing the possible eigenvalues of the operator weighted by $|b_m|^2$, the square of the expansion coefficient. The bigger the contribution of an eigenfunction $\phi_m(x, t)$ of $\hat{A}$ to $\Psi(x, t)$ (larger $|b_m|^2$), the more probable it is that the outcome of an individual measurement will be a_m and the more a_m will influence the average value $\langle a \rangle$. There is no way to predict which of the a_m will be found in an individual measurement. However, if this same experiment is repeated many times on identical systems, the **average value** can be predicted with very high precision. It is important to realize that this is not a shortcoming of how the "identical" systems were prepared. These systems are identical in every way and there is no reason to believe that we have left something out that resulted in this probabilistic result. Worked-out problems illustrating these concepts will be presented in the next chapter.

To illustrate the preceding discussion, consider the three different normalized superposition wave functions shown in Figure 3.4. They are made of the normalized eigenfunctions $\phi_1(x)$, $\phi_2(x)$, and $\phi_3(x)$ of the operator $\hat{A}$ with eigenvalues a_1, $4a_1$, and $9a_1$, respectively. The superposition wave functions are the following combinations of $\phi_1(x)$, $\phi_2(x)$, and $\phi_3(x)$:

$$\psi_1(x) = \frac{\sqrt{11}}{4}\phi_1(x) + \frac{1}{4}\phi_2(x) + \frac{1}{2}\phi_3(x)$$

$$\psi_2(x) = \frac{1}{2}\phi_1(x) + \frac{1}{4}\phi_2(x) + \frac{\sqrt{11}}{4}\phi_3(x)$$

$$\psi_3(x) = \frac{1}{2}\phi_1(x) + \frac{\sqrt{11}}{4}\phi_2(x) + \frac{1}{4}\phi_3(x) \quad \text{(3.11)}$$

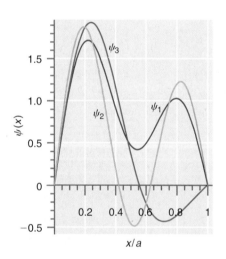

FIGURE 3.4

The three normalized wave functions $\psi_1(x)$, $\psi_2(x)$, and $\psi_3(x)$ (blue, yellow, and red curves, respectively) are defined over the interval $0 \leq x \leq a$. The amplitude of the wave functions is zero at both ends of the interval.

An individual measurement of the observable a gives only one of the values a_1, $4a_1$, and $9a_1$ regardless of which wave function describes the system. However, the probability of observing these values depends on the system wave function. For example, the probability of observing the value $9a_1$ is given by the square of the magnitude of the coefficient of $\phi_3(x)$ and is 1/4, 11/16, and 1/16, respectively, depending on whether the state is described by $\psi_1(x)$, $\psi_2(x)$, or $\psi_3(x)$.

You may have noticed that the postulate specified that the measurement was to be carried out only once each on a large number of identically prepared systems. What lies behind this requirement? We have just learned that the first measurement will give one of the eigenvalues of the operator corresponding to the observable being measured. We have also learned that we have no way to predict the outcome of a single measurement. What would you expect if a second measurement of the same observable were carried out on the same system? The experimentally established answer to this question is illustrated in Figure 3.3. In successive measurements on the same system, exactly the same result will be obtained that was obtained in the first experiment. If further successive measurements are carried out, all of the results will be the same. The probabilistic result is obtained only on the first measurement; after that, the result is deterministic.

How can this transition from a probabilistic to a **deterministic outcome** be understood? Note that the second and all successive results are exactly what would be expected if the system were in a particular eigenstate of the operator for which only one coefficient b_m is nonzero, namely, Ψ_{1s} or Ψ_{2s} or Ψ_{2p_x} or Ψ_{3s}, and not in the original superposition state $\Psi_{electronic} = b_1\Psi_{1s} + b_2\Psi_{2s} + b_3\Psi_{2p_x} + b_4\Psi_{3s}$. In fact, this is the key to understanding this very puzzling result. The act of carrying out a quantum mechanical measurement appears to convert the wave function of a system to an eigenfunction of the operator corresponding to the measured quantity! We are accustomed to thinking of our role in carrying out a measurement in classical mechanics as being passive. We simply note what the system is doing and it in turn takes no notice of us. The **measurement process** in quantum mechanics is radically different. In fact, the standard interpretation of quantum mechanics attributed to the school of Niels Bohr gives the measurement process a central role in the outcome of the experiment. This has vexed many scientists, most notably Albert Einstein. Applying this reasoning to the macroscopic world, he remarked to a colleague "Do you really think that the moon is not there when we are not looking at it?" However strange this may all seem, no one has devised an experiment to show that the view of the measurement process in quantum mechanics stated in this postulate is incorrect.

Assume now that the superposition state that describes the system is not known. This is generally the case for a real system. Can we determine the wave function from measurements like those shown in Figure 3.3? By measuring the frequency with which a particular eigenvalue is measured, the various $|b_m|^2$ can be determined. However, this only allows b_m to be determined to within a multiplicative factor $e^{i\theta}$ in which $0 < \theta < 2\pi$. Unfortunately, this does not provide enough information to reconstruct the wave function from experimental measurements. This is a general result; the wave function of a superposition state cannot be determined by any experimental means.

3.5 THE EVOLUTION IN TIME OF A QUANTUM MECHANICAL SYSTEM

POSTULATE 5

The evolution in time of a quantum mechanical system is governed by the time-dependent Schrödinger equation:

$$\hat{H}\Psi(x, t) = i\hbar\frac{\partial\Psi(x, t)}{\partial t} \qquad (3.12)$$

In this case, the total energy operator is given by $\hat{H} = (-\hbar^2/2m)(\partial^2/\partial x^2) + V(x, t)$. This looks like more familiar territory in that the equation has a unique solution for a set of given initial conditions. We call this behavior *deterministic* (like Newton's second law) in contrast to the probabilistic nature of Postulate 4. The fourth and fifth postulates are not contradictory. If a measurement is carried out at time t_0, Postulate 4 applies. If we ask what state the system will be in for a time $t_1 > t_0$, *without carrying out a measurement in this time interval*, Postulate 5 applies. If at time t_1, we carry out a measurement again, Postulate 4 will apply.

Note that for wave functions that are solutions of the time-independent Schrödinger equation, $\Psi(x, t) = \psi(x)e^{-i(E/\hbar)t}$. In this case, in solving the eigenvalue equation for any operator $\hat{A}$ that is not a function of time, we can write

$$\hat{A}(x)\Psi_n(x, t) = a_n\Psi_n(x, t)$$
$$\hat{A}(x)\psi_n(x)e^{-i(E/\hbar)t} = a_n\psi_n(x)e^{-i(E/\hbar)t} \quad \text{or}$$
$$\hat{A}(x)\psi_n(x) = a_n\psi_n(x) \qquad (3.13)$$

This means that eigenvalue equations can be written for $\hat{A}(x)$ using only the spatial part of the wave function $\psi(x)$, knowing that $\psi(x)$ and $\Psi(x, t)$ are related by $\Psi(x, t) = \psi(x)e^{-i(E/\hbar)t}$.

Vocabulary

average value

continuous function

deterministic outcome

eigenfunction

eigenvalue

expansion coefficient

expectation value

Hermitian operator

individual measurement

measurement process

normalization

observable

operator

orthonormal set

postulate

probabilistic outcome

single-valued function

superposition state

wave function

weighted average

4

Using Quantum Mechanics on Simple Systems

The framework described in Chapters 2 and 3 is used to solve two problems in a quantum mechanical framework that are familiar from classical mechanics. Both involve the motion of a particle on which no forces are acting. In the first case, the particle is not constrained. In the second, it is constrained to move within the confines of a box, but has no other forces acting on it. We find that unlike in classical mechanics, where the energy spectrum is continuous and the particle is equally likely to be found anywhere in the box, the quantum mechanical particle in the box has a discrete energy spectrum and has preferred positions that depend on the quantum mechanical state.

4.1 THE FREE PARTICLE

The simplest classical system imaginable is the free particle, a particle in a one-dimensional space on which no forces are acting. We begin with

$$F = ma = m\frac{d^2x}{dt^2} = 0 \qquad (4.1)$$

This differential equation can be solved to obtain

$$x = x_0 + v_0t \qquad (4.2)$$

Verify that this is a solution by substitution in Equation (4.1). The initial position x_0 and initial velocity v_0 arise from the constants of integration. To give them explicit values, the boundary conditions of the problem, namely, the initial position and velocity, must be known.

How is this problem solved using quantum mechanics? The condition that no forces can be acting on the particle means that the potential energy is constant and independent of x and t. Therefore, we use the time-independent Schrödinger equation in one dimension,

$$-\frac{\hbar^2}{2m}\frac{d^2\psi(x)}{dx^2} + V(x)\psi(x) = E\psi(x) \tag{4.3}$$

to solve for the dependence of the wave function $\psi(x)$ on x. Whenever the potential energy, $V(x)$, is constant, we can choose to make it zero because there is no fixed reference point for the zero of potential energy and only changes in this quantity are measurable. The Schrödinger equation for this problem reduces to

$$\frac{d^2\psi(x)}{dx^2} = -\frac{2m}{\hbar^2}E\psi(x) \tag{4.4}$$

In words, $\psi(x)$ is a function that can be differentiated twice to return the same function multiplied by a constant. Equation (4.4) has two solutions and the most appropriate form of these solutions (trigonometric or exponential) for our purposes are

$$\psi^+(x) = A_+e^{+i\sqrt{(2mE/\hbar^2)}x} = A_+e^{+ikx}$$

$$\psi^-(x) = A_-e^{-i\sqrt{(2mE/\hbar^2)}x} = A_-e^{-ikx} \tag{4.5}$$

in which the constants in the exponent have been combined using $k = 2\pi/\lambda = \sqrt{2mE/\hbar^2}$. We have been working with $\psi(x)$ rather than $\Psi(x, t)$. To obtain $\Psi(x, t)$, these two solutions are multiplied by $e^{-i(E/\hbar)t}$ or equivalently $e^{-i\omega t}$, where the relation $E = \hbar\omega$ has been used.

These solutions are plane waves, one moving to the right (positive x direction), the other moving to the left (negative x direction). The eigenvalues for the total energy can be found by substituting the wave functions of Equation (4.5) into Equation (4.4). For both solutions, $E = \hbar^2 k^2/2m$. Using the de Broglie relation (Section 1.5) to relate k with v, $k = mv/\hbar = \sqrt{2mE/\hbar^2}$. Because k is a constant, these wave functions represent waves moving at a constant velocity determined by their initial velocity. Therefore, the quantum mechanical solution of this problem contains the same information as the classical particle problem, namely, motion with a constant velocity. One other important similarity between the classical and quantum mechanical free particle is that both can take on all values of energy, because k is a continuous variable. The quantum mechanical free particle has a continuous energy spectrum. Why is this the case? We will learn the answer to this question in the next section of this chapter.

Of course, because a plane wave is not localized in space, we cannot speak of its position as was done for the particle. However, the **probability** of finding the particle in an interval of length dx can be calculated. The free-particle wave functions cannot be normalized over the interval $-\infty < x < \infty$, but if x is restricted to the interval $-L \leq x \leq L$ where L can be very large, the probability to find the particle described by $\psi^+(x)$ at position x in the interval dx is

$$P(x)\,dx = \frac{\psi^{+*}(x)\psi^+(x)\,dx}{\displaystyle\int_{-L}^{L}\psi^{+*}(x)\psi^+(x)\,dx} = \frac{A_+A_+e^{-ikx}e^{+ikx}\,dx}{A_+A_+\displaystyle\int_{-L}^{L}e^{-ikx}e^{+ikx}\,dx} = \frac{dx}{2L} \tag{4.6}$$

The same result is found for $\psi^-(x)$. The coefficients A_+ and A_- cancel because they appear in both the numerator and the denominator. Surprisingly, $P(x)\,dx$ is independent of x. This result states that the particle is equally likely to be anywhere in the interval, which is equivalent to saying that nothing is known about the position of the particle. As will be shown in Chapter 6, this result is linked to the fact that the momentum of the particle has been precisely specified to have the values $\hbar k$ and $-\hbar k$ for the wave functions $\psi^+(x) = A_+\exp\left[+i\sqrt{(2mE/\hbar^2)}\right]x$ and $\psi^-(x) = A_-\exp\left[-i\sqrt{(2mE/\hbar^2)}\right]x$, respectively. You can verify that the eigenfunctions of the total energy operator are also eigenfunctions of the momentum operator by applying the momentum operator to these total energy eigenfunctions.

4.2 THE PARTICLE IN A ONE-DIMENSIONAL BOX

The next case to be considered is the particle confined to a box. To keep the mathematics simple, the box is one dimensional; that is, it is the one-dimensional analog of a single atom moving freely in a cube that has impenetrable walls. Two- and three-dimensional boxes are dealt with in Section 4.3 and in the problems at the end of this chapter. The impenetrable walls are modeled by making the potential energy infinite outside of a region of width a. The potential is depicted in Figure 4.1.

$$V(x) = 0, \quad \text{for } a > x > 0$$
$$V(x) = \infty, \quad \text{for } x \geq a, x \leq 0 \tag{4.7}$$

How does this change in the potential affect the eigenfunctions that were obtained for the free particle? To answer this question, the Schrödinger equation is written in the following form:

$$\frac{d^2\psi(x)}{dx^2} = \frac{2m}{\hbar^2}[V(x) - E]\psi(x) \tag{4.8}$$

Outside of the box, where the potential energy is infinite, the second derivative of the wave function would be infinite if $\psi(x)$ were not zero for all x values outside the box. Because $d^2\psi(x)/dx^2$ must exist and be well behaved, $\psi(x)$ must be zero everywhere outside of the box including $x = 0$ and $x = a$. Moreover, because the wave function must be single valued at the boundaries of the box in order to be continuous,

$$\psi(0) = \psi(a) = 0 \tag{4.9}$$

Equation (4.9) lists **boundary conditions** that any well-behaved wave function for the one-dimensional box must satisfy.

Inside the box, where $V(x) = 0$, the Schrödinger equation is identical to that for a free particle [Equation (4.4)], so the solutions must be the same. For ease in applying the boundary conditions, the solution is written in a form equivalent to that of Equation (4.5):

$$\psi(x) = A \sin kx + B \cos kx \tag{4.10}$$

Now the boundary conditions given by Equation (4.9) are applied. Putting the values $x = 0$ and $x = a$ in Equation (4.10), we obtain

$$\psi(0) = 0 + B = 0$$
$$\psi(a) = A \sin ka = 0 \tag{4.11}$$

The first condition can only be satisfied by the condition that $B = 0$. The second condition can be satisfied if either $A = 0$ or if $ka = n\pi$ with n being an integer. Setting A equal to zero would mean that the wave function is always zero, which is unacceptable because then there is no particle in the interval. Therefore, we conclude that

$$\psi_n(x) = A \sin\left(\frac{n\pi x}{a}\right) \tag{4.12}$$

The requirement that $ka = n\pi$ will turn out to have important consequences for the energy spectrum of the particle in the box. The preceding discussion shows that acceptable wave functions for this problem must have the form

$$\psi_n(x) = A \sin\left(\frac{n\pi x}{a}\right), \quad \text{for } n = 1, 2, 3, 4, \ldots \tag{4.13}$$

Each different value of n corresponds to a different eigenfunction. To use the language learned in discussing operators, we have found the infinite set of eigenfunctions of the total energy operator for the potential energy defined by Equation (4.7).

the **Chemistry place**™

4.1 The Classical Particle in a Box

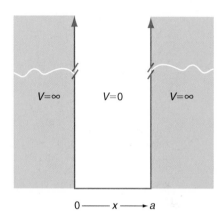

FIGURE 4.1

The potential described by Equation (4.7) is depicted. Because the particle is confined to the range $0 \leq x \leq a$, we say that it is confined to a one-dimensional box.

Note the undefined constant A in these equations. This constant can be determined by normalization, that is, by realizing that $\psi^*(x)\psi(x)\,dx$ represents the probability of finding the particle in the interval of width dx centered at x. Because the probability of finding the particle somewhere in the entire interval is one,

$$\int_0^a \psi^*(x)\,\psi(x)\,dx = A^*A\int_0^a \sin^2\left(\frac{n\pi x}{a}\right)dx = 1 \qquad (4.14)$$

This integral is evaluated using the standard integral

$$\int \sin^2(by)\,dy = \frac{y}{2} - \frac{\sin(2by)}{4b}$$

resulting in $A = \sqrt{2/a}$, so the normalized eigenfunctions are

$$\psi_n(x) = \sqrt{\frac{2}{a}}\sin\left(\frac{n\pi x}{a}\right) \qquad (4.15)$$

What are the energy eigenvalues that go with these eigenfunctions? Applying the total energy operator to the eigenfunctions will give back the eigenfunction multiplied by the eigenvalue. We find that

$$-\frac{\hbar^2}{2m}\frac{d^2\psi_n(x)}{dx^2} = \frac{\hbar^2}{2m}\left(\frac{n\pi}{a}\right)^2\sqrt{\frac{2}{a}}\sin\left(\frac{n\pi x}{a}\right) \qquad (4.16)$$

Because

$$-\frac{\hbar^2}{2m}\frac{d^2\psi_n(x)}{dx^2} = E_n\psi_n(x)$$

the following result is obtained:

$$E_n = \frac{\hbar^2}{2m}\left(\frac{n\pi}{a}\right)^2 = \frac{h^2n^2}{8ma^2}, \quad \text{for } n = 1, 2, 3, \ldots \qquad (4.17)$$

An important difference is seen when this result is compared to that obtained for the free particle. The energy for the particle in the box can only take on discrete values, and we say that the energy of the particle in the box is **quantized** and the integer n is a **quantum number**. Another important result of this calculation is that the lowest allowed energy is greater than zero. The particle has a nonzero minimum energy, known as a **zero point energy**.

Why are quite different results obtained for the free and the confined particle? A comparison of these two problems reveals that quantization entered through the confinement of the particle. Because the particle is confined to the box, the amplitude of all allowed wave functions must be zero everywhere outside the box. By considering the limit $a \to \infty$, the confinement condition is removed. Example Problem 4.1 shows that the discrete energy spectrum becomes continuous in this limit.

The lowest four energy levels for the particle in the box are shown in Figure 4.2 superimposed on an energy versus distance diagram. The eigenfunctions are also shown in this figure. Keep in mind that the time-independent part of the wave function is graphed. The full wave function is obtained by multiplying the wave functions shown in Figure 4.2 by $e^{-i(E/\hbar)t}$. If this is done, you will see that the variation of the total wave function with time is exactly what was shown in Figure 2.4 for a **standing wave**, if the real and imaginary parts of $\Psi(x, t)$ are considered separately. This result turns out to be general: the wave function for a stationary state is a standing and not a **traveling wave**. This result becomes clear by considering the boundary conditions for the particle in the box which state that the amplitude of the wave function is zero at the ends of the box for all times. A standing wave has nodes that are at fixed distances independent of time, whereas the nodes move in time for a traveling wave. For this reason, the boundary conditions cannot be satisfied for a traveling wave.

4.2 Energy Levels for the Particle in a Box

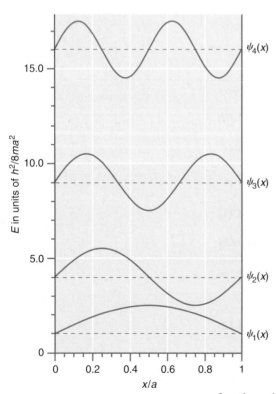

FIGURE 4.2

The first few eigenfunctions for the particle in a box are shown together with the corresponding energy eigenvalues. The energy scale is shown on the left. The wave function amplitude is shown on the right with the zero for each level indicated by the dashed line.

The particle in the box is also useful for showing that the quantization of the energy ultimately has its origin in the coupling of wave properties and boundary conditions. In moving from $\psi_n(x)$ to $\psi_{n+1}(x)$, the number of half-wavelengths, and therefore the number of **nodes**, has been increased by one. There is no way to add anything other than an integral number of half-wavelengths and still have nodes at the ends of the box. Therefore, the **wave vector** k will increase in discrete increments rather than continuously in going from one stationary state to another. Because

$$k = \frac{2\pi}{\lambda} = \frac{p}{\hbar} = \frac{\sqrt{2mE}}{\hbar}$$

the allowed energies, E, also increase in jumps rather than in a continuous fashion as in classical mechanics. Thinking in this way also helps to understand the origin of the zero point energy. Because $E = h^2/2m\lambda^2$, zero energy corresponds to an infinite wavelength. But the longest wavelength that has standing-wave nodes at the ends of the box is $\lambda = 2a$. Substituting this value in the equation for E gives exactly the zero point energy. Note that the zero point energy approaches zero as a approaches infinity. In this limit, the particle becomes free.

Looking at Equation (4.17), which shows the dependence of the total energy eigenvalues on the quantum number n, it is not immediately obvious that the energy spectrum will become continuous in the **classical limit** of very large n because the spacing between adjacent levels increases with n. This issue is addressed in Example Problem 4.1.

The total energy is one example of an observable that can be calculated once the eigenfunctions of the time-independent Schrödinger equation are known. Another observable that comes directly from solving this equation is the quantum mechanical analogue of position. Recall that the probability of finding the particle in any interval of width dx in the one-dimensional box is given by $\psi^*(x)\psi(x)\,dx$. The **probability density** $\psi^*(x)\psi(x)$ at a given point is shown in Figure 4.3 for the first few eigenfunctions.

How can these results be understood? Looking back at the discussion of waves in Chapter 2, recall that to ask for the position of a wave is not meaningful because the wave is not localized at a point. Wave-particle duality modifies the classical picture of being able to specify the location of a particle. Figure 4.3 shows the probability density of finding the particle in the vicinity of a given value of x rather than the position of that particle. We see that the probability of finding the particle outside of the box is zero, but that the probability of finding the particle within an interval dx in the box depends on the position and the wave function of the particle. Although $|\psi(x)|^2$ can be zero at nodal positions, $\int_{x-\Delta x}^{x+\Delta x} \psi^*(x')\psi(x')\,dx'$ is never zero for a finite interval Δx inside the box. This means that there is no finite length interval inside the box in which the particle is not found. For the ground state, it is much more likely that the particle is found near the center of the box than at the edges. A classical particle would be found with the same probability everywhere. Does this mean that quantum mechanics and classical mechanics are in conflict? No, because we need to consider large values of n to compare with the classical limit. However, a feature in Figure 4.3 that appears hard to understand is the oscillations in $\psi^2(x)$. They will not disappear for large n; they will just be spaced more closely together. Because there are no such oscillations for the classical case, we need to make the quantum oscillations disappear for very large n.

The way to understand the convergence to the classical limit is to consider the measurement process. Any measurement has a certain resolution that averages data over the resolution range. What effect will the limited resolution have on a measurement like probability? The result is shown in Figure 4.4. The probability density $\psi^2(x)$ is shown for the 1st, 30th, and 50th eigenstates of the particle in the box for three different limits of resolution. The probability density for the ground state is unaffected by including a resolution limit. However, as the resolution of the measurement decreases, we see that the probability density for the 50th state is beginning to approach the classical behavior of a constant probability everywhere. The classical limit is closer to $n = 1 \times 10^{10}$ rather than 50 for $E \approx kT$ at realistic temperatures and box dimensions on the order of

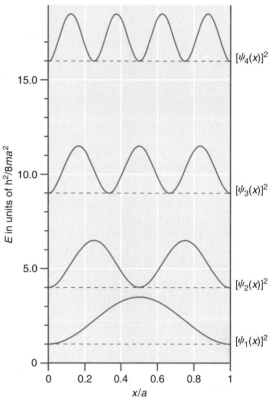

FIGURE 4.3

The square of the magnitude of the wave function, or probability density, is shown as a function of distance together with the corresponding energy eigenvalues. The energy scale is shown on the left. The square of the wave function amplitude is shown on the right with the zero for each level indicated by the dashed line.

4.3 Probability of Finding the Particle in a Given Interval

Higher resolution ⟶ Lower resolution

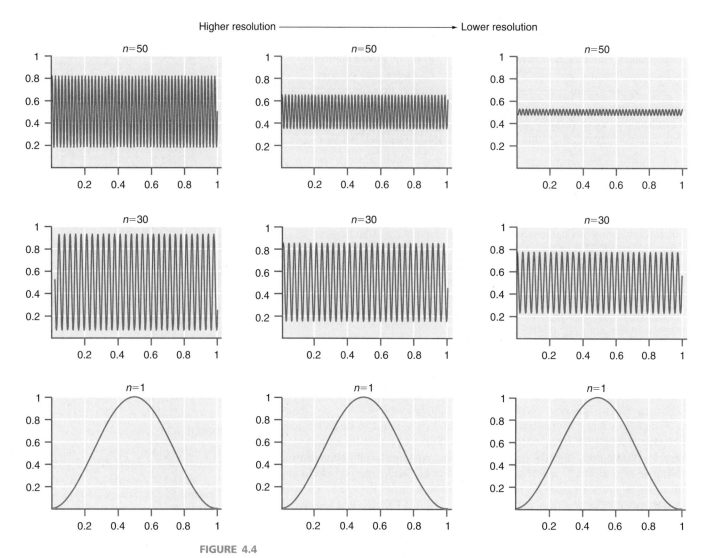

FIGURE 4.4

The three columns in this figure each show $\psi_n^2(x)/[\psi_1^2(x)]_{\max}$ as a function of x/a for $n = 1, 30$, and 50 (from bottom to top). In going from left to right, the data have been convoluted with an instrument function that averages the data over an increasingly wider range. Note that the probability of finding the particle in an interval dx becomes increasingly independent of position as n increases for lower resolution.

centimeters. You can see that the difference between the quantum and classical results disappears as n becomes large.

This first attempt to apply quantum rather than classical mechanics to two familiar problems has led to several useful insights. By representing a wave-particle as a wave, familiar questions that can be asked in classical mechanics become inappropriate. An example is "Where is the particle at time t_0?" The appropriate question in quantum mechanics is "What is the probability of finding the particle at time t_0 in an interval of length dx centered at the position x_0?" For the free particle, we found that the relationship between momentum and energy is the same as in classical mechanics and that there are no restrictions on the allowed energy. Restricting the motion of a particle to a finite region on the order of its wavelength has a significant effect on many observables associated with the particle. We saw that the origin of the effect is the requirement that the amplitude of the wave function be zero at the ends of the box for all times. This requirement changes the eigenfunctions of the Schrödinger equation from the traveling waves of the free particle to standing waves. Only discrete values of the particle momentum are allowed because of the condition $ka = n\pi$, $n = 1, 2, 3, \ldots$. Because $E = \hbar^2 k^2/2m$, the particle can only have certain values for the energy, and these values are determined by the dimensions of the box. Wave-particle duality also leads to a nonuniform probability for finding the particle in the box.

EXAMPLE PROBLEM 4.1

From the formula given for the energy levels for the particle in the box, $E_n = h^2 n^2 / 8ma^2$ for $n = 1, 2, 3, \ldots$, we can see that the spacing between adjacent levels increases with n. This appears to indicate that the energy spectrum does not become continuous for large n, which must be the case for the quantum mechanical result to be identical to the classical result in the high-energy limit. A better way to look at the spacing between levels is to form the ratio $(E_{n+1} - E_n)/E_n$. Form this ratio and show that $\Delta E/E$ becomes a smaller fraction of the energy as $n \to \infty$. This result shows that the energy spectrum becomes continuous for large n.

Solution

$$\frac{E_{n+1} - E_n}{E_n} = \left(\frac{h^2 [(n+1)^2 - n^2]/8ma^2}{h^2 n^2 / 8ma^2} \right) = \frac{2n+1}{n^2}$$

which approaches zero as $n \to \infty$. Both the level spacing and the energy increase with n, but the energy increases faster (as n^2), making the energy spectrum appear to be continuous as $n \to \infty$.

4.3 TWO- AND THREE-DIMENSIONAL BOXES

The one-dimensional box is a useful model system because the conceptual simplicity allows the focus to be on the quantum mechanics rather than on the mathematics. As will be shown in Chapter 5, the particle in the box model is very useful for understanding a number of real-world physical systems. The extension of the formalism developed for the one-dimensional problem to two and three dimensions has several aspects that are of use in understanding topics such as the rotation of molecules and the electronic structure of atoms, which cannot be reduced to one-dimensional problems.

Our focus here is on the three-dimensional box because the reduction in dimensionality from three to two is straightforward. The potential energy is given by

$$V(x, y, z) = 0 \quad \text{for } 0 < x < a; \quad 0 < y < b; \quad 0 < z < c$$
$$= \infty \quad \text{otherwise} \tag{4.18}$$

As before, the amplitude of the eigenfunctions of the total energy operator is identically zero outside the box. Inside the box, the Schrödinger equation can be written as

$$-\frac{\hbar^2}{2m} \left(\frac{\partial^2}{\partial x^2} + \frac{\partial^2}{\partial y^2} + \frac{\partial^2}{\partial z^2} \right) \psi(x, y, z) = E\psi(x, y, z) \tag{4.19}$$

This differential equation is solved assuming that $\psi(x, y, z)$ has the form

$$\psi(x, y, z) = X(x)Y(y)Z(z) \tag{4.20}$$

in which $\psi(x, y, z)$ is the product of three functions, each of which depends on only one of the variables. The assumption is valid in this case because $V(x, y, z)$ is independent of $x, y,$ and z inside the box. It is also valid for a potential of the form $V(x, y, z) = V_x(x) + V_y(y) + V_z(z)$. Substituting Equation (4.20) in Equation (4.19), we obtain

$$-\frac{\hbar^2}{2m} \left(Y(y)Z(z)\frac{d^2 X(x)}{dx^2} + X(x)Z(z)\frac{d^2 Y(y)}{dy^2} + X(x)Y(y)\frac{d^2 Z(z)}{dz^2} \right)$$
$$= EX(x)Y(y)Z(z) \tag{4.21}$$

Note that Equation (4.21) no longer contains partial derivatives because each of the three functions X, Y, and Z depends on only one variable. Dividing by the product $X(x)Y(y)Z(z)$ results in

$$-\frac{\hbar^2}{2m}\left[\frac{1}{X(x)}\frac{d^2X(x)}{dx^2} + \frac{1}{Y(y)}\frac{d^2Y(y)}{dy^2} + \frac{1}{Z(z)}\frac{d^2Z(z)}{dz^2}\right] = E \quad \textbf{(4.22)}$$

The form of this equation shows that E can be viewed as having independent contributions from the three coordinates, $E = E_x + E_y + E_z$, and the original differential equation in three variables reduces to three differential equations, each in one variable:

$$-\frac{\hbar^2}{2m}\frac{d^2X(x)}{dx^2} = E_x X(x); \quad -\frac{\hbar^2}{2m}\frac{d^2Y(y)}{dy^2} = E_y Y(y); \quad -\frac{\hbar^2}{2m}\frac{d^2Z(z)}{dz^2} = E_z Z(z)$$

$$\textbf{(4.23)}$$

Each of these equations has the same form as the equation that was solved for the one-dimensional problem. Therefore, the total energy eigenfunctions have the form

$$\psi_{n_x n_y n_z}(x, y, z) = N \sin\frac{n_x \pi x}{a} \sin\frac{n_y \pi y}{b} \sin\frac{n_z \pi z}{c} \quad \textbf{(4.24)}$$

and the total energy has the form

$$E = \frac{h^2}{8m}\left(\frac{n_x^2}{a^2} + \frac{n_y^2}{b^2} + \frac{n_z^2}{c^2}\right) \quad \textbf{(4.25)}$$

This is a general result. *If the total energy can be written as a sum of independent terms corresponding to different degrees of freedom, then the wave function is a product of individual terms, each corresponding to one of the degrees of freedom.*

Because this is a three-dimensional problem, the eigenfunctions depend on three quantum numbers. Because more than one set of the three quantum numbers may have the same energy [for example, (1, 2, 1), (2, 1, 1), and (1, 1, 2) if $a = b = c$], several distinct eigenfunctions of the total energy operator may have the same energy. In this case, we say that the energy level is **degenerate**, and the number of states that have the same energy is the **degeneracy** of the level.

What form do ψ and E take for the two-dimensional box? How many quantum numbers are needed to characterize ψ and E for the two-dimensional problem? Additional issues related to the functional form, degeneracy, and normalization of the total energy eigenfunctions are covered in the end-of-chapter problems.

We have made a considerable effort to understand the particle in the box, because this model is very useful in understanding properties that can be measured for real systems. Some of these systems will be discussed in Chapter 5. However, we first return to the postulates introduced in Chapter 3, now that the Schrödinger equation has been solved for an interesting system.

the Chemistry place

4.4 Eigenfunctions for the Two-Dimensional Box

4.4 USING THE POSTULATES TO UNDERSTAND THE PARTICLE IN THE BOX AND VICE VERSA

Because of its simplicity, the **particle in a box** is an excellent teaching tool for learning how to apply quantum mechanics to a relatively simple system. In this section, each of the postulates is applied to this problem using the eigenvalues and eigenfunctions calculated earlier. We begin with the first postulate.

POSTULATE 1:

The state of a quantum mechanical system is completely specified by a wave function $\Psi(x, t)$. The probability that a particle will be found at time t in a spatial interval of width dx centered at x_0 is given by $\Psi^*(x_0, t)\Psi(x_0, t)\, dx$.

This postulate states that all the information that can ever be obtained about the system is contained in the wave function. At this point it is useful to review the distinction between a wave function and an eigenfunction. A wave function is any mathematically well-behaved function that satisfies the boundary conditions and that can be normalized to allow a meaningful definition of probability. An eigenfunction must satisfy these and one more criterion. A wave function is an eigenfunction of an operator $\hat{A}$ only if it satisfies the relationship $\hat{A}\psi_n(x) = a_n\psi_n(x)$. These criteria are illustrated in Example Problem 4.2.

[handwritten margin note: difference between wave function and eigenfunction]

EXAMPLE PROBLEM 4.2

Consider the function $\psi(x) = c \sin(\pi x/a) + d \sin(2\pi x/a)$.

a. Is $\psi(x)$ an acceptable wave function for the particle in the box?

b. Is $\psi(x)$ an eigenfunction of the total energy operator, $\hat{H}$?

c. Is $\psi(x)$ normalized?

Solution

a. If $\psi(x)$ is to be an acceptable wave function, it must satisfy the boundary conditions $\psi(x) = 0$ at $x = 0$ and $x = a$. The first and second derivatives of $\psi(x)$ must also be well-behaved functions between $x = 0$ and $x = a$. This is the case for $\psi(x)$. We conclude that $\psi(x) = c \sin(\pi x/a) + d \sin(2\pi x/a)$ is an acceptable wave function for the particle in the box.

b. Although $\psi(x) = c \sin(\pi x/a) + d \sin(2\pi x/a)$ may be an acceptable wave function, it need not be an eigenfunction of a given operator. To see if $\psi(x)$ is an eigenfunction of the total energy operator, the operator is applied to the function:

$$-\frac{\hbar^2}{2m}\frac{d^2}{dx^2}\left(c \sin\left(\frac{\pi x}{a}\right) + d \sin\left(\frac{2\pi x}{a}\right)\right)$$

$$= \frac{\hbar^2\pi^2}{2ma^2}\left(c \sin\left(\frac{\pi x}{a}\right) + 4d \sin\left(\frac{2\pi x}{a}\right)\right)$$

The result of this operation is not $\psi(x)$ multiplied by a constant. Therefore, $\psi(x)$ is not an eigenfunction of the total energy operator.

c. To see if $\psi(x)$ is normalized, the following integral is evaluated:

$$\int_0^a \left|c \sin\left(\frac{\pi x}{a}\right) + d \sin\left(\frac{2\pi x}{a}\right)\right|^2 dx$$

$$= \int_0^a \left[c^*c \sin^2\left(\frac{\pi x}{a}\right) + d^*d \sin^2\left(\frac{2\pi x}{a}\right) + (cd^* + c^*d)\sin\left(\frac{\pi x}{a}\right)\sin\left(\frac{2\pi x}{a}\right)\right] dx$$

$$= \int_0^a |c|^2 \sin^2\left(\frac{\pi x}{a}\right) dx + \int_0^a |d|^2 \sin^2\left(\frac{2\pi x}{a}\right) dx$$

$$+ \int_0^a (cd^* + c^*d) \sin\left(\frac{\pi x}{a}\right) \sin\left(\frac{2\pi x}{a}\right) dx$$

Using the standard integral $\int \sin^2(by)dy = y/2 - (1/4b)\sin(2by)$ and recognizing that the third integral is zero because all sin nx functions with different n are orthogonal,

$$\int_0^a \left[|c|^2 \sin^2\left(\frac{\pi x}{a}\right) + |d|^2 \sin^2\left(\frac{2\pi x}{a}\right)\right] dx$$

$$= |c|^2\left[\frac{a}{2} - \frac{a(\sin 2\pi - \sin 0)}{4\pi}\right] + |d|^2\left[\frac{a}{2} - \frac{a(\sin 4\pi - \sin 0)}{8\pi}\right]$$

$$= \frac{a}{2}\left(|c|^2 + |d|^2\right)$$

Therefore, $\psi(x)$ is not normalized, but the function

$$\sqrt{\frac{2}{a}}\left[c\,\sin\left(\frac{\pi x}{a}\right) + d\,\sin\left(\frac{2\pi x}{a}\right)\right]$$

is normalized for the condition that $|c|^2 + |d|^2 = 1$.

Note that a superposition wave function has a more complicated dependence on time than does an eigenfunction of the total energy operator. For instance, $\psi(x,t)$ for the wave function under consideration is given by

$$\Psi(x,t) = \sqrt{\frac{2}{a}}\left[ce^{-iE_1 t/\hbar}\sin\left(\frac{\pi x}{a}\right) + de^{-iE_2 t/\hbar}\sin\left(\frac{2\pi x}{a}\right)\right] \neq \psi(x)f(t)$$

This wave function cannot be written as a product of a function of x and a function of t. It is not a standing wave and does not describe a state whose properties are, in general, independent of time.

4.5 Acceptable Wave Functions for the Particle in a Box

All of the particle in the box eigenfunctions, $\psi_n(x) = \sqrt{2/a}\,\sin(n\pi x/a)$, for $n = 1, 2, 3, \ldots$ are normalized, meaning that the total probability of finding the particle somewhere between $x = 0$ and $x = a$ is one. In other words, the particle is somewhere in the box. We cannot predict with certainty the outcome of a single measurement in which the position of the particle is determined, because these eigenfunctions of the total energy operator are not eigenfunctions of the position operator. In Chapter 6, we will discuss why the eigenvalues of $\hat{H}$ and $\hat{x}$ cannot be determined simultaneously. We can, however, predict the average value determined in a large number of independent measurements of the particle position. This is equivalent to asking for the probability density of finding the particle at a given position. The formula for calculating this probability is stated in the first postulate. The total probability of finding the particle in a finite length interval is obtained by integrating the probability density, as shown in Example Problem 4.3.

EXAMPLE PROBLEM 4.3

What is the probability, P, of finding the particle in the central third of the box if it is in its ground state?

Solution

For the ground state, $\psi_1(x) = \sqrt{2/a}\,\sin(\pi x/a)$. From the postulate, P is the sum of all the probabilities of finding the particle in intervals of width dx within the central third of the box. This probability is given by the integral

$$P = \frac{2}{a}\int_{a/3}^{2a/3}\sin^2\left(\frac{\pi x}{a}\right)dx$$

Solving this integral as in Example Problem 4.2,

$$P = \frac{2}{a}\left[\frac{a}{6} - \frac{a}{4\pi}\left(\sin\frac{4\pi}{3} - \sin\frac{2\pi}{3}\right)\right] = 0.609$$

Although we cannot predict the outcome of a single measurement, we can predict that for 60.9% of a large number of individual measurements, the particle is found in the central third of the box. What is the probability of finding a classical particle in this interval?

Postulate 2 is a recipe for associating classical observables with quantum mechanical operators and need not be considered further. Postulates 3 and 4 are best understood by considering them together.

POSTULATE 3:

In any single measurement of the observable that corresponds to the operator $\hat{A}$, the only values that will ever be measured are the eigenvalues of that operator.

POSTULATE 4:

If the system is in a state described by the wave function $\Psi(x, t)$, and the value of the observable a is measured once each on many identically prepared systems, the average value of all of these measurements is given by

$$\langle a \rangle = \frac{\int_{-\infty}^{\infty} \Psi^*(x,t) \hat{A} \Psi(x,t) \, dx}{\int_{-\infty}^{\infty} \Psi^*(x,t) \Psi(x,t) \, dx} \tag{4.26}$$

The wave function for particle in its ground state is $\psi(x) = \sqrt{2/a} \sin(\pi x/a)$, which is a normalized eigenfunction of the total energy operator. Applying the operator to this wave function returns the function multiplied by the constant . This is the value of the energy that is determined in any single measurement and, therefore, it is also the average of all values for the energy that are measured on many particles prepared in the same state.

Now consider a measurement of the total energy for a case in which the wave function of the system is not an eigenfunction of this operator. As you convinced yourself in Example Problem 4.2, the normalized wave function

$$\psi(x) = \sqrt{\frac{2}{a}} \left(c \sin\frac{\pi x}{a} + d \sin\frac{2\pi x}{a} \right)$$

where $|c|^2 + |d|^2 = 1$ is not an eigenfunction of $\hat{H}$. Postulate 4 says that the average value of the energy for a large number of identical measurements on a system whose state is described by a normalized wave function is

$$\langle E \rangle = \int_0^a \psi^*(x) \left[-\frac{\hbar^2}{2m} \frac{d^2}{dx^2} + V(x) \right] \psi(x) \, dx \tag{4.27}$$

We now substitute the expression for $\psi(x)$ into Equation (4.27):

$$\langle E \rangle = \frac{2}{a} \int_0^a \left(c^* \sin\frac{\pi x}{a} + d^* \sin\frac{2\pi x}{a} \right) \left[-\frac{\hbar^2}{2m} \frac{d^2}{dx^2} \right] \left(c \sin\frac{\pi x}{a} + d \sin\frac{2\pi x}{a} \right) \, dx \tag{4.28}$$

Multiplying out the terms in the brackets, and recognizing that each of the individual terms in the parentheses is an eigenfunction of the operator, $\langle E \rangle$ reduces to

$$\langle E \rangle = \frac{2}{a} \left[|c|^2 E_1 \int_0^a \sin^2\frac{\pi x}{a} \, dx + |d|^2 E_2 \int_0^a \sin^2\frac{2\pi x}{a} \, dx \right]$$

$$+ \frac{2}{a} \left[c^* d E_2 \int_0^a \sin\frac{\pi x}{a} \sin\frac{2\pi x}{a} \, dx + d^* c E_1 \int_0^a \sin\frac{\pi x}{a} \sin\frac{2\pi x}{a} \, dx \right] \tag{4.29}$$

We know the value of each of the first two integrals is $a/2$ from our efforts to normalize the functions. Each of the last two integrals is identically zero because the sine functions with different arguments are mutually orthogonal. Therefore, the result of these calculations is

$$\langle E \rangle = |c|^2 E_1 + |d|^2 E_2 \tag{4.30}$$

4.6 Expectation Values for E, p, and x for a Superposition Wave Function

where $E_n = n^2h^2/8ma^2$. Because $|c|^2 + |d|^2 = 1$, $\langle E \rangle$ is a weighted average of E_1 and E_2. As seen in Example Problem 4.2, the superposition wave function does not describe a stationary state, and the average values of observables such as $\langle p \rangle$ and $\langle x \rangle$ are functions of time as shown in Problem W4.6. However, the average energy is independent of time because the energy is conserved.

Note that this result is exactly what was derived for a more general case in discussing Postulate 4 (see Chapter 3). Now let's discuss in more detail what will be obtained for an individual measurement of the total energy and relate it to the result that was just derived for the average of many individual measurements. Postulate 3 says that in an individual measurement, only one of the eigenvalues of the operator can be measured. In this case, it means that only one of the infinite set of E_n given by $E_n = n^2h^2/8ma^2$, $n = 1, 2, 3, \ldots$, is a possible result of an individual measurement. What is the likelihood that the value E_2 will be measured? Postulate 4 gives a recipe for answering this question. It tells us to expand the system wave function in the complete set of functions that are the eigenfunctions of the operator of interest. The probability that an individual measurement will give E_n is given by the square of the expansion coefficient of that eigenfunction in the expression for the wave function. In the particular case under consideration, the wave function can be written as follows:

$$\psi(x) = c\psi_1(x) + d\psi_2(x) + 0 \times (\psi_3(x) + \ldots + \psi_n(x) + \ldots) \quad \textbf{(4.31)}$$

in which it has been made explicit that the coefficients of all the eigenfunctions other than $\psi_1(x)$ and $\psi_2(x)$ are zero. Therefore, given the wave function for the system, individual measurements on identically prepared systems will never give anything other than E_1 or E_2. The probability of obtaining E_1 is c^2 and the probability of obtaining E_2 is d^2. From this result, it is clear that the average value for the energy determined from a large number of measurements is $c^2E_1 + d^2E_2$.

A more detailed discussion of causality in quantum mechanics would lead us to a number of conclusions that differ significantly from our experience with classical mechanics. For instance, it is not possible to predict whether E_1 or E_2 would be measured in an individual measurement any more than the outcome of a single throw of a die can be predicted. However, if the energy is measured again on the same system (rather than carrying out a second measurement on an identically prepared system), the same result will be obtained as in the initial measurement. This conclusion also holds for all subsequent measurements. This last result is particularly intriguing because it suggests that through the measurement process, the system has been forced into an eigenfunction of the operator corresponding to the quantity being measured.

Now consider a measurement of the momentum or the position. As shown earlier, we need to know the wave function that describes the system to carry out such a calculation. For this calculation, assume that the system is in one of the eigenstates of the total energy operator, for which $\psi_n(x) = \sqrt{2/a} \sin(n\pi x/a)$. From the second postulate, and Table 3.1, the quantum mechanical operator associated with momentum is $-i\hbar(d/dx)$. Although $\psi(x)$ is an eigenfunction of the total energy operator, it is not clear if it is an eigenfunction of the momentum operator. Verify that it is not an eigenfunction of this operator by operating on the wave function with the momentum operator. We will return to this result in Chapter 6, but we first proceed in applying the postulates. Postulate 4 defines how the average value of the momentum obtained in a large number of individual measurements on an identically prepared system can be calculated. The result is given by

$$\langle p \rangle = \int_0^a \psi^*(x) \hat{p} \psi(x) \, dx$$

$$= \frac{2}{a} \int_0^a \sin\left(\frac{n\pi x}{a}\right)\left[-i\hbar\frac{d}{dx}\sin\left(\frac{n\pi x}{a}\right)\right] dx \quad \textbf{(4.32)}$$

$$= \frac{-2i\hbar n\pi}{a^2} \int_0^a \sin\left(\frac{n\pi x}{a}\right)\cos\left(\frac{n\pi x}{a}\right) dx = \frac{-i\hbar}{a}[\sin^2 n\pi - \sin^2 0] = 0$$

Note that the result is the same for all values of n. We know that the energy of the lowest state is greater than zero and that all the energy is in the form of kinetic energy. Because $E = p^2/2m \neq 0$, the magnitude of p must be greater than zero for an individual measurement. How can the result that the average value of the momentum is zero be understood?

Keep in mind that, classically, the particle is bouncing back and forth between the two walls of the one-dimensional box with a constant velocity. Therefore, it is equally likely that the particle is moving in the $+x$ and $-x$ directions and that its momentum is positive or negative. For this reason, the average momentum is zero. This result holds up in a quantum mechanical picture. However, a major difference exists between the quantum and classical pictures. In classical mechanics, the magnitude of the momentum of the particle is known to be $\sqrt{2m\,E_{kin}}$ exactly. In quantum mechanics, a consequence of confining the particle to a box of length a is that an uncertainty has been introduced in its momentum that is proportional to $1/a$. This issue will be discussed in depth in Chapter 6. The calculation for the average value of position is carried out in Example Problem 4.4.

EXAMPLE PROBLEM 4.4

Assume that a particle is confined to a box of length a, and that the system wave function is $\psi(x) = \sqrt{2/a}\,\sin(\pi x/a)$.

a. Is this state an eigenfunction of the position operator?

b. Calculate the average value of the position $\langle x \rangle$ that would be obtained for a large number of measurements. Explain your result.

Solution

a. The position operator $\hat{x} = x$. Because $x\,\psi(x) = \sqrt{2/a}\,x\,\sin(\pi x/a) \neq c\psi(x)$, where c is a constant, the wave function is not an eigenfunction of the position operator.

b. The expectation value is calculated using the fourth postulate:

$$\langle x \rangle = \frac{2}{a}\int_0^a \left\{\sin\left(\frac{\pi x}{a}\right)\right\} x \sin\left(\frac{\pi x}{a}\right)\,dx = \frac{2}{a}\int_0^a x \left\{\sin\left(\frac{\pi x}{a}\right)\right\}^2\,dx$$

Using the standard integral $\displaystyle\int x(\sin bx)^2\,dx = \frac{x^2}{4} - \frac{\cos 2bx}{8b^2} - \frac{x\sin 2bx}{4b}$

$$\langle x \rangle = \frac{2}{a}\left[\frac{x^2}{4} - \frac{\cos\left(\dfrac{2\pi x}{a}\right)}{8\left(\dfrac{\pi}{a}\right)^2} - \frac{x\sin\left(\dfrac{2\pi x}{a}\right)}{4\left(\dfrac{\pi}{a}\right)}\right]_0^a$$

$$= \frac{2}{a}\left[\left(\frac{a^2}{4} - \frac{a^2}{8\pi^2} - 0\right) + \frac{a^2}{8\pi^2}\right] = \frac{a}{2}$$

The average position is midway in the box. This is exactly what we would expect, because the particle is equally likely to be in each half of the box.

Vocabulary

boundary condition	particle in a box	standing waves
classical limit	probability	traveling waves
degeneracy	probability density	wave vector
degenerate	quantized	zero point energy
node	quantum number	

Questions on Concepts

Q4.1   We set the potential energy in the particle in the box equal to zero and justified it by saying that there is no absolute scale for potential energy. Is this also true for kinetic energy?

Q4.2   Discuss why a quantum mechanical particle in a box has a zero point energy in terms of its wavelength.

Q4.3   How does an expectation value for an observable differ from an average of all possible eigenvalues?

Q4.4   Is the probability distribution for a free particle consistent with a purely particle picture, a purely wave picture, or both?

Q4.5   Why is it not possible to normalize the free-particle wave functions over the whole range of motion of the particle?

Q4.6   The probability density for a particle in a box is an oscillatory function even for very large energies. Explain how the classical limit of a constant probability density that is independent of position is achieved.

Q4.7   Explain using words, rather than equations, why if $V(x, y, z) \neq V_x(x) + V_y(y) + V_z(z)$, the total energy eigenfunctions cannot be written in the form $\psi(x, y, z) = X(x)Y(y)Z(z)$.

Q4.8   Can a guitar string be in a superposition of states or is such a superposition only possible for a quantum mechanical system?

Q4.9   Show that for the particle in the box total energy eigenfunctions, $\psi_n(x) = \sqrt{2/a}\, \sin(n\pi x/a)$, $\psi(x)$ is a continuous function at the edges of the box. Is $d\psi/dx$ a continuous function of x at the edges of the box?

Q4.10   Why are standing-wave solutions for the free particle not compatible with the classical result $x = x_0 + v_0 t$?

Q4.11   What is the difference between probability and probability density?

Q4.12   Why are traveling-wave solutions for the particle in the box not compatible with the boundary conditions?

Q4.13   Can the particles in a one-dimensional box, a square two-dimensional box, and a cubic three-dimensional box all have degenerate energy levels?

Q4.14   Invoke wave-particle duality to address the following question: How does a particle get through a node in a wave function to get to the other side of the box?

Q4.15   Why is the zero point energy lower for a He atom in a box than for an electron?

Problems

Problem numbers in **red** indicate that the solution to the problem is given in the *Student's Solutions Manual*.

P4.1   This problem explores under what conditions the classical limit is reached for a macroscopic cubic box of edge length a. An argon atom of average translational energy 3/2 kT is confined in a cubic box of volume $V = 0.500 \text{ m}^3$ at 298 K. Use the result from Equation (4.25) for the dependence of the energy levels on a and on the quantum numbers n_x, n_y, and n_z.

a. What is the value of the "reduced quantum number" $\alpha = \sqrt{n_x^2 + n_y^2 + n_z^2}$ for $T = 298$ K?

b. What is the energy separation between the levels α and $\alpha + 1$? (*Hint*: Subtract $E_{\alpha+1}$ from E_α *before* plugging in numbers.)

c. Calculate the ratio $(E_{\alpha+1} - E_\alpha)/kT$ and use your result to conclude whether a classical or quantum mechanical description is appropriate for the particle.

P4.2   Calculate the expectation value $\langle x^2 \rangle$ for a particle in the state $n = 3$ moving in a one-dimensional box of length a. Is $\langle x^2 \rangle = \langle x \rangle^2$? Explain your answer.

P4.3   Normalize the total energy eigenfunctions for the three-dimensional box in the interval $0 \leq x \leq a$, $0 \leq y \leq b, 0 \leq z \leq c$.

P4.4   Is the superposition wave function for the free particle an eigenfunction of the momentum operator? Is it an eigenfunction of the total energy operator? Explain your result.

P4.5   Suppose that the wave function for a system can be written as

$$\psi(x) = \frac{1}{2}\phi_1(x) + \frac{1}{4}\phi_2(x) + \frac{3 + \sqrt{2}i}{4}\phi_3(x)$$

and that $\phi_1(x), \phi_2(x)$, and $\phi_3(x)$ are normalized eigenfunctions of the operator $\hat{E}_{kinetic}$ with eigenvalues E_1, $3E_1$, and $7E_1$, respectively.

a. Verify that $\psi(x)$ is normalized.

b. What are the possible values that you could obtain in measuring the kinetic energy on identically prepared systems?

c. What is the probability of measuring each of these eigenvalues?

d. What is the average value of $E_{kinetic}$ that you would obtain from a large number of measurements?

P4.6   Consider a free particle moving in one dimension whose probability of moving in the positive x direction is three times that for moving in the negative x direction. Give as much information as you can about the wave function of the particle.

P4.7   Consider the contour plots of Problem P4.19.

a. What are the most likely area or areas $\Delta x \Delta y$ to find the particle for each of the eigenfunctions of $\hat{H}$ depicted in plots a–f?

b. For the one-dimensional box, the nodes are points. What form do the nodes take for the two-dimensional box? Where are the nodes located in plots a–f? How many nodes are there in each contour plot?

P4.8 Evaluate the normalization integral for the eigenfunctions of $\hat{H}$ for the particle in the box $\psi_n(x) = A \sin(n\pi x/a)$ using the trigonometric identity $\sin^2 y = (1 - \cos 2y)/2$.

P4.9 The function $\psi(x) = A(x/a)^2[1 - (x/a)]$ is an acceptable wave function for the particle in the one-dimensional infinite depth box of length a. Calculate the normalization constant A and the expectation values $\langle x \rangle$ and $\langle x^2 \rangle$.

P4.10 What is the solution of the time-dependent Schrödinger equation $\Psi(x, t)$ for the total energy eigenfunction $\psi_4(x) = \sqrt{2/a} \sin(4\pi x/a)$ in the particle in the box model? Write $\omega = E/\hbar$ explicitly in terms of the parameters of the problem.

P4.11 Derive an equation for the probability that a particle characterized by the quantum number n is in the first 40.% $(0 \le x \le 0.40a)$ of an infinite depth box. Show that this probability approaches the classical limit as $n \to \infty$.

P4.12 It is useful to consider the result for the energy eigenvalues for the one-dimensional box $E_n = h^2 n^2/8ma^2$, $n = 1, 2, 3, \ldots$ as a function of n, m, and a.

a. By what factor do you need to change the box length to decrease the zero point energy by a factor of 275 for a fixed value of m?

b. By what factor would you have to change n for fixed values of a and m to increase the energy by a factor of 600.?

c. By what factor would you have to increase a at constant n to have the zero point energies of a He atom be equal to the zero point energy of a proton in the box?

P4.13 Show that the energy eigenvalues for the free particle, $E = \hbar^2 k^2/2m$, are consistent with the classical result $E = (1/2)mv^2$.

P4.14 a. Show by substitution into Equation (4.19) that the eigenfunctions of $\hat{H}$ for a box with lengths along the x, y, and z directions of a, b, and c, respectively, are

$$\psi_{n_x, n_y, n_z}(x, y, z) = N \sin\left(\frac{n_x \pi x}{a}\right) \sin\left(\frac{n_y \pi y}{b}\right) \sin\left(\frac{n_z \pi z}{c}\right)$$

b. Obtain an expression for E_{n_x, n_y, n_z} in terms of n_x, n_y, n_z, and a, b, and c.

P4.15 Calculate the wavelength of the light emitted when an electron in a one-dimensional box of length 3.0 nm makes a transition from the $n = 5$ state to the $n = 4$ state.

P4.16 A bowling ball has a weight of 15 lb and the length of the lane is approximately 60. feet. Treat the ball in the lane as a one-dimensional box. What quantum number corresponds to a velocity of 5.0 miles per hour?

P4.17 Using your result from P4.19, how many energy levels does a particle of mass m in a two-dimensional box of edge length a have with $E \le 25h^2/8ma^2$? What is the degeneracy of each level?

P4.18 Are the eigenfunctions of $\hat{H}$ for the particle in the one-dimensional box also eigenfunctions of the momentum operator $\hat{p}_x$? Calculate the average value of p_x for the case $n = 3$. Repeat your calculation for $n = 5$ and, from these two results, suggest an expression valid for all values of n. How does your result compare with the prediction based on classical physics?

P4.19 For a particle in a two-dimensional box, the total energy eigenfunctions are

$$\psi_{n_x n_y}(x, y) = N \sin\frac{n_x \pi x}{a} \sin\frac{n_y \pi y}{b}$$

a. Obtain an expression for E_{n_x, n_y} in terms of n_x, n_y, a, and b by substituting this wave function into the two-dimensional analog of Equation (4.19).

b. Contour plots of several eigenfunctions are shown here. The x and y directions of the box lie along the horizontal and vertical directions, respectively. The amplitude has been displayed as a gradation in colors. Regions of positive and negative amplitude are indicated. Identify the values of the quantum numbers n_x and n_y for plots a–f.

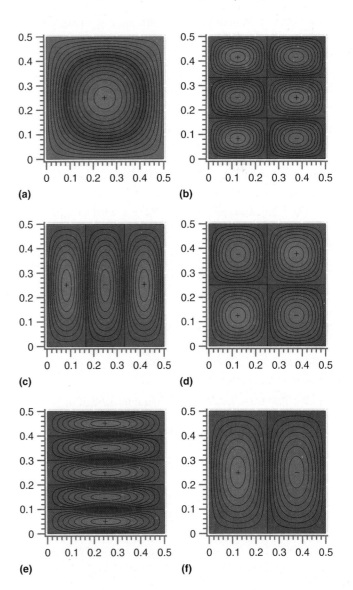

(a)

(b)

(c)

(d)

(e)

(f)

P4.20 Calculate (a) the zero point energy of a CO molecule in a one-dimensional box of length 1.00 cm and (b) the ratio of the zero point energy to kT at 300. K.

P4.21 Normalize the total energy eigenfunction for the rectangular two-dimensional box,

$$\psi_{n_x, n_y}(x, y) = N \sin\left(\frac{n_x \pi x}{a}\right) \sin\left(\frac{n_y \pi y}{b}\right)$$

in the interval $0 \le x \le a, 0 \le y \le b$.

P4.22 Generally, the quantization of translational motion is not significant for atoms because of their mass. However, this conclusion depends on the dimensions of the space to which they are confined. Zeolites are structures with small pores that we describe by a cube with edge length 1 nm. Calculate the energy of a H_2 molecule with $n_x = n_y = n_z = 10$. Compare this energy to kT at $T = 300$. K. Is a classical or a quantum description appropriate?

P4.23 Are the eigenfunctions of $\hat{H}$ for the particle in the one-dimensional box also eigenfunctions of the position operator $\hat{x}$? Calculate the average value of x for the case where $n = 3$. Explain your result by comparing it with what you would expect for a classical particle. Repeat your calculation for $n = 5$ and, from these two results, suggest an expression valid for all values of n. How does your result compare with the prediction based on classical physics?

P4.24 What are the energies of the lowest five energy levels in a three-dimensional box with $a = b = c$? What is the degeneracy of each level?

P4.25 In discussing the Boltzmann distribution in Chapter 2, we used the symbols g_i and g_j to indicate the degeneracies of the energy levels i and j. By degeneracy, we mean the number of distinct quantum states (different quantum numbers) all of which have the same energy.

a. Using your answer to Problem P4.19a, what is the degeneracy of the energy level $5h^2/8ma^2$ for the square two-dimensional box of edge length a?

b. Using your answer to Problem P4.14b, what is the degeneracy of the energy level $9h^2/8ma^2$ for a three-dimensional cubic box of edge length a?

P4.26 Show by examining the position of the nodes that $Re[A_+ e^{i(kx - \omega t)}]$ and $Re[A_- e^{i(-kx - \omega t)}]$ represent plane waves moving in the positive and negative x directions, respectively. The notation $Re[\ ]$ refers to the real part of the function in the brackets.

P4.27 Two wave functions are distinguishable if they lead to a different probability density. Which of the following wave functions are distinguishable from $\sin kx$?

a. $(e^{ikx} - e^{-ikx})/2$

b. $e^{i\theta} \sin kx$, θ a constant

c. $\cos(kx - \pi/2)$

d. $i \cos(kx + \pi/2)(\sin \theta + i \cos \theta)\left(-\dfrac{\sqrt{2}}{2} + i\dfrac{\sqrt{2}}{2}\right)$,

   θ a constant

P4.28 Is the superposition wave function $\psi(x) = \sqrt{2/a}[\sin(n\pi x/a) + \sin(m\pi x/a)]$ an eigenfunction of the total energy operator for the particle in the box?

P4.29 The smallest observed frequency for a transition between states of an electron in a one-dimensional box is 3.5×10^{14} s^{-1}. What is the length of the box?

P4.30 Are the total energy eigenfunctions for the free particle in one dimension, $\psi^+(x) = A_+ e^{+i\sqrt{(2mE/\hbar^2)}x}$ and $\psi^-(x) = A_- e^{-i\sqrt{(2mE/\hbar^2)}x}$, eigenfunctions of the one-dimensional linear momentum operator? If so, what are the eigenvalues?

P4.31 Use the eigenfunction $\psi(x) = A' e^{+ikx} + B' e^{-ikx}$ rather than $\psi(x) = A \sin kx + B \cos kx$ to apply the boundary conditions for the particle in the box.

a. How do the boundary conditions restrict the acceptable choices for A' and B' and for k?

b. Do these two functions give different probability densities if each is normalized?

P4.32 Consider a particle in a one-dimensional box defined by $V(x) = 0, a > x > 0$ and $V(x) = \infty, x \ge a, x \le 0$. Explain why each of the following unnormalized functions is or is not an acceptable wave function based on criteria such as being consistent with the boundary conditions, and with the association of $\psi^*(x)\psi(x)\,dx$ with probability.

a. $A \cos \dfrac{n\pi x}{a}$

b. $B(x + x^2)$

c. $C x^3(x - a)$

d. $\dfrac{D}{\sin(n\pi x/a)}$

P4.33 Use your result from Problem P4.19 and make an energy level diagram for the first five energy levels of a square two-dimensional box of edge length b. Indicate which of the energy levels are degenerate and the degeneracy of these levels.

P4.34 Calculate the probability that a particle in a one-dimensional box of length a is found between $0.18a$ and $0.22a$ when it is described by the following wave functions:

a. $\sqrt{\dfrac{2}{a}} \sin\left(\dfrac{\pi x}{a}\right)$

b. $\sqrt{\dfrac{2}{a}} \sin\left(\dfrac{5\pi x}{a}\right)$

What would you expect for a classical particle? Compare your results in the two cases with the classical result.

Web-Based Simulations, Animations, and Problems

W4.1   The motion of a classical particle in a box potential is simulated. The particle energy and the potential in the two halves of the box are varied using sliders. The kinetic energy is displayed as a function of the position x, and the result of measuring the probability of detecting the particle at x is displayed as a density plot. The student is asked to use the information gathered to explain the motion of the particle.

W4.2   Wave functions for $n = 1$–5 are shown for the particle in the infinite depth box and the energy levels are calculated. Sliders are used to vary the box length and the mass of the particle. The student is asked questions that clarify the relationship between the level energy, the mass, and the box length.

W4.3   The probability is calculated for finding a particle in the infinite depth box in the interval $0 \rightarrow 0.1a$, $0.1a \rightarrow 0.2a$, $\ldots$, $0.9a \rightarrow 1.0a$ for $n = 1$, $n = 2$, and $n = 50$. The student is asked to explain these results.

W4.4   Contour plots are generated for the total energy eigenfunctions of the particle in the two-dimensional infinite depth box,

$$\psi_{n_x n_y}(x, y) = N \sin \frac{n_x \pi x}{a} \sin \frac{n_y \pi y}{b}$$

The student is asked questions about the nodal structure of these eigenfunctions and asked to assign quantum numbers n_x and n_y to each contour plot.

W4.5   The student is asked to determine if the normalized wave function

$$\psi(x) = \sqrt{\frac{105}{a^7}} x^2 (x - a)^2$$

is an acceptable wave function for the particle in the infinite depth box based on graphs of $\psi(x)$ and $d\psi(x)/dx$ as a function of x. The wave function $\psi(x)$ is expanded in eigenfunctions of the total energy operator. The student is asked to determine the probability of observing certain values of the total energy in a measurement on the system.

W4.6   The normalized wave function,

$$\Psi(x, t) = \sqrt{\frac{2}{a}} \left[c e^{-iE_1 t/\hbar} \sin\left(\frac{\pi x}{a}\right) + d e^{-iE_2 t/\hbar} \sin\left(\frac{2\pi x}{a}\right) \right]$$

with $|c|^2 + |d|^2 = 1$ is a superposition of the ground state and first excited state for the particle in the infinite depth box. Simulations are carried out to determine if $\langle E \rangle$, $\langle p \rangle$, and $\langle x \rangle$ are independent of time for this superposition state.

5

The Particle in the Box and the Real World

Why have we spent so much time trying to understand the quantum mechanical particle in a box? The particle in a box is a simple model that can be used to explore concepts such as why core electrons are not involved in chemical bonds, the stabilizing effect of delocalized π electrons in aromatic molecules, and the ability of metals to conduct electrons. It also provides a framework for understanding the tunneling of quantum mechanical particles through (not over!) barriers and size quantization, both of which find applications in quantum wells and quantum dots.

5.1 THE PARTICLE IN THE FINITE DEPTH BOX

Before applying the particle in a box model to the "real world," the box must be modified to make it more realistic. This is done by letting the box have a finite depth, which allows the particle to escape. This modification is necessary to model problems such as the ionization of an atom. The potential is defined by

$$V(x) = 0, \quad \text{for } -a/2 < x < a/2$$
$$V(x) = V_0, \quad \text{for } x \geq a/2, x \leq -a/2 \tag{5.1}$$

The origin of the x coordinate has been changed from one end of the box (Chapter 4) to the center of the box to simplify the mathematics of solving the Schrödinger equation. The shift of the origin changes the functional form of the total energy eigenfunctions, as you will see in the end-of-chapter problems. However, it has no physical consequences in that eigenvalues and graphs of the eigenfunctions superimposed on the potential are identical for both choices of the point $x = 0$.

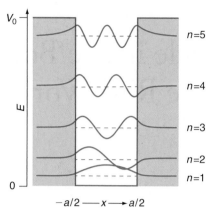

FIGURE 5.1

Eigenfunctions and allowed energy levels are shown for an electron in a well of depth $V_0 = 1.20 \times 10^{-18}$ J and width 1.00×10^{-9} m.

5.1 Energy Eigenfunctions and Eigenvalues for a Finite Depth Box

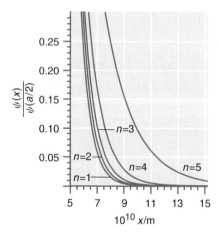

FIGURE 5.2

Decrease of the amplitude of the eigenfunctions as a function of distance from the center of the box. All eigenfunctions have been normalized to the value unity at $x = 5 \times 10^{-10}$ m for purposes of comparison.

How do the eigenfunctions and eigenvalues for the Schrödinger equation for the **finite depth box** differ from those for the infinitely deep potential? For $E > V(x)$ (inside the box), the eigenfunctions have the oscillatory behavior that was exhibited for the infinitely deep box. However, because $V_0 < \infty$, the reasoning following Equation (4.8) no longer holds; the amplitude of the eigenfunctions need not be zero at the ends of the box. For $E < V(x)$ (outside of the box), the eigenfunctions decay exponentially with distance from the box as we will show below. These two regions are considered separately. Inside the box, $V(x) = 0$, and

$$\frac{d^2\psi(x)}{dx^2} = -\frac{2mE}{\hbar^2}\psi(x) \tag{5.2}$$

Outside of the box, the Schrödinger equation has the form

$$\frac{d^2\psi(x)}{dx^2} = \frac{2m(V_0 - E)}{\hbar^2}\psi(x) \tag{5.3}$$

The difference in sign on the right-hand side makes a big difference in the eigenfunctions! Inside the box, the solutions have the same general form as discussed in Chapter 4, but outside the box, they have the form

$$\psi(x) = A\,e^{-\kappa x} + B e^{+\kappa x} \quad \text{for } \infty \ge x \ge a/2 \quad \text{and}$$

$$\psi(x) = A'\,e^{-\kappa x} + B'e^{+\kappa x} \quad \text{for } -\infty \le x \le -a/2$$

$$\text{where} \quad \kappa = \sqrt{\frac{2m(V_0 - E)}{\hbar^2}} \tag{5.4}$$

Convince yourself that the functions of Equation (5.4) are solutions to Equation (5.3). The coefficients (A, B and A', B') are different on each side of the box. Because $\psi(x)$ must remain finite for very large positive and negative values of x, $B = A' = 0$. By matching the wave functions and their derivatives at the boundaries of the three regions of the potential and imposing a normalization condition, the Schrödinger equation can be solved for the eigenfunctions and eigenvalues in the potential for given values of m, a, and V_0. The details of the solution are left to the end-of-chapter problems. The allowed energy levels and the corresponding eigenfunctions for a finite depth potential are shown in Figure 5.1. The yellow areas correspond to the region for which $E_{potential} > E_{total}$. Because $E_{total} = E_{kinetic} + E_{potential}$, $E_{kinetic} < 0$ in this region. For a particle, $E_{kinetic} = p^2/2m$ and a negative value for $E_{kinetic}$ implies that the momentum is imaginary. For this reason, $E_{kinetic} < 0$ defines what is called the **classically forbidden region**.

Two major differences in the solutions between the finite and the infinite depth box are immediately apparent. First, the potential has only a finite number of total energy eigenvalues, which correspond to bound states. The number depends on m, a, and V_0. Second, the amplitude of the wave function does not go to zero at the edge of the box. We explore the consequences of this second difference further when discussing tunneling. As seen in Figure 5.2, the falloff of the wave function inside the barrier is not the same for all eigenfunctions: $\psi(x)$ falls off most rapidly with distance for the most strongly bound state ($V_0 \gg E$) in the potential and most slowly for the least strongly bound state in the potential ($V_0 \sim E$). Convince yourself that Equation (5.4) predicts this trend.

5.2 DIFFERENCES IN OVERLAP BETWEEN CORE AND VALENCE ELECTRONS

Figure 5.2 shows that weakly bound states have wave functions that leak quite strongly into the region outside of the box. What are the consequences of this behavior? Take this potential as a crude model for electrons in an atom. Strongly bound levels correspond to

core electrons and weakly bound levels correspond to **valence electrons**. What happens when a second atom is placed close enough to the first atom that a chemical bond is formed? The results in Figure 5.3 show that the falloff of the wave functions for the weakly bound states in the box is gradual enough that both wave functions have a nonzero amplitude in the region between the wells. *These wave functions have a significant overlap.* Note that this is not the case for the strongly bound levels; these energy eigenfunctions have a small overlap.

We conclude that a correlation exists between the nonzero overlap required for chemical bond formation and the position of the energy level in the potential. This is our first application of the particle in the box model. It provides an understanding of why chemical bonds involve the least strongly bound, or valence, electrons and not the more strongly bound, or core, electrons. We will have more to say on this topic when the chemical bond is discussed in Chapter 12.

5.3 PI ELECTRONS IN CONJUGATED MOLECULES CAN BE TREATED AS MOVING FREELY IN A BOX

The absorption of light in the visible and ultraviolet (UV) part of the electromagnetic spectrum in molecules is a result of the excitation of electrons from occupied to unoccupied energy levels. If the electrons are delocalized as in an organic molecule with a **π-bonded network**, the maximum in the absorption spectrum shifts from the UV into the visible range. The greater the degree of **delocalization**, the more the absorption maximum shifts toward the red end of the visible spectrum. The energy levels for such a conjugated system can be described quite well with a one-dimensional particle in a box model. The series of dyes, 1,4-diphenyl–1,3-butadiene, 1,6-diphenyl–1,3,5-hexatriene, and 1,8-diphenyl–1,3,5,7-octatetraene consist of a planar backbone of alternating C—C and C=C bonds and have phenyl groups attached to the ends. The phenyl groups serve the purpose of decreasing the volatility of the compound. The π-bonded network does not include the phenyl groups, but does include the terminal carbon–phenyl group bond length. Only the π-bonded electrons are modeled using the particle in the box. Because each energy level can be occupied by two electrons, the HOMO corresponds to $n = 2$, 3, and 4 for the series of molecules considered.

The longest wavelength at which light is absorbed occurs when one of the electrons in the highest occupied energy level (HOMO) is promoted to the lowest lying unoccupied level (LUMO). As Equation (4.17) shows, the energy level spacing depends on the length of the π-bonded network. For 1,4-diphenyl–1,3-butadiene, 1,6-diphenyl–1,3,5-hexatriene, and 1,8-diphenyl–1,3,5,7-octatetraene, the maximum wavelength at which absorption occurs is 345, 375, and 390 nm, respectively. From these data, and taking into account the quantum numbers corresponding to the highest occupied and lowest unoccupied levels, the apparent network length can be calculated. We demonstrate the calculation for 1,6-diphenyl–1,3,5-hexatriene, for which the HOMO–LUMO transition corresponds to $n_i = 3 \rightarrow n_f = 4$ as indicated in Figure 5.4.

$$a = \sqrt{\frac{(n_f^2 - n_i^2)h^2}{8m\Delta E}} = \sqrt{\frac{(n_f^2 - n_i^2)h\lambda_{max}}{8mc}}$$

$$= \sqrt{\frac{(4^2 - 3^2)(6.626 \times 10^{-34} \text{ J s})(375 \times 10^{-9} \text{ m})}{8(9.11 \times 10^{-31} \text{ kg})(2.998 \times 10^8 \text{ m s}^{-1})}}$$

$$= 892 \text{ pm} \tag{5.5}$$

The apparent and calculated network length has been compared for each of the three molecules by B. D. Anderson [*J. Chemical Education* 74 (1997) 985]. Values are shown in Table 5.1. The agreement is reasonably good, given the simplicity of the model. Most importantly, the model correctly predicts that because λ is proportional to

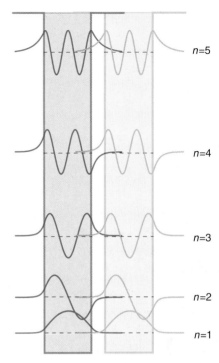

FIGURE 5.3

Overlap of wave functions from two closely spaced finite depth wells. The vertical scale has been expanded relative to Figure 5.1 to better display the overlap.

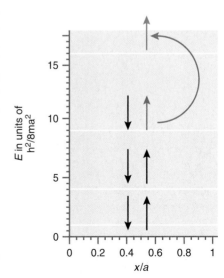

FIGURE 5.4

The HOMO to LUMO transition is shown for the particle in the box model of the π electrons in 1,6-diphenyl–1,3,5-hexatriene.

TABLE 5.1   CALCULATED NETWORK LENGTH FOR CONJUGATED MOLECULES

Compound	Apparent Network Length (pm)	Calculated Network Length (pm)
1,4-Diphenyl–1,3-butadiene	723	695
1,6-Diphenyl–1,3,5-hexatriene	892	973
1,8-Diphenyl–1,3,5,7-octatetraene	1030	1251

a^2, shorter π-bonded networks show absorption at smaller wavelengths. This trend is confirmed by experiment.

For 1,6-diphenyl–1,3,5-hexatriene in the **ground state**, the HOMO corresponds to $n = 3$. Does this mean that in a large number of molecules there will be very few molecules for which the $n = 4$ level is occupied at 300. K? This question can be answered with the help of the Boltzmann distribution.

The energy difference between the two levels is given by

$$\Delta E = \frac{h^2(n_f^2 - n_i^2)}{8ma^2} = \frac{7 \times (6.626 \times 10^{-34}\,\text{J s})^2}{8 \times 9.109 \times 10^{-31}\,\text{kg} \times (973 \times 10^{-12}\,\text{m})^2}$$

$$= 4.45 \times 10^{-19}\,\text{J} \tag{5.6}$$

Because there are two quantum states for each value of n, $g_4 = g_3 = 2$. The ratio of the population in the $n = 4$ level to that in the $n = 3$ level is given by

$$\frac{n_4}{n_3} = \frac{g_4}{g_3}e^{-\Delta E/kT} = \exp\left[-\frac{4.45 \times 10^{-19}\,\text{J}}{(1.381 \times 10^{-23}\,\text{J K}^{-1} \times 300.\,\text{K}}\right]$$

$$= 2.1 \times 10^{-47} \tag{5.7}$$

Therefore, the $n = 3 \rightarrow n = 4$ transition cannot be achieved by the exchange of translational energy in the collision between molecules at 300. K, and essentially all molecules are in their electronic ground state.

5.4 WHY DOES SODIUM CONDUCT ELECTRICITY AND WHY IS DIAMOND AN INSULATOR?

As discussed earlier, valence electrons on adjacent atoms in a molecule or a solid can have an appreciable overlap. This means that the electrons can "hop" from one atom to the next. Consider Na, which has one valence electron per atom. If two Na atoms are bonded to form a dimer, the valence level that was localized on each atom will be delocalized over both atoms as is illustrated in Figure 5.5. Now add additional Na atoms to form a one-dimensional Na crystal. A crystalline metal can be thought of as a box with a periodic corrugated potential at the bottom. To illustrate the relationship to a box model, the potential of a one-dimensional periodic array of Na^+ potentials arising from the atomic cores at lattice sites is shown in Figure 5.6. Because the Na 3s valence electrons can be found with equal probability at any Na atom, one electron per atom is delocalized over the whole metal sample. This is exactly the model of the particle in the box.

FIGURE 5.5

At large distances, the valence level on each Na atom is localized on that atom. When they are brought close enough together to form the dimer, the barrier between them is lowered, and the level is delocalized over both atoms. The quantity x_e represents the bond length of the dimer.

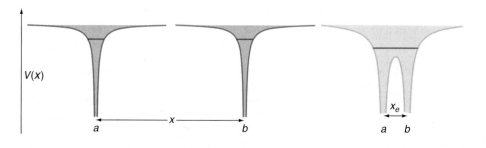

The potential of Figure 5.6 can be idealized to a box as shown in Figure 5.7. This box differs from the simple boxes discussed earlier in an essential way. There are many atoms in the atomic chain under consideration (large a), such that the energy levels for the delocalized electrons are very closely spaced in what is called the conduction band as discussed below. What is the energy-level spacing for the delocalized electrons in a 1.00-cm-long box? About 2×10^7 atoms will fit into the box. If each atom donates one electron to the band, you can easily show that at the highest filled level,

$$E_{n+1} - E_n = \frac{(n+1)^2 h^2}{8ma^2} - \frac{n^2 h^2}{8ma^2} = \frac{h^2}{8ma^2}(2n+1)$$

$$= (2n+1)(6.02 \times 10^{-34} \text{ J}) \qquad \textbf{(5.8)}$$

This spacing between levels is at most only $\approx 10^{-6}\, kT$ for $T = 300.$ K and therefore the energy spectrum is essentially continuous. All energies within the range bounded by the bottom of the red shaded area in Figure 5.7 for low energies and the dashed line for high energies are accessible. This set of continuous energy levels is referred to as an **energy band**. The band shown in Figure 5.7a extends up to the dashed line, beyond which there are no allowed energy levels until the energy has increased by ΔE. An energy range, ΔE, in which there are no allowed states is called a **band gap** (See Figure 5.15.). For Na, not all available states in the band are filled, as shown in Figure 5.7. The range of energies between the top of the red area and the dashed line corresponds to unfilled **conduction band** states. The fact that the band is only partially filled is critical in making Na an **electrical conductor**, as explained later.

What happens when an electrical potential is applied between the two ends of the box? The field gives rise to a gradient of potential energy along the box superimposed on the original potential as shown in Figure 5.7b. The unoccupied states on the side of the metal with the more positive electrical potential have a lower energy than the occupied states with the more negative electrical potential. This makes it energetically favorable for the electrons to move toward the end of the box with the more positive voltage as shown in Figure 5.7c. This flow of electrons through the metal is the current that flows through the "wire." It occurs because of the **overlap of wave functions** on adjacent atoms, which leads to hopping, and because the energy levels are so close together that they form a continuous energy spectrum.

What makes diamond an **insulator** in this picture? Bands are separated from one another by band gaps, in which there are no allowed eigenfunctions of the total energy

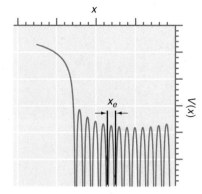

FIGURE 5.6

The potential energy resulting from a one-dimensional periodic array of Na$^+$ ions. One valence electron per Na is delocalized over this box. The quantity x_e represents the lattice spacing.

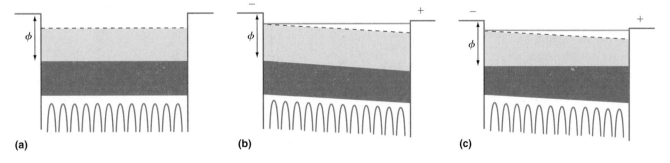

(a)   **(b)**   **(c)**

FIGURE 5.7

Idealization of a metal in the particle in the box model. The horizontal scale is greatly expanded to show the periodic potential. Actually, more than 10^7 atoms will fit into a 1-cm-long box. The red shaded band shows the range of energies filled by the valence electrons of the individual atoms. The highest energy that can be occupied in this band is indicated by the dashed line. The energy required to remove an electron from the highest occupied state is the *work function*, ϕ. **(a)** The metal without an applied potential. **(b)** The effect on the energy levels of applying an electric field. **(c)** The response of the metal to the change in the energy levels induced by the electric field. The thin solid line at the top of the band in parts (b) and (c) indicates where the energy of the highest level would lie in the absence of an electric field.

operator. In diamond, all quantum states in the band accessible to the delocalized valence electrons are filled. The highest energy filled band in semiconductors and insulators is called the **valence band**. In Figure 5.7, this corresponds to extending the red area up to the dashed line. As the energy increases, a range is encountered in which there are no allowed states of the system until the valence band of allowed energy levels is reached. This means that, although we could draw diagrams just like the upper two panels of Figure 5.7 for diamond, the system cannot respond as shown in the lower panel. There are no unoccupied states in the valence band that can be used to transport electrons through the crystal. Therefore, diamond is an insulator. **Semiconductors** also have a band gap separating the fully occupied valence and the empty conduction band. However, in semiconductors, the band gap is smaller than for insulators, allowing them to become conductors at elevated temperatures. The band structure of solids is discussed in more detail in Chapter 14 after the chemical bond has been discussed.

5.5 TUNNELING THROUGH A BARRIER

Consider a particle with energy E that is confined to a very large box. Within this box, a barrier of height V_0 separates two regions in which $E < V_0$. Classically, the particle will not pass the barrier region because it has insufficient energy to get over the barrier. This situation looks quite different in quantum mechanics. As long as the particle that leaks into the barrier sees a potential energy such that $V_0 > E$, its wave function will decay exponentially with distance. However, something surprising happens if the barrier is thin, meaning that $V_0 > E$ only over a distance comparable to the particle wavelength. The particle can escape *through* the barrier even though it does not have sufficient energy to go *over* the barrier. This effect, which is depicted in Figure 5.8, is known as **tunneling**. We briefly describe tunneling here and leave the details of the calculations for the end-of-chapter problems.

To investigate tunneling, we focus on the region near the edge of the box. The finite depth box is modified by making it wider and letting the barrier have a finite thickness on the right-hand side. The potential is now described by

$$V(x) = 0, \quad \text{for } x < 0$$
$$V(x) = V_0, \quad \text{for } 0 \le x \le a$$
$$V(x) = 0, \quad \text{for } x > a \qquad \qquad \textbf{(5.9)}$$

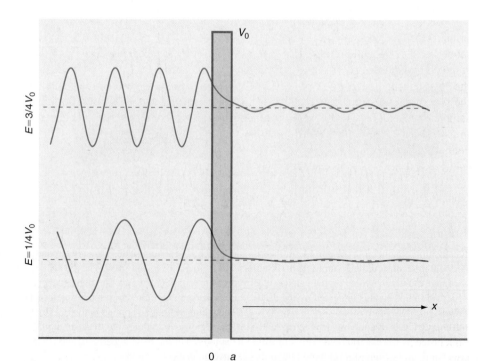

FIGURE 5.8

Waves corresponding to the indicated energy are incident from the left on a barrier of height $V_0 = 1.60 \times 10^{-19}$ J and width 9.00×10^{-10} m.

The incident wave functions are shown to the left of the barrier in Figure 5.8. If the **barrier width**, a, is small enough that $\psi(x)$ has not decayed to a negligibly small value at $x = a$, the wave function in the region $x > a$ has a finite amplitude. Because $V(x) = 0$ for $x > a$, the wave function is a traveling wave. This means that the particle has a finite probability of escaping from the well, although its energy is less than the height of the barrier. An exponentially decaying wave function is shown inside the barrier and the incident and transmitted wave functions are shown in the other two regions.

We see that tunneling is much more likely for particles with energies near the top of the barrier. This is a direct consequence of the degree to which the wave function in the barrier falls off with distance as $e^{-\kappa x}$, in which $1/\kappa$, called the **decay length**, is given by $\sqrt{\hbar^2/2m(V_0 - E)}$.

5.2 Tunneling through a Barrier

5.6 THE SCANNING TUNNELING MICROSCOPE

Researchers did not know that particles could tunnel through a barrier until the advent of quantum mechanics. In the early 1980s, the tunneling of electrons between two solids was used to develop an atomic resolution microscope. Gerd Binnig and Heinrich Rohrer received a Nobel Prize for the invention of the **scanning tunneling microscope (STM)** in 1986.

The STM allows the imaging of solid surfaces with atomic resolution with a surprisingly minimal mechanical complexity. The STM and a closely related device called the atomic force microscope (AFM) have been successfully used to study phenomena at atomic and near atomic resolution in a wide variety of areas including chemistry, physics, biology, and engineering. The invention of the STM and AFM played a significant role in enabling the development of nanotechnology. The essential elements of an STM are a sharp metallic tip and a conducting sample over which the tip is scanned to create an image of the sample surface. In an STM, the barrier between these two conductors is usually vacuum or air, and electrons are made to tunnel across this barrier as discussed later. As might be expected, the barrier width needs to be on the order of atomic dimensions to observe tunneling. Electrons with an energy of typically 5 eV are used to tunnel from the metal tip to the surface. This energy corresponds to the **work function**, as well as to the barrier height $V_0 - E$ in Figure 5.8. The decay length $\sqrt{\hbar^2/2m(V_0 - E)}$ for such an electron in the barrier is about 0.1 nm. Therefore, if the tip and sample are brought to within a nanometer of one another, electron tunneling will be observed between them.

How does a scanning tunneling microscope work? We address this question first in principle and then from a practical point of view. Because the particle in a box is a good model for the conduction of electrons in the metal solid, the tip and surface can be represented by boxes as shown in Figure 5.9. For convenience, the part of the box below the lowest energy that can be occupied by the core electrons has been omitted, and only the part of the box immediately adjacent to the tip–sample gap is shown. The tip and sample in general have different work functions as indicated. If they are not connected in an external circuit, their energy diagrams line up as in Figure 5.9a. When they are connected in an external circuit, charge flows between the tip and sample until the highest occupied level is the same everywhere as shown in Figure 5.9b.

Tunneling takes place at constant energy, which in Figure 5.9 corresponds to the horizontal dashed line. However, for the configuration shown in Figure 5.9b, there is no empty state on the sample into which an electron from the tip can tunnel. To allow tunneling to occur, a small $(0.01 - 1 \text{ V})$ electrical potential is placed between the two metals. This raises the highest filled energy level of the tip relative to that of the sample. Now tunneling of electrons can take place from tip to sample, resulting in a net current flow.

Up until now, we have discussed a tunneling junction, not a microscope. Figure 5.10 shows how an STM functions in an imaging mode. A radius of curvature of 100 nm at the apex of the tip is routinely achievable by electrolytically etching a metal wire. The sample could be a single crystal whose structure is to be investigated at an atomic scale. This junction is shown on an atomic scale in the bottom part of Figure 5.10. No matter how blunt the tip is, one atom is closer to the surface than all the others. At a tunneling

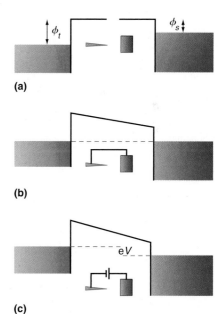

(a)

(b)

(c)

FIGURE 5.9

(a) If the conducting tip and surface are electrically isolated from one another, their energy diagrams line up. **(b)** If they are connected by a wire in an external circuit, charge flows from the lower work function material into the higher work function material until the highest occupied states have the same energy in both materials. **(c)** By applying a voltage V between the two materials, the highest occupied levels have an offset of energy eV. This allows tunneling to occur from left to right. The subscripts t and s refer to tip and surface.

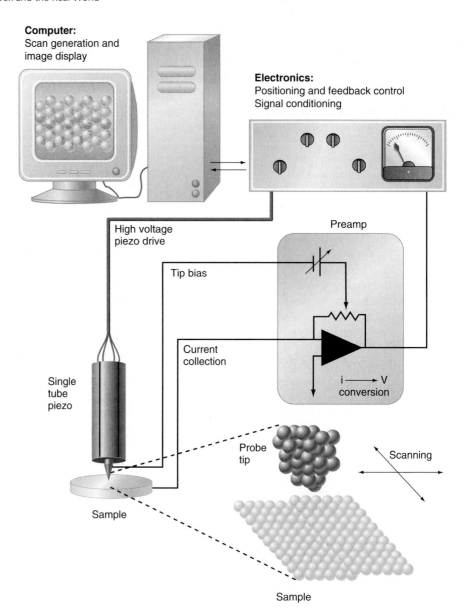

Computer:
Scan generation and
image display

Electronics:
Positioning and feedback control
Signal conditioning

Preamp

High voltage
piezo drive

Tip bias

Current
collection

Single
tube
piezo

i ⟶ V
conversion

Probe
tip

Scanning

Sample

Sample

FIGURE 5.10

Schematic representation of a scanning
tunneling microscope.

[Courtesy of Kevin Johnson, University of
Washington thesis, 1991.]

gap distance of about 0.5 nm, the tunneling current decreases by an order of magnitude
for every 0.1 nm that the gap is increased. Therefore, the next atoms back from the apex
of the tip make a negligible contribution to the tunneling current and the whole tip acts
like a single atom for tunneling.

The tip is mounted on a segmented tubular scanner made of a piezoelectric material
that changes its length in response to an applied voltage. In this way the tip can be
brought close to the surface by applying a voltage to the piezoelectric tube. Assume that
we have managed to bring the tip within tunneling range of the surface. On the magnified scale shown in Figure 5.10, the individual atoms in the tip and surface are seen at a
tip–surface spacing of about 0.5 nm. Keep in mind that the wave functions for the tunneling electrons in the tip decay rapidly in the region between tip and sample, as shown
earlier in Figure 5.2. If the tip is directly over a surface atom, the amplitude of the wave
function is large at the surface atom, and the tunneling current is high. If the tip is
between surface atoms, the amplitude of the wave functions is smaller and the tunneling
current will be lower. To scan over the surface, different voltages are applied to the
four segmented electrodes on the piezo tube. This allows a topographical image of
the surface to be obtained. Because the tunneling current varies exponentially with the
tip–surface distance, the microscope provides a very high sensitivity to changes in the
height of the surface that occur on an atomic scale.

In this abbreviated description, some details have been glossed over. The current is
usually kept constant as the tip is scanned over the surface using a feedback circuit to
keep the tip–surface distance constant. This is done by changing the voltage to the piezo

tube electrodes as the tip scans over the surface. Additionally, a vibrational isolation system is required to prevent the tip from crashing into the surface as a result of vibrations always present in a laboratory. Figure 5.11 provides an example of the detail that can be seen with a scanning tunneling microscope. The individual planes, which are stacked together to make the silicon crystal, and the 0.3-nm height change between planes are clearly seen. Defects in the crystal structure are also clearly resolved. Researchers are using this microscope in many new applications aimed at understanding the structure of solid surfaces and modifying surfaces atom by atom.

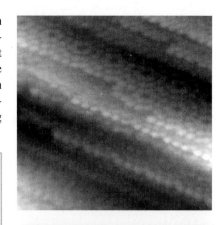

EXAMPLE PROBLEM 5.1

As was found for the finite depth well, the wave function amplitude decays in the barrier according to $\psi(x) = A \exp[-\sqrt{2m(V_0 - E)/\hbar^2}\, x]$. This result will be used to calculate the sensitivity of the scanning tunneling microscope. Assume that the tunneling current through a barrier of width a is proportional to $|A|^2 \exp[-2\sqrt{2m(V_0 - E)/\hbar^2}\, a]$.

a. If $V_0 - E$ is 4.50 eV, how much larger would the current be for a barrier width of 0.20 nm than for 0.30 nm?

b. A friend suggests to you that a proton tunneling microscope would be equally effective as an electron tunneling microscope. For a 0.20-nm barrier width, by what factor is the tunneling current changed if protons are used instead of electrons?

Solution

a. Putting the numbers into the formula given, we obtain

$$\frac{I(a = 2.0 \times 10^{-10}\,\text{m})}{I(a = 3.0 \times 10^{-10}\,\text{m})} = \exp\left[-2\sqrt{\frac{2m(V_0 - E)}{\hbar^2}}\right.$$

$$\left.\times (2.0 \times 10^{-10}\,\text{m} - 3.0 \times 10^{-10}\,\text{m})\right]$$

$$= \exp\left[-2\sqrt{\frac{2 \times 9.109 \times 10^{-31}\,\text{kg} \times 4.50\,\text{eV} \times 1.602 \times 10^{-19}\,\text{J/eV}}{(1.055 \times 10^{-34}\,\text{J s})^2}}\right.$$

$$\left.\times (-1.0 \times 10^{-10}\,\text{m})\right]$$

$$= 8.8$$

Even a small distance change results in a substantial change in the tunneling current.

b. We find that the tunneling current for protons is appreciably smaller than that for electrons.

$$\frac{I(proton)}{I(electron)} = \frac{\exp\left[-2\sqrt{\dfrac{2m_{proton}(V_0 - E)}{\hbar^2}}\, a\right]}{\exp\left[-2\sqrt{\dfrac{2m_{electron}(V_0 - E)}{\hbar^2}}\, a\right]}$$

$$= \exp\left[-2\sqrt{\frac{2(V_0 - E)}{\hbar^2}}\left(\sqrt{m_{proton}} - \sqrt{m_{electron}}\right)a\right]$$

$$= \exp\left[-2\sqrt{\frac{2 \times 4.50\,\text{eV} \times 1.602 \times 10^{-19}\,\text{J/eV}}{(1.055 \times 10^{-34}\,\text{J s})^2}} \times \left(\sqrt{1.67 \times 10^{-27}\,\text{kg}} - \sqrt{9.11 \times 10^{-31}\,\text{kg}}\right)\right.$$

$$\left.\times 2.0 \times 10^{-10}\,\text{m}\right]$$

$$= 1.2 \times 10^{-79}$$

This result does not make the proton tunneling microscope look very promising.

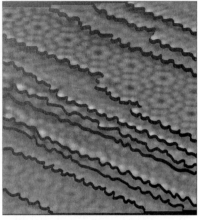

FIGURE 5.11

STM images of the (111) surface of Si. The upper image shows a 200.×200.-nm region with a high density of atomic steps, and the light dots correspond to individual Si atoms. The lower image shows how the image is related to the structure of parallel crystal planes separated by steps of one atom height. The step edges are shown as dark ribbons.

[Courtesy of Kevin Johnson, University of Washington thesis, 1991.]

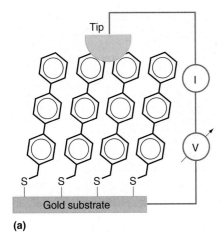

(a)

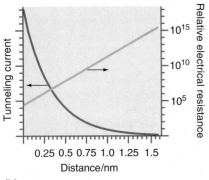

(b)

FIGURE 5.12

(a) An atomic force microscope tip is brought into contact with a self-assembled monolayer formed of molecules of a given length. The electrical current is measured as a function of the applied voltage. This allows the electrical resistance to be determined. **(b)** A plot of the resistance relative to that for zero length as a function of distance falls off exponentially (red curve). This relationship is also shown shown on a logarithmic scale (yellow curve).

Tunneling is also important in the transport of charge through a molecule, as demonstrated by the following discussion. How do you expect the electrical resistance of a single molecule to depend on its length? If the electron has to tunnel through the molecule to get from one end to the other, the resistance will increase exponentially with length because wave functions decay exponentially with distance inside a barrier. By means of atomic force microscopy (a close relative of scanning tunneling microscopy), the dependence of the electrical resistance of molecules on their length can be directly measured, as shown in Figure 5.12.

Using the technique of self-assembly, an ordered array of thiols is created on a conducting gold substrate. The vertical orientation shown in Figure 5.12 arises from the strong interaction between S and Au, and the long-range order arises from the repulsive interaction between one molecule and its neighbors. Using the positioning techniques discussed for the STM, an electrical contact is made with the adsorbed layer using a sharp tip, and the current is measured as a function of the applied voltage. This measurement allows the electrical resistance to be determined. Because the tip is in contact with more than one molecule, the measured resistance is that of several molecules in parallel. The electrical resistance is measured for molecules that are identical apart from their length. The resistance increases exponentially with length, showing that electrons must tunnel through the molecule to get from one end to the other.

In an example of more direct chemical interest, it has been shown by examining (donor)–bridge–(acceptor) molecules of different bridge lengths that electron tunneling is an important mechanism in redox reactions. Electron transfer can induce redox reactions, even if the donor and acceptor are separated by ~1 nm. By contrast, reactions that require covalent bond formation are not effective if the reaction partners are spaced that far apart.

5.7 TUNNELING IN CHEMICAL REACTIONS

Most chemical reactions are thermally activated; they proceed faster as the temperature of the reaction mixture is increased. This behavior is typical of reactions for which an energy barrier must be overcome in order to transform reactants into products. This barrier is referred to as the **activation energy** for the reaction. By increasing the temperature of the reactants, the fraction that has an energy that exceeds the activation energy is increased, allowing the reaction to proceed.

Tunneling provides another mechanism to convert reactants to products that does not require an increase in energy of the reactants for the reaction to proceed. It is well known that hydrogen transfer reactions can involve tunneling. An example is the reaction $R_1OH + R_2O^- \longrightarrow R_2OH + R_1O^-$, where R_1 and R_2 are two different organic groups. The test for tunneling in this case is to substitute deuterium for hydrogen. If the reaction is thermally activated, the change in reaction rate is small and can be attributed to the different ground-state vibrational frequency of $-OH$ and $-OD$ bonds (see Chapter 8). However, if tunneling occurs, the rate decreases greatly because the tunneling rate depends exponentially on the decay length $\sqrt{\hbar^2/2m(V_0 - E)}$. However, it is not widely appreciated that tunneling can be important for heavier atoms such as C and O. A report by Zuev *et al.* [*Science* 299 (2003), 867] shows that rate of the ring expansion reaction depicted in Figure 5.13 is faster than the predicted thermally activated rate by the factor 10^{152} at 10 K! This increase is due to the tunneling pathway. Because the tunneling rate depends exponentially on the product $2m(V_0 - E)$, heavy atom tunneling is only appreciable if $(V_0 - E)$ is very small. However, in a number of reactions, particularly in the fields of chemical catalysis and enzymology, this condition is met.

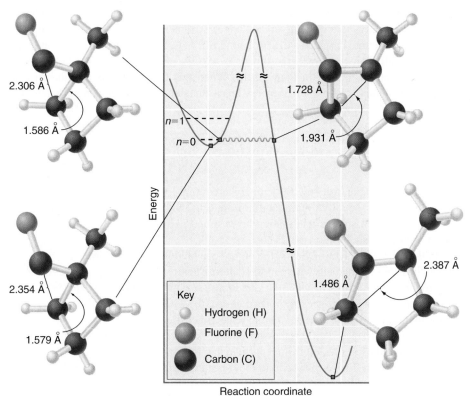

FIGURE 5.13

The structures of four species along the reaction path from reactant to product are shown together with a schematic energy diagram. The reaction occurs not by surmounting the barrier, but by tunneling through the barrier at the energy indicated by the wavy line.

[Adapted with permission from *Science* and Weston Borden, University of Washington. Copyright 2003, AAAS.]

SUPPLEMENTAL

5.8 QUANTUM WELLS AND QUANTUM DOTS

Just as not all atoms have the same ionization energy, not all solids have the same work function. The width and energetic position of the bands of allowed states are also not the same. These facts can be used to engineer some very useful devices. One good example is a device called a **quantum well structure**. Gallium arsenide is a widely used semiconductor in microelectronics applications. $Al_\alpha Ga_{1-\alpha}As$ is a substitutional alloy in which some of the Ga atoms are replaced by Al atoms. It can be combined with GaAs to form crystalline **heterostructures** that consist of alternating layers of GaAs and $Al_\alpha Ga_{1-\alpha}As$. Both substances are semiconductors that, like insulators, have a fully occupied energy band derived from their valence electrons. The fully occupied band is referred to as the valence band. As the energy increases, a band gap evolves that has no states, followed by an empty band that can be occupied by electrons, called the **conduction band**. However, there are only enough electrons in the electrically neutral crystal to fill the valence band. This is analogous to the H atom in which the $1s$ state is occupied and the $2s$ state is empty. This band structure is shown in Figure 5.14.

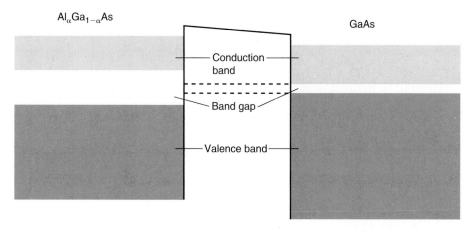

FIGURE 5.14

Schematic representation of relative positions of the bands in GaAs and $Al_\alpha Ga_{1-\alpha}As$ connected in an external circuit as in Figure 5.9b (not to scale). Note that the smaller band gap in GaAs lines up with the center of the larger band gap in $Al_\alpha Ga_{1-\alpha}As$.

By means of a technique called molecular beam epitaxy in which materials are slowly evaporated onto a growing crystal under extremely low pressures, one can grow a crystalline structure in which a 0.1- to 1-nm layer of GaAs is sandwiched between two macroscopically thick (several micrometers) $Al_\alpha Ga_{1-\alpha}As$ layers. Such a heterostructure is depicted on the left in Figure 5.15. When this GaAs layer is considered as a three-dimensional (3D) box, it has energy levels that depend on three quantum numbers because this is a 3D problem:

$$E_{n_x n_y n_z} = \frac{h^2}{8m} \left(\frac{n_z^2 + n_y^2}{b^2} + \frac{n_x^2}{a^2} \right) \qquad (5.10)$$

The length b is on the order of 1000. nm, whereas a is 1 to 10. nm. Therefore, the energy spectrum is essentially continuous in n_z and n_y, but discrete in n_x. What does the band-gap region in such an alternating layer structure look like? This can be deduced from Figure 5.14 and is shown in Figure 5.15.

In this very thin layer of GaAs, the empty conduction band has lower energy states in the GaAs region than elsewhere in the heterostructure. The $Al_\alpha Ga_{1-\alpha}As$ layers have macroscopic dimensions in all three directions, so that the particle in the box states form a continuous energy spectrum. By contrast, the GaAs layer has relatively large dimensions parallel to the layer, but atomic scale dimensions along the x direction perpendicular to the interface between the substances. Along this direction, the quantization conditions are those expected from a particle in a finite well, leading to discrete energy levels as shown in Figure 5.15. Along the other two directions, the energy-level spectrum is continuous. By choosing this unusual geometry for the box, the system has a continuous energy spectrum along the y and z directions, and a discrete energy spectrum along the x direction. As discussed later, it is possible to selectively change the discrete energy spectrum.

This is certainly a novel structure, and it is also useful because it can be made to function as a very efficient laser. In the ground state, the valence band is fully occupied and the conduction band is empty. The lowest energy excitation from the valence band into the conduction band lies in the GaAs layer. Therefore, it is possible to efficiently excite these transitions by putting an amount of energy into the system that is equal to or larger than the band-gap energy in GaAs, but less than the band-gap energy in $Al_\alpha Ga_{1-\alpha}As$. When the system decays to the ground state, a photon is emitted with frequency $\nu = \Delta E/h$, in which ΔE is the difference in energy between the excited energy level in the conduction and the empty states in the valence band. A laser of this type has two advantages over more conventional solid-state lasers. The first is that such lasers can be very efficient in producing photons. The second is that the energy levels in the GaAs layer can be changed by varying the layer thickness, as predicted by Equation

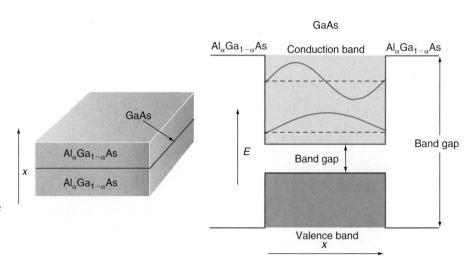

FIGURE 5.15

Schematic depiction of the heterostructure (left) and the band and band-gap structure in the immediate vicinity of the GaAs layer (right). Not to scale.

(5.10). This allows for tuning of the laser frequency through a limited range. Devices based on the principles outlined here are called *quantum well devices.*

The technique used to manufacture heterostructures like those just discussed is molecular beam epitaxy (MBE). Because the materials must be deposited in a very high vacuum, MBE is an expensive technique. New techniques involving size-controlled crystallization in solution offer a less expensive way to synthesize nanoscale particles. Such techniques can produce crystalline spherical particles of compound semiconductors such as CdSe with uniform diameters in the range of 1 to 10. nm. This results in the energy levels being quantized in all three directions and opens up new possibilities for these structures, which are called **quantum dots**. Quantum dots have a band-gap energy that strongly depends on their diameter for the reasons discussed earlier.

Assume that all states below the band gap are filled and all states above the band gap of width E_{bg} are empty in the ground state of the quantum dot, making it a semiconductor. Transitions from states below to those above the band gap can occur through absorption of visible light. Subsequently, the electron in the excited state can drop to an empty state below the band gap, emitting a photon in a process called fluorescence with a wavelength $\lambda = hc/E_{bg}$. Because the energy levels and E_{bg} depend on the length b, λ also depends on b. This property is illustrated in Figure 5.16a. For CdSe quantum dots, the emission wavelength increases from 450 nm (blue light) to 650 nm (red light) as the dot diameter increases from 2 to 8 nm. Figure 5.16b shows another important property of quantum dots. Although they absorb light over a wide range of wavelengths, they emit light in a much smaller range of wavelengths. This occurs because electrons excited from occupied states just below the band gap to states well above the band gap in absorption lose energy and relax to states just above the band gap before emitting light in fluorescence. The absorbed light is reemitted in a narrow frequency range determined by the band-gap energy of the semiconducting quantum dot.

Quantum dots are currently being used in bioanalytical methods. The usefulness of these quantum dots is their ability to act as tags for biologically interesting substrates

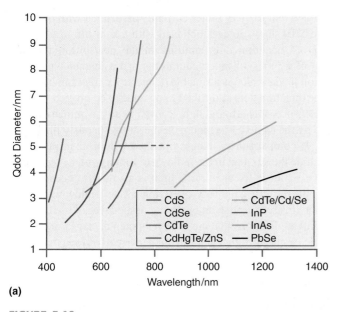

(a)

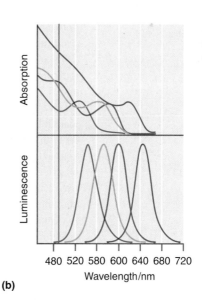

(b)

FIGURE 5.16

(a) The dependence of the wavelength of the light emitted in a transition from just above to just below the band gap is shown as a function of the quantum dot diameter for a number of materials. (b) The top panel shows the absorption spectrum of four CdSe/ZnS quantum dots of different diameters, and the bottom panel shows the corresponding emission spectrum. Note that absorption occurs over a much larger range of wavelengths than emission. The vertical bar indicates the wavelength of a 488-nm argon ion laser which can be used to excite electrons from below to above the band gap for all four diameters. Using this laser ensures that absorption and emission occur at distinctly different wavelengths.

[From X. Michalet et al., *Science*, 307, 538 (2005).]

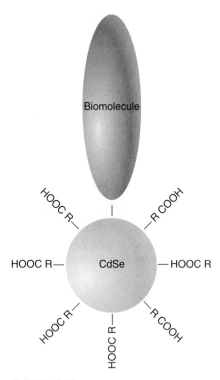

FIGURE 5.17

A CdSe quantum dot can be made soluble in an aqueous solution by coating it with a single molecular layer of an organic acid. When tethered to a biomolecule of interest, it can be used as a fluorescent tag to locate the biomolecule when the biomolecule is bound to a receptor in a heterogeneous environment such as a cell.

such as proteins, as shown in Figure 5.17. By functionalizing such quantum dots with an appropriate molecular layer, they can be made soluble in aqueous solutions and tethered to the protein of interest.

The following example illustrates the usefulness of a protein with a fluorescent label. After letting the tagged proteins enter a cell and attach to their receptors, the cell is illuminated with light and the quantum dots act as point sources of fluorescent light whose location can be imaged using optical microscopy. Because the light used for excitation and the fluorescent light have different wavelengths, it is easy to distinguish between them using optical filters. The same excitation wavelength can be used for quantum dots of different size, so that several different ligand–receptor combinations can be probed simultaneously if the individual ligands are tethered to quantum dots of differing diameter. It might appear that the number of possible different fluorescent tags is limited by the overlap in the wavelengths at which they fluoresce. However, one can also tether different combinations of a few different quantum dots to a protein, creating a barcode. For instance, the intensity versus wavelength distribution of the fluorescent signal from a tagged protein to which two 1-nm, one 3-nm, and two 5-nm quantum dots have been attached is different from all other distinct possible permutations of five quantum dots. This analysis method, which is based on size quantization, offers new analytical techniques for measuring the spatial distribution of molecules in inherently heterogeneous biological environments.

Because a quantum dot absorbs strongly over a wide range of wavelengths, but fluoresces in a narrow range of wavelengths, it can be used as an internal light source for imaging the interior of semitransparent specimens. Figure 5.18 shows an image obtained by projecting the capillary structure of adipose tissue in a 250-μm-thick specimen surrounding a surgically exposed ovary of a living mouse on a plane [Larson *et al., Science* 300 (2003), 1434]. Additionally, the rate of blood flow and the differences in systolic and diastolic pressure can be directly observed in these experiments. It is not possible to obtain such images with X-ray-based techniques, because of the absence of a contrast mechanism.

The usefulness of quantum dots in imaging tissues in vivo has significant potential applications in surgery provided that toxicity issues currently associated with quantum dots can be resolved. Figure 5.19 shows images obtained with a near-infrared camera resulting from the injection of quantum dots emitting in the near-infrared region (840–860 nm) into the paw of a mouse. The quantum dots had a diameter ~15 nm including the functionalization layer, which is a suitable diameter for trapping in lymph nodes. Fluorescence from the quantum dots allow the lymph node to be imaged through the overlying tissue layers. Because near-infrared light is invisible to the human eye and visible light is not registered by the camera, the mouse can be simultaneously illuminated with both types of light. A comparison of images obtained with near-infrared and visible light can be used to guide the surgeon in the removal of tumors and to verify that all affected tissues have been removed. Near-infrared light is also useful because this wavelength minimizes the absorption of light by overlying tissues, allowing tissues containing quantum dots to be imaged through an overlying tissue layer of 6–10 cm.

Quantum dots have several applications that are in developmental stages. It may be possible to use them to couple electrical signal amplification, currently based on charge

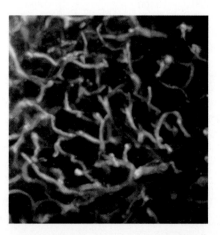

FIGURE 5.18

This 160 μm × 160 μm image was obtained by projecting the capillary structure in a 250-μm-thick specimen of adipose tissue in the skin of a living mouse using CdSe quantum dots that fluoresce at 550 nm.

[Reproduced with permission from Larson *et al., Science*, Vol 300, 30 May 2003, © 2003 American Association for the Advancement of Science.]

a

Pre-injection
autofluorescence

Color video 5 min
post-injection

NIR fluorescence
5 min post-injection

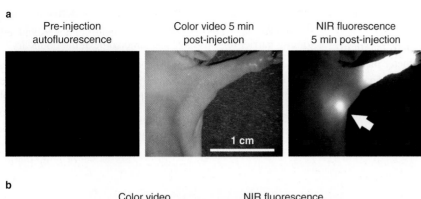

1 cm

b

Color video

NIR fluorescence

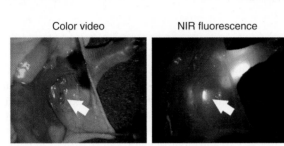

FIGURE 5.19

Quantum dots were injected into the paw of a mouse. **(a)** The middle panel shows a video image taken 5 minutes after the quantum dots were introduced. The right panel is a fluorescence image that shows localization of the quantum dots in the lymph node. The left panel shows that without the quantum dots, no fluorescence is observed. **(b)** Surgery after injection of a chemical mapping agent that is known to localize in lymph nodes confirms that the quantum dots are localized in the lymph nodes.
[From S. Kim *et al.*, *Nature Biotechnology* 22, 93 (2004).]

carrier conduction in semiconductors, with light amplification, in an application known as optoelectronics. Additionally, the reduced dimensions of quantum dots utilized as wavelength-tunable lasers allow them to be integrated into conventional silicon-based microelectronics.

Vocabulary

activation energy
band gap
barrier width
classically forbidden region
conduction band
core electrons
decay length
delocalization

electrical conductor
energy band
finite depth box
ground state
heterostructure
insulator
overlap of wave functions
π-bonded network

quantum dot
quantum well structure
scanning tunneling microscope (STM)
semiconductor
tunneling
valence band
valence electrons
work function

Questions on Concepts

Q5.1   Why is it necessary to apply a bias voltage between the tip and surface in a scanning tunneling microscope?

Q5.2   The amplitude of the wave on the right side of the barrier in Figure 5.8 is much smaller than that of the wave incident on the barrier. What happened to the "rest of the wave"?

Q5.3   Why isn't a tunneling current observed in an STM when the tip and the surface are 1 mm apart?

Q5.4   Redraw Figure 5.7 for an insulator.

Q5.5   Explain how it is possible to create a three-dimensional electron conductor that has a continuous energy spectrum in two dimensions and a discrete energy spectrum in the third dimension.

Q5.6   Explain, without using equations, why tunneling is more likely for the particle with $E = 3/4V_0$ than for $E = 1/4V_0$ in Figure 5.8.

Q5.7   What is the advantage of using quantum dots that fluoresce in the near infrared for surgical applications?

Q5.8   The overlap between wave functions can either be constructive or destructive, just as for waves. Can you distinguish between constructive and destructive overlap for the various energy levels in Figure 5.3?

Q5.9   Explain how you can use size-quantized quantum dots to create a protein with a barcode that can be read using light.

Q5.10   An STM can also be operated in a mode in which electrons tunnel from the surface into the tip. Use Figure 5.9

to explain how you would change the experimental setup to reverse the tunneling current.

Q5.11 For CdSe quantum dots, the emission wavelength increases from 450. nm to 650. nm as the dot diameter increases from 2 to 8 nm. Calculate the band gap energy for these two particle diameters.

Q5.12 Why is it necessary to functionalize CdSe quantum dots with groups such as organic acids to make them useful in bioanalytical applications?

Q5.13 Why must the amplitudes of the first derivatives of the energy eigenfunctions in the finite depth box and in the adjoining barrier regions have the same value at the boundary?

Q5.14 Why must the amplitudes of the energy eigenfunctions in the finite depth box and in the adjoining barrier regions have the same value at the boundary?

Q5.15 Explain how a quantum dot can absorb light over a range of wavelengths and emit light over a much smaller range of wavelengths.

Problems

P5.1 In this problem, you will calculate the transmission probability through the barrier illustrated in Figure 5.8. We first go through the mathematics leading to the solution. You will then carry out further calculations.

The domain in which the calculation is carried out is divided into three regions for which the potentials are

$$V(x) = 0 \quad \text{for } x \le 0 \qquad \text{Region I}$$
$$V(x) = V_0 \quad \text{for } 0 < x < a \qquad \text{Region II}$$
$$V(x) = 0 \quad \text{for } x \ge a \qquad \text{Region III}$$

The spatial part of the wave functions must have the following form in the three regions if $E < V_0$:

$$\psi(x) = A\exp\left[+i\sqrt{\frac{2mE}{\hbar^2}}\,x\right] + B\exp\left[-i\sqrt{\frac{2mE}{\hbar^2}}\,x\right]$$
$$= Ae^{+ikx} + Be^{-ikx} \qquad \text{Region I}$$

$$\psi(x) = C\exp\left[-\sqrt{\frac{2m(V_0 - E)}{\hbar^2}}\,x\right]$$
$$+ D\exp\left[+\sqrt{\frac{2m(V_0 - E)}{\hbar^2}}\,x\right]$$
$$= Ce^{-\kappa x} + De^{+\kappa x} \qquad \text{Region II}$$

$$\psi(x) = F\exp\left[+i\sqrt{\frac{2mE}{\hbar^2}}\,x\right] + G\exp\left[-i\sqrt{\frac{2mE}{\hbar^2}}\,x\right]$$
$$= Fe^{+ikx} + Ge^{-ikx} \qquad \text{Region III}$$

Assume that the wave approaches the barrier from the negative x direction. The coefficient B cannot be set equal to zero because $Be^{-i\sqrt{(2mE/\hbar^2)}\,x}$ represents reflection from the barrier. However, G can be set equal to zero because there is no wave incident on the barrier from the positive x direction.

a. The wave functions and their derivatives must be continuous at $x = 0$ and $x = a$. Show that the coefficients must satisfy the following conditions:

$$A + B = C + D \qquad Ce^{-\kappa a} + De^{+\kappa a} = Fe^{+ika}$$

$$A - B = -\frac{i\kappa}{k}(-C + D) \quad -Ce^{-\kappa a} + De^{+\kappa a} = \frac{ik}{\kappa}Fe^{+ika}$$

b. Because the transmission probability is given by $|F/A|^2$, it is useful to manipulate these equations to get a relationship between F and A. By adding and subtracting the first pair of equations, A and B can be expressed in terms of C and D. The second pair of equations can be combined in the same way to give equations for D and C in terms of F. Show that

$$D = \frac{ik\,e^{+ika} + \kappa e^{+ika}}{2\kappa\,e^{+\kappa a}}F$$

$$C = \frac{-ik\,e^{+ika} + \kappa e^{+ika}}{2\kappa\,e^{-\kappa a}}F, \text{ and}$$

$$A = \frac{(ik - \kappa)C + (ik + \kappa)D}{2ik}$$

c. Substitute these results for C and D in terms of F into

$$A = \frac{(ik - \kappa)C + (ik + \kappa)D}{2ik}$$

to relate A and F. Show that

$$2ikA = \frac{e^{+ika}}{2\kappa}[(ik - \kappa)(-ik + \kappa)e^{+\kappa a}$$
$$+ (ik + \kappa)(ik + \kappa)e^{-\kappa a}]F$$

d. Using the hyperbolic trigonometric functions

$$\sinh x = \frac{e^x - e^{-x}}{2} \text{ and } \cosh x = \frac{e^x + e^{-x}}{2}$$

and the relationship $\cosh^2 x - \sinh^2 x = 1$, show that

$$\left|\frac{F}{A}\right|^2 = \frac{16(\kappa k)^2}{16(\kappa k)^2 + (4(k^2 - \kappa^2)^2 + 16(\kappa k)^2)\sinh^2(\kappa a)}$$

$$= \frac{1}{1 + [(k^2 + \kappa^2)^2 \sinh^2(\kappa a)]/4(\kappa k)^2}$$

e. Plot the transmission probability for an electron as a function of energy for $V_0 = 1.6 \times 10^{-19}$ J and $a = 9.0 \times 10^{-10}$ m up to an energy of 8×10^{-19} J. At what energy is the tunneling probability 0.1? At what energy is the tunneling probability 0.02?

f. Plot the transmission probability for an electron of energy 0.50×10^{-19} J as a function of the barrier width for $V_0 = 1.6 \times 10^{-19}$ J between 2×10^{-10}

and 8×10^{-10} m. At what barrier width is the transmission probability 0.2?

P5.2 Semiconductors can become conductive if their temperature is raised sufficiently to populate the (empty) conduction band from the highest filled levels in the valence band. The ratio of the populations in the highest level of the conduction band to that of the lowest level in the valence band is

$$\frac{n_{conduction}}{n_{valence}} = \frac{g_{conduction}}{g_{valence}} e^{-\Delta E/kT}$$

where ΔE is the band gap, which is 0.661 eV for Ge and 5.5 eV for diamond. Assume for simplicity that the ratio of the degeneracies is one and that the semiconductor becomes sufficiently conductive when

$$\frac{n_{conduction}}{n_{valence}} = 1.0 \times 10^{-6}$$

At what temperatures will germanium and diamond become sufficiently conductive? Given that diamond sublimates near 3000 K, could you heat diamond enough to make it conductive and not sublimate it?

P5.3 For the π-network of β-carotene modeled using the particle in the box, the position-dependent probability density of finding 1 of the 22 electrons is given by

$$P_n(x) = |\psi_n(x)|^2 = \frac{2}{a} \sin^2\left(\frac{n\pi x}{a}\right)$$

The quantum number n in this equation is determined by the energy level of the electron under consideration. As we saw in Chapter 4, this function is strongly position dependent. The question addressed in this problem is as follows: Would you also expect the total probability density defined by $P_{total}(x) = \sum_n |\psi_n(x)|^2$ to be strongly position dependent? The sum is over all the electrons in the π-network.

a. Calculate the total probability density $P_{total}(x) = \sum_n |\psi_n(x)|^2$ using the box length $a = 29.0$ nm and plot your results as a function of x. Does $P_{total}(x)$ have the same value near the ends and at the middle of the molecule?

b. Determine $\Delta P_{total}(x)/\langle P_{total}(x)\rangle$, where $\Delta P_{total}(x)$ is the peak-to-peak amplitude of $P_{total}(x)$ in the interval between 12.0 and 16.0 nm.

c. Compare the result of part (b) with what you would obtain for an electron in the highest occupied energy level.

d. What value would you expect for $P_{total}(x)$ if the electrons were uniformly distributed over the molecule? How does this value compare with your result from part (a)?

P5.4 Calculate the energy levels of the π-network in hexatriene, C_6H_8, using the particle in the box model. To calculate the box length, assume that the molecule is linear and use the values 135 and 154 pm for C=C and C—C bonds. What is the wavelength of light required to induce a

transition from the ground state to the first excited state? How does this compare with the experimentally observed value of 240 nm? What does the comparison made suggest to you about estimating the length of the π-network by adding bond lengths for this molecule?

P5.5 Calculate the energy levels of the π-network in octatetraene, C_8H_{10}, using the particle in the box model. To calculate the box length, assume that the molecule is linear and use the values 135 and 154 pm for C=C and C—C bonds. What is the wavelength of light required to induce a transition from the ground state to the first excited state?

P5.6 The maximum safe current in a copper wire with a diameter of 3.0 mm is about 20. amperes. In an STM, a current of 7.5×10^{-10} A passes from the tip to the surface in a filament of diameter ~1.5 nm. Compare the current density in the copper wire with that in the STM.

P5.7 In this problem, you will solve for the total energy eigenfunctions and eigenvalues for an electron in a finite depth box. We first go through the calculation for the box parameters used in Figure 5.1. You will then carry out the calculation for a different set of parameters.

We describe the potential in this way:

$$V(x) = V_0 \quad \text{for } x \le -\frac{a}{2} \qquad \text{Region I}$$

$$V(x) = 0 \quad \text{for } -\frac{a}{2} < x < \frac{a}{2} \quad \text{Region II}$$

$$V(x) = V_0 \quad \text{for } x \ge \frac{a}{2} \qquad \text{Region III}$$

The eigenfunctions must have the following form in these three regions:

$$\psi(x) = B\exp\left[+\sqrt{\frac{2m(V_0 - E)}{\hbar^2}}\,x\right]$$
$$+ B'\exp\left[-\sqrt{\frac{2m(V_0 - E)}{\hbar^2}}\,x\right]$$
$$= Be^{+\kappa x} + B'e^{-\kappa x} \qquad \text{Region I}$$

$$\psi(x) = C\sin\sqrt{\frac{2mE}{\hbar^2}}\,x + D\cos\sqrt{\frac{2mE}{\hbar^2}}\,x$$
$$= C\sin kx + D\cos kx \qquad \text{Region II}$$

$$\psi(x) = A\exp\left[-\sqrt{\frac{2m(V_0 - E)}{\hbar^2}}\,x\right]$$
$$+ A'\exp\left[+\sqrt{\frac{2m(V_0 - E)}{\hbar^2}}\,x\right]$$
$$= Ae^{-\kappa x} + A'e^{+\kappa x} \qquad \text{Region III}$$

So that the wave functions remain finite at large positive and negative values of x, $A' = B' = 0$. An additional condition must also be satisfied. To arrive at physically meaningful solutions for the eigenfunctions, the wave functions in the separate regions must have the same amplitude and derivatives at the values of $x = a/2$ and $x = -a/2$ bounding the regions. This restricts the possible values for the coefficients A, B, C, and D. Show that applying these conditions gives the following equations:

$$Be^{-\kappa(a/2)} = -C\sin k\frac{a}{2} + D\cos k\frac{a}{2}$$

$$B\kappa e^{-\kappa(a/2)} = Ck\cos k\frac{a}{2} + Dk\sin k\frac{a}{2}$$

$$Ae^{-\kappa(a/2)} = C\sin k\frac{a}{2} + D\cos k\frac{a}{2}$$

$$-A\kappa e^{-\kappa(a/2)} = Ck\cos k\frac{a}{2} - Dk\sin k\frac{a}{2}$$

These two pairs of equations differ on the right side only by the sign of one term. We can obtain a set of equations that contain fewer coefficients by adding and subtracting each pair of equations to give

$$(A + B)e^{-\kappa(a/2)} = 2D\cos\left(k\frac{a}{2}\right)$$

$$(A - B)e^{-\kappa(a/2)} = 2C\sin\left(k\frac{a}{2}\right)$$

$$(A + B)\kappa e^{-\kappa(a/2)} = 2Dk\sin\left(k\frac{a}{2}\right)$$

$$-(A - B)\kappa e^{-\kappa(a/2)} = 2Ck\cos\left(k\frac{a}{2}\right)$$

At this point we notice that by dividing the equations in each pair, the coefficients can be eliminated to give

$$\kappa = k\tan\left(k\frac{a}{2}\right) \quad\text{or}\quad \sqrt{\frac{2m(V_0 - E)}{\hbar^2}}$$

$$= \sqrt{\frac{2mE}{\hbar^2}}\tan\left(\sqrt{\frac{2mE}{\hbar^2}}\frac{a}{2}\right) \quad\text{and}$$

$$-\kappa = k\cot an\left(k\frac{a}{2}\right) \quad\text{or}\quad -\sqrt{\frac{2m(V_0 - E)}{\hbar^2}}$$

$$= \sqrt{\frac{2mE}{\hbar^2}}\cot\left(\sqrt{\frac{2mE}{\hbar^2}}\frac{a}{2}\right)$$

Multiplying these equations on both sides by $a/2$ gives dimensionless parameters and the final equations are

$$\sqrt{\frac{m(V_0 - E)a^2}{2\hbar^2}} = \sqrt{\frac{mEa^2}{2\hbar^2}}\tan\sqrt{\frac{mEa^2}{2\hbar^2}} \quad\text{and}$$

$$-\sqrt{\frac{m(V_0 - E)a^2}{2\hbar^2}} = \sqrt{\frac{mEa^2}{2\hbar^2}}\cot\sqrt{\frac{mEa^2}{2\hbar^2}}$$

The allowed energy values E must satisfy these equations. They can be obtained by graphing the two sides of each equation against E. The intersections of the two curves are the allowed energy eigenvalues. For the parameters in the caption of Figure 5.1, $V_0 = 1.20 \times 10^{-18}$ J and $a = 1.00 \times 10^{-9}$ m, the following two graphs are obtained:

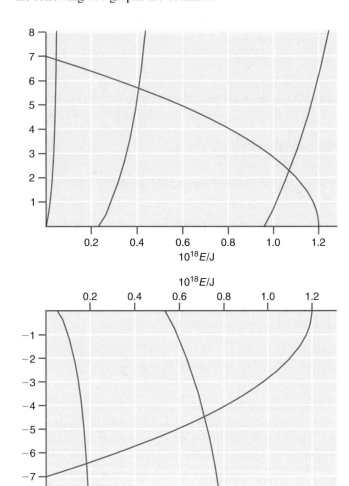

The five allowed energy levels are at 4.61×10^{-20}, 4.09×10^{-19}, and 1.07×10^{-18} J (top figure), and 1.84×10^{-19}, and 7.13×10^{-19} J (bottom figure).

a. Given these values, calculate λ for each energy level. Is the relation $\lambda = 2a/n$ (for n an integer) that arose from the calculations on the infinitely deep box still valid? Compare the values with the corresponding energy level in the infinitely deep box. Explain why the differences arise.

b. Repeat this calculation for $V_0 = 5.00 \times 10^{-19}$ J and $a = 0.900 \times 10^{-9}$ m. Do you think that there will be fewer or more bound states than for the problem just worked out? How many allowed energy levels are there for this well depth and what is the energy corresponding to each level?

Computational Problems

More detailed instructions on carrying out this calculation using Spartan Physical Chemistry are found on the book website at *www.chemplace.com*.

C5.1 Build (a) ethylene, (b) the trans conformation for 1,3 butadiene, and (c) all trans hexatriene and calculate the ground-state (singlet) energy of these molecules using the B3LYP method with the 6-311+G** basis set. Repeat your calculation for the triplet state, which corresponds to the excitation of a π electron from the highest filled energy level to the lowest unoccupied energy level. Use a nonplanar input geometry for the triplet states. Compare the energy difference from these calculations to literature values of the maximum in the UV-visible absorption spectrum.

Web-Based Simulations, Animations, and Problems

W5.1 The Schrödinger equation is solved numerically for the particle in the finite height box. Using the condition that the wave function must approach zero amplitude in the classically forbidden region, the energy levels are determined for a fixed particle mass, box depth, and box length. The particle mass and energy and the box depth and length are varied with sliders to demonstrate how the number of bound states varies with these parameters.

W5.2 The Schrödinger equation is solved numerically to calculate the tunneling probability for a particle through a thin finite barrier. Sliders are used to vary the barrier width and height and the particle energy and mass. The dependence of the tunneling probability on these variables is investigated.

6

Commuting and Noncommuting Operators and the Surprising Consequences of Entanglement

Classical physics predicts that there is no limit to the amount of information (observables) that can be known about a system at a given instant of time. This is not the case in quantum mechanics. Two observables can be known simultaneously only if the outcome of the measurements is independent of the order in which they are conducted. An uncertainty relation limits the degree to which observables of noncommuting operators can be known simultaneously. Although this result is counterintuitive from a classical perspective, the Stern–Gerlach experiment clearly demonstrates that this prediction of quantum mechanics is obeyed at the atomic level. Because a quantum state can be a superposition of individual states, two particles can be entangled. Entanglement is the basis of both teleportation and quantum computing.

6.1 COMMUTATION RELATIONS

In classical mechanics, a system under consideration can in principle be described completely. For instance, for a mass falling in a gravitational field, its position, momentum, kinetic energy, and potential energy can be determined simultaneously at any point on its trajectory. The uncertainty in the measurements is seemingly only limited by the capabilities of the measurement technique. All of these observables (and many more) can be known simultaneously. This is not generally true from a quantum mechanical perspective. In the quantum world, some observables can be known simultaneously with high accuracy, but others have a fundamental uncertainty that cannot be removed through more sensitive measurement techniques. However, as will be shown later, in the classical limit of very large quantum numbers, the fundamental uncertainty for these observables is less than the uncertainty associated with experimental techniques.

The values of two different observables a and b, which correspond to the operators $\hat{A}$ and $\hat{B}$, can be simultaneously determined only if the measurement process used does not change the state of the system. Otherwise, the system on which the two measurements is carried out is not the same. Let $\psi_n(x)$ be the wave function that characterizes the system. How can the measurements of the observables corresponding to the operators $\hat{A}$ and $\hat{B}$ be described? Carrying out a measurement of the observables corresponding first to the operator $\hat{A}$ and subsequently to the operator $\hat{B}$ is equivalent to evaluating $\hat{B}[\hat{A}\psi_n(x)]$. If $\psi_n(x)$ is an eigenfunction of $\hat{A}$, then $\hat{B}[\hat{A}\psi_n(x)] = \alpha_n\hat{B}\psi_n(x)$. The only case in which the second measurement does not change the state of the system is if $\psi_n(x)$ is also an eigenfunction of $\hat{B}$. In this case, $\hat{B}[\hat{A}\psi_n(x)] = \beta_n\alpha_n\psi_n(x)$. Reversing the order of the two operations gives $\hat{A}[\hat{B}\psi_n(x)] = \alpha_n\beta_n\psi_n(x)$. Because the eigenvalues β_n and α_n are simply constants, $\beta_n\alpha_n\psi_n(x) = \alpha_n\beta_n\psi_n(x)$ and, therefore, $\hat{B}[\hat{A}\psi_n(x)] = \hat{A}[\hat{B}\psi_n(x)]$.

We have just shown that the act of measurement changes the state of the system unless the system wave function is an eigenfunction of the two different operators. Therefore, this is a condition for being able to simultaneously know the observables corresponding to these operators. How can one know if two operators have a common set of eigenfunctions? The example just discussed suggests a simple test that can be applied. Only if

$$\hat{A}[\hat{B}\,f(x)] - \hat{B}[\hat{A}\,f(x)] = 0 \tag{6.1}$$

for $f(x)$, an arbitrary function, will $\hat{A}$ and $\hat{B}$ have a common set of eigenfunctions, and only then can the corresponding observables be known simultaneously.

If two operators have a common set of eigenfunctions, we say that they **commute**. The difference $\hat{A}[\hat{B}\,f(x)] - \hat{B}[\hat{A}\,f(x)]$ is abbreviated $[\hat{A}, \hat{B}]f(x)$ and the expression in the square brackets is called the **commutator** of the operators $\hat{A}$ and $\hat{B}$. If the value of the commutator is not zero for an arbitrary function $f(x)$, the corresponding observables cannot be determined simultaneously and exactly. We will have more to say about what is meant by *exactly* later in this chapter.

EXAMPLE PROBLEM 6.1

Determine whether the momentum and (a) the kinetic energy and (b) the total energy can be known simultaneously.

Solution

To solve these problems, we determine whether two operators $\hat{A}$ and $\hat{B}$ commute by evaluating the commutator $\hat{A}[\hat{B}\,f(x)] - \hat{B}[\hat{A}\,f(x)]$. If the commutator is zero, the two observables can be determined simultaneously and exactly.

a. For momentum and kinetic energy, we evaluate

$$-i\hbar\frac{d}{dx}\left(-\frac{\hbar^2}{2m}\frac{d^2}{dx^2}\right)f(x) - \left(-\frac{\hbar^2}{2m}\frac{d^2}{dx^2}\right)\left(-i\hbar\frac{d}{dx}\right)f(x)$$

In calculating the third derivative, it does not matter if the function is first differentiated twice and then once or the other way around. Therefore, the momentum and the kinetic energy can be determined simultaneously and exactly.

b. For momentum and total energy, we evaluate

$$-i\hbar\frac{d}{dx}\left(-\frac{\hbar^2}{2m}\frac{d^2}{dx^2} + V(x)\right)f(x) - \left(-\frac{\hbar^2}{2m}\frac{d^2}{dx^2} + V(x)\right)\left(-i\hbar\frac{d}{dx}\right)f(x)$$

Because the kinetic energy and momentum operators commute, per part (a), this expression is equal to

$$-i\hbar\frac{d}{dx}(V(x)f(x)) + i\hbar V(x)\frac{d}{dx}f(x)$$

$$= -i\hbar V(x)\frac{d}{dx}f(x) - i\hbar f(x)\frac{d}{dx}V(x) + i\hbar V(x)\frac{d}{dx}f(x)$$

$$= -i\hbar f(x)\frac{d}{dx}V(x)$$

We conclude the following:

$$\left[V(x), -i\hbar\frac{d}{dx}\right] = -i\hbar\frac{d}{dx}V(x) \neq 0$$

Therefore, the momentum and the total energy cannot be known simultaneously and exactly. Note that the arbitrary function $f(x)$ is not present in the final expression for the commutator. Note also that the momentum and the total energy can be known simultaneously if $[dV(x)]/dx = 0$. This corresponds to a constant potential energy for all values of x, in other words, the free particle of Section 4.1.

Now apply the formalism just discussed to the particle in the box in its lowest energy state. In Chapter 4, we found that although the wave function is an eigenfunction of the total energy operator, it is not an eigenfunction of the momentum operator. Therefore, these two operators do not commute. If the total energy of the particle is measured, the value $E = h^2/8ma^2$ is obtained. If the average momentum is subsequently determined from a number of individual measurements, the result is $\langle p_x \rangle = 0$. This result merely states that it is equally likely that positive and negative values will be obtained. There is no way of knowing what the magnitude and sign of the momentum will be for an individual measurement. Because the energy is known precisely, nothing is known about the momentum. This result is consistent with the fact that the two operators do not commute.

6.2 THE STERN–GERLACH EXPERIMENT

Consider now a real experiment in a simple quantum mechanical framework that illustrates some of the concepts discussed in the preceding section in more concrete terms. This experiment also illustrates how quantum mechanical concepts of measurement arose out of analyzing results obtained in the laboratory. In this experiment, a beam of silver atoms having a well-defined direction is passed through a magnetic field that has a constant value in the x–y plane and varies linearly with the z coordinate, which is chosen to be perpendicular to the path of the atoms. We say that the magnetic field has a gradient in the z direction. An atomic beam of silver atoms can be made in a vacuum system by heating solid silver in an oven to a temperature at which the vapor pressure of Ag is in the range of 10^{-2} torr. Letting the atoms escape through a series of collimating apertures in the wall of the oven results in a beam of Ag atoms, all traveling in the same direction, which is chosen to be the y direction. The atoms pass through the magnetic field and are detected some distance beyond the magnet. The torque acting on the individual Ag atoms is depicted in Figure 6.1, and the **Stern–Gerlach experiment** is shown schematically in Figure 6.2.

Silver atoms have a single unpaired electron that has an intrinsic magnetic moment. We return to the consequences of this fact later. The magnetic moment is associated with what is called the *electron spin*, although the picture of a spherical electron spinning around an axis through its center is incorrect. It turns out that the spin emerges naturally in relativistic quantum mechanics. (View these remarks as an aside because none of this was known at the time the Stern–Gerlach experiment was conducted.)

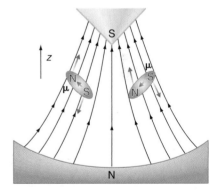

FIGURE 6.1

The effect of an inhomogeneous magnetic field on magnetic dipoles is to orient and deflect them in opposite directions, depending on the sign of the component of the magnetic moment along the z direction.

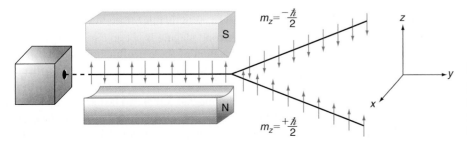

FIGURE 6.2

Schematic representation of the Stern–Gerlach experiment. The inhomogeneous magnetic field separates the beam into two, and only two, components.

Because each atom has a magnetic moment associated with the unpaired electron, the atom is deflected in the z direction as it passes through the inhomogeneous magnetic field. The atom is not deflected along the x and y directions, because the magnetic field is constant along these directions. What outcome is expected in this experiment? Consider the classical system of a beam of magnetic dipoles. We expect that the magnetic dipoles are randomly oriented in space and that only their z component is affected by the magnet. Because the z component takes on all possible values between $+|\boldsymbol{\mu}|$ and $-|\boldsymbol{\mu}|$, where $\boldsymbol{\mu}$ is the magnetic moment of the atom, the silver atoms will be equally distributed along a range of z values at the detector. The z values can be predicted from the geometry of the experiment and the strength of the field gradient if the magnetic moment is known.

What are the results of the experiment? Silver atoms are deflected only in the z direction, but only two z values are observed. One corresponds to an upward deflection and the other to a downward deflection of the same magnitude. *What conclusions can be drawn from this experiment?* We conclude that the operator called "measure the z component of the magnetic moment," denoted by $\hat{A}$, has only two eigenfunctions with eigenvalues that are equal in magnitude, but opposite in sign. We call the two eigenfunctions α and β and assume that they are normalized. Because the experiment shows that these two eigenfunctions form a complete set (only two deflection angles are observed), any acceptable wave function can be written as a linear combination of α and β. Therefore, the initial normalized wave function that describes a single silver atom is

$$\psi = c_1\alpha + c_2\beta \quad \text{with } |c_1|^2 + |c_2|^2 = 1 \tag{6.2}$$

We cannot specify the values of c_1 and c_2, because they refer to individual measurements, and only the total number of silver atoms in the two deflected beams at the detector is measured. However, after the beam has passed through the magnet for a time, the relative number of Ag atoms that was deflected upward and downward can be measured. This ratio is one, and therefore $|c_1|^2_{average} = |c_2|^2_{average} = 1/2$. The averaging is over all the atoms that have landed on the detector.

Now carry this experiment a step further. We follow the path of the downwardly deflected atoms, which have the wave function $\psi = \alpha$ and deflect them once again. However, this time the magnet has been turned 90° so that the magnetic field gradient is in the x direction. Note that now there is an inhomogeneity in the x direction, such that the atoms are separated along this direction. The operator is now "measure the x component of the magnetic moment," which is denoted $\hat{B}$. The experiment shows that this operator also has two and only two eigenfunctions that we call δ and γ. They have the same eigenvalues as α and β, respectively. If the relative number of Ag atoms deflected in the $+x$ and $-x$ directions is measured, the ratio is determined to be one. We conclude that the wave function prior to entering the second magnet was

$$\psi = c_3\delta + c_4\gamma \quad \text{with } |c_3|^2 + |c_4|^2 = 1 \tag{6.3}$$

As before, note that $|c_3|^2_{average} = |c_4|^2_{average} = 1/2$.

Now comes the punch line. We ask the question "Do the operators $\hat{A}$ and $\hat{B}$ commute?" This question is answered by repeating the first measurement, to see if the state of the system has been changed by carrying out the second measurement. Experimentally, a third magnet that has the same alignment as the first magnet is added. This third magnet acts on one of the two separated beams that have emerged from the second magnet, as shown in Figure 6.3. If the operators commute, a single downwardly deflected beam of Ag atoms corresponding to $\psi = \alpha$ will be observed. If they do not commute, the wave function for the atoms entering the third magnet will no longer be an eigenfunction of $\hat{A}$ and two beams will be observed. Why is this? If the wave function that describes the Ag atom emerging from the second magnet is not an eigenfunction of $\hat{A}$, it can be still represented as a linear combination of the eigenfunctions of $\hat{A}$. A state whose wave function is a linear combination of α and β will give rise to two deflected beams of Ag atoms.

The result of the experiment with the third magnet is that two beams emerge, just as was seen from the first magnet! *We conclude that the operators $\hat{A}$, "measure the z component of the magnetic moment," and $\hat{B}$, "measure the x component of the magnetic mo-*

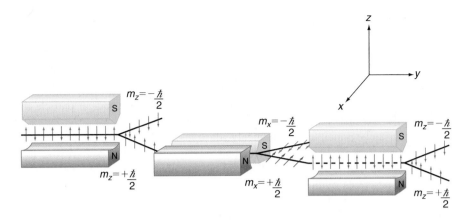

FIGURE 6.3

One of the beams exiting from the first magnet has been passed through a second magnet rotated by 90°. Again the beam is split into two components. The third magnet gives a result that is different than what would have been expected from classical physics.

ment," *do not commute*. This means that a silver atom does not simultaneously have well-defined values for both μ_z and μ_x. This is, of course, not the conclusion you would have reached by applying classical mechanics to a classical magnetic moment. The experiment is a good illustration of how the quantum mechanical postulates arose from consideration of the outcomes of experiments.

Because the magnetic moment and the angular momentum of a charged particle differ only by a multiplicative constant, we have also shown that the operators for the individual components of the angular momentum vector do not commute. The consequences of this result will be discussed in Chapter 7.

6.2.1 The History of the Stern–Gerlach Experiment

This classic experiment, carried out in 1921, was designed to distinguish between the quantum mechanical model of the atom proposed by Niels Bohr and classical planetary models. A silver beam generated by an oven in a vacuum chamber was collimated by two narrow slits of 0.03-mm width. The beam passed through an inhomogeneous magnet 3.5 cm in length and impinged on a glass plate. After about an hour of operation, the plate was removed and examined visually. Only about one atomic layer of Ag was deposited on the plate in this time, making the detection of the spatial distribution of the silver atoms very difficult. The key to their successful detection was that both Stern and Gerlach smoked cheap cigars with a high sulfur content. The sulfur-containing smoke reacted with the Ag atoms, producing Ag_2S, which was clearly visible under a microscope, even though the amount deposited was less than 10^{-7} mol. Upon successful completion of the experiment, Gerlach sent Bohr the following postcard, which shows the result obtained without the magnetic field (left) and with the magnetic field (right). The

Courtesy of the Niels Bohr Archive, Copenhagen.

splitting of the beam into two distinct components is clearly visible. The handwritten notes explain the experiment and congratulate Bohr, saying that the results confirm his theory.

Although the results did not confirm the classical model of the atom, the agreement with the Bohr model turned out to be fortuitous and incorrect. Several years later, researchers discovered that the electron has an intrinsic angular moment (spin). This angular moment—and not a magnetic moment produced by electrons orbiting around the nucleus—is the basis for the deflection observed. A more detailed account of this experiment can be found in an article by B. Friedrich and D. Herschbach in the December 2003 issue of *Physics Today*.

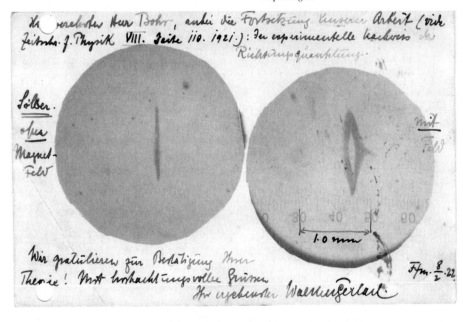

6.3 THE HEISENBERG UNCERTAINTY PRINCIPLE

The best-known case of noncommuting operators concerns position and momentum and is associated with the **Heisenberg uncertainty principle**. This principle quantifies the uncertainty in the position and momentum of a quantum mechanical particle that arises from the fact that $[\hat{x}, \hat{p}_x] \neq 0$.

The uncertainty principle can be nicely illustrated with the free particle. As discussed in Section 4.1, the free-particle total energy eigenfunctions have the form $\Psi(x, t) = A \exp i(kx - \omega t - \phi)$. What can be said about the position and momentum of states described by this wave function? It is convenient to set $\phi = 0$ and $t = 0$ so that we can focus on the spatial variation of $\psi(x)$.

By operating on this wave function with the momentum operator, it can be easily shown that it is an eigenfunction of the momentum operator with the eigenvalue $\hbar k = hk/2\pi$. To discuss probability, this wave function must be normalized. As shown in Section 4.1, a plane wave cannot be normalized over an interval that is infinite, but it can be normalized over the finite interval $-L \leq x \leq L$:

$$\int_{-L}^{L} A^*\psi^*(x) A\psi(x)\, dx = 1$$

$$A^*A \int_{-L}^{L} e^{-ikx} e^{ikx}\, dx = 1 \tag{6.4}$$

$$|A| = \frac{1}{\sqrt{2L}}$$

Now that the function is normalized, we calculate the probability of finding the particle near $x = x_0$:

$$P(x_0)\, dx = \psi^*(x_0) \psi(x_0)\, dx \tag{6.5}$$

We see that the probability is $P(x)\, dx = dx/2L$ independent of position. This means that it is equally probable that the particle will be found anywhere. Now let the interval length L become arbitrarily large. The probability of finding the particle within the interval dx centered at $x = x_0$ approaches zero! *We conclude that if a particle is prepared in a state in which the momentum is exactly known, then its position is completely unknown.* It turns out that if a particle is prepared such that its position is exactly known (the wave function is an eigenfunction of the position operator), then its momentum is completely unknown.

This result is completely at variance with expectations based on classical mechanics, because a simultaneous knowledge of position and momentum is essential to calculating trajectories of particles subject to forces. How can this counterintuitive result be understood? Note that a single plane wave with a precisely specified wavelength has been used to represent the particle. This means that the momentum of the particle is known precisely. As a result, nothing is known about the position of the particle.

The uncertainty in position arises because the momentum is precisely known. Is it possible to construct a wave function for which the momentum is not precisely known? Will such a wave function give more information about the position of the particle than the plane wave $\Psi(x, t) = A \exp i(kx - \omega t - \phi)$ does? These questions can be answered by constructing a wave function that is a superposition of several plane waves and then examining its properties. Consider the superposition of plane waves of very similar wave vectors given by

$$\psi(x) = \frac{1}{2} A\, e^{ik_0 x} + \frac{1}{2} A \sum_{n=-m}^{n=m} e^{i(k_0 + n\Delta k)x}, \quad \text{with } \Delta k \ll k \tag{6.6}$$

Convince yourself that this superposition wave function is not an eigenfunction of the momentum operator. The upper portion of Figure 6.4 shows the real part of each of the 21 individual terms in an interval of approximately seven wavelengths about an arbitrarily

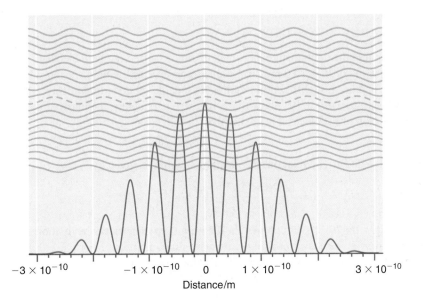

FIGURE 6.4

The top part of the figure shows 21 waves, each of which has zero amplitude outside the range of distances shown. They have been displaced vertically for purposes of display. The bottom part shows the probability density $\psi^*(x)\psi(x)$ resulting from adding all 21 waves. The wave vector k_0 has the value $7.00 \times 10^{10} \text{ m}^{-1}$.

chosen zero of distance for $m = 10$. We also choose to make the amplitude of the wave function zero outside of the range of distances shown. This ensures that the particle is somewhere in the interval.

How does the amplitude of $\psi(x)$ vary over the interval? At $x = 0$, all 21 waves constructively interfere, but at $x = \pm 3.14 \times 10^{-10}$ m, they undergo destructive interference. Consequently, the wave function, which is a superposition of these waves, has a maximum amplitude at $x = 0$ and a value of zero at $x = \pm 3.14 \times 10^{-10}$ m. The amplitude oscillates about zero at intermediate values of x. How does the probability density vary over the interval? Evaluating Equation (6.6) for $m = 10$ and forming $|\psi(x)|^2$, the function shown in the lower part of Figure 6.4 is obtained. Because $|\psi(x)|^2$ is strongly peaked at the center of the interval, by superposing these 21 waves, we see that the particle has been localized. The oscillations shown in Figure 6.4 are a result of having taken only 21 terms in the superposition. They would disappear, leaving a broad smooth curve which is the envelope of the red curve, if an infinite number of waves of intermediate wavelengths had been included in the superposition.

What does this calculation show? Because $\psi(x)$ is not an eigenfunction of the momentum operator, an uncertainty is connected with the momentum of the particle. In going from a single plane wave to the superposition function $\psi(x)$, the uncertainty in momentum has increased. As the curve for $|\psi(x)|^2$ in Figure 6.4 shows, increasing the uncertainty in momentum has decreased the uncertainty in position. Such a superposition wave function is referred to as a **wave packet** because it has wave character, but is localized to a finite interval.

Because 21 waves of differing momentum have been superposed to construct the wave function, the momentum is no longer exactly known. Can we make this statement more quantitative? The value of p is known fairly well if $\Delta k \ll k_0$ because an individual measurement of the momentum for a state described by $\psi(x)$ gives values in the following range:

$$\hbar(k_0 - m\Delta k) \leq p \leq \hbar(k_0 + m\Delta k) \qquad \textbf{(6.7)}$$

Comparing the results just obtained for $\psi(x)$ with those for a single plane wave of precisely determined momentum allows the following conclusion to be made: *as a result of the superposition of many plane waves, the position of the particle is no longer completely unknown, and the momentum of the particle is no longer exactly known.* Figure 6.4 shows that the approximate position of the particle can be known as long as an uncertainty in its momentum can be tolerated. The lesson of this discussion is that both position and momentum cannot be known *exactly and simultaneously* in quantum mechanics. We must accept a trade-off between the uncertainty in p and that of x. This result was quantified by Heisenberg in his famous uncertainty principle:

$$\Delta p \, \Delta x \geq \frac{\hbar}{2} \qquad \textbf{(6.8)}$$

6.1 The Heisenberg Uncertainty Principle

EXAMPLE PROBLEM 6.2

Assume that the double-slit experiment could be carried out with electrons using a slit spacing of $b = 10.0$ nm. To be able to observe diffraction, we choose $\lambda = b$, and because diffraction requires reasonably monochromatic radiation, we choose $\Delta p / p = 0.01$. Show that with these parameters, the uncertainty in the position of the electron is greater than the slit spacing b.

Solution

Using the de Broglie relation, the mean momentum is given by

$$\langle p \rangle = \frac{h}{\lambda} = \frac{6.626 \times 10^{-34} \text{ J s}}{1.00 \times 10^{-8} \text{ m}} = 6.626 \times 10^{-26} \text{ kg m s}^{-1}$$

and $\Delta p = 6.626 \times 10^{-28}$ kg m s^{-1}. The minimum uncertainty in position is given by

$$\Delta x = \frac{\hbar}{2\Delta p} = \frac{1.055 \times 10^{-34} \text{ J s}}{2 \times 6.626 \times 10^{-28} \text{ kg m s}^{-1}} = 7.96 \times 10^{-8} \text{ m}$$

which is greater than the slit spacing. Note that the concept of an electron trajectory is not well defined under these conditions. This offers an explanation for the observation that the electron appears to go through both slits simultaneously!

6.2 Wave Packets and the Uncertainty Principle

If the right-hand side of the inequality were equal to zero instead of $\hbar/2$, then it would be possible to know both the position and momentum exactly. This is not the case and, therefore, it is not possible to calculate a trajectory of a quantum mechanical particle exactly. The trajectory of a particle for which the momentum and energy are exactly known is not a well-defined concept in quantum mechanics. However, you can get a good approximation for a "trajectory" in quantum mechanical systems by using wave packets.

What is the practical effect of the uncertainty principle? Does this mean that you have no idea what trajectories the electrons in your TV picture tube will follow or where a baseball thrown by a pitcher will pass a waiting batter? As mentioned earlier, this gets down to what is meant by *exact*. An exact trajectory could be calculated if $\hbar$ were equal to zero, rather than being a small number. However, because $\hbar$ is a very small number, the uncertainty principle does not affect the calculation of the trajectories of baseballs, rockets, or other macroscopic objects. Although the uncertainty principle holds for both electrons and for baseballs, the effect is so small that it is not detectable for large masses.

EXAMPLE PROBLEM 6.3

The electrons in a TV picture tube have an energy of about 1.00×10^4 eV. If $\Delta p / p = 0.01$ in the direction of the electron trajectory for this case, calculate the minimum uncertainty in the position that defines where the electrons land on the phosphor in the picture tube.

Solution

Using the relation $\langle p \rangle = \sqrt{2mE}$, the momentum is calculated as follows:

$$\langle p \rangle = \sqrt{2 \times 9.109 \times 10^{-31} \text{ kg} \times 1.00 \times 10^4 \text{ eV} \times 1.602 \times 10^{-19} \text{ J/eV}}$$
$$= 5.41 \times 10^{-23} \text{ kg m s}^{-1}$$

Proceeding as in Example Problem 6.2,

$$\Delta x = \frac{\hbar}{2\Delta p} = \frac{1.055 \times 10^{-34} \text{ J s}}{2 \times 5.41 \times 10^{-25} \text{ kg m s}^{-1}} = 9.8 \times 10^{-11} \text{ m}$$

This distance is much smaller than could be measured and, therefore, the uncertainty principle has no effect in this instance.

EXAMPLE PROBLEM 6.4

An (over)educated baseball player tries to convince his manager that he cannot hit a 100 mile per hour (26.8 m s^{-1}) baseball that has a mass of 140 g and relative momentum uncertainty of 1% because the uncertainty principle does not allow him to estimate its position within 0.1 mm. Is his argument valid?

Solution

The momentum is calculated using the following equation:

$p = mv = 0.140 \text{ kg} \times 26.8 \text{ m s}^{-1} = 3.75 \text{ kg m s}^{-1}$, and $\Delta p = 0.0375 \text{ kg m s}^{-1}$

Substituting in the uncertainty principle,

$$\Delta x = \frac{\hbar}{2\Delta p} = \frac{1.055 \times 10^{-34} \text{ J s}}{2 \times 0.0375 \text{ kg m s}^{-1}} = 1.41 \times 10^{-33} \text{ m}$$

The uncertainty is not zero, but it is well below the experimental sensitivity. Sorry, back to the minor leagues.

This result—that it is not possible to know the exact values of two observables simultaneously—is not restricted to position and momentum. It applies to any two observables whose corresponding operators do not commute. Energy and time are another example of two observables that are linked by an uncertainty principle. The energy of the H atom with the electron in the $1s$ state can only be known to high accuracy because it has a very long lifetime. This is the case because there is no lower state to which it can decay. Excited states that rapidly decay to the ground state have an uncertainty in their energy. Evaluation of the commutator is the means used to test whether any two observables can be determined simultaneously and exactly.

SUPPLEMENTAL

6.4 THE HEISENBERG UNCERTAINTY PRINCIPLE EXPRESSED IN TERMS OF STANDARD DEVIATIONS

This section addresses the topic of how to use the Heisenberg uncertainty principle in a quantitative fashion. This inequality can be written in the form

$$\sigma_x \sigma_p \geq \frac{\hbar}{2} \qquad \textbf{(6.9)}$$

In this equation, σ_p and σ_x are the standard deviations that would be obtained by analyzing the distribution of a large number of measured values of position and momentum. The **standard deviations**, σ_p and σ_x, are related to observables by the relations

$$\sigma_p^2 = \langle p^2 \rangle - \langle p \rangle^2 \quad \text{and} \quad \sigma_x^2 = \langle x^2 \rangle - \langle x \rangle^2 \qquad \textbf{(6.10)}$$

where σ_p^2 is called the variance.

EXAMPLE PROBLEM 6.5

Starting with the definition for the standard deviation, $\sigma_x = \sqrt{1/N \sum_{i=1}^{N} (x_i - \langle x \rangle)^2}$, derive the expression for σ_x^2 in Equation (6.10).

Solution

$$\sigma_x^2 = \frac{1}{N} \sum_{i=1}^{N} (x_i - \langle x \rangle)^2 = \frac{1}{N} \sum_{i=1}^{N} (x_i^2 - 2x_i \langle x \rangle + \langle x \rangle^2)$$

$$= \langle x^2 \rangle - 2\langle x \rangle \langle x \rangle + \langle x \rangle^2$$

$$= \langle x^2 \rangle - \langle x \rangle^2$$

The fourth postulate of quantum mechanics tells how to calculate these observables from the normalized wave functions:

$$\langle p^2 \rangle = \int \psi^*(x)\hat{p}^2\psi(x)\,dx \quad \text{and}$$

$$\langle p \rangle^2 = \left(\int \psi^*(x)\hat{p}\psi(x)\,dx \right)^2$$

Similarly,

$$\langle x^2 \rangle = \int \psi^*(x)\hat{x}^2\psi(x)\,dx \quad \text{and}$$

$$\langle x \rangle^2 = \left(\int \psi^*(x)\hat{x}\psi(x)\,dx \right)^2 \tag{6.11}$$

To illustrate how to use the Heisenberg uncertainty principle, we carry out a calculation for σ_p and σ_x using the particle in the box as an example. The normalized wave functions are given by $\psi_n(x) = \sqrt{2/a}\,\sin(n\pi x/a)$ and the operators needed are $\hat{p} = -i\hbar(\partial/\partial x)$ and $\hat{x} = x$.

Using the standard integrals

$$\int x\sin^2 ax\,dx = \frac{x^2}{4} - \frac{1}{4a}x\sin 2ax - \frac{1}{8a^2}\cos 2ax \quad \text{and}$$

$$\int x^2\sin^2 ax\,dx = \frac{1}{6}x^3 - \left(\frac{1}{4a}x^2 - \frac{1}{8a^3} \right)\sin 2ax - \frac{1}{4a^2}x\cos 2ax$$

it is found that

$$\langle x \rangle = \int_0^a \sqrt{\frac{2}{a}}\sin\left(\frac{n\pi x}{a}\right)x\sqrt{\frac{2}{a}}\sin\left(\frac{n\pi x}{a}\right)dx = \frac{2}{a}\int_0^a x\sin^2\left(\frac{n\pi x}{a}\right)dx = \frac{1}{2}a$$

$$\langle x^2 \rangle = \int_0^a \sqrt{\frac{2}{a}}\sin\left(\frac{n\pi x}{a}\right)x^2\sqrt{\frac{2}{a}}\sin\left(\frac{n\pi x}{a}\right)dx = \frac{2}{a}\int_0^a x^2\sin^2\left(\frac{n\pi x}{a}\right)dx$$

$$= a^2\left(\frac{1}{3} - \frac{1}{2\pi^2 n^2} \right)$$

$$\langle p \rangle = \int_0^a \sqrt{\frac{2}{a}}\sin\left(\frac{n\pi x}{a}\right)\left(-i\hbar\frac{\partial}{\partial x}\sqrt{\frac{2}{a}}\sin\left(\frac{n\pi x}{a}\right)\right)dx$$

$$= -i\hbar\frac{2\pi n}{a^2}\int_0^a \sin\left(\frac{n\pi x}{a}\right)\cos\left(\frac{n\pi x}{a}\right)dx = 0$$

$$\langle p^2 \rangle = \int_0^a \sqrt{\frac{2}{a}}\sin\left(\frac{n\pi x}{a}\right)\left(-\hbar^2\frac{\partial^2}{\partial x^2}\sqrt{\frac{2}{a}}\sin\left(\frac{n\pi x}{a}\right)\right)dx$$

$$= \frac{2\pi^2 n^2\hbar^2}{a^3}\int_0^a \sin^2\left(\frac{n\pi x}{a}\right)dx = \frac{n^2\pi^2\hbar^2}{a^2}$$

With these results, σ_p becomes

$$\sigma_p = \sqrt{\frac{n^2\pi^2\hbar^2}{a^2}} = \frac{n\pi\hbar}{a} \quad \text{and} \quad \sigma_x = a\sqrt{\left(\frac{1}{12} - \frac{1}{2\pi^2 n^2} \right)} \tag{6.12}$$

Next, these results are verified as being compatible with the uncertainty principle for $n = 1$:

$$\sigma_p \sigma_x = \frac{n\pi\hbar}{a}\sqrt{a^2\left(\frac{1}{12} - \frac{1}{2\pi^2 n^2}\right)} = \hbar\sqrt{\left(\frac{\pi^2 n^2}{12} - \frac{1}{2}\right)}$$

$$= 0.57\hbar > \frac{\hbar}{2} \quad \text{for } n = 1 \qquad \textbf{(6.13)}$$

Because this function has its minimum value for $n = 1$, the uncertainty principle is satisfied for all values of n.

In evaluating a quantum mechanical result, it is useful to make sure that it converges to the classical result as $n \to \infty$. To do so, the relative uncertainties in x and p are evaluated. The quantity $\sqrt{\langle p^2 \rangle}$ rather than $\langle p \rangle$ is used for this calculation because $\langle p \rangle = 0$. The following result is obtained:

$$\frac{\sigma_x}{\langle x \rangle} = \frac{a\sqrt{\left(\frac{1}{12} - \frac{1}{2\pi^2 n^2}\right)}}{a/2} = \sqrt{\frac{1}{3} - \frac{2}{\pi^2 n^2}} \quad \text{and} \quad \frac{\sigma_p}{\sqrt{\langle p^2 \rangle}} = \frac{n\pi\hbar/a}{n\pi\hbar/a} = 1 \quad \textbf{(6.14)}$$

The interesting result is obtained that the relative uncertainty $\sigma_x/\langle x \rangle$ increases as $n \to \infty$. How can this result be understood? Looking back at the probability density in Figure 6.4, we see that the particle is most likely to be found near the center of the box for $n = 1$, whereas it is equally likely to be anywhere in the box for large n. The fact that the ground-state particle is more confined than the classical particle is at first surprising, but is consistent with the discussion in Chapter 4.

The result that the relative uncertainty in momentum is independent of the momentum is counterintuitive because in the classical limit, the uncertainty in the momentum is expected to be negligible. It turns out that the result for $\sigma_p/\sqrt{\langle p^2 \rangle}$ in Equation (6.14) is misleading because there are two values of p for a given value of p^2. The variance calculated earlier is characteristic of the set of the two p values, and what we want to know is $\sigma_p/\sqrt{\langle p^2 \rangle}$ for each value of p individually. How can the desired result be obtained?

The result is obtained by expanding the eigenfunctions $\psi_n(x)$ in the eigenfunctions of the momentum operator. In a fashion similar to that used to generate the data in Figure 6.4, we ask what values of k and what relative amplitudes A_k are required to represent the wave functions

$$\psi_n(x) = \sqrt{\frac{2}{a}}\sin\left(\frac{n\pi x}{a}\right), \quad \text{for } n = 1, 2, 3, 4, \ldots \quad a > x > 0 \quad \text{and}$$

$$\psi_n(x) = 0, \quad \text{for } 0 \geq x, \quad x \geq a \qquad \textbf{(6.15)}$$

in the form

$$\psi_n(x) = \sum_{k=-\infty}^{k=\infty} A_k e^{ikx} \qquad \textbf{(6.16)}$$

Expressing the eigenfunctions in this way allows the probability density of observing a particular value of p for a particle whose wave function is an eigenfunction of the total energy operator to be calculated. As outlined in the discussion of the fourth postulate in Chapter 3, the probability density of measuring a given momentum is proportional to $A_k^* A_k$. This quantity is shown as a function of k for several values of n in Figure 6.5, where, for $n = 101$, the result looks quite classical, in that the observed values are sharply peaked at the two classically predicted values $p = \pm\sqrt{2mE}$. However, as n becomes smaller, quantum effects become much clearer. The most probable values of p are still given by $p = \pm\sqrt{2mE}$ for $n = 5$ and 15, but subsidiary maxima are seen, and the width of the peaks (which is a measure of the uncertainty in p) is substantial. For $n = 1$, the distribution is peaked at $p = 0$, rather than the classical values. For this lowest energy state, quantum and classical mechanics give very different results.

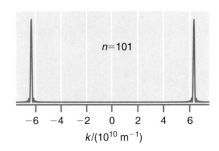

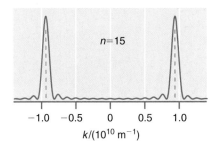

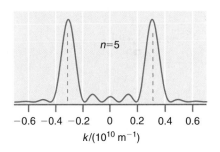

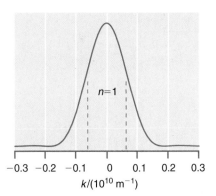

FIGURE 6.5

The relative probability density of observing a particular value of k, $A_k^* A_k$ (vertical axis), is graphed versus k for a 5.00-nm-long box for several values of n. The dashed lines for $n = 1$, 5, and 15 show the classically expected values $p = \pm\sqrt{2mE}$.

6.3 Expanding the Total Energy Eigenfunctions in Eigenfunctions of the Momentum Operator

Figure 6.5 demonstrates that the relative uncertainty $\sigma_p/\sqrt{\langle p^2 \rangle}$ decreases as p increases. You will explore this issue more quantitatively in the end-of-chapter problems. The counterintuitive result of Equation (6.14)—that the relative uncertainty in the momentum is constant—is an artifact of characterizing the distribution consisting of two widely separated peaks by one variance, rather than looking at each of the peaks individually.

S U P P L E M E N T A L

6.5 A THOUGHT EXPERIMENT USING A PARTICLE IN A THREE-DIMENSIONAL BOX

Think of the following experiment: A particle is put in an opaque box, and the top is securely fastened. From the outside, a partition is slid into the box, dividing it into two equal leak-tight volumes. This partition allows the initial box to be separated into two separate leak-tight boxes, each with half the volume. These two boxes are separated by sending one of them to the moon. Finally, an observer opens one of the boxes. The observer finds that the box he has opened is either empty or that it contains the particle. From the viewpoint of classical mechanics, this is a straightforward experiment. If the box that was opened is empty, then that half of the box was empty when the partition was initially inserted. What does this problem look like from a quantum mechanical point of view? The individual steps are illustrated in Figure 6.6.

Initially, we know only that the particle is somewhere in the box before the partition is inserted. Because it exhibits wave-particle duality, the position of the particle cannot be determined exactly. If two eigenstates of the position operator, ψ_{left} and ψ_{right}, are defined, then the initial wave function is given by

$$\psi = a\psi_{left} + b\psi_{right}, \quad \text{with } |a|^2 + |b|^2 = 1 \qquad \textbf{(6.17)}$$

In the figure, it has been assumed that $a = b$. The square of the wave function is nonzero everywhere in the box, and goes to zero at the walls. When the partition is inserted, what we have just said is again true, except that now the wave function also goes to zero along the partition. Classically, the particle is either in the left- or the right-hand side of the combined box, although it is not known which of these possibilities applies.

From a quantum mechanical perspective, such a definitive statement cannot be made. We can merely say that there is an equal probability of finding the particle in each of the two parts of the original box. Therefore, when the two halves of the box are separated, the integral of the square of the wave function is one-half in each of the smaller boxes.

Now the box is opened. This is equivalent to applying the position operator to the wave function of Equation (6.17). According to the discussion in Chapter 3, the wave function becomes either ψ_{left} or ψ_{right}. We do not know which of these will be the final wave function of the system, but we do know that in a large number of measurements, the probability of finding it on the left is a^2. Assume the case shown in the top part of Figure 6.6 in which the particle is found in the left box. *In that case, the integral of the square of the wave function in that box instantaneously changes from 0.5 to 1.0 at the moment we look into the box, and the integral of the square of the wave function in the other box drops from 0.5 to zero!* Because this result does not depend on the distance of separation between the boxes, this distance can be made large enough that the boxes are not coupled by a physical force. Even so, the one box "knows" instantaneously what has been learned about the other box. This is the interpretation of quantum mechanics attributed to the Copenhagen school of Niels Bohr, which gives the act of measurement a central role in the outcome of an experiment. Nearly 80 years after the formulation of quantum theory, the search for an "observer-free" theory has not yet led to a widely accepted alternative to the interpretation of the Copenhagen school.

Before you dismiss this scenario as unrealistic, and accept the classical view that the particle really is in one part of the box *or* the other, have another look at Figure 3.3. The results shown there demonstrate clearly that the outcome of an experiment on *identically prepared* quantum mechanical systems is inherently probabilistic. Therefore, the wave

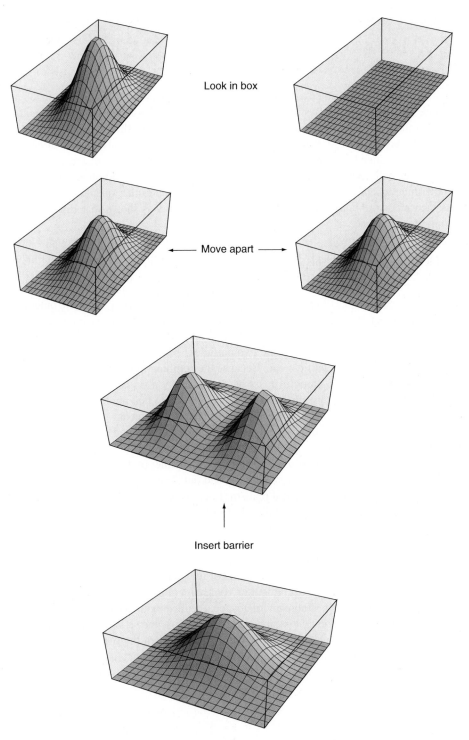

Look in box

← Move apart →

Insert barrier

FIGURE 6.6

Thought experiment using a particle in a box. The square of the wave function is plotted along the x and y coordinates of the box.

function for an individual system must be formulated in such a way that it includes all possible outcomes of an experiment. This means that, in general, it describes a superposition state. The result that measurements on identically prepared systems lead to different outcomes has been amply documented by experiments at the atomic level, and this precludes the certainty in the classical assertion that the particle really is in one part of the box *or* the other. Where does the classical limit appear in this case? For instance, one might ask why the motion of a human being is not described by the Schrödinger equation rather than Newton's second law if every atom in our body is described by quantum mechanics. This topic is an active area of research, and the current view is that the superposition wave function of a macroscopic system is unstable because of interactions with the environment. The superposition state decays very rapidly to a single term. This decay has the consequence that the strange behavior characteristic of quantum mechanical superposition states is no longer observed in large "classical systems."

6.6 ENTANGLED STATES, TELEPORTATION, AND QUANTUM COMPUTERS

Erwin Schrödinger first noted a prediction of quantum mechanics that was very much at variance with classical physics. It is that two quantum particles can be coupled in such a way that their properties are no longer independent of one another no matter how far apart they may be. We say that the particles are **entangled**. This consequence of entanglement was pointed out by Einstein and called a "spooky action at a distance" to indicate what he believed to be a serious flaw in quantum mechanics. Definitive experiments to determine whether entanglement was real or an indication that quantum mechanics was incorrect were not possible until the 1970s, when it was shown that Einstein was wrong in this instance.

Consider the following example of entanglement. A particle with no magnetic moment decays giving two identical particles whose z component of the magnetic moment (which we call m_z) can take on the values $\pm 1/2$. Each of these particles is sent through a Stern–Gerlach analyzer as described in Section 6.2. A series of measurements of m_z for particle one gives $\pm 1/2$ in a random pattern; it is not possible to predict the outcome of a single measurement. However, because angular momentum is conserved, if one of the particles is found to have m_z equal to $+1/2(\uparrow)$, the other must have $m_z = -1/2(\downarrow)$. There are only two possibilities for the two particles, $\uparrow\downarrow$ or $\downarrow\uparrow$ where the left arrow in each case indicates m_z for particle one. Because the combinations $\uparrow\downarrow$ or $\downarrow\uparrow$ occur with equal probability, the two particles must be described by a single superposition wave function which we write schematically as $\uparrow\downarrow + \downarrow\uparrow$. Note that neither particle can be described by its own wave function as a result of the entanglement.

This result implies that the second particle has no well-defined value of m_z until a measurement is carried out on the first particle. Because the roles of particles one and two can be reversed, quantum mechanics tells us that neither of the particles has a well-defined value of m_z until a measurement is carried out. This result violates a basic principle of classical physics called local realism. **Local realism** asserts that: (1) Measured results correspond to elements of reality. For example, if I determine that your hair is black, according to local realism, your hair was black before I make the measurement and is black regardless of whether a measurement is ever made. (2) Measured results are independent of any action that might be taken at a distant location at the same time. If you have an identical twin on the other side of the planet, a measurement of her hair color has no influence on a measurement of your hair color made at the same time.

The experiment described above violates local realism both because that there is no value for m_z until a measurement is carried out and because the m_z values of the two particles remain coupled no matter how far apart they are when the first measurement is made. An experiment that illustrates this surprising result is depicted in Figure 6.7. Two entangled photons are passed through optical fibers to locations spaced 10 km apart.

FIGURE 6.7

The spatial distribution of the light intensity for photon 2 shows diffraction pattern even though it has not passed through a slit. This result arises because photons 1 and 2 are entangled.

Source: D. V. Strekalov *et al.*, *Physical Review Letters*, 74 (1995) 3600.

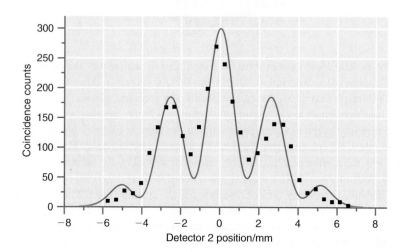

Photon 1 is passed through a double slit and exhibits a diffraction pattern. If the profile of the light intensity corresponding to photon 2 is determined, it corresponds to that of a photon that has passed through a double slit, even though it hasn't! If you and your identical twin were quantum mechanically entangled, neither your hair color nor that of your twin would be known before a measurement was made. Any possible hair color would be equally likely to be determined for you in a measurement, and your twin would be found to have the same hair color.

Does entanglement suggest that information can be transmitted instantaneously over an arbitrarily large distance? To answer this question, we consider how information about a system can be transmitted to a distant location, first for a classical system and then for a quantum mechanical system. Classically, a copy of the original information or object is created at the distant location. A classical system can be copied as often as desired and the accuracy of the copy is limited only by the quality of the tools used. In principle, the copies can be so well made that they are indistinguishable from the original. The speed with which information is transferred is limited by the speed of light. By contrast, the information needed to make a copy of a quantum mechanical system cannot be obtained, because it is impossible to determine the state of the system exactly by measurement. If the system wave function is given by

$$\psi = \sum_m b_m \phi_m \qquad (6.20)$$

in which the ϕ_m are the eigenfunctions of an appropriate quantum mechanical operator, experiments can only determine the absolute magnitudes $|b_m|^2$. This is not enough information to determine the wave function. Therefore, the information needed to make a copy is not available. Making a copy of a quantum mechanical system is also in violation of the Heisenberg uncertainty principle. If a copy could be made, one could easily measure the momentum of one of the copies and measure the position of the other copy. If this were possible, both the momentum and position could be known simultaneously.

Given these limitations of knowledge of quantum mechanical systems, how can a quantum mechanical system be transported to a distant location, and how is this transfer related to entanglement? Consider the following experiment described by Anton Zeilinger in *Scientific American*, April 2000, in which a photon at one location was recreated at a second location. Although photons were used in this experiment, there is no reason in principle why atoms or molecules could not be transferred from one location to another in the same way.

Bob and Alice are at distant locations and share an entangled photon pair, of which Bob has photon B and Alice has photon A as shown in Figure 6.8. Each of them carefully stores his or her photon so that the entanglement is maintained. At a later time, Alice has another photon that we call X, which she would like to send to Bob. How can this be done? She cannot measure the polarization state directly and send this information to Bob, because the act of measurement would change the state of the photon. Instead, she entangles X and A.

What are the consequences of the entanglement of A and X on B? We know that whatever state X has, A must have the orthogonal state. If X is vertically (horizontally) polarized, then A must be horizontally (vertically) polarized. However, the same logic must apply to A and B because they are also entangled. Whatever state A has, B must have the orthogonal state. If the state of B is orthogonal to that of A and the state of A is orthogonal to that of X, *then the states of B and X must be identical*. This follows from the fact that there are only two possible eigenfunctions of the polarization operator.

What has been accomplished by this experiment? Photon B acquires the original polarization of Alice's photon X and is therefore identical in every way to the original state of X. However, the state of X has been irreversibly changed at Alice's location, because in order to know that photons A and X have been successfully entangled, Alice has to pass both her photons through a detector. Therefore, the properties of X have been changed at Alice's location and transferred to Bob's location. This process is called **teleportation**, defined as the transfer of a quantum state from one location to another. Note that the uncertainty principle has not been violated because the photon has been teleported rather than copied.

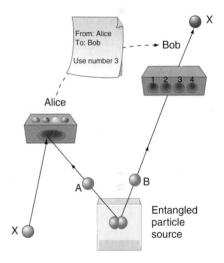

FIGURE 6.8

Teleportation of photon X from Alice to Bob. Note the classical communication channel that Alice uses to communicate the outcome of her measurement on A and X to Bob.

Maintaining the entanglement of pairs A and B and A and X is the crucial ingredient of teleportation. Neither Bob nor Alice knows the state of X at the start or the end of the experiment. This is the case because neither of them has measured the state of the photon directly. Had they done so, the state of the photon would have been irreversibly changed. It is only because they did not determine the state of the photon that the recreation of photon X at Bob's location was possible.

If the preceding outcome were the only possible outcome of Alice's entanglement of A and X, the transmission of information from Alice to Bob would be instantaneous, regardless of the distance between them. Therefore, it would be faster than the speed of light. Unfortunately, it turns out that Alice's entanglement of A and X has four possible outcomes, which we won't discuss other than to say that each is equally probable in the entanglement of an individual photon pair. Although there is no way to predict which of the four outcomes will occur, Alice has detectors that will tell her *after the fact* which outcome occurred.

In each of these outcomes, the entanglement of A and X is transferred to B, but in three of the four, Bob must carry out an operation on B, such as to rotate its polarization by a fixed angle, in order to make B identical to X. How does this affect what Bob knows about B? Without knowing which of the four outcomes Alice detected, Bob doesn't know how his photon has been transformed. Only if Alice sends him the result of her measurement does Bob know what he must do to B to make it indistinguishable from X. It is the need for this additional information that limits the speed of quantum information transfer through teleportation. Although the state of Bob's photon B is instantaneously transformed as Alice entangles A and X, he cannot interpret his results without additional information from her. Because Alice's information must be sent to Bob using conventional methods such as phone, fax, or e-mail, the overall process of teleportation is limited by the speed of light. Although the state of entangled particles changes instantaneously, information transmission utilizing entanglement cannot proceed faster than the speed of light.

In principle, the same technique could be used to teleport an atom, a molecule, or even an organism. The primary requirement is that it must be possible to create entangled pairs of the object to be teleported. The initial experiment was carried out with photons because experimental methods to entangle photons are available. As discussed earlier, entangled states are fragile and can decay to a single eigenfunction of the operator rapidly through interactions with the environment. This is especially true of systems containing a large number of atoms. However, it has been possible to entangle atoms, and it seems within reach to entangle small molecules.

Entanglement has a further interesting application. It provides the basis for the **quantum computer**, which currently exists only as a concept. Such a computer would be far more powerful than the largest supercomputers currently available. How does a quantum computer differ from a classical computer? In a classical computer, information is stored in bits. A **bit** generally takes the form of a macroscopic object like a wire or a memory element that can be described in terms of a property such as a voltage. Two different ranges of voltage are used to represent the numbers 0 and 1. Within this binary system, an n bit memory can have 2^n possible states that range between 00000...00 and 11111. . .11. Information such as text and images can be stored in the form of such states. Mathematical or logical operations can be represented as transformations between such memory states. Logic gates operate on binary strings to carry out mathematical operations. Software provides an instruction set to route the data through the logic gates that are the heart of the computer hardware. This is the basis on which classical computers operate.

The quantum analog of the bit, in which two numbers characterize the entity, is called a **qubit**, which has the property that it is *simultaneously* a linear combination of 0 and 1, rather than being either 1 or 0. The concept of a qubit is best explained using Figure 6.9.

A single photon is incident from the top left on a beam splitter, which has a probability of 0.5 for reflection and for transmission of a photon. We assign a photon with this direction the value 0. Just as for a particle incident on a double slit, each photon follows *both* pathways of reflection and transmission and reflection, rather than being either reflected or transmitted. The two mirrors are used to combine the two pathways at a second beam splitter where each incident photon again follows both transmission

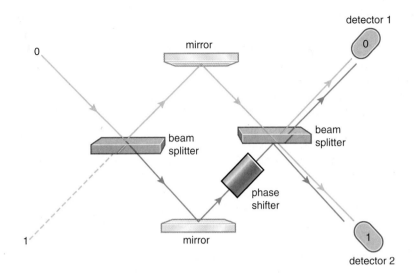

FIGURE 6.9

A combination of beam splitters and mirrors with a phase shifter is used to illustrate how a qubit can be generated. Partial reflection occurs because a semi-transparent silver layer is evaporated onto a glass substrate. The reflecting layer is on the top of the left beam splitter and on the bottom of the right beam splitter.

and reflection pathways. On the final part of the path to the detectors, the initial reflection and transmission pathways are combined and interference of the two beams occurs. To understand what the detectors register, three simple rules must be followed: (1) If a photon is reflected at an interface for which the refractive index behind the mirror is larger than in front of the mirror, a phase shift of π (half a wavelength) occurs. (2) If a photon is reflected at an interface for which the refractive index behind the mirror is smaller than in front of the mirror, no phase shift occurs. (3) In passing through the higher index glass making up the beam splitters, a phase shift ϕ occurs which is proportional to the path length.

We first remove the phase shifter. Along the red pathway leading to detector 1, phase shifts of π, π, and 2ϕ occur for a total of $2\pi + 2\phi$. Along the green pathway also leading to detector 1, phase shifts of ϕ, π, and ϕ occur for a total of $\pi + 2\phi$. Therefore, the two pathways are out of phase by π, meaning that destructive interference occurs and no signal is registered at detector 1. We next carry out the same analysis for detector 2. Along the red pathway leading to detector 2, phase shifts of π, π, and ϕ occur for a total of $2\pi + \phi$. Along the green pathway also leading to detector 1, phase shifts of ϕ, π, and π occur for a total of $2\pi + \phi$. There is no phase difference between the two pathways. Therefore, constructive interference occurs and a signal is registered at detector 1. Had we considered the incident photon indicated by the dashed path given the value 1, we would have found that a signal is registered at detector 1, but not at detector 2. This device forms a NOT gate, because it makes the transformations $0 \rightarrow 1$ and $1 \rightarrow 0$.

If a phase shifter in the form of a piece of glass of variable thickness is inserted into one leg of the interferometer, the relative phase of the two pathways can be changed to any value from 0 to π. This means that the incoming signals 0 and 1 are transformed into a superposition of 0 and 1. In this way, a bit has been transformed into a qubit.

The advantage of a qubit over a bit can be illustrated with the following example. Only one of eight numbers can be stored in a conventional three-bit array. By contrast, qubits can be entangled with one another so that a 3-qubit entangled array is in a superposition of all eight possible binary strings of length 3. More generally, an M-qubit entangled array is in a superposition of all 2^M possible binary strings of length M. If this input signal can be processed using quantum gates without destroying the entanglement, 2^M simultaneous calculations ($\sim 10^{30}$ for $M = 100$) can be done in parallel by an M-qubit quantum computer.

There is a significant limiting factor in such a calculation. Although 2^M simultaneous calculations could be carried out, only one of them will be registered at the output through the collapse of the superposition wave function. Furthermore, it is not possible to know which of the 2^M calculations corresponds to the output. Fortunately, algorithms can be devised for which the final result depends logically on all of the 2^M intermediate results. In particular, Shor's algorithm would allow a quantum computer to factor large numbers orders of magnitude faster than a conventional computer. Although this may seem like an esoteric task, factorization is the primary method of encoding sensitive data sent over the Internet. Therefore, quantum computers are expected to have a major role in communications technology if their theoretically achievable potential can be realized.

Three major hurdles must be overcome to construct a quantum computer: the entanglement of real qubits, the maintenance of entanglement over a long enough time to allow calculations to be carried out, and the extraction of the desired result from the superposition of all possible outcomes. Ions trapped in electromagnetic fields and nuclear spins on different atoms in a molecule have been successfully entangled, and are useful models for quantum computers. However, the ultimate goal is a solid-state device that is compatible with current microelectronic technology. Entanglement can best be maintained in systems that are isolated from the environment, but one cannot exchange input and output with a totally isolated system. Therefore, overcoming this hurdle will involve a compromise between ease of access and the lifetime of the entangled array. The third hurdle can be overcome for algorithms for which the logical output contains information from all of the 2^M possible pathways. This limits the applicability of a quantum computer, but the enormous increase in computational power achievable in principle for suitable applications relative to a conventional computer justifies the resources invested in its creation.

Vocabulary

bit

commutator

commute

entangled

Heisenberg uncertainty principle

local realism

quantum computer

qubit

standard deviation

Stern–Gerlach experiment

teleportation

wave packet

Questions on Concepts

Q6.1 Why does the Stern–Gerlach experiment show that the operator "measure the z component of the magnetic moment of an Ag atom" has only two eigenfunctions with eigenvalues that have the same magnitude and opposite sign?

Q6.2 Have a closer look at Equation (6.6) and Figure 6.4. How would Figure 6.4 change if m increases? Generalize your conclusion to make a statement of how well the momentum is known if the position is known exactly.

Q6.3 Why is maintaining the entanglement of pairs A and B and A and X the crucial ingredient of teleportation?

Q6.4 Why is it not possible to reconstruct the wave function of a quantum mechanical superposition state from experiments?

Q6.5 Why does the relative uncertainty in x for the particle in the box increase as $n \rightarrow \infty$?

Q6.6 Why is the statistical concept of variance a good measure of uncertainty in a quantum mechanical measurement?

Q6.7 Derive a relationship between $[\hat{A}, \hat{B}]$ and $[\hat{B}, \hat{A}]$.

Q6.8 How does our study of the eigenfunctions for the particle in the box let us conclude that the position uncertainty has its minimum value for $n = 1$?

Q6.9 What is the difference between a bit and a qubit?

Q6.10 How does the Heisenberg uncertainty principle allow us to conclude that it is not possible to make exact copies of quantum mechanical objects?

Q6.11 Which result of the Stern–Gerlach experiment allows us to conclude that the operators for the z and x components of the magnetic moment do not commute?

Q6.12 Why is it impossible to determine the state of a quantum mechanical system exactly by measurement?

Q6.13 Why is the motion of a human being not described by the Schrödinger equation rather than Newton's second law if every atom in our body is described by quantum mechanics?

Q6.14 Explain the following statement: if $\hbar = 0$, it would be possible to measure the position and momentum of a particle exactly and simultaneously.

Q6.15 Why is $\sqrt{\langle p^2 \rangle}$ rather than $\langle p \rangle$ used to calculate the relative uncertainty for the particle in the box?

Problems

Problem numbers in **red** indicate that the solution to the problem is given in the *Student's Solutions Manual*.

P6.1 In this problem, we consider the calculations for σ_p and σ_x for the particle in the box shown in Figure 6.5 in more detail. In particular, we want to determine how the absolute uncertainty Δp_x and the relative uncertainty $\Delta p_x/p_x$ of a single peak corresponding to either the most probable positive or negative momentum depends on the quantum number n.

a. First we must relate k and p_x. From $E = p_x^2/2m$ and $E = n^2h^2/8ma^2$, show that $p_x = nh/2a$.

b. Use the result from part (a) together with the relation linking the length of the box and the allowed wavelengths to obtain $p_x = \hbar k$.

c. Relate Δp_x and $\Delta p_x/p_x$ with k and Δk.

d. The following graph shows $|A_k|^2$ versus $k\text{-}k_{peak}$. By plotting the results of Figure 6.5 in this way, all peaks appear at the same value of the abscissa. Successive curves have been shifted upward to avoid overlap. Use the width of the $|A_k|^2$ peak at half height as a measure of Δk. What can you conclude from this graph about the dependence of Δp_x on n?

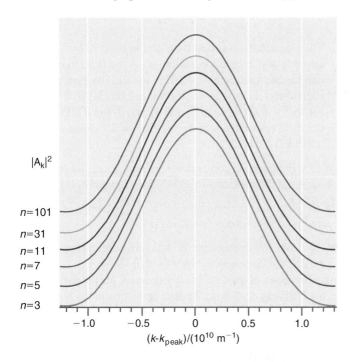

e. The following graph shows $|A_k|^2$ versus k/n for $n = 3$, $n = 5$, $n = 7$, $n = 11$, $n = 31$, and $n = 101$. Use the width of the $|A_k|^2$ peak at half height as a measure of $\Delta k/n$. Using the graphs, determine the dependence of $\Delta p_x/p_x$ on n. One way to do this is to assume that the width depends on n like $(\Delta p_x/p_x) = n^\alpha$ where α is a constant to be determined. If this relationship holds, a plot of $\ln(\Delta p_x/p_x)$ versus $\ln n$ will be linear and the slope will give the constant α.

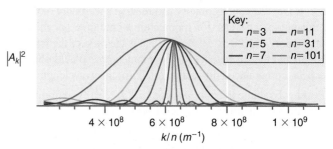

P6.2 Consider the results of Figure 6.5 more quantitatively. Describe the values of x and k by $x \pm \Delta x$ and $k_0 \pm \Delta k$. Evaluate Δx from the zero of distance to the point at which the envelope of $\psi^*(x)\psi(x)$ is reduced to one-half of its peak value. Evaluate Δk from $\Delta k = |1/2(k_0 - k_{min})|$ where k_0 is the average wavevector of the set of 21 waves (11th of 21) and k_{min} corresponds to the 21st of the 21 waves. Is your estimated value of $\Delta p \, \Delta x = \hbar \, \Delta k \, \Delta x$ in reasonable agreement with the Heisenberg uncertainty principle?

P6.3 Evaluate the commutator $[d/dr, 1/r]$ by applying the operators to an arbitrary function $f(r)$.

P6.4 Show

a. that $\psi(x) = e^{-x^2/2}$ is an eigenfunction of $\hat{A} = x^2 - \partial^2/\partial x^2$; and

b. that $\hat{B}\psi(x)$ (where $\hat{B} = x - \partial/\partial x$) is another eigenfunction of $\hat{A}$.

P6.5 Another important uncertainty principle is encountered in time-dependent systems. It relates the lifetime of a state Δt with the measured spread in the photon energy ΔE associated with the decay of this state to a stationary state of the system. "Derive" the relation $\Delta E \, \Delta t \geq \hbar/2$ in the following steps.

a. Starting from $E = p_x^2/2m$ and $\Delta E = (dE/dp_x)\Delta p_x$, show that $\Delta E = v_x \Delta p_x$.

b. Using $v_x = \Delta x/\Delta t$, show that $\Delta E \, \Delta t = \Delta p_x \Delta x \geq \hbar/2$.

c. Estimate the width of a spectral line originating from the decay of a state of lifetime 1.0×10^{-9} s and 1.0×10^{-11} s in inverse seconds and inverse centimeters.

P6.6 Evaluate the commutator $[\hat{x}(\partial/\partial y), \hat{y}]$ by applying the operators to an arbitrary function $f(x,y)$.

P6.7 Evaluate $[\hat{A}, \hat{B}]$ if $\hat{A} = x^2 - d^2/dx^2$ and $\hat{B} = x - d/dx$.

P6.8 Consider the entangled wave function for two photons,

$$\psi_{12} = \frac{1}{\sqrt{2}}(\psi_1(H)\psi_2(V) + \psi_1(V)\psi_2(H))$$

Assume that the polarization operator $\hat{P}_i$ has the properties $\hat{P}_i\psi_i(H) = -\psi_i(H)$ and $\hat{P}_i\psi_i(V) = +\psi_i(V)$ where $i = 1$ or $i = 2$.

a. Show that ψ_{12} is not an eigenfunction of $\hat{P}_1$ or $\hat{P}_2$.

b. Show that each of the two terms in ψ_{12} is an eigenfunction of the polarization operator $\hat{P}_1$.

c. What is the average value of the polarization P_1 that you will measure on identically prepared systems?

P6.9 Evaluate the commutator $[\hat{p}_x, \hat{p}_x^2]$ by applying the operators to an arbitrary function $f(x)$.

P6.10 Revisit the double-slit experiment of Example Problem 6.2. Using the same geometry and relative uncertainty in the momentum, what electron momentum would give a position uncertainty of 1.00×10^{-9} m? What is the ratio of the wavelength and the slit spacing for this momentum? Would you expect a pronounced diffraction effect for this wavelength?

P6.11 Evaluate the commutator $[d^2/dx^2, x]$ by applying the operators to an arbitrary function $f(x)$.

P6.12 Revisit the TV picture tube of Example Problem 6.3. Keeping all other parameters the same, what electron energy would result in a position uncertainty of 1.00×10^{-8} m along the direction of motion?

P6.13 Evaluate the commutator $[(d^2/dx^2) - x, (d/dx) + x^2]$ by applying the operators to an arbitrary function $f(x)$.

P6.14 If the wave function describing a system is not an eigenfunction of the operator $\hat{B}$, measurements on identically prepared systems will give different results. The variance of this set of results is defined in error analysis as $\sigma_B^2 = \langle (B - \langle B \rangle)^2 \rangle$, where B is the value of the observable in a single measurement and $\langle B \rangle$ is the average of all measurements. Using the definition of average value from the quantum mechanical postulates, $\langle A \rangle = \int \psi^*(x) \hat{A} \psi(x) \, dx$, show that $\sigma_B^2 = \langle B^2 \rangle - \langle B \rangle^2$.

P6.15 Apply the Heisenberg uncertainty principle to estimate the zero point energy for the particle in the box.

a. First justify the assumption that $\Delta x \leq a$ and that, as a result, $\Delta p \geq \hbar/2a$. Justify the statement that, if $\Delta p \geq 0$, we cannot know that $E = p^2/2m$ is identically zero.

b. Make this application more quantitative. Assume that $\Delta x = 0.50a$ and $\Delta p = 0.50p$ where p is the momentum in the lowest energy state. Calculate the total energy of this state based on these assumptions and compare your result with the ground-state energy for the particle in the box.

c. Compare your estimates for Δp and Δx with the more rigorously derived uncertainties σ_p and σ_x of Equation (6.13).

P6.16 Evaluate the commutator $[d/dx, x^2]$ by applying the operators to an arbitrary function $f(x)$.

P6.17 Evaluate the commutator $[\hat{x}, \hat{p}_x]$ by applying the operators to an arbitrary function $f(x)$. What value does the commutator $[\hat{p}_x, \hat{x}]$ have?

P6.18 In this problem, you will carry out the calculations that describe the Stern–Gerlach experiment shown in Figure 6.2. Classically, a magnetic dipole $\boldsymbol{\mu}$ has the potential energy $E = -\boldsymbol{\mu} \cdot \mathbf{B}$. If the field has a gradient in the z direction, the magnetic moment will experience a force, leading it to be deflected in the z direction. Because classically $\boldsymbol{\mu}$ can take on any value in the range $-|\boldsymbol{\mu}| \leq \mu_z \leq |\boldsymbol{\mu}|$, a continuous range of positive and negative z deflections of a beam along the y direction will be observed. From a quantum mechanical perspective, the forces are the same as in the classi-

cal picture, but μ_z can only take on a discrete set of values. Therefore, the incident beam will be split into a discrete set of beams that have different deflections in the z direction.

a. The geometry of the experiment is shown here. In the region of the magnet indicated by d_1, the Ag atom experiences a constant force. It continues its motion in the force-free region indicated by d_2.

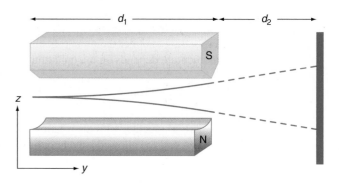

If the force inside the magnet is F_z, show that $|z| = 1/2(F_z/m_{Ag})t_1^2 + t_2 v_z(t_1)$. The flight times t_1 and t_2 correspond to the regions d_1 and d_2.

b. Show that assuming a small deflection,

$$|z| = F_z \left(\frac{d_1 d_2 + \frac{1}{2} d_1^2}{m_{Ag} v_y^2} \right)$$

c. The magnetic moment of the electron is given by $|\boldsymbol{\mu}| = g_S \mu_B/2$. In this equation, μ_B is the Bohr magneton and has the value 9.274×10^{-24} J/tesla. The gyromagnetic ratio of the electron g_S has the value 2.00231. If $\partial B_z/\partial z = 1000.$ tesla m^{-1}, and d_1 and d_2 are 0.200 and 0.250 m, respectively, and $v_y = 500.$ m s^{-1}, what values of z will be observed?

P6.19 Evaluate the commutator $[(d/dx) - x, (d/dx) + x]$ by applying the operators to an arbitrary function $f(x)$.

P6.20 Evaluate the commutator $[\hat{x}, \hat{p}_x^2]$ by applying the operators to an arbitrary function $f(x)$.

P6.21 What is wrong with the following argument? We know that the functions $\psi_n(x) = \sqrt{2/a} \sin(n\pi x/a)$ are eigenfunctions of the total energy operator for the particle in the infinitely deep box. We also know that in the box, $E = p_x^2/2m + V(x) = p_x^2/2m$. Therefore, the operator for E_{total} is proportional to the operator for p_x^2. Because the operators for p_x^2 and p_x commute as you demonstrated in Problem P6.8, the functions $\psi_n(x) = \sqrt{2/a} \sin(n\pi x/a)$ are eigenfunctions of both the total energy and momentum operators.

P6.22 For linear operators A, B, and C, show that $[\hat{A}, \hat{B}\hat{C}] = [\hat{A}, \hat{B}]\hat{C} + \hat{B}[\hat{A}, \hat{C}]$.

P6.23 The muzzle velocity of a rifle bullet is about 900. m s^{-1} along the direction of motion. If the bullet weighs 30. g, and the uncertainty in its momentum is 0.10%, how accurately can the position of the bullet be measured along the direction of motion?

Web-Based Simulations, Animations, and Problems

W6.1   The simulation of particle diffraction from a single slit is used to illustrate the dependence between the uncertainty in the position and momentum. The slit width and particle velocity are varied using sliders.

W6.2   The Heisenberg uncertainty principle states that $\Delta p \Delta x > \hbar/2$. In an experiment, it is more likely that λ is varied rather than p, where λ is the de Broglie wavelength of the particle. The relationship between Δx and $\Delta \lambda$ will be determined using a simulation. Δx will be measured as a function of $\Delta \lambda$ at a constant value of λ, and as a function of λ for a constant value of $\Delta \lambda$.

W6.3   The uncertainty in momentum will be determined for the total energy eigenfunctions for the particle in the infinite depth box for several values of the quantum number n. The function describing the distribution in k,

$$g_n(k) = \frac{1}{\sqrt{2\pi}} \int_{-\infty}^{\infty} f_n(x)\, e^{-ikx}\, dx = \frac{1}{\sqrt{2\pi}} \int_{0}^{a} \sin\frac{n\pi x}{a}\, e^{-ikx}\, dx$$

will be determined. The values of k for which this function has maxima will be compared with that expected for a classical particle of momentum $p = \sqrt{2mE}$. The width in k of the function $g_n(k)$ on n will be investigated.

A Quantum Mechanical Model for the Vibration and Rotation of Molecules

A molecule has translational, vibrational, and rotational types of motion. Each of these can be separately described by its own energy spectrum and energy eigenfunctions. As shown in Chapter 4, the particle in the box is a useful model for exploring the translational degree of freedom. In this chapter, quantum mechanics is used to study the vibration and rotation of a diatomic molecule. We first consider the vibrational degree of freedom, modeled by the harmonic oscillator. Like the particle in the box, the quantum mechanical harmonic oscillator has a discrete energy spectrum. We then formulate and solve a quantum mechanical model for rotational motion. This model provides a basis for understanding the orbital motion of electrons around the nucleus of an atom as well as the rotation of a molecule about its principal axes.

7.1 SOLVING THE SCHRÖDINGER EQUATION FOR THE QUANTUM MECHANICAL HARMONIC OSCILLATOR

The discussion of the free particle and the particle in the box in Chapters 4 and 5 was useful for understanding how **translational motion** in various potentials is described in the context of wave-particle duality. In applying quantum mechanics to molecules, two other types of motion that molecules can undergo require discussion: **vibration** and **rotation**. A quantum mechanical model for vibration is formulated in this section and for rotation in Sections 7.2 and 7.3. A review of the classical analogues of these quantum mechanical models can be found in the optional review sections, Sections 7.6 and 7.7. You may want to read through these sections before proceeding further.

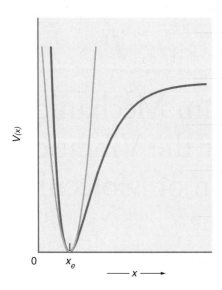

FIGURE 7.1

Potential energy, $V(x)$, as a function of the bond length, x, for a diatomic molecule. The zero of energy is chosen to be the bottom of the potential. The red curve depicts a realistic potential in which the molecule dissociates for large values of x. The yellow curve shows a harmonic potential, $V(x) = (1/2)kx^2$, which is a good approximation to the realistic potential near the bottom of the well. x_e represents the equilibrium bond length.

7.1 The Classical Harmonic Oscillator

The simplest vibrational motion that can be imagined is that which occurs in a diatomic molecule. Vibration involves a displacement of the atoms from their equilibrium positions, which is dictated by the chemical bond length between the atoms. The energy needed to stretch the chemical bond can be described by a simple potential function such as that shown in Figure 7.1. The existence of a stable chemical bond implies that a minimum energy exists at some internuclear distance, called the **bond length**. The position of atoms in a molecule is dynamic rather than static. Think of the chemical bond as a spring rather than a rigid bar connecting the two atoms. Thermal energy increases the vibrational amplitude of the atoms about their equilibrium positions, but does not change the vibrational frequency to a good approximation. The potential becomes steeply repulsive at short distances as the electron clouds of the atoms interpenetrate. It levels out at large distances because the overlap of electrons between the atoms required for chemical bond formation falls to zero.

The exact form of $V(x)$ as a function of x depends on the molecule under consideration. However, as will be shown in Chapter 8, only the lowest one or two vibrational energy levels are occupied for most molecules for $T \sim 300$ K. Therefore, it is a good approximation to say that the functional form of the potential energy near the equilibrium bond length can be approximated by the harmonic potential

$$V(x) = \frac{1}{2}kx^2 \tag{7.1}$$

In Equation (7.1), k is the **force constant** and is not to be confused with the wave vector or the Boltzmann constant. For weakly bound molecules or high temperatures, the more realistic Morse potential (red curve in Figure 7.1) discussed in Section 8.3 should be used. Vibration is studied in the center of mass coordinates because only the relative motion of the atoms is of interest. Transformation to these coordinates means that rather than having two atoms of mass m_1 and m_2 oscillating about their center of mass, we consider the mathematically equivalent problem of a single reduced mass $\mu = (m_1 m_2)/(m_1 + m_2)$ tethered to a wall of infinite mass with a spring of force constant, k (see Optional Section 7.6).

We expect the wave-particle of mass μ vibrating around its equilibrium distance to be described by a set of wave functions $\psi_n(x)$. To find these wave functions and the corresponding allowed energies in the vibrational motion, the Schrödinger equation with the appropriate potential energy function must be solved:

$$-\frac{\hbar^2}{2\mu}\frac{d^2\psi_n(x)}{dx^2} + \frac{kx^2}{2}\psi_n(x) = E_n\psi_n(x) \tag{7.2}$$

The solution of this second-order differential equation was well known in the mathematical literature from other contexts well before the development of quantum mechanics. We simply state that the normalized wave functions are

$$\psi_n(x) = A_n H_n(\alpha^{1/2}x)e^{-\alpha x^2/2}, \quad \text{for } n = 0, 1, 2, \ldots \tag{7.3}$$

EXAMPLE PROBLEM 7.1

Show that the function $e^{-\beta x^2}$ satisfies the Schrödinger equation for the quantum harmonic oscillator. What conditions does this place on β? What is E?

Solution

$$-\frac{\hbar^2}{2\mu}\frac{d^2\psi_n(x)}{dx^2} + V(x)\psi_n(x) = E_n\psi_n(x)$$

$$-\frac{\hbar^2}{2\mu}\frac{d^2\left(e^{-\beta x^2}\right)}{dx^2} + V(x)\left(e^{-\beta x^2}\right) = -\frac{\hbar^2}{2\mu}\frac{d\left(-2\beta x e^{-\beta x^2}\right)}{dx} + \frac{1}{2}kx^2\left(e^{-\beta x^2}\right)$$

$$= -\frac{\hbar^2}{2\mu}\left(-2\beta\, e^{-\beta x^2}\right) + \frac{\hbar^2}{2\mu}\left(-4\beta^2 x^2 e^{-\beta x^2}\right)$$

$$+ \frac{1}{2}kx^2\left(e^{-\beta x^2}\right)$$

The function is an eigenfunction of the total energy operator only if the last two terms cancel:

$$\hat{H}_{total}\, e^{-\beta x^2} = \frac{\hbar^2 \beta}{\mu} e^{-\beta x^2} \quad \text{if } \beta^2 = \frac{1}{4}\frac{k\mu}{\hbar^2}$$

Finally, $E = \dfrac{\hbar^2 \beta}{\mu} = \dfrac{\hbar^2}{\mu}\sqrt{\dfrac{1}{4}\dfrac{k\mu}{\hbar^2}} = \dfrac{\hbar}{2}\sqrt{\dfrac{k}{\mu}}$

In the preceding equation, several constants have been combined to give $\alpha = \sqrt{k\mu/\hbar^2}$, and the normalization constant A_n is given by

$$A_n = \frac{1}{\sqrt{2^n n!}}\left(\frac{\alpha}{\pi}\right)^{1/4} \tag{7.4}$$

The solution is written in this manner because the set of functions $H_n(\alpha^{1/2}x)$ is well known in mathematics as **Hermite polynomials**. The first few eigenfunctions $\psi_n(x)$ are given by

$$\psi_0(x) = \left(\frac{\alpha}{\pi}\right)^{1/4} e^{-(1/2)\alpha x^2}$$

$$\psi_1(x) = \left(\frac{4\alpha^3}{\pi}\right)^{1/4} x e^{-(1/2)\alpha x^2}$$

$$\psi_2(x) = \left(\frac{\alpha}{4\pi}\right)^{1/4} (2\alpha x^2 - 1) e^{-(1/2)\alpha x^2}$$

$$\psi_3(x) = \left(\frac{\alpha^3}{9\pi}\right)^{1/4} (2\alpha x^3 - 3x) e^{-(1/2)\alpha x^2} \tag{7.5}$$

where $\psi_0, \psi_2, \psi_4, \ldots$ are even functions of x, $[\psi(x) = \psi(-x)]$, whereas $\psi_1, \psi_3, \psi_5, \ldots$ are odd functions of x $[\psi(x) = -\psi(-x)]$.

A necessary boundary condition is that the amplitude of the wave functions remains finite at large values of x. As for the particle in the box, this boundary condition gives rise to quantization. In this case, the quantization condition is not easy to derive. However, it can be shown that the amplitude of the wave functions approaches zero for large x values only if the following condition is met:

the
Chemistry place

7.2 Energy Levels and Eigenfunctions for the Harmonic Oscillator

$$E_n = \hbar\sqrt{\frac{k}{\mu}}\left(n + \frac{1}{2}\right) = h\nu\left(n + \frac{1}{2}\right) \quad \text{with } n = 0, 1, 2, 3, \ldots \tag{7.6}$$

Once again, we see that the imposition of boundary conditions has led to a discrete energy spectrum. Unlike the classical analogue, the energy stored in the quantum mechanical harmonic oscillator can only take on certain values. As for the particle in the box, the lowest state accessible to the system still has a nonzero energy, referred to as a **zero point energy**. The **frequency of oscillation** is given by

$$\nu = \frac{1}{2\pi}\sqrt{\frac{k}{\mu}} \tag{7.7}$$

EXAMPLE PROBLEM 7.2

a. Is $\psi_1(x) = (4\alpha^3/\pi)^{1/4} x e^{-(1/2)\alpha x^2}$ an eigenfunction of the kinetic energy operator? Is it an eigenfunction of the potential energy operator?

b. What are the average values of the kinetic and potential energies for a quantum mechanical oscillator in this state?

Solution

a. As discussed in Chapter 6, neither the potential energy operator nor the kinetic energy operator commutes with the total energy operator. Therefore, because $\psi_1(x) = (4\alpha^3/\pi)^{1/4}xe^{-(1/2)\alpha x^2}$ is an eigenfunction of the total energy operator, it is not an eigenfunction of the potential or kinetic energy operators.

b. The fourth postulate states how the average value of an observable can be calculated. Because

$$\hat{E}_{potential}(x) = V(x) \quad \text{and} \quad \hat{E}_{kinetic}(x) = -\frac{\hbar^2}{2\mu}\frac{d^2}{dx^2}$$

then

$$\langle E_{potential}\rangle = \int \psi_1^*(x)\, V(x)\psi_1(x)\, dx$$

$$= \int_{-\infty}^{\infty}\left(\frac{4\alpha^3}{\pi}\right)^{1/4}xe^{-(1/2)\alpha x^2}\left(\frac{1}{2}kx^2\right)\left(\frac{4\alpha^3}{\pi}\right)^{1/4}xe^{-(1/2)\alpha x^2}\, dx$$

$$= \frac{1}{2}k\left(\frac{4\alpha^3}{\pi}\right)^{1/2}\int_{-\infty}^{\infty}x^4e^{-\alpha x^2}\, dx = k\left(\frac{4\alpha^3}{\pi}\right)^{1/2}\int_{0}^{\infty}x^4e^{-\alpha x^2}\, dx$$

The limits can be changed as indicated in the last integral because the integrand is an even function of x. To obtain the solution, the following standard integral found in the Math Supplement is used:

$$\int_{0}^{\infty}x^{2n}e^{-ax^2}dx = \frac{1\times 3\times 5\cdots(2n-1)}{2^{n+1}a^n}\sqrt{\frac{\pi}{a}}$$

The calculated values for the average potential and kinetic energy are

$$\langle E_{potential}\rangle = \frac{1}{2}k\left(\frac{4\alpha^3}{\pi}\right)^{1/2}\left(\sqrt{\frac{\pi}{\alpha}}\right)\frac{3}{4\alpha^2}$$

$$= \frac{3k}{4\alpha} = \frac{3}{4}\hbar\sqrt{\frac{k}{\mu}}$$

$$\langle E_{kinetic}\rangle = \int \psi_1^*(x)\left(-\frac{\hbar^2}{2\mu}\frac{d^2}{dx^2}\right)\psi_1(x)\, dx$$

$$= \int_{-\infty}^{\infty}\left(\frac{4\alpha^3}{\pi}\right)^{1/4}xe^{-(1/2)\alpha x^2}\left(-\frac{\hbar^2}{2\mu}\frac{d^2}{dx^2}\right)\left(\frac{4\alpha^3}{\pi}\right)^{1/4}xe^{-(1/2)\alpha x^2}\, dx$$

$$= -\frac{\hbar^2}{2\mu}\left(\frac{4\alpha^3}{\pi}\right)^{1/2}\int_{-\infty}^{\infty}(\alpha^2 x^4 - 3\alpha x^2)e^{-\alpha x^2}\, dx$$

$$= -\frac{\hbar^2}{\mu}\left(\frac{4\alpha^3}{\pi}\right)^{1/2}\int_{0}^{\infty}(\alpha^2 x^4 - 3\alpha x^2)e^{-\alpha x^2}\, dx$$

$$= -\frac{\hbar^2}{2\mu}\left(\frac{4\alpha^3}{\pi}\right)^{1/2}\left(\alpha^2\left[\sqrt{\frac{\pi}{\alpha}}\frac{3}{4\alpha^2}\right] - 3\alpha\left[\sqrt{\frac{\pi}{\alpha}}\frac{1}{2\alpha}\right]\right)$$

$$= \frac{3}{4}\frac{\hbar^2\alpha}{\mu} = \frac{3}{4}\hbar\sqrt{\frac{k}{\mu}}$$

Note that just as for the classical harmonic oscillator (see Optional Review Section 7.6), the average values of the kinetic and potential energies are equal. When the kinetic energy has its maximum value, the potential energy is zero and vice versa. In general, we find that for the nth state,

$$\langle E_{kinetic,n}\rangle = \langle E_{potential,n}\rangle = \frac{\hbar}{2}\sqrt{\frac{k}{\mu}}\left(n + \frac{1}{2}\right)$$

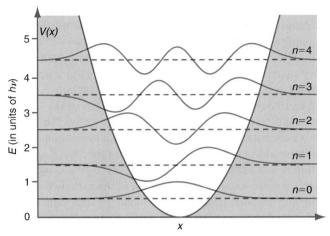

FIGURE 7.2

The first few eigenfunctions of the quantum mechanical harmonic oscillator are plotted together with the potential function. The energy scale is shown on the left. The amplitude of the eigenfunctions is shown superimposed on the energy level, with the zero of amplitude for the eigenfunctions indicated by the dashed lines. The yellow area indicates the classically forbidden region for which $E < V$.

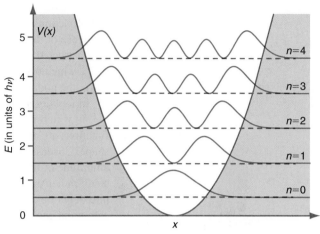

FIGURE 7.3

The square of the first few eigenfunctions of the quantum mechanical harmonic oscillator (the probability density) is superimposed on the energy spectrum and plotted together with the potential function. The yellow area indicates the classically forbidden region.

As was done for the particle in the box, it is useful to plot $\psi(x)$ and $\psi^2(x)$ against x. They are shown superimposed on the potential energy function in Figures 7.2 and 7.3.

It is instructive to compare the quantum mechanical with the classical results. In quantum mechanics the value of x cannot be known if the system is in an eigenstate of the total energy operator, because these two operators do not commute. This issue arose earlier in considering the particle in the box. What can one say about x, the amplitude of the vibration?

Only the probability of the vibrational amplitude having a particular value of x within an interval dx can be calculated, and this probability is given by $\psi^2(x)\,dx$. For the classical harmonic oscillator, the probability of finding a particular value of x within the interval dx can also be calculated. Because the probability density varies inversely with the velocity, its maximum values are found at the turning points and its minimum value is found at $x = 0$. To visualize this behavior, imagine a frictionless ball rolling on a parabolic track under the influence of gravity. The ball moves fastest at the lowest point on the track, and stops momentarily as it reverses its direction at the highest points on either side of the track. Figure 7.4 shows a comparison of $\psi^2_{12}(x)$ and the probability density of finding a particular amplitude for a classical oscillator with the same total energy as a function of x. A large quantum number has been used for comparison because in the limit of high energies (very large quantum numbers), classical and quantum mechanics give the same result.

The main difference between the classical and quantum mechanical results are the oscillations in $\psi^2_{12}(x)$, which are absent in the classical result. However, in calculating the probability of finding the value x for the oscillation amplitude in the interval Δx, it is necessary to evaluate

$$\int_{-\Delta x/2}^{\Delta x/2} \psi^2(x)\,dx$$

rather than the probability density $\psi^2(x)$. For large quantum numbers, the interval Δx in which the probability is calculated will be large in comparison to the distance between neighboring oscillations in $\psi^2(x)$. Therefore, the oscillations in the probability density $\psi^2(x)$ are averaged out, and the quantum and classical results agree well. The argument is the same as that used in calculating the results for the particle in the box shown in Figure 4.4.

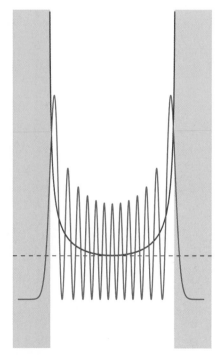

FIGURE 7.4

The calculated probability density for the vibrational amplitude is shown for the 12th eigenstate of the quantum mechanical oscillator (red curve). The classical result is shown by the blue curve. The yellow area indicates the classically forbidden region.

We have been working with the time-independent Schrödinger equation, whose eigenfunctions allow the probability density to be calculated. To describe the time dependence of the oscillation amplitude, the total wave function, $\Psi_n(x, t) = e^{-i\omega t}\psi_n(x)$, is constructed. The spatial amplitude shown in Figure 7.2 is modulated by the factor $e^{-i\omega t}$, which has a frequency given by $\omega = \sqrt{k/\mu}$. Because $\Psi_n(x, t)$ is a standing wave, the nodal positions shown in Figures 7.2 and 7.3 do not move with time.

In looking at Figures 7.2 and 7.3, several similarities are seen with Figures 4.2 and 4.3, in which the equivalent results were shown for the particle in the box. The eigenfunctions are again standing waves, but they are now in a box with a more complicated shape. Successive eigenfunctions add one more oscillation within the "box," and the amplitude of the wave function is small at the edge of the "box." (The reason why it is small rather than zero follows the same lines as the discussion of the particle in the finite depth box in Chapter 5.) The quantum mechanical harmonic oscillator also has a zero point energy, meaning that the lowest possible energy state still has vibrational energy. The origin of this zero point energy is similar to that for the particle in the box. By attaching a spring to the particle, its motion has been constrained. As the spring is made stiffer (larger k), the particle is more constrained and the zero point energy increases. This is the same trend observed for the particle in the box as the length is decreased.

Note, however, that important differences exist in the two problems that are a result of the more complicated shape of the harmonic oscillator "box." Although the wave functions show oscillatory behavior, they are no longer represented by simple sine functions. Because the classical probability density is not independent of x (see Figure 7.4), simple sine functions do not lead to the correct classical limit. The energy spacing is the same between adjacent energy levels; that is, it does not increase with the quantum number as was the case for the particle in the box. These differences show the sensitivity of the eigenfunctions and eigenvalues to the functional form of the potential.

7.3 Probability of Finding the Oscillator in the Classically Forbidden Region

7.2 SOLVING THE SCHRÖDINGER EQUATION FOR ROTATION IN TWO DIMENSIONS

Quantum mechanical models were developed for translation in Chapter 4 and for vibration in Section 7.1. We now consider rotation to complete the description of the fundamental types of motion available to a molecule. To a good approximation, the three types of motion—translation, vibration, and rotation—can be dealt with independently. This treatment is exact rather than approximate (1) if the translational part of the total energy operator depends only on the translational coordinates of the center of mass, (2) if the rotational part depends only on the angular coordinates of the center of mass, and (3) if the vibrational part depends only on the internal coordinates of the molecule. This condition cannot be exactly satisfied, because the types of motion are not totally decoupled. For example, the average bond length of a rapidly rotating molecule is slightly longer than for a molecule that is not rotating because of the centrifugal forces acting on the atoms. However, although the coupling between the types of motion can be measured using sensitive spectroscopic techniques, the coupling is small for most molecules.

Neglecting this coupling, the total energy operator can be written as a sum of individual operators for the types of motion for the molecule:

$$\hat{H}_{total} = \hat{H}_{trans}(r_{cm}) + \hat{H}_{vib}(\tau_{internal}) + \hat{H}_{rot}(\theta_{cm}, \phi_{cm}) \qquad (7.8)$$

In this equation, r_{cm}, θ_{cm}, and ϕ_{cm} refer to the spatial coordinates of the center of mass in spherical coordinates (see the Math Supplement, Appendix A). The symbol $\tau_{internal}$ refers collectively to the vibrational amplitudes of all atoms in the molecule around their equilibrium position. Because different variables appear in $\hat{H}_{trans}(r_{cm})$, $\hat{H}_{vib}(\tau_{internal})$, and $\hat{H}_{rot}(\theta_{cm}, \phi_{cm})$, it is possible to solve the Schrödinger equation for each type of motion separately. In this approximation, the total energy is given by the sum of the individual contributions, $E_{total} = E_{translational}(r_{cm}) + E_{vibrational}(\tau_{internal}) + E_{rotational}(\theta_{cm}, \phi_{cm})$, and the system wave function is a product of the eigenfunctions for the three types of motion:

$$\psi_{total} = \psi_{trans}(r_{cm})\psi_{vib}(\tau_{internal})\psi_{rot}(\theta_{cm}, \phi_{cm}) \qquad (7.9)$$

Because the wave function is a product of individual terms that depend on different variables, what has been accomplished in Equation (7.9) is a **separation of variables**.

Whereas for a diatomic molecule, translation can be considered in one to three independent dimensions, and vibration in one dimension, rotation requires at least a two-dimensional description. We restrict our considerations to diatomic molecules because the motion is easy to visualize. However, the process outlined next can be generalized to any molecule if several angular coordinates are included. Rotation in a two-dimensional space is discussed first because the mathematical formalism needed to describe such a problem is less complicated. The formalism is extended in Section 7.3 to rotation in three dimensions.

Rotation in two dimensions occurs only in a constrained geometry. An example is a molecule adsorbed on a smooth surface. Consider a diatomic molecule with masses m_1 and m_2 and a fixed bond length r_0 freely rotating in the $x-y$ plane. Because the bond length is assumed to remain constant as the molecule rotates, this model is often referred to as the **rigid rotor**. By transforming to the center of mass coordinate system, this problem becomes equivalent to a single reduced mass $\mu = (m_1 m_2)/(m_1 + m_2)$ rotating in the $x-y$ plane on a ring of radius r_0. This geometry is referred to as a particle on a ring.

EXAMPLE PROBLEM 7.3

The bond length for $H^{19}F$ is 91.68×10^{-12} m. Where does the axis of rotation intersect the molecular axis?

Solution

The position of the center of mass is given by $x_{cm} = (m_H x_H + m_F x_F)/(m_H + m_F)$. We choose the origin of our coordinate system to be at the F atom, so $x_F = 0$ and $x_H = 91.68 \times 10^{-12}$ m. Substituting $m_F = 18.9984$ amu and $m_H = 1.008$ amu, we find that $x_{cm} = 4.62 \times 10^{-12}$ m. Therefore $x_F = 4.62 \times 10^{-12}$ m and $x_H = 87.06 \times 10^{-12}$ m. We see that the axis of rotation is very close to the F atom. This effect is even more pronounced for HI or HCl.

Because it has been assumed that the particle experiences no hindrance to rotation, the potential energy is constant everywhere. Therefore, we can conveniently set $V(x, y) = 0$ everywhere without affecting the eigenfunctions of the total energy operator. The Schrödinger equation in Cartesian coordinates for this problem is

$$-\frac{\hbar^2}{2\mu}\left(\frac{\partial^2 \psi(x, y)}{\partial x^2} + \frac{\partial^2 \psi(x, y)}{\partial y^2}\right)_{r=r_0} = E\psi(x, y) \qquad (7.10)$$

The subscript after the bracket makes it clear that the radius is constant. Although Equation (7.10) is correct, it is always best to choose a coordinate system that reflects the symmetry of the system being considered. In this case, two-dimensional polar coordinates with the variables r and ϕ are the logical choice. In these coordinates, with r fixed at r_0, the operator $(\partial^2/\partial x^2) + (\partial^2/\partial y^2)$ becomes $(1/r_0^2)(\partial^2/\partial\phi^2)$. Therefore, the Schrödinger equation takes the simple form

$$-\frac{\hbar^2}{2\mu r_0^2}\frac{d^2\Phi(\phi)}{d\phi^2} = E\Phi(\phi) \qquad (7.11)$$

where the eigenfunction $\Phi(\phi)$ depends only on the angle ϕ. We have changed the symbol for the wave function to emphasize the change in the variables. This equation has the same form as the Schrödinger equation for a free particle, which was solved in Chapter 4. You should verify that the two linearly independent solutions to this equation are

$$\Phi_+(\phi) = A_{+\phi}e^{im_l\phi} \quad \text{and} \quad \Phi_-(\phi) = A_{-\phi}e^{-im_l\phi} \qquad (7.12)$$

The two solutions correspond to counterclockwise and clockwise rotation.

EXAMPLE PROBLEM 7.4

Determine the normalization constant $A_{+\phi}$ in Equation (7.12).

Solution

The variable ϕ can take on values between 0 and 2π. The following result is obtained:

$$\int_0^{2\pi} \Phi_{m_l}^*(\phi)\, \Phi_{m_l}(\phi)\, d\phi = 1$$

$$(A_{+\phi})^2 \int_0^{2\pi} e^{-im_l\phi} e^{im_l\phi}\, d\phi = (A_{+\phi})^2 \int_0^{2\pi} d\phi = 1$$

$$A_{+\phi} = \frac{1}{\sqrt{2\pi}}$$

Convince yourself that $A_{-\phi}$ has the same value.

To obtain solutions of the Schrödinger equation that describe this physical problem, it is necessary to introduce the boundary condition $\Phi(\phi + 2\pi) = \Phi(\phi)$. This condition states that there is no way to distinguish the particle that has rotated n times around the circle from one that has rotated $n + 1$ times around the circle. Without this condition, the probability density would have multiple values for ϕ and $\phi + 2n\pi$, as shown in Figure 7.5, which is unacceptable. Applying the single-value condition to the eigenfunction, $e^{im_l[\phi+2\pi]} = e^{im_l\phi}$ or $e^{2\pi im_l} = 1$. Using Euler's relation, this expression is equivalent to

$$\cos 2\pi m_l + i \sin 2\pi m_l = 1 \tag{7.13}$$

To satisfy this condition, m_l must equal $0, \pm 1, \pm 2, \pm 3, \ldots$. The boundary condition generates the quantization rules for the quantum number m_l. The motivation for using the subscript l on the quantum number m will become clear when rotation in three dimensions is considered.

What do these eigenfunctions look like? Because they are complex functions of the angle ϕ, only the real part of the function is shown in Figure 7.6. The imaginary part is identical in shape, but is shifted in phase by the angle $\pi/2$. Note that, as for the particle in the box and the harmonic oscillator, the lowest energy state has no nodes, and the number of nodes, which is twice the quantum number, increases with m_l.

Putting the eigenfunctions back into Equation (7.11) allows the corresponding eigenvalues, E_{m_l}, to be calculated. The energy-level spectrum is discrete and is given by

$$E_{m_l} = \frac{\hbar^2 m_l^2}{2\mu r_0^2} = \frac{\hbar^2 m_l^2}{2I} \quad \text{for } m_l = 0, \pm 1, \pm 2, \pm 3, \ldots \tag{7.14}$$

In Equation (7.14), $I = \mu r_0^2$ is the moment of inertia. Note that states with $+m_l$ and $-m_l$ have the same energy although the wave functions corresponding to these states are orthogonal to one another. We say that the energy levels with $m_l \neq 0$ are *twofold degenerate*.

The origin of the energy quantization is again a boundary condition. In this case, imagine the ring as a box of length 2π defined by the variable ϕ. The boundary condition given in Equation (7.13) states that an integral number of wavelengths must fit into this "box." Keep in mind that for a classical rigid rotor,

$$E = \frac{|\mathbf{l}|^2}{2\mu r_0^2} = \frac{|\mathbf{l}|^2}{2I} = \frac{1}{2} I \omega^2$$

where, throughout this chapter, the symbol $\mathbf{l}$ is used for the angular momentum vector, $|\mathbf{l}|$ for its magnitude, and $\hat{l}$ for the angular momentum operator. The same relationship

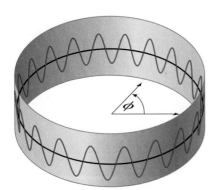

$m_l = \pm\text{integer}$

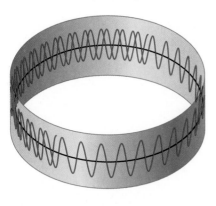

$m_l \neq \pm\text{integer}$

FIGURE 7.5

If the condition $m_l = $ integer is not met, the wave function does not have the same value for $\phi + 2\pi$ as for ϕ. The real part of the wave function is plotted as a function of ϕ in each case.

also holds for the quantum mechanical rigid rotor, with the association $\omega = m_l \hbar / I$. Therefore, the quantization of energy means that only a discrete set of rotational frequencies is allowed.

One aspect of the eigenvalues for free rotation in two dimensions is different from what was encountered with the particle in the box or the harmonic oscillator: no zero point energy is associated with free rotational motion; $E_{m_l} = 0$ when $m_l = 0$. Why is this the case? A zero point energy appears only if the potential confines the motion to a limited region. In free rotation, there is no confinement and no zero point energy. Of course, a diatomic molecule also vibrates. Therefore, the rotating molecule has a zero point energy associated with this degree of freedom.

The angular momentum can also be calculated for the two-dimensional rigid rotor. For rotation in the $x-y$ plane, the angular momentum vector lies on the z axis. The angular momentum operator in these coordinates takes the simple form $\hat{l}_z = -i\hbar(\partial/\partial\phi)$. Applying this operator to an eigenfunction,

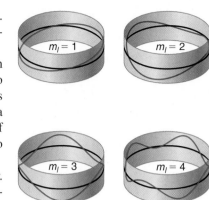

$$\hat{l}_z \Phi_+(\phi) = \frac{-i\hbar}{\sqrt{2\pi}} \frac{d\,e^{im_l\phi}}{d\phi} = \frac{m_l\hbar}{\sqrt{2\pi}} e^{im_l\phi} = m_l\hbar\,\Phi_+(\phi) \qquad \textbf{(7.15)}$$

A similar equation can be written for $\Phi_-(\phi)$. This result shows that the angular momentum is quantized. We see that $\Phi_+(\phi)$ and $\Phi_-(\phi)$ are eigenfunctions of both the total energy and the angular momentum operators for the two-dimensional rigid rotor. As we will see, this is not the case for rotation in three dimensions. Because the angular momentum has the values $+\hbar m_l$ and $-\hbar m_l$, Equation (7.14) can be written in the following form:

$$E_{m_l} = \frac{\hbar^2 m_l^2}{2I} = \frac{|\mathbf{l}|^2}{2I}$$

just as in classical mechanics.

What can be said about the angle of the molecular axis with respect to a fixed direction in the $x-y$ plane? We know that the probability of finding a particular angle ϕ in the interval $d\phi$ is

$$P(\phi)\,d\phi = \Phi^*(\phi)\Phi(\phi)\,d\phi = \left(\frac{1}{\sqrt{2\pi}}\right)^2 e^{\pm im_l\phi} e^{\mp im_l\phi}\,d\phi = \frac{d\phi}{2\pi} \qquad \textbf{(7.16)}$$

The probability of finding the particle in a given interval $d\phi$ is the same for all values of ϕ. Just as for the position of a free particle whose linear momentum is precisely defined, nothing is known about the angular position of the molecule whose angular momentum is precisely defined. The origin of this result is that the operators $\hat{\phi}$ and $\hat{l}_z$ do not commute, just as $\hat{x}$ and $\hat{p}_x$ do not commute.

FIGURE 7.6

The real part of the second through seventh eigenfunctions for the rigid rotor with rotation confined to a plane is plotted as a function of ϕ. In the center of mass coordinates, this problem is equivalent to the particle on a ring. What does the first eigenfunction look like?

7.3 SOLVING THE SCHRÖDINGER EQUATION FOR ROTATION IN THREE DIMENSIONS

In the case just considered, the motion has been constrained to two dimensions. Now imagine the more familiar case of a diatomic molecule freely rotating in three-dimensional space. This problem is not more difficult, but the mathematics is more cumbersome than the two-dimensional case just considered. Again, we transform to the center of mass coordinate system, and the rotational motion is transformed to the motion of a particle on the surface of a sphere of radius r_0. As before, it is advantageous to express the kinetic and potential energy operators in an appropriate coordinate system, which in this case is spherical coordinates. Because there is no hindrance to rotation, the potential energy is zero. In this coordinate system, which is depicted in Figure 2.5, the Schrödinger equation turns out to be

$$-\frac{\hbar^2}{2\mu r_0^2}\left[\frac{1}{\sin\theta}\frac{\partial}{\partial\theta}\left(\sin\theta\frac{\partial Y(\theta,\phi)}{\partial\theta}\right) + \frac{1}{\sin^2\theta}\frac{\partial^2 Y(\theta,\phi)}{\partial\phi^2}\right] = EY(\theta,\phi) \qquad \textbf{(7.17)}$$

Figure 2.5 defines the relationship between x, y, and z in Cartesian coordinates, and r, θ, and ϕ in spherical coordinates.

Our task is to find the eigenfunctions $Y(\theta, \phi)$ and the corresponding eigenvalues that are the solutions of this equation. Although the solution of this partial differential equation is not discussed in detail here, the first few steps are outlined because they provide some important physical insights. Combining constants in the form

$$\beta = \frac{2\mu r_0^2 E}{\hbar^2} \qquad (7.18)$$

multiplying through on the left by $\sin^2 \theta$, and rearranging this equation results in Equation (7.19)

$$\sin \theta \frac{\partial}{\partial \theta} \left(\sin \theta \frac{\partial Y(\theta, \phi)}{\partial \theta} \right) + [\beta \sin^2 \theta] Y(\theta, \phi) = -\frac{\partial^2 Y(\theta, \phi)}{\partial \phi^2} \qquad (7.19)$$

On the right side of the equation, the differentiation is with respect to ϕ only. On the left side of the equation, the differentiation is with respect to θ only. If this equation is to hold for all ϕ and θ, $Y(\theta, \phi)$ must be the product of two functions, each of which depends on only one of the two independent variables:

$$Y(\theta, \phi) = \Theta(\theta)\Phi(\phi) \qquad (7.20)$$

The fact that the function $Y(\theta, \phi)$ can be written as a product of two functions, each of which depends on only one of the two variables, offers a major simplification in solving the differential equation, Equation (7.17). This separation of variables has also been invoked previously in solving the three-dimensional particle in the box problem in Section 4.3, and in separating the translational, vibrational, and rotational types of motion in Section 7.2.

The functions $Y(\theta, \phi)$ are known as the **spherical harmonic functions** and are discussed in detail later in this chapter. Substituting Equation (7.20) into Equation (7.19) and dividing through by $\Theta(\theta)\Phi(\phi)$, we obtain

$$\frac{1}{\Theta(\theta)} \sin \theta \frac{d}{d\theta} \left(\sin \theta \frac{d\Theta(\theta)}{d\theta} \right) + \beta \sin^2 \theta = -\frac{1}{\Phi(\phi)} \frac{d^2 \Phi(\phi)}{d\phi^2} \qquad (7.21)$$

Note that this equation no longer contains partial derivatives. Because each side of the equation depends on only one of the variables for all values of the variables, it must be true that the equality is satisfied by a constant:

$$\frac{1}{\Theta(\theta)} \sin \theta \frac{d}{d\theta} \left(\sin \theta \frac{d\Theta(\theta)}{d\theta} \right) + \beta \sin^2 \theta = m_l^2 \quad \text{and}$$

$$\frac{1}{\Phi(\phi)} \frac{d^2 \Phi(\phi)}{d\phi^2} = -m_l^2 \qquad (7.22)$$

Looking back at the differential equation for rotation in two dimensions, it is clear why the constant is written in this way. The solutions for the second equation can be obtained immediately, because the same equation was solved for the molecule rotating in two dimensions:

$$\Phi_+(\phi) = A_{+\phi} e^{im_l \phi} \quad \text{and} \quad \Phi_-(\phi) = A_{-\phi} e^{-im_l \phi}, \quad \text{for } m_l = 0, 1, 2, 3, \ldots \qquad (7.23)$$

where the part of $Y(\theta, \phi)$ that depends on ϕ is associated with the quantum number m_l.

The first equation in Equations (7.22) allows the part of $Y(\theta, \phi)$ that depends on θ to be determined. It can be solved to give a set of eigenfunctions and their corresponding eigenvalues. Rather than work through the solution, the results are summarized with a focus on the eigenvalues. A discussion of the spherical harmonics is postponed until later in this chapter. Two boundary conditions must be satisfied to solve Equations (7.22). To make sure that the functions $Y(\theta, \phi)$ are single-valued functions of θ and ϕ and that the amplitude of these functions remains finite everywhere, the following conditions must be met. We state rather than derive these conditions:

$$\beta = l(l+1), \quad \text{for } l = 0, 1, 2, 3, \ldots \quad \text{and}$$

$$m_l = -l, -(l-1), -(l-2), \ldots, 0, \ldots, (l-2), (l-1), l \qquad (7.24)$$

Both l and m_l must be integers. Note that l and m_l are the quantum numbers for the 3D rigid rotor; to emphasize this result, the spherical harmonic functions are written in the form

$$Y(\theta, \phi) = Y_l^{m_l}(\theta, \phi) = \Theta_l^{m_l}(\theta)\Phi_{m_l}(\phi) \qquad (7.25)$$

The function $\Theta_l^{m_l}(\theta)$ is associated with both quantum numbers l and m_l, and the function $\Phi_{m_l}(\phi)$ is associated only with the quantum number m_l. The values of the quantum numbers l and m_l are dependent on one another. For a given value of l, there are $2l + 1$ different values of m_l ranging from $-l$ to $+l$. We next consider the origin of these quantum numbers more closely.

Why are there two quantum numbers for rotation in three dimensions, whereas there is only one for rotation in two dimensions? The answer is related to the dimensionality of the problem. For rotation in two dimensions, r was held constant. Therefore, ϕ is the only variable in the problem and there is only one boundary condition. For rotation in three dimensions, r is again held constant and, therefore, only the two boundary conditions on θ and ϕ generate quantum numbers. For the same reason, the particle in the one-dimensional box is characterized by a single quantum number, whereas three quantum numbers are required to characterize the particle in the three-dimensional box.

What observables of the rotating molecule are associated with the quantum numbers l and m_l? From the equation

$$\beta = \frac{2\mu r_0^2 E}{\hbar^2} = \frac{2I}{\hbar^2}E$$

the energy eigenvalues for rotation in three dimensions can be obtained. This shows that the quantum number l is associated with the total energy observable,

$$E_l = \frac{\hbar^2}{2I}l(l + 1), \quad \text{for } l = 0, 1, 2, 3, \ldots \qquad (7.26)$$

and that the total energy eigenfunctions $Y_l^{m_l}(\theta, \phi)$ satisfy the eigenvalue equation

$$\hat{H}_{total} Y_l^{m_l}(\theta, \phi) = \frac{\hbar^2}{2I}l(l + 1)Y_l^{m_l}(\theta, \phi), \quad \text{for } l = 0, 1, 2, 3, \ldots \qquad (7.27)$$

Note that the rotational energy values are quantized and that, once again, each quantization condition arises through a boundary condition. Note that the energy levels depend differently on the quantum number than the energy levels for rotation in two dimensions for which

$$E_{m_l} = \frac{\hbar^2 m_l^2}{2\mu r_0^2} = \frac{\hbar^2 m_l^2}{2I}, \quad \text{for } m_l = 0, \pm 1, \pm 2, \pm 3, \ldots$$

For rotation in three dimensions the energy depends on the quantum number l, but not on m_l. Why is this the case? As will be shown later, the quantum number m_l determines the z component of the vector $\mathbf{l}$. Because $E_{total} = |\mathbf{l}|^2/2\mu r_0^2$, the energy of rotation depends only on the magnitude of the angular momentum and not its direction. Therefore, all $2l + 1$ total energy eigenfunctions that have the same l value, but different m_l values, have the same energy. This means that the **degeneracy** of each energy level is $2l + 1$. Recall that for rotation in two dimensions, the degeneracy of each energy level is two, except for the $m_l = 0$ level, which is nondegenerate.

7.4 THE QUANTIZATION OF ANGULAR MOMENTUM

We continue our discussion of three-dimensional rotation, although now it is discussed in the context of **angular momentum** rather than energy as was done earlier. Why is angular momentum important in quantum chemistry? Perhaps the best way to understand why is to consider a familiar example from introductory chemistry, namely, the s,

p, and d **orbitals** associated with atoms of the periodic table. This notation will be discussed in more detail in Chapter 9. For now, assume that the bonding behavior of s, p, and d electrons is quite different. Why is an s orbital spherically symmetrical, whereas a p orbital has a dumbbell structure? Why are three energetically degenerate p orbitals directed along the x, y, and z directions? The origin of these chemically important properties is related to the particular value of l or m_l associated with these orbitals.

As discussed earlier, the spherical harmonic functions $Y_l^{m_l}(\theta, \phi)$, are eigenfunctions of the total energy operator for a molecule that rotates freely in three dimensions. Are these functions also eigenfunctions of other operators of interest to us? Because the potential energy is zero for a free rotor, the total energy stored in rotational motion is given by $E_{total} = |\mathbf{l}|^2/2I$, in which $\mathbf{l}$ is the angular momentum and $I = \mu r_0^2$. Note that E_{total} and $|\mathbf{l}^2|$ differ only by the constant $1/2I$. Therefore, the corresponding operators $\hat{H}_{total}$ and $\hat{l}^2$ also satisfy this relationship. Because they differ only by a multiplicative constant, these two operators commute with one another and have a common set of eigenfunctions. Furthermore, because E_{total} is quantized, it can be concluded that $|\mathbf{l}^2|$ is also quantized. Using the proportionality of E_{total} and $|\mathbf{l}^2|$, the eigenvalue equation for the operator $\hat{l}^2$ can immediately be written from Equation (7.27):

$$\hat{l}^2 Y_l^{m_l}(\theta, \phi) = \hbar^2 l(l + 1) Y_l^{m_l}(\theta, \phi) \qquad (7.28)$$

The notation explicitly shows that the quantum numbers l and m_l are defining indices for the eigenfunctions of $\hat{H}_{total}$ and $\hat{l}^2$. It also states that the eigenvalues for $\hat{l}^2$ are given by $\hbar^2 l(l + 1)$, which means that the magnitude of the angular momentum takes on the quantized values $|\mathbf{l}| = \hbar\sqrt{l(l + 1)}$.

Note that it is $\hat{l}^2$ and not $\hat{l}$ that commutes with $\hat{H}_{total}$. We now focus our attention on the angular momentum $\mathbf{l}$ and the corresponding operator $\hat{l}^2$. How many components does $\mathbf{l}$ have? For rotation in the x–y plane, the angular momentum vector has only a single component that lies on the z axis. For rotation in three dimensions, the angular momentum vector has the three components l_x, l_y, and l_z, which are obtained from the vector cross product $\mathbf{l} = \mathbf{r} \times \mathbf{p}$. See Optional Review Section 7.7 and the Math Supplement for a more detailed discussion of the cross product and angular motion. As might be expected from the discussion of the Stern–Gerlach experiment in Chapter 6, the operators $\hat{l}_x$, $\hat{l}_y$, and $\hat{l}_z$ do not commute.

As you will see when working the end-of-chapter problems, the operators $\hat{l}_x$, $\hat{l}_y$, and $\hat{l}_z$ have the following form in Cartesian coordinates:

$$\hat{l}_x = -i\hbar\left(y\frac{\partial}{\partial z} - z\frac{\partial}{\partial y} \right)$$

$$\hat{l}_y = -i\hbar\left(z\frac{\partial}{\partial x} - x\frac{\partial}{\partial z} \right)$$

$$\hat{l}_z = -i\hbar\left(x\frac{\partial}{\partial y} - y\frac{\partial}{\partial x} \right) \qquad (7.29)$$

Although not derived here, the operators have the following form in spherical coordinates:

$$\hat{l}_x = -i\hbar\left(-\sin\phi\frac{\partial}{\partial\theta} - \cot\theta\cos\phi\frac{\partial}{\partial\phi} \right)$$

$$\hat{l}_y = -i\hbar\left(\cos\phi\frac{\partial}{\partial\theta} - \cot\theta\sin\phi\frac{\partial}{\partial\phi} \right)$$

$$\hat{l}_z = -i\hbar\left(\frac{\partial}{\partial\phi} \right) \qquad (7.30)$$

As you will verify in the end-of-chapter problems for the operators in Cartesian coordinates, the commutators relating the operators $\hat{l}_x$, $\hat{l}_y$, and $\hat{l}_z$ are given by

$$[\hat{l}_x, \hat{l}_y] = i\hbar\hat{l}_z$$

$$[\hat{l}_y, \hat{l}_z] = i\hbar\hat{l}_x$$

$$[\hat{l}_z, \hat{l}_x] = i\hbar\hat{l}_y \qquad (7.31)$$

Note that the order of the commutator is important, that is, $[\hat{l}_x, \hat{l}_y] = -[\hat{l}_y, \hat{l}_x]$.

What are the consequences of the fact that the operators corresponding to the components of the angular momentum do not commute with one another? Because the commutators are not zero, the direction of the angular momentum vector cannot be specified for rotation in three dimensions. To do so, it would be necessary to know all three components simultaneously, which would require that the three commutators in Equation (7.31) are zero. Given that $\hat{l}_x$, $\hat{l}_y$, and $\hat{l}_z$ do not commute, what can be known about the components of the angular momentum for a molecule whose wave function is an eigenfunction of the total energy operator?

To answer this question, we look more closely at the operators for the individual components of the angular momentum. In spherical coordinates, $\hat{l}_x$ and $\hat{l}_y$ depend on both θ and ϕ, but as Equation (7.30) shows, $\hat{l}_z$ depends only on ϕ. As shown earlier, the spherical harmonics, $Y_l^{m_l}(\theta, \phi) = \Theta_l^{m_l}(\theta)\Phi_{m_l}(\phi)$, are eigenfunctions of the total energy operator and of $\hat{l}^2$. We now show that the spherical harmonics are also eigenfunctions of $\hat{l}_z$. Applying $\hat{l}_z$ to the functions $Y_l^{m_l}(\theta, \phi)$, we obtain

$$\hat{l}_z(Y_l^{m_l}(\theta, \phi)) = \Theta(\theta)\left[-i\hbar \frac{\partial}{\partial \phi}\left(\frac{1}{\sqrt{2\pi}} e^{\pm i m_l \phi}\right)\right] = \pm m_l \hbar \Theta(\theta)\Phi(\phi),$$

for $m_l = 0, \pm 1, \pm 2, \pm 3, \ldots, \pm l$

(7.32)

showing that the $Y_l^{m_l}(\theta, \phi)$ are eigenfunctions of $\hat{l}_z$. What can we conclude from Equation (7.32)? Because the spherical harmonics are eigenfunctions of both $\hat{l}^2$ and $\hat{l}_z$, both the magnitude of $|\mathbf{l}|$ and its z component can be known simultaneously. In other words, one can know the length of the vector $\mathbf{l}$ and one of its components, but it is not possible to simultaneously know the other two components of $\mathbf{l}$.

Why has $\hat{l}_z$ rather than $\hat{l}_x$ or $\hat{l}_y$ been singled out, and what makes the z component special? There is nothing special about the z direction, and one could have just as easily chosen another direction. The way in which the variables are defined in spherical coordinates makes $\hat{l}_z$ take on a simple form. Therefore, when a direction is chosen, it is convenient to make it the z direction. The essence of the preceding discussion is that one can know the magnitude of $\mathbf{l}$ and only one of its components simultaneously. The consequences of the different commutation relations among $\hat{H}$, $\hat{l}^2$, $\hat{l}_x$, $\hat{l}_y$, and $\hat{l}_z$ are explored in Supplemental Section 7.8, which deals with spatial quantization.

7.5 THE SPHERICAL HARMONIC FUNCTIONS

Until now, only the eigenvalues for $\hat{l}^2$, $\hat{H}$, and $\hat{l}_z$ for rotation in three dimensions have been discussed. We now discuss the spherical harmonic functions, $Y_l^{m_l}(\theta, \phi)$, which are the eigenfunctions common to these three operators. They are listed here for the first few values of l and m_l:

$$Y_0^0(\theta, \phi) = \frac{1}{(4\pi)^{1/2}}$$

$$Y_1^0(\theta, \phi) = \left(\frac{3}{4\pi}\right)^{1/2} \cos\theta$$

$$Y_1^{\pm 1}(\theta, \phi) = \left(\frac{3}{8\pi}\right)^{1/2} \sin\theta \, e^{\pm i\phi}$$

$$Y_2^0(\theta, \phi) = \left(\frac{5}{16\pi}\right)^{1/2} (3\cos^2\theta - 1)$$

$$Y_2^{\pm 1}(\theta, \phi) = \left(\frac{15}{8\pi}\right)^{1/2} \sin\theta \cos\theta \, e^{\pm i\phi}$$

$$Y_2^{\pm 2}(\theta, \phi) = \left(\frac{15}{32\pi}\right)^{1/2} \sin^2\theta \, e^{\pm 2i\phi}$$

(7.33)

As seen earlier in Equation (7.12), the ϕ dependence is a simple exponential function. The θ dependence enters as a polynomial in $\sin\theta$ and $\cos\theta$. The numerical factor in front of these functions ensures that they are normalized over the intervals $0 \leq \theta \leq \pi$ and $0 \leq \phi \leq 2\pi$. Because the spherical harmonics are eigenfunctions of the time-independent Schrödinger equation, they represent standing waves on the surface of a sphere, in which the nodal positions are independent of time.

For $l = 0$, the eigenfunction is equal to a constant that is determined by the normalization condition. What does this mean? Remember that the square of the wave function gives the probability density for finding the particle at the coordinates θ and ϕ within the interval $d\theta$ and $d\phi$. These coordinates specify the angle defining the internuclear axis in a diatomic molecule. The fact that the wave function is independent of θ and ϕ means that any orientation of the internuclear axis in the rotation of a molecule is equally likely. This must be the case for a state in which the angular momentum is zero. A net angular momentum, corresponding to $l > 0$, requires the wave function and the probability density distribution to not have spherical symmetry.

These functions are complex unless $m_l = 0$. Graphing complex functions requires double the number of dimensions as for real functions, so that it is customary to instead form appropriate linear combinations of the $Y_l^{m_l}(\theta, \phi)$ values to generate real functions. These functions, which still form an orthonormal set, are given in the following equations. Equation (7.34) lists the p functions, and Equation (7.35) lists the d functions. We recognize the abbreviations in connection with the orbital designations for the hydrogen atom. As shown in Chapter 9, the functions shown in Figures 7.7 and 7.8 appear in the

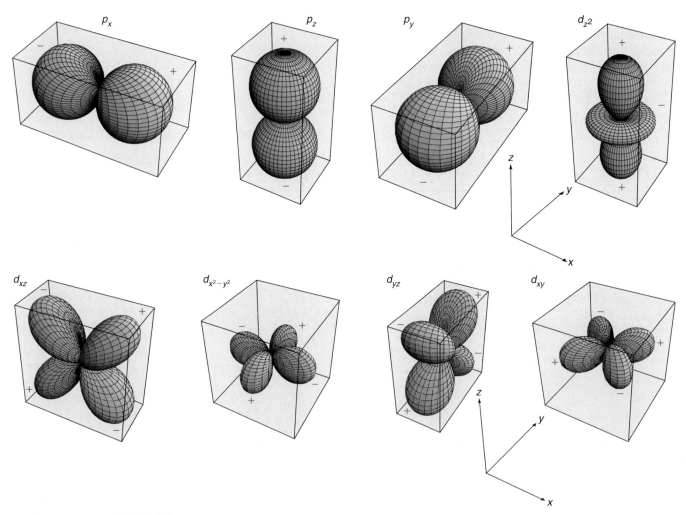

FIGURE 7.7

3D perspective plots of the p and d linear combinations of the spherical harmonics. The plots show three-dimensional surfaces in which the relationship of the angles θ and ϕ to the Cartesian axes is defined in Figure 2.5. The distance from the origin to a point on the surface, (θ, ϕ), represents the absolute magnitude of the functions defined by Equations (7.34) and (7.35). The sign of the functions in the different lobes is indicated by plus and minus signs.

solutions of the Schrödinger equation for the hydrogen atom. Because of this, they merit more discussion.

$$p_x = \frac{1}{\sqrt{2}}(Y_1^1 + Y_1^{-1}) = \sqrt{\frac{3}{4\pi}}\sin\theta\cos\phi$$

$$p_y = \frac{1}{\sqrt{2}\,i}(Y_1^1 - Y_1^{-1}) = \sqrt{\frac{3}{4\pi}}\sin\theta\sin\phi \qquad (7.34)$$

$$p_z = Y_1^0 = \sqrt{\frac{3}{4\pi}}\cos\theta$$

$$d_{z^2} = Y_2^0 = \sqrt{\frac{5}{16\pi}}(3\cos^2\theta - 1)$$

$$d_{xz} = \frac{1}{\sqrt{2}}(Y_2^1 + Y_2^{-1}) = \sqrt{\frac{15}{4\pi}}\sin\theta\cos\theta\cos\phi$$

$$d_{yz} = \frac{1}{\sqrt{2}\,i}(Y_2^1 - Y_2^{-1}) = \sqrt{\frac{15}{4\pi}}\sin\theta\cos\theta\sin\phi \qquad (7.35)$$

$$d_{x^2-y^2} = \frac{1}{\sqrt{2}}(Y_2^2 + Y_2^{-2}) = \sqrt{\frac{15}{16\pi}}\sin^2\theta\cos 2\phi$$

$$d_{xy} = \frac{1}{\sqrt{2}\,i}(Y_2^2 - Y_2^{-2}) = \sqrt{\frac{15}{16\pi}}\sin^2\theta\sin 2\phi$$

These functions depend on two variables, θ and ϕ, and the way in which they are named refers them back to Cartesian coordinates. In graphing the functions, spherical coordinates have been used, whereby the radial coordinate is used to display the magnitude of the amplitude, $r = |f(\theta, \phi)|$. All the functions generate lobular patterns in which the amplitude of the function in a lobe is either positive or negative. These signs are indicated in the plots.

The p functions form a set of three mutually perpendicular dumbbell structures. The wave function has the same amplitude, but a different sign in the two lobes, and each function has a nodal plane passing through the origin. Four of the five d functions have a more complex four-lobed shape with nodal planes separating lobes in which the function has opposite signs. Because l is larger for the d than for the p functions, more nodes are seen in both angles. As for the particle in the box wave functions, an increase in the number of nodes corresponds to an increase in the energy of the quantum state. For the particle in the box, an increase in the number of nodes over a fixed interval corresponds to a shorter wavelength and, through the de Broglie relation, to a higher linear momentum. For the rigid rotor, the analogy is exact, but to the angular momentum instead of the linear momentum. We return to the spherical harmonic functions when discussing the orbitals for the H atom in Chapter 9.

Up to this point, questions have been asked about the energy and the momentum. What can be learned about the angular orientation of the internuclear axis for the rotating molecule? This information is given by the probability density, defined by the first postulate as the square of the magnitude of the wave function. The probability density for the p and d functions is very similar in shape to the wave function amplitude shown in Figure 7.7, although the amplitude in all lobes is positive. Taking the p_z plot as an example, Figure 7.7 shows that the maximum amplitude of $|Y_1^0|^2$ is found along the positive and negative z axis. A point on the z axis corresponds to the probability density for finding the molecular axis parallel to the z axis.

An alternate graphical representation can be used that recognizes that spherical harmonics can be used to represent waves on the surface of a sphere. This can be done by displaying the amplitude of the desired function on the sphere at the location θ, ϕ using a color scale. This is done in Figure 7.9, where the square of the amplitude of the p_z and p_y functions is plotted as a color scale on the surface of a sphere. Black and red regions correspond to high and low probability densities, respectively. For the p_z function, there

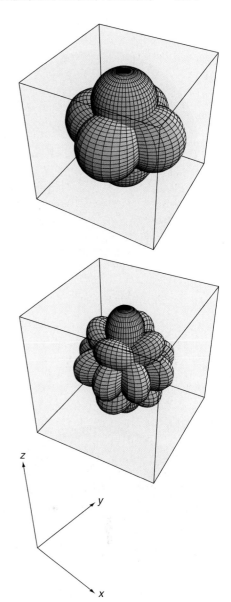

FIGURE 7.8

3D perspective plots show the three p and the five d linear combinations of the spherical harmonics superimposed. The convention used in displaying the functions is explained in the text and in the caption for Figure 7.7.

p_z

z

y

p_y

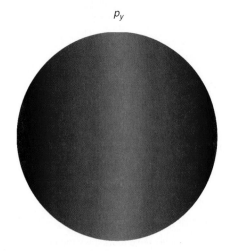

FIGURE 7.9

The absolute magnitude of the amplitude of the p_z and p_y functions is plotted on the surface of a unit sphere. Black and red regions correspond to high and low probability densities, respectively.

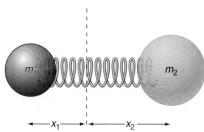

m_1 m_2

x_1 x_2

FIGURE 7.10

Two unequal masses are shown connected by a spring of force constant k. The intersection of the vertical line with the spring indicates the center of mass.

is a much higher probability density of finding the particle near the z axis than in the $z = 0$ plane. This means that the molecular axis is much more likely to be parallel to the z axis than to lie in the $x-y$ plane. For a state whose wave function is p_y, the internuclear axis is much more likely to be parallel to the y axis than to lie in the $y = 0$ plane. This is consistent with the angular orientation of the maxima of these functions shown in Figure 7.7. Why isn't the probability density more sharply peaked in a small angular region near the z or y axis? If the wave function is the p_z function, E_{total}, $|\mathbf{l}^2|$, and l_z *are well defined.* However, the operators for the angular coordinates ϕ and θ do not commute with the operators for E_{total}, $|\mathbf{l}^2|$, and l_z. As a consequence, the angular position coordinates are not known exactly and only average values can be determined for these observables.

OPTIONAL REVIEW

7.6 THE CLASSICAL HARMONIC OSCILLATOR

The harmonic oscillator is reviewed from the perspective of classical mechanics in this section. Consider two masses that are connected by a coiled spring. When at rest, the spring is at its equilibrium length. If the masses are pushed together, the spring is compressed, and if the masses are pulled apart, the spring is extended. In each case, the spring resists any attempt to move the masses away from their equilibrium positions. If the deviation of the spacing between the masses from its rest position is denoted by x, then

$$x = [x_1(t) + x_2(t)] - [x_1 + x_2]_{equilibrium} \tag{7.36}$$

Positive and negative values of x correspond to stretching and compression of the spring, respectively, as shown in Figure 7.10. Experimentally, it is found that to double x, the force exerted on the system must be doubled. This means that a linear relationship exists between the force and x given by

$$F = -k\,x \tag{7.37}$$

In this equation, k is called the spring constant. The negative sign shows that the force and the displacement are in opposite directions.

Before undertaking a mathematical analysis of the **harmonic oscillator**, make a mental image of what happens when the spring is either stretched or compressed and then let go. In either case, the force that the spring exerts on each of the masses will be in the direction opposite that of the applied force. As soon as the spacing of the masses reaches its equilibrium distance, the direction of the spring's force will change. This causes the direction of motion to reverse and an initial stretch becomes a compression and vice versa. In the absence of dissipative forces, the particles continue through alternate half cycles of being farther apart and closer together than their equilibrium distance. This system exhibits **oscillatory behavior**.

You should note one other feature of the system at this point. Although the masses move in opposite directions, the magnitudes of their displacements are not equal if their masses are unequal. This makes it hard to develop a simple picture of the time evolution of this system. However, somewhere between the masses is a point that does not move as the oscillatory behavior proceeds. This is called the center of mass, and a transformation to **center of mass coordinates** gives us a simpler description of the oscillatory motion of the harmonic oscillator.

Before going on, it is also useful to summarize what information is needed to describe the harmonic oscillator and what information can be derived using classical mechanics. The oscillator is described by the two masses, m_1 and m_2, and the force constant, k, which allows the force, F, acting on each of the masses to be calculated. To solve Newton's second law of motion, two independent pieces of information are needed that describe the state of the system at a given initial time. The value of x and the kinetic energy of the oscillator at a given time will do. From this information, the positions, x_1 and x_2, the velocities, $\mathbf{v}_1$ and $\mathbf{v}_2$, and the kinetic and potential energies of each mass can be determined as a function of time. This is more information than necessary, because we are more interested in the potential and kinetic energies associated with the entire oscillator as a unit than with the values for each of the masses separately. This is the main reason for working with the center of mass coordinates.

In the center of mass coordinates, the physical picture of the system changes from two masses connected by a spring of force constant k to a single mass, called the reduced mass, μ, connected by a spring of the same force constant to an immovable wall. Why is this transformation used? We do this because only the *relative* motion of these two masses with respect to one another and not their individual motions is of interest. This change of coordinates also reduces the description of the periodic motion to a single coordinate.

The location of the center of mass, x_{cm}, and the reduced mass, μ, are given by the equations

$$x_{cm} = \frac{m_1 x_1 + m_2 x_2}{m_1 + m_2} \tag{7.38}$$

and

$$\mu = \frac{m_1 m_2}{m_1 + m_2} \tag{7.39}$$

We now use Newton's second law of motion to investigate the dynamics of the harmonic oscillator. Because the motion is in one dimension, the scalar magnitude of the force and acceleration can be used in what follows. Recall that the variable x denotes the deviation of the spring extension from its equilibrium position. Starting with

$$F = \mu a = \mu \frac{d^2 x}{dt^2} \tag{7.40}$$

and using Equation (7.37) for the force, the differential equation

$$\mu \frac{d^2 x}{dt^2} + kx = 0 \tag{7.41}$$

describes the time dependence of the distance between the masses relative to its equilibrium value.

The general solution to this differential equation is

$$x(t) = c_1 e^{+i\sqrt{(k/\mu)}t} + c_2 e^{-i\sqrt{(k/\mu)}t} \tag{7.42}$$

in which c_1 and c_2 are arbitrary coefficients. At this point the **Euler formula**, $r e^{\pm i\theta} = r\cos\theta \pm ir\sin\theta$, is used to recast Equation (7.42) in the form

$$x(t) = c_1\left(\cos\sqrt{\frac{k}{\mu}}t + i\sin\sqrt{\frac{k}{\mu}}t\right) + c_2\left(\cos\sqrt{\frac{k}{\mu}}t - i\sin\sqrt{\frac{k}{\mu}}t\right) \tag{7.43}$$

The last equation can be further simplified to

$$x(t) = b_1\cos\sqrt{\frac{k}{\mu}}t + b_2\sin\sqrt{\frac{k}{\mu}}t \tag{7.44}$$

with $b_1 = c_1 + c_2$ and $b_2 = i(c_1 - c_2)$. This is the general solution of the differential equation with no restrictions on b_1 and b_2. Because the amplitude of oscillation is real, a boundary condition is imposed which requires that b_1 and b_2 be real. The general solution contains two constants of integration that can be determined for a specific solution through the boundary conditions, $x(0)$, and $v(0) = [dx(t)/dt]_{t=0}$. For instance, if $x(0) = 0$ and $v(0) = v_0$, then

$$x(0) = b_1\cos\sqrt{\frac{k}{\mu}} \times 0 = b_1 = 0$$

$$v(0) = \left(\frac{dx(t)}{dt}\right)_{t=0} = b_2\sqrt{\frac{k}{\mu}}\cos\sqrt{\frac{k}{\mu}} \times 0 = b_2\sqrt{\frac{k}{\mu}} \quad \text{and}$$

$$b_2 = \sqrt{\frac{\mu}{k}}v_0 \tag{7.45}$$

The specific solution takes the form

$$x(t) = \sqrt{\frac{\mu}{k}} \, v_0 \sin \sqrt{\frac{k}{\mu}} t \tag{7.46}$$

Note that only the second term in Equation (7.44) remains. This is because we arbitrarily choose $x(0) = 0$ and $v(0) = v_0$. Other boundary conditions could lead to solutions in which both b_1 and b_2 are nonzero.

Because the sine and cosine functions are periodic functions of the variable t, x exhibits oscillatory motion. The period of oscillation, T, is defined by the relation

$$\sqrt{\frac{k}{\mu}} \, (t + T) - \sqrt{\frac{k}{\mu}} t = 2\pi \tag{7.47}$$

giving

$$T = 2\pi \sqrt{\frac{\mu}{k}} \tag{7.48}$$

The inverse of T is called the frequency, ν:

$$\nu = \frac{1}{2\pi} \sqrt{\frac{k}{\mu}} \tag{7.49}$$

These definitions of ν and T allow x to be written in the form

$$x(t) = b_1 \cos 2\pi \frac{t}{T} + b_2 \sin 2\pi \frac{t}{T} \tag{7.50}$$

Often, the angular frequency, $\omega = 2\pi\nu$, is introduced, giving

$$x(t) = b_1 \cos \omega t + b_2 \sin \omega t, \quad \text{or equivalently,} \quad x(t) = A \sin(\omega t + \alpha) \tag{7.51}$$

where the phase shift, α, is explored later in Example Problem 7.5. The oscillatory periodic motion of the harmonic oscillator is depicted in Figure 7.11.

With a mathematical description of the motion, your mental picture of the oscillatory behavior can be tested. Because the potential energy, $E_{potential}$, and the kinetic energy, $E_{kinetic}$, of the oscillator are related to the magnitude of **x** and **v** by the equations

$$E_{potential} = \frac{1}{2} k x^2 \quad \text{and} \quad E_{kinetic} = \frac{1}{2} \mu v^2 \tag{7.52}$$

$E_{potential}$ and $E_{kinetic}$ can be expressed in terms of x, as is done in Example Problem 7.5.

Visualize the essential features of the harmonic oscillator in terms of the potential and kinetic energies. Energy can be pumped into a harmonic oscillator at rest by stretching or compressing the spring. When energy is no longer being added, the masses oscillate around their equilibrium positions. This oscillatory motion continues indefinitely in the absence of energy dissipation such as that resulting from friction. The maximum displacement from the equilibrium position depends on the force constant and the amount of energy taken up. It can be expressed in terms of the position and the velocity at $t = 0$. The kinetic and potential energies also oscillate with time. The harmonic oscillator can have any positive value for the total energy. As the amount of energy increases, its maximum amplitude of vibration and its maximum velocity increase. Because there are no constraints on what value of the energy is allowed, the classical harmonic oscillator has a **continuous energy spectrum**.

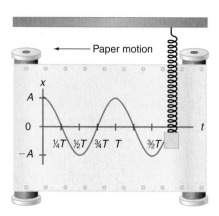

FIGURE 7.11

The periodic motion of a harmonic oscillator is revealed if the vertical motion is displayed on a chart recorder.

EXAMPLE PROBLEM 7.5

For a harmonic oscillator described by $x(t) = A \sin(\omega t + \alpha)$, $\omega = (k/\mu)^{1/2}$, answer the following questions.

a. What are the units of A? What role does α have in this equation?

b. Graph the kinetic and potential energies given by the following equations as a function of time:

$$E_{kinetic} = \frac{1}{2} m v^2 \quad \text{and} \quad E_{potential} = \frac{1}{2} k x^2$$

c. Show that the sum of the kinetic and potential energies is independent of time.

Solution

a. Because $x(t)$ has the units of length and the sine function is dimensionless, A must have the units of length. The quantity α sets the value of x at $t = 0$, because $x(0) = A\sin(\alpha)$.

b. We begin by expressing the kinetic and potential energies in terms of $x(t)$:

$$E_{kinetic} = \frac{1}{2}\mu v^2 = \frac{1}{2}\mu\left(\frac{dx}{dt}\right)^2$$

$$= \frac{\mu}{2}(A\omega\cos(\omega t + \alpha))^2$$

$$= \frac{1}{2}\mu\omega^2 A^2 \cos^2(\omega t + \alpha)$$

$$E_{potential} = \frac{1}{2}kx^2 = \frac{1}{2}kA^2\sin^2(\omega t + \alpha) \text{ because } \omega = \sqrt{\frac{k}{\mu}} \text{ and } k = \mu\omega^2$$

$$= \frac{1}{2}\mu\omega^2 A^2 \sin^2(\omega t + \alpha)$$

In the following figure, the energy is expressed in increments of $(1/2)\mu\omega^2 A^2$ and we have arbitrarily chosen $\alpha = \pi/6$. Note that the kinetic and potential energies are out of phase. Why is this the case?

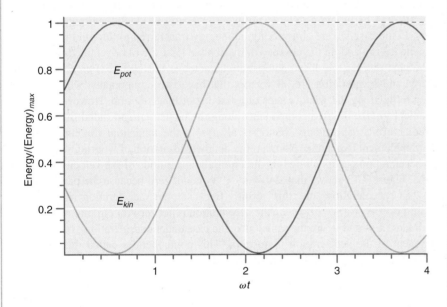

c. The black dashed line in the preceding figure is the sum of the kinetic and potential energies, which is a constant. This can be verified algebraically by adding the expressions for $E_{kinetic}$ and $E_{potential}$:

$$E_{total} = \frac{1}{2}\mu\omega^2 A^2 \cos^2(\omega t + \alpha) + \frac{1}{2}\mu\omega^2 A^2 \sin^2(\omega t + \alpha)$$

$$= \frac{1}{2}\mu\omega^2 A^2$$

Note that the sum of the kinetic and potential energies is independent of time, as must be the case, because no energy is added to the system after the initial stretching of the spring.

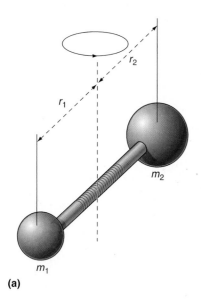

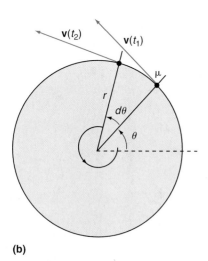

(b)

FIGURE 7.12

(a) The rigid rotor consists of two masses separated by a fixed distance. The dashed vertical line is the axis of rotation. It is perpendicular to the plane of rotation and passes through the center of mass. **(b)** The rigid rotor in the center of mass coordinates is a single particle of reduced mass μ rotating on a ring of radius equal to the bond length. The position and velocity of the reduced mass μ are shown at two times.

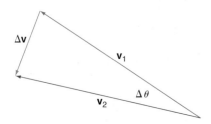

FIGURE 7.13

Vector diagram of $\mathbf{v}_1$, $\mathbf{v}_2$, and $\Delta\mathbf{v}$ from Figure 7.12.

OPTIONAL REVIEW

7.7 ANGULAR MOTION AND THE CLASSICAL RIGID ROTOR

The harmonic oscillator is a good example of linear motion. In this system, the vectors for the velocity, momentum, and acceleration are all parallel to the direction of motion. However, not all motion is linear, making it necessary to analyze the motion induced if the applied force is not along the initial direction of motion. Why is rotational motion of interest to chemists? Energy can be taken up by a molecule in any of several ways. The first of these is translational kinetic energy, which is associated with the collective motion of all atoms in the molecule, or with the center of mass. A second way to store energy, in the form of vibrational energy, was just discussed. It was shown that vibrational energy is both kinetic and potential and can be taken up by stretching bonds within a molecule. Now set the molecule spinning in addition to having it vibrate and undergo translational motion. Additional energy is taken up in this collective rotational motion. The rigid rotor is a simple example of angular motion. It is a good model for thinking about rotation of a diatomic molecule. The term *rigid* stems from the assumption that the rotational motion does not result in a stretching of the bond.

Consider the rigid rotor shown in Figure 7.12. The axis of rotation is perpendicular to the plane of rotation and passes through the center of mass. The distance of the individual masses from the center of mass are indicated. As for the harmonic oscillator, it is convenient to view the motion of the rigid rotor in the center of mass coordinates because only the relative motion of the two masses is of interest. In these coordinates, the dumbbell is equivalent to a single mass $\mu = (m_1 m_2)/(m_1 + m_2)$ rotating in a plane at a distance from a fixed axis equal to the bond length.

In the rotating system under consideration, no force opposes the rotation. For this reason, potential energy cannot be stored in the motion of the rigid rotor. This means that unlike the harmonic oscillator, no interconversion of kinetic and potential energy occurs. In the case we are considering, all energy transferred to the rigid rotor appears as kinetic energy and, in the absence of dissipative losses, will be retained indefinitely.

We next discuss the observables that characterize this system. Force, momentum, velocity, and acceleration are all vectors that have two components, which could be chosen to lie along the x and y axes of a fixed coordinate system. However, it is more convenient to take the two components along the tangential and radial directions. For a circular orbit, the velocity vector is always in the tangential direction. (See the Math Supplement, Appendix A for a more detailed discussion of working with vectors.) If no acceleration occurs along this direction, the magnitude of the velocity is constant in time. Figure 7.13 shows that $\Delta\mathbf{v} = \mathbf{v}_2 - \mathbf{v}_1$ is not zero because the particle experiences an acceleration on this orbit. Because the acceleration is given by $\mathbf{a} = \lim(\mathbf{v}_2 - \mathbf{v}_1)/(t_2 - t_1)_{\Delta t \to 0}$ the acceleration is not zero for circular motion. If the magnitudes of $\mathbf{v}_1$ and $\mathbf{v}_2$ are the same (as in the case under consideration), only the radial component of the acceleration is nonzero. This component is called the **centripetal acceleration**, $a_{centripetal}$, and has the magnitude

$$a_{centripetal}(t) = \frac{|\mathbf{v}(t)|^2}{r} \qquad (7.53)$$

In circular motion, the total accumulated rotation angle, θ, is analogous to the distance variable in linear motion. The angle is typically measured in **radians**. A radian is the angle for which the arc length is equal to the radius. This definition means that radians are related to degrees by 2π radians $= 360°$ or one radian $= 57.3°$. **Angular velocity** and **angular acceleration**, which are analogous to $\mathbf{v}$ and $\mathbf{a}$ in linear motion, are defined by

$$|\boldsymbol{\omega}| = \frac{d\theta}{dt} \quad \text{and} \quad \alpha = \frac{d|\boldsymbol{\omega}|}{dt} = \frac{d^2\theta}{dt^2} \qquad (7.54)$$

The directions of both $\boldsymbol{\omega}$ and $d\boldsymbol{\omega}/dt$ are determined by the right-hand rule, and point along the axis of rotation. The application of the right-hand rule in determining the direction of ω is illustrated in Figure 7.14. The angular acceleration is nonzero if the particle is not moving at constant speed on its circular orbit. Keep in mind that ω is a vector perpendicular to the plane of rotation. Because the velocity is also defined by the expression

$$\mathbf{v} = \frac{\Delta \mathbf{s}}{\Delta t} = \frac{\mathbf{r}\Delta \boldsymbol{\theta}}{\Delta t} \quad \text{in the limit as} \quad \Delta t \to 0, \quad \mathbf{v} = \frac{\mathbf{r}d\theta}{dt} = r\boldsymbol{\omega} \qquad (7.55)$$

the magnitudes of the angular and linear velocities can be related. In the case under consideration, the acceleration along the direction of motion is zero, and the expression for $d\theta/dt$ in Equation (7.54) can be integrated to obtain

$$\theta = \theta_0 + \omega t \qquad (7.56)$$

You can show that for a constant acceleration along the direction of motion,

$$\omega = \omega_0 + \alpha t$$

$$\theta = \theta_0 + \omega_0 t + \frac{1}{2}\alpha t^2 \qquad (7.57)$$

The kinetic energy can be expressed in the form

$$E_{kinetic} = \frac{1}{2}\mu v^2 = \frac{1}{2}\mu r^2\omega^2 = \frac{1}{2}I\omega^2 \qquad (7.58)$$

The quantity μr^2 is called the **moment of inertia** and given the symbol I. With this definition, the kinetic energy takes on a form similar to that in linear motion with the moment of inertia and the angular velocity taking on the role of the mass and linear velocity.

To develop a relationship similar to $\mathbf{F} = m\mathbf{a} = d\mathbf{p}/dt$ for angular motion, the angular momentum, $\mathbf{l}$, is defined by

$$\mathbf{l} = \mathbf{r} \times \mathbf{p} \qquad (7.59)$$

in which $\times$ indicates the vector cross product between $\mathbf{r}$ and $\mathbf{p}$. The use of the right-hand rule to determine the orientation of $\mathbf{l}$ relative to $\mathbf{r}$ and $\mathbf{p}$ is shown in Figure 7.15.

The magnitude of $\mathbf{l}$ is given by

$$l = pr \sin\phi = \mu v r \sin\phi \qquad (7.60)$$

in which ϕ is the angle between the vectors $\mathbf{r}$ and $\mathbf{p}$. For circular motion, $\mathbf{r}$ and $\mathbf{p}$ are perpendicular so that $l = mvr$. The equation $E = p^2/2m$ and the definition of angular momentum can be used to express the kinetic energy in terms of l:

$$E = \frac{p^2}{2\mu} = \frac{l^2}{2\mu r^2} = \frac{l^2}{2I} \qquad (7.61)$$

Classical mechanics does not place any restrictions on the direction or magnitude of $\mathbf{l}$. As for any observable for a classical system, the magnitude of $\mathbf{l}$ can change by an incrementally small amount. Therefore, any amount of energy can be stored in the rigid rotor and an increase in the energy appears as an increase in the angular frequency. Because the amount of energy can be increased by an infinitesimally small amount, the classical rigid rotor has a continuous energy spectrum, just like the classical harmonic oscillator.

SUPPLEMENTAL

7.8 SPATIAL QUANTIZATION

The fact that the operators $\hat{H}$, $\hat{l}^2$, and $\hat{l}_z$ commute whereas $\hat{l}_x$, $\hat{l}_y$, and $\hat{l}_z$ do not commute with one another states that the energy, the magnitude of the angular momentum vector, and the value of any one of its components can be known simultaneously, but

FIGURE 7.14

The right-hand rule is used to determine the direction of the angular velocity vector.

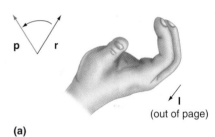

(a)

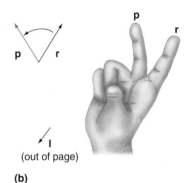

(b)

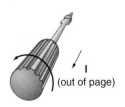

(c)

FIGURE 7.15

The right-hand rule is used to determine the cross product between two vectors.

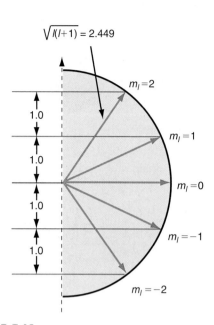

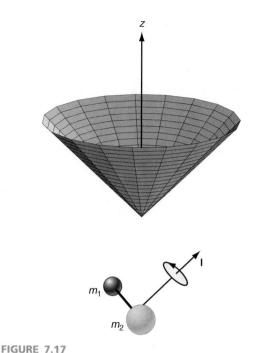

FIGURE 7.16

Possible orientations of the angular momentum vector $|\mathbf{l}| = \sqrt{l(l+1)}\,\hbar$ and $l_z = m_l\hbar$, $l \geq |m_l|$ for $l = 2$. The lengths indicated are in units of $\hbar$.

FIGURE 7.17

Example of an angular momentum vector for which only l^2, l_z, and $l_x^2 + l_y^2$ are known. In this case, $l = 2$ and $m_l = +2$. The right side of the figure illustrates a classical rigid rotor for which the angular momentum vector has the same l_z component.

that the other two components of the angular momentum cannot be known. Contrast this with classical mechanics in which all three components of an angular momentum vector can be specified simultaneously. In that case, both the length of the vector and its direction can be known.

We summarize what can be known about the angular momentum vector associated with a molecule rotating in three dimensions pictorially. In doing so, classical and quantum mechanical descriptions are mixed. For this reason, the following is a **semiclassical** description. The one component that is known is chosen to be along the z direction. In Figure 7.16, we show what can be known about $\mathbf{l}$ and l_z. The magnitude of $\mathbf{l}$ is $\sqrt{l(l+1)}\,\hbar$ and that of $l_z = m_l\hbar$. The vector $\mathbf{l}$ cannot lie on the z axis because $|\sqrt{m_l(m_l+1)}| > |m_l|$ so that $|m_l| \leq l$. From another point of view, $\mathbf{l}$ cannot lie on the z axis because the commutators in Equation (7.31) are not zero. If $\mathbf{l}$ did lie on the z axis, then l_x and l_y would both be zero and, therefore, all three components of the vector $\mathbf{l}$ could be known simultaneously. The fact that only $|\mathbf{l}|$ and one of its components can be known simultaneously is a direct manifestation of the fact that the operators $\hat{l}_x$, $\hat{l}_y$, and $\hat{l}_z$ do not commute with one another.

Although the picture in Figure 7.16 is useful, it does not depict $\mathbf{l}$ as a three-dimensional vector. We modify this figure to take the three-dimensional nature of $\mathbf{l}$ into account in Figure 7.17 for the case where $l = 2$ and $m_l = 2$. The vector $\mathbf{l}$ is depicted as a line on the surface of the cone beginning at its apex. The magnitude of $\mathbf{l}$ and its projection on the z axis are known exactly and can be determined from the figure. However, the components of the angular momentum vector along the x and y axes, l_x and l_y, cannot be known exactly and simultaneously. All that is knowable about them is that $l^2 - l_z^2 = l_x^2 + l_y^2 = l(l+1)\hbar^2 - m_l^2\hbar^2$. This equation defines the circle terminating the cone at its open end. Figure 7.17 depicts all that can be known simultaneously about the components of the angular momentum. To give a more physical picture to Figure 7.17, a classical rigid rotor for which the z component of the angular momentum vector is the same as for the quantum mechanical case is also shown. Do not take this comparison literally, because the rotor can be depicted as shown only because all three components of the angular momentum can be known simultaneously. This is not possible for a quantum mechanical rigid rotor.

Figure 7.18 combines the information about all possible values of m_l consistent with $l = 2$ in one figure. Such a depiction is often referred to as a **vector model of angular momentum**. Only the orientations of **l** for which the vector lies on one of the cones are allowed. A surprising result emerges from these considerations. Not only are the possible magnitudes of the angular momentum quantized, but the vector can only have certain orientations in space! This result is referred to as **spatial quantization**.

What is the analogous situation in classical mechanics? Because l_x, l_y, and l_z can be known simultaneously for a classical system, and because their values are not quantized, the possible orientations of **l** map out a continuous spherical surface. The contrast between classical and quantum mechanical behavior is clearly evident! It is also apparent how quantum and classical results merge for high energies (large quantum numbers). For a given l value, there are $2l + 1$ conical surfaces on a vector diagram like that shown in Figure 7.18. For large values of l, the individual cones are so close together that they merge into a sphere, and the angular momentum vector no longer exhibits spatial quantization.

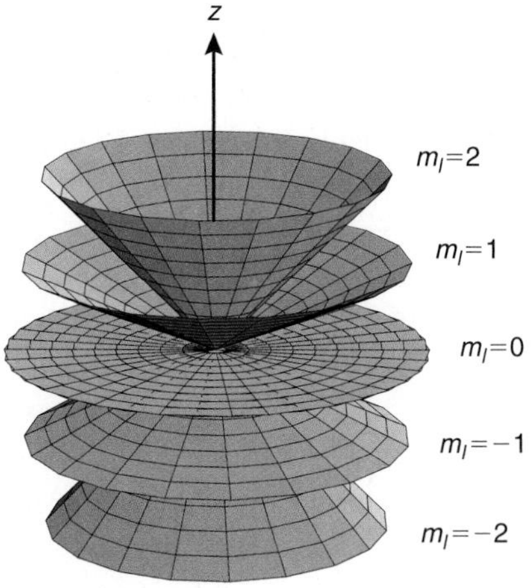

FIGURE 7.18

All possible orientations of an angular momentum vector with $l = 2$. The z component of the angular momentum is shown in units of $\hbar$.

EXAMPLE PROBLEM 7.6

How many cones of the type shown in Figure 7.18 will there be for $l = 1000.$? What is the closest allowed angle between **l** and the z axis?

Solution

There will be $2l + 1$ or 2001 cones. The smallest allowed angle is for $l_z = 1000.\hbar$, and is given by

$$\cos \phi = \frac{l}{\sqrt{l(l + 1)}} = \frac{1000.}{1000.50}$$

$$\phi = 0.03 \text{ radians} = 1.7°$$

Vocabulary

angular acceleration	force constant	rotation
angular momentum	frequency of oscillation	semiclassical
angular velocity	harmonic oscillator	separation of variables
bond length	Hermite polynomial	spatial quantization
center of mass coordinates	moment of inertia	spherical harmonic functions
centripetal acceleration	orbital	translational motion
continuous energy spectrum	oscillatory behavior	vector model of angular momentum
degeneracy	radians	vibration
Euler formula	rigid rotor	zero point energy

Questions on Concepts

Q7.1 Why is the probability of finding the harmonic oscillator at its maximum extension or compression larger than that for finding it at its rest position?

Q7.2 Why does the energy of a rotating molecule depend on l, but not on m_l?

Q7.3 Are the real functions listed in Equations (7.34) and (7.35) eigenfunctions of $\hat{l}_z$? Justify your answer.

Q7.4 Spatial quantization was discussed in Supplemental Section 7.8. Suppose that we have a gas consisting of atoms, each of which has a nonzero angular momentum. Are all of their angular momentum vectors aligned?

Q7.5 Does the average length of a harmonic oscillator depend on its energy? Answer this question by referring to Figure 7.1. The average length is the midpoint of the horizontal line connecting the two parts of $V(x)$.

Q7.6   Why can the angular momentum vector lie on the z axis for two-dimensional rotation in the $x-y$ plane but not for rotation in three-dimensional space?

Q7.7   What is the functional dependence of the total energy of the quantum harmonic oscillator on the position variable x?

Q7.8   Explain in words why the amplitude of the total energy eigenfunctions for the quantum mechanical harmonic oscillator increases with $|x|$ as shown in Figure 7.4.

Q7.9   Why is it possible to write the total energy eigenfunctions for rotation in three dimensions in the form $Y(\theta, \phi) = \Theta(\theta)\Phi(\phi)$?

Q7.10   The two linearly independent total energy eigenfunctions for rotation in two dimensions are

$$\Phi_+(\phi) = \frac{1}{\sqrt{2\pi}}e^{im_l\phi} \quad \text{and} \quad \Phi_-(\phi) = \frac{1}{\sqrt{2\pi}}e^{-im_l\phi}$$

What is the difference in motion for these two solutions? Explain your answer.

Q7.11   Why is only one quantum number needed to characterize the eigenfunctions for rotation in two dimensions, whereas two quantum numbers are needed to characterize the eigenfunctions for rotation in three dimensions?

Q7.12   What makes the z direction special such that $\hat{l}^2$, $\hat{H}$, and $\hat{l}_z$ commute, whereas $\hat{l}_y$, $\hat{l}_z$, and $\hat{l}_x$ do not commute?

Q7.13   How are the spherical harmonics combined to form real p and d functions? What is the advantage in doing so?

Q7.14   Does the bond length of a real molecule depend on its energy? Answer this question by referring to Figure 7.1. The bond length is the midpoint of the horizontal line connecting the two parts of $V(x)$.

Q7.15   The zero point energy of the particle in the box goes to zero as the length of the box approaches infinity. What is the appropriate analogue for the quantum harmonic oscillator?

Problems

Problem numbers in **red** indicate that the solution to the problem is given in the *Student's Solutions Manual*.

P7.1   A gas-phase $^1H^{35}Cl$ molecule, with a bond length of 127.5 pm, rotates in a three-dimensional space.

a. Calculate the zero point energy associated with this rotation.

b. What is the smallest quantum of energy that can be absorbed by this molecule in a rotational excitation?

P7.2   In this problem you will derive the commutator $[\hat{l}_x, \hat{l}_y] = i\hbar\hat{l}_z$.

a. The angular momentum vector in three dimensions has the form $\mathbf{l} = \mathbf{i}l_x + \mathbf{j}l_y + \mathbf{k}l_z$ where the unit vectors in the x, y, and z directions are denoted by $\mathbf{i}$, $\mathbf{j}$, and $\mathbf{k}$. Determine l_x, l_y, and l_z by expanding the 3×3 cross product $\mathbf{l} = \mathbf{r} \times \mathbf{p}$. The vectors $\mathbf{r}$ and $\mathbf{p}$ are given by $\mathbf{r} = \mathbf{i}x + \mathbf{j}y + \mathbf{k}z$ and $\mathbf{p} = \mathbf{i}p_x + \mathbf{j}p_y + \mathbf{k}p_z$.

b. Substitute the operators for position and momentum in your expressions for l_x and l_y. Always write the position operator to the left of the momentum operator in a simple product of the two.

c. Show that $[\hat{l}_x, \hat{l}_y] = i\hbar\hat{l}_z$.

P7.3   In discussing molecular rotation, the quantum number J is used rather than l. Using the Boltzmann distribution, calculate n_J/n_0 for $^1H^{19}F$ for $J = 0, 5, 10,$ and 20 at $T = 650.$ K. Does n_J/n_0 go through a maximum as J increases? If so, what can you say about the value of J corresponding to the maximum?

P7.4   Draw a picture (to scale) showing all angular momentum cones consistent with $l = 5$. Calculate the half angles for each of the cones.

P7.5 $^1H^{127}I$ has a force constant of 314 N m^{-1} and a bond length of 160.92 pm. Calculate the frequency of the light corresponding to the lowest energy pure vibrational and pure rotational transitions. In what regions of the electromagnetic spectrum do the transitions lie?

P7.6   The wave functions p_x and d_{xz} are linear combinations of the spherical harmonic functions, which are eigenfunctions of the operators $\hat{H}$, $\hat{l}^2$, and $\hat{l}_z$ for rotation in three dimensions. The combinations have been chosen to yield real functions. Are these functions still eigenfunctions of $\hat{l}_z$? Answer this question by applying the operator to the functions.

P7.7   At what values of θ does $Y_2^0(\theta, \phi) = (5/16\pi)^{1/2}$ $(3\cos^2\theta - 1)$ have nodes? Are the nodes points, lines, planes, or other surfaces?

P7.8   The vibrational frequency for $^{19}F_2$ expressed in wave numbers is 916.64 cm^{-1}. What is the force constant associated with the F-F bond? How much would a classical spring with this force constant be elongated if a mass of 2.50 kg were attached to it? Use the gravitational acceleration on Earth at sea level for this problem.

P7.9   In discussing molecular rotation, the quantum number J is used rather than l. Calculate E_{rot}/kT for $^1H^{19}F$ for $J = 0$, 5, 10, and 20 at 298 K. For which of these values of J is $E_{rot}/kT \geq 10.$?

P7.10   Show by carrying out the necessary integration that the eigenfunctions of the Schrödinger equation for rotation in two dimensions,

$$\frac{1}{\sqrt{2\pi}}e^{im_l\phi} \quad \text{and} \quad \frac{1}{\sqrt{2\pi}}e^{in_l\phi}, \quad m_l \neq n_l$$

are orthogonal.

P7.11 Evaluate the average of the square of the linear momentum of the quantum harmonic oscillator, $\langle p_x^2 \rangle$, for the ground state ($n = 0$) and first two excited states ($n = 1$ and $n = 2$). Use the hint about evaluating integrals in Problem P7.12.

P7.12 Show by carrying out the appropriate integration that the total energy eigenfunctions for the harmonic oscillator $\psi_0(x) = (\alpha/\pi)^{1/4}e^{-(1/2)\alpha x^2}$ and $\psi_2(x) = (\alpha/4\pi)^{1/4}(2\alpha x^2 - 1)e^{-(1/2)\alpha x^2}$ are orthogonal over the interval $-\infty < x < \infty$ and that $\psi_2(x)$ is normalized over the same interval. In evaluating integrals of this type,

$$\int_{-\infty}^{\infty} f(x)\, dx = 0 \text{ if } f(x) \text{ is an odd function of } x \text{ and}$$

$$\int_{-\infty}^{\infty} f(x)\, dx = 2\int_{0}^{\infty} f(x)\, dx \text{ if } f(x) \text{ is an even function of } x.$$

P7.13 Two 2.50-g masses are attached by a spring with a force constant of $k = 725$ kg s^{-2}. Calculate the zero point energy of the system and compare it with the thermal energy kT at 298 K. If the zero point energy were converted to translational energy, what would be the speed of the masses?

P7.14 Calculate the frequency and wavelength of the radiation absorbed when a quantum harmonic oscillator with a frequency of 5.58×10^{13} s^{-1} makes a transition from the $n = 3$ to the $n = 4$ state.

P7.15 Evaluate the average kinetic and potential energies, $\langle E_{kinetic} \rangle$ and $\langle E_{potential} \rangle$, for the ground state ($n = 0$) of the harmonic oscillator by carrying out the appropriate integrations.

P7.16 The vibrational frequency of $^1H^{19}F$ is 1.24×10^{14} s^{-1}. Calculate the force constant of the molecule. How large a mass would be required to stretch a classical spring with this force constant by 1.00 cm? Use the gravitational acceleration on Earth at sea level for this problem.

P7.17 Use $\sqrt{\langle x^2 \rangle}$ as calculated in Problem P18.32 as a measure of the vibrational amplitude for a molecule. What fraction is $\sqrt{\langle x^2 \rangle}$ of the 141.4-pm bond length of the $^1H^{81}Br$ molecule for $n = 0$, 1, and 2? The force constant for the $^1H^{81}Br$ molecule is 412 N m^{-1}.

P7.18 A coin with a mass of 5.67 g suspended on a rubber band has a vibrational frequency of 3.00 s^{-1}. Calculate (a) the force constant of the rubber band; (b) the zero point energy; (c) the total vibrational energy if the maximum displacement is 0.500 cm; and (d) the vibrational quantum number corresponding to the energy in part (c).

P7.19 Calculate the position of the center of mass of (a) $H^{35}Cl$, which has a bond length of 127.5 pm; and (b) $^{12}C^{16}O$, which has a bond length of 112.8 pm.

P7.20 Show that the function $Y_2^0(\theta, \phi) = (5/16\pi)^{1/2}(3\cos^2\theta - 1)$ is normalized over the interval $0 \leq \theta \leq \pi$ and $0 \leq \phi \leq 2\pi$.

P7.21 Is it possible to simultaneously know the angular orientation of a molecule rotating in a two-dimensional space and its angular momentum? Answer this question by evaluating the commutator $[\phi, -i\hbar(\partial/\partial\phi)]$.

P7.22 The force constant for the D_2 molecule is 577 N m^{-1}. Calculate the vibrational zero point energy of this molecule. If this amount of energy were converted to translational energy, how fast would the molecule be moving? Compare this speed to the root mean square speed from the kinetic gas theory, $|\mathbf{v}|_{rms} = \sqrt{3kT/m}$ for $T = 300.$ K.

P7.23 The force constant for a $H^{35}Cl$ molecule is 516 N m^{-1}.

a. Calculate the zero point vibrational energy for this molecule for a harmonic potential.

b. Calculate the light frequency needed to excite this molecule from the ground state to the first excited state.

P7.24 In discussing molecular rotation, the quantum number J is used rather than l. At 300. K, most molecules are in their ground vibrational state. Is this also true for their rotational degree of freedom? Calculate $n_{J=2}/n_{J=0}$ and $n_{J=15}/n_{J=0}$ for the $H^{19}F$ molecule. Make sure that you take the degeneracy of the levels into account.

P7.25 A $^1H^{127}I$ molecule, with a bond length of 160.92 pm, absorbed on a surface rotates in two dimensions.

a. Calculate the zero point energy associated with this rotation.

b. What is the smallest quantum of energy that can be absorbed by this molecule in a rotational excitation?

P7.26 Using your results for Problems P7.11, 29, 31, and 32, calculate the uncertainties in the position and momentum $\sigma_p^2 = \langle p^2 \rangle - \langle p \rangle^2$ and $\sigma_x^2 = \langle x^2 \rangle - \langle x \rangle^2$ for the ground state ($n = 0$) and first two excited states ($n = 1$ and $n = 2$) of the quantum harmonic oscillator. Compare your results with the predictions of the Heisenberg uncertainty principle.

P7.27 Evaluate the average kinetic and potential energies, $\langle E_{kinetic} \rangle$ and $\langle E_{potential} \rangle$, for the second excited state ($n = 2$) of the harmonic oscillator by carrying out the appropriate integrations.

P7.28 By substituting in the Schrödinger equation for the harmonic oscillator, show that the ground-state vibrational wave function is an eigenfunction of the total energy operator. Determine the energy eigenvalue.

P7.29 Evaluate the average linear momentum of the quantum harmonic oscillator, $\langle p_x \rangle$, for the ground state ($n = 0$) and first two excited states ($n = 1$ and $n = 2$). Use the hint about evaluating integrals in Problem P7.12.

P7.30 By substituting in the Schrödinger equation for rotation in three dimensions, show that the rotational wave function $(5/16\pi)^{1/2}(3\cos^2\theta - 1)$ is an eigenfunction of the total energy operator. Determine the energy eigenvalue.

P7.31 Evaluate the average of the square of the vibrational amplitude of the quantum harmonic oscillator about its equilibrium value, $\langle x^2 \rangle$, for the ground state ($n = 0$) and first two excited states ($n = 1$ and $n = 2$). Use the hint about evaluating integrals in Problem P7.12.

P7.32 Evaluate the average vibrational amplitude of the quantum harmonic oscillator about its equilibrium value, $\langle x \rangle$, for the ground state ($n = 0$) and first two excited states ($n = 1$ and $n = 2$). Use the hint about evaluating integrals given in Problem P7.12.

P7.33 For molecular rotation, the symbol J rather than l is used as the quantum number for angular momentum. A $^{1}H^{19}F$ molecule has the rotational quantum number $J = 10$ and vibrational quantum number $n = 0$.

a. Calculate the rotational and vibrational energy of the molecule. Compare each of these energies with kT at 300. K.

b. Calculate the period for vibration and rotation. How many times does the molecule rotate during one vibrational period?

P7.34 Calculate the first five energy levels for a $^{12}C^{16}O$ molecule, which has a bond length of 112.8 pm, (a) if it rotates freely in three dimensions and (b) if it is absorbed on a surface and forced to rotate in two dimensions.

P7.35 Calculate the constants b_1 and b_2 in Equation (7.44) for the condition $x(0) = x_{max}$, the maximum extension of the oscillator. What is $v(0)$ for this condition?

Web-Based Simulations, Animations, and Problems

W7.1 The motion of a particle in a harmonic potential is investigated, and the particle energy and force constant k are varied using sliders. The potential and kinetic energy are displayed as a function of the position x, and the result of measuring the probability of detecting the particle at x is displayed as a density plot. The student is asked to use the information gathered to explain the motion of the particle.

W7.2 The allowed energy levels for the harmonic oscillator are determined by numerical integration of the Schrödinger equation, starting in the classically forbidden region to the left of the potential. The criterion that the energy is an eigenvalue for the problem is that the wave function decays to zero in the classically forbidden region to the right of the potential. The zero point energy is determined for different values of k. The results are graphed to obtain a functional relationship between the zero point energy and k.

W7.3 The probability of finding the harmonic oscillator in the classically forbidden region, P_n, is calculated. The student generates a set of values for P_n for $n = 0, 1, 2, ..., 20$ and graphs them.

8

The Vibrational and Rotational Spectroscopy of Diatomic Molecules

Chemists have a wide range of spectroscopic techniques available to them. With these techniques, unknown molecules can be identified, bond lengths can be measured, and the force constants associated with chemical bonds can be determined. Spectroscopic techniques are based on transitions that occur between different energy states of molecules when they interact with electromagnetic radiation. In this chapter, we describe how light interacts with molecules to induce transitions between states. In particular, we discuss the absorption of electromagnetic radiation in the infrared and microwave regions of the spectrum. Light of these wavelengths induces transitions between eigenstates of vibrational and rotational energy.

8.1 AN INTRODUCTION TO SPECTROSCOPY

The various forms of **spectroscopy** are among the most powerful tools that chemists have at their disposal to probe the world at an atomic and molecular level. In this chapter, we begin a discussion of molecular spectroscopy that will be taken up again in later chapters. Atomic spectroscopy will be discussed separately in Chapter 11. The information that is accessible through molecular spectroscopy includes bond lengths (rotational spectroscopy) and the vibrational frequencies of molecules (vibrational spectroscopy). In addition, the allowed energy levels for electrons in molecules can be determined with electronic spectroscopy, which is discussed in Chapter 14. This spectroscopic information is crucial for a deeper understanding of the chemical bonding and the reactivity of molecules. In all spectroscopies, atoms or molecules absorb electromagnetic radiation and undergo transitions between allowed quantum states.

In the spectroscopies of interest to chemists, the atoms or molecules are confined in a container that is partly transparent to electromagnetic radiation. In most experiments, the attenuation or enhancement of the incident radiation resulting from absorption or

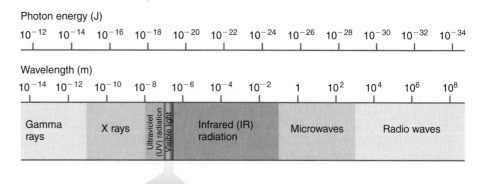

Photon energy (J)

Wavelength (m)

Gamma rays | X rays | Ultraviolet (UV) radiation | Visible light | Infrared (IR) radiation | Microwaves | Radio waves

Visible light

400 nm   700 nm

FIGURE 8.1

The electromagnetic spectrum depicted on a logarithmic wavelength scale.

the **Chemistry place™**

8.1 Energy Levels and Emission Spectra

emission of radiation is measured as a function of the incident wavelength or frequency. Because quantum mechanical systems have a discrete energy spectrum, an absorption or emission spectrum consists of individual peaks, each of which is associated with a transition between two allowed energy levels of the system. As we show later in Supplemental Section 8.9, the frequency at which energy is absorbed or emitted is related to the energy levels involved in the transitions by

$$h\nu = |E_2 - E_1| \tag{8.1}$$

The photon energy that is used in chemical spectroscopies spans more than sixteen orders of magnitude in going from the radio frequency to the X-ray region. This is an indication of the very different energy-level spacings probed by these techniques. The energy-level spacing is smallest in nuclear magnetic resonance (NMR) spectroscopy, which is discussed in Chapter 17, and largest for electronic spectroscopy. Transitions between rotational and vibrational energy levels are intermediate between these two extremes, with rotational energy levels being more closely spaced than vibrational energy levels. The electromagnetic spectrum is depicted schematically in Figure 8.1. An examination of this figure shows that visible light is a very small part of this spectrum.

The spectral regions associated with various spectroscopies are shown in Table 8.1. Spectroscopists commonly use the quantity **wave number**, $\widetilde{\nu} = 1/\lambda$, which has units of inverse centimeters, rather than the wavelength λ or frequency ν to designate spectral transitions for historical reasons. The relationship between ν and $\widetilde{\nu}$ is given by $\nu = \widetilde{\nu}c$, where c is the speed of light. It is important to use consistent units when calculating the energy difference between states associated with a frequency in units of inverse seconds and in units of inverse centimeters. Equation (8.1) expressed for both units is $|E_2 - E_1| = h\nu = hc\widetilde{\nu}$.

The fact that atoms and molecules possess a set of discrete energy levels is an essential feature of all spectroscopies. If all molecules had a continuous energy spectrum, it

TABLE 8.1   IMPORTANT SPECTROSCOPIES AND THEIR SPECTRAL RANGE

Spectral Range	λ (m)	ν (Hz)	$\widetilde{\nu}$ (cm^{-1})	Energy (J)	Spectroscopy
Radio	>0.1	$<3 \times 10^9$	>0.1	$<2 \times 10^{-24}$	NMR
Microwave	$0.001 - 0.1$	$3 \times 10^9 - 3 \times 10^{11}$	$0.1 - 10$	$2 \times 10^{-24} - 2 \times 10^{-22}$	Rotational
Infrared	$7 \times 10^{-7} - 1 \times 10^{-3}$	$3 \times 10^{11} - 4 \times 10^{14}$	$10 - 1 \times 10^4$	$2 \times 10^{-22} - 3 \times 10^{-19}$	Vibrational
Visible	$4 \times 10^{-7} - 7 \times 10^{-7}$	$4 \times 10^{14} - 7 \times 10^{14}$	$1 \times 10^4 - 3 \times 10^4$	$3 \times 10^{-19} - 5 \times 10^{-19}$	Electronic
Ultraviolet	$1 \times 10^{-8} - 4 \times 10^{-7}$	$7 \times 10^{14} - 3 \times 10^{16}$	$3 \times 10^4 - 1 \times 10^6$	$5 \times 10^{-19} - 2 \times 10^{-17}$	Electronic

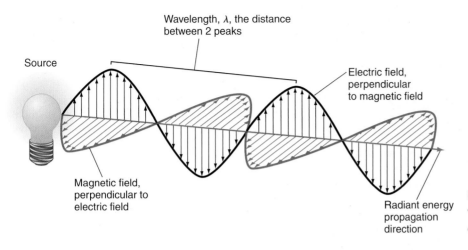

FIGURE 8.2

The electric and magnetic fields associated with a traveling light wave.

would be very difficult to distinguish them on the basis of their absorption spectra. However, as discussed in Section 8.4, not all transitions between arbitrarily chosen states occur. **Selection rules** tell us which transitions will be experimentally observed. Because spectroscopies involve transitions between quantum states, we must first describe how electromagnetic radiation interacts with molecules.

We begin with a qualitative description of energy transfer from the electromagnetic field to a molecule leading to vibrational excitation. Light is an electromagnetic traveling wave that has perpendicular magnetic and electric field components as shown in Figure 8.2. Consider the effect of a time-dependent electric field on a classical dipolar diatomic "molecule" constrained to move in one dimension. Such a molecule is depicted in Figure 8.3. If the spring were replaced by a rigid rod, the molecule could not take up energy from the field. However, the spring allows the two masses to oscillate about their equilibrium distance, thereby generating a periodically varying dipole moment. If the electric field and oscillation of the dipole moment have the same frequency and if they are in phase, the molecule can absorb energy from the field. For a classical "molecule," any amount of energy can be taken up and the absorption spectrum is continuous.

For a real quantum mechanical molecule, the interaction with the electromagnetic field is similar. The electric field acts on a dipole moment within the molecule that can be of two types: permanent and dynamic. Polar molecules like HCl have a **permanent dipole moment**. As molecules vibrate, an additional induced **dynamic dipole moment** can be generated. How does the dynamic dipole arise? The magnitude of the dipole moment depends on the bond length and the degree to which charge is transferred from H to Cl. In turn, the charge transfer depends on the overlap of the electron densities of the two atoms which in turn depends on the internuclear distance. As the molecule vibrates, its dipole moment changes because of these effects, generating a dynamic dipole moment. Because the vibrational amplitude is a small fraction of the bond distance, the dynamic dipole moment is generally small compared to the permanent dipole moment.

As will be seen in the next section, it is the dynamic rather than the permanent dipole moment that determines if a molecule will absorb energy in the infrared region. By contrast, it is the permanent dipole moment that determines if a molecule will undergo rotational transitions by absorbing energy in the microwave region. Homonuclear diatomic molecules have neither permanent nor dynamic dipole moments and cannot absorb infrared radiation. However, vibrational spectroscopy on these molecules can be carried out using the Raman effect as discussed in Section 8.8.

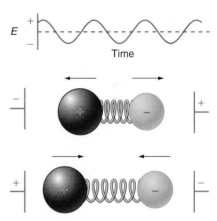

FIGURE 8.3

Schematic of the interaction of a classical harmonic oscillator constrained to move in one dimension under the influence of an electric field. The sinusoidally varying electric field shown at the top of the figure is applied between a pair of capacitor plates. The arrows indicate the direction of force on each of the two charged masses. If the phases of the field and vibration are as shown and the frequencies are equal, the oscillator will absorb energy in both the stretching and compression half cycles.

8.2 ABSORPTION, SPONTANEOUS EMISSION, AND STIMULATED EMISSION

We now move from a classical picture to a quantum mechanical description involving discrete energy levels. The basic processes by which photon-assisted transitions between energy levels occur are **absorption**, **spontaneous emission**, and **stimulated emission**. For simplicity, only transitions in a two-level system are considered as shown in Figure 8.4.

In absorption, the incident photon induces a transition to a higher level, and in emission, a photon is emitted as an excited state relaxes to one of lower energy. Absorption

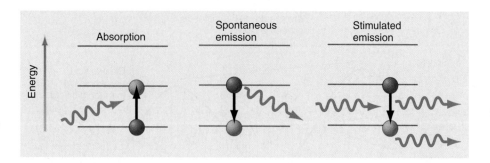

FIGURE 8.4

The three basic processes by which photon-assisted transitions occur. Orange and red filled circles indicate empty and occupied levels, respectively.

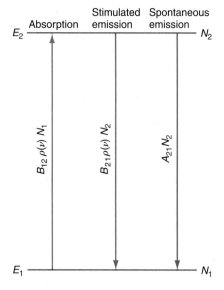

FIGURE 8.5

The rate at which transitions occur between two levels. It is in each case proportional to the product of the appropriate rate coefficient A_{21}, B_{12}, or B_{21} for the process and the population in the originating state, N_1 or N_2. For absorption and stimulated emission, the rate is additionally proportional to the radiation density, $\rho(\nu)$.

and stimulated emission are initiated by a photon incident on the molecule of interest. As the name implies, spontaneous emission is a random event and its rate is related to the lifetime of the excited state. These three processes are not independent in a system at equilibrium, as can be seen by considering Figure 8.5. At equilibrium, the overall transition rate from level 1 to 2 must be the same as that from 2 to 1. This means that

$$B_{12}\rho(\nu)N_1 = B_{21}\rho(\nu)N_2 + A_{21}N_2 \tag{8.2}$$

Whereas spontaneous emission is independent of the radiation density at a given frequency, $\rho(\nu)$, the rates of absorption and stimulated emission are directly proportional to $\rho(\nu)$. The proportionality constants for the three processes are A_{21}, B_{12}, and B_{21}, respectively. Each of these rates is directly proportional to the number of molecules (N_1 or N_2) in the state from which the transition originates. This means that unless the lower state is populated, a signal will not be observed in an absorption experiment. Similarly, unless the upper state is populated, a signal will not be observed in an emission experiment.

The appropriate function to use for $\rho(\nu)$ in the equilibrium equation, Equation (8.2), is the blackbody spectral density function of Equation (1.7), because $\rho(\nu)$ is the distribution of frequencies at equilibrium for a given temperature. Following this reasoning, Einstein concluded that

$$B_{12} = B_{21} \quad \text{and} \quad \frac{A_{21}}{B_{21}} = \frac{16\pi^2\hbar\nu^3}{c^3} \tag{8.3}$$

This result is derived in Example Problem 8.1.

EXAMPLE PROBLEM 8.1

Derive the equations $B_{12} = B_{21}$ and $A_{21}/B_{21} = 16\pi^2\hbar\nu^3/c^3$ using these two pieces of information: (1) the overall rate of transition between levels 1 and 2 (see Figure 8.5) is zero at equilibrium, and (2) the ratio of N_2 to N_1 is governed by the Boltzmann distribution.

Solution

The rate of transitions from level 1 to level 2 is equal and opposite to the transitions from level 2 to level 1. This gives the equation $B_{12}\rho(\nu)N_1 = B_{21}\rho(\nu)N_2 + A_{21}N_2$. The Boltzmann distribution function states that

$$\frac{N_2}{N_1} = \frac{g_2}{g_1}e^{-h\nu/kT}$$

In this case, $g_2 = g_1$. These two equations can be solved for $\rho(\nu)$, giving $\rho(\nu) = A_{21}/(B_{12}e^{h\nu/kT} - B_{21})$. As Planck showed, $\rho(\nu)$ has the form shown in Equation (1.7) so that

$$\rho(\nu) = \frac{A_{21}}{B_{12}e^{h\nu/kT} - B_{21}} = \frac{8\pi h\nu^3}{c^3}\frac{1}{e^{h\nu/kT} - 1}$$

For these two expressions to be equal, $B_{12} = B_{21}$ and $A_{21}/B_{21} = 8\pi h\nu^3/c^3 = 16\pi^2\hbar\nu^3/c^3$.

Spontaneous emission and stimulated emission differ in an important respect. Spontaneous emission is a completely random process, and the emitted photons are incoherent, by which we mean that their phases are random. In stimulated emission, the phase and direction of propagation are the same as that of the incident photon. This is referred to as coherent photon emission. A lightbulb is an **incoherent photon source**. The phase relation between individual photons is random, and because the propagation direction of the photons is also random, the intensity of the source falls off as the square of the distance. A laser is a **coherent source** of radiation. All photons are in phase, and because they have the same propagation direction, the divergence of the beam is very small. This explains why a laser beam that is reflected from the moon still has a measurable intensity when it returns to Earth. We will have more to say about lasers when atomic spectroscopy is discussed in Chapter 11.

8.3 AN INTRODUCTION TO VIBRATIONAL SPECTROSCOPY

We now have a framework with which we can discuss spectroscopy as a chemical tool. Two features have enabled vibrational spectroscopy to achieve the importance that it has as a tool in chemistry. The first is that the vibrational frequency depends primarily on the identity of the two vibrating atoms on either end of the bond and to a much lesser degree on the presence of atoms farther away from the bond. This property generates characteristic frequencies for atoms joined by a bond known as **group frequencies**. We discuss group frequencies further in Section 8.5. The second feature is that a particular vibrational mode in a molecule has only one characteristic frequency of appreciable intensity. We discuss this feature next.

In any spectroscopy, transitions occur from one energy level to another. As discussed in Section 8.2, the energy level from which the transition originates must be occupied in order to generate a spectral signal. Which of the infinite set of vibrational levels has a substantial probability of being occupied? Table 8.2 shows the number of diatomic molecules in the first excited vibrational state (N_1) relative to those in the ground state (N_0) at 300 and 1000 K. The calculations have been carried out using the Boltzmann distribution. We see that nearly all the molecules in a macroscopic sample are in their ground vibrational state at room temperature because $N_1/N_0 \ll 1$. Even at 1000 K, N_1/N_0 is very small except for Br_2. This means that for these molecules, absorption of light at the characteristic frequency will occur predominantly from molecules in the $n = 0$ state. What final states are possible? As shown in the next section, for absorption by a quantum mechanical harmonic oscillator, $\Delta n = n_{final} - n_{initial} = +1$. Because only the $n = 0$ state has an appreciable population, with few exceptions only the $n = 0 \rightarrow n = 1$ transition is observed in vibrational spectroscopy.

TABLE 8.2   VIBRATIONAL STATE POPULATIONS FOR SELECTED DIATOMIC MOLECULES

Molecule	$\tilde{\nu}$ (cm^{-1})	ν (s^{-1})	N_1/N_0 for 300. K	N_1/N_0 for 1000. K
H–H	4400	1.32×10^{14}	6.88×10^{-10}	1.78×10^{-3}
H–F	4138	1.24×10^{14}	2.42×10^{-9}	2.60×10^{-3}
H–Br	2649	7.94×10^{13}	3.05×10^{-6}	2.21×10^{-2}
N–N	2358	7.07×10^{13}	1.23×10^{-5}	3.36×10^{-2}
C–O	2170	6.51×10^{13}	3.03×10^{-5}	4.41×10^{-2}
Br–Br	323	9.68×10^{12}	0.213	0.628

EXAMPLE PROBLEM 8.2

A strong absorption of infrared radiation is observed for $^1H^{35}Cl$ at 2991 cm^{-1}.

- Calculate the force constant, k, for this molecule.
- By what factor do you expect this frequency to shift if deuterium is substituted for hydrogen in this molecule? The force constant is unaffected by this substitution.

Solution

a. We first write $\Delta E = h\nu = hc/\lambda = \hbar\sqrt{k/\mu}$. Solving for k,

$$k = 4\pi^2\left(\frac{c}{\lambda}\right)^2 \mu$$

$$= 4\pi^2(2.998 \times 10^8 \text{ m s}^{-1})^2 \left(\frac{2991}{\text{cm}} \times \frac{100 \text{ cm}}{1 \text{ m}}\right)^2 \frac{(1.008)(34.969) \text{ amu}}{35.977}$$

$$\times \left(\frac{1.661 \times 10^{-27} \text{ kg}}{1 \text{ amu}}\right)$$

$$= 516.3 \text{ N m}^{-1}$$

b. $\dfrac{\nu_{DCl}}{\nu_{HCl}} = \sqrt{\dfrac{\mu_{HCl}}{\mu_{DCl}}} = \sqrt{\dfrac{m_H m_{Cl}(m_D + m_{Cl})}{m_D m_{Cl}(m_H + m_{Cl})}} = \sqrt{\left(\dfrac{1.0078}{2.0140}\right)\left(\dfrac{36.983}{35.977}\right)}$

$$= 0.717$$

The vibrational frequency for DCl is lower by a substantial amount. Would the shift be as great if ^{37}Cl were substituted for ^{35}Cl? The fact that vibrational frequencies are so strongly shifted by isotopic substitution of deuterium for hydrogen makes infrared spectroscopy a valuable tool for determining the presence of hydrogen atoms in molecules.

8.2 The Morse Potential

Note that the high sensitivity available in modern instrumentation to carry out vibrational spectroscopy does make it possible in favorable cases to see vibrational transitions originating from the $n = 0$ state for which $\Delta n = +2, +3, \ldots$. These **overtone** transitions are much weaker than the $\Delta n = +1$ absorption, but are possible because the selection rule $\Delta n = +1$ is not rigorously obeyed for an anharmonic potential, as discussed later. This more advanced topic is explored in Problem P8.22 at the end of the chapter.

The overtone transitions are useful because they allow us to determine the degree to which real molecular potentials differ from the simple **harmonic potential**, $V(x) = (1/2)kx^2$. To a good approximation, a realistic **anharmonic potential** can be described in analytical form by the **Morse potential**:

$$V(x) = D_e\left[1 - e^{-\alpha(x-x_e)}\right]^2 \tag{8.4}$$

in which D_e is the dissociation energy relative to the bottom of the potential and $\alpha = \sqrt{k/2D_e}$. The force constant, k, for the Morse potential is defined by $k = (d^2V/dx^2)_{x=x_e}$. Note that this definition for k is also valid for the harmonic potential. The **bond energy** D_0 is defined with respect to the lowest allowed level, rather than to the bottom of the potential, as shown in Figure 8.6.

The energy levels for this potential are given by

$$E_n = h\nu\left(n + \frac{1}{2}\right) - \frac{(h\nu)^2}{4D_e}\left(n + \frac{1}{2}\right)^2 \tag{8.5}$$

The second term gives the anharmonic correction to the energy levels. Measurements of the frequencies of the overtone vibrations allow the parameter D_e in the Morse potential to be determined for a specific molecule. This provides a useful method for determining the details of the interaction potential in a molecule.

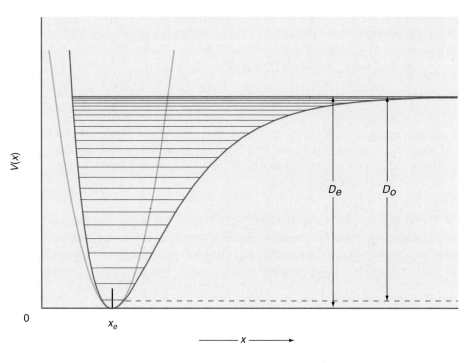

FIGURE 8.6

Morse potential, $V(x)$ (red curve), as a function of the bond length, x, for HCl, using the parameters from Example Problem 8.3. The zero of energy is chosen to be the bottom of the potential. The yellow curve shows a harmonic potential, which is a good approximation to the Morse potential near the bottom of the well. The horizontal lines indicate allowed energy levels in the Morse potential. D_e and D_0 represent the bond energies defined with respect to the bottom of the potential and the lowest state, respectively, and x_e is the equilibrium bond length.

EXAMPLE PROBLEM 8.3

The Morse potential can be used to model dissociation as illustrated in this example. The $^1H^{35}Cl$ molecule can be described by a Morse potential with $D_e = 7.41 \times 10^{-19}$ J. The force constant k for this molecule is 516.3 N m^{-1} and $\nu = 8.97 \times 10^{13}$ s^{-1}. Calculate the number of allowed vibrational states in this potential and the bond energy for the $^1H^{35}Cl$ molecule.

Solution

We solve the equation

$$E_n = h\nu\left(n + \frac{1}{2}\right) - \frac{(h\nu)^2}{4D_e}\left(n + \frac{1}{2}\right)^2 = D_e$$

to obtain the highest value of n consistent with the potential. Using the parameters given earlier, we obtain the following equation for n:

$$-1.1918 \times 10^{-21} n^2 + 5.8243 \times 10^{-20} n + 2.942 \times 10^{-20} = 7.41 \times 10^{-19}$$

Both solutions to this quadratic equation give $n = 24.4$, so we conclude that the potential has 24 allowed levels. If the left side of the equation is graphed versus n, we obtain the results shown in the following figure.

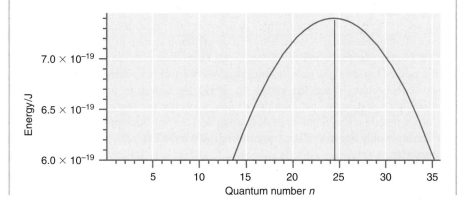

Note that E_n decreases for $n > 25$. This is mathematically correct, but unphysical because for $n > 24$, the molecule has a continuous energy spectrum, and Equation (8.5) is no longer valid.

The bond energy, D_0, is not D_e, but $D_e - E_0$ where

$$E_0 = \frac{h\nu}{2} - \frac{(h\nu)^2}{16D_e}$$

from Equation (8.5), because the molecule has a zero point vibrational energy. Using the parameters given earlier, the bond energy is 7.11×10^{-19} J. The Morse and harmonic potentials as well as the allowed energy levels for this molecule are shown in Figure 8.6.

The material-dependent parameters that determine the frequencies observed in vibrational spectroscopy for diatomic molecules are the force constant, k, and the reduced mass, μ. The corresponding parameters for rotational spectroscopy (see Section 8.6) are the rotational constant, B, and the bond length, r or x_e. These parameters, along with the bond energy D_0, are listed in Table 8.3 for selected molecules. The quantities B and $\tilde{\nu}$ are expressed in units of inverse centimeters.

TABLE 8.3 VALUES OF MOLECULAR CONSTANTS FOR SELECTED DIATOMIC MOLECULES

	$\tilde{\nu}\,(\text{cm}^{-1})$	$\nu\,(\text{s}^{-1})$	$x_e\,(\text{pm})$	$k\,(\text{N m}^{-1})$	$B\,(\text{cm}^{-1})$	$D_0\,(\text{kJ mol}^{-1})$	$D_0\,(\text{J molecule}^{-1})$
H_2	4401	1.32×10^{14}	74.14	575	60.853	436	7.24×10^{-19}
D_2	3115	9.33×10^{13}	74.15	577	30.444	443	7.36×10^{-19}
$^1\text{H}^{81}\text{Br}$	2649	7.94×10^{13}	141.4	412	8.4649	366	6.08×10^{-19}
$^1\text{H}^{35}\text{Cl}$	2991	8.97×10^{13}	127.5	516	10.5934	432	7.17×10^{-19}
$^1\text{H}^{19}\text{F}$	4138	1.24×10^{14}	91.68	966	20.9557	570	9.46×10^{-19}
$^1\text{H}^{127}\text{I}$	2309	6.92×10^{13}	160.92	314	6.4264	298	4.95×10^{-19}
$^{35}\text{Cl}_2$	559.7	1.68×10^{13}	198.8	323	0.2440	243	4.03×10^{-19}
$^{79}\text{Br}_2$	325.3	9.75×10^{12}	228.1	246	0.082107	194	3.22×10^{-19}
$^{19}\text{F}_2$	916.6	2.75×10^{13}	141.2	470	0.89019	159	2.64×10^{-19}
$^{127}\text{I}_2$	214.5	6.43×10^{12}	266.6	172	0.03737	152	2.52×10^{-19}
$^{14}\text{N}_2$	2359	7.07×10^{13}	109.8	2295	1.99824	945	1.57×10^{-18}
$^{16}\text{O}_2$	1580.	4.74×10^{13}	120.8	1177	1.44563	498	8.27×10^{-19}
$^{12}\text{C}^{16}\text{O}$	2170.	2.56×10^{13}	112.8	1902	1.9313	1076	1.79×10^{-18}

Source: Lide, D. R., Ed., *CRC Handbook of Chemistry and Physics*, 83rd Edition. CRC Press, Boca Raton, FL, 2003.

8.4 THE ORIGIN OF SELECTION RULES

Every spectroscopy has selection rules that govern the transitions that can occur between different states of a system. This is a great simplification in the interpretation of spectra, because far fewer transitions occur than would be the case if there were no selection rules. How do these selection rules arise? We use the general model for transitions between states discussed earlier to derive the selection rules for vibrational spectroscopy based on the quantum mechanical harmonic oscillator.

As discussed later in Supplemental Section 8.9, the transition probability from state n to state m is only nonzero if the **transition dipole moment**, μ_x^{mn}, satisfies the following condition:

$$\mu_x^{mn} = \int \psi_m^*(x)\mu_x(x)\psi_n(x)\,dx \neq 0 \tag{8.6}$$

In this equation, x is the spatial variable and μ_x is the dipole moment along the electric field direction, which we take to be the x axis.

In the following discussion, we show how selection rules for vibrational excitation arise from Equation (8.6). As discussed in Section 8.1, the dipole moment μ_x will change slightly as the molecule vibrates. Because the amplitude of vibration x is an oscillatory function of t, the molecule will have a time-dependent dynamic dipole moment. We take this into account by expanding μ_x in a Taylor series about the equilibrium bond length. Because x is the amplitude of vibration, the equilibrium bond length corresponds to $x = 0$:

$$\mu_x(x(t)) = \mu_{0x} + x(t)\left(\frac{d\mu_x}{dx}\right)_{x=0} + \ldots \tag{8.7}$$

in which the values of μ_{0x} and $(d\mu_x/dx)_{x=0}$ depend on the molecule under consideration. Note that because $x = x(t)$, $\mu_x(x(t))$ is a function of time. The first term in Equation (8.7) is the permanent dipole moment, and the second term is the dynamic dipole moment. As we saw earlier, for absorption experiments, it is reasonable to assume that only the $n = 0$ state is populated. Using Equation (7.5), which gives explicit expressions for the eigenfunctions m:

$$\mu_x^{m0} = A_m A_0 \mu_{0x} \int_{-\infty}^{\infty} H_m(\alpha^{1/2}x)\, H_0(\alpha^{1/2}x)e^{-\alpha x^2}\, dx$$

$$+ A_m A_0 \left[\left(\frac{d\mu_x}{dx}\right)_{x=0}\right] \int_{-\infty}^{\infty} H_m(\alpha^{1/2}x)\, x\, H_0(\alpha^{1/2}x)e^{-\alpha x^2}\, dx \tag{8.8}$$

The first integral is zero because different eigenfunctions are orthogonal. To solve the second integral, we need to use the specific functional form of $H_m(\alpha^{1/2}x)$. However, because the integration is over the symmetric interval $-\infty < x < \infty$, this integral is zero if the integrand is an odd function of x. As Equation (7.5) shows, the Hermite polynomials $H_m(\alpha^{1/2}x)$ are odd functions of x if m is odd and even functions of x if m is even. The term $x\, H_0(\alpha^{1/2}x)e^{-\alpha x^2}$ in the integrand is an odd function of x and, therefore, μ_x^{m0} is zero if $H_m(\alpha^{1/2}x)$ is even. This simplifies the problem because only transitions of the type

$$n = 0 \rightarrow m = 2b + 1, \quad \text{for } b = 0, 1, 2, \ldots \tag{8.9}$$

can have nonzero values for μ_x^{m0}.

Do all the transitions indicated in Equation (8.9) lead to nonzero values of μ_x^{m0}? To answer this question, the integrand $H_m(\alpha^{1/2}x)\, x\, H_0(\alpha^{1/2}x)\, e^{-\alpha x^2}$ is graphed against x for the transitions $n = 0 \rightarrow m = 1$, $n = 0 \rightarrow m = 3$, and $n = 0 \rightarrow m = 5$ in Figure 8.7. Whereas the integrand is positive everywhere for the $n = 0 \rightarrow n = 1$ transition, the areas above the dashed line exactly cancel those below the line for the $n = 0 \rightarrow m = 3$ and $n = 0 \rightarrow m = 5$ transitions, showing that the value of $\mu_x^{mn} = 0$. Therefore, $\mu_x^{m0} \neq 0$ only for the first of the three transitions under consideration. It can be shown more generally that in the dipole approximation, the selection rule for absorption is $\Delta n = +1$, and for emission, it is $\Delta n = -1$. We have derived this selection rule for the particular case of vibrational spectroscopy, and selection rules are different for different spectroscopies. However, within the dipole approximation, the selection rules for any spectroscopy are calculated using Equation (8.6) and the appropriate total energy eigenfunctions.

Note that because we found that the first integral in Equation (8.8) was zero, the absence or presence of a permanent dipole moment μ_{0x} is not relevant for the absorption of infrared radiation. For vibrational excitation to occur, the dynamic dipole moment must be nonzero. Because of this condition, homonuclear diatomic molecules do not absorb light in the infrared. This has important consequences for our environment. The temperature of the Earth is determined primarily by an energy balance between visible and ultraviolet (UV) radiation absorbed from the sun and infrared radiation emitted by the planet. The molecules N_2, O_2, and H_2, which have a zero permanent and transient dipole moment, together with the rare gases make up 99.93% of the atmosphere. These gases do not absorb the infrared radiation emitted by the Earth. Therefore, almost all of the emitted infrared radiation passes through the atmosphere and escapes into space. By

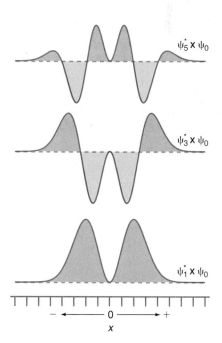

FIGURE 8.7

The integrand $H_m(\alpha^{1/2}x) \times H_0(\alpha^{1/2}x)e^{-\alpha x^2}$ is graphed as a function of x for the transitions $n = 0 \rightarrow m = 1$, $n = 0 \rightarrow m = 3$, and $n = 0 \rightarrow m = 5$. The dashed line shows the zero level for each graph.

contrast, greenhouse gases such as CO_2, NO_x, and hydrocarbons absorb the infrared radiation emitted by the Earth and radiate a portion of it back to the Earth. The result is an increase in the Earth's temperature and global warming. However, as you will conclude in Problem P8.15 at the end of the chapter, not all the vibrational modes of CO_2 are infrared active.

8.5 INFRARED ABSORPTION SPECTROSCOPY

The most basic result of quantum mechanics is that atoms and molecules possess a discrete energy spectrum and that energy can only be absorbed or emitted in amounts that correspond to the difference between two energy levels. Because the energy spectrum for each chemical species is unique, the allowed transitions between these levels provide a "fingerprint" for that species. Using such a fingerprint to identify and quantify the species is a primary role of all chemical spectroscopies. For a known molecule, the vibrational spectrum can also be used to determine the symmetry of the molecule and the force constants associated with the characteristic vibrations.

In absorption spectroscopies in general, electromagnetic radiation from a source of the appropriate wavelength is incident on a sample that is confined in a cell. The chemical species in the sample undergo transitions that are allowed by the appropriate selection rules among rotational, vibrational, or electronic states. The incident light of intensity $I_0(\lambda)$ is attenuated in passing a distance dl through the sample as described by the differential form of the **Beer-Lambert law** in which M is the concentration of the absorber; $I(\lambda)$ is the intensity of the transmitted light leaving the cell. Units of moles per liter are commonly used for M in liquid solutions, and partial pressure is used for gas mixtures:

$$dI(\lambda) = -\varepsilon(\lambda)MI(\lambda)\,dl \qquad (8.10)$$

This equation can be integrated to give

$$\frac{I(\lambda)}{I_0(\lambda)} = e^{-\varepsilon(\lambda)Ml} \qquad (8.11)$$

The information on the discrete energy spectrum of the chemical species in the cell is contained in the wavelength dependence of the **molar absorption coefficient**, $\varepsilon(\lambda)$. It is evident that the strength of the absorption is proportional to $I(\lambda)/I_0(\lambda)$, which increases with $\varepsilon(\lambda)$, M, and with path length l. Because $\varepsilon(\lambda)$ is a function of the wavelength, absorption spectroscopy experiments typically consist of the elements shown in Figure 8.8. In the most basic form of this spectroscopy, a **monochromator** is used to separate the broadband radiation from the source into its constituent wavelengths. After passing through the sample, the transmitted light impinges on the detector. With this setup, only one wavelength can be measured at a time. This form of absorption spectroscopy is unnecessarily time consuming in comparison with Fourier transform techniques, which are discussed in Supplemental Section 8.7.

Light source

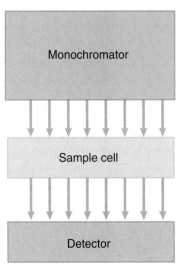

FIGURE 8.8

In an absorption experiment, the dependence of the sample absorption on wavelength is determined. A monochromator is used to filter out a particular wavelength from the broadband light source.

EXAMPLE PROBLEM 8.4

The molar absorption coefficient $\varepsilon(\lambda)$ for ethane is 40 $(\text{cm bar})^{-1}$ at a wavelength of 12 μm. Calculate $I(\lambda)/I_0(\lambda)$ in a 1.0-cm-long absorption cell if ethane is present at a contamination level of 2.0 ppm in one bar of air. What cell length is required to make $I(\lambda)/I_0(\lambda) = 0.90$?

Solution

Using $\dfrac{I(\lambda)}{I_0(\lambda)} = e^{-\varepsilon(\lambda)Ml}$

$\dfrac{I(\lambda)}{I_0(\lambda)} = \exp\left\{-[40.(\text{cm bar})^{-1}(2.0 \times 10^{-6}\,\text{bar})(1.0\,\text{cm})]\right\} = 0.9992 \approx 1.0$

This result shows that for this cell length, light absorption is difficult to detect.

Rearranging the Beer–Lambert equation, we have

$$l = -\frac{1}{M\varepsilon(\lambda)}\ln\left(\frac{I(\lambda)}{I_0(\lambda)}\right) = -\frac{1}{40.(\text{cm bar})^{-1}(2.0 \times 10^{-6}\text{ bar})}\ln(0.90)$$

$$= 1.3 \times 10^3 \text{ cm}$$

Path lengths of this order are possible in sample cells in which the light undergoes multiple reflections from mirrors outside of the cell. Even much longer path lengths are possible in cavity ringdown spectroscopy. In this method, the absorption cell is mounted between two highly focusing mirrors with a reflectivity greater than 99.99%. Because of the many reflections that take place between the mirrors without appreciable attenuation of the light, the effective length of the cell is very large. The detection sensitivity to molecules such as NO_2 is less than 10 parts per billion using this technique.

How does $\varepsilon(\lambda)$ depend on the wavelength or frequency? We know that for a harmonic oscillator, $\nu = (1/2\pi)\sqrt{k/\mu}$ so that the masses of the atoms and the force constant of the bond determine the resonant frequency. Now consider a molecule such as

$$\begin{array}{c}O\\\parallel\\R-C-R'\end{array}$$

The vibrational frequency of the C and O atoms in the carbonyl group is determined by the force constant for the $C{=}O$ bond. This force constant is primarily determined by the chemical bond between these atoms and to a much lesser degree by the adjacent R and R′ groups. For this reason, the carbonyl group has a characteristic frequency at which it absorbs infrared radiation that varies in a narrow range for different molecules. These group frequencies are very valuable in determining the structure of molecules, and an illustrative set is shown in Table 8.4.

TABLE 8.4   SELECTED GROUP FREQUENCIES

Group	Frequency (cm^{-1})	Group	Frequency (cm^{-1})
O—H stretch	3450–3650	C=O stretch	1650–1750
N—H stretch	3300–3500	C=C stretch	1620–1680
C—H stretch	2800–3000	C—C stretch	1200–1300
C—H bend	1450–1480	C—Cl stretch	600–800

We have shown that a diatomic molecule has a single vibrational peak of appreciable intensity, because the overtone frequencies have a low intensity. How many vibrational peaks are observed for larger molecules in an infrared absorption experiment? A molecule consisting of n atoms has three translational degrees of freedom, and two or three rotational degrees of freedom depending on whether it is a linear or nonlinear molecule. The remaining $3n - 6$ (nonlinear molecule) or $3n - 5$ (linear) degrees of freedom are vibrational modes. For example, benzene has 30 vibrational modes. However, some of these modes have the same frequency (they are degenerate in energy), so that benzene has only 20 distinct vibrational frequencies.

We now examine some experimental data. Vibrational spectra for gas-phase CO and CH_4 are shown in Figure 8.9. Because CO and CH_4 are linear and nonlinear molecules, we expect one and nine vibrational modes, respectively. However, the spectrum for CH_4 shows two rather than nine peaks that we might associate with vibrational transitions. We also see several unexpected broad peaks in the CH_4 spectrum. The single peak in the CO spectrum is much broader than would be expected for a vibrational peak, and it has a deep minimum at the central frequency.

These spectra look different than expected for two reasons. The broadening in the CO absorption peak and the broad envelopes of additional peaks for CH_4 result from transitions between different rotational energy states that occur simultaneously with the

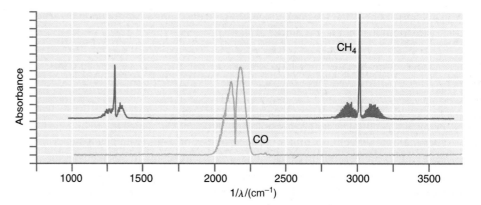

FIGURE 8.9

Infrared absorption spectra of gaseous CO and CH$_4$. The curves are offset vertically for clarity.

$n = 0 \rightarrow n = 1$ transition between vibrational energy levels. We discuss transitions between rotational energy levels and analyze a high-resolution infrared absorption spectrum for a diatomic molecule in some detail in Section 8.6. At this point we simply note that absorption of infrared radiation results in both rotational and vibrational rather than just vibrational transitions.

The second unexpected feature in Figure 8.9 is that two and not nine peaks are observed in the CH$_4$ spectrum. Why is this? To discuss the vibrational modes of polyatomic species in more detail, the information about molecular symmetry and group theory discussed in Chapter 16 is needed. At this point, we simply state the results. In applying group theory to the CH$_4$ molecule, the 1306-cm^{-1} peak can be associated with three degenerate C—H bending modes, and the 3020-cm^{-1} peak can be associated with three degenerate C—H stretching modes. This still leaves three vibrational modes unaccounted for. Again applying group theory to the CH$_4$ molecule, one finds that these modes are symmetric and do not satisfy the condition $d\mu_x/dx \neq 0$. Therefore, they are infrared inactive. However, all modes for CO and CH$_4$ are active in Raman spectroscopy, which we discuss in Supplemental Section 8.8.

Of the 30 vibrational modes for benzene, four peaks (corresponding to 7 of the 30 modes) are observed in infrared spectroscopy, and seven peaks (corresponding to 12 of the 30 modes) are observed in Raman spectroscopy. None of the frequencies is observed in both Raman and infrared spectroscopy. Eleven vibrational modes are neither infrared nor Raman active.

Although the discussion to this point might lead us to believe that each bonded pair of atoms in a molecule vibrates independently of the others, this is not the case. For example, we might think that the linear CO$_2$ molecule has a single C=O stretching frequency, because the two C=O bonds are equivalent. However, experiments show that this molecule has two distinct C=O stretching frequencies. Why is this the case? Although the two bonds are equivalent, the vibrational motion in these bonds is not independent. When one C=O bond vibrates, the atomic positions and electron distribution throughout the molecule are changed, thereby influencing the other C=O bond. In other words, we can view the CO$_2$ molecule as consisting of two coupled harmonic oscillators. In the center of mass coordinates, each of the two C=O groups is modeled as a mass coupled to a wall by a spring with force constant k_1 (see Section 7.1). We model the coupling as a second spring with force constant k_2 that connects the two oscillators. The model is depicted in Figure 8.10.

FIGURE 8.10

The coupled oscillator model of CO$_2$ is shown. The vertical dashed lines show the equilibrium positions. The symmetric vibrational mode is shown in the upper part of the figure, and the antisymmetric (or assymetric) vibrational mode is shown in the lower part of the figure.

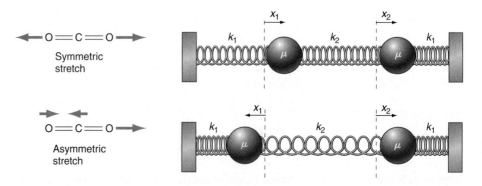

This coupled system has two vibrational frequencies: the symmetrical and antisymmetric modes. In the symmetrical mode, the vibrational amplitude is equal in both magnitude and sign for the individual oscillators. In this case, the C atom does not move. This is equivalent to the coupling spring in Figure 8.10 having the same length during the whole vibrational period. Therefore, the vibrational frequency is unaffected by the coupling and is given by

$$\nu_{symmetric} = \frac{1}{2\pi}\sqrt{\frac{k_1}{\mu}} \tag{8.12}$$

In the antisymmetric mode, the C atom does move. This is equivalent to the vibrational amplitude being equal in magnitude and opposite in sign for the individual oscillators. In this mode, the spring representing the coupling is doubly stretched, once by each of the oscillators. The resulting force on the reduced mass representing each oscillator is

$$F = -(k_1 + 2k_2)x \tag{8.13}$$

and the resulting frequency of this antisymmetric mode is

$$\nu_{antisymmetric} = \frac{1}{2\pi}\sqrt{\frac{k_1 + 2k_2}{\mu}} \tag{8.14}$$

We see that the C=O bond coupling gives rise to two different vibrational stretching frequencies and that the antisymmetric mode has the higher frequency as illustrated in Figure 8.11 for H_2O.

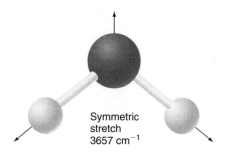
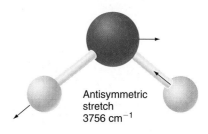

Symmetric stretch 3657 cm^{-1}

Antisymmetric stretch 3756 cm^{-1}

FIGURE 8.11

The two O—H stretching modes for the H_2O molecule are shown. Note that neither of them corresponds to the stretching of a single localized O—H bond. This is the case because, although equivalent, the two bonds are coupled, rather than independent.

8.6 ROTATIONAL SPECTROSCOPY

As for the harmonic oscillator, a selection rule governs the absorption of electromagnetic energy for a molecule to change its rotational energy, namely, $\Delta J = J_{final} - J_{initial} = \pm 1$. Although we do not derive this selection rule, Example Problem 8.5 shows that it holds for a specific case.

Note that we have just changed the symbol for the angular momentum quantum number from l to J. The quantum number l is usually used for orbital angular momentum (for example, the electron orbiting around the nucleus), and J is usually used for rotating molecules.

the **Chemistry place**™

8.3 Normal Modes for H_2O
8.4 Normal Modes for CO_2
8.5 Normal Modes for NH_3
8.6 Normal Modes for Formaldehyde

EXAMPLE PROBLEM 8.5

Using the following total energy eigenfunctions for the three-dimensional rigid rotor, show that the $J = 0 \to J = 1$ transition is allowed, and that the $J = 0 \to J = 2$ transition is forbidden:

$$Y_0^0(\theta, \phi) = \frac{1}{(4\pi)^{1/2}}$$

$$Y_1^0(\theta, \phi) = \left(\frac{3}{4\pi}\right)^{1/2}\cos\theta$$

$$Y_2^0(\theta, \phi) = \left(\frac{5}{16\pi}\right)^{1/2}(3\cos^2\theta - 1)$$

The notation $Y_J^{M_J}$ is used for the preceding functions.

Solution

Assuming the electromagnetic field to lie along the z axis, $\mu_z = \mu\cos\theta$, and the transition dipole moment takes the form

$$\mu_z^{J0} = \mu\int_0^{2\pi} d\phi \int_0^{\pi} Y_J^0(\theta, \phi)(\cos\theta)Y_0^0(\theta, \phi)\sin\theta\, d\theta$$

For the $J = 0 \rightarrow J = 1$ transition,

$$\mu_z^{10} = \mu \frac{\sqrt{3}}{4\pi} \int_0^{2\pi} d\phi \int_0^\pi \cos^2\theta \sin\theta \, d\theta = \mu \frac{\sqrt{3}}{2}\left[-\frac{\cos^3\theta}{3}\right]_{\theta=0}^{\theta=\pi} = \frac{\mu\sqrt{3}}{3} \neq 0$$

For the $J = 0 \rightarrow J = 2$ transition,

$$\mu_z^{20} = \mu \frac{\sqrt{5}}{8\pi} \int_0^{2\pi} d\phi \int_0^\pi (3\cos^2\theta - 1)\cos\theta \sin\theta \, d\theta$$

$$= \mu \frac{\sqrt{5}}{8\pi}\left[-\frac{3\cos^4\theta}{4} + \frac{\cos^2\theta}{2}\right]_{\theta=0}^{\theta=\pi}$$

$$= \mu \frac{\sqrt{5}}{8\pi}\left[-\frac{1}{4} + \frac{1}{4}\right] = 0$$

The preceding calculations show that the $J = 0 \rightarrow J = 1$ transition is allowed and that the $J = 0 \rightarrow J = 2$ transition is forbidden.

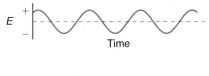

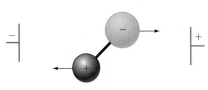

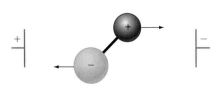

FIGURE 8.12

The interaction of a rigid rotor with an electric field. Imagine the sinusoidally varying electric field shown at the top of the figure applied between a pair of capacitor plates. The arrows indicate the direction of force on each of the two charged masses. If the frequencies of the field and rotation are equal, the rotor will absorb energy from the electric field.

In discussing vibrational spectroscopy, we learned that a molecule must have a nonzero dynamic dipole moment to absorb infrared radiation. By contrast, a molecule must have a permanent dipole moment to absorb energy in the microwave frequency range in which rotational transitions occur. As was the case for vibrational spectroscopy, the dominant interaction with the electric field is through the dipole moment. This is shown schematically in Figure 8.12.

The dependence of the rotational energy on the quantum number is given by

$$E = \frac{\hbar^2}{2\mu r_0^2}J(J + 1) = \frac{h^2}{8\pi^2\mu r_0^2}J(J + 1) = hcBJ(J + 1) \tag{8.15}$$

In this equation, the constants specific to a molecule are combined in the so-called **rotational constant** $B = h/(8\pi^2 c\mu r_0^2)$. The factor c is included in B so that it has the units of inverse centimeters rather than inverse seconds. The energy levels and transitions allowed by the selection rule $\Delta J = J_{final} - J_{initial} = \pm 1$ as well as a simulated rotational spectrum are shown in Figure 8.13.

We can calculate the energy corresponding to rotational transitions for $\Delta J = +1$ and $\Delta J = -1$. $\Delta J = +1$ corresponds to absorption and $\Delta J = -1$ corresponds to emission of a photon. In the following equations, J is the quantum number of the state from which the transition occurs.

$$\Delta E = E(J_{final}) - E(J_{initial})$$

for $\Delta J = +1$

$$\Delta E_+ = \frac{\hbar^2}{2\mu r_0^2}(J + 1)(J + 2) - \frac{\hbar^2}{2\mu r_0^2}J(J + 1)$$

$$= \frac{\hbar^2}{2\mu r_0^2}\left[(J^2 + 3J + 2) - (J^2 + J)\right] = \frac{\hbar^2}{\mu r_0^2}(J + 1)$$

$$= 2hcB(J + 1) \quad \text{and for} \quad \Delta J = -1$$

$$\Delta E_- = \frac{\hbar^2}{2\mu r_0^2}(J - 1)J - \frac{\hbar^2}{2\mu r_0^2}J(J + 1)$$

$$= \frac{\hbar^2}{2\mu r_0^2}\left[(J^2 - J) - (J^2 + J)\right] = -\frac{\hbar^2}{\mu r_0^2}J = -2hcBJ \tag{8.16}$$

Note that $|\Delta E_+| \neq |\Delta E_-|$ because the energy levels are not equally spaced. We see that the larger the J value of the originating energy level, the more energetic the photon must be to promote excitation to the next highest energy level. Because the rotational energy does not depend on m_J, each energy level is $2J+1$-fold degenerate.

EXAMPLE PROBLEM 8.6

Because of the very high precision of frequency measurements, bond lengths can be determined with a correspondingly high precision, as illustrated in this example. From the rotational microwave spectrum of $^1H^{35}Cl$, we find that $B = 10.59342$ cm^{-1}. Given that the masses of ^1H and ^{35}Cl are 1.0078250 and 34.9688527 amu, respectively, determine the bond length of the $^1H^{35}Cl$ molecule.

Solution

$$B = \frac{h}{8\pi^2 \mu c r_0^2}$$

$$r_0 = \sqrt{\frac{h}{8\pi^2 \mu c B}}$$

$$= \sqrt{\frac{6.6260755 \times 10^{-34}\,\text{J s}}{8\pi^2 c \left(\frac{(1.0078250)(34.9688527)\,\text{amu}}{1.0078250 + 34.9688527}\right)(1.66054 \times 10^{-27}\,\text{kg amu}^{-1})(10.59342\,\text{cm}^{-1})}}$$

$$= 1.274553 \times 10^{-10}\,\text{m}$$

The structure of a rotational spectrum becomes more apparent when we consider the energy-level spacing in more detail. Table 8.5 shows the frequencies needed to excite various transitions consistent with the selection rule $\Delta J = J_{final} - J_{initial} = +1$. Each of these transitions can lead to absorption of electromagnetic radiation. We see that for successive initial values of J, the ΔE associated with the transition increases in such a way that the difference between these $\Delta \nu$, which we call $\Delta(\Delta \nu)$, is constant. This means that the spectrum for a molecule immersed in a microwave field with a broad range of frequencies shows a series of equally spaced lines, separated in frequency by $2cB$ as seen in Figure 8.13.

How many absorption peaks will be observed? For vibrational spectroscopy, we expect only one intense peak for the following reasons. The energy-level spacing between adjacent levels is the same for all values of the quantum number in the harmonic approximation so that given the selection rule $\Delta n = +1$, all transitions have the same frequency. Also, in general only the $n = 0$ energy level has a significant population so that even taking anharmonicity into account will not generate additional peaks originating from peaks with $n > 0$. However, the situation is different for rotational transitions. Note that because the rotational energy levels are not equally spaced in energy, different transitions give rise to separate peaks. Additionally, $\Delta E_{rotation} \ll kT$ under most conditions so that many rotational energy levels will be populated. Therefore, many peaks are generally observed in a rotational spectrum.

TABLE 8.5	FREQUENCIES NEEDED TO EXCITE VARIOUS ROTATIONAL TRANSITIONS	
$J \to J'$	$\Delta \nu$	$\Delta(\Delta \nu)$
$0 \to 1$	$2cB$	$2cB$
$1 \to 2$	$4cB$	$2cB$
$2 \to 3$	$6cB$	$2cB$
$3 \to 4$	$8cB$	$2cB$
$4 \to 5$	$10cB$	$2cB$

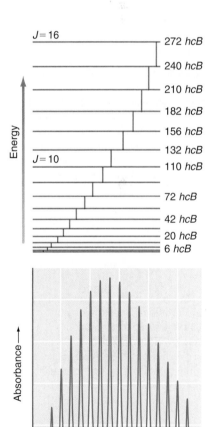

FIGURE 8.13

The energy levels for a rigid rotor are shown in the top panel and the spectrum observed through absorption of microwave radiation is shown in the bottom panel. The allowed transitions between levels are shown as vertical bars.

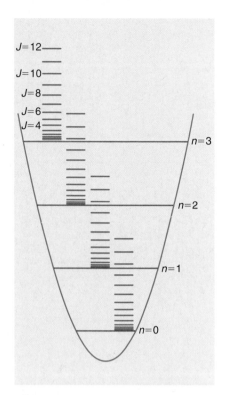

FIGURE 8.14

Schematic representation of rotational and vibrational levels. Each vibrational level has a set of rotational levels associated with it. Therefore, vibrational transitions usually also involve rotational transitions. The rotational levels are shown on an expanded energy scale and are much more closely spaced for real molecules.

8.7 Rotational Spectroscopy of Diatomic Molecules

Up to this point, we have considered rotation and vibration separately. In the microwave region of the electromagnetic spectrum, the photon energy is sufficient to excite rotational transitions, but not to excite vibrational transitions. However, this is not the case for infrared radiation. Diatomic molecules that absorb infrared radiation can make transitions in which both n and J change according to the selection rules $\Delta n = +1$ and $\Delta J = \pm 1$. Therefore, an infrared absorption spectrum contains both vibrational and rotational transitions. What does a rotational-vibrational spectrum look like? To answer this question, first consider the relative photon energies associated with rotational and vibrational excitation. The energy levels for both degrees of freedom are indicated schematically in Figure 8.14. The ratio of the smallest value of ΔE in a rotational transition to that in a vibrational transition is

$$\frac{\Delta E_{rot}}{\Delta E_{vib}} = \frac{\hbar^2/\mu r_0^2}{\hbar\sqrt{k/\mu}} = \frac{\hbar}{r_0^2\sqrt{k\mu}} \tag{8.17}$$

This ratio is molecule specific, but we consider two extremes. For H_2 and I_2, $\Delta E_{rot}/\Delta E_{vib}$ is 0.028 and 0.00034, respectively where ΔE_{rot} is for the $J = 0 \rightarrow J = 1$ transition. In both cases, there are many rotational levels between adjacent vibrational levels. Large moments of inertia (large atomic masses and/or long bonds) and large force constants (strong bonds) lead to a smaller value of $\Delta E_{rot}/\Delta E_{vib}$. It is largely the difference in the **moment of inertia** $I = \mu r_0^2$ that makes the ratio so different for I_2 and H_2.

On the basis of the previous discussion, what will be seen in an infrared absorption experiment on a diatomic molecule in which both rotational and vibrational transitions occur? As discussed in Section 8.3, the dominant vibrational transition is $n = 0 \rightarrow n = 1$. All transitions must now satisfy two selection rules, $\Delta n = +1$ and $\Delta J = \pm 1$. As discussed earlier, a vibrational-rotational spectrum will exhibit many different rotational transitions. What can one predict about the relative intensities of the peaks? Recall that the intensity of a spectral line in an absorption experiment is determined by the number of molecules in the energy level from which the transition originates. (This rule holds as long as the upper state population is small compared to the lower state population.) How many molecules are there in states for a given value of J relative to the number in the ground state for which $J = 0$? This ratio can be calculated using the Boltzmann distribution:

$$\frac{n_J}{n_0} = \frac{g_J}{g_0}e^{-(E_J - E_0)/kT} = (2J + 1)e^{-\hbar^2 J(J+1)/(2\,IkT)} \tag{8.18}$$

The term in front of the exponential gives the degeneracy of the energy level corresponding to J. It generally dominates n_J/n_0 for small J and sufficiently large T. However, as J increases, the exponential term causes n_J/n_0 to decrease rapidly with increasing J. For molecules with a large moment of inertia, the exponential term does not dominate until J is quite large. As a result, many rotational energy levels are occupied and this behavior is seen for CO in Figure 8.15. Because many levels are occupied, a large number of peaks are observed in a rotational spectrum. For a molecule with a small moment of inertia, the rotational levels can be far enough apart that few rotational states are populated. This behavior is shown in Figure 8.15 for HD. At 100 K, only the $J = 0, 1,$ and 2 states have an appreciable population. Increasing the temperature raises this upper value of J to approximately 4 and 7 for 300 and 700 K, respectively. The corresponding J values for CO are 13, 23, and 33.

Therefore, as long as the ratio n_J/n_0 increases with J, the intensity of the spectral peaks originating from states with those J values will increase. Beyond the J values for which n_J/n_0 increases, the intensity of the peaks decreases.

A simulated rotational-vibrational infrared absorption spectrum for HCl is shown in Figure 8.16. Such a spectrum consists of two nearly symmetric parts. The higher frequency part of the spectrum corresponds to transitions in which $\Delta J = +1$ and is called the **R branch**. The lower frequency part of the spectrum corresponds to transitions in which $\Delta J = -1$ and is called the **P branch**. Note that the gap in the center of the spectrum corresponds to $\Delta J = 0$, which is a forbidden transition in the dipole approximation for a linear molecule. Without going into more detail, note that Raman spectroscopy (see

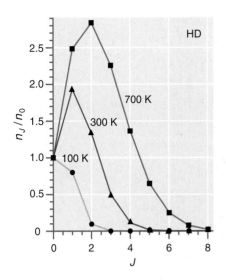

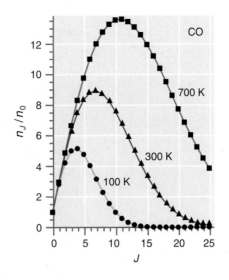

FIGURE 8.15

The number of molecules in energy levels corresponding to the quantum number J relative to the number in the ground state is shown as a function of J for two molecules at three different temperatures.

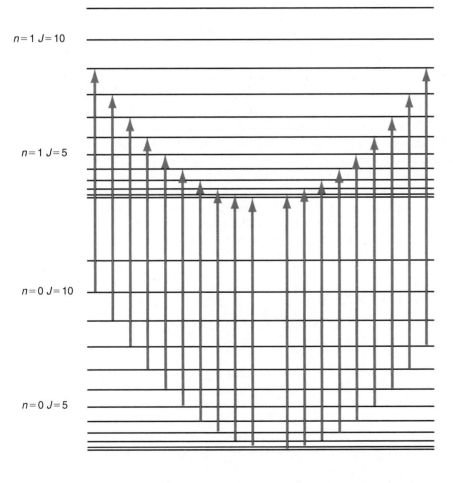

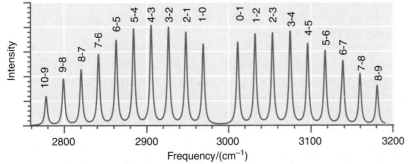

FIGURE 8.16

Simulated 300 K infrared absorption spectrum and energy diagram for $H^{35}Cl$. The two indices above the peak refer to the initial (first) and final (second) J values. The region of the spectrum with $\Delta J = +1$ (higher frequency) is called the R branch, and the region of the spectrum with $\Delta J = -1$ (lower frequency) is called the P branch. The energy levels in the upper part of the figure are not drawn to scale.

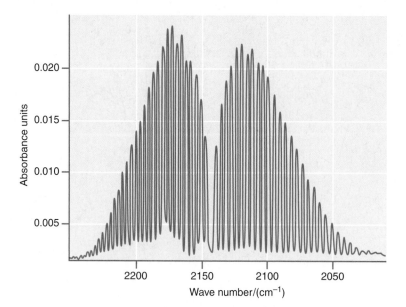

FIGURE 8.17

A high-resolution spectrum is shown for CO in which the P and R branches are resolved into the individual rotational transitions.

8.8 Rotational-Vibrational Spectroscopy of Diatomic Molecules

Supplemental Section 8.8) also shows both rotational and vibrational transitions. However, the selection rules are different. For rotational Raman spectra, the selection rule is $\Delta J = 0, \pm 2$, and not $\Delta J = \pm 1$ as it is for infrared absorption spectra.

Based on this discussion of rotational-vibrational spectroscopy and the results shown in Figures 8.15 and 8.16, it is useful to revisit the infrared spectra of CO and CH_4 shown in Figure 8.9. The broad unresolved peaks seen for CO between 2000 and 2250 cm^{-1} are the P and R branches corresponding to rotational-vibrational excitations. The minimum near 2150 cm^{-1} corresponds to the forbidden $\Delta J = 0$ transition. The broad and only partially resolved peaks for CH_4 seen around the sharp peaks centered near 1300 and 3000 cm^{-1} are again the P and R branches. The $\Delta J = 0$ transition is allowed for methane and is the reason why the sharp central peaks are observed in the methane spectrum seen in Figure 8.9. To demonstrate the origin of the broad CO peaks in Figure 8.9, a high-resolution infrared absorption spectrum for this molecule is shown in Figure 8.17. It is apparent that the envelopes of the P and R branches in this figure correspond to the broad unresolved peaks in Figure 8.9.

SUPPLEMENTAL

8.7 FOURIER TRANSFORM INFRARED SPECTROSCOPY

How are infrared absorption spectra obtained in practice? We now turn to a discussion of Fourier transform infrared (FTIR) spectroscopy, which is the most widely used technique for obtaining vibrational absorption spectra. FTIR spectroscopy improves on the schematic absorption experiment shown in Figure 8.8 by eliminating the monochromator and by using a broadband blackbody radiation source. By simultaneously analyzing the absorption throughout the spectral range of the light source, it achieves a **multiplex advantage** that is equivalent to carrying out many single-wavelength experiments in parallel. This technique allows a spectrum to be obtained in a short time and has led to a revolution in the field of vibrational spectroscopy. We describe how **FTIR spectroscopy** works in this section.

The multiplex advantage in FTIR is gained by using a **Michelson interferometer** to determine the frequencies at which radiation is absorbed by molecules. A schematic drawing of this instrument is shown in Figure 8.18. The functioning of a Michelson interferometer is first explained by analyzing its effect on monochromatic radiation. An incoming traveling plane wave of amplitude $A_0 \exp i(kx - \omega t)$ and intensity I_0 impinges on a beam splitter S that both transmits and reflects 50% of the incident light. Each of these two waves is reflected back from a mirror (M$_1$ or M$_2$) and is incident on the beam splitter S. The wave that is reflected back from the movable mirror M$_2$ and

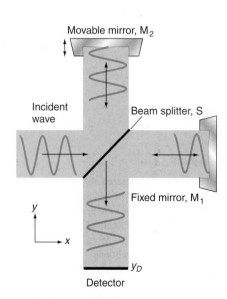

FIGURE 8.18

Schematic diagram of a Michelson interferometer.

transmitted by S interferes with the wave that is reflected from the fixed mirror M_1 and reflected from S. The recombined wave resulting from this interference travels in the negative y direction and has an amplitude at the detector plane $y_- = y_D$ given by

$$A(t) = \frac{A_0}{\sqrt{2}}\left[\exp i(ky_D - \omega t) + \exp i(k[y_D + \Delta d(t)] - \omega t)\right]$$

$$= \frac{A_0}{\sqrt{2}}\left(1 + e^{i\delta(t)}\right)\exp i(ky_D - \omega t) \qquad \textbf{(8.19)}$$

The phase difference $\delta(t)$ results from the path difference that the two interfering waves have traveled, Δd. It arises because mirrors M_1 and M_2 are not equidistant from the beam splitter:

$$\delta(t) = \frac{2\pi}{\lambda}(2SM_1 - 2SM_2) = \frac{2\pi}{\lambda}\Delta d(t) \qquad \textbf{(8.20)}$$

In this equation, SM_1 and SM_2 are the distances between the beam splitter and mirrors 1 and 2, respectively. The intensity of the resultant wave at the detector plane, I, is proportional to the product $A(t)A^*(t)$:

$$I(t) = \frac{I_0}{2}(1 + \cos\delta(t)) = \frac{I_0}{2}\left(1 + \cos\frac{2\pi\Delta d(t)}{\lambda}\right) \qquad \textbf{(8.21)}$$

where $I_0 = A_0^2$. The intensity varies periodically with distance as mirror M_2 is moved toward the beam splitter. Whenever $\Delta d = n\lambda$, the interference is constructive and the maximum intensity is transmitted to the detector. Whenever $\Delta d = (2n + 1)(\lambda/2)$, the interference is destructive and the wave is fully reflected back into the source.

The signal measured by the detector is called an **interferogram** because it results from the interference of the two waves. In this case the interferogram is described by a single sine wave, so that a frequency analysis of the intensity gives a single frequency corresponding to the incident plane wave. The output of the interferometer for a single incident frequency is shown in Figure 8.19. This simple example illustrates how the frequency of the radiation that enters the interferometer can be determined from the experimentally obtained interferogram.

We now consider the more interesting case encountered when the incident wave is composed of a number of different frequencies. This case describes a realistic situation in which a blackbody source of infrared light passes through a sample and enters the interferometer. Only certain frequencies from the source are absorbed by vibrational excitations of the molecules. The interferometer sees the blackbody distribution of frequencies of the source from which certain frequencies have been attenuated through absorption. What can we expect for the case of several incident frequencies? We can write the amplitude of the wave resulting from the interference of the two reflections from mirrors M_1 and M_2 as follows:

$$A(t) = \sum_j \frac{A_j}{\sqrt{2}}\left(1 + \exp\left[i\left(\frac{2\pi\Delta d(t)}{\lambda_j}\right)\right]\right)\exp\left[i\left(\frac{2\pi}{\lambda_j}y_D - \omega_j t\right)\right] \qquad \textbf{(8.22)}$$

In this equation, the subscript j refers to the individual frequencies incident in the beam entering the interferometer. As you will see in the problems at the end of this chapter, if the mirror is moving with a velocity v, the measured intensity at the detector is

$$I(t) = \frac{1}{2}\sum_j I(\omega_j)\left(1 + \cos\left[\omega_j\frac{2v}{c}t\right]\right) \qquad \textbf{(8.23)}$$

The interferogram $I(t)$ is determined by the distribution of frequencies entering the interferometer. Figure 8.20b shows interferograms that result from the sample spectra shown in Figure 8.20a. In practice, the opposite path is followed, in which the measured interferogram is converted to a spectrum using Fourier transform techniques.

Note the contrast between the interferogram of Figure 8.20 and the interferogram in Figure 8.19. Whereas the interferogram in Figure 8.19 is calculated for a single frequency, a real spectral line has a finite width in frequency. The effect of this finite

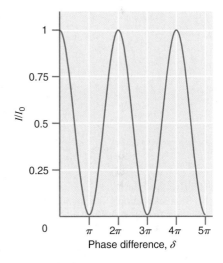

FIGURE 8.19
Intensity measured at the detector as a function of the phase difference in the two arms of a Michelson interferometer.

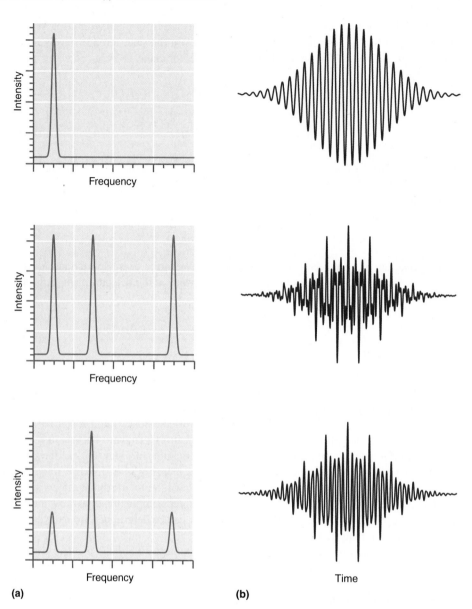

(a)

FIGURE 8.20
(a) Simulated sample spectra and **(b)** the resulting interferograms calculated using Equation (8.23).

width is to damp the amplitude of the interferogram for longer and shorter times relative to the central value of $t = 0$. For our purposes, it is sufficient to note that the interferograms for the different sample spectra are clearly different. Although the characteristic absorption frequencies cannot be obtained directly by inspection of the interferogram, they are readily apparent after the data have been Fourier transformed from the time domain into the frequency domain.

Because the information about absorption at all frequencies is determined simultaneously, an FTIR spectrum can be obtained quickly with high sensitivity. For example, the components of automobile gas exhaust are typically (percent by volume) N_2 (71%), CO_2 (18%), H_2O (9.2%), CO (0.85%), O_2 and noble gases (0.7%), NO_x (0.08%), and hydrocarbons (0.05%). The concentration of these components other than N_2, O_2, and the rare gases can be determined in well under a minute by recording a single FTIR spectrum.

SUPPLEMENTAL

8.8 RAMAN SPECTROSCOPY

As discussed in the previous sections, absorption of light in the infrared portion of the spectrum can lead to transitions between eigenstates of the vibrational and rotational energy. Another interaction between a molecule and an electromagnetic field can also lead to vibrational and rotational excitation. It is called the **Raman effect** after its discoverer and involves scattering of a photon by the molecule. You can think of scattering as the collision between a molecule and a photon in which energy and momentum are transferred between the two collision partners. Raman spectroscopy complements infrared absorption spectroscopy because it obeys different selection rules. For instance, the stretching mode in a homonuclear diatomic molecule is Raman active, but infrared inactive. The reasons for this difference will become clear after molecular symmetry and group theory are discussed in Chapter 16.

Consider a molecule with a characteristic vibrational frequency ν_{vib} in an electromagnetic field that has a time-dependent electric field given by

$$E = E_0 \cos 2\pi\nu t \qquad (8.24)$$

The electric field distorts the molecule slightly because the negative valence electrons and the positive nuclei and their tightly bound core electrons experience forces in opposite directions. This induces a time-dependent dipole moment of magnitude $\mu_{induced}(t)$ in the molecule of the same frequency as the field. The dipole moment is linearly proportional to the magnitude of the electric field, and the proportionality constant is the **polarizability**, α. The polarizability is an anisotropic quantity and its value depends on the direction of the electric field relative to the molecular axes:

$$\mu_{induced}(t) = \alpha E_0 \cos 2\pi\nu t \qquad (8.25)$$

The polarizability depends on the bond length $x_e + x(t)$, where x_e is the equilibrium value. The polarizability α can be expanded in a Taylor-Maclaurin series (see the Math Supplement, Appendix A) in which terms beyond the first order have been neglected:

$$\alpha(x) = \alpha(x_e) + x\left(\frac{d\alpha}{dx}\right)_{x=x_e} + \ldots \qquad (8.26)$$

Due to the vibration of the molecule, $x(t)$ is time dependent and is given by

$$x(t) = x_{max} \cos 2\pi\nu_{vib}t \qquad (8.27)$$

Combining this result with Equation (8.26), we can rewrite Equation (8.25) in the form

$$\mu_{induced}(t) = \alpha E = E_0 \cos 2\pi\nu t \left[\alpha(x_e) + \left[\left(\frac{d\alpha}{dx}\right)_{x=x_e}\right] x_{max} \cos 2\pi\nu_{vib}t \right] \qquad (8.28)$$

which can be simplified, using the trigonometric identity $\cos x \cos y = \frac{1}{2}[\cos(x-y) + \cos(x+y)]$ to

$$\mu_{induced}(t) = \alpha E = \alpha(x_e)E_0 \cos 2\pi\nu t$$
$$+ \left[\left(\frac{d\alpha}{dx}\right)_{x=x_e}\right] x_{max} E_0[\cos(2\pi\nu + 2\pi\nu_{vib})t + \cos(2\pi\nu - 2\pi\nu_{vib})t] \qquad (8.29)$$

The time-varying dipole moment radiates light of the same frequency as the dipole moment, and at the frequencies ν, $(\nu - \nu_{vib})$, and $(\nu + \nu_{vib})$. These three frequencies are referred to as the **Rayleigh**, **Stokes**, and **anti-Stokes frequencies**, respectively. We see that in addition to scattered light at the incident frequency, light will also be scattered at frequencies corresponding to vibrational excitation and de-excitation. Higher order terms in the expansion for the polarizability [Equation (8.26)] also lead to scattered light at the frequencies $\nu \pm 2\nu_{vib}, \nu \pm 3\nu_{vib}, \ldots$, but the scattered intensity at these frequencies is much weaker than at the primary frequencies.

Equation (8.29) illustrates that the intensity of the Stokes and anti-Stokes peaks is zero unless $d\alpha/dx \neq 0$. We conclude that for vibrational modes to be Raman active, the polarizability of the molecule must change as it vibrates. This condition is satisfied for many vibrational modes and, in particular, it is satisfied for the stretching vibration of a homonuclear molecule, although $d\mu_x/dx = 0$ for these molecules, making them infrared inactive. Not all vibrational modes that are infrared active are Raman active and vice versa. This is why infrared and Raman spectroscopies provide a valuable complement to one another.

A schematic picture of the scattering event in Raman spectroscopy on an energy scale is shown in Figure 8.21. This diagram is quite different from that considered

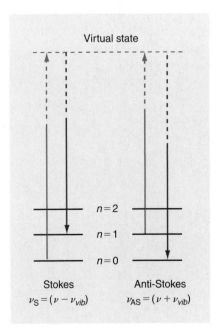

FIGURE 8.21
Schematic depiction of the Raman scattering event. The spectral peak resulting in vibrational excitation is called the Stokes peak, and the spectral peak originating from vibrational de-excitation is called the anti-Stokes peak.

earlier in depicting a transition between two states. The initial and final states are the $n = 0$ and $n = 1$ states at the bottom of the figure. To visualize the interaction of the molecule with the photon of energy $h\nu$, which is much greater than the vibrational energy spacing, we imagine the photon to be absorbed by the molecule, resulting in a much higher intermediate energy "state." This very short-lived "state" quickly decays to the final state. Whereas the initial and final states are eigenfunctions of the time-independent Schrödinger equation, the upper "state" in this energy diagram need not satisfy this condition. Therefore, it is referred to as a virtual state.

Are the intensities of the Stokes and anti-Stokes peaks equal? We know that their relative intensity is governed by the relative number of molecules in the originating states. For the Stokes line, the transition originates from the $n = 0$ state, whereas for the anti-Stokes line, the transition originates from the $n = 1$ state. Therefore, the relative intensity of the Stokes and anti-Stokes peaks can be calculated using the Boltzmann distribution:

$$\frac{I_{anti\text{-}Stokes}}{I_{Stokes}} = \frac{n_{excited}}{n_{ground}} = \frac{e^{-3h\nu/2kT}}{e^{-h\nu/2kT}} = e^{-h\nu/kT} \qquad \textbf{(8.30)}$$

For vibrations for which $\widetilde{\nu}$ is in the range of 1000 to 3000 cm^{-1}, $\frac{I_{anti\text{-}Stokes}}{I_{Stokes}}$ ranges between 8×10^{-3} and 5×10^{-7} at 300 K. This calculation shows that the intensities of the Stokes and anti-Stokes peaks will be quite different. In this discussion of the Raman effect, we have only considered vibrational transitions. However, just as for infrared absorption spectra, Raman spectra show peaks originating from both vibrational and rotational transitions.

Raman and infrared absorption spectroscopy are complementary and both can be used to study the vibration of molecules. Both techniques can be used to determine the identities of molecules in a complex mixture by comparing the observed spectral peaks with characteristic group frequencies. The most significant difference between these two spectroscopies is the light source needed to implement the technique. For infrared absorption spectroscopy, the light source is in the infrared. Because Raman spectroscopy is a scattering technique, the frequency of the light used need not match the frequency of the transition being studied. Therefore, a source in the visible part of the spectrum is generally used to study rotational and vibrational modes. This has several advantages over infrared sources. By shifting the vibrational spectrum from the infrared into the visible part of the spectrum, commonly available lasers can be used to obtain Raman spectra. Intense lasers are necessary because the probability for Raman scattering is generally on the order of 10^{-6} or less. Furthermore, shifting the frequency of the source from the infrared into the visible part of the spectrum can reduce interference with absorbing species that are not of primary interest. For instance, infrared spectra of aqueous solutions always contain strong water peaks that may mask other peaks of interest. By shifting the source frequency to the visible part of the spectrum, such interferences can be eliminated.

Another interesting application of the Raman effect is the Raman microscope or microprobe. Because Raman spectroscopy is done in the visible part of the light spectrum, it can be combined with optical microscopy to obtain spectroscopic information with a spatial resolution of better than 0.01 mm. An area in which this technique has proved particularly useful is as a nondestructive probe of the composition of gas inclusions such as CH_4, CO, H_2S, N_2, and O_2 in mineral samples. Raman microscopy has also been used in biopsy analyses to identify mineral particles in the lung tissues of silicosis victims and to analyze the composition of gallstones.

SUPPLEMENTAL

8.9 HOW DOES THE TRANSITION RATE BETWEEN STATES DEPEND ON FREQUENCY?

Now that we have some familiarity with the terms *absorption*, *spontaneous emission*, and *stimulated emission*, the frequency dependence of the interaction of molecules with light can be examined. Until now, we have only dealt with potential energy functions that are independent of time. In any spectroscopic method, transitions occur from one state to another. Transitions cannot be induced by a time-independent potential, because the eigenfunctions of the time-independent Schrödinger equation have a constant energy. We now outline how a time-dependent electromagnetic field of light incident on molecules with a discrete set of energy levels can induce transitions between these levels.

To make the mathematics more tractable, we consider a two-state system in which the states are denoted by 1 and 2. The system is described by the wave function of Equation (8.31) in which x represents the spatial coordinate and t represents time; here we assume that $E_2 > E_1$:

$$\Phi(x, t) = a_1(t)\psi_1(x)e^{-i(E_1/\hbar)t} + a_2(t)\psi_2(x)e^{-i(E_2/\hbar)t} \tag{8.31}$$

To simplify the equations that follow, we use the notation $\Psi_1 = \psi_1(x)e^{-i(E_1/\hbar)t}$ and $\Psi_2 = \psi_2(x)e^{-i(E_2/\hbar)t}$. At time $t = 0$, the system is in its ground state with $a_1 = 1$ and $a_2 = 0$. Both $\psi_1(x)$ and $\psi_2(x)$ are eigenfunctions of the total energy operator H_0, which does not depend on time, with eigenvalues E_1 and E_2, respectively.

When the light is turned on, the molecule interacts with the electric field of the light through its permanent or induced dipole moment, and the time-dependent potential energy is given by

$$\hat{H}_{\text{int}}(t) = -\boldsymbol{\mu} \cdot \mathbf{E} = -\mu_x E_0 \cos 2\pi\nu t = -\frac{\mu_x E_0}{2}\left(e^{2\pi i\nu t} + e^{-2\pi i\nu t}\right) \tag{8.32}$$

We have assumed that the electric field E_0 lies along the x axis. Writing the operator $\hat{H}_{int}(t)$ in this way is called the **dipole approximation**, because much smaller terms involving higher order multipoles are neglected. What change will the system undergo under the influence of the light? We can expect transitions from the ground state to the first excited state to occur. This means that a_1 will decrease and a_2 will increase with time.

Because this is a time-dependent system, we must solve the time-dependent Schrödinger equation $(\hat{H}_0 + \hat{H}_{int})\Phi(x, t) = i\hbar(\partial\Phi(x, t)/\partial t)$ to obtain a_1 and a_2 as functions of time. In this two-level system, Ψ_1 and Ψ_2 form a complete set, and therefore the eigenfunctions of the operator $\hat{H}_0 + \hat{H}_{int}$ must be a linear combination of Ψ_1 and Ψ_2:

$$\Phi(x, t) = a_1(t)\Psi_1 + a_2(t)\Psi_2 \tag{8.33}$$

Substituting Equation (8.33) in the time-dependent Schrödinger equation we obtain

$$a_1(t)\hat{H}_0\Psi_1 + a_2(t)\hat{H}_0\Psi_2 + a_1(t)\,\hat{H}_{int}\Psi_1 + a_2(t)\,\hat{H}_{int}\Psi_2$$

$$= i\hbar\Psi_1\frac{da_1(t)}{dt} + i\hbar\Psi_2\frac{da_2(t)}{dt} + i\hbar a_1(t)\frac{d\Psi_1(t)}{dt} + i\hbar a_2(t)\frac{d\Psi_2(t)}{dt} \tag{8.34}$$

This equation can be simplified by evaluating $i\hbar a_1(t)\dfrac{d\Psi_1(t)}{dt} + i\hbar a_2(t)\dfrac{d\Psi_2(t)}{dt}$:

$$i\hbar a_1(t)\frac{d\Psi_1(t)}{dt} + i\hbar a_2(t)\frac{d\Psi_2(t)}{dt} = i\hbar a_1(t)\psi_1(x)\frac{de^{-i(E_1/\hbar)t}}{dt} + i\hbar a_2(t)\psi_2(x)\frac{de^{-i(E_2/\hbar)t}}{dt}$$

$$= a_1(t)E_1\psi_1(x) + a_2(t)E_2\psi_2(x) \tag{8.35}$$

Because $\hat{H}_0\Psi_1 = E_1\Psi_1$ and $\hat{H}_0\Psi_2 = E_2\Psi_2$, $i\hbar a_1(t)\dfrac{d\Psi_1(t)}{dt} + i\hbar a_2(t)\dfrac{d\Psi_2(t)}{dt}$ cancels out $a_1(t)\hat{H}_0\Psi_1 + a_2(t)\hat{H}_0\Psi_2$ on the left side of Equation (8.34) and this equation takes the simpler form

$$a_1(t)\hat{H}_{int}\Psi_1 + a_2(t)\hat{H}_{int}\Psi_2 = i\hbar\Psi_1\frac{da_1(t)}{dt} + i\hbar\Psi_2\frac{da_2(t)}{dt} \tag{8.36}$$

We now multiply on the left by Ψ_2^* and integrate over the spatial coordinate, x, to obtain

$$a_1(t)\int \Psi_2^*\hat{H}_{int}\Psi_1\,dx + a_2(t)\int \Psi_2^*\hat{H}_{int}\Psi_2\,dx$$

$$= i\hbar\frac{da_1(t)}{dt}\int \Psi_2^*\Psi_1\,dx + i\hbar\frac{da_2(t)}{dt}\int \Psi_2^*\Psi_2\,dx \tag{8.37}$$

Because Ψ_1 and Ψ_2 are orthonormal, this equation can be simplified to

$$i\hbar \frac{da_2(t)}{dt} = a_1(t) \int \Psi_2^* \hat{H}_{int} \Psi_1 \, dx + a_2(t) \int \Psi_2^* \hat{H}_{int} \Psi_2 \, dx \qquad (8.38)$$

The mathematics can be simplified if only changes in the coefficients $a_1(t)$ and $a_2(t)$ for small values of t are considered. In this limit, we can replace $a_1(t)$ and $a_2(t)$ on the right side of this equation by their initial values, $a_1(t) = 1$ and $a_2(t) = 0$. Therefore, only one term remains on the right side of Equation (8.38). It turns out that imposing this limit does not affect the general conclusions drawn next. We also replace Ψ_1 and Ψ_2 by the more complete notations $\Psi_1 = \psi_1(x)e^{-i(E_1/\hbar)t}$ and $\Psi_2 = \psi_2(x)e^{-i(E_2/\hbar)t}$. After doing so, the following equations are obtained:

$$i\hbar \frac{da_2(t)}{dt} = \exp\left[\frac{i}{\hbar}(E_2 - E_1)t\right] \int \psi_2^*(x) H_{int} \psi_1(x) \, dx$$

$$i\hbar \frac{da_2(t)}{dt} = -\frac{E_0}{2} \exp\left[\frac{i}{\hbar}(E_2 - E_1)t\right] (\exp[2\pi i\nu t] + \exp[-2\pi i\nu t])$$

$$\times \int \psi_2^*(x) \mu_x \psi_1(x) \, dx$$

$$i\hbar \frac{da_2(t)}{dt} = -\frac{E_0}{2}\left(\exp\left[\frac{i}{\hbar}(E_2 - E_1 + h\nu)t\right] + \exp\left[\frac{i}{\hbar}(E_2 - E_1 - h\nu)t\right]\right)$$

$$\times \int \psi_2^*(x) \mu_x \psi_1(x) \, dx$$

$$i\hbar \frac{da_2(t)}{dt} = -\mu_x^{21} \frac{E_0}{2}\left(\exp\left[\frac{i}{\hbar}(E_2 - E_1 + h\nu)t\right] + \exp\left[\frac{i}{\hbar}(E_2 - E_1 - h\nu)t\right]\right) \quad (8.39)$$

In the last equation we have introduced the transition dipole moment, μ_x^{21}, defined by $\mu_x^{21} = \int \psi_2^*(x) \mu_x \psi_1(x) \, dx$. The transition dipole moment is important because it generates the selection rules for any spectroscopy, as discussed in Section 8.4. Next, the last equation in Equations (8.39) is integrated with respect to time, using the dummy variable t',

$$a_2(t) = \frac{i}{\hbar} \mu_x^{21} \frac{E_0}{2} \int_0^t \left(e^{\frac{i}{\hbar}(E_2 - E_1 + h\nu)t'} + e^{\frac{i}{\hbar}(E_2 - E_1 - h\nu)t'}\right) dt' \qquad (8.40)$$

to obtain $a_2(t)$:

$$a_2(t) = \mu_x^{21} \frac{E_0}{2}\left(\frac{-1 + e^{\frac{i}{\hbar}(E_2 - E_1 + h\nu)t}}{E_2 - E_1 + h\nu} + \frac{-1 + e^{-\frac{i}{\hbar}(E_2 - E_1 - h\nu)t}}{E_2 - E_1 - h\nu}\right) \qquad (8.41)$$

This expression looks complicated, but it contains a great deal of useful information that can be extracted fairly easily. Most importantly, it is seen that $a_2(t) = 0$ for all times unless $\mu_z^{21} \neq 0$. Next, we look at the terms in the parentheses. The numerator in each of the terms is an oscillating function of time. The period of oscillation approaches zero in the first and second terms as $E_1 - E_2 \to h\nu$ and $E_2 - E_1 \to h\nu$, respectively. In these limits, the denominator approaches zero. These are the conditions that lead a_2 to grow rapidly with time. The second term corresponds to absorption of a photon because we have chosen $E_2 > E_1$. The first term corresponds to stimulated emission of a photon. Stimulated emission is of importance in understanding lasers and this process is discussed in more detail in Chapter 11. However, because the current topic is absorption, we focus on the narrow range of energy around $E_2 - E_1 = h\nu$ in which only the absorption peak appears. The behavior of $a_2(t)$ at the resonance is not easy to discern, because $\lim a_2(t) = 0/0$ as $E_2 - E_1 \to h\nu$. We use L'Hôpital's rule,

$$\lim\left[\frac{f(x)}{g(x)}\right]_{x \to 0} = \lim\left[\frac{df(x)/dx}{dg(x)/dx}\right]_{x \to 0} \qquad (8.42)$$

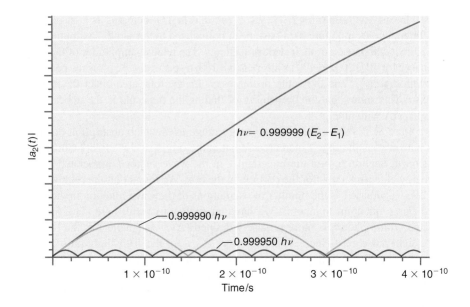

FIGURE 8.22
The change in the magnitude of $a_2(t)$ with time for three photon energies. The curves are calculated for $E_2 - E_1 = 5.00 \times 10^{-19}$ J.

which in this case takes the form

$$\lim\left[\frac{-1 + \exp\left[-\frac{i}{\hbar}(E_2 - E_1 - h\nu)t\right]}{(E_2 - E_1 - h\nu)}\right]_{E_2-E_1-h\nu\to0}$$

$$= \lim\left[\frac{\left[d\left(-1 + \exp\left[-\frac{i}{\hbar}\left(E_2 - E_1 - h\nu\right)t\right]\right)\right]/d(E_2 - E_1 - h\nu)}{d(E_2 - E_1 - h\nu)/d(E_2 - E_1 - h\nu)}\right]_{E_2-E_1-h\nu\to0}$$

$$= -\frac{it}{\hbar}\left[\exp\left(\frac{-i}{\hbar}[E_2 - E_1 - h\nu]t\right)\right]_{E_2-E_1-h\nu=0} = -\frac{it}{\hbar} \quad (8.43)$$

The important result that emerges from this calculation is that at the resonance condition $E_2 - E_1 = h\nu$, the magnitude of $a_2(t)$ increases linearly with t. How does $a_2(t)$ change with t near but not at the resonance condition? We can get this information if $a_2(t)$ is graphed versus t for the $h\nu$ values near the resonance as shown in Figure 8.22.

Figure 8.22 shows that $a_2(t)$ increases nearly linearly with time for small values of t if $E_2 - E_1$ is extremely close to $h\nu$. This means that the probability of finding the atom or molecule in the excited state increases with time. However, for photon energies that deviate even by 1 ppm from this limit, $a_2(t)$ will oscillate and remain small. The oscillations will be more frequent and smaller in amplitude the more $h\nu$ differs from $E_2 - E_1$. The probability of finding the atom or molecule in the excited state remains small if $h\nu$ differs even slightly from $E_2 - E_1$, and the atom or molecule is unable to take up energy from the electromagnetic field. *We conclude that the rate of transition from the ground to the excited state is appreciable only if $h\nu$ is essentially equal to $E_2 - E_1$.*

Our final goal is to find an expression for $a_2^*(t)a_2(t)$ that represents the probability of finding the molecule in the excited state with energy E_2 after it has been exposed to the light for the time t. We leave this part of the derivation for the end-of-chapter problems and simply state the result here:

$$a_2^*(t)a_2(t) = E_0^2[\mu_x^{21}]^2\frac{\sin^2[(E_2 - E_1 - h\nu)t/2\hbar]}{(E_2 - E_1 - h\nu)^2} \quad (8.44)$$

Figure 8.23 shows a graph of $a_2^*(t)a_2(t)$ against $(E_2 - E_1 - h\nu)/(E_2 - E_1)$ for 40, 120, and 400 ps. As expected, the probability of finding the molecule in the excited state is sharply peaked if the photon energy satisfies the condition $E_2 - E_1 = h\nu$. Because $a_2(t)$

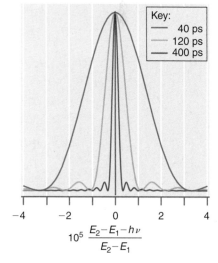

FIGURE 8.23
Graph of $a_2^*(t)a_2(t)$ against $(E_2 - E_1 - h\nu)/(E_2 - E_1)$. In the calculations $E_2 - E_1 = 5.00 \times 10^{-19}$ J. The range of $E_2 - E_1 - h\nu$ shown in the graph is only 80 parts per million of $E_2 - E_1$. The broadest and narrowest resonances are observed for $t = 40$ and 400 ps, respectively. The intermediate resonance is for $t = 120$ ps. All curves have been normalized to the same amplitude so that the peak widths can be directly compared.

increases linearly with time for $E_2 - E_1 = h\nu$, $a_2^*(t)a_2(t)$ increases as t^2 at resonance. The different curves in Figure 8.23 have been normalized to the same maximum value to allow a direct comparison of their widths in energy. The relative amplitudes of $a_2^*(t)a_2(t)$ at resonance for 40, 120, and 400 ps are 1, 9, and 100, respectively. Because the peak height varies with time as t^2 and the width decreases as $1/t$, the total area under the resonance varies as t. This shows that the probability of finding the molecule in the upper state increases linearly with time.

As Figure 8.23 shows, the photon energy range over which absorption occurs becomes narrower as the time t increases. What is the origin of this effect? Consider the discussion in Section 6.4 on the uncertainty in the wave vector k associated with the total energy eigenfunctions for the particle in the box. We found that the relative uncertainty $\Delta k/k$ decreased as the number of oscillations of the wave function in the box increased (large quantum numbers). We have an analogous result here. The frequency associated with a photon is more precisely defined ($\Delta\nu/\nu$ is smaller) as the time over which a light pulse exists increases. Therefore, $\Delta\nu/\nu$ for the 40-ps light pulse is larger than that for the 400-ps light pulse. For this reason, the range of energy over which transitions occur is larger for the 40-ps pulse than for the 400-ps pulse.

The probability density $a_2^*(t)a_2(t)$ is closely related to that observed in an absorption spectrum. How is the broadening that was just discussed related to the linewidth observed in an experimentally determined spectrum? To answer this question, we must distinguish between an intrinsic and a measured linewidth. By intrinsic linewidth, we mean the linewidth that would be measured if the spectrometer were perfect. However, a real spectrometer is defined by an instrument function, which is the output of the spectrometer for a very narrow spectral peak. The observed spectrum results from the convolution of the instrument function with the intrinsic linewidth.

Based on theoretical calculations, the intrinsic linewidth for vibrational spectra is less than $\sim 10^{-3}$ cm^{-1}. This is very small compared with the resolution of conventional infrared spectrometers, which is typically no better than 0.1 cm^{-1}. Therefore, the width of peaks in a spectrum is generally determined by the instrumental function as shown in the top panel of Figure 8.24, and gives no information about the intrinsic linewidth. However, peaks that are broader than the instrument function are obtained if a sample contains many different local environments for the entity generating the peak. For example, the O—H stretching region in an infrared spectrum in liquid water is very broad. This is the case because of the many different local geometries that arise from hydrogen bonding between H_2O molecules, and each of them gives rise to a slightly different O—H stretching frequency. This effect is referred to as **inhomogeneous broadening** and is illustrated in the bottom panel of Figure 8.24.

FIGURE 8.24

An experimental spectrum, shown on the right, arises from the convolution (indicated by the symbol *) of the instrument function and the intrinsic linewidth of the transition. For a homogeneous sample (top panel) the width of the spectrum is generally determined by the instrument function. For an inhomogeneous sample, the intrinsic linewidth is the sum of the linewidths of the many different local environments. For inhomogeneous samples, the width of the spectrum can be determined by the intrinsic linewidth, rather than the instrument function.

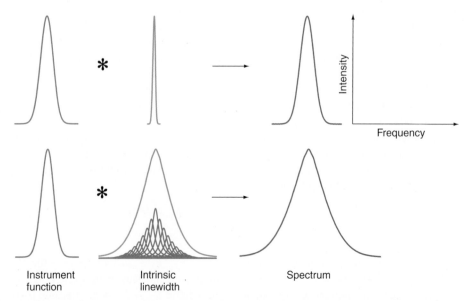

Vocabulary

absorption

anharmonic potential

anti–Stokes frequency

Beer–Lambert law

bond energy

coherent source

dipole approximation

dynamic dipole moment

FTIR spectroscopy

group frequencies

harmonic potential

incoherent photon source

inhomogeneous broadening

interferogram

Michelson interferometer

molar absorption coefficient

moment of inertia

monochromator

Morse potential

multiplex advantage

overtone

P branch

permanent dipole moment

polarizability

R branch

Raman spectroscopy

Rayleigh frequency

rotational constant

selection rule

spectroscopy

spontaneous emission

stimulated emission

Stokes frequency

transition dipole moment

Questions on Concepts

Q8.1 Why would you observe a pure rotational spectrum in the microwave region and a rotational-vibrational spectrum rather than a pure vibrational spectrum in the infrared region?

Q8.2 Solids generally expand as the temperature increases. Such an expansion results from an increase in the bond length between adjacent atoms as the vibrational amplitude increases. Will a harmonic potential lead to thermal expansion? Will a Morse potential lead to thermal expansion?

Q8.3 How can you observe vibrational transitions in Raman spectroscopy using visible light lasers where the photon energy is much larger than the vibrational energy spacing?

Q8.4 A molecule in an excited state can decay to the ground state either by stimulated emission or spontaneous emission. Use the Einstein coefficients to predict how the relative probability of these processes changes as the frequency of the transition doubles.

Q8.5 In Figure 8.15, n_J/n_0 increases initially with J for all three temperatures for CO, but only for the two highest temperatures for HD. Explain this difference.

Q8.6 What is the difference between the transition dipole moment and the dynamic dipole moment?

Q8.7 Nitrogen and oxygen do not absorb infrared radiation and are therefore not greenhouse gases. Why is this the case?

Q8.8 Does the initial excitation in Raman spectroscopy take place to a stationary state of the system? Explain your answer.

Q8.9 What feature of the Morse potential makes it suitable for modeling dissociation of a diatomic molecule?

Q8.10 If the rotational levels of a diatomic molecule were equally spaced and the selection rule remained unchanged, how would the appearance of the rotational-vibrational spectrum in Figure 8.16 change?

Q8.11 If a spectral peak is broadened, can you always conclude that the excited state has a short lifetime?

Q8.12 What is the difference between a permanent and a dynamic dipole moment?

Q8.13 What is the explanation for the absence of a peak in the rotational-vibrational spectrum near 3000 cm^{-1} in Figure 8.14?

Q8.14 What is the advantage in acquiring a vibrational spectrum using a FTIR spectrometer over a spectrometer in which the absorption is measured separately at each wavelength?

Q8.15 The number of molecules in a given energy level is proportional to $e^{-\Delta E/kT}$ where ΔE is the difference in energy between the level in question and the ground state. How is it possible that a higher lying rotational energy level can have a higher population than the ground state?

Problems

Problem numbers in **red** indicate that the solution to the problem is given in the *Student's Solutions Manual*.

P8.1 The $^1H^{81}Br$ molecule can be described by a Morse potential with $D_e = 7.70 \times 10^{-19}$ J. The force constant k for this molecule is 412 N m^{-1} and $\nu = 7.94 \times 10^{13}$ s^{-1}.

a. Calculate the lowest four energy levels for a Morse potential using the formula

$$E_n = h\nu\left(n + \frac{1}{2}\right) - \frac{(h\nu)^2}{4D_e}\left(n + \frac{1}{2}\right)^2$$

b. Calculate the fundamental frequency ν_0 corresponding to the transition $n = 0 \rightarrow n = 1$ and the frequencies of the first three overtone vibrations. How large would the relative error be if you assume that the first three overtone frequencies are $2\nu_0$, $3\nu_0$, and $4\nu_0$?

P8.2 The infrared spectrum of $^7Li^{35}Cl$ has an intense line at 643 cm^{-1}. Calculate the force constant and period of vibration of this molecule.

P8.3 Purification of water for drinking using UV light is a viable way to provide potable water in many areas of the world. Experimentally, the decrease in UV light of wavelength 250 nm follows the empirical relation $I/I_0 = e^{-\varepsilon' l}$ where l is the distance that the light passed through the water and ε' is an effective absorption coefficient. $\varepsilon' = 0.070$ cm^{-1} for pure water and 0.30 cm^{-1} for water exiting a wastewater treatment plant. What distance corresponds to a decrease in I of 10.% from its incident value for a) pure water and b) waste water?

P8.4 A simulated infrared absorption spectrum of a gas-phase organic compound is shown in the following figure. Use the characteristic group frequencies listed in Section 8.6 to decide whether this compound is more likely to be Cl_2CO, $(CH_3)_2CO$, CH_3OH, CH_3COOH, CH_3CN, CCl_4, or C_3H_8. Explain your reasoning.

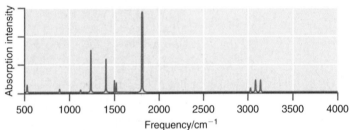

P8.5 The molecules $^{16}O^{12}C^{32}S$ and $^{16}O^{12}C^{34}S$ have values for $h/8\pi^2 I$ of 6081.490×10^6 s^{-1} and 5932.816×10^6 s^{-1}, respectively. Calculate the C—O and C—S bond distances.

P8.6 A simulated infrared absorption spectrum of a gas-phase organic compound is shown in the following figure. Use the characteristic group frequencies listed in Section 8.6 to decide whether this compound is more likely to be Cl_2CO, $(CH_3)_2CO$, CH_3OH, CH_3COOH, CH_3CN, CCl_4, or C_3H_8. Explain your reasoning.

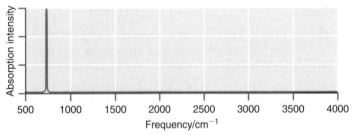

P8.7 The rotational constant for $^2D^{19}F$ determined from microwave spectroscopy is 11.007 cm^{-1}. The atomic masses of ^{19}F and ^2D are 18.9984032 and 2.0141018 amu, respectively. Calculate the bond length in $^2D^{19}F$ to the maximum number of significant figures consistent with this information.

P8.8 An infrared absorption spectrum of an organic compound is shown in the following figure. Use the characteristic group frequencies listed in Section 8.6 to decide whether this compound is more likely to be ethyl amine, pentanol, or acetone.

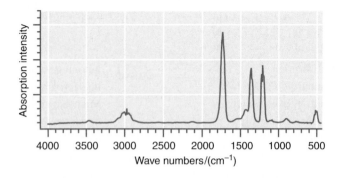

P8.9 Calculate the zero point energies for H^{35}Cl and D^{35}Cl. Compare the difference in the zero point energies to kT at 300. K.

P8.10 Write an expression for the moment of inertia of the acetylene molecule in terms of the bond distances. Does this molecule have a pure rotational spectrum?

P8.11 Show that the selection rule for the two-dimensional rotor in the dipole approximation is $\Delta m_l = \pm 1$. Use $A_{+\phi}e^{im_1\phi}$ and $A'_{+\phi}e^{im_2\phi}$ for the initial and final states of the rotor and $\mu\cos\phi$ as the dipole moment element.

P8.12 Following Example Problem 8.5, show that the $J = 1 \rightarrow J = 2$ rotational transition is allowed.

P8.13 Selection rules in the dipole approximation are determined by the integral $\mu_x^{mn} = \int \psi_m^*(\tau)\mu_x(\tau)\psi_n(\tau)\,d\tau$. If this integral is nonzero, the transition will be observed in an absorption spectrum. If the integral is zero, the transition is "forbidden" in the dipole approximation. It actually occurs with low probability because the dipole approximation is not exact. Consider the particle in the one-dimensional box and set $\mu_x = -ex$.

a. Calculate μ_x^{12} and μ_x^{13} in the dipole approximation. Can you see a pattern and discern a selection rule? You may need to evaluate a few more integrals of the type μ_x^{1m}. The standard integral

$$\int x\sin\frac{\pi x}{a}\sin\frac{n\pi x}{a}\,dx$$

$$= \frac{1}{2}\left(\frac{a^2\cos\frac{(n-1)\pi x}{a}}{(n-1)^2\pi^2} + \frac{ax\sin\frac{(n-1)\pi x}{a}}{(n-1)\pi}\right)$$

$$- \frac{1}{2}\left(\frac{a^2\cos\frac{(n+1)\pi x}{a}}{(n+1)^2\pi^2} + \frac{ax\sin\frac{(n+1)\pi x}{a}}{(n+1)\pi}\right)$$

is useful for solving this problem.

b. Determine the ratio μ_x^{12}/μ_x^{14}. On the basis of your result, would you modify the selection rule that you determined in part a?

P8.14 The bond length of $^{23}Na^{19}F$ is 192.6 pm. Calculate the value of B and the spacing between lines in the pure rotational spectrum of this molecule in units of s^{-1}.

P8.15 Calculating the motion of individual atoms in the vibrational modes of molecules (called normal modes) is an advanced topic. Given the normal modes shown in the following figure, decide which of the normal modes of CO_2 and H_2O have a nonzero dynamical dipole moment and are therefore infrared active. The motion of the atoms in the second of the two doubly degenerate bend modes for CO_2 is identical to the first, but is perpendicular to the plane of the page.

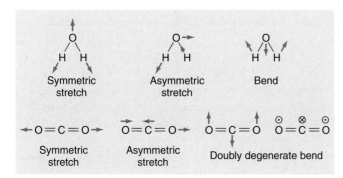

P8.16 The force constants for H_2 and Br_2 are 575 and 246 N m^{-1}, respectively. Calculate the ratio of the vibrational state populations n_1/n_0 and n_2/n_0 at $T = 300.$ and at 1000. K.

P8.17 The rigid rotor model can be improved by recognizing that in a realistic anharmonic potential, the bond length increases with the vibrational quantum number n. Therefore, the rotational constant depends on n, and it can be shown that $B_n = B - (n + 1/2)\alpha$, where B is the rigid rotor value. The constant α can be obtained from experimental spectra. For $^1H^{127}I$, $B = 6.551$ cm^{-1} and $\alpha = 0.183$ cm^{-1}. Using this more accurate formula for B_n, calculate the bond length for HI in the ground state and for $n = 3$.

P8.18 Greenhouse gases generated from human activity absorb infrared radiation from the Earth and keep it from being dispersed outside our atmosphere. This is a major cause of global warming. Compare the path length required to absorb 99% of the Earth's radiation near a wavelength of 7 μm for CH_3CCl_3 $[\varepsilon(\lambda) = 1.8$ (cm atm)$^{-1}]$ and the chlorofluorocarbon CFC-14 $[\varepsilon(\lambda) = 4.1 \times 10^3$ (cm atm)$^{-1}]$ assuming that each of these gases has a partial pressure of 2.0×10^{-6} bar.

P8.19 Show that the Morse potential approaches the harmonic potential for small values of the vibrational amplitude. (*Hint*: Expand the Morse potential in a Taylor-Maclaurin series.)

P8.20 The rotational constant for $^{127}I^{79}Br$ determined from microwave spectroscopy is 0.1141619 cm^{-1}. The atomic masses of ^{127}I and ^{79}Br are 126.904473 and 78.918336 amu, respectively. Calculate the bond length in $^{127}I^{79}Br$ to the maximum number of significant figures consistent with this information.

P8.21 A simulated infrared absorption spectrum of a gas-phase organic compound is shown in the following figure. Use the characteristic group frequencies listed in Section 8.6 to decide whether this compound is more likely to be Cl_2CO, $(CH_3)_2CO$, CH_3OH, CH_3COOH, CH_3CN, CCl_4, or C_3H_8. Explain your reasoning.

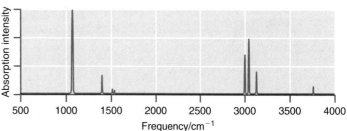

P8.22 Overtone transitions in vibrational absorption spectra for which $\Delta n = +2, +3, \ldots$ are forbidden for the harmonic potential $V = (1/2)kx^2$ because $\mu_x^{mn} = 0$ for $|m - n| \neq 1$ as shown in Section 8.4. However, overtone transitions are allowed for the more realistic anharmonic potential. In this problem, you will explore how the selection rule is modified by including anharmonic terms in the potential. We do so in an indirect manner by including additional terms in the expansion of the dipole moment $\mu_x(x) = \mu_{0x} + x(d\mu_x/dx)_{r_e} + \ldots$, but assuming that the harmonic oscillator total energy eigenfunctions are still valid. This approximation is valid if the anharmonic correction to the harmonic potential is small. You will show that including the next term in the expansion of the dipole moment, which is proportional to x^2, makes the transitions $\Delta n = \pm 2$ allowed.

a. Show that Equation (8.8) becomes

$$\mu_x^{m0} = A_m A_0 \mu_{0x} \int_{-\infty}^{\infty} H_m(\alpha^{1/2}x) H_0(\alpha^{1/2}x)e^{-\alpha x^2} dx$$

$$+ A_m A_0 \left(\frac{d\mu_x}{dx}\right)_{x=0} \int_{-\infty}^{\infty} H_m(\alpha^{1/2}x) x H_0(\alpha^{1/2}x)e^{-\alpha x^2} dx$$

$$+ \frac{A_m A_0}{2!} \left(\frac{d^2\mu_x}{dx^2}\right)_{x=0} \int_{-\infty}^{\infty} H_m(\alpha^{1/2}x) x^2 H_0(\alpha^{1/2}x)e^{-\alpha x^2} dx$$

b. Evaluate the effect of adding the additional term to μ_x^{mn}. You will need the recursion relationship $\alpha^{1/2}x H_n(\alpha^{1/2}x) = n H_{n-1}(\alpha^{1/2}x) + \frac{1}{2}H_{n+1}(\alpha^{1/2}x)$.

c. Show that both the transitions $n = 0 \rightarrow n = 1$ and $n = 0 \rightarrow n = 2$ are allowed in this case.

P8.23 The fundamental vibrational frequencies for $^1H^{19}F$ and $^2D^{19}F$ are 4138.52 and 2998.25 cm^{-1}, respectively, and D_e for both molecules is 5.86 eV. What is the difference in the bond energy of the two molecules?

P8.24 A simulated infrared absorption spectrum of a gas-phase organic compound is shown in the following figure. Use the characteristic group frequencies listed in Section 8.6 to decide whether this compound is more likely to be Cl_2CO,

$(CH_3)_2CO$, CH_3OH, CH_3COOH, CH_3CN, CCl_4, or C_3H_8. Explain your reasoning.

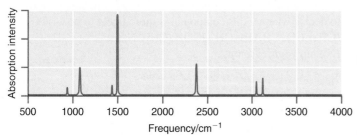

P8.25 Isotopic substitution is used to identify characteristic groups in an unknown compound using vibrational spectroscopy. Consider the $C{=}C$ bond in ethene ($^{12}C_2{}^1H_4$). By what factor would the frequency change if deuterium were substituted for all the hydrogen atoms? Treat the H and D atoms as being rigidly attached to the carbon.

P8.26 A simulated infrared absorption spectrum of a gas-phase organic compound is shown in the following figure. Use the characteristic group frequencies listed in Section 8.6 to decide whether this compound is more likely to be Cl_2CO, $(CH_3)_2CO$, CH_3OH, CH_3COOH, CH_3CN, CCl_4, or C_3H_8. Explain your reasoning.

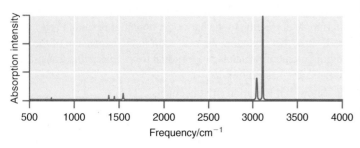

P8.27 Fill in the missing step in the derivation that led to the calculation of the spectral line shape in Figure 8.22. Starting from

$$a_2(t) = \mu_x^{21}\frac{E_0}{2}\left(\frac{1 - e^{\frac{i}{\hbar}(E_2-E_1+h\nu)t}}{E_2 - E_1 + h\nu} + \frac{1 - e^{-\frac{i}{\hbar}(E_2-E_1-h\nu)t}}{E_2 - E_1 - h\nu}\right)$$

and neglecting the first term in the parentheses, show that

$$a_2^*(t)a_2(t) = E_0^2[\mu_x^{21}]^2\frac{\sin^2[(E_2 - E_1 - h\nu)t/2\hbar]}{(E_2 - E_1 - h\nu)^2}$$

P8.28 The force constant to $^{79}Br^{35}Cl$ is 282 N m^{-1}. Calculate the vibrational frequency and zero point energy of this molecule.

P8.29 Because the intensity of a transition to first order is proportional to the population of the originating state, the J value for which the maximum intensity is observed in a rotational-vibrational spectrum is not generally $J = 0$. Treat J in the equation

$$\frac{n_J}{n_0} = \frac{g_J}{g_0}e^{-(E_J-E_0)/kT} = (2J + 1)e^{-\hbar^2J(J+1)/(2IkT)}$$

as a continuous variable.

a. Show that

$$\frac{d\left(\dfrac{n_J}{n_0}\right)}{dJ} = 2e^{-\hbar^2J(J+1)/(2IkT)} - \frac{(2J + 1)^2\hbar^2}{2IkT}e^{-\hbar^2J(J+1)/(2IkT)}$$

b. Show that setting $d(n_J/n_0)/dJ = 0$ gives the equation

$$2 - \frac{(2J_{max} + 1)^2\hbar^2}{2IkT} = 0$$

c. Show that the solution of this quadratic equation is

$$J_{max} = \frac{1}{2}\left[\sqrt{\frac{4IkT}{\hbar^2}} - 1\right]$$

In this problem, we assume that the intensity of the individual peaks is solely determined by the population in the originating state and that it does not depend on the initial and final J values.

P8.30 A strong absorption band in the infrared region of the electromagnetic spectrum is observed at $\tilde{\nu} = 2170$ cm^{-1} for $^{12}C^{16}O$. Assuming that the harmonic potential applies, calculate the fundamental frequency ν in units of inverse seconds, the vibrational period in seconds, and the zero point energy for the molecule in joules and electron-volts.

P8.31 The spacing between lines in the pure rotational spectrum of $^7Li^1H$ is 4.505×10^{11} s^{-1}. Calculate the bond length of this molecule.

P8.32 In this problem, you will derive the equations used to explain the Michelson interferometer for incident light of a single frequency.

a. Show that the expression

$$A(t) = \frac{A_0}{\sqrt{2}}(1 + e^{i\delta(t)})\exp i(ky_D - \omega t)$$

represents the sum of two waves of the form $A_0/\sqrt{2}\exp i(kx - \omega t)$, one of which is phase shifted by the amount $\delta(t)$ evaluated at the position y_D.

b. Show using the definition $I(t) = A(t)A^*(t)$, that $I(t) = I_0/2(1 + \cos\delta(t))$.

c. Expressing $\delta(t)$ in terms of $\Delta d(t)$, show that

$$I(t) = \frac{I_0}{2}\left(1 + \cos\frac{2\pi\,\Delta d(t)}{\lambda}\right)$$

d. Expressing $\Delta d(t)$ in terms of the mirror velocity v, show that

$$I(t) = \frac{I_0}{2}\left(1 + \cos\frac{2v}{c}\omega t\right)$$

P8.33 Calculate the moment of inertia, the magnitude of the rotational angular momentum, and the energy in the $J = 1$ rotational state for $^1H_2$ in which the bond length of $^1H_2$ is 74.6 pm. The atomic mass of ^1H is 1.007825 amu.

P8.34 A simulated infrared absorption spectrum of a gas-phase organic compound is shown in the following figure. Use the characteristic group frequencies listed in Section 8.6 to decide whether this compound is more likely to be Cl_2CO, $(CH_3)_2CO$, CH_3OH, CH_3COOH, CH_3CN, CCl_4, or C_3H_8. Explain your reasoning.

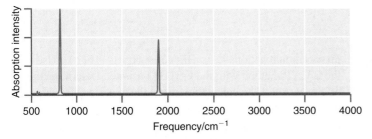

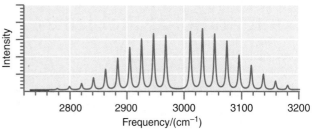

P8.35 A measurement of the vibrational energy levels of $^{12}C^{16}O$ gives the relationship

$$\tilde{\nu}(n) = 2170.21\left(n + \frac{1}{2}\right)cm^{-1} - 13.461\left(n + \frac{1}{2}\right)^2 cm^{-1}$$

where n is the vibrational quantum number. The fundamental vibrational frequency is $\tilde{\nu}_0 = 2170.21\ cm^{-1}$. From these data, calculate the depth D_e of the Morse potential for $^{12}C^{16}O$. Calculate the bond energy of the molecule.

P8.36 Using the formula for the energy levels for the Morse potential,

$$E_n = h\nu\left(n + \frac{1}{2}\right) - \frac{(h\nu)^2}{4D_e}\left(n + \frac{1}{2}\right)^2$$

show that the energy spacing between adjacent levels is given by

$$E_{n+1} - E_n = h\nu - \frac{(h\nu)^2}{2D_e}(1 - n)$$

For $^1H^{35}Cl$, $D_e = 7.41 \times 10^{-19}\ J$ and $\nu = 8.97 \times 10^{13}\ s^{-1}$. As n increases, the energy difference between adjacent levels decreases. Calculate the smallest value of n for which $E_{n+1} - E_n < 0.5(E_1 - E_0)$.

P8.37 In Problem P8.29 you obtained the result

$$J_{max} = 1/2[\sqrt{4IkT/\hbar^2} - 1]$$

Using this result, estimate T for the simulated $^1H^{35}Cl$ rotational spectra shown in the following figure. Give realistic estimates of the precision with which you can determine T from the spectra. In generating the simulation, we assumed that the intensity of the individual peaks is solely determined by the population in the originating state and that it does not depend on the initial and final J values.

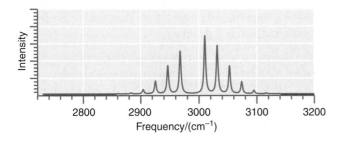

P8.38 50.% of the 190 nm wavelength light incident on a 10.0-mm-thick piece of fused silica quartz glass passes through the glass and the remainder is absorbed. What percentage of the light will pass through a 20.0-mm-thick piece of the same glass?

P8.39 Calculate the moment of inertia, the magnitude of the angular momentum, and the energy in the $J = 1$ rotational state for $^{16}O_2$ in which the bond length is 120.8 pm. The atomic mass of ^{16}O is 15.99491 amu.

P8.40 A simulated infrared absorption spectrum of a gas-phase organic compound is shown in the following figure. Use the characteristic group frequencies listed in Section 8.6 to decide whether this compound is more likely to be Cl_2CO, $(CH_3)_2CO$, CH_3OH, CH_3COOH, CH_3CN, CCl_4, or C_3H_8. Explain your reasoning.

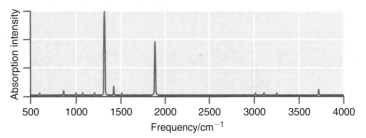

P8.41 An infrared absorption spectrum of an organic compound is shown in the following figure. Use the characteristic group frequencies listed in Section 8.6 to decide whether this compound is more likely to be hexene, hexane, or hexanol.

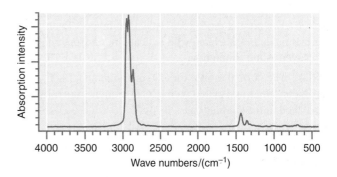

Computational Problems

More detailed instructions on carrying out these calculations using Spartan Physical Chemistry are found on the book website at *www.chemplace.com*.

C8.1   Build structures for the gas-phase (a) hydrogen fluoride ($^1H^{19}F$), (b) hydrogen chloride ($^1H^{35}Cl$), (c) carbon monoxide ($^{12}C^{16}O$, and (d) sodium chloride ($^{23}Na^{35}Cl$) molecules. (For Spartan, these are the default isotopic masses.) Calculate the equilibrium geometry and the IR spectrum using the B3LYP method with the 6-311+G** basis set.

a. Compare your result for the vibrational frequency with the experimental value listed in Table 8.3. What is the relative error in the calculation?

b. Calculate the force constant from the vibrational frequency and reduced mass. Determine the relative error using the experimental value in Table 8.3.

c. Calculate the values for the rotational constant, B, using the calculated bond length. Determine the relative error using the experimental value in Table 8.3.

C8.2   Calculate the bond energy in gaseous (a) hydrogen fluoride ($^1H^{19}F$), (b) hydrogen chloride ($^1H^{35}Cl$), (c) carbon monoxide ($^{12}C^{16}O$), and (d) sodium chloride ($^{23}Na^{35}Cl$) molecules by comparing the total energies of the species in the dissociation reactions [e.g., $HF(g) \rightarrow H(g) + F(g)$]. Use the B3LYP method with the 6-31G* basis set. Determine the relative error of the calculation using the experimental value in Table 8.3.

C8.3   Build structures for the gas-phase (a) NF_3, (b) PCl, and (c) SO_3 molecules. Calculate the equilibrium geometry and the IR spectrum using the B3LYP method with the 6-31G* basis set. Animate the vibrational normal modes and classify them as symmetrical stretch, symmetrical deformation, degenerate stretch, and degenerate deformation.

C8.4   Build structures for the gas-phase (a) F_2CO, (b) Cl_2CO, and (c) O_2NF molecules of the structural form X_2YZ. Calculate the equilibrium geometry and the IR spectrum using the B3LYP method with the 6-311+G** basis set. Animate the vibrational normal modes and classify them as Y-Z stretch, YX_2 scissors, antisymmetric X-Y stretch, YX_2 rock, and Y-X_2 wag.

C8.5   Build structures for the bent gas-phase (a) HOF, (b) ClOO, and (c) HSO molecules of the structural form XYZ. Calculate the equilibrium geometry and the IR spectrum using the B3LYP method with the 6-31G* basis set. Animate the vibrational normal modes and classify them as Y-Z stretch, X-Y stretch, and X-Y-Z bend.

Web-Based Simulations, Animations, and Problems

W8.1   A pair of emission spectra, one from an unknown (hypothetical) atom and one resulting from the electron energy levels entered using sliders, is displayed. The student adjusts the displayed energy levels in order to replicate the atomic spectrum and, hence, determine the actual electron energy levels in the atom.

W8.2   The number of allowed energy levels in a Morse potential is determined for variable values of the vibrational frequency and the well depth.

W8.3   The normal modes for H_2O are animated. Each normal mode is associated with a local motion from a list displayed in the simulation.

W8.4   The normal modes for CO_2 are animated. Each normal mode is associated with a local motion from a list displayed in the simulation.

W8.5   The normal modes for NH_3 are animated. Each normal mode is associated with a local motion from a list displayed in the simulation.

W8.6   The normal modes for formaldehyde are animated. Each normal mode is associated with a local motion from a list displayed in the simulation.

W8.7   Simulated rotational (microwave) spectra are generated for one or more of the diatomic molecules $^{12}C^{16}O$, $^1H^{19}F$, $^1H^{35}Cl$, $^1H^{79}Br$, and $^1H^{127}I$. Using a slider, the temperature is varied. The J value corresponding to the maximum intensity peak is determined and compared with the prediction from the formula

$$J_{max} = \frac{1}{2}\left[\sqrt{\frac{4I\,kT}{\hbar^2}} - 1\right]$$

The number of peaks that have an intensity greater than half of that for the largest peak is determined at different temperatures. The frequencies of the peaks are then used to generate the rotational constants B and α_e.

W8.8   Simulated rotational-vibrational (infrared) spectra are generated for one or more diatomic molecules including $^{12}C^{16}O$, $^1H^{19}F$, $^1H^{35}Cl$, $^1H^{79}Br$, or $^1H^{127}I$ for predetermined temperatures. The frequencies of the peaks are then used to generate the rotational constants B and α_e, and the force constant k.

9

The Hydrogen Atom

Classical physics is unable to explain the stability of atoms. In a classical picture, the electrons radiate energy because they undergo accelerated motion as they orbit the positively charged nucleus. As a consequence, the electrons fall into the nucleus. This is clearly incorrect! In this chapter, we solve the Schrödinger equation for the motion of an electron in a spherically symmetric Coulomb potential. To emphasize the similarities and differences between quantum mechanical and classical models, a comparison is made between the quantum mechanical picture of the hydrogen atom and the popularly depicted shell picture of the atom.

9.1 FORMULATING THE SCHRÖDINGER EQUATION

After having applied quantum mechanics to a number of simple problems, we turn to one of the triumphs of quantum mechanics: the understanding of atomic structure and spectroscopy. As discussed in Chapter 10, for atoms with more than one electron, the Schrödinger equation must be solved numerically. However, for the hydrogen atom, the Schrödinger equation can be solved analytically, and many of the results we obtain from that solution can be generalized to many-electron atoms.

To set the stage historically, experiments by Rutherford had established that the positive charge associated with an atom was localized at the center of the atom and that the electrons were spread out over a large volume (relative to nuclear dimensions) centered at the nucleus. The **shell model** in which the electrons are confined in spherical shells centered at the nucleus had a major flaw when viewed from the vantage point of classical physics. An electron orbiting around the nucleus undergoes accelerated motion and radiates energy. Therefore, it will eventually fall into the nucleus. Atoms are not stable according to classical mechanics. The challenge for quantum mechanics was to provide a framework within which the stability of atoms could be understood.

We picture the hydrogen atom as made up of an electron moving about a proton located at the origin of the coordinate system. The two particles attract one another and the interaction potential is given by a simple **Coulomb potential**:

$$V(\mathbf{r}) = -\frac{e^2}{4\pi\varepsilon_0 |\mathbf{r}|} = -\frac{e^2}{4\pi\varepsilon_0 r} \tag{9.1}$$

In this equation, e is the electron charge, m_e is the electron mass, and ε_0 is the permittivity of free space. In the text that follows, we abbreviate the magnitude of the vector $\mathbf{r}$ as r. Because the potential is spherically symmetrical, we choose spherical polar coordinates (r, θ, ϕ) to formulate the Schrödinger equation for this problem. In doing so, it takes on the formidable form

$$-\frac{\hbar^2}{2m_e}\left[\frac{1}{r^2}\frac{\partial}{\partial r}\left(r^2\frac{\partial\psi(r,\theta,\phi)}{\partial r}\right) + \frac{1}{r^2\sin\theta}\frac{\partial}{\partial\theta}\left(\sin\theta\frac{\partial\psi(r,\theta,\phi)}{\partial\theta}\right) + \frac{1}{r^2\sin^2\theta}\frac{\partial^2\psi(r,\theta,\phi)}{\partial\phi^2}\right]$$

$$-\frac{e^2}{4\pi\varepsilon_0 r}\psi(r,\theta,\phi) = E\psi(r,\theta,\phi) \tag{9.2}$$

9.2 SOLVING THE SCHRÖDINGER EQUATION FOR THE HYDROGEN ATOM

Because $V(r)$ depends only on r and not on the angles θ and ϕ, we can achieve a **separation of variables**, as discussed in Section 7.4, and write the wave function as a product of three functions, each of which depends on only one of the variables:

$$\psi(r,\theta,\phi) = R(r)\Theta(\theta)\Phi(\phi) \tag{9.3}$$

This simplifies the solution of the partial differential equation greatly. We also recognize that, apart from constants, the angular part of Equation (9.2) is the operator $\hat{l}^2$. Therefore, the angular part of $\psi(r,\theta,\phi)$ is the product $\Theta(\theta)\Phi(\phi)$ that we encountered in solving the Schrödinger equation for the rigid rotor, namely, the spherical harmonic functions. Therefore, the only part of $\psi(r,\theta,\phi)$ that remains unknown is the radial function $R(r)$.

Equation (9.2) can be reduced to a radial equation in the following way. Substituting the product function $\psi(r,\theta,\phi) = R(r)\Theta(\theta)\Phi(\phi)$ into Equation (9.2), and taking out those parts not affected by the partial derivative in front of each term, we obtain

$$-\frac{\hbar^2}{2m_e r^2}\Theta(\theta)\Phi(\phi)\frac{d}{dr}\left[r^2\frac{dR(r)}{dr}\right] + \frac{1}{2m_e r^2}R(r)\hat{l}^2\Theta(\theta)\Phi(\phi)$$

$$-\Theta(\theta)\Phi(\phi)\left[\frac{e^2}{4\pi\varepsilon_0 r}\right]R(r) = ER(r)\Theta(\theta)\Phi(\phi) \tag{9.4}$$

where the operator $\hat{l}^2$ has been defined in Section 7.4. But we know that $\hat{l}^2\Theta(\theta)\Phi(\phi) = \hbar^2 l(l+1)\Theta(\theta)\Phi(\phi)$. Putting this result into Equation (9.4), and canceling the product $\Theta(\theta)\Phi(\phi)$ that appears in each term, a differential equation is obtained for $R(r)$:

$$-\frac{\hbar^2}{2m_e r^2}\frac{d}{dr}\left[r^2\frac{dR(r)}{dr}\right] + \left[\frac{\hbar^2 l(l+1)}{2m_e r^2} - \frac{e^2}{4\pi\varepsilon_0 r}\right]R(r) = ER(r) \tag{9.5}$$

Before continuing, we summarize the preceding discussion. The Schrödinger equation was formulated for the hydrogen atom. It differs from the rigid rotor problem, where r has a fixed value, in that the potential is not zero; instead, it depends inversely on r. Because the potential is not dependent on the angular coordinates, the solutions to the Schrödinger equation for θ and ϕ are the same as those obtained for the rigid rotor. In the rigid rotor, r was fixed at a constant value that is appropriate for a diatomic molecule with a stiff bond. For the electron–proton distance in the hydrogen atom, this is clearly not appropriate, and the wave function will depend on r. We have been able to separate out the dependence of the wave function on the radial coordinate, r, from that on the angles θ and ϕ. We now take a closer look at the eigenvalues and eigenfunctions for Equation (9.5).

Note that the second term on the left-hand side of Equation (9.5) can be viewed as an **effective potential**, $V_{eff}(r)$. It is made up of the **centripetal potential**, which varies as $+1/r^2$, and the Coulomb potential, which varies as $-1/r$:

$$V_{eff}(r) = \frac{\hbar^2 l(l+1)}{2m_e r^2} - \frac{e^2}{4\pi\varepsilon_0 r} \qquad (9.6)$$

Each of the terms that contribute to $V_{eff}(r)$ and their sums are graphed as a function of distance in Figure 9.1.

Because the first term is repulsive and varies more rapidly with r than the Coulomb potential, it dominates at small distances if $l \neq 0$. Both terms approach zero for large values of r. The resultant potential is repulsive at short distances for $l > 0$ and is more repulsive the greater the value of l. The net result of this repulsive centripetal potential is to force the electrons in orbitals with $l > 0$ (looking ahead, p, d, and f electrons) farther from the nucleus than s electrons for which $l = 0$.

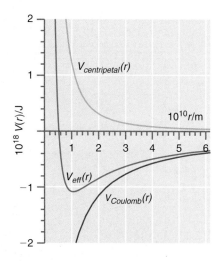

FIGURE 9.1

The individual contributions to the effective potential and their sum are plotted as a function of distance. The centripetal potential used is for $l = 1$; larger values of l make the effective potential more repulsive at small r.

9.3 EIGENVALUES AND EIGENFUNCTIONS FOR THE TOTAL ENERGY

Equation (9.5) can be solved using standard mathematical methods, so we concern ourselves only with the results. Note that the energy, E, only appears in the radial equation and not in the angular equation. Because only one variable is involved in this equation, the energy is expected to depend on a single quantum number. The quantization condition that results from the restriction that $R(r)$ be well behaved at large values of r $[R(r) \to 0$ as $r \to \infty]$ is

$$E_n = -\frac{m_e e^4}{8\varepsilon_0^2 h^2 n^2}, \quad \text{for } n = 1, 2, 3, 4, \ldots \qquad (9.7)$$

This formula is usually simplified by combining a number of constants in the form $a_0 = \varepsilon_0 h^2 / \pi m_e e^2$. The quantity a_0 has the value 0.529 Å and is called the **Bohr radius**. Use of this definition leads to the following formula:

$$E_n = -\frac{e^2}{8\pi\varepsilon_0 a_0 n^2} = -\frac{2.179 \times 10^{-18}\,\text{J}}{n^2} = -\frac{13.60\,\text{eV}}{n^2} \quad n = 1, 2, 3, 4, \ldots \quad (9.8)$$

Note that E_n goes to zero as $n \to \infty$. As previously emphasized, the zero of energy is a matter of convention rather than being a quantity that can be determined. As n approaches infinity, the electron is on average farther and farther from the nucleus, and the zero of energy corresponds to the electron at infinite separation from the nucleus. All negative energies correspond to bound states of the electron in the Coulomb potential.

As has been done previously for the particle in the box and the harmonic oscillator, the energy eigenvalues can be superimposed on a potential energy diagram, as shown in Figure 9.2. The potential forms a "box" that acts to confine the particle. This box has a peculiar form in that it is infinitely deep at the center of the atom, and the depth falls off inversely with distance from the proton. Figure 9.2 shows that the two lowest energy levels have an appreciable separation in energy and that the separation for adjacent energy levels becomes rapidly smaller as $n \to \infty$. All states for which $5 < n < \infty$ have energies in the narrow range between $\sim -1 \times 10^{-19}$ J and zero. Although this seems

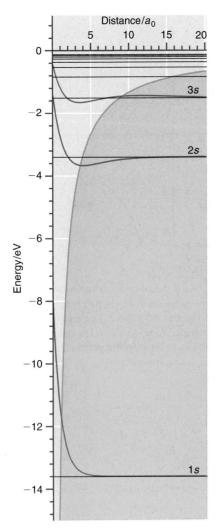

FIGURE 9.2

The border of the yellow classically forbidden region is the Coulomb potential, which is shown together with E_n for $n = 1$ through $n = 10$ and the 1s, 2s, and 3s wave functions.

strange at first, it is exactly what is expected based on the results for the particle in the box. Because of the shape of the potential, the H atom box is very narrow for the first few energy eigenstates, but becomes very wide for large n. The particle in the box formula [Equation (4.17)] predicts that the energy spacing varies as the inverse of the square of box length. This is the trend seen in Figure 9.2. Note also that the wave functions penetrate into the classically forbidden region just as for the particle in the finite depth box and the harmonic oscillator.

Although the energy depends on a single quantum number, n, the eigenfunctions $\psi(r, \theta, \phi)$ are associated with three quantum numbers because three boundary conditions arise in a three-dimensional problem. The other two quantum numbers are l and m_l, which arise from the angular coordinates. As for the rigid rotor, these quantum numbers are not independent. Their relationship is given by

$$n = 1, 2, 3, 4, \ldots$$

$$l = 0, 1, 2, 3, \ldots, n - 1$$

$$m_l = 0, \pm 1, \pm 2, \pm 3, \ldots \pm l \qquad \textbf{(9.9)}$$

The relationship between l and m_l was discussed in Section 7.5. Although we do not present a justification of the relationship between n and l here, all the conditions in Equation (9.9) emerge naturally out of the boundary conditions in the solution of the differential equations.

The radial functions, $R(r)$, are products of an exponential function with a polynomial in the dimensionless variable r/a_0. Their functional form depends on the quantum numbers n and l. The first few radial functions $R_{nl}(r)$ are as follows:

$$n = 1, l = 0 \quad R_{10}(r) = 2 \left(\frac{1}{a_0} \right)^{3/2} e^{-r/a_0}$$

$$n = 2, l = 0 \quad R_{20}(r) = \frac{1}{\sqrt{8}} \left(\frac{1}{a_0} \right)^{3/2} \left(2 - \frac{r}{a_0} \right) e^{-r/2a_0}$$

$$n = 2, l = 1 \quad R_{21}(r) = \frac{1}{\sqrt{24}} \left(\frac{1}{a_0} \right)^{3/2} \frac{r}{a_0} e^{-r/2a_0}$$

$$n = 3, l = 0 \quad R_{30}(r) = \frac{2}{81\sqrt{3}} \left(\frac{1}{a_0} \right)^{3/2} \left(27 - 18\frac{r}{a_0} + 2\frac{r^2}{a_0^2} \right) e^{-r/3a_0}$$

$$n = 3, l = 1 \quad R_{31}(r) = \frac{4}{81\sqrt{6}} \left(\frac{1}{a_0} \right)^{3/2} \left(6\frac{r}{a_0} - \frac{r^2}{a_0^2} \right) e^{-r/3a_0}$$

$$n = 3, l = 2 \quad R_{32}(r) = \frac{4}{81\sqrt{30}} \left(\frac{1}{a_0} \right)^{3/2} \frac{r^2}{a_0^2} e^{-r/3a_0}$$

To form the hydrogen atom eigenfunctions, we combine $R_{nl}(r)$ with the spherical harmonics and list here the first few of the infinite set of normalized wave functions $\psi(r, \theta, \phi) = R(r)\Theta(\theta)\Phi(\phi)$ for the hydrogen atom. Note that, in general, the eigenfunctions depend on r, θ, and ϕ, but are not functions of θ and ϕ for $l = 0$. The quantum numbers are associated with the wave functions using the notation ψ_{nlm_l}:

$$n = 1, l = 0, m_l = 0 \quad \psi_{100}(r) = \frac{1}{\sqrt{\pi}} \left(\frac{1}{a_0} \right)^{3/2} e^{-r/a_0}$$

$$n = 2, l = 0, m_l = 0 \quad \psi_{200}(r) = \frac{1}{4\sqrt{2\pi}} \left(\frac{1}{a_0} \right)^{3/2} \left(2 - \frac{r}{a_0} \right) e^{-r/2a_0}$$

$$n = 2, l = 1, m_l = 0 \quad \psi_{210}(r, \theta, \phi) = \frac{1}{4\sqrt{2\pi}} \left(\frac{1}{a_0} \right)^{3/2} \frac{r}{a_0} e^{-r/2a_0} \cos\theta$$

$$n = 2, l = 1, m_l = \pm 1 \quad \psi_{21\pm1}(r, \theta, \phi) = \frac{1}{8\sqrt{\pi}} \left(\frac{1}{a_0}\right)^{3/2} \frac{r}{a_0} e^{-r/2a_0} \sin\theta \, e^{\pm i\phi}$$

$$n = 3, l = 0, m_l = 0 \quad \psi_{300}(r) = \frac{1}{81\sqrt{3\pi}} \left(\frac{1}{a_0}\right)^{3/2} \left(27 - 18\frac{r}{a_0} + 2\frac{r^2}{a_0^2}\right) e^{-r/3a_0}$$

$$n = 3, l = 1, m_l = 0 \quad \psi_{310}(r, \theta, \phi) = \frac{1}{81} \left(\frac{2}{\pi}\right)^{1/2} \left(\frac{1}{a_0}\right)^{3/2} \left(6\frac{r}{a_0} - \frac{r^2}{a_0^2}\right) e^{-r/3a_0} \cos\theta$$

$$n = 3, l = 1, m_l = \pm 1 \quad \psi_{31\pm1}(r, \theta, \phi) = \frac{1}{81\sqrt{\pi}} \left(\frac{1}{a_0}\right)^{3/2} \left(6\frac{r}{a_0} - \frac{r^2}{a_0^2}\right) e^{-r/3a_0} \sin\theta \, e^{\pm i\phi}$$

$$n = 3, l = 2, m_l = 0 \quad \psi_{320}(r, \theta, \phi) = \frac{1}{81\sqrt{6\pi}} \left(\frac{1}{a_0}\right)^{3/2} \frac{r^2}{a_0^2} e^{-r/3a_0}(3\cos^2\theta - 1)$$

$$n = 3, l = 2, m_l = \pm 1 \quad \psi_{32\pm1}(r, \theta, \phi) = \frac{1}{81\sqrt{\pi}} \left(\frac{1}{a_0}\right)^{3/2} \frac{r^2}{a_0^2} e^{-r/3a_0} \sin\theta \cos\theta \, e^{\pm i\phi}$$

$$n = 3, l = 2, m_l = \pm 2 \quad \psi_{32\pm2}(r, \theta, \phi) = \frac{1}{162\sqrt{\pi}} \left(\frac{1}{a_0}\right)^{3/2} \frac{r^2}{a_0^2} e^{-r/3a_0} \sin^2\theta \, e^{\pm 2i\phi}$$

These functions are referred to both as the H atom eigenfunctions and the H atom **orbitals**. A shorthand notation for the quantum numbers is to give the numerical value of n followed by a symbol indicating the values of l and m_l. The letters s, p, d, and f are used to denote $l = 0, 1, 2,$ and 3, respectively, and $\psi_{100}(r)$ is referred to as the $1s$ orbital or wave function and all three wave functions with $n = 2$ and $l = 1$ are referred to as $2p$ orbitals. The wave functions are real functions when $m_l = 0$, and complex functions otherwise. The angular and radial portions of the wave functions have nodes that are discussed in more detail later in this chapter. These functions have been normalized in keeping with the association between probability density and $\psi(r, \theta, \phi)$ stated in the first postulate (see Chapter 3).

EXAMPLE PROBLEM 9.1

Normalize the functions $e^{-r/2a_0}$ and $(r/a_0)e^{-r/2a_0} \sin\theta \, e^{+i\phi}$ in three-dimensional spherical coordinates.

Solution

In general, a wave function $\psi(\tau)$ is normalized by multiplying it by a constant N defined by $N^2 \int \psi^*(\tau)\psi(\tau) \, d\tau = 1$. In three-dimensional spherical coordinates, $d\tau = r^2 \sin\theta \, dr \, d\theta \, d\phi$, as discussed in Section 2.6. The normalization integral becomes $N^2 \int_0^\pi \sin\theta \, d\theta \int_0^{2\pi} d\phi \int_0^\infty \psi^*(r, \theta, \phi) \psi(r, \theta, \phi) r^2 \, dr = 1$.

For the first function,

$$N^2 \int\limits_0^\pi \sin\theta \, d\theta \int\limits_0^{2\pi} d\phi \int\limits_0^\infty e^{-r/2a_0} e^{-r/2a_0} r^2 \, dr = 1$$

We use the standard integral

$$\int\limits_0^\infty x^n e^{-ax} \, dx = \frac{n!}{a^{n+1}}$$

Integrating over the angles θ and ϕ, we obtain $4\pi N^2 \int_0^\infty e^{-r/2a_0} e^{-r/2a_0} r^2 \, dr = 1$. Evaluating the integral over r,

$$4\pi N^2 \frac{2!}{1/a_0^3} = 1 \quad \text{or} \quad N = \frac{1}{2\sqrt{2\pi}} \left(\frac{1}{a_0}\right)^{3/2}$$

For the second function,

$$N^2 \int_0^\pi \sin\theta\, d\theta \int_0^{2\pi} d\phi \int_0^\infty \left(\frac{r}{a_0} e^{-r/2a_0} \sin\theta\, e^{-i\phi}\right)\left(\frac{r}{a_0} e^{-r/2a_0} \sin\theta\, e^{+i\phi}\right) r^2\, dr = 1$$

This simplifies to

$$N^2 \int_0^\pi \sin^3\theta\, d\theta \int_0^{2\pi} d\phi \int_0^\infty \left(\frac{r}{a_0}\right)^2 e^{-r/a_0} r^2\, dr = 1$$

Integrating over the angles θ and ϕ using the result $\int_0^\pi \sin^3\theta\, d\theta = 4/3$, we obtain

$$\frac{8\pi}{3} N^2 \int_0^\infty \left(\frac{r}{a_0}\right)^2 e^{-r/a_0} r^2\, dr = 1$$

Using the same standard integral as in the first part of the problem,

$$\frac{8\pi}{3} N^2 \frac{1}{a_0^2}\left(\frac{4!}{1/a_0^5}\right) = 1 \quad \text{or} \quad N = \frac{1}{8\sqrt{\pi}}\left(\frac{1}{a_0}\right)^{3/2}$$

Each eigenfunction listed here describes a separate state of the hydrogen atom. However, as we have seen, the energy depends only on the quantum number n. Therefore, all states with the same value for n, but different values for l and m_l, have the same energy and we say that the energy levels are degenerate. Using the formulas given in Equation (9.9), we can see that the **degeneracy** of a given level is n^2. Therefore, the $n = 2$ level has a fourfold degeneracy and the $n = 3$ level has a ninefold degeneracy.

The angular part of each hydrogen atom total energy eigenfunction is a spherical harmonic function. As discussed in Section 7.8, these functions are complex unless $m_l = 0$. To facilitate making graphs, it is useful to form combinations of those hydrogen orbitals $\psi_{nlm_l}(r, \theta, \phi)$ for which $m_l \neq 0$ are real functions of r, θ, and ϕ. As discussed in Section 7.5, this is done by forming linear combinations of $\psi_{nlm_l}(r, \theta, \phi)$ and $\psi_{nl-m_l}(r, \theta, \phi)$. The first few of these combinations, resulting in the $2p$, $3p$, and $3d$ orbitals, are shown here:

$$\psi_{2p_x}(r, \theta, \phi) = \frac{1}{4\sqrt{2\pi}}\left(\frac{1}{a_0}\right)^{3/2}\frac{r}{a_0} e^{-r/2a_0} \sin\theta \cos\phi$$

$$\psi_{2p_y}(r, \theta, \phi) = \frac{1}{4\sqrt{2\pi}}\left(\frac{1}{a_0}\right)^{3/2}\frac{r}{a_0} e^{-r/2a_0} \sin\theta \sin\phi$$

$$\psi_{2p_z}(r, \theta, \phi) = \frac{1}{4\sqrt{2\pi}}\left(\frac{1}{a_0}\right)^{3/2}\frac{r}{a_0} e^{-r/2a_0} \cos\theta$$

$$\psi_{3p_x}(r, \theta, \phi) = \frac{\sqrt{2}}{81\sqrt{\pi}}\left(\frac{1}{a_0}\right)^{3/2}\left(6\frac{r}{a_0} - \frac{r^2}{a_0^2}\right) e^{-r/3a_0} \sin\theta \cos\phi$$

$$\psi_{3p_y}(r, \theta, \phi) = \frac{\sqrt{2}}{81\sqrt{\pi}}\left(\frac{1}{a_0}\right)^{3/2}\left(6\frac{r}{a_0} - \frac{r^2}{a_0^2}\right) e^{-r/3a_0} \sin\theta \sin\phi$$

$$\psi_{3p_z}(r, \theta, \phi) = \frac{\sqrt{2}}{81\sqrt{\pi}}\left(\frac{1}{a_0}\right)^{3/2}\left(6\frac{r}{a_0} - \frac{r^2}{a_0^2}\right) e^{-r/3a_0} \cos\theta$$

$$\psi_{3d_{z^2}}(r, \theta, \phi) = \frac{1}{81\sqrt{6\pi}}\left(\frac{1}{a_0}\right)^{3/2}\frac{r^2}{a_0^2} e^{-r/3a_0}(3\cos^2\theta - 1)$$

$$\psi_{3d_{xz}}(r, \theta, \phi) = \frac{\sqrt{2}}{81\sqrt{\pi}}\left(\frac{1}{a_0}\right)^{3/2}\frac{r^2}{a_0^2}e^{-r/3a_0}\sin\theta\cos\theta\cos\phi$$

$$\psi_{3d_{yz}}(r, \theta, \phi) = \frac{\sqrt{2}}{81\sqrt{\pi}}\left(\frac{1}{a_0}\right)^{3/2}\frac{r^2}{a_0^2}e^{-r/3a_0}\sin\theta\cos\theta\sin\phi$$

$$\psi_{3d_{x^2-y^2}}(r, \theta, \phi) = \frac{1}{81\sqrt{2\pi}}\left(\frac{1}{a_0}\right)^{3/2}\frac{r^2}{a_0^2}e^{-r/3a_0}\sin^2\theta\cos 2\phi$$

$$\psi_{3d_{xy}}(r, \theta, \phi) = \frac{1}{81\sqrt{2\pi}}\left(\frac{1}{a_0}\right)^{3/2}\frac{r^2}{a_0^2}e^{-r/3a_0}\sin^2\theta\sin 2\phi$$

When is it appropriate to use these functions as opposed to the complex functions $\psi_{nlm_l}(r)$? The real functions are more useful in visualizing chemical bonds, so those will generally be used throughout this book. However, both representations are useful in different applications, and we note that although the real functions are eigenfunctions of $\hat{H}$ and $\hat{l}^2$, they are not eigenfunctions of $\hat{l}_z$.

The challenge we posed for quantum mechanics at the beginning of this chapter was to provide an understanding for the stability of atoms. By verifying that there is a set of eigenfunctions and eigenvalues of the time-independent Schrödinger equation for a system consisting of a proton and an electron, we have demonstrated that there are states whose energy is independent of time. Because the energy eigenvalues are all negative numbers, all of these states are more stable than the reference state of zero energy that corresponds to the proton and electron separated by an infinite distance. Because $n \geq 1$, the energy cannot approach $-\infty$, corresponding to the electron falling into the nucleus. These results show that when the wave nature of the electron is taken into account, the H atom is stable.

As with any new theory, the true test is consistency with experimental data. Although the wave functions are not directly observable, we know that the spectral lines from a hydrogen arc lamp measured as early as 1885 must involve transitions between two stable states of the hydrogen atom. Therefore, the frequencies measured by the early experimentalists in emission spectra must be given by

$$\nu = \left|\frac{1}{h}\left(E_{initial} - E_{final}\right)\right| \qquad (9.10)$$

In a more exact treatment, the origin of the coordinate system describing the H atom is placed at the center of mass of the proton and electron rather than at the position of the proton. Using Equations (9.7) with the reduced mass of the atom in place of m_e and (9.10), quantum theory predicts that the frequencies of all the spectral lines are given by

$$\nu = \left|\frac{\mu e^4}{8\varepsilon_0^2 h^3}\left(\frac{1}{n_{initial}^2} - \frac{1}{n_{final}^2}\right)\right| \qquad (9.11)$$

where μ is the reduced mass of the atom. Spectroscopists commonly refer to spectral lines in units of wave numbers. Rather than reporting values of ν, they use the units $\tilde{\nu} = \nu/c = 1/\lambda$. The combination of constants $m_e e^4/8\varepsilon_0^2 h^3 c$ is called the **Rydberg constant**. It has the units of energy and is equal to $2.1798736 \times 10^{-18}$ J, which corresponds to $109,737.31534$ cm^{-1}. The reduced mass for H is 0.05% less than m_e.

Equation (9.11) quantitatively predicts all observed spectral lines for the hydrogen atom. It also correctly predicts the very small shifts in frequency observed for the isotopes of hydrogen, which have slightly different reduced masses. The agreement between theory and experiment verifies that the quantum mechanical model for the hydrogen atom is valid and accurate. We discuss the selection rules for transitions between electronic states in atoms in Chapter 11. Some of these transitions are shown superimposed on a set of energy levels in Figure 9.3.

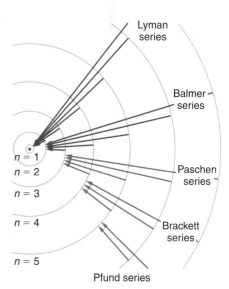

FIGURE 9.3

Energy-level diagram for the hydrogen atom showing the allowed transitions for $n < 6$. Because for $E > 0$, the energy levels are continuous, the absorption spectrum will be continuous above an energy that depends on the initial n value. The different sets of transitions are named after the scientists who first investigated them.

EXAMPLE PROBLEM 9.2

Consider an excited state of the H atom with the electron in the 2s orbital.

a. Is the wave function that describes this state,

$$\psi_{200}(r) = \frac{1}{\sqrt{32\pi}}\left(\frac{1}{a_0}\right)^{3/2}\left(2 - \frac{r}{a_0}\right)e^{-r/2a_0}$$

an eigenfunction of the kinetic energy? Of the potential energy?

b. Calculate the average values of the kinetic and potential energies for an atom described by this wave function.

Solution

a. We know that this function is an eigenfunction of the total energy operator because it is a solution of the Schrödinger equation. You can convince yourself that the total energy operator does not commute with either the kinetic energy operator or the potential energy operator by extending the discussion of Example Problem 9.1. Therefore, this wave function cannot be an eigenfunction of either of these operators.

b. The average value of the kinetic energy is given by

$$\langle E_{kinetic}\rangle = \int \psi^*(\tau)\,\hat{E}_{kinetic}\,\psi(\tau)\,d\tau$$

$$= -\frac{\hbar^2}{2m_e}\frac{1}{32\pi a_0^3}\int_0^{2\pi} d\phi \int_0^{\pi} \sin\theta\, d\theta \int_0^{\infty}\left(2 - \frac{r}{a_0}\right) \times$$

$$e^{-r/2a_0}\left(\frac{1}{r^2}\frac{d}{dr}\left[r^2\frac{d}{dr}\left\{\left(2 - \frac{r}{a_0}\right)e^{-r/2a_0}\right\}\right]\right)r^2\,dr$$

$$= -\frac{\hbar^2}{2m_e}\frac{1}{8a_0^3}\int_0^{\infty}\left(2 - \frac{r}{a_0}\right)e^{-r/2a_0}\left(-\frac{e^{-r/2a_0}}{4a_0^3 r}\right)(16a_0^2 - 10a_0 r + r^2)r^2\,dr$$

$$= -\frac{\hbar^2}{2m_e}\frac{1}{8a_0^3}\left(\frac{9}{a_0^2}\int_0^{\infty}r^2 e^{-r/a_0}\,dr - \frac{8}{a_0}\int_0^{\infty}r e^{-r/a_0}\,dr - \frac{3}{a_0^3}\int_0^{\infty}r^3 e^{-r/a_0}\,dr\right.$$

$$\left. + \frac{1}{4a_0^4}\int_0^{\infty}r^4 e^{-r/a_0}\,dr\right)$$

We use the standard integral, $\int_0^{\infty} x^n e^{-ax}\,dx = n!/a^{n+1}$:

$$\langle E_{kinetic}\rangle = \hbar^2/8m_e a_0^2$$

Using the relationship $a_0 = \varepsilon_0 h^2/\pi m_e e^2$,

$$\langle E_{kinetic}\rangle = \frac{e^2}{32\pi\varepsilon_0 a_0} = -E_n, \quad \text{for } n = 2$$

The average potential energy is given by

$$\langle E_{potential}\rangle = \int \psi^*(\tau)\,\hat{E}_{potential}\,\psi(\tau)\,d\tau$$

$$= -\frac{e^2}{4\pi\varepsilon_0}\frac{1}{32\pi a_0^3}\int_0^{2\pi} d\phi \int_0^{\pi} \sin\theta\, d\theta \int_0^{\infty}\left[\left(2 - \frac{r}{a_0}\right)e^{-r/2a_0}\right] \times$$

$$\left(\frac{1}{r}\right)\left[\left(2 - \frac{r}{a_0}\right)e^{-r/2a_0}\right]r^2\,dr$$

$$= -\frac{e^2}{4\pi\varepsilon_0}\frac{1}{8a_0^3}\left(4\int_0^{\infty}r e^{-r/a_0}\,dr - \frac{4}{a_0}\int_0^{\infty}r^2 e^{-r/a_0}\,dr + \frac{1}{a_0^2}\int_0^{\infty}r^3 e^{-r/a_0}\,dr\right)$$

$$= -\frac{e^2}{4\pi\varepsilon_0}\frac{1}{8a_0^3}(2a_0^2)$$

$$= -\frac{e^2}{16\pi\varepsilon_0 a_0} = 2\,E_n \quad \text{for } n = 2$$

We see that $\langle E_{potential}\rangle = 2\langle E_{total}\rangle$ and $\langle E_{potential}\rangle = -2\langle E_{kinetic}\rangle$. The relationship of the kinetic and potential energies is a specific example of the **virial theorem** and holds for any system in which the potential is Coulombic.

9.4 THE HYDROGEN ATOM ORBITALS

We now turn to the total energy eigenfunctions (or orbitals) of the hydrogen atom. What insight can be gained from them? Recall the early quantum mechanics shell model of atoms proposed by Niels Bohr. It depicted electrons as orbiting around the nucleus and associated orbits of small radius with more negative energies. Only certain orbits were allowed in order to give rise to a discrete energy spectrum. This model was discarded because defining orbits exactly is inconsistent with the Heisenberg uncertainty principle. The model postulated by Schrödinger and other pioneers of quantum theory replaced knowledge of the location of the electron in the hydrogen atom with knowledge of the probability of finding it in a small volume element at a specific location. As we have seen in considering the particle in the box and the harmonic oscillator, this probability is proportional to $\psi^*(r, \theta, \phi)\psi(r, \theta, \phi)\,d\tau$.

To what extent does the exact quantum mechanical solution resemble the shell model? To answer this question, information must be extracted from the H atom orbitals. A new concept, the radial distribution function, is introduced for this purpose. We begin our discussion by focusing on the wave functions $\psi_{nlm_l}(r, \theta, \phi)$. Next we discuss what can be learned about the probability of finding the electron in a particular region in space, $\psi_{nlm_l}^2(r, \theta, \phi)\,r^2\sin\theta\,dr\,d\theta\,d\phi$. Finally, we define the radial distribution function and look at the similarities and differences between quantum mechanical and shell models of the hydrogen atom.

The initial step is to look at the ground-state (lowest energy state) wave function for the hydrogen atom, and to find a good way to visualize this function. Because

$$\psi_{100}(r) = \frac{1}{\sqrt{\pi}}\left(\frac{1}{a_0}\right)^{3/2}e^{-r/a_0}$$

is a function of the three spatial coordinates x, y, and z, we need a four-dimensional space to plot ψ_{100} as a function of all its variables. Because such a space is not readily available, the number of variables will be reduced. The dimensionality of the representation can be reduced by evaluating $r = \sqrt{x^2 + y^2 + z^2}$ in one of the x–y, x–z, or y–z planes by setting the third coordinate equal to zero. Three common ways of depicting $\psi_{100}(r)$ are shown in Figures 9.4 through 9.6. In Figure 9.4a, a three-dimensional plot of $\psi_{100}(r)$

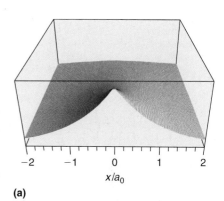

(a)

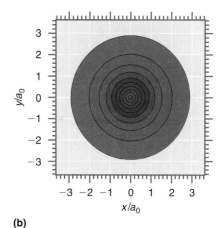

(b)

FIGURE 9.4

(a) 3D perspective and **(b)** contour plot of $\psi_{100}(r)$. Red and blue contours correspond to the most positive and least positive values of the wave function, respectively.

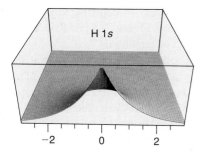
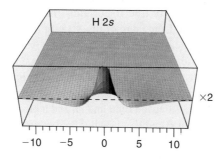
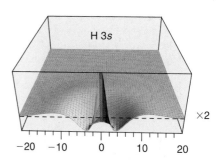

FIGURE 9.5

Three-dimensional perspective plots of the $1s$, $2s$, and $3s$ orbitals. The dashed lines indicate the zero of amplitude for the wave functions. The "×2" refers to the fact that the amplitude of the wave function has been multiplied by 2 to make the subsidiary maxima apparent. The horizontal axis shows radial distance in units of a_0.

FIGURE 9.6

Plot of $a_0^{3/2} R(r)$ versus r/a_0 for the first few H atomic orbitals.

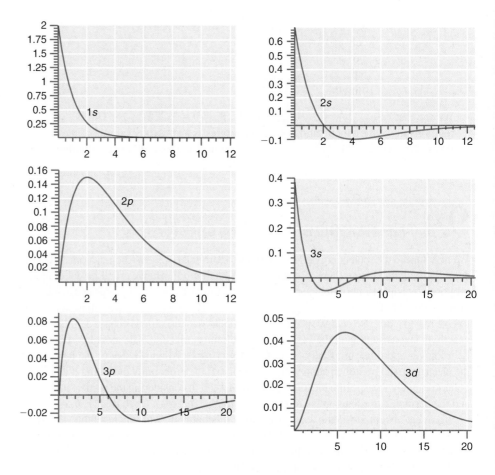

evaluated in the $x-y$ half-plane $(z = 0, y \geq 0)$ is shown in perspective. Although it is difficult to extract quantitative information from such a plot directly, it allows a good visualization of the function. We clearly see that the wave function has its maximum value at $r = 0$ (the nuclear position) and that it falls off rapidly with increasing distance from the nucleus.

More quantitative information is available in a contour plot shown in Figure 9.4b in which $\psi_{100}(r)$ is evaluated in the $x-y$ plane from a vantage point on the z axis. In this case, the outermost contour represents 10% of the maximum value, and successive contours are spaced at equal intervals. The shading also indicates the value of the function, with red and blue representing high and low values, respectively. This way of depicting $\psi_{100}(r)$ is more quantitative than that of Figure 9.4a in that we can recognize that the contours of constant amplitude are circles and that the contour spacing becomes smaller as r approaches zero. A third useful representation is to show the value of the function with two variables set equal to zero. This represents a cut through $\psi_{100}(r)$ in a plane perpendicular to the $x-y$ plane. Because ψ_{100} is independent of the angular coordinates, the same result is obtained for all planes containing the z axis. As we will see later, this is only true for orbitals for which $l = 0$. This way of depicting $\psi_{100}(r)$ is shown as the front edge of the 3D plot in Figure 9.4a and in Figures 9.5 and 9.6. All of these graphical representations contain exactly the same information.

Because $\psi_{100}(r)$ is a function of the single variable, r, the function can be graphed directly. However, it is important to keep in mind that r is a three-dimensional function of x, y, and z. For $l = 0$, the wave function does not depend on θ and ϕ. For $l > 0$, the wave function does depend on θ and ϕ, and a plot of the amplitude of a wave function versus r assumes that θ and ϕ are being held constant at values that need to be specified. These values generally correspond to a maximum in the angular part of the wave function.

In Figure 9.6, the radial wave function amplitude, $R(r)$, is graphed versus r. What should we expect having solved the particle in the box and harmonic oscillator problems? Because the eigenfunctions of the Schrödinger equation are standing waves, the solutions should be oscillating functions that have nodes. There should be no nodes in the ground state, and the number of nodes should increase as the quantum number increases.

First consider the eigenfunctions with $l = 0$, namely, the 1s, 2s, and 3s orbitals. From Figure 9.5, we clearly see that ψ_{100} has no nodes as expected. The 2s and 3s orbitals have one and two nodes, respectively. Because these nodes correspond to constant values of r, they are spherical **nodal surfaces**, rather than the nodal points previously encountered for one-dimensional potentials.

Now consider the eigenfunctions with $l > 0$. Why do the 2p and 3d functions in Figure 9.6 appear not to have nodal surfaces? This is related to the fact that the function is graphed for particular values of θ and ϕ. To see the nodes, the angular part of these eigenfunctions must be displayed.

Whereas the spherically symmetric s orbitals are equally well represented by the three forms of graphics described earlier, the p and d orbitals can best be visualized with a contour plot analogous to that of Figure 9.4b. Contour plots for the $2p_y$, $3p_y$, $3d_{xy}$, and $3d_{z^2}$ wave functions are shown in Figure 9.7. This nomenclature was defined in Section 9.3. We can now see that the $2p_y$ wave function has a nodal plane defined by $y = 0$; however, it appears in the angular rather than the radial part of the wave function. It can be shown that the radial part of the energy eigenfunctions has $n - l - 1$ nodal surfaces, not counting the one at infinity. There are l nodal surfaces in the angular part of the energy eigenfunctions, making a total of $n - 1$ nodes, just as was obtained for the particle in the box and the harmonic oscillator. As can be seen in Figure 9.7, the $3p_y$ wave function has a second nodal surface in addition to the nodal plane at $y = 0$. This second node comes from the radial part of the energy eigenfunction and is a spherical surface. The d orbitals have a more complex nodal structure that can include spheres, planes, and cones. The $3d_{xy}$ orbital has two nodal planes that intersect in the z axis. The $3d_{z^2}$ orbital has two conical nodal surfaces, whose axis of rotation is the z axis.

EXAMPLE PROBLEM 9.3

Locate the nodal surfaces in

$$\psi_{310}(r, \theta, \phi) = \frac{1}{81}\left(\frac{2}{\pi}\right)^{1/2}\left(\frac{1}{a_0}\right)^{3/2}\left(6\frac{r}{a_0} - \frac{r^2}{a_0^2}\right)e^{-r/3a_0}\cos\theta$$

Solution

We consider the angular and radial nodal surfaces separately. The angular part, $\cos\theta$, is zero for $\theta = \pi/2$. In three-dimensional space, this corresponds to the plane $z = 0$.

The radial part of the equations is zero for finite values of r/a_0 for $(6r/a_0 - r^2/a_0^2) = 0$.

This occurs at $r = 0$ and at $r = 6a_0$. The first value is a point in three-dimensional space and the second is a spherical surface. This wave function has one angular and one radial node. In general, an orbital characterized by n and l has l angular nodes and $n - l - 1$ radial nodes.

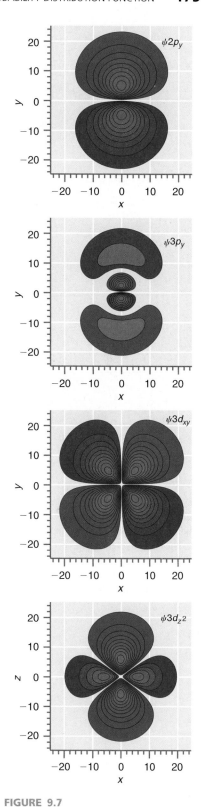

FIGURE 9.7

Contour plots for the orbitals indicated. The colors red and blue indicate the most positive and least positive value of the wave function amplitude, respectively. Distances are in units of a_0.

9.5 THE RADIAL PROBABILITY DISTRIBUTION FUNCTION

To continue the discussion of the similarities and differences between the quantum mechanical and shell models of the hydrogen atom, let us see what information can be obtained from $\psi^2_{nlm_l}(r, \theta, \phi)$, which is the probability density of finding the electron at a particular point in space. We again consider the s orbitals and the p and d orbitals separately. We first show $\psi^2_{n00}(r, \theta, \phi)$ as a 3D graphic in Figure 9.8 for $n = 1$, 2, and 3. Subsidiary maxima are seen in addition to the main maximum at $r = 0$. Figure 9.9 shows a graph of $R^2_{nlm_l}(r)$ as a function of r.

Now consider $\psi^2_{nlm_l}(r, \theta, \phi)$ for $l > 0$. As expected from the effect of the centripetal potential (see Figure 9.1), the electron is pushed away from the nucleus, so that

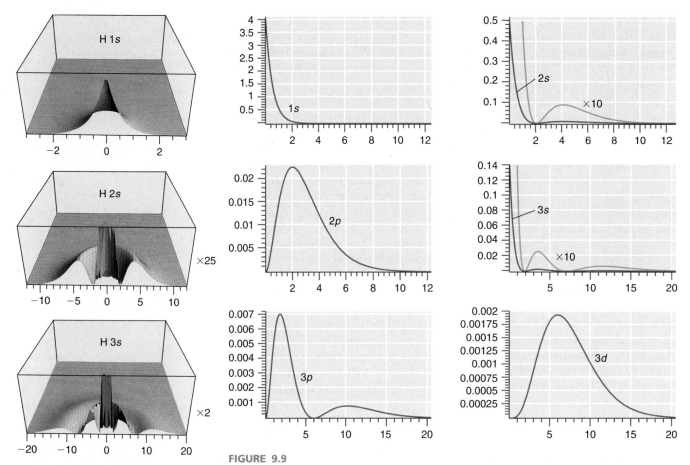

FIGURE 9.8

3D perspective plots of the square of the wave functions for the orbitals indicated. The numbers on the axes are in units of a_0. The "×25" refers to the fact that the amplitude of the wave function has been multiplied by 25 to make the subsidiary maxima apparent.

FIGURE 9.9

Plot of $a_0^3 R^2(r)$ versus r/a_0 for the first few H atomic orbitals. The numbers on the horizontal axis are in units of a_0.

$\psi_{nlm_l}^2(r, \theta, \phi)$ goes to zero as r approaches zero. Because a nonzero angular momentum is associated with these states, $\psi_{nlm_l}^2(r, \theta, \phi)$ is not spherically symmetric. All of this makes sense in terms of our picture of p and d orbitals.

EXAMPLE PROBLEM 9.4

a. At what point does the probability density for the electron in a 2s orbital have its maximum value?

b. Assume that the nuclear diameter for H is 2×10^{-15} m. Using this assumption, calculate the total probability of finding the electron in the nucleus if it occupies the 2s orbital.

Solution

a. The point at which $\psi^*(\tau)\psi(\tau) \, d\tau$ and, therefore, $\psi(\tau)$ has its greatest value is found from the wave function:

$$\psi_{200}(r) = \frac{1}{\sqrt{32\pi}} \left(\frac{1}{a_0}\right)^{3/2} \left(2 - \frac{r}{a_0}\right) e^{-r/2a_0}$$

which has its maximum value at $r = 0$, or at the nucleus as seen in Figure 9.6.

b. The result obtained in part (a) seems unphysical, but is a consequence of wave-particle duality in describing electrons. It is really only a problem if the total probability of finding the electron within the nucleus is significant. This probability is given by

$$P = \frac{1}{32\pi} \left(\frac{1}{a_0}\right)^3 \int_0^{2\pi} d\phi \int_0^{\pi} \sin\theta \, d\theta \int_0^{r_{nucleus}} r^2 \left(2 - \frac{r}{a_0}\right)^2 e^{-r/a_0} \, dr$$

Because $r_{nucleus} \ll a_0$, we can evaluate the integrand by assuming that $(2 - r/a_0)^2 \, e^{-r/a_0} \sim 2$ over the interval $0 \le r \le r_{nucleus}$:

$$P = \frac{1}{32\pi} \left(\frac{1}{a_0}\right)^3 4\pi \left[\left(2 - \frac{r_{nucleus}}{a_0}\right)^2 e^{-r_{nucleus}/a_0}\right] \int_0^{r_{nucleus}} r^2 \, dr$$

$$= \frac{1}{32\pi} \left(\frac{1}{a_0}\right)^3 \left[\left(2 - \frac{r_{nucleus}}{a_0}\right)^2 e^{-r_{nucleus}/a_0}\right] \frac{4\pi}{3} r_{nucleus}^3$$

Because $2 - (r_{nucleus}/a_0) \approx 2$ and $e^{-r_{nucleus}/a_0} \approx 1$,

$$P = \frac{1}{6}\left(\frac{r_{nucleus}}{a_0}\right)^3 = 9.0 \times 10^{-15}$$

Because this probability is vanishingly small, even though the wave function has its maximum amplitude at the nucleus, the probability of finding the electron in the nucleus is essentially zero.

At this point, we ask a different question involving probability. What is the most probable distance from the nucleus at which the electron will be found? For the $1s$, $2s$, and $3s$ orbitals, the maximum probability density is at the nucleus. This result seems to predict that the most likely orbit for the electron has a radius of zero. Clearly, we are missing something, because this result is inconsistent with a shell model. It turns out that we are not asking the right question. The probability as calculated from the probability density is correct, but it gives the likelihood of finding the particle in the vicinity of a *particular point* for a given value of r, θ, and ϕ. Why is this not the information we are looking for? Imagine that you know a planet has a circular orbit and you want to determine the radius of the orbit. To do so, you must find the planet. If you looked at only one point on a spherical shell of a given radius for different values of the radius, you would be unlikely to find the planet. To find the planet, you need to look everywhere on a shell of a given radius simultaneously.

How do we apply this reasoning to finding the electron on the hydrogen atom? The question we need to ask is "What is the probability of finding the electron at a particular value of r, regardless of the values of θ and ϕ?" This probability is obtained by integrating the probability density $\psi_{nlm_l}^2(r, \theta, \phi) \, r^2 \sin\theta \, dr \, d\theta \, d\phi$ over all values of θ and ϕ. This gives the probability of finding the electron in a spherical shell of radius r and thickness dr rather than the probability of finding the electron near a given point on the spherical shell of thickness dr with the particular coordinates r_0, θ_0, ϕ_0. For example, for the $1s$ orbital the probability of finding the electron in a spherical shell of radius r and thickness dr is

$$P(r) \, dr = \frac{1}{\pi a_0^3} \int_0^{2\pi} d\phi \int_0^{\pi} \sin\theta \, d\theta \, r^2 e^{-2r/a_0} \, dr$$

$$= \frac{4}{a_0^3} r^2 e^{-2r/a_0} \, dr \tag{9.12}$$

EXAMPLE PROBLEM 9.5

Consider an excited hydrogen atom with the electron in the $2s$ orbital.

a. Calculate the probability of finding the electron in the volume about a point defined by

$$5.22 \times 10^{-10} \text{ m} \le r \le 5.26 \times 10^{-10} \text{ m}, \frac{\pi}{2} - 0.01 \le \phi \le \frac{\pi}{2} + 0.01,$$

$$\frac{\pi}{2} - 0.01 \le \theta \le \frac{\pi}{2} + 0.01$$

b. Calculate the probability of finding the electron in the spherical shell defined by

$$5.22 \times 10^{-10} \text{ m} \le r \le 5.26 \times 10^{-10} \text{ m}$$

Solution

a. We numerically solve the integral $\iiint \psi^*(r, \theta, \phi)\psi(r, \theta, \phi)\, r^2 \sin\theta\, dr\, d\theta\, d\phi$. The result is

$$P = \frac{1}{32\pi}\left(\frac{1}{a_0}\right)^3 \int_{\frac{\pi}{2}-0.01}^{\frac{\pi}{2}+0.01} d\phi \int_{\frac{\pi}{2}-0.01}^{\frac{\pi}{2}+0.01} \sin\theta\, d\theta \int_{5.22\times10^{-10}}^{5.26\times10^{-10}} r^2\left(2 - \frac{r}{a_0}\right)^2 e^{-r/a_0}\, dr$$

$$= 0.0099 \times 6.76 \times 10^{30} \times 0.020 \times 0.0199 \times 3.43 \times 10^{-33} = 9.20 \times 10^{-8}$$

b. In this case, we integrate over all values of the angles:

$$P = \frac{1}{32\pi}\left(\frac{1}{a_0}\right)^3 \int_0^{2\pi} d\phi \int_0^{\pi} \sin\theta\, d\theta \int_{5.22\times10^{-10}}^{5.26\times10^{-10}} r^2\left(2 - \frac{r}{a_0}\right)^2 e^{-r/a_0}\, dr$$

$$= 0.0099 \times 6.76 \times 10^{30} \times 4\pi \times 3.43 \times 10^{-33} = 2.89 \times 10^{-3}$$

This probability is greater than that calculated in part (a) by a factor of 3.2×10^4 because we have integrated the probability density over the whole spherical shell of thickness 4×10^{-12} m.

Because the integration of the probability density over the angles θ and ϕ amounts to an averaging of $\psi(r, \theta, \phi)^2$ over all angles, it is most meaningful for the s orbitals whose amplitudes are independent of the angular coordinates. However, to arrive at a uniform definition for all orbitals regardless of their l values, a new function is defined that is called the **radial distribution function**, $P(r)$, in terms of the radial function, $R(r)$:

$$P(r)\, dr = r^2[R(r)]^2\, dr \tag{9.13}$$

The radial distribution is the probability function of choice to determine the most likely radius to find the electron for a given orbital. Understanding the difference between the radial distribution function, $P(r)\, dr$, and the probability, $\psi^*(r)\psi(r)\, r^2 \sin\theta\, dr\, d\theta\, d\phi$, is very important in making sense of the hydrogen atom wave functions.

EXAMPLE PROBLEM 9.6

Calculate the maxima in the radial probability distribution for the $2s$ orbital. What is the most probable distance from the nucleus for an electron in this orbital? Are there subsidiary maxima?

Solution

The radial distribution function is

$$P(r) = r^2 R^2(r) = \frac{1}{8}\left(\frac{1}{a_0}\right)^3 r^2 \left(2 - \frac{r}{a_0}\right)^2 e^{-r/a_0}$$

To find the maxima, we plot $P(r)$ and

$$\frac{dP(r)}{dr} = \frac{r}{8a_0^6}(8a_0^3 - 16a_0^2 r + 8a_0 r^2 - r^3)e^{-r/a_0}$$

versus r/a_0 and look for the nodes in this function. These functions are plotted as a function of r/a_0 in the following figure:

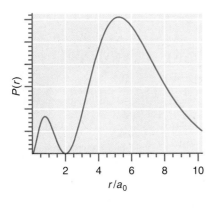
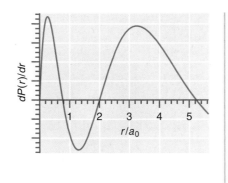

We see that the principal maximum in $P(r)$ is at $5.24\ a_0$. This corresponds to the most probable distance of a $2s$ electron from the nucleus. The subsidiary maximum is at $0.76\ a_0$. The minimum is at $2\ a_0$.

The resulting radial distribution function only depends on r, and not on θ and ϕ. Therefore, we can display $P(r)\ dr$ versus r in a graph as shown in Figure 9.10.

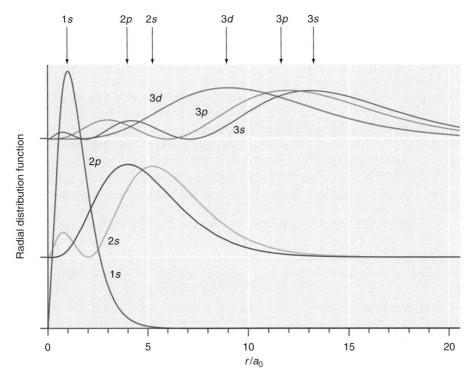

FIGURE 9.10

Plot of $r^2 a_0^3 R^2(r)$ versus r/a_0 for the first few H atomic orbitals. The curves for $n = 2$ and $n = 3$ have been displaced vertically as indicated. The position of the principal maxima for each orbital is indicated by an arrow.

9.6 THE VALIDITY OF THE SHELL MODEL OF AN ATOM

What can we conclude from Figure 9.10 regarding a shell model for the hydrogen atom? By now, we have become accustomed to the idea of wave-particle duality. Waves are not sharply localized, so that a shell model like that shown in Figure 9.11 with electrons as point masses orbiting around the nucleus is not viable in quantum mechanics. If there are some remnants of a shell model in the hydrogen atom, there is a greater likelihood of finding the electron at some distance from the nucleus than others.

The quantum mechanical analogue of the shell model can be generated in the following way. Imagine that 3D images of the shell model for hydrogen with the electron

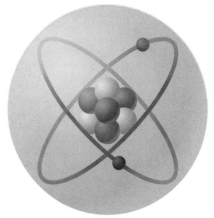

FIGURE 9.11

A shell model of an atom with electrons moving on spherical shells around the nucleus.

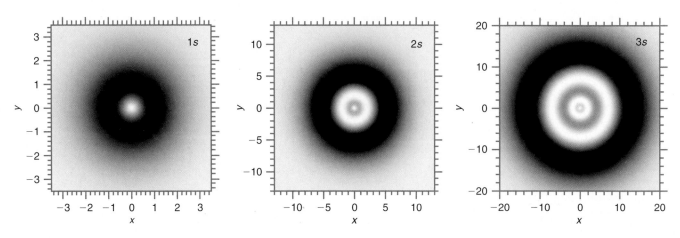

FIGURE 9.12

The radial probability distribution evaluated for $z = 0$ is plotted in the $x–y$ plane with lengths in units of a_0. Darker regions correspond to greater values of the function. The sharp circle in a classical shell model becomes a broad ring in a quantum mechanical model over which the probability of finding the electron varies. Less intense subsidiary rings are also observed for the $2s$ and $3s$ orbitals.

in the $1s$, $2s$, or $3s$ levels were taken at a large number of random times. A cut through the resulting images at the $z = 0$ plane would reveal sharply defined circles with a different radius for each orbital. The quantum mechanical analogue of this process is depicted in Figure 9.12. The principal maxima seen in Figure 9.10 are the source of the darkest rings in each part of Figure 9.12. The rings are broad in comparison to the sharp circle of the classical model. The subsidiary maxima seen in Figure 9.10 appear as less intense rings for the $2s$ and $3s$ orbitals.

The radial distribution function gives results that are more in keeping with our intuition and with a shell model than what we saw in the plots for the probability density $\psi^*(r, \theta, \phi)\psi(r, \theta, \phi)$. For the $1s$ orbital, the radial distribution function is peaked at a value of a_0. However, the peak has a considerable width, whereas a shell model would give a sharp peak of nearly zero width. This contradiction is reminiscent of our discussion of the double-slit diffraction experiment. Because wave-particle duality is well established, it is not useful to formulate models that are purely particle-like or purely wave-like. The broadening of the orbital shell over what we would expect in a particle picture is a direct manifestation of the wave nature of the electron, and the existence of an orbit is what we would expect in a particle picture. Both aspects of wave-particle duality are evident.

It is useful to summarize the main features that appear in Figures 9.10 and 9.12 for the radial probability distribution. We see broad maxima that move to greater values of r as n increases. This means that the electron is on average farther away from the nucleus for large n. From Equation (9.8), as n increases, the electron is less strongly bound. Both of these results are consistent with that expected from the Coulomb potential. However, we also see nodes and subsidiary maxima in the radial distribution function. How can these features be explained? Nodes are always present in standing waves, and eigenfunctions of the time-independent Schrödinger equation are standing waves. The nodes are directly analogous to the nodes observed for the particle in the box wave functions and are a manifestation of wave-particle duality. The subsidiary maxima are another manifestation of the wave character of the electron and occur whenever wave interference occurs. Recall that such subsidiary maxima are also observed in diffraction experiments. It is tempting to assign orbital radii to the H atomic orbitals with values corresponding to the positions of the principal maxima. The maxima are indicated by arrows in Figure 9.10. However, this amounts to reducing a complex function to a single number and is unwise.

Vocabulary

Bohr radius	effective potential	Rydberg constant
centripetal potential	nodal surface	separation of variables
Coulomb potential	orbital	shell model
degeneracy	radial distribution function	virial theorem

Questions on Concepts

Q9.1 What possible geometrical forms can the nodes in the angular function for p and d orbitals in the H atom have? What possible geometrical forms can the nodes in the radial function for s, p, and d orbitals in the H atom have?

Q9.2 What transition gives rise to the highest frequency spectral line in the Lyman series?

Q9.3 Is it always true that the probability of finding the electron in the H atom is greater in the interval $r - dr < r < r + dr$ than in the interval $r - dr < r < r + dr$ $\quad \theta - d\theta < \theta < \theta + d\theta$ $\phi - d\phi < \phi < \phi + d\phi$?

Q9.4 Why are the total energy eigenfunctions for the H atom not eigenfunctions of the kinetic energy?

Q9.5 How do the results shown in Figure 9.10 differ from the predictions of the Bohr model of the H atom?

Q9.6 What effect does the centripetal potential have in determining the maximum in the radial function for the 3s, 3p, and 3d orbitals?

Q9.7 How does the effective potential differ for p and d electrons?

Q9.8 Why does the centripetal potential dominate the effective potential for small values of r?

Q9.9 If the probability density of finding the electron in the 1s orbital in the H atom has its maximum value for $r = 0$, does this mean that the proton and electron are located at the same point in space?

Q9.10 Explain the different degree to which the 1s, 2s, and 3s total energy eigenfunctions penetrate into the classically forbidden region.

Q9.11 What are the units of the H atom total energy eigenfunctions? Why is $a_0^{3/2} R(r)$ graphed in Figure 9.6 rather than $R(r)$?

Q9.12 Why is the radial probability function rather than $\psi^*(r)\psi(r) r^2 \sin\theta \, dr \, d\theta \, d\phi$ the best measure of the probability of finding the electron at a distance r from the nucleus?

Q9.13 Use an analogy with the particle in the box to explain why the energy levels for the H atom are more closely spaced as n increases.

Q9.14 Explain why the radial distribution function rather than the square of the magnitude of the wave function should be used to make a comparison with the shell model of the atom.

Q9.15 What is the difference between an angular and a radial node? How can you distinguish the two types of nodes in a contour diagram such as Figure 9.7?

Problems

Problem numbers in **red** indicate that the solution to the problem is given in the *Student's Solutions Manual*.

P9.1 Calculate the wave number corresponding to the most and least energetic spectral lines in the Lyman, Balmer, and Paschen series for the hydrogen atom.

P9.2 Show that the function $(r/a_0)e^{-r/2a_0}$ is a solution of the following differential equation for $l = 1$

$$-\frac{\hbar^2}{2m_e r^2} \frac{d}{dr}\left[r^2 \frac{dR(r)}{dr} \right]$$
$$+ \left[\frac{\hbar^2 l(l+1)}{2m_e r^2} - \frac{e^2}{4\pi\varepsilon_0 r} \right] R(r) = ER(r)$$

What is the eigenvalue? Using this result, what is the value for the principal quantum number n for this function?

P9.3 Determine the probability of finding the electron in the region for which the ψ_{320} wavefunction is negative (the toroidal region).

P9.4 Calculate the expectation value for the potential energy of the H atom with the electron in the 1s orbital. Compare your result with the total energy.

P9.5 Calculate the probability that the 1s electron for H will be found between $r = 0$ and $r = a_0$.

P9.6 Using the result of Problem P9.13, calculate the probability of finding the electron in the 1s state outside a sphere of radius $0.5a_0$, $3a_0$, and $5a_0$.

P9.7 Calculate the distance from the nucleus for which the radial distribution function for the 2p orbital has its main and subsidiary maxima.

P9.8 Calculate the mean value of the radius $\langle r \rangle$ at which you would find the electron if the H atom wave function is $\psi_{100}(r)$.

P9.9 Calculate the expectation value for the kinetic energy of the H atom with the electron in the 2s orbital. Compare your result with the total energy.

P9.10 Ions with a single electron such as He^+, Li^{2+}, and Be^{3+} are described by the H atom wave functions with Z/a_0 substituted for $1/a_0$, where Z is the nuclear charge. The 1s wave function becomes $\psi(r) = 1/\sqrt{\pi}(Z/a_0)^{3/2}e^{-Zr/a_0}$. Using this result, calculate the total energy for the 1s state in H, He^+, Li^{2+}, and Be^{3+} by substitution in the Schrödinger equation.

P9.11 Ions with a single electron such as He^+, Li^{2+}, and Be^{3+} are described by the H atom wave functions with Z/a_0 substituted for $1/a_0$, where Z is the nuclear charge. The 1s wave function becomes $\psi(r) = 1/\sqrt{\pi}(Z/a_0)^{3/2}e^{-Zr/a_0}$.

Using this result, compare the mean value of the radius $\langle r \rangle$ at which you would find the $1s$ electron in H, He$^+$, Li^{2+}, and Be^{3+}.

P9.12   As the principal quantum number n increases, the electron is more likely to be found far from the nucleus. It can be shown that for H and for ions with only one electron such as He$^+$,

$$\langle r \rangle_{nl} = \frac{n^2 a_0}{Z}\left[1 + \frac{1}{2}\left(1 - \frac{l(l+1)}{n^2} \right) \right]$$

Calculate the value of n for an s state in the hydrogen atom such that $\langle r \rangle = 1000. \, a_0$. Round up to the nearest integer. What is the ionization energy of the H atom in this state in electron-volts? Compare your answer with the ionization energy of the H atom in the ground state.

P9.13   In this problem, you will calculate the probability density of finding an electron within a sphere of radius r for the H atom in its ground state.

a. Show using integration by parts, $\int u \, dv = uv - \int v \, du$, that $\int r^2 e^{-r/\alpha} \, dr = e^{-r/\alpha}(-2\alpha^3 - 2\alpha^2 r - \alpha r^2)$.

b. Using this result, show that the probability density of finding the electron within a sphere of radius r for the hydrogen atom in its ground state is

$$1 - e^{-2r/a_0} - \frac{2r}{a_0}\left(1 + \frac{r}{a_0} \right)e^{-2r/a_0}$$

c. Evaluate this probability density for $r = 0.10 \, a_0$, $r = 1.0 \, a_0$, and $r = 4.0 \, a_0$.

P9.14   Calculate the expectation value $\langle r - \langle r \rangle \rangle^2$ if the H atom wave function is $\psi_{100}(r)$.

P9.15   In spherical coordinates, $z = r \cos \theta$. Calculate $\langle z \rangle$ and $\langle z^2 \rangle$ for the H atom in its ground state. Without doing the calculation, what would you expect for $\langle x \rangle$ and $\langle y \rangle$, and $\langle x^2 \rangle$ and $\langle y^2 \rangle$? Why?

P9.16   The force acting between the electron and the proton in the H atom is given by $F = -e^2/4\pi\varepsilon_0 r^2$. Calculate the expectation value $\langle F \rangle$ for the $1s$ and $2p_z$ states of the H atom in terms of e, ε_0, and a_0.

P9.17   The d orbitals have the nomenclature d_{z^2}, d_{xy}, d_{xz}, d_{yz}, and $d_{x^2-y^2}$. Show how the d orbital

$$\psi_{3d_{yz}}(r, \theta, \phi) = \frac{\sqrt{2}}{81\sqrt{\pi}}\left(\frac{1}{a_0} \right)^{3/2} \frac{r^2}{a_0^2} e^{-r/3a_0} \sin \theta \cos \theta \sin \phi$$

can be written in the form $yzF(r)$.

P9.18   Calculate the expectation value of the moment of inertia of the H atom in the $2s$ and $2p_z$ states in terms of μ and a_0.

P9.19   The energy levels for ions with a single electron such as He$^+$, Li^{2+}, and Be^{3+} are given by $E_n = -Z^2 e^2/8\pi\varepsilon_0 a_0 n^2$, $n = 1, 2, 3, 4, \ldots$. Calculate the ionization energies of H, He$^+$, Li^{2+}, and Be^{3+} in their ground states in units of electron-volts (eV).

P9.20   Calculate the mean value of the radius $\langle r \rangle$ at which you would find the electron if the H atom wave function is $\psi_{210}(r, \theta, \phi)$.

P9.21   The total energy eigenvalues for the hydrogen atom are given by $E_n = -e^2/8\pi\varepsilon_0 a_0 n^2$, $n = 1, 2, 3, 4, \ldots$, and the

three quantum numbers associated with the total energy eigenfunctions are related by $n = 1, 2, 3, 4, \ldots$; $l = 0, 1, 2, 3, \ldots, n - 1$; and $m_l = 0, \pm 1, \pm 2, \pm 3, \ldots \pm l$. Using the nomenclature ψ_{nlm_l}, list all eigenfunctions that have the following total energy eigenvalues:

a. $E = -\dfrac{e^2}{32\pi\,\varepsilon_0 a_0}$ b. $E = -\dfrac{e^2}{72\pi\,\varepsilon_0 a_0}$

c. $E = -\dfrac{e^2}{128\pi\,\varepsilon_0 a_0}$

d. What is the degeneracy of each of these energy levels?

P9.22   Locate the radial and angular nodes in the H orbitals $\psi_{3p_x}(r, \theta, \phi)$ and $\psi_{3p_z}(r, \theta, \phi)$.

P9.23   Calculate the average value of the kinetic and potential energies for the H atom in its ground state.

P9.24   Show by substitution that $\psi_{100}(r, \theta, \phi) = 1/\sqrt{\pi}(1/a_0)^{3/2}e^{-r/a_0}$ is a solution of

$$-\frac{\hbar^2}{2m_e}\left[\frac{1}{r^2}\frac{\partial}{\partial r}\left(r^2 \frac{\partial\psi(r,\theta,\phi)}{\partial r} \right) + \frac{1}{r^2 \sin\theta}\frac{\partial}{\partial\theta}\left(\sin\theta \frac{\partial\psi(r,\theta,\phi)}{\partial\theta} \right) \right.$$
$$\left. + \frac{1}{r^2 \sin^2\theta}\frac{\partial^2\psi(r,\theta,\phi)}{\partial\phi^2} \right]$$
$$-\frac{e^2}{4\pi\varepsilon_0 r}\psi(r,\theta,\phi) = E\psi(r,\theta,\phi)$$

What is the eigenvalue for total energy? Use the relation $a_0 = \varepsilon_0 h^2/(\pi m_e e^2)$.

P9.25   Show that the total energy eigenfunctions $\psi_{100}(r)$ and $\psi_{200}(r)$ are orthogonal.

P9.26   As will be discussed in Chapter 10, core electrons shield valence electrons so that they experience an effective nuclear charge Z_{eff} rather than the full nuclear charge. Given that the first ionization energy of Li is 5.39 eV, use the formula in Problem P9.19 to estimate the effective nuclear charge experienced by the $2s$ electron in Li.

P9.27   Is the total energy wave function

$$\psi_{310}(r, \theta, \phi) = \frac{1}{81}\left(\frac{2}{\pi} \right)^{1/2}\left(\frac{1}{a_0} \right)^{3/2}\left(6\frac{r}{a_0} - \frac{r^2}{a_0^2} \right)e^{-r/3a_0} \cos \theta$$

an eigenfunction of any other operators? If so, which ones? What are the eigenvalues?

P9.28   Show that the total energy eigenfunctions $\psi_{210}(r, \theta, \phi)$ and $\psi_{211}(r, \theta, \phi)$ are orthogonal. Do you have to integrate over all three variables to show that the functions are orthogonal?

P9.29   Calculate $\langle r \rangle$ and the most probable value of r for the H atom in its ground state. Explain why they differ with a drawing.

P9.30   How many radial and angular nodes are there in the following H orbitals?

a. $\psi_{2p_x}(r, \theta, \phi)$

b. $\psi_{2s}(r)$

c. $\psi_{3d_{xz}}(r, \theta, \phi)$

d. $\psi_{3d_{x^2-y^2}}(r, \theta, \phi)$

10

Many-Electron Atoms

The Schrödinger equation cannot be solved analytically for atoms containing more than one electron because of the electron–electron repulsion term in the potential energy. Instead, approximate numerical methods can be used to obtain the eigenfunctions and eigenvalues of the Schrödinger equation for many-electron atoms. Having more than one electron in an atom also raises new issues that we have not considered, including the indistinguishability of electrons, the electron spin, and the interaction between orbital and spin magnetic moments. The Hartree–Fock method provides a way to calculate total energies and orbital energies for many-electron atoms in the limit that the motion of individual electrons is assumed to be uncorrelated.

10.1 HELIUM: THE SMALLEST MANY-ELECTRON ATOM

The Schrödinger equation for the hydrogen atom can be solved analytically because this atom has only one electron. The complexity of solving the Schrödinger equation for systems that have more than one electron can be illustrated using the He atom. Centering the coordinate system at the nucleus and neglecting the kinetic energy of the nucleus, the Schrödinger equation takes the form

$$\left(-\frac{\hbar^2}{2m_{e1}} \nabla_{e1}^2 - \frac{\hbar^2}{2m_{e2}} \nabla_{e2}^2 - \frac{2e^2}{4\pi\varepsilon_0 r_1} - \frac{2e^2}{4\pi\varepsilon_0 r_2} + \frac{e^2}{4\pi\varepsilon_0 r_{12}} \right) \psi(\mathbf{r}_1, \mathbf{r}_2)$$

$$= E\psi(\mathbf{r}_1, \mathbf{r}_2) \tag{10.1}$$

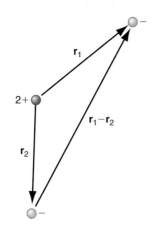

FIGURE 10.1

The top image shows the proton and two electrons that need to be considered for correlated electron motion. The bottom image shows that if the position of electron 2 is averaged over its orbit, electron 1 sees a spherically symmetric charge distribution due to the proton and electron 2.

In this equation, $r_1 = |\mathbf{r}_1|$ and $r_2 = |\mathbf{r}_2|$ are the distances of electrons 1 and 2 from the nucleus, $r_{12} = |\mathbf{r}_1 - \mathbf{r}_2|$, and ∇_{e1}^2 is shorthand for

$$\frac{1}{r_1^2}\frac{\partial}{\partial r_1}\left(r_1^2\frac{\partial}{\partial r_1}\right) + \frac{1}{r_1^2\sin^2\theta_1}\frac{\partial^2}{\partial\phi_1^2} + \frac{1}{r_1^2\sin\theta_1}\frac{\partial}{\partial\theta_1}\left(\sin\theta_1\frac{\partial}{\partial\theta_1}\right)$$

This is the part of the operator that is associated with the kinetic energy of electron 1, expressed in spherical coordinates. The last three terms in the operator are the potential energy operators for the electron–nucleus attraction and the electron–electron repulsion. The variables $r_1 = |\mathbf{r}_1|$, $r_2 = |\mathbf{r}_2|$, and $r_{12} = |\mathbf{r}_1 - \mathbf{r}_2|$ are shown in Figure 10.1.

The eigenfunctions of the Schrödinger equation depend on the coordinates of both electrons. If this formalism is applied to argon, each eigenfunction depends simultaneously on the coordinates of 18 electrons! Although true from a rigorous mathematical perspective, we also know that electrons in different atomic orbitals have quite different properties. For instance, valence electrons are involved in chemical bonds, and core electrons are not. Therefore, it seems reasonable to express a many-electron eigenfunction in terms of individual electron orbitals, each of which depends only on the coordinates of one electron. This is called the **orbital approximation** in which the many-electron eigenfunctions of the Schrödinger equation are expressed as a product of one-electron orbitals:

$$\psi(\mathbf{r}_1, \mathbf{r}_2, \ldots, \mathbf{r}_n) = \phi_1(\mathbf{r}_1)\phi_2(\mathbf{r}_2)\ldots\phi_n(\mathbf{r}_n) \tag{10.2}$$

This is not equivalent to saying that all of the electrons are independent of one another because, as we will see, the functional form for each $\phi_n(\mathbf{r}_n)$ is influenced by all the other electrons. The one-electron orbitals $\phi_n(\mathbf{r}_n)$ turn out to be quite similar to the functions $\psi_{nlm_l}(r, \theta, \phi)$ obtained for the hydrogen atom in Chapter 9, and they are labeled with indices such as $1s$, $2p$, and $3d$. Each of the $\phi_n(\mathbf{r}_n)$ is associated with a one-electron **orbital energy ε_n.**

The orbital approximation allows an n-electron Schrödinger equation to be written as n one-electron Schrödinger equations, one for each electron. However, a further problem arises in solving these n equations. Because of the form of the **electron–electron repulsion** term, $e^2/4\pi\varepsilon_0 r_{12}$, the operator $\hat{H}$ no longer has spherical symmetry. Therefore, the Schrödinger equation cannot be solved analytically, and numerical methods must be used. For these methods to be effective, further approximations beyond the orbital approximation have to be made. Perhaps the most serious of these approximations is that one cannot easily include what electrons do naturally in a many-electron atom, namely, stay out of each other's way by undergoing a correlated motion. Whereas **electron correlation** ensures that the repulsion among electrons is minimized, tractable numerical methods to solve the Schrödinger equation assume that the electrons move independently of one another. As discussed in Chapter 15 corrections can be made that largely eliminate the errors generated through this assumption.

A schematic illustration of how a neglect of electron correlation simplifies solving the Schrödinger equation is shown in Figure 10.1 for the He atom. We know from introductory chemistry that both electrons occupy what we call the $1s$ *orbital*, implying that the wave functions are similar to

$$\frac{1}{\sqrt{\pi}}\left(\frac{\zeta}{a_0}\right)^{3/2} e^{-\zeta r/a_0}$$

Zeta (ζ) is the **effective nuclear charge** felt by the electron. The importance of ζ in determining chemical properties is discussed later in this chapter. If the assumption is made that the motion of electrons 1 and 2 is uncorrelated, electron 1 can interact with the nucleus and the spatially averaged charge distribution arising from electron 2. This spatially averaged charge distribution is determined by $\phi^*(\mathbf{r}_2)\phi(\mathbf{r}_2)$. Think of electron 2 as being smeared out in a distribution that is spherically symmetrical about the nucleus, with a negative charge in the volume element $d\tau$ equal to $-e\phi^*(\mathbf{r}_2)\phi(\mathbf{r}_2)\,d\tau$.

The advantage of this approximation becomes apparent in Figure 10.1, because the effective charge distribution that electron 1 experiences is spherically symmetrical.

Because the potential energy V depends only on r, each one-electron wave function can be written as a product of radial and angular functions, $\phi(\mathbf{r}) = \phi(r, \theta, \phi) = R(r)\Theta(\theta)\Phi(\phi)$. Although the radial functions differ from those for the hydrogen atom, the angular functions are the same so that the $s, p, d, f, \ldots$ nomenclature used for the hydrogen atom also applies to many-electron atoms.

This quick look at the helium atom illustrates the approach that we take in the rest of this chapter. The Schrödinger equation is solved for many-electron atoms by approximating the true wave function by products of orbitals, each of which depends only on the coordinates of one electron. This approximation reduces the n-electron Schrödinger equation to n one-electron Schrödinger equations. The set of n equations is solved to obtain the one-electron energies and orbitals ε_i and ϕ_i. The solutions are approximate because of the orbital approximation and because electron correlation is neglected. However, before this approach is implemented, two important concepts must be introduced, namely, electron spin and the indistinguishability of electrons.

10.2 INTRODUCING ELECTRON SPIN

Electron spin plays an important part in formulating the Schrödinger equation for many-electron atoms. In discussing the Stern–Gerlach experiment in Chapter 6, we focused on commutation relations rather than the other surprising result of this experiment, which is that two and only two deflected beams are observed. In order for a silver atom to be deflected in an inhomogeneous field, it must have a magnetic moment and an associated angular momentum. What is the origin of this moment? An electric current passing through a loop of wire produces a magnetic field and, therefore, the loop has a magnetic moment. An electron in an orbit around a nucleus for which $l > 0$ has a nonzero angular momentum because of the nonspherical electron charge distribution. However, Ag has a closed-shell configuration plus a single $5s$ valence electron. A closed shell has a spherical electron charge distribution and no net angular momentum. Therefore, the magnetic moment must be associated with the $5s$ electron which has no orbital angular momentum. If this electron has an intrinsic angular momentum, which we call s, it will be split into $2s + 1$ components in passing through the magnet. The fact that two components are observed in the Stern–Gerlach experiment shows that $s = 1/2$. Therefore, there is a z component of angular momentum $s_z = m_s\hbar = \pm\hbar/2$ associated with the $5s$ electron. The origin of this effect cannot be an orbital angular momentum because for an s electron, $l = 0$, and because orbital angular momentum comes in quanta twice that size. This intrinsic electron spin angular momentum is a vector called $\mathbf{s}$ and its z component is called s_z to distinguish it from orbital angular momentum. The term *intrinsic* refers to the fact that the spin is independent of the environment of the electron. The use of the term *spin* implies that the electron is spinning about an axis. Although the nomenclature is appealing, there is no physical basis for this association.

How does the existence of spin change what has been discussed up to now? As we show later, each of the orbitals in a many-electron atom can be doubly occupied; one electron has $m_s = +1/2$, and the other has $m_s = -1/2$. This adds a fourth quantum number to the H atom eigenfunctions that is now labeled $\psi_{nlm_lm_s}(r, \theta, \phi)$. Because electron spin is an intrinsic property of the electron, it does not depend on the spatial variables r, θ, and ϕ.

How can this additional quantum number be incorporated in the formalism described for the hydrogen atom? This can be done by defining spin wave functions called α and β, which are eigenfunctions of the spin angular momentum operators $\hat{s}^2$ and $\hat{s}_z$. Because all angular momentum operators have the same properties, the spin operators follow the commutation rules listed in Equation (7.31). As for the orbital angular momentum, only the magnitude of the spin angular momentum and one of its components can be known simultaneously. The spin operators $\hat{s}^2$ and $\hat{s}_z$ have the following properties:

$$\hat{s}^2\alpha = \hbar^2 s\,(s + 1)\alpha = \frac{\hbar^2}{2}\left(\frac{1}{2} + 1\right)\alpha$$

$$\hat{s}^2\beta = \hbar^2 s(s+1)\beta = \frac{\hbar^2}{2}\left(\frac{1}{2}+1\right)\beta$$

$$\hat{s}_z\alpha = m_s\hbar\alpha = \frac{\hbar}{2}\alpha, \quad \hat{s}_z\beta = m_s\hbar\beta = -\frac{\hbar}{2}\beta$$

$$\int \alpha^*\beta\,d\sigma = \int \beta^*\alpha\,d\sigma = 0$$

$$\int \alpha^*\alpha\,d\sigma = \int \beta^*\beta\,d\sigma = 1 \tag{10.3}$$

In these equations, σ is called the spin variable. It is not a spatial variable and the "integration" over σ exists only formally so that we can define orthogonality. The H atom eigenfunctions are redefined by multiplying them by α and β and including a quantum number for spin. For example, the H atom $1s$ eigenfunctions take the form

$$\psi_{100\frac{1}{2}}(r) = \frac{1}{\sqrt{\pi}}\left(\frac{1}{a_0}\right)^{3/2}e^{-r/a_0}\alpha \quad \text{and}$$

$$\psi_{100\frac{1}{2}}(r) = \frac{1}{\sqrt{\pi}}\left(\frac{1}{a_0}\right)^{3/2}e^{-r/a_0}\beta \tag{10.4}$$

The eigenfunctions remain orthonormal because with this formalism

$$\iiint \psi^*_{100\frac{1}{2}}(r,\sigma)\psi_{100-\frac{1}{2}}(r,\sigma)\,dx\,dy\,dz\,d\sigma$$

$$= \iiint \psi^*_{100}(r)\psi_{100}(r)\,dx\,dy\,dz\int\alpha^*\beta\,d\sigma = 0$$

and

$$\iiint \psi^*_{100\frac{1}{2}}(r,\sigma)\psi_{100\frac{1}{2}}(r,\sigma)\,dx\,dy\,dz\,d\sigma$$

$$= \iiint \psi^*_{100}(r)\psi_{100}(r)\,dx\,dy\,dz\int\alpha^*\alpha\,d\sigma = 1 \tag{10.5}$$

These two eigenfunctions have the same energy because the total energy operator of Equation (10.1) does not depend on the spin coordinates. Having discussed how to include electron spin in a wave function, we now take on the issue of keeping track of electrons in a many-electron atom.

10.3 WAVE FUNCTIONS MUST REFLECT THE INDISTINGUISHABILITY OF ELECTRONS

In discussing He in Section 10.1, the electrons were numbered 1 and 2. Macroscopic objects can be distinguished from one another, but in an atom, we have no way to distinguish between any two electrons. This fact needs to be taken into account in the formulation of a wave function. How can **indistinguishability** be introduced into the orbital approximation? Consider an n-electron wave function written as the product of n one-electron wave functions which we describe using the notation $\psi(1,2,\ldots,n) = \psi(r_1\theta_1\phi_1\sigma_1, r_2\theta_2\phi_2\sigma_2, \ldots, r_n\theta_n\phi_n\sigma_n)$. The position variables are suppressed in favor of keeping track of the electrons. How does indistinguishability affect how the wave function is written? We know that the wave function itself is not an observable, but the square of the magnitude of the wave function is proportional to the electron density and is an observable. Because the two electrons in He are indistinguishable, no observable of the system can be changed if the electron labels 1 and 2 are interchanged. Therefore, $\psi^2(1,2) = \psi^2(2,1)$. This equation can be satisfied either by $\psi(1,2) = \psi(2,1)$ or $\psi(1,2) = -\psi(2,1)$. We refer to the wave function as being a **symmetric wave function** if $\psi(1,2) = \psi(2,1)$ or an **antisymmetric wave function** if $\psi(1,2) = -\psi(2,1)$. For a ground-state He atom, examples of symmetric and antisymmetric wave functions are as follows:

$$\psi_{symmetric}(1,2) = \phi_{1s}(1)\alpha(1)\phi_{1s}(2)\beta(2) + \phi_{1s}(2)\alpha(2)\phi_{1s}(1)\beta(1) \ \text{ and}$$

$$\psi_{antisymmetric}(1,2) = \phi_{1s}(1)\alpha(1)\phi_{1s}(2)\beta(2) - \phi_{1s}(2)\alpha(2)\phi_{1s}(1)\beta(1) \quad \textbf{(10.6)}$$

where $\phi(1) = \phi(\mathbf{r}_1)$. Wolfgang Pauli showed that only an antisymmetric wave function is allowed for electrons, a result that can be formulated as a further fundamental postulate of quantum mechanics.

POSTULATE 6:

Wave functions describing a many-electron system must change sign (be antisymmetric) under the exchange of any two electrons.

This postulate is also known as the **Pauli exclusion principle**. This principle states that different product wave functions of the type $\psi(1, 2, 3, \ldots, n) = \phi_1(1)\phi_2(2)\ldots\phi_n(n)$ must be combined such that the resulting wave function changes sign when any two electrons are interchanged. A combination of such terms is required, because a single-product wave function cannot be made antisymmetric in the interchange of two electrons. For example, $\phi_{1s}(1)\alpha(1)\phi_{1s}(2)\beta(2) \neq -\phi_{1s}(2)\alpha(2)\phi_{1s}(1)\beta(1)$.

How can antisymmetric wave functions be constructed? Fortunately, there is a simple way to do so using determinants. They are known as **Slater determinants** and have the form

$$\psi(1,2,3, \ldots, n) = \frac{1}{\sqrt{n!}} \begin{vmatrix} \phi_1(1)\alpha(1) & \phi_1(1)\beta(1) & \ldots & \phi_m(1)\beta(1) \\ \phi_1(2)\alpha(2) & \phi_1(2)\beta(2) & \ldots & \phi_m(2)\beta(2) \\ \ldots & \ldots & \ldots & \ldots \\ \phi_1(n)\alpha(n) & \phi_1(n)\beta(n) & \ldots & \phi_m(n)\beta(n) \end{vmatrix} \quad \textbf{(10.7)}$$

where $m = n/2$ if n is even and $m = (n + 1)/2$ if n is odd. The one-electron orbitals in which the n electrons are sequentially filled are listed going across each row with one row for each electron. The factor in front of the determinant takes care of the normalization if the one-electron orbitals are individually normalized. The Slater determinant is simply a recipe for constructing an antisymmetric wave function, and none of the individual entries in the determinant has a separate reality. For the ground state of He, the antisymmetric wave function is the 2×2 determinant:

$$\psi(1,2) = \frac{1}{\sqrt{2}} \begin{vmatrix} 1s(1)\alpha(1) & 1s(1)\beta(1) \\ 1s(2)\alpha(2) & 1s(2)\beta(2) \end{vmatrix}$$

$$= \frac{1}{\sqrt{2}}[1s(1)\alpha(1)1s(2)\beta(2) - 1s(1)\beta(1)1s(2)\alpha(2)] \quad \textbf{(10.8)}$$

The shorthand notation $\phi_{1s}(1)\alpha(1) = \phi_{100+\frac{1}{2}}(r_1, \theta_1, \phi_1, \sigma_1) = 1s(1)\alpha(1)$ has been used in the preceding determinant.

Determinants are used in constructing antisymmetric wave functions because their value automatically changes sign when two rows (which refer to individual electrons) are interchanged. This can easily be verified by comparing the values of the following determinants:

$$\begin{vmatrix} 3 & 6 \\ 4 & 2 \end{vmatrix} \ \text{ and } \ \begin{vmatrix} 4 & 2 \\ 3 & 6 \end{vmatrix}$$

Writing the wave function as a determinant also demonstrates another formulation of the Pauli exclusion principle. The value of a determinant is zero if two rows are identical. *This is equivalent to saying that the wave function is zero if all quantum numbers of any two electrons are the same.* Example Problem 10.1 illustrates how to work with determinants. Further information on determinants can be found in the Math Supplement (see Appendix A).

EXAMPLE PROBLEM 10.1

Consider the determinant

$$\begin{vmatrix} 3 & 1 & 5 \\ 4 & -2 & 1 \\ 3 & 2 & 7 \end{vmatrix}$$

a. Evaluate the determinant by expanding it in the cofactors of the first row.

b. Show that the value of the related determinant

$$\begin{vmatrix} 4 & -2 & 1 \\ 4 & -2 & 1 \\ 3 & 2 & 7 \end{vmatrix}$$

in which the first two rows are identical, is zero.

c. Show that exchanging the first two rows changes the sign of the value of the determinant.

Solution

a. The value of a 2 × 2 determinant

$$\begin{vmatrix} a & b \\ c & d \end{vmatrix} = ad - bc$$

We reduce a higher order determinant to a 2 × 2 determinant by expanding it in the cofactors of a row or column (see the Math Supplement). Any row or column can be used for this reduction, and all will yield the same result. The cofactor of an element a_{ij}, where i is the index of the row and j is the index of the column, is the $(n - 1) \times (n - 1)$ determinant that is left by ignoring the elements in the ith row and in the jth column. In our case, we reduce the 3 × 3 determinant to a sum of 2 × 2 determinants by adding the first row cofactors, each of which is multiplied by $(-1)^{i+j} a_{ij}$. For the given determinant,

$$\begin{vmatrix} 3 & 1 & 5 \\ 4 & -2 & 1 \\ 3 & 2 & 7 \end{vmatrix} = 3(-1)^{1+1}\begin{vmatrix} -2 & 1 \\ 2 & 7 \end{vmatrix} + 1(-1)^{1+2}\begin{vmatrix} 4 & 1 \\ 3 & 7 \end{vmatrix} + 5(-1)^{1+3}\begin{vmatrix} 4 & -2 \\ 3 & 2 \end{vmatrix}$$

$$= 3(-14 - 2) - 1(28 - 3) + 5(8 + 6) = -3$$

b.
$$\begin{vmatrix} 4 & -2 & 1 \\ 4 & -2 & 1 \\ 3 & 2 & 7 \end{vmatrix} = 4(-1)^{1+1}\begin{vmatrix} -2 & 1 \\ 2 & 7 \end{vmatrix} + (-2)(-1)^{1+2}\begin{vmatrix} 4 & 1 \\ 3 & 7 \end{vmatrix}$$

$$+ 1(-1)^{1+3}\begin{vmatrix} 4 & -2 \\ 3 & 2 \end{vmatrix}$$

$$= 4(-14 - 2) + 2(28 - 3) + 1(8 + 6) = 0$$

c.
$$\begin{vmatrix} 4 & -2 & 1 \\ 3 & 1 & 5 \\ 3 & 2 & 7 \end{vmatrix} = 4(-1)^{1+1}\begin{vmatrix} 1 & 5 \\ 2 & 7 \end{vmatrix} - 2(-1)^{1+2}\begin{vmatrix} 3 & 5 \\ 3 & 7 \end{vmatrix}$$

$$+ 1(-1)^{1+3}\begin{vmatrix} 3 & 1 \\ 3 & 2 \end{vmatrix}$$

$$= 4(7 - 10) + 2(21 - 15) + 1(6 - 3) = +3$$

For ground-state helium, both electrons have the same values of n, l, and m_l, but the values of m_s are $+1/2$ for one electron and $-1/2$ for the other. We now describe the way in which electrons are assigned to orbitals by a configuration. A **configuration** specifies the values of n and l for each electron. For example, the configuration for ground-state He is $1s^2$ and that for ground state F is $1s^2 2s^2 p^5$. The quantum numbers m_l and m_s are not specified in a configuration. Describing the quantum state of an atom requires this information, as will be discussed in Chapter 11.

EXAMPLE PROBLEM 10.2

This problem illustrates how determinantal wave functions can be associated with putting α and β spins in a set of orbitals. The first excited state of the helium atom can be described by the configuration $1s^1 2s^1$. However, four different spin orientations are consistent with this notation, as shown pictorially below. Don't take these pictures too literally, because they imply that one electron can be distinguished from another.

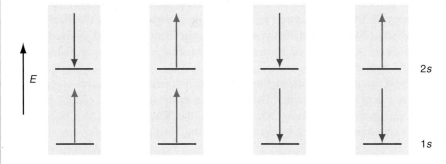

Keep in mind that the ↑ and ↓ notation commonly used for α and β spins is shorthand for the more accurate vector model depiction discussed in Chapter 7 and shown here:

Write determinantal wave functions that correspond to these pictures.

Solution

$$\psi_1(1,2) = \frac{1}{\sqrt{2}}\begin{vmatrix} 1s(1)\alpha(1) & 2s(1)\beta(1) \\ 1s(2)\alpha(2) & 2s(2)\beta(2) \end{vmatrix} \qquad \psi_2(1,2) = \frac{1}{\sqrt{2}}\begin{vmatrix} 1s(1)\alpha(1) & 2s(1)\alpha(1) \\ 1s(2)\alpha(2) & 2s(2)\alpha(2) \end{vmatrix}$$

$$\psi_3(1,2) = \frac{1}{\sqrt{2}}\begin{vmatrix} 1s(1)\beta(1) & 2s(1)\beta(1) \\ 1s(2)\beta(2) & 2s(2)\beta(2) \end{vmatrix} \qquad \psi_4(1,2) = \frac{1}{\sqrt{2}}\begin{vmatrix} 1s(1)\beta(1) & 2s(1)\alpha(1) \\ 1s(2)\beta(2) & 2s(2)\alpha(2) \end{vmatrix}$$

The neutral atom that has three electrons is Li. If the third electron is put in the $1s$ orbital, the determinantal wave function

$$\psi(1,2,3) = \frac{1}{\sqrt{3!}}\begin{vmatrix} 1s(1)\alpha(1) & 1s(1)\beta(1) & 1s(1)\alpha(1) \\ 1s(2)\alpha(2) & 1s(2)\beta(2) & 1s(2)\alpha(2) \\ 1s(3)\alpha(3) & 1s(3)\beta(3) & 1s(3)\alpha(3) \end{vmatrix}$$

is obtained, where the third electron can have either α or β spin. However, the first and third columns in this determinant are identical, so that $\psi(1,2,3) = 0$. Therefore, the third electron must go into the next higher energy orbital with $n = 2$. *This example shows that the Pauli exclusion principle requires that each orbital have a maximum occupancy of two electrons.* The configuration of ground-state Li is $1s^2 2s^1$. For $n = 2$,

l can take on the value 0 with the only possible m_l value of 0, or 1 with the possible m_l values of 0, +1, and −1. Each of the possible sets of n, l, and m_l can be combined with $m_s = \pm 1/2$. Therefore, there are eight different sets of quantum numbers for $n = 2$. The set of orbitals with the same values of n and l comprises a **subshell**, and the set of orbitals with the same n value comprises a **shell**. The connotation of a shell is demonstrated pictorially in Figure 9.12.

10.4 USING THE VARIATIONAL METHOD TO SOLVE THE SCHRÖDINGER EQUATION

In Section 10.1, we concluded that electron–electron repulsion terms in the total energy operator for many-electron atoms preclude an analytical solution to the Schrödinger equation. However, numerical methods are available for calculating one-electron energies and the orbitals ε_i and ϕ_i that include electron–electron repulsion. The goal is to obtain as good an approximation as possible to the eigenfunctions and eigenvalues of total energy for the many-electron atom. Only one of these methods, the **Hartree–Fock self-consistent field method** combined with the **variational method**, is discussed here. Other methods that go beyond Hartree–Fock by including electron correlation are discussed in Chapter 15.

We next discuss the variational method, which is frequently used in computational chemistry calculations. Consider a system in its ground state with energy E_0 and the corresponding eigenfunction ψ_0, which satisfies the equation $\hat{H}\psi_0 = E_0\psi_0$. Multiplying this expression on the left by ψ_0^* and integrating results in the following equation:

$$E_0 = \frac{\int \psi_0^* \hat{H} \psi_0 \, d\tau}{\int \psi_0^* \psi_0 \, d\tau} \tag{10.9}$$

The denominator takes into account that the wave function may not be normalized. For a many-electron atom, the total energy operator can be formulated, but the exact total energy eigenfunctions are unknown. How can the energy be calculated in this case? The **variational theorem** states that no matter what approximate wave function Φ is substituted for the ground-state eigenfunction in Equation (10.9), the energy is always greater than or equal to the true energy. Expressed mathematically, the theorem says that

$$E = \frac{\int \Phi^* \hat{H} \Phi \, d\tau}{\int \Phi^* \Phi \, d\tau} \geq E_0 \tag{10.10}$$

The proof of this theorem is included as an end-of-chapter problem. How can this method be implemented to obtain good approximate wave functions and energies? We parameterize the **trial wave function** Φ and find the optimal values for the parameters by minimizing the energy with respect to each parameter. This procedure gives the best energy that can be obtained for that particular choice of a trial wave function. The better the choice made for the trial function, the closer the calculated energy will be to the true energy.

We illustrate this formalism using the particle in the box as a specific example. Any trial function used must satisfy a number of general conditions (single valued, normalizable, the function and its first derivative are continuous) and also the boundary condition that the wave function goes to zero at the ends of the box. We use the trial function of Equation (10.11) to approximate the ground-state wave function. Convince yourself that this wave function satisfies the criteria just listed. This function contains a single parameter α that is used to minimize the energy:

$$\Phi(x) = \left(\frac{x}{a} - \frac{x^3}{a^3}\right) + \alpha\left(\frac{x^5}{a^5} - \frac{1}{2}\left(\frac{x^7}{a^7} + \frac{x^9}{a^9}\right)\right), \quad 0 < x < a \tag{10.11}$$

We first calculate the energy for $\alpha = 0$ and obtain

$$E = \frac{-\dfrac{\hbar^2}{2m}\displaystyle\int_0^a \left(\dfrac{x}{a} - \dfrac{x^3}{a^3}\right)\dfrac{d^2}{dx^2}\left(\dfrac{x}{a} - \dfrac{x^3}{a^3}\right)dx}{\displaystyle\int_0^a \left(\dfrac{x}{a} - \dfrac{x^3}{a^3}\right)^2 dx} = 0.133\frac{h^2}{ma^2} \qquad \textbf{(10.12)}$$

Because the trial function is not the exact ground-state wave function, the energy is higher than the exact value $E_0 = 0.125(h^2/ma^2)$. How similar is the trial function to the ground-state eigenfunction? A comparison between the exact solution and the trial function with $\alpha = 0$ is shown in Figure 10.2a.

To find the optimal value for α, E is first expressed in terms of h, m, a, and α, and then minimized with respect to α. The energy E is given by

$$E = \frac{-\dfrac{\hbar^2}{2m}\displaystyle\int_0^a \left[\begin{array}{l}\left[\left(\dfrac{x}{a} - \dfrac{x^3}{a^3}\right) + \alpha\left(\dfrac{x^5}{a^5} - \dfrac{1}{2}\left(\dfrac{x^7}{a^7} + \dfrac{x^9}{a^9}\right)\right)\right] \times \\[3mm] \dfrac{d^2}{dx^2}\left[\left(\dfrac{x}{a} - \dfrac{x^3}{a^3}\right) + \alpha\left(\dfrac{x^5}{a^5} - \dfrac{1}{2}\left(\dfrac{x^7}{a^7} + \dfrac{x^9}{a^9}\right)\right)\right]\end{array}\right]dx}{\displaystyle\int_0^a \left[\left(\dfrac{x}{a} - \dfrac{x^3}{a^3}\right) + \alpha\left(\dfrac{x^5}{a^5} - \dfrac{1}{2}\left(\dfrac{x^7}{a^7} + \dfrac{x^9}{a^9}\right)\right)\right]^2 dx} \qquad \textbf{(10.13)}$$

Carrying out this integration gives E in terms of $\hbar$, m, a, and α:

$$E = \frac{\hbar^2}{2ma^2}\frac{\left(\dfrac{4}{5} + \dfrac{116\alpha}{231} + \dfrac{40247\alpha^2}{109395}\right)}{\left(\dfrac{8}{105} + \dfrac{8\alpha}{273} + \dfrac{1514\alpha^2}{230945}\right)} \qquad \textbf{(10.14)}$$

To minimize the energy with respect to the variational parameter, we differentiate this function with respect to α, set the resulting equation equal to zero, and solve for α. The solutions are $\alpha = -5.74$ and $\alpha = -0.345$. The second of these solutions corresponds to the minimum in E. Substituting this value in Equation (10.14) gives $E = 0.127(h^2/ma^2)$, which is very close to the true value of $0.125(h^2/ma^2)$. The optimized trial function is shown in Figure 10.2b. We can see that, by choosing the optimal value of α, $E \rightarrow E_0$ and $\Phi \rightarrow \psi_0$. No better value for the energy can be obtained with this particular choice of a trial wave function, and this illustrates a limitation of the variational method. The "best" energy obtained depends on the choice of the trial function. For example, a lower energy is obtained if a function of the type $\Phi(x) = x^\alpha(a - x)^\alpha$ is minimized with respect to α. This example shows how the variational method can be implemented by optimizing approximate solutions to the Schrödinger equation.

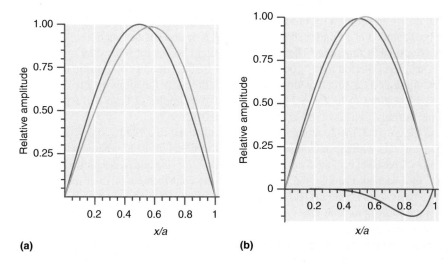

(a)   (b)

FIGURE 10.2

Exact (red curve) and approximate (yellow curve) wave functions for the ground state of the particle in the box. **(a)** The approximate wave function contains only the first term in Equation (10.11). **(b)** The optimal approximate wave function contains both terms of Equation (10.11). The blue curve shows the contribution of the second term in Equation (10.11) to the approximate wave function.

10.5 THE HARTREE–FOCK SELF-CONSISTENT FIELD METHOD

We now return to the problem at hand, namely, solving the Schrödinger equation for many-electron atoms. The starting point is to use the orbital approximation and to take the Pauli exclusion principle into account. Antisymmetry of the wave function with respect to electron exchange is accomplished by expressing the wave function as a Slater determinant

$$\psi(1,2,3,\ldots,n) = \frac{1}{\sqrt{n!}}\begin{vmatrix} \phi_1(1)\alpha(1) & \phi_1(1)\beta(1) & \ldots & \phi_m(1)\beta(1) \\ \phi_1(2)\alpha(2) & \phi_1(2)\beta(2) & \ldots & \phi_m(2)\beta(2) \\ \ldots & \ldots & \ldots & \ldots \\ \phi_1(n)\alpha(n) & \phi_1(n)\beta(n) & \ldots & \phi_m(n)\beta(n) \end{vmatrix} \quad (10.15)$$

in which the individual entries $\phi_j(k)$ are modified H atom orbitals as described later. The Hartree–Fock method is a prescription for finding the single Slater determinant that gives the lowest energy for the ground-state atom in the absence of electron correlation. (More correctly, configurations with more than one unpaired electron require more than one Slater determinant.)

As for the helium atom discussed in Section 10.1, it is assumed that the electrons are uncorrelated and that a particular electron feels the spatially averaged electron charge distribution of the remaining $n - 1$ electrons. These approximations reduce the n-electron Schrödinger equation to n one-electron Schrödinger equations which have the form

$$\left(-\frac{\hbar^2}{2m}\nabla_i^2 + V_i^{eff}(\mathbf{r}_i)\right)\phi_i(\mathbf{r}_i) = \varepsilon_i\phi_i(\mathbf{r}_i), \quad i = 1,\ldots,n \quad (10.16)$$

in which the effective potential energy felt by the first electron, $V_1^{eff}(\mathbf{r}_1)$, takes into account the electron-nuclear attraction, and the repulsion between electron 1 and all other electrons. The Hartree–Fock method allows the best (in a variational sense) one-electron orbitals $\phi_i(\mathbf{r}_i)$ and the corresponding orbital energies ε_i to be calculated.

Because of the neglect of electron correlation, the effective potential is spherically symmetrical and, therefore, the angular part of the wave functions is identical to the solutions for the hydrogen atom. *This means that the s, p, d, ... orbital nomenclature derived for the hydrogen atom remains intact for the one-electron orbitals for all atoms.* What remains to be found are solutions to the radial part of the Schrödinger equation.

To optimize the radial part of the determinantal wave function, the variational method outlined in Section 10.4 is used. What functions should be used for the individual entries $\phi_j(r)$ in the determinant? Each $\phi_j(r)$ is expressed as a linear combination of suitable **basis functions** $f_i(r)$ as shown in Equation (10.17).

$$\phi_j(r) = \sum_{i=1}^{m} c_i f_i(r) \quad (10.17)$$

What do we mean by a set of suitable functions? Recall that a well-behaved function can be expanded in a Fourier series as a sum of sine and cosine functions, which in this context are basis functions. There are many other choices for individual members of a basis set. The criterion for a "good" basis set is that the number of terms in the sum representing $\phi_j(r)$ is as small as possible and that the basis functions enable the Hartree–Fock calculations to be carried out rapidly. Two examples of basis set expansions for atomic orbitals are shown in Figures 10.3 and 10.4.

In Figure 10.3, the 2p atomic orbital of Ne obtained in a Hartree–Fock calculation is shown together with the individual contributions to Equation 10.17 where each member of the $m = 4$ basis set is of the form $f_i(r) = N_i r \exp[-\zeta_i r/a_0]$ and N_i is a normalization constant. In a second example, the H 1s AO and the contributions of each member of the $m = 3$ basis set to Equation (10.17) are shown in Figure 10.4, where the basis set functions are of the form $f_i(r) = N_i \exp[-\zeta_i(r/a_0)^2]$ (Gaussian functions). In both cases, the coefficients c_i in Equation 10.17 are used as variational parameters to optimize $\phi_j(r)$ and the ζ_i values are optimized separately. Although the Gaussian functions do not represent the H 1s

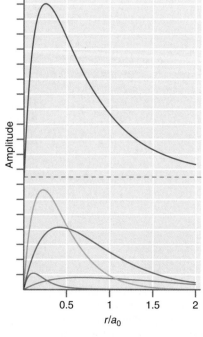

FIGURE 10.3

The top curve shows the radial function for the 2p orbital in Ne determined in a Hartree–Fock calculation. It has been shifted upward for clarity. The bottom four curves are the individual terms in the four-element basis set.

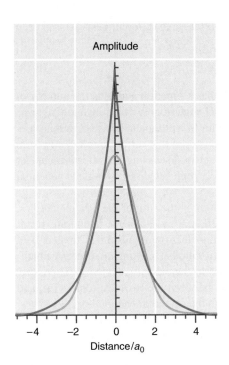

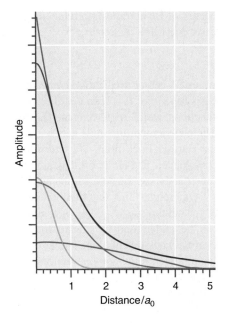

FIGURE 10.4

The left panel shows a fit to a H $1s$ orbital with a single Gaussian function. The agreement is not good. The right panel shows a best fit (dark blue curve) using a basis set of three Gaussian functions. Except very near the nucleus, the fit is very good.

function accurately near the nucleus, they are well suited to Hartree–Fock calculations and are the most widely used basis functions in computational chemistry. (See Chapter 15 for a more detailed discussion of Gaussian basis functions.)

The preceding discussion describes the input to a calculation of the orbital energies, but there is a problem in proceeding with the calculation. To solve the Schrödinger equation for electron 1, $V_1^{eff}(\mathbf{r}_1)$ must be known, and this means that we must know the functional form of all the other orbitals $\phi_2(\mathbf{r}_2)$, $\phi_3(\mathbf{r}_3)$, $\ldots$, $\phi_n(\mathbf{r}_n)$. This is also the case for the remaining $n-1$ electrons. In other words, the answers must be known in order to solve the problem.

The way out of this quandary is to use an iterative approach. A reasonable guess is made for an initial set of $\phi_j(k)$. Using these orbitals, an effective potential is calculated, and the energy and improved orbital functions, $\phi_j'(k)$, for each of the n electrons are calculated. The $\phi_j'(k)$ are used to calculate a new effective potential, which is used to calculate a further improved set of orbitals, $\phi_j''(k)$, and this procedure is repeated for all electrons until the solutions for the energies and orbitals are self-consistent, meaning that they do not change significantly in a further iteration. This procedure, coupled with the variational method in optimizing the parameters in the orbitals, is very effective in giving the best one-electron orbitals and energies available for a many-electron atom in the absence of electron correlation. More accurate calculations that include electron correlation are discussed in Chapter 15.

The accuracy of a Hartree–Fock calculation depends primarily on the size of the basis set. This dependence is illustrated in Table 10.1 in which the calculated total

TABLE 10.1   TOTAL ENERGY AND $1s$ ORBITAL ENERGY FOR He FOR THREE DIFFERENT BASIS SETS USED TO REPRESENT THE $1s$ ORBITAL

Number of Basis Functions, m	Exponents, ζ_i	Total Energy of He, ε_{total} (eV)	$1s$ Orbital Energy, ε_{1s} (eV)	$\varepsilon_{total} - 2\varepsilon_{1s}$ (eV)
5	1.41714, 2.37682, 4.39628, 6.52699, 7.94252	−77.8703	−24.9787	−27.9129
2	2.91093, 1.45363	−77.8701	−24.9787	−27.9133
1	1.68750	−77.4887	−24.3945	−28.6998

The data is taken from E. Clementi and C. Roetti, *Atomic Data and Nuclear Data Tables*, 14 (1974), 177.

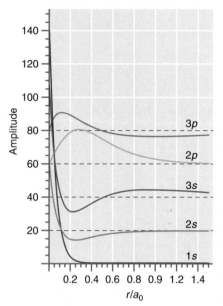

FIGURE 10.5

Hartree–Fock radial functions are shown for Ar. The curves are offset vertically to allow individual functions to be compared.

[Calculated from data in E. Clementi and C. Roetti, *Atomic Data and Nuclear Data Tables*, 14 (1974), 177.]

energy of He and the $1s$ orbital energy are shown for three different basis sets. In each case, $\phi_{1s}(r)$ has the form

$$\phi_{1s}(r) = \sum_{i=1}^{m} c_i N_i e^{-\zeta_i r/a_0} \tag{10.18}$$

where N_i is a normalization constant for the ith basis function and m is the number of basis functions. It is seen that there is almost no change in going from two to five basis functions, which represents the Hartree–Fock limit of a complete basis set in this case. The one element or single zeta basis set gives an energy that differs significantly from the Hartree–Fock limit. We return to this basis set in discussing the effective nuclear charge later. The He $1s$ orbital cannot be accurately represented by a single exponential function as was the case for the hydrogen atom.

One might think that the total energy of an atom is the sum of the orbital energies, or for helium, $\varepsilon_{total} = 2\varepsilon_{1s}$. As shown in Table 10.1, $\varepsilon_{total} - 2\varepsilon_{1s} < 0$, and this result can be understood by considering how electron–electron repulsion is treated in a Hartree–Fock calculation. The $1s$ orbital energy is calculated using an effective potential in which repulsion between the two electrons in the orbital is included. Therefore, assuming that $\varepsilon_{total} = 2\varepsilon_{1s}$ counts the repulsion between the two electrons twice, and gives a value for ε_{total} which is more positive than the true total energy.

Radial functions for Ar in the Hartree–Fock limit of a large basis set are shown in Figure 10.5. It is seen that they have the same nodal structure as the orbitals for the hydrogen atom.

The Hartree–Fock radial functions can be used to obtain the radial probability distribution for many-electron atoms from

$$P(r) = \sum_i n_i r^2 R_i(r) \tag{10.19}$$

where $R_i(r)$ is the radial function corresponding to the ith subshell, for example $2s$, $3p$, or $4d$, and n_i is the number of electrons in the subshell. $P(r)$ is shown for Ne, Ar, and Kr in Figure 10.6. Note that the radial distribution exhibits a number of maxima, one for each occupied shell and that the contributions from different shells overlap. The width of $P(r)$ for a given shell increases with n; it is smallest for $n = 1$, and largest for $n = 4$.

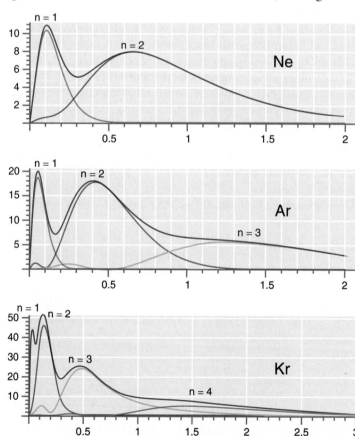

FIGURE 10.6

Radial distribution functions calculated from Hartree–Fock wave functions are shown for Ne, Ar, and Kr. The colored curves show the contributions from the individual shells and the dark blue curve shows the total radial distribution function.

[Calculated from data in E. Clementi and C. Roetti, *Atomic Data and Nuclear Data Tables*, 14 (1974), 177.]

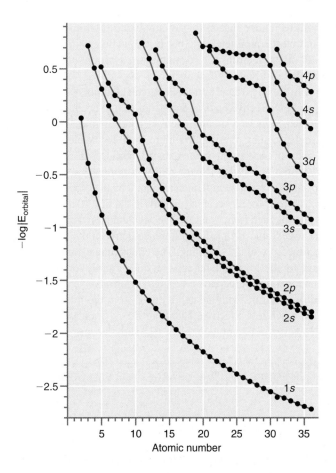

FIGURE 10.7

The one-electron orbital energies obtained from Hartree–Fock calculations are shown on a logarithmic scale for the first 36 elements.

[The data are taken from E. Clementi and C. Roetti, *Atomic Data and Nuclear Data Tables,* 14 (1974), 177.]

Hartree–Fock orbital energies ε_i are shown in Figure 10.7 for the first 36 elements in the periodic table of the elements. An important result of these calculations is that the ε_i for many-electron atoms depend on both the principal quantum number n and on the angular momentum quantum number l. Within a shell of principal quantum number n, $\varepsilon_{ns} < \varepsilon_{np} < \varepsilon_{nd} < \ldots$. This was not the case for the H atom. This result can be understood by considering the radial distribution functions for Kr shown in Figure 10.8. As discussed in Chapter 9, this function gives the probability of finding an electron at a given distance from the nucleus. The subsidiary maximum near $r = a_0$ in the $3s$ radial distribution function indicates that there is a higher probability of finding the $3s$ electron close to the nucleus than is the case for the $3p$ and $3d$ electrons. The potential energy associated with the attraction between the nucleus and the electron falls off as $1/r$, so that its magnitude increases substantially as the electron comes closer to the nucleus. As a result, the $3s$ electron is bound more strongly to the nucleus and, therefore, the orbital energy is more negative than for the $3p$ and $3d$ electrons. The same argument can be used to understand why the $3p$ orbital energy is lower than that for the $3d$ orbital. Figure 10.5 also shows that the energy of a given orbital decreases strongly with the atomic number. This is a result of the increase in the attractive force between the nucleus and an electron as the charge on the nucleus increases.

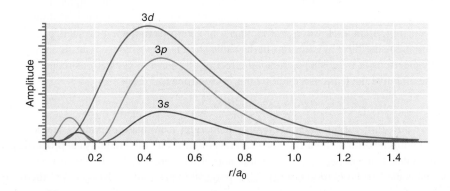

FIGURE 10.8

The contributions of the $3s$, $3p$, and $3d$ subshells to the radial distribution function of krypton obtained from Hartree–Fock calculations are shown.

[Calculated from data in E. Clementi and C. Roetti, *Atomic Data and Nuclear Data Tables,* 14 (1974), 177.]

It is important to realize that ε_i for a many-electron atom depends on the electron configuration and on the atomic charge because ε_i is determined in part by the average distribution of all other electrons. For example, the Hartree–Fock limiting value for ε_{1s} is -67.4 eV for neutral Li and -76.0 eV for Li^+.

A further useful result from Hartree–Fock calculations are values for the effective nuclear charge, ζ. The effective nuclear charge takes into account that an electron farther from the nucleus experiences a smaller nuclear charge than that experienced by an inner electron. This can be seen by referring to Figure 10.1. To the electron in question, it looks as though the nuclear charge has been reduced because of the presence of the other smeared-out electrons. This effect is particularly important for valence electrons and we say that they are *shielded* from the full nuclear charge by the core electrons closer to the nucleus. Table 10.2 shows ζ for all occupied orbitals in the first 10 atoms in the periodic table. The zeta values are obtained from a Hartree–Fock calculation using the single zeta basis set discussed earlier (See Table 10.1.). The difference between the true and effective nuclear charge is a direct measure of the shielding. The effective nuclear charge is nearly equal to the nuclear charge for the $1s$ orbital, but falls off quite rapidly for the outermost electron as the principal quantum number increases. Whereas electrons of smaller n value are quite effective in **shielding** electrons with greater n values from the full nuclear change, those in the same shell are much less effective. Therefore $Z - \zeta$ increases as we move across the periodic table. However, as Example Problem 10.3 shows, some subtle effects are involved.

TABLE 10.2 EFFECTIVE NUCLEAR CHARGES FOR SELECTED ATOMS

	H (1)							He (2)
$1s$	1.00							1.69

	Li (3)	Be (4)	B (5)	C (6)	N (7)	O (8)	F (9)	Ne (10)
$1s$	2.69	3.68	4.68	5.67	6.66	7.66	8.65	9.64
$2s$	1.28	1.91	2.58	3.22	3.85	4.49	5.13	5.76
$2p$			2.42	3.14	3.83	4.45	5.10	5.76

EXAMPLE PROBLEM 10.3

The effective nuclear charge seen by a $2s$ electron in Li is 1.28. We might expect this number to be 1.0 rather than 1.28. Why is ζ larger than 1? Similarly, explain the effective nuclear charge seen by a $2s$ electron in carbon.

Solution

The effective nuclear charge seen by a $2s$ electron in Li will be only 1.0 if all the charge associated with the $1s$ electrons is located between the nucleus and the $2s$ shell. As Figure 9.10 shows, a significant fraction of the charge is located farther from the nucleus than the $2s$ shell, and some of the charge is quite close to the nucleus. Therefore, the effective nuclear charge seen by the $2s$ electrons is reduced by a number smaller than 2. On the basis of the argument presented for Li, we expect the shielding by the $1s$ electrons in carbon to be incomplete and we might expect the effective nuclear charge felt by the $2s$ electrons in carbon to be more than 4. However, carbon has four electrons in the $n = 2$ shell, and although shielding by electrons in the same shell is less effective than shielding by electrons in inner shells, the total effect of all four $n = 2$ electrons reduces the effective nuclear charge felt by the $2s$ electrons to 3.22.

We now turn our attention to the orbital energies ε_i. What observables can be associated with the orbital energies? The most meaningful link of ε_i to physical properties is to the ionization energy. To a reasonable approximation, $-\varepsilon_i$ for the highest occupied orbital is the first **ionization energy**. This association is known as **Koopmans' theorem**

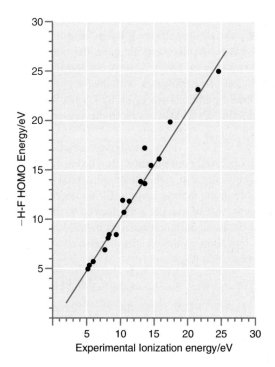

FIGURE 10.9

The negative of the highest occupied Hartree–Fock orbital is graphed against experimentally determined first ionization energies of the first 18 elements. If the two values were identical, all points would lie on the red line.

in the "frozen core" limit in which it is assumed that the electron distribution in the atom is not affected by the removal of an electron in the ionization event. Figure 10.9 shows that the agreement between the experimentally determined first ionization energy and the highest occupied orbital energy is quite good.

By analogy, $-\varepsilon_i$ for the lowest unoccupied orbital should give the **electron affinity** for a particular atom. However, Hartree–Fock electron affinity calculations are much less accurate than ionization energies. The electron affinity for F based on $-\varepsilon_i$ for the lowest unoccupied orbital is negative. This result predicts that the F^- ion is less stable than the neutral F atom, contrary to experiment. A better estimate of the electron affinity of F is obtained by comparing the total energies of F and F^-. This gives a value for the electron affinity of 0.013 eV which is still much smaller than the experimental value of 3.34 eV. More accurate calculations, including electron correlation as discussed in Chapter 15 are necessary to obtain accurate results for the ionization energy and electron affinity of atoms.

The electron configuration of most atoms can be obtained by using Figure 10.10, which shows the order in which the atomic orbitals are generally filled based on the orbital energy sequence of Figure 10.7. Filling orbitals in this sequential order is known as the **Aufbau principle**, and it is often asserted that the relative order of orbital energies explains the electron configurations of the atoms in the periodic table. However, this assertion is not always true.

To illustrate this point, consider the known configurations of the first transition series shown in Table 10.3. Figure 10.7 shows that the 4s orbital is lower in energy than

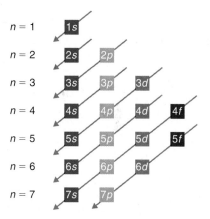

FIGURE 10.10

The order in which orbitals in many-electron atoms are filled for most atoms is described by the gray lines, starting from the top of the figure. Twelve of the 40 transition elements show departures from this order.

TABLE 10.3 CONFIGURATIONS FOR FOURTH ROW ATOMS

Nuclear Charge	Element	Electron Configuration	Nuclear Charge	Element	Electron Configuration
19	K	$[Ar]4s^1$	25	Mn	$[Ar]4s^23d^5$
20	Ca	$[Ar]4s^2$	26	Fe	$[Ar]4s^23d^6$
21	Sc	$[Ar]4s^23d^1$	27	Co	$[Ar]4s^23d^7$
22	Ti	$[Ar]4s^23d^2$	28	Ni	$[Ar]4s^23d^8$
23	V	$[Ar]4s^23d^3$	29	Cu	$[Ar]4s^13d^{10}$
24	Cr	$[Ar]4s^13d^5$	30	Zn	$[Ar]4s^23d^{10}$

the 3*d* orbital for K and Ca, but that the order is reversed for higher atomic numbers. Is the order in which the *s* and *d* subshells are filled in the 4th period explained by the relative energy of the orbitals? If this were the case, the configuration $[Ar]4s^03d^n$ with $n = 3, \ldots, 10$ would be predicted for the sequence scandium-nickel. [Ar] is an abbreviation for the configuration of Ar. However, with the exception of Cr and Cu, the experimentally determined configurations are given by $[Ar]4s^23d^n$, with $n = 1, \ldots, 10$ for the sequence scandium-zinc. Cr and Cu have a single 4*s* electron because a half-filled or filled d shell lowers the energy of an atom.

As has been shown by L. G. Vanquickenbourne *et al.* [*J. Chemical Education* 71 (1994), 469–471], the observed configurations can be explained if the total energies of the various possible configurations rather than the orbital energies are compared. We show that it is favorable in the neutral atom to fill the *s* orbital before the *d* orbital by considering the energetic cost of moving a 4*s* electron to the 3*d* orbital. The difference in the total energy of the two configurations is a balance between the orbital energies and the electrostatic repulsion of the electrons involved in the promotion. ΔE for a $4s^23d^n \rightarrow 4s^13d^{n+1}$ promotion is given by

$$\Delta E(4s \rightarrow 3d) \cong (\varepsilon_{3d} - \varepsilon_{4s}) + [E_{repulsive}(3d, 3d) - E_{repulsive}(3d, 4s)] \quad \textbf{(10.20)}$$

The second term in Equation (10.20) represents the difference in the repulsive energies of the two configurations. What is the sign of the second term? Figure 10.6 shows that the distance corresponding to the principal maxima in the radial probability distribution for a typical many-electron atom follows the order $3s > 3p > 3d$. We conclude that the *d* electrons are more localized than the *s* electrons, and therefore the repulsive energies follow the order $E_{repulsive}(3d, 3d) > E_{repulsive}(3d, 4s) > E_{repulsive}(4s, 4s)$. Therefore the sign of the second term in Equation 10.20 is positive. For this transition metal series, the magnitude of the repulsive term is greater than the magnitude of the difference in the orbital energies. Therefore, even though $(\varepsilon_{3d} - \varepsilon_{4s}) < 0$ for scandium, the promotion $4s^23d^1 \rightarrow 4s^03d^3$ does not occur because $(\varepsilon_{3d} - \varepsilon_{4s}) + [E_{repulsive}(3d, 3d) - E_{repulsive}(3d, 4s)] > 0$ and is larger than $|(\varepsilon_{3d} - \varepsilon_{4s})|$. The energy lowering from promotion to the lower orbital energy is more than offset by the energy increase resulting from electron repulsion. Therefore, Sc has the configuration $[Ar]4s^23d^1$ rather than $[Ar]4s^03d^3$.

These calculations also explain the seemingly anomalous configurations for the doubly charged positive ions in the sequence scandium-zinc, which are $[Ar]4s^03d^n$ with $n = 1, \ldots, 10$. The removal of two electrons significantly increases the effective nuclear charge felt by the remaining electrons. As a result, both ε_{4s} and ε_{3d} are lowered substantially, but ε_{3d} is lowered more. Therefore, $\varepsilon_{3d} - \varepsilon_{4s}$ becomes more negative. For the doubly charged ions, the magnitude of the repulsive term is less than the magnitude of the difference in the orbital energies. As a consequence, the doubly ionized configurations are those that would be predicted by filling the lower lying 3*d* orbital before the 4*s* orbital.

Recall that Hartree–Fock calculations neglect electron correlation. Therefore, the total energy is larger than the true energy by an amount called the **correlation energy**. For example, the correlation energy for He is 110 kJ mol^{-1}. This amount increases somewhat faster than the number of electrons in the atom. Although the correlation energy is a small percentage of the total energy of the atom and decreases with the atomic number (1.4% for He and 0.1% for K), it presents a problem in the application of Hartree–Fock calculations to chemical reactions for the following reason. In chemical reactions, we are not interested in the total energies of the reactants and products, but rather in the difference between the Gibbs energy and enthalpy of the reactants and products. These changes are on the order of 100 kJ mol^{-1}, so that errors in quantum chemical calculations resulting from the neglect of the electron correlation can lead to significant errors in thermodynamic calculations. However, the neglect of correlation is often less serious than might be expected. The resulting error in the total energy is often similar for the reactants and products if the number of unpaired electrons is the same for reactants and products. For such reactions, the neglect of electron correlation largely cancels in thermodynamic calculations. Additionally, the coordinated work of many quantum chemists over decades has led to computational methods that go beyond Hartree–Fock by including electron correlation. These advances make it possible to calculate thermodynamic functions and activation energies for many reactions for which it would be very difficult to obtain experimental data. These computational methods will be discussed in Chapter 15.

10.6 UNDERSTANDING TRENDS IN THE PERIODIC TABLE FROM HARTREE–FOCK CALCULATIONS

We briefly summarize the main results of Hartree–Fock calculations for atoms:

- The orbital energy depends on both n and l. Within a shell of principal quantum number n, $\varepsilon_{ns} < \varepsilon_{np} < \varepsilon_{nd} < \ldots.$
- Electrons in a many-electron atom are shielded from the full nuclear charge by other electrons. Shielding can be modeled in terms of an effective nuclear charge. Core electrons are more effective in shielding outer electrons than electrons in the same shell.
- The ground-state configuration for an atom results from a balance between orbital energies and electron–electron repulsion.

In addition to the orbital energies, two parameters that can be calculated using the Hartree–Fock method are very useful in understanding chemical trends in the periodic table. They are the atomic radius and the electronegativity. Values for atomic radii are obtained by calculating the radius of the sphere that contains ~90% of the electron charge. This radius is determined by the effective charge felt by valence shell electrons.

The degree to which atoms accept or donate electrons to other atoms in a reaction is closely related to the first ionization energy and the electron affinity, which we associate with the HOMO and LUMO orbitals. For example, the energy of these orbitals allows us to predict whether the ionic NaCl species is better described by Na^+Cl^- or Na^-Cl^+. Formation of Na^+ and Cl^- ions at infinite separation requires

$$\Delta E = E^{Na}_{ionization} - E^{Cl}_{electron\ affinity} = 5.14\ eV - 3.61\ eV = 1.53\ eV \qquad \textbf{(10.21)}$$

Formation of oppositely charged ions requires

$$\Delta E = E^{Cl}_{ionization} - E^{Na}_{electron\ affinity} = 12.97\ eV - 0.55\ eV = 12.42\ eV \qquad \textbf{(10.22)}$$

In each case, additional energy is gained by bringing the ions together. Clearly the formation of Na^+Cl^- is favored over Na^-Cl^+. The concept of **electronegativity**, which is given the symbol χ, quantifies this tendency of atoms to either accept or donate electrons to another atom in a chemical bond. Because the noble gases in group VIII do not form chemical bonds (with very few exceptions), they are not generally assigned values of χ.

Several definitions of electronegativity (which has no units) exist, but all lead to similar results when scaled to the same numerical range. For instance, χ as defined by Mulliken is given by

$$\chi = 0.187(IE + EA) + 0.17 \qquad \textbf{(10.23)}$$

where IE is the first ionization energy and EA is the electron affinity. It is basically the average of the first ionization energy and the electron affinity with the parameters 0.187 and 0.17 chosen to optimize the correlation with the earlier electronegativity scale of Pauling which is based on bond energies. The Mulliken definition of χ can be understood using Figure 10.11.

Assume that an atom with a small ionization energy and electron affinity (A) forms a bond with an atom that has a larger ionization energy and electron affinity (B). Partial charge transfer from A to B lowers the energy of the system and is therefore favored over the reverse process which increases the energy of the system. Chemical bonds between atoms with large differences in χ have a strong ionic character because significant electron transfer occurs. Chemical bonds between atoms that have similar χ values are largely covalent, because the driving force for electron transfer is small, and valence electrons are shared nearly equally by the atoms.

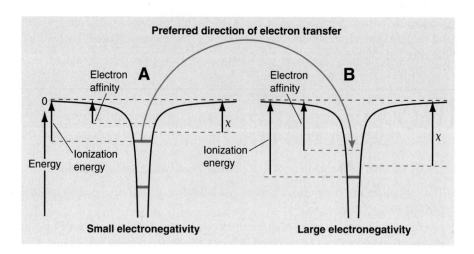

FIGURE 10.11

The energy of the molecule AB is lowered if electron charge is transferred from A to B rather than from B to A.

Figure 10.12 compares values for the atomic radius, first ionization energy, and χ as a function of atomic number up to $Z = 55$. This range spans one period in which only the $1s$ orbital is filled, two short periods in which only s and p orbitals are filled, and two longer periods in which d orbitals are also filled. Beginning with the covalent radius, we see the trends predicted from calculated ζ values for the valence electrons which increase in going across a period and down a group as shown for the main group elements in Figure 10.13.

The radii decrease continuously in going across a period, but increase abruptly as n increases by one in moving to the next period. Moving down a group of the periodic table, the radius increases with n because ζ increases more slowly with the nuclear charge than in moving across a period. Small radii are coupled with large ζ, and this combination leads to a large ionization energy. Therefore, changes in the ionization energy follow the opposite trend to that for the atomic radii. The ionization energy falls in moving down a column, because the atomic radius increases more rapidly than ζ

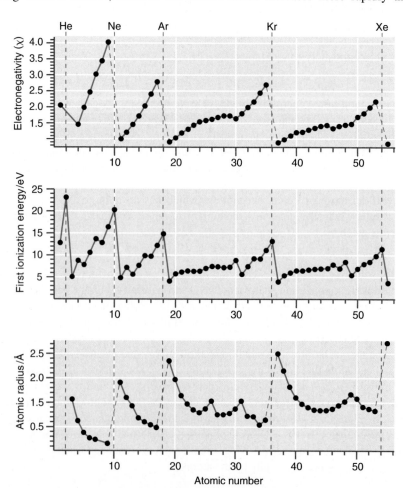

FIGURE 10.12

The electronegativity, first ionization energy, and covalent atomic radius are plotted as a function of the atomic number for the first 55 elements. Dashed vertical lines mark the completion of each period.

H 1s 1							He 1s 1.69
Li 2s 1.28	Be 2s 1.91	B 2s 2.58 2p 2.42	C 2s 3.22 2p 3.14	N 2s 3.85 2p 3.83	O 2s 4.49 2p 4.45	F 2s 5.13 2p 5.10	Ne 2s 5.76 2p 5.76
Na 3s 2.51	Mg 3s 3.31	Al 3s 4.12 3p 4.07	Si 3s 4.90 3p 4.29	P 3s 5.64 3p 4.89	S 3s 6.37 3p 5.48	Cl 3s 7.07 3p 6.12	Ar 3s 7.76 3p 6.76
K 4s 3.50	Ca 4s 4.40	Ga 4s 7.07 4p 6.22	Ge 4s 8.04 4p 6.78	As 4s 8.94 4p 7.45	Se 4s 9.76 4p 8.29	Br 4s 10.55 4p 9.03	Kr 4s 11.32 4p 9.77
Rb 5s 4.98	Sr 5s 6.07	In 5s 9.51 5p 8.47	Sn 5s 10.63 5p 9.10	Sb 5s 10.61 5p 9.99	Te 5s 12.54 5p 10.81	I 5s 13.40 5p 11.61	Xe 5s 14.22 5p 12.42

FIGURE 10.13

Effective nuclear charges are shown for valence shell electrons of main group elements in the first five periods in the periodic table.

increases. The electronegativity follows the same pattern as the ionization energy because in general the ionization energy is larger than the electron affinity.

Vocabulary

antiparallel spins
antisymmetric wave function
Aufbau principle
basis functions
configuration
correlation energy
effective nuclear charge
electron affinity
electron correlation
electron–electron repulsion
electron spin

electronegativity
Hartree–Fock self-consistent field
   method
indistinguishability
ionization energy
Koopmans' theorem
orbital approximation
orbital energy
paired electrons
parallel spins
Pauli exclusion principle

shell
shielding
Slater determinant
subshell
symmetric wave function
total angular momentum
trial wave function
unpaired electrons
variational method
variational theorem

Questions on Concepts

Q10.1 Why does the effective nuclear charge for the 1s orbital increase by 0.99 in going from oxygen to fluorine, but it only increases by 0.65 for the 2p orbital?

Q10.2 There are more electrons in the $n = 4$ shell than for the $n = 3$ shell in krypton. However, the peak in the radial distribution in Figure 10.6 is smaller for the $n = 4$ shell than for the $n = 3$ shell. Explain this fact.

Q10.3 How is the effective nuclear charge related to the size of the basis set in a Hartree–Fock calculation?

Q10.4 The angular functions, $\Theta(\theta)\Phi(\phi)$, for the one-electron Hartree–Fock orbitals are the same as for the hydrogen atom, and the radial functions and radial probability functions are similar to those for the hydrogen atom. For this question, assume that the latter two functions are identical to those for the hydrogen atom. The following figure shows (a) a contour plot in the $x-y$ plane with the y axis being the vertical axis, (b) the radial function, and (c) the radial probability distribution for a one-electron orbital. Identify the orbital ($2s$, $4d_{xz}$, and so on).

(a)

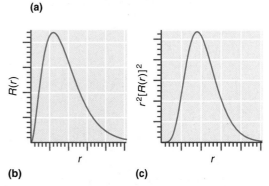

(b)   (c)

Q10.5 What is the functional dependence of the 1s orbital energy on Z in Figure 10.7? Check your answer against a few data points.

Q10.6 See Question Q10.4.

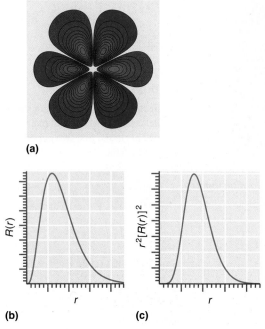

(a)

(b)   (c)

Q10.7 Explain why shielding is more effective by electrons in a shell of lower principal quantum number than by electrons having the same principal quantum number.

Q10.8 Are the elements of a basis set observable in an experiment? Explain your reasoning.

Q10.9 Show using an example that the following two formulations of the Pauli exclusion principle are equivalent:

a. Wave functions describing a many-electron system must change sign under the exchange of any two electrons.

b. No two electrons may have the same values for all four quantum numbers.

Q10.10 See Question Q10.4.

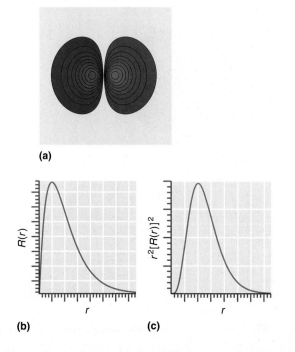

(a)

(b)   (c)

Q10.11 See Question Q10.4.

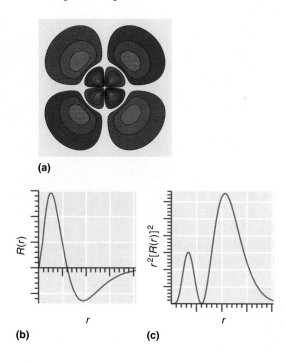

(a)

(b)   (c)

Q10.12 Why is the total energy of a many-electron atom not equal to the sum of the orbital energies for each electron?

Q10.13 See Question Q10.4.

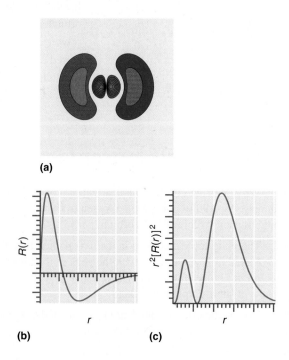

(a)

(b)   (c)

Q10.14   See Question Q10.4.

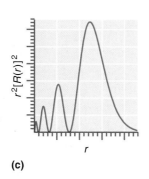

(a)

(b)   (c)

Q10.15   See Question Q10.4.

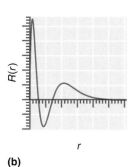

(a)

(b)   (c)

Q10.16   Show that the Slater determinant formalism automatically incorporates the Pauli exclusion principle by evaluating the He ground-state wave function of Equation (10.8), giving both electrons the same quantum numbers.

Q10.17   Is there a physical reality associated with the individual entries of a Slater determinant?

Q10.18   See Question Q10.4.

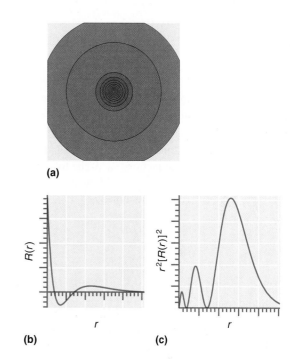

(a)

(b)   (c)

Q10.19   How can you tell if one basis set is better than another in calculating the total energy of an atom?

Q10.20   Why is the $s, p, d, \ldots$ nomenclature derived for the H atom also valid for many-electron atoms?

Q10.21   Would the trial wave function

$$\Phi(x) = \left(\frac{x}{a} - \frac{x^3}{a^3}\right) + \alpha\left(\frac{x^5}{a^5} - \frac{1}{2}\left(\frac{x^7}{a^7}\right)\right), \quad 0 < x < a$$

have been a suitable choice for the calculations carried out in Section 10.4? Justify your answer.

Problems

Problem numbers in **red** indicate that the solution to the problem is given in the *Student's Solutions Manual*.

P10.1   Is $\psi(1, 2) = 1s(1)\alpha(1)\,1s(2)\beta(2) + 1s(2)\alpha(2)1s(1)\beta(1)$ an eigenfunction of the operator $\hat{S}_z$? If so, what is its eigenvalue M_S?

P10.2   Calculate the angles that a spin angular momentum vector for an individual electron can make with the z axis.

P10.3   In this problem we represent the spin eigenfunctions and operators as vectors and matrices.

a. The spin eigenfunctions are often represented as the column vectors

$$\alpha = \begin{pmatrix} 1 \\ 0 \end{pmatrix} \quad \text{and} \quad \beta = \begin{pmatrix} 0 \\ 1 \end{pmatrix}$$

Show that α and β are orthogonal using this representation.

b. If the spin angular momentum operators are represented by the matrices

$$\hat{s}_x = \frac{\hbar}{2}\begin{pmatrix} 0 & 1 \\ 1 & 0 \end{pmatrix}, \hat{s}_y = \frac{\hbar}{2}\begin{pmatrix} 0 & -i \\ i & 0 \end{pmatrix}, \hat{s}_z = \frac{\hbar}{2}\begin{pmatrix} 1 & 0 \\ 0 & -1 \end{pmatrix}$$

show that the commutation rule $[\hat{s}_x, \hat{s}_y] = i\hbar\hat{s}_z$ holds.

c. Show that

$$\hat{s}^2 = \hat{s}_x^2 + \hat{s}_y^2 + \hat{s}_z^2 = \frac{\hbar^2}{4}\begin{pmatrix} 3 & 0 \\ 0 & 3 \end{pmatrix}$$

d. Show that α and β are eigenfunctions of $\hat{s}_z$ and $\hat{s}^2$. What are the eigenvalues?

e. Show that α and β are not eigenfunctions of $\hat{s}_x$ and $\hat{s}_y$.

P10.4 In this problem you will prove that the ground-state energy for a system obtained using the variational method is greater than the true energy.

a. The approximate wave function Φ can be expanded in the true (but unknown) eigenfunctions ψ_n of the total energy operator in the form $\Phi = \sum_n c_n\psi_n$. Show that by substituting $\Phi = \sum_n c_n\psi_n$ in the equation

$$E = \frac{\displaystyle\int \Phi^*\hat{H}\Phi\,d\tau}{\displaystyle\int \Phi^*\Phi\,d\tau}$$

you obtain the result

$$E = \frac{\displaystyle\sum_n\sum_m \int (c_n^*\psi_n^*)\hat{H}(c_m\psi_m)\,d\tau}{\displaystyle\sum_n\sum_m \int (c_n^*\psi_n^*)(c_m\psi_m)\,d\tau}$$

b. Because the ψ_n are eigenfunctions of $\hat{H}$, they are orthogonal and $\hat{H}\psi_n = E_n\psi_n$. Show that this information allows us to simplify the expression for E from part (a) to

$$E = \frac{\displaystyle\sum_m E_m c_m^* c_m}{\displaystyle\sum_m c_m^* c_m}$$

c. Arrange the terms in the summation such that the first energy is the true ground-state energy E_0 and the energy increases with the summation index m. Why can you conclude that $E - E_0 \geq 0$?

P10.5 In this problem you will show that the charge density of the filled $n = 2, l = 1$ subshell is spherically symmetrical and that therefore $\mathbf{L} = 0$. The angular distribution of the electron charge is simply the sum of the squares of the magnitude of the angular part of the wave functions for $l = 1$ and $m_l = -1, 0,$ and 1.

a. Given that the angular part of these wave functions is

$$Y_1^0(\theta, \phi) = \left(\frac{3}{4\pi}\right)^{1/2}\cos\theta$$

$$Y_1^1(\theta, \phi) = \left(\frac{3}{8\pi}\right)^{1/2}\sin\theta\,e^{i\phi}$$

$$Y_1^{-1}(\theta, \phi) = \left(\frac{3}{8\pi}\right)^{1/2}\sin\theta\,e^{-i\phi}$$

write an expression for $|Y_1^0(\theta, \phi)|^2 + |Y_1^1(\theta, \phi)|^2 + |Y_1^{-1}(\theta, \phi)|^2$.

b. Show that $|Y_1^0(\theta, \phi)|^2 + |Y_1^1(\theta, \phi)|^2 + |Y_1^{-1}(\theta, \phi)|^2$ does not depend on θ and ϕ.

c. Why does this result show that the charge density for the filled $n = 2, l = 1$ subshell is spherically symmetrical?

P10.6 The operator for the square of the total spin of two electrons is $\hat{S}_{total}^2 = (\hat{S}_1 + \hat{S}_2)^2 = \hat{S}_1^2 + \hat{S}_2^2 + 2(\hat{S}_{1x}\hat{S}_{2x} + \hat{S}_{1y}\hat{S}_{2y} + \hat{S}_{1z}\hat{S}_{2z})$. Given that

$$\hat{S}_x\alpha = \frac{\hbar}{2}\beta, \quad \hat{S}_y\alpha = \frac{i\hbar}{2}\beta, \quad \hat{S}_z\alpha = \frac{\hbar}{2}\alpha,$$

$$\hat{S}_x\beta = \frac{\hbar}{2}\alpha, \quad \hat{S}_y\beta = \frac{i\hbar}{2}\alpha, \quad \hat{S}_z\beta = \frac{\hbar}{2}\beta,$$

show that $\alpha(1)\alpha(2)$ and $\beta(1)\beta(2)$ are eigenfunctions of the operator $\hat{S}_{total}^2$. What is the eigenvalue in each case?

P10.7 Show that the functions $[\alpha(1)\beta(2) + \beta(1)\alpha(2)]/\sqrt{2}$ and $[\alpha(1)\beta(2) - \beta(1)\alpha(2)]/\sqrt{2}$ are eigenfunctions of $\hat{S}_{total}^2$. What is the eigenvalue in each case?

P10.8 In this problem, you will use the variational method to find the optimal $1s$ wave function for the hydrogen atom starting from the trial function $\Phi(r) = e^{-\alpha r}$ with α as the variational parameter. You will minimize

$$E(\alpha) = \frac{\displaystyle\int \Phi^*\hat{H}\Phi\,d\tau}{\displaystyle\int \Phi^*\Phi\,d\tau}$$

with respect to α.

a. Show that

$$\hat{H}\Phi = -\frac{\hbar^2}{2m_e}\frac{1}{r^2}\frac{\partial}{\partial r}\left(r^2\frac{\partial\Phi(r)}{\partial r}\right) - \frac{e^2}{4\pi\varepsilon_0 r}\Phi(r)$$

$$= \frac{\alpha\hbar^2}{2m_e r^2}(2r - \alpha r^2)e^{-\alpha r} - \frac{e^2}{4\pi\varepsilon_0 r}e^{-\alpha r}$$

b. Obtain the result $\int \Phi^*\hat{H}\Phi\,d\tau = 4\pi\int_0^\infty r^2\Phi^*\hat{H}\Phi\,dr = \pi\hbar^2/2m_e\alpha - e^2/4\varepsilon_0\alpha^2$ using the standard integrals in the Math Supplement.

c. Show that $\int \Phi^*\Phi\,d\tau = 4\pi\int_0^\infty r^2\Phi^*\Phi\,dr = \pi/\alpha^3$ using the standard integrals in the Math Supplement.

d. You now have the result $E(\alpha) = \hbar^2\alpha^2/2m_e - e^2\alpha/4\pi\varepsilon_0$. Minimize this function with respect to α and obtain the optimal value of α.

e. Is $E(\alpha_{optimal})$ equal to or greater than the true energy? Why?

P10.9 You have commissioned a measurement of the second ionization energy from two independent research teams. You find that they do not agree and decide to plot the data together with known values of the first ionization energy. The results are shown here:

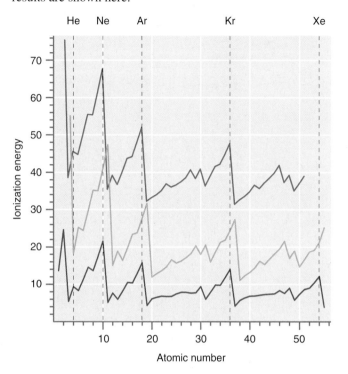

The lowest curve is for the first ionization energy and the upper two curves are the results for the second ionization energy from the two research teams. The uppermost curve has been shifted vertically to avoid an overlap with the other new data set. On the basis of your knowledge of the periodic table, you suddenly know which of the two sets of data is correct and the error that one of the teams of researchers made. Which data set is correct? Explain your reasoning.

P10.10 Classify the following functions as symmetric, antisymmetric, or neither in the exchange of electrons 1 and 2:

a. $[1s(1)2s(2) + 2s(1)1s(2)] \times$
$[\alpha(1)\beta(2) - \beta(1)\alpha(2)]$

b. $[1s(1)2s(2) + 2s(1)1s(2)]\alpha(1)\alpha(2)$

c. $[1s(1)2s(2) + 2s(1)1s(2)][\alpha(1)\beta(2) + \beta(1)\alpha(2)]$

d. $[1s(1)2s(2) - 2s(1)1s(2)][\alpha(1)\beta(2) + \beta(1)\alpha(2)]$

e. $[1s(1)2s(2) + 2s(1)1s(2)] \times$
$[\alpha(1)\beta(2) - \beta(1)\alpha(2) + \alpha(1)\alpha(2)]$

P10.11 Write the Slater determinant for the ground-state configuration of Be.

P10.12 The exact energy of a ground state He atom is -79.01 eV. Calculate the correlation energy and the ratio of the correlation energy to the total energy for He using the results in Table 10.1.

Computational Problems

More detailed instructions on carrying out these calculations using Spartan Physical Chemistry are found on the book website at *www.chemplace.com*. Gaussian basis sets are discussed in Chapter 15.

C10.1 Calculate the total energy and $1s$ orbital energy for Ne using the Hartree–Fock method and the (a) 3-21G, (b) 6-31G*, and (c) 6-311+G** basis sets. Note the number of basis functions used in the calculations. Calculate the relative error of your result compared with the Hartree–Fock limit shown in Table 10.1 for each basis set. Rank the basis sets in terms of their approach to the Hartree–Fock limit for the total energy.

C10.2 Calculate the total energy and $4s$ orbital energy for K using the Hartree–Fock method and the (a) 3-21G and (b) 6-31G* basis sets. Note the number of basis functions used in the calculations. Calculate the percentage deviation from the Hartree–Fock limits which are -16245.7 eV for the total energy and -3.996 eV for the $4s$ orbital energy. Rank the basis sets in terms of their approach to the Hartree–Fock limit for the total energy. What percentage error in the Hartree–Fock limit to the total energy corresponds to a typical reaction enthalpy change of 100 kJ mol^{-1}?

C10.3 Calculate the ionization energy for (a) Li, (b) F, (c) S, (d) Cl, and (e) Ne using the Hartree–Fock method and the 6-311+G** basis set. Carry out the calculation in two different ways: (a) Use Koopmans' theorem and (b) compare the total energy of the neutral and singly ionized atom. Compare your answers with literature values.

C10.4 Calculate the electron affinity for (a) Li, (b) F, (c) S, and (d) Cl using the Hartree–Fock method and the 6-311+G** basis set by comparing the total energy of the neutral and singly ionized atom. Compare your answers with literature values.

C10.5 Using your results from C10.3 and C10.4, calculate the Mulliken electronegativity for (a) Li, (b) F, (c) S, (d) Cl. Compare your results with literature values.

C10.6 To assess the accuracy of the Hartree–Fock method for calculating energy changes in reactions, calculate the total energy change for the reaction $CH_3OH \rightarrow CH_3 + OH$ by calculating the difference in the total energy of reactants and products (ΔU) using the Hartree–Fock method and the 6-31G* basis set. Compare your result with a calculation using the B3LYP method and the same basis set and with the experimental value of 410. kJ mol^{-1}. As discussed in Chapter 15, the B3LYP method takes electron correlation into account. What percentage error in the Hartree–Fock total energy for CH_3OH would account for the difference between the calculated and experimental value of ΔU?

11

Quantum States for Many-Electron Atoms and Atomic Spectroscopy

Having more than one electron in an atom raises the issues of the indistinguishability of electrons, the electron spin, and the interaction between orbital and spin magnetic moments. Taking these issues into consideration leads to a new set of quantum numbers for the states of many-electron atoms, and the grouping of these states into levels and terms. Atomic spectroscopies give information on the discrete energy levels of an atom and provide the basis for understanding the coupling of the individual spin and orbital angular momentum vectors in a many-electron atom. Because the discrete energy levels for atoms differ, atomic spectroscopies give information on the identity and concentration of atoms in a sample. For this reason, atomic spectroscopies are widely used in analytical chemistry. The discrete energy spectra of atoms and the difference in the rates of transition between quantum states can be used to construct lasers that provide an intense and coherent source of monochromatic radiation. Atomic spectroscopies can also provide elemental identification specific to the first few atomic layers of a solid. The reactions of electronically excited atoms can differ dramatically from their ground-state counterparts, as evidenced by reactions in the Earth's atmosphere.

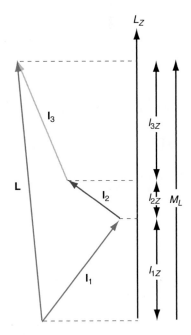

FIGURE 11.1

The sum of three classical angular momentum vectors is depicted. Whereas it is necessary to know the direction of each vector to calculate $\mathbf{L}$, this is not necessary to calculate M_L. As discussed in Section 7.8, each angular momentum vector would need to be represented by a cone to be consistent with the commutation relations among $\hat{L}_x$, $\hat{L}_y$, and $\hat{L}_z$.

11.1 GOOD QUANTUM NUMBERS, TERMS, LEVELS, AND STATES

How are quantum numbers assigned to many-electron atoms? The quantum numbers n, l, m_l, and m_s that were used to characterize total energy eigenfunctions for the H atom are associated with the eigenvalues of the operators $\hat{H}$, $\hat{l}^2$, $\hat{l}_z$, and $\hat{s}_z$. It can be shown that the eigenvalues of a given operator are independent of time only if the operator commutes with $\hat{H}$. The H atom quantum numbers are **good quantum numbers** because the set of operators $\hat{l}^2$, $\hat{l}_z$, $\hat{s}^2$, and $\hat{s}_z$ commutes with the total energy operator $\hat{H}$. Operators that generate good quantum numbers are of particular interest to us in obtaining the values of time-independent observables for atoms and molecules.

However, n, l, m_l, and m_s are not good quantum numbers for any many-electron atom or ion. Therefore, another set of quantum numbers must be found whose corresponding operators do commute with $\hat{H}$. Our primary focus is on a model that adequately describes atoms with $Z < 40$, and we extend this model to atoms for which $Z > 40$ in Section 11.3, where the reason for this restriction on the value of the atomic number is explained. Good quantum numbers are generated by forming vector sums of the electron orbital and spin angular momenta separately, $\mathbf{L}$ and $\mathbf{S}$, which have the z components M_L and M_S, respectively. Only electrons in unfilled subshells contribute to these sums:

$$\mathbf{L} = \sum_i \mathbf{l}_i, \qquad \mathbf{S} = \sum_i \mathbf{s}_i \qquad \qquad \textbf{(11.1)}$$

As discussed in Chapter 7 for $\mathbf{l}$, the magnitudes of $\mathbf{L}$ and $\mathbf{S}$ are $\sqrt{L(L+1)}\hbar$ and $\sqrt{S(S+1)}\hbar$, respectively.

Figure 11.1 illustrates vector addition in classical physics. Note that in order to carry out the vector summations, all three vector components must be known. However, it follows from the commutation rules between $\hat{l}_x$, $\hat{l}_y$, and $\hat{l}_z$ [see Equation (7.31)] that only the length of an angular momentum vector and one of its components can be known in quantum mechanics. This means that the summation shown in Figure 11.1 cannot actually be carried out. By contrast, it is easy to form the sum M_L because the components l_{zi} add as scalars, $M_L = \sum_i l_{zi}$. As we will see later, it is sufficient to know M_L and M_S in order to determine the good quantum numbers L and S.

We next discuss many electron atom operators $\hat{L}^2$, $\hat{L}_z$, $\hat{S}^2$, and $\hat{S}_z$, which are formed from one electron operators. These operators commute with $\hat{H}$ for a many-electron atom with $Z < 40$. The capitalized form of the operators refers to the resultant for all electrons in unfilled subshells of the atom. These operators are defined by

$$\hat{S}_z = \sum_i \hat{s}_{z,i} \quad \text{and} \quad \hat{S}^2 = \left(\sum_i \hat{s}_i \right)^2$$

$$\hat{L}_z = \sum_i \hat{l}_{z,i} \quad \text{and} \quad \hat{L}^2 = \left(\sum_i \hat{l}_i \right)^2 \qquad \textbf{(11.2)}$$

in which the index i refers to the individual electrons. The good quantum numbers for many-electron atoms for $Z < 40$ are L, S, M_L, and M_S.

As can be inferred from Equation (11.2), the calculation for $\hat{S}^2$ is somewhat complex and is not discussed here. By contrast, $\hat{S}_z$ can be calculated easily as shown in Example Problem 11.1.

FIGURE 11.2

The top line shows the level of approximation, the second shows the group of states that are degenerate in energy, and the bottom line shows the good quantum numbers in each level of approximation.

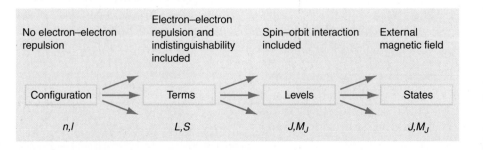

EXAMPLE PROBLEM 11.1

Is $\psi(1, 2) = 1s(1)\alpha(1)1s(2)\beta(2) - 1s(2)\alpha(2)1s(1)\beta(1)$ an eigenfunction of the operator $\hat{S}_z$? If so, what is its eigenvalue M_S?

Solution

$$\hat{S}_z = \hat{s}_z(1) + \hat{s}_z(2) \text{ where } \hat{s}_z(i) \text{ acts only on electron } i$$

$$\hat{S}_z\psi(1, 2) = (\hat{s}_z(1) + \hat{s}_z(2))\psi(1, 2)$$

$$= (\hat{s}_z(1) + \hat{s}_z(2))[1s(1)\alpha(1)1s(2)\beta(2) - 1s(2)\alpha(2)1s(1)\beta(1)]$$

$$= (\hat{s}_z(1))[1s(1)\alpha(1)1s(2)\beta(2) - 1s(2)\alpha(2)1s(1)\beta(1)]$$

$$+ (\hat{s}_z(2))[1s(1)\alpha(1)1s(2)\beta(2) - 1s(2)\alpha(2)1s(1)\beta(1)]$$

$$= \frac{\hbar}{2}[1s(1)\alpha(1)1s(2)\beta(2)] + \frac{\hbar}{2}[1s(2)\alpha(2)1s(1)\beta(1)]$$

$$- \frac{\hbar}{2}[1s(1)\alpha(1)1s(2)\beta(2)] - \frac{\hbar}{2}[1s(2)\alpha(2)1s(1)\beta(1)]$$

$$= \left(\frac{\hbar}{2} - \frac{\hbar}{2}\right)[1s(1)\alpha(1)1s(2)\beta(2) - 1s(2)\alpha(2)1s(1)\beta(1)] = 0 \times \psi(1, 2)$$

This result shows that the wave function is an eigenfunction of $\hat{S}_z$ with $M_S = 0$.

Although a configuration is a very useful way to describe the electronic structure of atoms, it does not completely specify the quantum state of a many-electron atom because it is based on the one-electron quantum numbers n and l. Taking electron–electron repulsion into account and invoking the Pauli exclusion principle splits a configuration into terms. A **term** is a group of states that has the same L and S values. Describing the states of many-electron atoms by terms is appropriate for atoms with a nuclear charge of $Z < 40$ because L and S are "good enough" quantum numbers for these atoms, meaning that the difference in energy between levels in a term is very small compared to the energy separation of the terms. Levels will be discussed in Section 11.3.

11.2 THE ENERGY OF A CONFIGURATION DEPENDS ON BOTH ORBITAL AND SPIN ANGULAR MOMENTUM

As proved in Supplemental Section 11.11, the energy of an atom depends on the value of the quantum number S. If an atom has at least two unpaired electrons (electrons in orbitals that are singly occupied), then the atom can have more than one value for S. Consider the excited state of He with the configuration $1s^1 2s^1$. Because both electrons have $l = 0$, $|\mathbf{L}| = 0$. We next show that there are two different values of $|\mathbf{S}|$ consistent with the $1s^1 2s^1$ configuration and formulate antisymmetric wave functions for each value of S.

Recall that an individual electron can be characterized by a spin angular momentum vector $\mathbf{s}$ of magnitude $|\mathbf{s}| = \sqrt{s(s + 1)}\hbar$, where the quantum number s can only have the single value $s = 1/2$. The vector $\mathbf{s}$ has $2s + 1 = 2$ possible orientations with the z component $s_z = \pm 1/2\hbar$. We say that two spins can only be **parallel**, $\alpha(1)\alpha(2)$ and $\beta(1)\beta(2)$, or **antiparallel**, $\alpha(1)\beta(2)$ and $\beta(1)\alpha(2)$.

By looking at Figure 11.3, convince yourself that adding the scalar components m_s for the two electrons in each of the four possible combinations gives the values $M_S = m_{s1} + m_{s2} = 0$ twice, as well as $M_S = m_{s1} + m_{s2} = +1$ and -1. Surprisingly, the possible

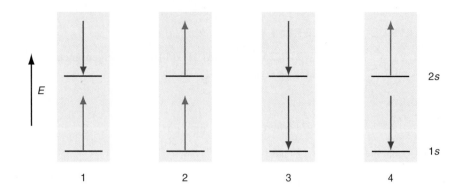

FIGURE 11.3

Possible alignment of the spins in the He configuration $1s^1 2s^1$. An upward-pointing arrow corresponds to $m_s = +1/2$ and a downward-pointing arrow corresponds to $m_s = -1/2$.

values of S for He in the $1s^1 2s^1$ configuration can be deduced using only this information about M_S. We know that $S \geq |M_S|$ because the spin angular momentum follows the same rules as the orbital angular momentum. Because there is no value for $M_S > 1$ among these four possible spin combinations, $M_S = \pm 1$ is only consistent with $S = 1$. Because M_S takes on all integral values between $+S$ and $-S$, the $S = 1$ group must include $M_S = 0, +1,$ and -1. This accounts for three of the four values of M_S listed above. The one remaining combination has $M_S = 0$, which is only consistent with $S = 0$.

We have just shown that three of the four possible spin combinations are characterized by $S = 1$ with $M_S = \pm 1$ and 0 and that the fourth has $S = 0$ with $M_S = 0$. Because of the number of possible M_s values, the $S = 0$ spin combination is called a **singlet** and the $S = 1$ spin combination is called a **triplet**. Singlet and triplet states are encountered frequently in chemistry, and are associated with **paired** and **unpaired electrons**, respectively.

Now that we know the S values for the four spin combinations, we can write antisymmetric wave functions for He $1s^1 2s^1$ that are eigenfunctions of $\hat{S}^2$ with $S = 0$ and $S = 1$.

$$S = 0 \quad \psi_{singlet} = \frac{1}{\sqrt{2}}[1s(1)2s(2) + 2s(1)1s(2)]\frac{1}{\sqrt{2}}[\alpha(1)\beta(2) - \beta(1)\alpha(2)]$$

$$S = 1 \quad \psi_{triplet} = \frac{1}{\sqrt{2}}[1s(1)2s(2) - 2s(1)1s(2)]$$

$$\begin{cases} \alpha(1)\alpha(2) \quad \text{or} \\ \beta(1)\beta(2) \quad \text{or} \\ \frac{1}{\sqrt{2}}[\alpha(1)\beta(2) + \beta(1)\alpha(2)] \end{cases} \tag{11.3}$$

For the wave functions that describe the three different states for the triplet, $S = 1$ and $|\mathbf{S}| = \sqrt{2}\hbar$, and (from top to bottom) $M_S = 1, -1,$ and 0. The singlet consists of a single state with $S = 0$ and $M_S = 0$. Note that the antisymmetry of the total wave function is achieved by making the spatial part symmetric and the spin part antisymmetric for the singlet wave function, and the other way around for the triplet wave functions.

The vector model of angular momentum can be used to depict singlet and triplet states, as shown in Figure 11.4. Although the individual spins cannot be located on the cones, their motion is coupled so that $S = 0$ for the singlet state. For a triplet state, there is a similar coordinated precession, but in this case, the vectors add rather than cancel and $S = 1$. Because $S = 1$, there must be three different cones corresponding to $M_S = -1, 0,$ and 1.

We next make it plausible that the total energy for many-electron atoms also depends on $|\mathbf{L}|$. In the Hartree–Fock self-consistent field method, the actual positions of the electrons are approximated by their average positions. This results in a spherically symmetric charge distribution for closed subshells. As discussed in Chapter 10, this approximation greatly simplifies the calculations of orbital energies and wave functions for many-electron atoms. However, by looking at the angular part of the hydrogen atom wave functions (Figure 9.7), you can see that if l is not zero (for example, the $2p$

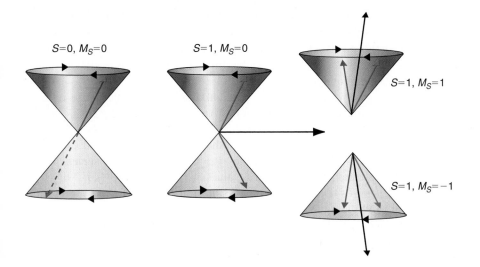

FIGURE 11.4

Vector model of the singlet and triplet states. The individual spin angular momentum vectors and their vector sum **S** (black arrow) are shown for the triplet states. For the singlet state (left image), $|\mathbf{S}| = 0$ and $M_S = 0$. The dashed arrow in the left image indicates that the vector on the yellow cone is on the opposite side of the cone from the vector on the blue cone.

electrons in carbon), the electron probability distribution is not spherically symmetrical. Electrons in states characterized by $l = 1$ that have different values of m_l ($-1, 0,$ or $+1$) have different orientations of the same spatial probability distributions. Two such electrons, therefore, have different repulsive interactions depending on their m_l values. By looking at Figure 7.7, you can see that two electrons in the p_x orbital repel each other more strongly than if one of the electrons is in the p_x orbital and the other is in the p_z or p_y orbital.

Because the m_s value constrains the choices for m_l for electrons in the same orbital through the Pauli principle, the repulsive interactions between these electrons are determined by both l and s. Recall that a configuration specifies only the n and l values for the electrons and not the m_l and m_s values. For many atoms, the configuration does not completely define the quantum state. When is this the case and how does the angular momentum affect the orbital energies of the atom? As you will verify in the end-of-chapter problems, only partially filled subshells contribute to **L** and **S**. Under what conditions do the values of m_l and m_s for a given configuration lead to different spatial distributions of electrons and therefore to a different electron–electron repulsion? This occurs when there are at least two electrons in the valence shell and when there are multiple possible choices in m_l and m_s for these electrons consistent with the Pauli principle and the configuration. This is not the case for the ground states of the rare gases, the alkali metals, the alkaline earth metals, group III, and the halogens. Atoms in all of these groups have either a filled shell or subshell, or only one electron or one electron fewer than the maximum number of electrons in a subshell. None of these atoms has more than one unpaired electron in its ground state and all are uniquely described by their configuration. However, the ground states for carbon, nitrogen, and oxygen are not completely described by a configuration. Several quantum states, all of which are consistent with the configuration, have significantly different values for the total energy as well as different chemical reactivities.

For atoms with $Z < 40$, the total energy is essentially independent of M_S and M_L. Therefore, a group of different quantum states that have the same values for L and S but different values of M_L and M_S is degenerate in energy. Such a group of states is called a term, and the L and S values for the term are indicated by the **term symbol** $^{(2S+1)}L$. Terms with $L = 0, 1, 2, 3, 4, \ldots$ are given the symbols S, P, D, F, G, ..., respectively. Because there are $2L + 1$ quantum states (different M_L values) for a given value of L and the $2S + 1$ states (different M_S values) for a given value of S, a term will include $(2L + 1)(2S + 1)$ quantum states, all of which have the same energy to a good approximation. This is the **degeneracy of a term**. The superscript $2S + 1$ is called the **multiplicity**, and the words *singlet* and *triplet* refer to $2S + 1 = 1$ and 3, respectively. Extending this formalism, $2S + 1 = 2$ and 4 are associated with doublets and quartets. For a filled subshell or shell,

$$M_L = \sum_i m_{li} = M_S = \sum_i m_{si} = 0 \qquad \textbf{(11.4)}$$

and $M_L = 0$ and $M_S = 0$ are only consistent with $L = 0$ and $S = 0$. Therefore atoms with a filled valence subshell or shell are all characterized by the term ^1S. Note that the term symbols do not depend on the principal quantum number of the valence shell. Carbon, which has the $1s^2 2s^2 2p^2$ configuration, has the same set of terms as silicon, which has the $1s^2 2s^2 2p^6 3s^2 3p^2$ configuration.

How are terms generated for a given configuration? The simplest case is for a configuration with singly occupied subshells. An example is C $1s^2 2s^2 2p^1 3d^1$, in which an electron has been promoted from the $2p$ to the $3d$ orbital. Only the $2p$ and $3d$ electrons need be considered, because the other electrons are in filled subshells. The possible values of L and S are given by the **Clebsch–Gordon series**. When applied to the two-electron case, allowed L values are given by $l_1 + l_2, l_1 + l_2 - 1, l_1 + l_2 - 2, \ldots, |l_1 - l_2|$. Using the same rule, the allowed S values are $s_1 - s_2$ and $s_1 + s_2$. For our example, $l_1 = 1$, $l_2 = 2$, and $s_1 = s_2 = 1/2$. Therefore, L can have the values 3, 2, and 1, and S can have the values 1 and 0. We conclude that the $1s^2 2s^2 2p^1 3d^1$ configuration generates ^3F, ^3D, ^3P, ^1F, ^1D, and ^1P terms. The degeneracy of these terms, $(2L + 1)(2S + 1)$, is 21, 15, 9, 7, 5, and 3, respectively, which corresponds to a total of 60 quantum states. Looking back at the configuration, the $2p$ electron can have $m_l = \pm 1$ and 0 and $m_s = \pm 1/2$. This gives six possible combinations of m_l and m_s. The $3d$ electron can have $m_l = \pm 1$, ± 2, and 0, and $m_s = \pm 1/2$. This gives 10 possible combinations of m_l and m_s. Because any combination for the $2p$ electron can be used with any combination of the $3d$ electron, there are a total of $6 \times 10 = 60$ combinations of m_l and m_s consistent with the $1s^2 2s^2 2p^1 3d^1$ configuration. These combinations generate the 60 states that belong to the ^3F, ^3D, ^3P, ^1F, ^1D, and ^1P terms.

The same method can be extended to more than two electrons by first calculating L and S for two electrons and adding in the remaining electrons one by one. For example, consider the L values associated with the C $1s^2 2s^1 2p^1 3p^1 3d^1$ configuration. Combining the $2s$ and $2p$ electrons gives only $L = 1$. Combining this L value with the $3p$ electron gives 2, 1, and 0. Combining these values with the $3d$ electron gives possible L values of 4, 3, 2, 1, and 0. The maximum value of S is $n/2$, where n is the number of different singly filled subshells. The minimum value of S is 0 if n is even, and $1/2$ if n is odd. For our example, the possible S values are 2, 1, and 0. Which terms are generated by these values of L and S?

Assigning terms to a configuration is more complicated if subshells have more than one electron, because the Pauli exclusion principle must be obeyed. To illustrate such a case, consider the ground state of carbon, which has the configuration $1s^2 2s^2 2p^2$. We need only consider the $2p$ electrons. Because m_l can have any of the values $-1, 0,$ or $+1$, and m_s can have the values $+1/2$ and $-1/2$ for p electrons, six combinations of the quantum numbers m_s and m_l for the first electron are possible. The second electron will have one fewer possible combination because of the Pauli principle. This appears to give a total of $6 \times 5 = 30$ combinations of quantum numbers for the two electrons. However, this assumes that the electrons are distinguishable which overcounts the possible number of combinations by a factor of 2. Taking this into account, there are 15 possible quantum states of the carbon atom consistent with the configuration $1s^2 2s^2 2p^2$, which are shown schematically in Figure 11.5.

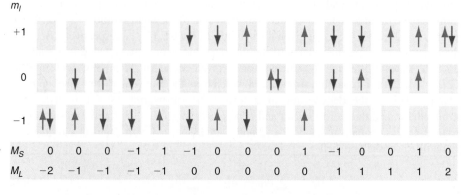

FIGURE 11.5

The different ways in which two electrons can be placed in p orbitals is shown. Upward- and downward-pointing arrows correspond to $m_s = +1/2$ and $m_s = -1/2$, respectively, and M_S and M_L are the scalar sums of the m_s and m_l, respectively.

m_l															
+1				↓	↓	↑		↑	↓	↓	↑	↑	↑↓
0	↓	↑	↓	↑			↑↓		↓	↑	↓	↑	
−1	↑↓	↑	↓	↓	↑	↓	↑	↓		↑				
M_S	0	0	0	−1	1	−1	0	0	0	1	−1	0	0	1	0
M_L	−2	−1	−1	−1	−1	0	0	0	0	0	1	1	1	1	2

TABLE 11.1   STATES AND TERMS FOR THE np^2 CONFIGURATION

m_{l1}	m_{l2}	$M_L = m_{l1} + m_{l2}$	m_{s1}	m_{s2}	$M_S = m_{s1} + m_{s2}$	Term
-1	-1	-2	$1/2$	$-1/2$	0	^1D
			$-1/2$	$-1/2$	-1	^3P
0	-1	-1	$-1/2$	$1/2$	0	$^1D, {}^3P$
			$1/2$	$-1/2$	0	
			$1/2$	$1/2$	1	^3P
0	0	0	$1/2$	$-1/2$	0	
			$-1/2$	$1/2$	0	$^1D, {}^3P, {}^1S$
			$1/2$	$-1/2$	0	
1	-1	0	$-1/2$	$-1/2$	-1	^3P
			$1/2$	$1/2$	1	^3P
			$-1/2$	$-1/2$	-1	^3P
1	0	1	$-1/2$	$1/2$	0	$^1D, {}^3P$
			$1/2$	$-1/2$	0	
			$1/2$	$1/2$	1	^3P
1	1	2	$1/2$	$-1/2$	0	^1D

To determine the possible terms consistent with a p^2 configuration, it is convenient to display the information in Figure 11.5 in tabular form, as shown in Table 11.1. In setting up Table 11.1, we have relied only on the z components of the vectors m_{si} and m_{li}. Using these components, $M_S = \sum_i m_{si}$ and $M_L = \sum_i m_{li}$ can be easily calculated because no vector addition is involved. To derive terms from this table, it is necessary to determine what values for L and S are consistent with the tabulated M_S and M_L values. How can this be done knowing only M_L and M_S? We first determine which values of L and S are consistent with the entries for M_L and M_S in the table given that $-S \le M_S \le +S$ and $-L \le M_L \le +L$. A good way to start is to look at the highest value for $|M_L|$ first. This requires careful bookkeeping.

The top and bottom entries in the table have the largest M_L values of -2 and $+2$, respectively. They must belong to a term with $L = 2$ (a D term), because $|M_L|$ can be no greater than L. All states with M_L values of -2 and $+2$ have $M_S = 0$ because the set of quantum numbers for each electron must differ. Stated differently, because $m_{l1} = m_{l2}, m_{s1} \ne m_{s2}$, and therefore $M_S = 0$. We conclude that $S = 0, 2S + 1 = 1$, and the D term must be ^1D. This term has $(2S + 1)(2L + 1) = 5$ states associated with it. It includes states with $M_L = -2, -1, 0, +1$, and $+2$, all of which have $M_S = 0$. These 5 states are mentally removed from the table, which leaves us with 10 states. Of those remaining, the next highest value of $|M_L|$ is $+1$, which must belong to a P term. Because there is a combination with $M_L = 1$ and $M_S = 1$, the P term must be ^3P. This term has $(2S + 1)(2L + 1) = 9$ states associated with it and by mentally removing these 9 states from the table, a single state is left with $M_L = M_S = 0$. This is a complete ^1S term. By a process of elimination, we have found that the 15 combinations of m_l and m_s consistent with the configuration $1s^2 2s^2 2p^2$ separate into ^1D, ^1S, and ^3P terms. This conclusion is true for any np^2 configuration. Because the ^1D, ^1S, and ^3P terms have 5, 1, and 9 states associated with them, a total of 15 states are associated with the terms of the $1s^2 2s^2 2p^2$ configuration just as for the classification scheme based on the individual quantum numbers n, l, m_s, and m_l.

EXAMPLE PROBLEM 11.2

What terms result from the configuration ns^1d^1? How many quantum states are associated with each term?

Solution

Because the electrons are not in the same subshell, the Pauli principle does not limit the combinations of m_l and m_s. Using the guidelines formulated earlier, $S_{min} = 1/2 - 1/2 = 0$, $S_{max} = 1/2 + 1/2 = 1$, $L_{min} = 2 - 0 = 2$, and $L_{max} = 2 + 0 = 2$. Therefore, the terms that arise from the configuration ns^1d^1 are ^3D and ^1D. Table 11.2 shows how these terms arise from the individual quantum numbers. In setting up the table, we have relied only on the z components of the vectors m_{si} and m_{li}. Using these components, $M_S = \sum_i m_{si}$ and $M_L = \sum_i m_{li}$ can be easily calculated because no vector addition is involved. Because each term has $(2S + 1)(2L + 1)$ states, the ^3D term consists of 15 states, and the ^1D term consists of 5 states as shown in Table 11.2.

TABLE 11.2 STATES AND TERMS FOR THE ns^1d^1 CONFIGURATION

m_{l1}	m_{l2}	$M_L = m_{l1} + m_{l2}$	m_{s1}	m_{s2}	$M_S = m_{s1} + m_{s2}$	Term
0	−2	−2	−1/2	−1/2	−1	^3D
			−1/2	1/2	0	$^1D, {}^3D$
			1/2	−1/2	0	$^1D, {}^3D$
			1/2	1/2	1	^3D
0	−1	−1	−1/2	−1/2	−1	^3D
			−1/2	1/2	0	$^1D, {}^3D$
			1/2	−1/2	0	$^1D, {}^3D$
			1/2	1/2	1	^3D
0	0	0	−1/2	−1/2	−1	^3D
			−1/2	1/2	0	$^1D, {}^3D$
			1/2	−1/2	0	$^1D, {}^3D$
			1/2	1/2	1	^3D
0	1	1	−1/2	−1/2	−1	^3D
			−1/2	1/2	0	$^1D, {}^3D$
			1/2	−1/2	0	$^1D, {}^3D$
			1/2	1/2	1	^3D
0	2	2	−1/2	−1/2	−1	^3D
			−1/2	1/2	0	$^1D, {}^3D$
			1/2	−1/2	0	$^1D, {}^3D$
			1/2	1/2	1	^3D

The preceding discussion has demonstrated how to generate the terms associated with a particular configuration. The same procedure can be followed for any configuration and a few examples are shown in Table 11.3 for electrons in the same shell. The numbers in parentheses behind the term symbol indicate the number of different terms of that type that belong to the configuration. A simplifying feature in generating terms is that the same results are obtained for a given number of electrons or "missing electrons" (sometimes called holes) in a subshell. For example, d^1 and d^9 configurations

TABLE 11.3 POSSIBLE TERMS FOR INDICATED CONFIGURATIONS

Electron Configuration	Term Symbol
s^1	^2S
p^1, p^5	^2P
p^2, p^4	$^1S, {}^1D, {}^3P$
p^3	$^2P, {}^2D, {}^4S$
d^1, d^9	^2D
d^2, d^8	$^1S, {}^1D, {}^1G, {}^3P, {}^3F$
d^3, d^7	$^4F, {}^4P, {}^2H, {}^2G, {}^2F, {}^2D\,(2), {}^2P$
d^4, d^6	$^5D, {}^3H, {}^3G, {}^3F\,(2), {}^3D, {}^3P\,(2), {}^1I, {}^1G\,(2), {}^1F, {}^1D\,(2),$ $^1S\,(2)$
d^5	$^6S, {}^4G, {}^4F, {}^4D, {}^4P, {}^2I, {}^2H, {}^2G\,(2), {}^2F\,(2), {}^2D\,(3), {}^2P, {}^2S$

result in the same terms. Note that configurations with a single electron or hole in the unfilled shell or subshell give only a single term as discussed earlier. In filled shells or subshells, $M_L = M_S = 0$ because m_l and m_s take on all possible values between their maximum positive and negative values. For this reason the term symbol for s^2, p^6, and d^{10} is ^1S.

EXAMPLE PROBLEM 11.3

How many states are consistent with a d^2 configuration? What L values result from this configuration?

Solution

The first electron can have any of the m_l values ±2, ±1, and 0, and either of the m_s values $\pm 1/2$. This gives 10 combinations. The second electron can have 9 combinations and the total number of combinations for both electrons is $10 \times 9 = 90$. However, because the electrons are not distinguishable, we must divide this number by two and obtain 45 states. Using the formula $L = l_1 + l_2, l_1 + l_2 - 1, \ldots, |l_1 - l_2|$, we conclude that L values of 4, 3, 2, 1, and 0 are allowed. Therefore, this configuration gives rise to G, F, D, P, and S terms. Table 11.3 shows that the allowed terms for these L values are $^1S, {}^1D, {}^1G, {}^3P,$ and ^3F. The degeneracy of each term is given by $(2L + 1)(2S + 1)$, and is 1, 5, 9, 9, and 21, respectively. Therefore, the d^2 configuration gives rise to 45 distinct quantum states, just as was calculated based on the possible combinations of m_l and m_s.

The relative energy of the different terms has not been discussed yet. From the examination of a large body of spectroscopic data, Friedrich Hund deduced **Hund's rules**, which state that for a given configuration:

RULE 1:

The lowest energy term is that which has the greatest spin multiplicity. For example, the ^3P term of an np^2 configuration is lower in energy than the ^1D and ^1S terms.

RULE 2:

For terms that have the same spin multiplicity, the term with the greatest orbital angular momentum lies lowest in energy. For example, the ^1D term of an np^2 configuration is lower in energy than the ^1S term.

Hund's rules predict that in placing electrons in one-electron orbitals, the number of unpaired electrons should be maximized. This is why Cr has the configuration $[Ar]4s^1 3d^5$ rather than $[Ar]4s^2 3d^4$. Hund's rules imply that the energetic consequences of electron–electron repulsion are greater for spin than for orbital angular momentum. As we will see in Section 11.10, atoms in quantum states described by different terms can have substantially different chemical reactivity.

Although some care is needed to establish the terms that belong to a particular configuration such as p^n or d^n, it is straightforward to predict the lowest energy term among the possible terms using the following recipe. Create boxes, one for each of the possible values of m_l. Place the electrons specified by the configuration in the boxes in such a way that $M_L = \sum_i m_{li}$ is maximized and that the number of unpaired spins is maximized. L and S for the lowest energy term are given by $L = M_{L,\,max} = \sum_i m_{li}$ and $S = M_{S,\,max} = \sum_i m_{si}$. This procedure is illustrated for the p^2 and d^6 configurations in Example Problem 11.4.

EXAMPLE PROBLEM 11.4

Determine the lowest energy term for the p^2 and d^6 configurations.

Solution

The placement of the electrons is as shown here:

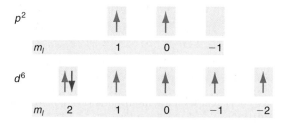

For the p^2 configuration, $M_{L,\,max} = 1$ and $M_{S,\,max} = 1$. Therefore, the lowest energy term is ^3P. For the d^6 configuration, $M_{L,\,max} = 2$ and $M_{S,\,max} = 2$. Therefore, the lowest energy term is ^5D. It is important to realize that this procedure only provides a recipe for finding the lowest energy term. The picture used in the recipe has no basis in reality, because no association of a term with particular values of m_s and m_l can be made.

11.3 SPIN-ORBIT COUPLING BREAKS UP A TERM INTO LEVELS

Up until now, we have said that all states in a term have the same energy. This is a good approximation for atoms with $Z < 40$. However, even for these atoms, the terms are split into closely spaced levels. What is this splitting due to? We know that electrons have nonzero magnetic moments if $L > 0$ and $S > 0$. The separate spin and orbital magnetic moments can interact through **spin-orbit coupling**, just as two bar magnets interact. As a result of this interaction, the total energy operator contains an extra term proportional to $\mathbf{L} \cdot \mathbf{S}$. Under these conditions, the operators $\hat{L}^2$, $\hat{L}_z$, $\hat{S}^2$, and $\hat{S}_z$ no longer commute with $\hat{H}$, but the operators $\hat{J}^2$ and $\hat{J}_z$ where $\mathbf{J}$ is the **total angular momentum** defined by

$$\mathbf{J} = \mathbf{L} + \mathbf{S} \tag{11.5}$$

do commute with $\hat{H}$. If the coupling is sufficiently large as in atoms for which $Z > 40$, the only good quantum numbers are J and M_J, the projection of J on the z axis. The magnitude of $\mathbf{J}$ can take on all values given by $J = L + S, L + S - 1, L + S - 2, \ldots,$ $|L - S|$ and M_J can take on all values between zero and J that differ by one.

For example, the ^3P term has J values of 2, 1, and 0. All quantum states with the same J value have the same energy and belong to the same **level**. The additional quantum number J is included in the nomenclature for a level as a subscript in the form $^{(2S+1)}L_J$. In counting states, $2J + 1$ states with different M_J values are associated with each J value. This gives five states associated with $^3P_2$, three states associated with $^3P_1$, and one state associated with $^3P_0$. The total of nine states in the three levels is the same as the number of states in the ^3P term, as deduced from the formula $(2L + 1)(2S + 1)$.

Taking spin-orbit coupling into account gives Hund's third rule:

RULE 3:

The order in energy of levels in a term is given by the following:

- If the unfilled subshell is exactly or more than half full, the level with the highest J value has the lowest energy.
- If the unfilled subshell is less than half full, the level with the lowest J value has the lowest energy.

Therefore, the $^3P_0$ level has the lowest energy for an np^2 configuration. The $^3P_2$ level has the lowest energy for an np^4 configuration, which describes O.

In a magnetic field, states with the same J, but different M_J, have different energies. For atoms with $Z < 40$, this energy splitting is less than the energy separation between levels, which is in turn less than the energy separation between terms. However, all of these effects are observable in spectroscopies as shown for carbon in Figure 11.6, and many of them have practical implications in analytical chemistry. Clearly the energy levels of many-electron atoms have a higher level of complexity than those for the hydrogen atom. This complicates the understanding of the electronic structure of atoms. It also gives more detailed information about atoms through spectroscopic experiments that can be used to better understand the quantum mechanics of many-electron atoms.

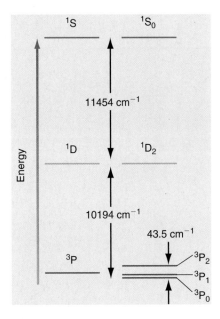

FIGURE 11.6

The energy of the carbon atom assuming a spherically symmetric electron distribution would give a single energy for a configuration. Taking the dependence of the electron repulsion on the directions of L and S into account splits the configuration into terms of different energy as shown. Taking the coupling of L and S into account leads to a further splitting of the terms into levels according to the J values as shown on the right. The separation of the levels for the ^3P term has been multiplied by a factor of 25 to make it visible.

EXAMPLE PROBLEM 11.5

What values of J are consistent with the terms ^2P and ^3D? How many states with different values of M_J correspond to each?

Solution

The quantum number J can take on all values given by $J = L + S, L + S - 1, L + S - 2, \ldots, |L - S|$. For the ^2P term, $L = 1$ and $S = 1/2$. Therefore, J can have the values 3/2 and 1/2. There are $2J + 1$ values of M_J, or 4 and 2 states, respectively.

For the ^3D term, $L = 2$ and $S = 1$. Therefore, J can have the values 3, 2, and 1. There are $2J + 1$ values of M_J or 7, 5, and 3 states, respectively.

11.4 THE ESSENTIALS OF ATOMIC SPECTROSCOPY

With an understanding of the quantum states of many-electron atoms, we turn our attention to atomic spectroscopy. All spectroscopies involve the absorption or emission of electromagnetic radiation that induces transitions between states of a quantum mechanical system. In this chapter, we discuss transitions between electronic states in atoms. Whereas the energies involved in rotational and vibrational transitions are on the order of 1 and 10 kJ mol^{-1}, respectively, photon energies associated with electronic transitions are on the order of 200 to 1000 kJ mol^{-1}. Typically, such energies are associated with visible, UV, or X-ray photons.

The information on atomic energy levels discussed in previous sections is derived from atomic spectra. The interpretation of spectra requires knowledge of the selection rules for the spectroscopy being used. **Selection rules** can be derived based on the **dipole approximation** (Section 8.4). Although transitions that are forbidden in the dipole approximation may be allowed in a higher level theory, but the transitions are very weak. In Chapter 8, the dipole selection rule $\Delta n = \pm 1$ was derived for vibrational transitions, and it was stated without proof that the selection rule for rotational transitions in diatomic molecules is $\Delta J = \pm 1$. What selection rules apply for transitions between atomic levels? If the $\mathbf{L} - \mathbf{S}$ coupling scheme outlined in Section 11.2 applies (atomic numbers less than ~40), the dipole selection rules for atomic transitions are $\Delta l = \pm 1$, and $\Delta L = 0, \pm 1$, and $\Delta J = 0, \pm 1$. There is an additional selection rule, $\Delta S = 0$, for the spin angular momentum. Note that the first selection rule refers to the angular momentum of an electron involved in the transition, whereas the other rules refer to the vector sums for all electrons in the atom. Keep in mind that aside from the selection rule cited above, the quantum number J in this chapter refers to the total electron angular momentum and not to the rotational angular momentum.

Atomic spectroscopy is important in many practical applications such as analytical chemistry and lasers, which we discuss in this chapter. At a fundamental level, the relative energy of individual quantum states can be measured to high precision using spectroscopic techniques. An application of such high-precision measurements is the standard for the time unit of a second, which is based on a transition between states in the cesium atom that has the frequency $9192631770 \text{ s}^{-1}$.

Because the energy levels of the hydrogen atom can be written as

$$E_n = -\frac{m_e e^4}{8\varepsilon_0^2 h^2 n^2} \tag{11.6}$$

where n is the principal quantum number, the frequency for absorption lines in the hydrogen spectrum is given by

$$\tilde{\nu} = \frac{m_e e^4}{8\varepsilon_0^2 h^3 c}\left(\frac{1}{n_{initial}^2} - \frac{1}{n_{final}^2}\right) = R_H\left(\frac{1}{n_{initial}^2} - \frac{1}{n_{final}^2}\right) \tag{11.7}$$

where R_H is the Rydberg constant. The derivation of this formula was one of the early major triumphs of quantum mechanics. The Rydberg constant is one of the most precisely known fundamental constants, and it has the value $10973731.534 \text{ cm}^{-1}$. [See Problem 11.18 for a more exact equation for $\tilde{\nu}$ that is based on having the coordinate system centered at the center of mass rather than the nucleus.] The series of spectral lines associated with $n_{initial} = 1$ is called the Lyman series, and the series associated with $n_{initial} = 2, 3, 4,$ and 5 are called the Balmer, Paschen, Brackett, and Pfund series, respectively, after the spectroscopists who identified them.

EXAMPLE PROBLEM 11.6

The absorption spectrum of the hydrogen atom shows lines at 82,258; 97,491; 102,823; 105,290; and 106,631 cm^{-1}. There are no lower frequency lines in the spectrum. Use graphical methods to determine $n_{initial}$ and the ionization energy of the hydrogen atom in this state.

Solution

The knowledge that frequencies for transitions follow a formula like that of Equation (11.7) allows $n_{initial}$ and the ionization energy to be determined from a limited number of transitions between bound states. The plot of $\tilde{\nu}$ versus assumed values of $1/n_{final}^2$ has a slope of $-R_H$ and an intercept with the frequency axis of $R_H/n_{initial}^2$. However, both $n_{initial}$ and n_{final} are unknown, so that in plotting the data, n_{final} values have to be assigned to the observed frequencies. For the lowest energy transition, n_{final} is $n_{initial} + 1$. We try different combinations of n_{final} and $n_{initial}$ values to see if the slope and intercept are consistent with the expected values of $-R_H$ and $R_H/n_{initial}^2$. In this case, the sequence of spectral lines is assumed to correspond to $n_{final} = 2, 3, 4, 5$, and 6 for an assumed value of $n_{initial} = 1$; $n_{final} = 3, 4, 5, 6$, and 7 for an assumed value of $n_{initial} = 2$; and $n_{final} = 4, 5, 6, 7$, and 8 for an assumed value of $n_{initial} = 3$. The plots are shown in the following figure:

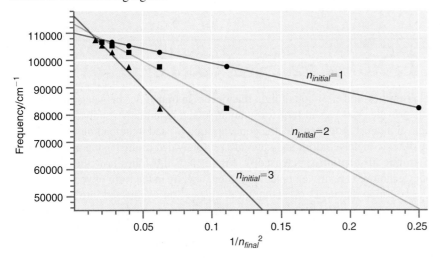

The slopes and intercepts calculated for these assumed values of $n_{initial}$ are:

Assumed $n_{initial}$	Slope (cm^{-1})	Intercept (cm^{-1})
1	-1.10×10^5	1.10×10^5
2	-2.71×10^5	1.13×10^5
3	-5.23×10^5	1.16×10^5

Because the slope is $-R_H$, for only one of the three assumed values, we conclude that $n_{initial} = 1$. The ionization energy of the hydrogen atom in this state is hcR_H, corresponding to $n_{final} \rightarrow \infty$, or 2.18×10^{-18} J. The appropriate number of significant figures for the slope and intercept is approximate in this example and must be based on an error analysis of the data.

Information from atomic spectra is generally displayed in a standard format called a **Grotrian diagram**. An example is shown in Figure 11.7 for He, for which $\mathbf{L} - \mathbf{S}$ coupling is a good model. The figure shows the configuration information next to the energy level and the configurations are arranged according to their energy and term symbols. The triplet and singlet states are shown in separate parts of the diagram because transitions between these states do not occur as a consequence of the $\Delta S = 0$

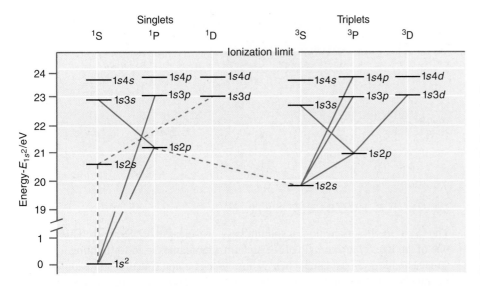

FIGURE 11.7

The ground and the first few excited states of the He atom are shown on an energy scale. All terms for which $l > 2$ and $n > 4$ have been omitted to simplify the presentation. The top horizontal line indicates the ionization energy of He. Below this energy, all states are discrete. Above this level, the energy spectrum is continuous. Several, but not all, allowed (solid lines) and forbidden (dashed lines) transitions are shown. Which selection rule do the forbidden transitions violate?

selection rule. The ^{3}P and ^{3}D He terms are split into levels with different J values, but because the spin-orbit interaction is so small for He, the splitting is not shown in Figure 11.7. For example, the ^{3}P$_0$ and ^{3}P$_2$ levels arising from the $1s2p$ configuration differ in energy by only 0.0006%.

11.5 ANALYTICAL TECHNIQUES BASED ON ATOMIC SPECTROSCOPY

The absorption and emission of light that occurs in transitions between different atomic levels provides a powerful tool for qualitative and quantitative analysis of samples of chemical interest. For example, the concentration of lead in human blood and the presence of toxic metals in drinking water are routinely determined using **atomic emission** and **atomic absorption spectroscopy**. Figure 11.8 illustrates how these two spectroscopic techniques are implemented. A sample, ideally in the form of very small droplets ($\sim 1-10\ \mu$m in diameter) of a solution or suspension, is injected into the heated zone of the spectrometer. The heated zone may take the form of a flame, an electrically heated graphite furnace, or a plasma arc source. The main requirement of the heated zone is that it must convert a portion of the molecules in the sample of interest into atoms in their ground and excited states.

We first discuss atomic emission spectroscopy. In this technique, the light emitted by excited-state atoms as they undergo transitions back down to the ground state is dispersed into its component wavelengths by a monochromator and the intensity of the radiation is measured as a function of wavelength. Because the emitted light intensity is proportional to the number of excited-state atoms and because the wavelengths at which emission occurs are characteristic for the atom, the technique can be used for both qualitative and quantitative analysis. Temperatures in the range of 1800 to 3500 K can be achieved in flames and carbon furnaces and up to 10,000 K can be reached in plasma arc sources. These high temperatures are required to produce sufficient excited-state atoms that emit light as demonstrated in Example Problem 11.7.

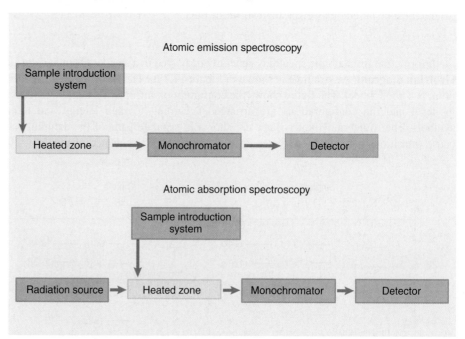

FIGURE 11.8

Schematic diagram of atomic emission and atomic absorption spectroscopies.

EXAMPLE PROBLEM 11.7

The ^{2}S$_{1/2}$ $\rightarrow$ ^{2}P$_{3/2}$ transition in sodium has a wavelength of 589.0 nm. This is one of the lines characteristic of the sodium vapor lamps used for lighting streets, and it gives the lamps their yellow-orange color. Calculate the ratio of the number of atoms in these two states at 1500., 2500., and 3500. K. The following figure is a Grotrian diagram for Na (not to scale) in which the transition of interest is shown as a blue line.

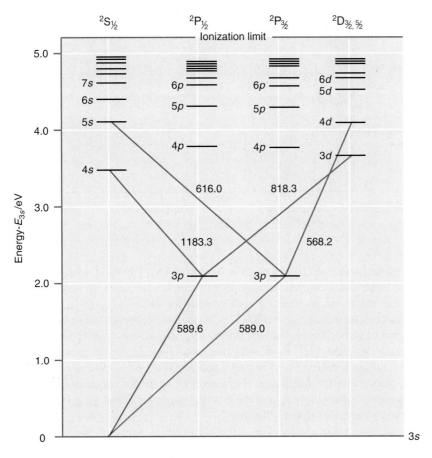

Solution

The ratio of atoms in the upper and lower levels is given by the Boltzmann distribution:

$$\frac{n_{upper}}{n_{lower}} = \frac{g_{upper}}{g_{lower}} e^{-(\varepsilon_{upper} - \varepsilon_{lower})/kT}$$

The degeneracies, g, are given by $2J + 1$, which is the number of states in each level:

$$g_{upper} = 2 \times \frac{3}{2} + 1 = 4 \quad \text{and} \quad g_{lower} = 2 \times \frac{1}{2} + 1 = 2$$

From the Boltzmann distribution,

$$\frac{n_{upper}}{n_{lower}} = \frac{g_{upper}}{g_{lower}} \exp\left(-hc/\lambda_{transition} kT\right)$$

$$= \frac{4}{2} \exp\left[-\frac{(6.626 \times 10^{-34} \text{ J s})(2.998 \times 10^{8} \text{ m s}^{-1})}{(589.0 \times 10^{-9} \text{ m})(1.381 \times 10^{-23} \text{ J K}^{-1}) \times T}\right]$$

$$= 2 \exp\left(-\frac{24421.6}{T}\right) = 1.699 \times 10^{-7} \text{ at } 1500. \text{ K}, 1.144 \times 10^{-4} \text{ at } 2500. \text{ K},$$

and 1.865×10^{-3} at 3500. K

As seen in the preceding example problem, the fraction of atoms in the excited state is quite small, but increases rapidly with temperature. The very high temperature plasma arc sources are widely used because they allow light emission from both more

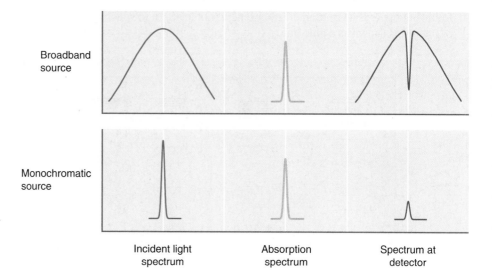

FIGURE 11.9

The intensity of light as a function of its frequency is shown at the entrance to the heated zone and at the detector for broadband and monochromatic sources. The absorption spectrum of the atom to be detected is shown in the middle column.

highly excited states and from ions to be observed. This greatly increases the sensitivity of the technique. However, because photons can be detected with very high efficiency, measurements can be obtained from systems for which n_{upper}/n_{lower} is quite small. For instance, a temperature of ~3000 K is reached in an oxygen-natural gas flame. If a small amount of NaCl is put into the flame, $n_{upper}/n_{lower} \sim 6 \times 10^{-4}$ for Na as shown in Example Problem 11.7. Even with this rather low degree of excitation, a bright yellow emission resulting from the 589.0 and 589.6-nm emission lines in the flame is clearly visible with the naked eye. The sensitivity of the technique can be greatly enhanced using photomultipliers, and spectral transitions for which $n_{upper}/n_{lower} < 10^{-10}$ are routinely used in analytical chemistry.

Atomic absorption spectroscopy differs from atomic emission spectroscopy in that light is passed through the heated zone and the absorption associated with transitions from the lower to the upper state is detected. Because this technique relies on the population of low-lying rather than highly excited atomic states, it has some advantages in sensitivity over atomic emission spectroscopy. It became a very widely used technique when researchers realized that the sensitivity would be greatly enhanced if the light source were nearly monochromatic with a wavelength centered at $\lambda_{transition}$. The advantage of this arrangement can be seen from Figure 11.9.

Only a small fraction of the broadband light that passes through the heated zone is absorbed in the transition of interest. To detect the absorption, the light needs to be dispersed with a grating and the intensity of the light must be measured as a function of frequency. Because the monochromatic source matches the transition both in frequency and in linewidth, detection is much easier. Only a simple monochromator is needed to remove background light before the light is focused on the detector. The key to the implementation of this technique was the development of hollow cathode gas discharge lamps that emit light at the characteristic frequencies of the cathode materials. By using an array of these relatively inexpensive lamps on a single spectrometer, analyses for a number of different elements of interest can be carried out.

The sensitivity of atomic emission and absorption spectroscopy depends on the element and ranges from $10^{-4}\,\mu g/mL$ for Mg to $10^{-2}\,\mu g/mL$ for Pt. These techniques are used in a wide variety of applications, including drinking water analysis and engine wear, by detecting trace amounts of abraded metals in lubricating oil.

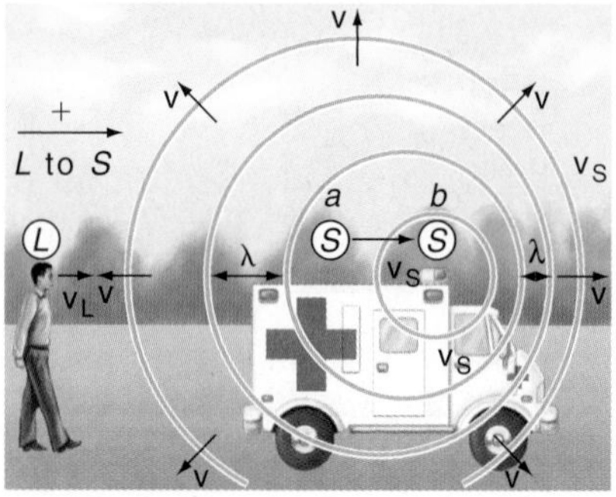

FIGURE 11.10

The frequency of light or sound at the position of the observer L depends on the relative velocity of the source S and observer.

11.6 THE DOPPLER EFFECT

A further application of atomic spectroscopy results from the **Doppler effect**. If a source is radiating light and moving relative to an observer, the observer sees a change in the frequency of the light as shown in Figure 11.10 for sound.

The shift in frequency is given by the formula

$$\omega = \omega_0 \sqrt{(1 \pm v_z/c)/(1 \mp v_z/c)} \tag{11.8}$$

In this formula, v_z is the velocity component in the observation direction, c is the speed of light, and ω_0 is the light frequency in the frame in which the source is stationary. The upper and lower signs refer to the object approaching and receding from the observer, respectively. Note that the frequency shift is positive for objects that are approaching (a so-called "blue shift") and negative for objects that are receding (a so-called "red shift"). The **Doppler shift** is used to measure the speed at which stars and other radiating astronomical objects are moving relative to the Earth.

EXAMPLE PROBLEM 11.8

A line in the Lyman emission series for atomic hydrogen $(n_{final} = 1)$, for which the wavelength is at 121.6 nm for an atom at rest, is seen for a particular quasar at 445.1 nm. Is the source approaching toward or receding from the observer? What is the magnitude of the velocity?

Solution

Because the frequency observed is less than that which would be observed for an atom at rest, the object is receding. The relative velocity is given by

$$\left(\frac{\omega}{\omega_0}\right)^2 = \left(1 - \frac{v_z}{c}\right)\Big/\left(1 + \frac{v_z}{c}\right), \text{ or }$$

$$\frac{v_z}{c} = \frac{1 - (\omega/\omega_0)^2}{1 + (\omega/\omega_0)^2} = \frac{1 - (\lambda_0/\lambda)^2}{1 + (\lambda_0/\lambda)^2}$$

$$= \frac{1 - (121.6/445.1)^2}{1 + (121.6/445.1)^2} = 0.8611; \quad v_z = 2.582 \times 10^8 \, \text{m s}^{-1}$$

For source velocities much less than the speed of light, the nonrelativistic formula

$$\omega = \omega_0 \left(1 \Big/ 1 \mp \frac{v_z}{c}\right) \tag{11.9}$$

applies. This formula is appropriate for a gas of atoms or molecules for which the distribution of speeds is given by the Maxwell–Boltzmann distribution. Because all velocity directions are equally represented for a particular speed, v_z has a large range for a gas at a given temperature, centered at $v_z = 0$. Therefore, the frequency is not shifted; instead, the spectral line is broadened. This is called **Doppler broadening**. Because atomic and molecular velocities are very small compared with the speed of light, the broadening of a line of frequency ω_0 is on the order of 1 part in 10^6. This effect is not as dramatic as the shift in frequency for the quasar, but still of importance in determining the linewidth of a laser, as we will see in the next section.

11.7 THE HELIUM-NEON LASER

In this section, we demonstrate the relevance of the basic principles discussed earlier to the functioning of a laser. We focus on the common He-Ne laser. To understand the He-Ne laser, the concepts of absorption, spontaneous emission, and stimulated emission introduced in Chapter 8 are used. All three processes obey the same selection rules: $\Delta l = \pm 1$ for an electron and $\Delta L = 0$ or ± 1 for an atom.

Spontaneous and stimulated emission differ in an important respect. **Spontaneous emission** is a completely random process, and the photons that are emitted are

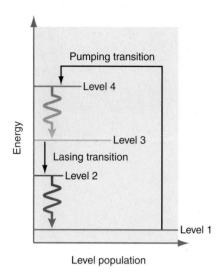

FIGURE 11.11

Schematic representation of a four-state laser. The energy is plotted vertically, and the level population is plotted horizontally.

incoherent, meaning that their phases are random. A light bulb is an **incoherent photon source**. The phase relation between individual photons is random, and because the propagation direction of the photons is random, the intensity of the source falls off as the square of the distance. In **stimulated emission**, the phase and direction of propagation are the same as that of the incident photon. This is referred to as coherent photon emission. A **laser** is a coherent source of radiation. All photons are in phase, and because they have the same propagation direction, the divergence of the beam is very small. This explains why a laser beam that is reflected from the moon still has a measurable intensity when it returns to the earth. This discussion makes it clear that a **coherent photon source** must be based on stimulated rather than spontaneous emission. However, $B_{12} = B_{21}$, as was shown in Section 8.2. Therefore, the rates of absorption and stimulated emission are equal for $N_1 = N_2$. Stimulated emission with an amplification of the light will only dominate over absorption if $N_2 > N_1$. This is called a **population inversion** because for equal level degeneracies, the higher energy state has the higher population at equilibrium. The key to making a practical laser is to create a stable population inversion. Although a population inversion is not possible under equilibrium conditions, it is possible to maintain such a distribution under steady-state conditions if the relative rates of the transitions between levels are appropriate. This is illustrated in Figure 11.11.

Figure 11.11 can be used to understand how the population inversion between the levels involved in the lasing transition is established and maintained. The lengths of the horizontal lines representing the levels are proportional to the level populations N_1 to N_4. The initial step involves creating a significant population in level 4 by transitions from level 1. This is accomplished by an external source, which for the He-Ne laser is an electrical discharge in a tube containing the gas mixture. Relaxation to level 3 can occur through spontaneous emission of a photon as indicated by the curvy arrow. Similarly, relaxation from level 2 to level 1 can also occur through spontaneous emission of a photon. If this second relaxation process is fast compared to the first, N_3 will be maintained at a higher level than N_2. In this way, a population inversion is established between levels two and three. The advantage of having the lasing transition between levels 3 and 2 rather than 2 and 1 is that N_2 can be kept low if relaxation to level 1 from level 2 is fast. It is not possible to keep N_1 at a low level because atoms in the ground state cannot decay to a lower state.

This discussion shows how a population inversion can be established. How can a continuous lasing transition based on stimulated emission be maintained? This is made possible by carrying out the process indicated in Figure 11.1 in an **optical resonator** as shown in Figure 11.12.

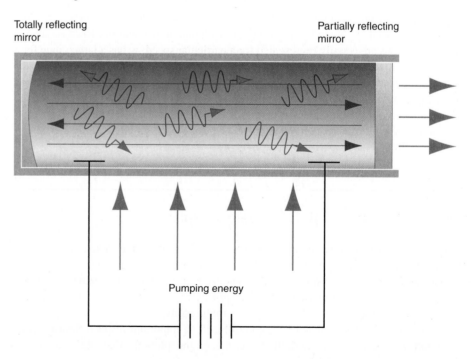

FIGURE 11.12

Schematic representation of a He-Ne laser operated as an optical resonator. The parallel lines in the resonator represent coherent stimulated emission that is amplified by the resonator, and the red waves represent incoherent spontaneous emission events.

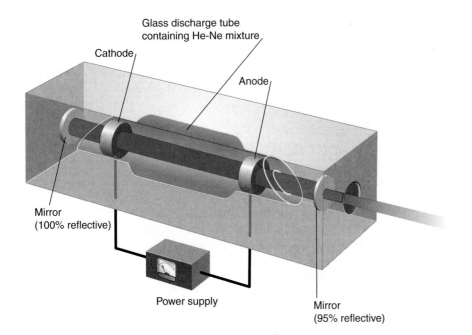

FIGURE 11.13

Schematic diagram of a He-Ne laser.

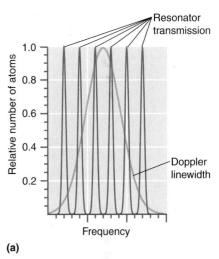

(a)

The He-Ne mixture is put into a glass tube with carefully aligned parallel mirrors on each end. Electrodes are inserted to maintain the electrical discharge that pumps level 4 from level 1. Light reflected back and forth in the optical cavity between the two mirrors interferes constructively only if $n\lambda = n(c/\nu) = 2d$, where d is the distance between mirrors and n is an integer. The next constructive interference occurs when $n \rightarrow n + 1$. The difference in frequency between these two modes is $\Delta\nu = c/2d$, which defines the bandwidth of the cavity. The number of modes that contribute to laser action is determined by two factors: the frequencies of the **resonator modes** and the width in frequency of the stimulated emission transition. The width of the transition is determined by Doppler broadening, which arises through the thermal motion of gas-phase atoms or molecules. A schematic diagram of a He-Ne laser, including the anode, cathode, and power supply needed to maintain the electrical discharge as well as the optical resonator, is shown in Figure 11.13.

Six of the possible resonator modes are indicated in Figure 11.14. The curve labeled "Doppler linewidth" gives the relative number of atoms in the resonator as a function of the frequency at which they emit light. The product of these two functions gives the relative intensities of the stimulated emission at the different frequencies supported by the resonator. This is shown in Figure 11.14b. Because of losses in the cavity, a laser transition can only be sustained if enough atoms in the cavity are in the excited state at a supported resonance. In Figure 11.14, only two resonator modes lead to a sufficient intensity to sustain the laser. The main effect of the optical resonator is to decrease the $\Delta\nu$ associated with the frequency of the lasing transition to less than the Doppler limit. Example Problem 11.9 shows how the number of supported modes varies with the gas temperature.

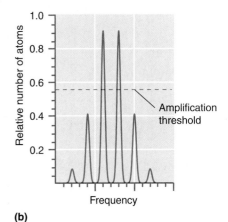

(b)

FIGURE 11.14

The linewidth of a transition in a He-Ne laser is Doppler broadened through the Maxwell–Boltzmann velocity distribution. **(a)** The resonator transmission decreases the linewidth of the lasing transition to less than the Doppler limit. **(b)** The amplification threshold further reduces the number of frequencies supported by the resonator.

EXAMPLE PROBLEM 11.9

The distribution function that describes the probability of finding a particular value of magnitude of the velocity along one dimension, v, in a gas at temperature T is given by

$$f(\mathrm{v})d\mathrm{v} = \sqrt{\frac{m}{2\pi kT}}\, e^{-m\mathrm{v}^2/2kT} d\mathrm{v}$$

This velocity distribution leads to the broadening of a laser line in frequency given by

$$I(\nu)d\nu = \sqrt{\frac{mc^2}{2\pi kT \nu_0^2}} \exp\left\{-\frac{mc^2}{2kT}\left(\frac{\nu - \nu_0}{\nu_0}\right)^2\right\} d\nu$$

The symbol c stands for the speed of light, and k is the Boltzmann constant. We next calculate the broadening of the 632.8 nm line in the He-Ne laser at a function of T.

a. Plot $I(\nu)$ for $T = 100.0$, 300.0, and 1000. K, using the mass appropriate for a Ne atom, and determine the width in frequency at half the maximum amplitude of $I(\nu)$ for each of the three temperatures.

b. Assuming that the amplification threshold is 50% of the maximum amplitude, how many modes could lead to amplification in a cavity of length 100. cm?

Solution

a. This function is of the form of a normal or Gaussian distribution given by

$$f_{\nu_0, \sigma}(\nu)\, d\nu = \frac{1}{\sigma\sqrt{2\pi}} e^{-(\nu-\nu_0)/2\sigma^2}\, d\nu$$

The full width at half height is 2.35 σ, or for this case, $2.35\sqrt{kT\nu_0^2/mc^2}$. This gives half widths of 7.554×10^8 s^{-1}, 1.308×10^9 s^{-1}, and 2.388×10^9 s^{-1} at temperatures of 100.0, 300.0, and 1000. K, respectively. The functions $I(\nu)$ are plotted here:

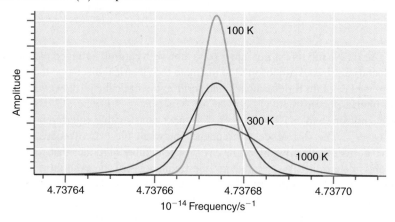

b. The frequency spacing between two modes is given by

$$\Delta\nu = \frac{c}{2d} = \frac{2.998 \times 10^8 \text{ m s}^{-1}}{2 \times 1.00 \text{ m}} = 1.50 \times 10^8 \text{ s}^{-1}$$

The width of the velocity distribution will support 5 modes at 100.0 K, 8 modes at 300.0 K, and 15 modes at 1000. K. The smaller Doppler broadening at low temperatures reduces the number of possible modes considerably.

11.1 Simulation of a Laser

By adding a further optical filter in the laser tube, it is possible to have only one mode enhanced by the resonator. This means that only the transition of interest is enhanced by multiple reflections in the optical resonator. In addition, all light of the correct frequency, but with a propagation direction that is not perpendicular to the mirrors, is not enhanced through multiple reflections. In this way, the resonator establishes a standing wave at the lasing frequency that has a propagation direction aligned along the laser tube axis. This causes further photons to be emitted from the lasing medium (He-Ne mixture) through stimulated emission. As discussed previously, these photons are exactly in phase with the photons that stimulate the emission and have the same propagation direction. These photons amplify the standing wave, which because of its greater intensity, causes even more stimulated emission. Allowing one of the end mirrors to be partially transmitting lets some of the light escape, and the result is a coherent, well-collimated laser beam.

To this point, the laser has been discussed at a schematic level. How does this discussion relate to the atomic energy levels of He and Ne? The relationship is shown in Figure 11.15. The electrical discharge in the laser tube produces electrons and positively charged ions. The electrons are accelerated in the electric field and can excite the He atoms from states in the ^{1}S term of the $1s^2$ configuration to states in the ^{1}S and ^{3}S terms of the $1s2s$ configuration. This is the **pumping transition** in the scheme shown in Figure 11.11. This

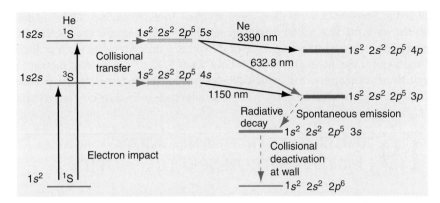

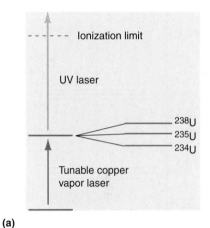

FIGURE 11.15

Transitions in the He-Ne laser. The slanted solid lines in the upper right side of the figure show three possible lasing transitions.

transition occurs through a collision rather than through the absorption of a photon and, therefore, the normal selection rules do not apply. Because the selection rules $\Delta S = 0$ and $\Delta l = \pm 1$ prohibit transitions to the ground state, these states are long lived. The excited He atoms efficiently transfer their energy through collisions to states in the $2p^5 5s$ and $2p^5 4s$ configurations of Ne. This creates a population inversion relative to Ne states in the $2p^5 4p$ and $2p^5 3p$ configuration. These levels are involved in the **lasing transition** through stimulated emission. Spontaneous emission to states in the $2p^5 3s$ configuration and collisional deactivation at the inner surface of the optical resonator depopulate the lower state of the lasing transitions and ensure that the population inversion is maintained. The initial excitation is to excited states of He, which consist of a single term. However, the excited-state configurations of Ne give rise to several terms (^3P and ^1P for $2p^5 4s$ and $2p^5 5s$, and ^3D, ^1D, ^3P, ^1P, ^3S, and ^1S for $2p^5 3p$ and $2p^5 4p$). The manifold of these states is indicated in the figure by thicker lines, indicating a range of energies.

Note that a number of wavelengths can lead to lasing transitions. Coating the mirrors in the optical resonator ensures that they are reflective only in the range of interest. The resonator is usually configured to support the 632.8-nm transition in the visible part of the spectrum. This corresponds to the red light usually seen from He-Ne lasers.

11.8 LASER ISOTOPE SEPARATION

A number of ways are available for separating atoms and molecules into their different isotopes. Separation by diffusion in the gas phase is possible because the speed of molecules depends on the molecular weight, M, as $M^{-1/2}$. To fabricate fuel rods for nuclear reactors, uranium fuel which contains the isotopes ^{234}U, ^{235}U, and ^{238}U must be enriched in the fissionable isotope ^{235}U. This has been done on a large scale by reacting uranium with fluorine to produce the gas-phase molecule UF_6. This gas is enriched in the ^{235}U isotope by centrifugation.

It is feasible, although not practical on an industrial scale, to create a much higher degree of enrichment by using selective laser ionization of the ^{235}U isotope. The principle is shown schematically in Figure 11.16. A tunable copper vapor laser with a very narrow linewidth is used to excite ground-state uranium atoms to an excited state involving the $7s$ electrons. As Figure 11.16b indicates, the electronic states of the different isotopes have slightly different energies, and the bandwidth of the laser is sufficiently small that only one isotope is excited. A second laser pulse is used to ionize the selectively excited isotope. The ions can be collected by electrostatic attraction to a metal electrode at an appropriate electrical potential. Because neither of the lasers produces photons of energy sufficient to ionize the uranium atoms directly, only those atoms selectively excited by the copper vapor laser by the first pulse are ionized by the second pulse.

Why do the different isotopes have slightly different atomic energy levels? The $-1/r$ Coulomb potential that attracts the electron to the nucleus is valid outside the nucleus, but the potential levels off inside the nucleus, which has a diameter of about 1.6×10^{-14} m or 3×10^{-4} a_0. The distance at which the Coulomb potential is no longer valid depends on the nuclear diameter and, therefore, on the number of neutrons in the nucleus. This effect is negligible for states with $l > 0$, because the effective potential

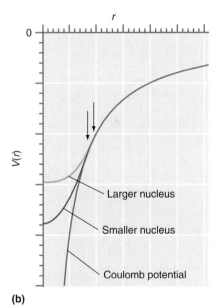

FIGURE 11.16

The principle of laser isotope excitation is shown schematically (not to scale). The electron-nucleus potential deviates from a Coulomb potential at the nuclear radius, which is indicated by the vertical arrows in part **(b)**. Because this distance depends on the nuclear volume, it depends on the number of neutrons in the nucleus for a given atomic number. **(a)** This very small variation in $V(r)$ for isotopes ^{234}U through ^{238}U gives rise to an energy splitting for states involving s electrons. By means of a combined two-photon excitation and ionization, this splitting can be utilized to selectively ionize a particular isotope.

discussed in Section 9.2 keeps these electrons away from the nucleus. Only *s* states, for which the wave function has its maximum amplitude at the center of the nucleus, exhibit energy-level splitting. The magnitude of the splitting depends on the nuclear diameter. The splitting for the uranium isotopes is only about 2×10^{-3} percent of the ionization energy. However, the very small bandwidth that is attainable in lasers allows selective excitation to a single level even in cases for which the energy-level spacing is very small.

11.9 AUGER ELECTRON AND X-RAY PHOTOELECTRON SPECTROSCOPIES

Most of the atomic spectroscopies that we have discussed have been illustrated with gas-phase examples. Another useful application of spectroscopic methods is in the analysis of the elemental composition of surfaces. This capability is important in such fields as corrosion and heterogeneous catalysis in which a chemical reaction takes place at the interface between a solid phase and a gaseous or liquid phase. Sampling of a surface in a way that is relevant to the process requires that the method be sensitive only to the first few atomic layers of the solid. The two spectroscopies described in this section satisfy this requirement for reasons to be discussed next. They are also applicable in other environments such as the gas phase.

Both of these spectroscopies involve the ejection of an electron from individual atoms in a solid and the measurement of the electron energy. To avoid energy losses due to collisions of the ejected electron with gas-phase molecules, the solid sample is examined in a vacuum chamber. If the electron has sufficient energy to escape from the solid into the vacuum, it will have a characteristic energy simply related to the energy level from which it originated. To escape into the vacuum, it must travel from its point of origination to the surface of the solid. This process is analogous to a gas-phase atom traveling through a gas. The atom will travel a certain distance (that depends on the gas pressure) before it collides with another atom. In the collision, it exchanges energy and momentum with its collision partner and thereby loses memory about its previous momentum and energy. Similarly, an electron generated in the solid traveling toward the surface suffers collisions with other electrons and loses memory of the energy levels from which it originated if its path is too long. Only those atoms within one **inelastic mean free path** of the surface eject electrons into the gas phase whose energy can be assigned to the atomic energy levels. Electrons emitted from other atoms simply contribute to the background signal. The mean free path for electrons depends on the energy of the electrons, but is relatively material independent. It has its minimum value of about 2 atomic layers near 40 eV and increases slowly to about 10 atomic layers at 1000 eV. Electrons in this energy range that have been ejected from atoms can provide information that is highly surface sensitive.

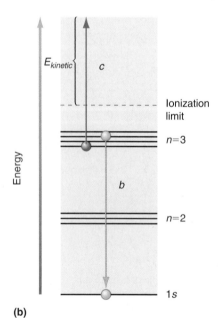

(a)

(b)

FIGURE 11.17

The principle of Auger electron spectroscopy is illustrated schematically. **(a)** A core level hole is formed by energy transfer from an incident photon or electron. **(b)** The core hole is filled through relaxation from a higher level and a third electron is emitted to conserve energy. The energy of the emitted electron can be measured and is characteristic of the particular element.

EXAMPLE PROBLEM 11.10

Upon impingement of X-rays from a laboratory source, titanium atoms near the surface of bulk TiO_2 emit electrons into a vacuum with energy of 790 eV. The finite mean free path of these electrons leads to an attenuation of the signal for Ti atoms beneath the surface according to $[I(d)]/[I(0)] = e^{-d/\lambda}$. In this equation, d is the distance to the surface and λ is the mean free path. If λ is 2.0 nm, what is the sensitivity of Ti atoms 10.0 nm below the surface relative to those at the surface?

Solution

Substituting in the equation $[I(d)]/[I(0)] = e^{-d/\lambda}$, we obtain
$\frac{I(100)}{I(0)} = e^{-10.0/2.0} = 6.7 \times 10^{-3}$. This result illustrates the surface
sensitivity of the technique.

Auger electron spectroscopy (AES) is schematically illustrated in Figure 11.17. An electron (or photon) ejects an electron from a low-lying level in an atom. This hole is quickly filled by a transition from a higher state. This event alone, however, does not

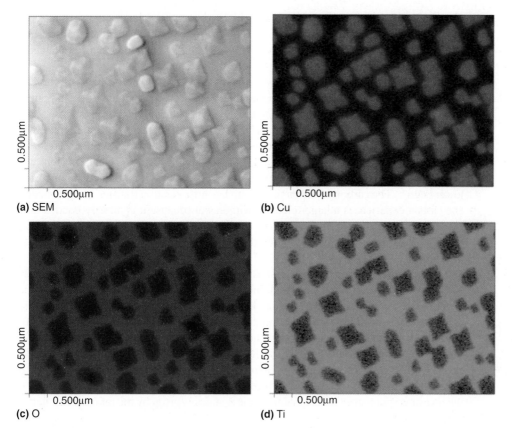

(a) SEM

(b) Cu

(c) O

(d) Ti

FIGURE 11.18

Scanning electron microscopy and scanning Auger spectroscopy images for copper, oxygen, and titanium are shown for a 0.5×0.5-μm area of a $SrTiO_3$ crystal surface on which Cu_2O nanodots have been deposited by Cu evaporation in an oxygen containing plasma. **(a)** The SEM image obtained without energy analysis shows structure, but gives no information on the elemental distribution. **(b)**, **(c)**, and **(d)** are obtained using energy analysis of backscattered electrons. Light and dark areas correspond to high and low values respective of Cu **(b)**, and O **(c)**, and Ti **(d)**. An analysis of the data shows that the light areas in (b) contain Cu and O, but no Ti, and that the dark areas in (d) contain no Ti.

[Photos (a), (b), and (d) courtesy of Pacific Northwest National Laboratory. Photo (c) courtesy of Liang *et al.*/Electrochemical Society Proceedings Volume 2001–5.]

conserve energy. Energy conservation is accomplished by the simultaneous ejection of a second electron into the gas phase. It is the kinetic energy of this electron that is measured. Although three different energy levels are involved in this spectroscopy, the signatures of different atoms are quite easy to distinguish. The main advantage of using electrons rather than photons to create the initial hole is to gain spatial resolution. Electron beams can be focused to a spot size on the order of 10–100 nm and, therefore, Auger spectroscopy with electron excitation is routinely used in many industries to map out elemental distribution at the surfaces of solids with very high lateral resolution.

Results using scanning Auger spectroscopy to study the growth of Cu_2O nanodots on the surface of a $SrTiO_3$ crystal are shown in Figure 11.18. The scanning electron microscopy (SEM) image shows the structure of the surface, but gives no information on the elemental composition. Scanning Auger images of the same area are shown for Cu, Ti, and O. These images show that Cu does not uniformly coat the surface, but instead forms three-dimensional crystallites. This conclusion can be drawn from the absence of Cu and the presence of Ti in the areas between the nanodots. The Cu_2O nanodots appear darker (lower O content) than the underlying $SrTiO_3$ surface in the oxygen image because Cu_2O has only one O for every two Cu cations, whereas the surface has three O in a formula unit containing two cations. These results show that the nanodot deposition process can be understood using a surface-sensitive spectroscopy.

X-ray photoelectron spectroscopy (XPS) is simpler than AES in that only one level is involved. A photon of energy is absorbed by an atom, and to conserve energy, an electron is ejected with kinetic energy

$$E_{kinetic} = h\nu - E_{binding} \qquad (11.10)$$

A small correction term that involves the work functions of the solid (defined in Chapter 1) and the detector has been omitted. A schematic picture of the process that gives rise to an ejected electron is shown in Figure 11.19. Currently, no off-the-shelf X-ray lasers are available, so that sources cannot be made with very small bandwidths. However, using monochromatized X-ray sources, distinctly different peaks are observed for substances in which the same atom is present in chemically nonequivalent environments. This

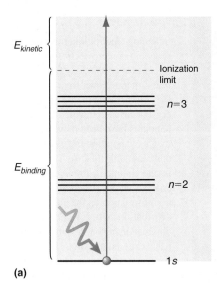

$E_{kinetic}$

Ionization limit

$n=3$

$E_{binding}$

$n=2$

$1s$

(a)

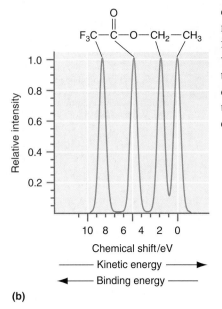

(b)

FIGURE 11.19

(a) Principle of X-ray photoelectron spectroscopy. **(b)** A spectrum exhibiting chemical shifts for the carbon $1s$ level of the ethyltrifluoroacetate molecule is shown.

FIGURE 11.20

XP spectra are shown for the deposition of Fe films on a crystalline MgO surface. Note the splitting of the peaks originating from the $2p$ core level as a result of spin-orbit interaction. Shake-up features originate when valence electrons are promoted to higher levels in the photoemission event. This promotion reduces the kinetic energy of the ejected electron.

[Graph courtesy of Scott A. Chambers/Pacific Northwest National Laboratory.]

chemical shift is also illustrated in Figure 11.19. A positive value for the chemical shift indicates a higher binding energy for the electron in the atom than would be measured for the free atom. The origin of the chemical shift can be understood in a simple model, although accurate calculations require a more detailed treatment.

Consider the different binding environment of the carbon atom in ethyltrifluoroacetate, whose structure is shown in Figure 11.19b. The carbon atom in the CF_3 group experiences a net electron withdrawal to the much more electronegative F atoms. Therefore, the $1s$ electrons lose some of the shielding effect they had from the $2p$ electrons and, as a result, the C $1s$ electron experiences a slightly greater nuclear charge. This leads to an increase in the binding energy, or a positive chemical shift. The carbon atom with double and single bonds to oxygen experiences an electron withdrawal, although to a lesser degree. The carbon of the methyl group has little electron transfer, and the methylene carbon experiences a larger electron withdrawal because it is directly bonded to an oxygen. Although these effects are small, they are easily measurable. Therefore, a photoelectron spectrum gives information on the oxidation state as well as on the identity of the element.

An example in which the surface sensitivity of XPS is used is shown in Figure 11.20. The growth of an iron film on a crystalline magnesium oxide surface is monitored under different conditions. The X-ray photon ejects an electron from the $2p$ level of Fe species near the surface, and the signal is dominated by those species within ~ 1 nm of the surface. The spin angular momentum **s** of the remaining electron in the Fe $2p$ level couples with its orbital angular momentum **l** to form the total angular momentum vector **j** with two possible values for the quantum number, $j = 1/2$ and $3/2$. These two states are of different energy and, therefore, two peaks are observed in the spectrum. The ratio of the measured photoemission signal from these states is given by the ratio of their degeneracy, or

$$\frac{I(2p_{3/2})}{I(2p_{1/2})} = \frac{2 \times \left(\frac{3}{2}\right) + 1}{2 \times \left(\frac{1}{2}\right) + 1} = 2$$

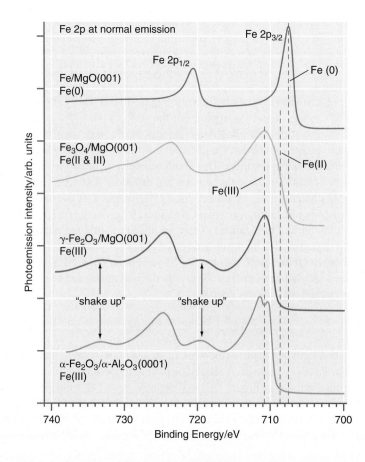

The binding energy corresponding to the Fe peaks clearly shows that deposition in vacuum leads to metallic or zero valent Fe species. Different crystalline phases of the iron oxide formed by exposing the film to oxygen gas while depositing iron have different ratios of Fe(II) and Fe(III).

11.10 SELECTIVE CHEMISTRY OF EXCITED STATES: O(^{3}P) AND O(^{1}D)

The interaction of sunlight with molecules in the atmosphere leads to an interconnected set of chemical reactions, which in part determines the composition of the Earth's atmosphere. Oxygen is a major species involved in these reactions. Solar radiation in the ultraviolet range governs the concentrations of $O\cdot$, O_2, and O_3 according to

$$O_2 + h\nu \longrightarrow O\cdot + O\cdot$$

$$O\cdot + O_2 + M \longrightarrow O_3$$

$$O_3 + h\nu \longrightarrow O\cdot + O_2$$

$$O\cdot + O_3 \longrightarrow 2O_2 \qquad\qquad (11.11)$$

where M is another molecule in the atmosphere that takes up the energy released in forming O_3. For wavelengths less than 315 nm, the dissociation of molecular oxygen occurs as

$$O_2 + h\nu \longrightarrow O\cdot(^3P) + O\cdot(^1D) \qquad\qquad (11.12)$$

and for longer wavelengths, dissociation to two ground-state ^3P oxygen atoms occurs. In this reaction, $\cdot O(^1D)$ has an excess energy of 190 kJ mol^{-1} relative to $\cdot O(^3P)$. This energy can be used to overcome an activation barrier to reaction. For example, the reaction

$$\cdot O(^3P) + H_2O \longrightarrow \cdot OH + \cdot OH \qquad\qquad (11.13)$$

is endothermic by 70 kJ mol^{-1}, whereas the reaction

$$\cdot O(^1D) + H_2O \longrightarrow \cdot OH + \cdot OH \qquad\qquad (11.14)$$

is exothermic by 120 kJ mol^{-1}. Because a radiative transition from $\cdot O(^1D)$ to $\cdot O(^3P)$ is forbidden by the selection rule $\Delta S = 0$, the $O(^1D)$ atoms are long lived and their concentration is predominantly depleted by reactions with other species.

The $\cdot O(^1D)$ atoms are primarily responsible for generating reactive hydroxyl and methyl radicals through the reactions

$$\cdot O(^1D) + H_2O \longrightarrow \cdot OH + \cdot OH$$

$$\cdot O(^1D) + CH_4 \longrightarrow \cdot OH + \cdot CH_3 \qquad\qquad (11.15)$$

As before, the reactivity for $\cdot O(^1D)$ is much higher than that for $\cdot O(^3P)$ largely because the excess energy in electronic excitation can be used to overcome the activation barrier for the reaction. $\cdot O(^1D)$ is also involved in generating the reactive NO intermediate from N_2O and $\cdot Cl$ from chlorofluorocarbons.

SUPPLEMENTAL

11.11 CONFIGURATIONS WITH PAIRED AND UNPAIRED ELECTRON SPINS DIFFER IN ENERGY

In this section, we show that the energy of a configuration depends on whether the spins are paired for a specific case, namely, the excited states of He with the configuration $1s^1 2s^1$. Antisymmetric wave functions for each value of S were formulated in Section 10.3.

In the following, the energy for the singlet and triplet states is calculated, and we prove the important result that the triplet state lies lower in energy than the singlet state.

The Schrödinger equation for the singlet wave function can be written as

$$\left(\hat{H}_1 + \hat{H}_2 + \frac{e^2}{4\pi\varepsilon_0 r_{12}}\right)\psi(1, 2) = E_{singlet}\psi(1, 2) \qquad (11.16)$$

where $\hat{H}_1$ and $\hat{H}_2$ are the total energy operators neglecting electron–electron repulsion and the subscripts refer to the electron involved. $\psi(1, 2)$ is the unknown exact wave function. The spin part of the wave function is not included because the total energy operator does not contain terms that depend on spin. Because we do not know the exact wave function, we approximate it by the simple singlet wave function of Equation 11.3. Keep in mind that the singlet wave function is not an eigenfunction of the total energy operator. To obtain the expectation value for the total energy using this approximate wave function, one multiplies on the left by the complex conjugate of the wave function and integrates over the spatial coordinates. Because the $1s$ and $2s$ functions are real, the function and its complex conjugate are identical.

$$E_{singlet} = \frac{1}{2}\iint [1s(1)2s(2) + 2s(1)1s(2)]\left(\hat{H}_1 + \hat{H}_2 + \frac{e^2}{4\pi\varepsilon_0 r_{12}}\right)$$
$$\times [1s(1)2s(2) + 2s(1)1s(2)]d\tau_1 d\tau_2 \qquad (11.17)$$

As you will see when you work the end-of-chapter problems, the two integrals arising from $\hat{H}_1$ and $\hat{H}_2$ give $E_{1s} + E_{2s}$ where $E_{1s} = -e^2/2\pi\varepsilon_0 a_0$ and $E_{2s} = -e^2/8\pi\varepsilon_0 a_0$ are the H atom eigenvalues for $\zeta = 2$. Using this result,

$$E_{singlet} = E_{1s} + E_{2s}$$
$$+ \frac{1}{2}\iint [1s(1)2s(2) + 2s(1)1s(2)]\left(\frac{e^2}{4\pi\varepsilon_0 r_{12}}\right)$$
$$\times [1s(1)2s(2) + 2s(1)1s(2)]d\tau_1 d\tau_2 \qquad (11.18)$$

The remaining integral can be simplified as you will also see in the end-of-chapter problems to yield $E_{singlet}$:

$$E_{singlet} = E_{1s} + E_{2s} + J_{12} + K_{12}, \text{ where} \qquad (11.19)$$

$$J_{12} = \frac{e^2}{8\pi\varepsilon_0}\iint [1s(1)]^2\left(\frac{1}{r_{12}}\right)[2s(2)]^2 d\tau_1 d\tau_2 \text{ and}$$

$$K_{12} = \frac{e^2}{8\pi\varepsilon_0}\iint [1s(1)2s(2)]\left(\frac{1}{r_{12}}\right)[1s(2)2s(1)]d\tau_1 d\tau_2 \qquad (11.20)$$

If the calculation is carried out for the triplet state, the corresponding result is

$$E_{triplet} = E_{1s} + E_{2s} + J_{12} - K_{12} \qquad (11.21)$$

Focus on the results rather than the mathematics. In the absence of the repulsive interaction between the two electrons, the total energy is simply $E_{1s} + E_{2s}$. Including the Coulomb repulsion between the electrons and making the wave function antisymmetric give rise to the additional terms J_{12} and K_{12}. The energy shift relative to $E_{1s} + E_{2s}$ is $J_{12} + K_{12}$ for the singlet state and $J_{12} - K_{12}$ for the triplet state. This shows that the triplet and singlet states of $He(1s^12s^1)$ have energies that differ by $2K_{12}$. Because all the terms that appear in the first integral of Equation (11.20) are positive, $J_{12} > 0$. It can also be shown that K_{12} is positive. *Therefore, it has been shown that the triplet state for the first excited state of He lies lower in energy than the singlet state. This is a general result.*

Looking back, this result is based on a purely mathematical argument. Can a physical meaning be attached to J_{12} and K_{12}? Imagine that electrons 1 and 2 were point charges. In that case, the integral J_{12} simplifies to $e^2/4\pi\varepsilon_0|\mathbf{r}_1 - \mathbf{r}_2|$. The electrons of He can be thought of as diffuse charge clouds. The integral J_{12} is simply the electrostatic interaction between the diffuse charge distributions $\rho(1)$ and $\rho(2)$ where

$\rho(2) = [2s(2)]^2 \, d\tau_2$ and $\rho(1) = [1s(1)]^2 \, d\tau_1$. Because J_{12} can be interpreted in this way, it is called the **Coulomb integral**. Unlike J_{12}, the integral K_{12} has no classical physical interpretation. The product $[1s(1)2s(2)] [1s(2)2s(1)] \, d\tau_1 d\tau_2$ does not fit the definition of charge because it does not have the form $|\psi(1)|^2 |\psi(2)|^2 \, d\tau_1 d\tau_2$. Because the electrons have been exchanged between the two parts of this product, K_{12} is referred to as the **exchange integral**. It has no classical analogue and arises from the fact that the singlet and triplet wave functions are written as a superposition of two terms in order to satisfy the Pauli exclusion principle.

The singlet and triplet wave functions also differ in the degree to which they include electron correlation. We know that electrons avoid one another because of their Coulomb repulsion. If we let electron 2 approach electron 1, the spatial part of the singlet wave function

$$\frac{1}{\sqrt{2}}[1s(1)2s(2) + 2s(1)1s(2)] \rightarrow \frac{1}{\sqrt{2}}[1s(1)2s(1) + 2s(1)1s(1)]$$

$$= \frac{2}{\sqrt{2}} 1s(1)2s(1)$$

because $r_2, \theta_2, \phi_2 \rightarrow r_1, \theta_1, \phi_1$, but the spatial part of the triplet wave function

$$\frac{1}{\sqrt{2}}[1s(1)2s(2) - 2s(1)1s(2)] \rightarrow \frac{1}{\sqrt{2}}[1s(1)2s(1) - 2s(1)1s(1)] = 0$$

This shows that the triplet wave function has a greater degree of electron correlation built into it than the singlet wave function, because the probability of finding both electrons in a given region falls to zero as the electrons approach one another.

Why is the energy of the triplet state lower than that of the singlet state? One might think that the electron–electron repulsion is lower in the triplet state because of the electron correlation and that this is the origin of the lower total energy. In fact, this is not correct. A more detailed analysis shows that the electron–electron repulsion is actually greater in the triplet than in the singlet state. However, on average the electrons are slightly closer to the nucleus in the triplet state. The increased electron–nucleus attraction outweighs the electron–electron repulsion and, therefore, the triplet state has a lower energy. Note that spin influences the energy even though the total energy operator does not contain any terms involving spin. Spin enters our calculation through the antisymmetrization required by the Pauli exclusion principle rather than through the total energy operator. *Generalizing this result, it can be concluded that for a given configuration, a state in which the spins are unpaired has a lower energy than a state in which the spins are paired.*

Vocabulary

antiparallel	good quantum number	resonator modes
atomic absorption spectroscopy	Grotrian diagram	selection rule
atomic emission spectroscopy	Hund's rules	singlet
Auger electron spectroscopy	incoherent photon source	spin-orbit coupling
chemical shift	inelastic mean free path	spontaneous emission
Clebsch–Gordon series	laser	stimulated emission
coherent photon source	lasing transition	term
Coulomb integral	level	term symbol
degeneracy of a term	multiplicity	total angular momentum
dipole approximation	optical resonator	triplet
Doppler broadening	paired electrons	unpaired electrons
Doppler effect	parallel spins	X-ray photoelectron spectroscopy
Doppler shift	population inversion	
exchange integral	pumping transition	

Questions on Concepts

Q11.1 Justify the statement that the Coulomb integral J defined in Equation (11.20) is positive by explicitly formulating the integral that describes the interaction between two negative classical charge clouds.

Q11.2 Without invoking equations, explain why the energy of the triplet state is lower than that of the singlet state for He in the $1s^1 2s^1$ configuration.

Q11.3 How can the width of a laser line be less than that determined by Doppler broadening?

Q11.4 Why is an electronically excited atom more reactive than the same ground-state atom?

Q11.5 Why is atomic absorption spectroscopy more sensitive in many applications than atomic emission spectroscopy?

Q11.6 Why does the Doppler effect lead to a shift in the wavelength of a star, but to a broadening of a transition in a gas?

Q11.7 Why are n, l, m_l, and m_s not good quantum numbers for many-electron atoms?

Q11.8 Write an equation giving the relationship between the Rydberg constant for H and for Li^{2+}.

Q11.9 Can the individual states in Table 11.1 be distinguished experimentally?

Q11.10 How is it possible to determine the L and S value of a term knowing only the M_L and M_S values of the states?

Q11.11 What is the origin of the chemical shift in XPS?

Q11.12 Why are two medium-energy photons rather than one high-energy photon used in laser isotope separation?

Q11.13 Why does one need to put a sample in a vacuum chamber to study it with XPS or AES?

Q11.14 Why is XPS a surface-sensitive technique?

Q11.15 Explain the direction of the chemical shifts for Fe(0), Fe(II), and Fe(III) in Figure 11.20.

Problems

Problem numbers in **red** indicate that the solution to the problem is given in the *Student's Solutions Manual*.

P11.1 The principal line in the emission spectrum of potassium is violet. On close examination, the line is seen to be a doublet with wavelengths of 393.366 and 396.847 nm. Explain the source of this doublet.

P11.2 The absorption spectrum of the hydrogen atom shows lines at 5334, 7804, 9145, 9953, and 10,478 cm^{-1}. There are no lower frequency lines in the spectrum. Use the graphical methods discussed in Example Problem 11.6 to determine $n_{initial}$ and the ionization energy of the hydrogen atom in this state. Assume values for $n_{initial}$ of 1, 2, and 3.

P11.3 Using Table 11.3, which lists the possible terms that arise from a given configuration, and Hund's rules, write the term symbols for the ground state of the atoms H through F in the form $^{(2S+1)}L_J$.

P11.4 In this problem, you will supply the missing steps in the derivation of the formula $E_{singlet} = E_{1s} + E_{2s} + J + K$ for the singlet level of the $1s^1 2s^1$ configuration of He.

a. Expand Equation (11.17) to obtain

$$E_{singlet} = \frac{1}{2} \iint \frac{[1s(1)2s(2) + 2s(1)1s(2)](\hat{H}_1)}{[1s(1)2s(2) + 2s(1)1s(2)]d\tau_1 d\tau_2}$$

$$+ \frac{1}{2} \iint \frac{[1s(1)2s(2) + 2s(1)1s(2)](\hat{H}_2)}{[1s(1)2s(2) + 2s(1)1s(2)]d\tau_1 d\tau_2}$$

$$+ \frac{1}{2} \iint [1s(1)2s(2) + 2s(1)1s(2)]$$

$$\left(\frac{e^2}{4\pi\varepsilon_0 |\mathbf{r}_1 - \mathbf{r}_2|} \right)$$

$$[1s(1)2s(2) + 2s(1)1s(2)]d\tau_1 d\tau_2$$

b. Starting from the equations $\hat{H}_i 1s(i) = E_{1s} 1s(i)$ and $\hat{H}_i 2s(i) = E_{2s} 2s(i)$, show that

$$E_{singlet} = E_{1s} + E_{2s}$$

$$+ \frac{1}{2} \iint [1s(1)2s(2) + 2s(1)1s(2)] \left(\frac{e^2}{4\pi\varepsilon_0 |\mathbf{r}_1 - \mathbf{r}_2|} \right)$$

$$[1s(1)2s(2) + 2s(1)1s(2)]d\tau_1 d\tau_2$$

c. Expand the previous equation using the definitions

$$J = \frac{e^2}{8\pi\varepsilon_0} \iint [1s(1)]^2 \left(\frac{1}{|\mathbf{r}_1 - \mathbf{r}_2|} \right) [2s(2)]^2 d\tau_1 d\tau_2 \text{ and}$$

$$K = \frac{e^2}{8\pi\varepsilon_0} \iint [1s(1)2s(2)] \left(\frac{1}{|\mathbf{r}_1 - \mathbf{r}_2|} \right) [1s(2)2s(1)]$$

$$d\tau_1 d\tau_2$$

to obtain the desired result, $E_{singlet} = E_{1s} + E_{2s} + J + K$.

P11.5 What J values are possible for a ^6H term? Calculate the number of states associated with each level and show that the total number of states is the same as that calculated from $(2S + 1)(2L + 1)$.

P11.6 Using Table 11.3, which lists the possible terms that arise from a given configuration, and Hund's rules, write the configurations and term symbols for the ground state of the ions F^- and Ca^{2+} in the form $^{(2S+1)}L_J$.

P11.7 The Doppler broadening in a gas can be expressed as $\Delta\nu = 2\nu_0/c \sqrt{2RT \ln 2/M}$, where M is the molar mass. For the sodium $3p\ ^2P_{3/2} \rightarrow 3s\ ^2S_{1/2}$ transition, $\nu_0 = 5.0933 \times 10^{14}\ s^{-1}$. Calculate $\Delta\nu$ and $\Delta\nu/\nu_0$ at 500.0 K.

P11.8 Calculate the transition dipole moment, $\mu_z^{mn} = \int \psi_m^*(\tau) \mu_z \psi_n(\tau) d\tau$ where $\mu_z = -er \cos\theta$ for a transition from the $1s$ level to the $2p_z$ level in H. Show that this transition is allowed. The integration is over r, θ, and ϕ. Use

$$\psi_{210}(r, \theta, \phi) = \frac{1}{\sqrt{32\pi}}\left(\frac{1}{a_0}\right)^{3/2}\frac{r}{a_0}e^{-r/2a_0}\cos\theta$$

for the $2p_z$ wave function.

P11.9 Consider the $1s\,np\ ^3P \rightarrow 1s\,nd\ ^3D$ transition in He. Draw an energy-level diagram, taking the spin-orbit coupling that splits terms into levels into account. Into how many levels does each term split? The selection rule for transitions in this case is $\Delta J = 0, \pm 1$. How many transitions will be observed in an absorption spectrum? Show the allowed transitions in your energy diagram.

P11.10 Atomic emission experiments of a mixture show a calcium line at 422.673 nm corresponding to a $^1P_1 \rightarrow ^1S_0$ transition and a doublet due to potassium $^2P_{3/2} \rightarrow ^2S_{1/2}$ and $^2P_{1/2} \rightarrow ^2S_{1/2}$ transitions at 764.494 and 769.901 nm, respectively.

a. Calculate the ratio g_{upper}/g_{lower} for each of these transitions.

b. Calculate n_{upper}/n_{lower} for a temperature of 1600°C for each transition.

P11.11 How many ways are there to place three electrons into an f subshell? What is the ground-state term for the f^3 configuration, and how many states are associated with this term? See Problem P11.36.

P11.12 Calculate the wavelengths of the first three lines of the Lyman, Balmer, and Paschen series, and the series limit (the shortest wavelength) for each series.

P11.13 The Lyman series in the hydrogen atom corresponds to transitions that originate from the $n = 1$ level in absorption or that terminate in the $n = 1$ level for emission. Calculate the energy, frequency (in inverse seconds and inverse centimeters), and wavelength of the least and most energetic transition in this series.

P11.14 The inelastic mean free path of electrons in a solid, λ, governs the surface sensitivity of techniques such as AES and XPS. The electrons generated below the surface must make their way to the surface without losing energy in order to give elemental and chemical shift information. An empirical expression for elements that give λ as a function of the kinetic energy of the electron generated in AES or XPS is $\lambda = 538E^{-2} + 0.41(lE)^{0.5}$. The units of λ are monolayers, E is the kinetic energy of the electron, and l is the monolayer thickness in nanometers. On the basis of this equation, what kinetic energy maximizes the surface sensitivity for a monolayer thickness of 0.3 nm? An equation solver would be helpful in obtaining the answer.

P11.15 The effective path length that an electron travels before being ejected into the vacuum is related to the depth below the surface at which it is generated and the exit angle by $d = \lambda \cos\theta$, where λ is the inelastic mean free path and θ is the angle between the surface normal and the exit direction.

a. Justify this equation based on a sketch of the path that an electron travels before exiting into the vacuum.

b. The XPS signal from a thin layer on a solid surface is given by $I = I_0(1 - e^{-d/\lambda\cos\theta})$, where I_0 is the signal

that would be obtained from an infinitely thick layer, and d and λ are defined in Problem P11.14. Calculate the ratio I/I_0 at $\theta = 0$ for $\lambda = 2d$. Calculate the exit angle required to increase I/I_0 to 0.50.

P11.16 List the allowed quantum numbers m_l and m_s for the following subshells and determine the maximum occupancy of the subshells:

a. $2p$ b. $3d$ c. $4f$ d. $5g$

P11.17 What are the levels that arise from a ^4F term? How many states are there in each level?

P11.18 As discussed in Chapter 9, in a more exact solution of the Schrödinger equation for the hydrogen atom, the coordinate system is placed at the center of mass of the atom rather than at the nucleus. In that case, the energy levels for a one-electron atom or ion of nuclear charge Z are given by

$$E_n = -\frac{Z^2\mu e^4}{32\pi^2\varepsilon_0^2\hbar^2 n^2}$$

where μ is the reduced mass of the atom. The masses of an electron, a proton, and a tritium (^3H or T) nucleus are given by 9.1094×10^{-31} kg, 1.6726×10^{-27} kg, and 5.0074×10^{-27} kg, respectively. Calculate the frequency of the $n = 1 \rightarrow n = 4$ transition in H and T to five significant figures. Which of the transitions, $1s \rightarrow 4s$, $1s \rightarrow 4p$, $1s \rightarrow 4d$, could the frequencies correspond to?

P11.19 Derive the ground-state term symbols for the following configurations:

a. s^1d^5 b. f^3 c. g^2

P11.20 Calculate the terms that can arise from the configuration $np^1n'p^1$, $n \neq n'$. Compare your results with those derived in the text for np^2. Which configuration has more terms and why?

P11.21 For a closed-shell atom, an antisymmetric wave function can be represented by a single Slater determinant. For an open-shell atom, more than one determinant is needed. Show that the wave function for the $M_S = 0$ triplet state of He $1s^1 2s^1$ is a linear combination of two of the Slater determinants of Example Problem 10.2. Which of the two are needed and what is the linear combination?

P11.22 Calculate the transition dipole moment, $\mu_z^{mn} = \int \psi_m^*(\tau)\mu_z\psi_n(\tau)\,d\tau$ where $\mu_z = -er\cos\theta$ for a transition from the $1s$ level to the $2s$ level in H. Show that this transition is forbidden. The integration is over r, θ, and ϕ.

P11.23 Use the transition frequencies shown in Example Problem 11.7 to calculate the energy (in joules and electron-volts) of the six levels relative to the $3s^2\ S_{1/2}$ level. State your answers with the correct number of significant figures.

P11.24 Derive the ground-state term symbols for the following atoms or ions:

a. H   b. F^- c. Na^+ d. Sc

P11.25 The spectrum of the hydrogen atom reflects the splitting of the $1s^2$ S and $2p^2$ P terms into levels. The energy difference between the levels in each term is much smaller than the difference in energy between the terms. Given this information, how many spectral lines are observed in the $1s^2$ S $\rightarrow$ $2p^2$ P transition? Are the frequencies of these transitions very similar or quite different?

P11.26 Using Table 11.3, which lists the possible terms that arise from a given configuration, and Hund's rules, write the term symbols for the ground state of the atoms K through Cu, excluding Cr, in the form $^{(2S+1)}L_J$.

P11.27 What atomic terms are possible for the following electron configurations? Which of the possible terms has the lowest energy?

a. ns^1np^1 b. ns^1nd^1 c. ns^2np^1 d. ns^1np^2

P11.28 Two angular momenta with quantum numbers $j_1 = 3/2$ and $j_2 = 5/2$ are added. What are the possible values of J for the resultant angular momentum states?

P11.29 Derive the ground-state term symbols for the following configurations:

a. d^2 b. f^9 c. f^{14}

P11.30 The first ionization potential of ground-state He is 24.6 eV. The wavelength of light associated with the $1s2p$ ^{1}P term is 58.44 nm. What is the ionization energy of the He atom in this excited state?

P11.31 In the Na absorption spectrum, the following transitions are observed:

$4p$ ^{2}P $\rightarrow$ $3s$ ^{2}S $\lambda = 330.26$ nm

$3p$ ^{2}P $\rightarrow$ $3s$ ^{2}S $\lambda = 589.593$ nm, 588.996 nm

$5s$ ^{2}S $\rightarrow$ $3p$ ^{2}P $\lambda = 616.073$ nm, 615.421 nm

Calculate the energies of the $4p$ ^{2}P and $5s$ ^{2}S states with respect to the $3s$ ^{2}S ground state.

P11.32 The Grotrian diagram in Figure 11.7 shows a number of allowed electronic transitions for He. Which of the following transitions shows multiple spectral peaks due to a splitting of terms into levels? How many peaks are observed in each case? Are any of the following transitions between energy levels forbidden by the selection rules?

a. $1s^2$ ^{1}S $\longrightarrow$ $1s2p$ ^{1}P

b. $1s2p^1$P $\longrightarrow$ $1s3s$ ^{1}S

c. $1s2s^3$S $\longrightarrow$ $1s2p$ ^{3}P

d. $1s2p^3$P $\longrightarrow$ $1s3d$ ^{3}D

P11.33 List the quantum numbers L and S that are consistent with the following terms:

a. ^{4}S   b. ^{4}G   c. ^{3}P   d. ^{2}D

P11.34 The transition Al[Ne]$(3s)^2(3p)^1 \longrightarrow$ Al[Ne] $(3s)^2(4s)^1$ has two lines given by $\widetilde{\nu} = 25354.8$ cm^{-1} and $\widetilde{\nu} = 25242.7$ cm^{-1}. The transition Al[Ne]$(3s)^2(3p)^1 \longrightarrow$ Al[Ne]$(3s)^2(3d)^1$ has three lines given by $\widetilde{\nu} = 32444.8$ cm^{-1}, $\widetilde{\nu} = 32334.0$ cm^{-1}, and $\widetilde{\nu} = 32332.7$ cm^{-1}. Sketch an energy-level diagram of the states involved and explain the source of all lines. [Hint: The lowest energy levels are P levels and the highest are D levels. The energy spacing between the D levels is less than for the P levels.]

P11.35 Given that the levels in the ^{3}P term for carbon have the relative energies (expressed in wave numbers) of $^3P_1 - ^3P_0 = 16.4$ cm^{-1} and $^3P_2 - ^3P_1 = 27.1$ cm^{-1}, calculate the ratio of the number of C atoms in the $^3P_2$ and $^3P_0$ levels at 200.0 and 1000. K.

P11.36 A general way to calculate the number of states that arise from a given configuration is as follows. Calculate the combinations of m_l and m_s for the first electron, and call that number n. The number of combinations used is the number of electrons, which we call m. The number of unused combinations is $n-m$. According to probability theory, the number of distinct permutations that arise from distributing the m electrons among the n combinations is $n!/[m!(n-m)!]$. For example, the number of states arising from a p^2 configuration is $6!/[2!4!] = 15$, which is the result obtained in Section 10.8. Using this formula, calculate the number of possible ways to place five electrons in a d subshell. What is the ground-state term for the d^5 configuration and how many states does the term include?

P11.37 The ground-state level for the phosphorus atom is $^4S_{3/2}$. List the possible values of L, M_L, S, M_S, J, and M_J consistent with this level.

P11.38 Derive the ground-state term symbols for the following atoms or ions:

a. F   b. Na   c. P

Web-Based Simulations, Animations, and Problems

W11.1 The individual processes of absorption, spontaneous emission, and stimulated emission are simulated in a two-level system. The level of pumping needed to sustain lasing is experimentally determined by comparing the population of the upper and lower levels in the lasing transition.

The Chemical Bond in Diatomic Molecules

The chemical bond is at the heart of chemistry. We begin with a qualitative molecular orbital model for chemical bonding using the H_2^+ molecule as an example. We show that H_2^+ is more stable than widely separated H and H^+ because of delocalization of the electron over the molecule and localization of the electron in the region between the two nuclei. The molecular orbital model provides a good understanding of the electronic structure of diatomic molecules and is used to understand the bond order, bond energy, and bond length of homonuclear diatomic molecules. The formalism is extended to describe bonding in strongly polar molecules such as HF.

12.1 THE SIMPLEST ONE-ELECTRON MOLECULE: H_2^+

Because the essence of chemistry is bonds between atoms, chemists need to have a firm understanding of the theory of the chemical bond. In this chapter, the origin of the chemical bond is explored using the H_2^+ molecule as an example. We then discuss chemical bonding in first and second row diatomic molecules. In Chapter 13, localized and delocalized bonding models will be used to understand and predict the shape of small molecules. The discussion in this chapter and Chapter 13 is largely qualitative in character. In Chapter 15, numerical methods for quantum chemical calculations on molecules are discussed. You may find it useful to work on Chapter 15 in parallel with Chapters 12 and 13.

A chemical bond is formed between two atoms if the energy of the molecule is lower than the energy of the separated atoms. What is it about the electron distribution around the nuclei that makes this happen? As discussed in Chapter 10, the introduction of a second electron vastly complicates the task of finding solutions to the Schrödinger

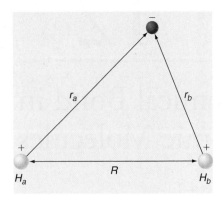

FIGURE 12.1

The two protons and the electron are shown at one instant in time. The quantities R, r_a, and r_b represent the distances between the charged particles.

equation for atoms. This is also true for molecules. For this reason, we initially focus our attention on the simplest molecule possible, the one-electron H_2^+ molecular ion, for which the Schrödinger equation can be solved exactly for a fixed internuclear distance. However, just as for atoms, the Schrödinger equation cannot be solved exactly for any molecule containing more than one electron. Therefore, we approach H_2^+ using an approximate model. This model gives considerable insight into chemical bonding and, most importantly, can be extended easily to other molecules.

We begin by setting up the Schrödinger equation for H_2^+. Figure 12.1 shows the relative positions of the two protons and the electron in H_2^+ at a particular instant in time. The total energy operator for this molecule has the form

$$\hat{H} = -\frac{\hbar^2}{2m_p}(\nabla_a^2 + \nabla_b^2) - \frac{\hbar^2}{2m_e}\nabla_e^2 - \frac{e^2}{4\pi\varepsilon_0}\left(\frac{1}{r_a} + \frac{1}{r_b}\right) + \frac{e^2}{4\pi\varepsilon_0}\frac{1}{R} \quad \textbf{(12.1)}$$

The first term is the kinetic energy operator for each of the nuclei, labeled a and b. The second term is the electron kinetic energy, the third term is the attractive Coulombic interaction between the electron and each of the nuclei, and the last term is the nuclear–nuclear repulsion.

The parts of $\hat{H}$ for the motion of the nuclei and the electrons, both of which appear in Equation (12.1), can be separated using the **Born–Oppenheimer approximation**, which is discussed in more detail in Chapter 14. Because the electron is lighter than the proton by a factor of nearly 2000, the electron charge quickly rearranges in response to the slower periodic motion of the nuclei in molecular vibrations. Because of the very different timescales for nuclear and electron motion, the two motions can be decoupled and we can solve the Schrödinger equation for a fixed nuclear separation and then calculate the energy of the molecule for that distance. If this procedure is repeated for many values of the internuclear separation, we can determine an energy function, $E(R)$.

From experimental results, we know that H_2^+ is a stable species, so that a solution of the Schrödinger equation for H_2^+ must give at least one bound state. We define the zero of energy as an H atom and an H^+ ion that are infinitely separated. Given this choice of the zero of energy, a stable molecule has a negative energy. The energy function $E(R)$ has a minimum value for a distance R_e, which is the equilibrium bond length.

We next discuss how to construct approximate wave functions for the H_2^+ molecule. Imagine slowly bringing together an H atom and an H^+ ion. At infinite separation, the electron is in a $1s$ orbital on either one nucleus or the other. However, as the internuclear distance approaches R_e, the potential energy wells for the two species overlap, and the barrier between them is lowered. Consequently, the electron can move back and forth between the Coulomb wells on the two nuclei. It is equally likely to be on nucleus a as on nucleus b so that the molecular wave function looks like the superposition of a $1s$ orbital on each nucleus. This model is shown pictorially in Figure 12.2.

FIGURE 12.2

The potential energy of the H_2^+ molecule is shown for two different values of R (red curves). At large distances, the electron will be localized in a $1s$ orbital either on nucleus a or b. However, at the equilibrium bond length R_e, the two Coulomb potentials overlap, allowing the electron to be delocalized over the whole molecule. The blue curve represents the amplitude of the atomic (top panel) and molecular (bottom panel) wave functions, and the solid horizontal lines represent the corresponding energy eigenvalues.

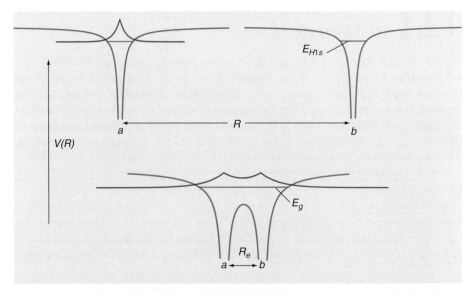

12.2 THE MOLECULAR WAVE FUNCTION FOR GROUND-STATE H_2^+

What does the electron distribution in H_2^+ look like? In answering this question, we consider the relative energies of two H atoms compared to the H_2 molecule. Two H atoms are more stable than the four infinitely separated charges by 2624 kJ mol^{-1}. The H_2 molecule is more stable than two infinitely separated H atoms by 436 kJ mol^{-1}. Therefore, the chemical bond lowers the total energy of the two protons and two electrons by only 17%. For chemical bonds in general, the **bond energy** is a small fraction of the total energy of the widely separated electrons and nuclei. This suggests that the charge distribution in a molecule is not very different from a superposition of the charge distribution of the individual atoms. It also suggests that the wave functions for molecules look like a linear superposition of the wave functions of the individual atoms. This is our starting point for the following discussion of the chemical bond in H_2^+.

On the basis of this reasoning, we approximate the **molecular wave function** for H_2^+ by

$$\psi = c_a \phi_{H1s_a} + c_b \phi_{H1s_b} \tag{12.2}$$

In this equation, ϕ_{H1s} is a H1s **atomic orbital (AO)**. To allow the electron distribution around each nucleus to change as the bond is formed, a **variational parameter**, ζ, is inserted in each AO:

$$\phi_{H1s} = \frac{1}{\sqrt{\pi}} \left(\frac{\zeta}{a_0} \right)^{3/2} e^{-\zeta r/a_0}$$

This parameter looks like an effective nuclear charge. You will see in the end-of-chapter problems that varying ζ allows the size of the orbital to change.

Note that in contrast to the AOs, the molecular wave function is delocalized over the molecule. This idea is expressed mathematically in Equation (12.2) by making the molecular orbital a linear combination of the 1s AO on nucleus a and the 1s AO on nucleus b. What are the values of the coefficients c_a and c_b? Recall that identical quantum mechanical particles are indistinguishable. Therefore, it is not meaningful to put a and b labels on the nuclei. Observables, and in particular the probability density $\psi^* \psi$, must not change when the two nuclei in a homonuclear diatomic molecule are interchanged. This requires that

$$|c_a| = |c_b| \quad \text{or} \quad c_a = \pm c_b \tag{12.3}$$

Although the signs of c_a and c_b can differ, the magnitude of the coefficients is the same.

Equation (12.3) enables us to generate two MOs from the two AOs:

$$\psi_g = c_g (\phi_{H1s_a} + \phi_{H1s_b})$$
$$\psi_u = c_u (\phi_{H1s_a} - \phi_{H1s_b}) \tag{12.4}$$

The wave functions for this homonuclear diatomic molecule are classified as g or u based on whether they change signs upon undergoing inversion through the center of the molecule. The subscripts g and u refer to the German words *gerade* and *ungerade* which can be translated as even and odd and are also referred to as **symmetric** and **antisymmetric**. If the origin of the coordinate system is placed at the center of the molecule, inversion corresponds to $\psi(x, y, z) \rightarrow \psi(-x, -y, -z)$. If this operation leaves the wave function unchanged, that is, $\psi(x, y, z) = \psi(-x, -y, -z)$, it has **g symmetry**. If $\psi(x, y, z) = -\psi(-x, -y, -z)$, the wave function has **u symmetry**. See Figures 12.5 and 12.11 for illustrations of g and u MOs. We will see later that only ψ_g describes a stable, chemically bonded H_2^+ molecule.

The values of c_g and c_u can be determined by normalizing ψ_u and ψ_g. Note that the integrals used in the normalization are over all three spatial coordinates. Normalization requires that

$$1 = \int c_g^* (\phi_{H1s_a}^* + \phi_{H1s_b}^*) c_g (\phi_{H1s_a} + \phi_{H1s_b}) \, d\tau$$

$$= c_g^2 \left(\int \phi_{H1s_a}^* \phi_{H1s_a} \, d\tau + \int \phi_{H1s_b}^* \phi_{H1s_b} \, d\tau + 2 \int \phi_{H1s_b}^* \phi_{H1s_a} \, d\tau \right) \tag{12.5}$$

FIGURE 12.3

The amplitude of two H1s atomic orbitals is shown along an axis connecting the atoms. The overlap is appreciable only for regions in which the amplitude of both AOs is significantly different from zero. Such a region is shown schematically in yellow. In reality, the overlap occurs in three dimensional space.

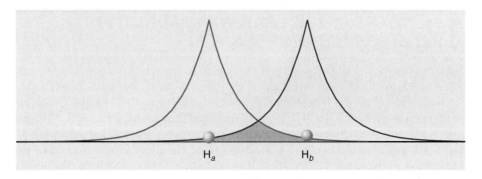

The first two integrals have the value one because the H_{1s} orbitals are normalized. The third integral is called the **overlap integral** and is abbreviated $S_{ab} = \int \phi^*_{H1s_b} \phi_{H1s_a} d\tau$. The overlap is a new concept that was not encountered in atomic systems. The meaning of S_{ab} is indicated pictorially in Figure 12.3. Carrying out the integration in the third integral in Equation (12.5), we obtain $1 = c_g^2(2 + 2S_{ab})$. The result for c_g is

$$c_g = \frac{1}{\sqrt{2 + 2S_{ab}}} \tag{12.6}$$

The coefficient c_u has a similar form, as you will see in the end-of-chapter problems.

If the AOs under the integral in Equation (12.5) were both centered on the same nucleus, the value of S_{ab} would be one. Contributions to S_{ab} arise only when both of the orbital amplitudes are significantly different from zero, which makes it a sensitive function of the internuclear separation. The value of S_{ab} for two $1s$ AOs is a number that lies between one as $R \to 0$ and zero as $R \to \infty$. The electron is delocalized over the molecule only if S_{ab} is significantly greater than zero.

12.3 THE ENERGY CORRESPONDING TO THE H_2^+ MOLECULAR WAVE FUNCTIONS ψ_g AND ψ_u

Keep in mind that the molecular wave functions in Equation (12.4) are approximate rather than exact eigenfunctions of the total energy operator of Equation (12.1). Therefore, we can only calculate the expectation value for the state corresponding to ψ_g:

$$E_g = \frac{\int \psi^*_g \hat{H} \psi_g \, d\tau}{\int \psi^*_g \psi_g \, d\tau}$$

$$= \frac{1}{2(1 + S_{ab})} \left(\int \phi^*_{H1s_a} \hat{H} \phi_{H1s_a} d\tau + \int \phi^*_{H1s_b} \hat{H} \phi_{H1s_b} d\tau \right.$$

$$\left. + \int \phi^*_{H1s_b} \hat{H} \phi_{H1s_a} d\tau + \int \phi^*_{H1s_a} \hat{H} \phi_{H1s_b} d\tau \right)$$

$$= \frac{H_{aa} + H_{ab}}{1 + S_{ab}} \tag{12.7}$$

In the last line of the preceding equation, the symbol H_{ij} is a shorthand notation for the integrals involving $\hat{H}$ and the AOs i and j as follows:

$$H_{ij} = \int \phi^*_i(\tau) \hat{H} \phi_j(\tau) \, d\tau \tag{12.8}$$

Equation (12.7) was simplified by setting $H_{aa} = H_{bb}$ because the two AOs are identical, and $H_{ab} = H_{ba}$ because both AOs and the operator $\hat{H}$ are real. More formally, $H_{ab} = H_{ba}$ because $\hat{H}$ is a Hermitian operator (see the discussion of Postulate 2 in Chapter 3).

What is the energy corresponding to ψ_u? In the end-of-chapter problems, you will see that

$$E_u = \frac{H_{aa} - H_{ab}}{1 - S_{ab}} \qquad (12.9)$$

Looking ahead, we will find that the total energy corresponding to ψ_g is lower than that corresponding to ψ_u, and that only ψ_g describes a stable H_2^+ molecule. To understand this difference, we must look in more detail at the integrals H_{aa} and H_{ab}.

To evaluate H_{aa}, we substitute the formula for the total energy operator from Equation (12.1), except for the nuclear kinetic energy that was removed in making the Born–Oppenheimer approximation:

$$H_{aa} = \int \phi_{H1s_a}^* \left(-\frac{\hbar^2}{2m} \nabla^2 - \frac{e^2}{4\pi\varepsilon_0 r_a} \right) \phi_{H1s_a} \, d\tau + \frac{e^2}{4\pi\varepsilon_0 R} \int \phi_{H1s_a}^* \phi_{H1s_a} \, d\tau$$

$$- \int \phi_{H1s_a}^* \left(\frac{e^2}{4\pi\varepsilon_0 r_b} \right) \phi_{H1s_a} \, d\tau \qquad (12.10)$$

Assume initially that $\zeta = 1$, in which case ϕ_{H1s_a} is an eigenfunction of the operator

$$-\frac{\hbar^2}{2m} \nabla^2 - \frac{e^2}{4\pi\varepsilon_0 r_a}$$

and the atomic wave functions are normalized, H_{aa} is given by

$$H_{aa} = E_{1s} + \frac{e^2}{4\pi\varepsilon_0 R} - J, \text{ where } J = \int \phi_{H1s_a}^* \left(\frac{e^2}{4\pi\varepsilon_0 r_b} \right) \phi_{H1s_a} \, d\tau \qquad (12.11)$$

Here, J represents the energy of interaction of the electron viewed as a negative diffuse charge cloud on atom a with the positively charged nucleus b. This result is exactly what would be calculated in classical electrostatics for a diffuse negative charge of density $\phi_{H1s_a}^* \phi_{H1s_a}$. What is the physical meaning of the energy H_{aa}? The quantity H_{aa} represents the total energy of an undisturbed hydrogen atom separated from a bare proton by the distance R. This energy is referred to as the nonbonded energy, and it is the reference energy against which the strength of the chemical bond is measured.

Next the energy $H_{ab} = H_{ba}$ is evaluated. Substituting as before, we find that

$$H_{ba} = \int \phi_{H1s_b}^* \left(-\frac{\hbar^2}{2m} \nabla^2 - \frac{e^2}{4\pi\varepsilon_0 r_a} \right) \phi_{H1s_a} \, d\tau + \frac{e^2}{4\pi\varepsilon_0 R} \int \phi_{H1s_b}^* \phi_{H1s_a} \, d\tau$$

$$- \int \phi_{H1s_b}^* \left(\frac{e^2}{4\pi\varepsilon_0 r_b} \right) \phi_{H1s_a} \, d\tau \qquad (12.12)$$

Evaluating the first integral gives

$$\int \phi_{H1s_b}^* \left(-\frac{\hbar^2}{2m} \nabla^2 - \frac{e^2}{4\pi\varepsilon_0 r_a} \right) \phi_{H1s_a} \, d\tau = E_{1s} \int \phi_{H1s_b}^* \phi_{H1s_a} \, d\tau = S_{ab} E_{1s}$$

Evaluation of the other integrals in Equation (12.12) gives

$$H_{ab} = S_{ab} \left(E_{1s} + \frac{e^2}{4\pi\varepsilon_0 R} \right) - K \quad \text{where } K = \int \phi_{H1s_b}^* \left(\frac{e^2}{4\pi\varepsilon_0 r_b} \right) \phi_{H1s_a} \, d\tau$$

In this model, K plays a central role in the lowering of the energy that leads to the formation of a bond. However, it has no simple physical interpretation. It is a consequence of the interference between the two atomic orbitals in the molecular wave function. Both J and K are positive, because all terms that appear in the integrals are positive over the entire range of the integration. It turns out that at the equilibrium distance $R = R_e$, both H_{aa} and H_{ab} are negative.

The differences ΔE_g and ΔE_u between the energy of the molecule in the states described by ψ_g and ψ_u and the energy of the hypothetical nonbonded state are calculated next. As shown in Example Problem 12.1,

$$\Delta E_g = E_g - H_{aa} = \frac{-K + S_{ab}J}{1 + S_{ab}} \quad \text{and} \quad \Delta E_u = E_u - H_{aa} = \frac{K - S_{ab}J}{1 - S_{ab}} \qquad (12.13)$$

EXAMPLE PROBLEM 12.1

Using Equations (12.11) and (12.12), show that the change in energy resulting from bond formation, $\Delta E_g = E_g - H_{aa}$ and $\Delta E_u = E_u - H_{aa}$, can be expressed in terms of J, K, and S_{ab} as

$$\Delta E_g = E_g - H_{aa} = \frac{-K + S_{ab}J}{1 + S_{ab}} \quad \text{and} \quad \Delta E_u = E_u - H_{aa} = \frac{K - S_{ab}J}{1 - S_{ab}}$$

Solution

Starting from

$$E_g = \frac{H_{aa} + H_{ab}}{1 + S_{ab}}$$

we have

$$E_g = \frac{H_{aa} + H_{ab}}{1 + S_{ab}} = H_{aa} + \frac{H_{aa} + H_{ab} - (1 + S_{ab})H_{aa}}{1 + S_{ab}} = H_{aa} + \frac{H_{ab} - S_{ab}H_{aa}}{1 + S_{ab}}$$

$$\Delta E_g = E_g - H_{aa} = \frac{H_{ab} - S_{ab}H_{aa}}{1 + S_{ab}}$$

$$\Delta E_g = \frac{S_{ab}\left(E_{1s} + \dfrac{e^2}{4\pi\varepsilon_0 R}\right) - K - S_{ab}\left(E_{1s} + \dfrac{e^2}{4\pi\varepsilon_0 R} - J\right)}{1 + S_{ab}} = \frac{-K + S_{ab}J}{1 + S_{ab}}$$

$$E_u = \frac{H_{aa} - H_{ab}}{1 - S_{ab}} = H_{aa} + \frac{H_{aa} - H_{ab} - (1 - S_{ab})H_{aa}}{1 - S_{ab}} = H_{aa} + \frac{-H_{ab} + S_{ab}H_{aa}}{1 - S_{ab}}$$

$$\Delta E_u = \frac{-S_{ab}\left(E_{1s} + \dfrac{e^2}{4\pi\varepsilon_0 R}\right) + K + S_{ab}\left(E_{1s} + \dfrac{e^2}{4\pi\varepsilon_0 R} - J\right)}{1 - S_{ab}} = \frac{K - S_{ab}J}{1 - S_{ab}}$$

Looking at the numerators in the expressions shown in Example Problem 12.1, we see that the energy difference ΔE is opposite in sign for ψ_g and ψ_u. A quantitative calculation at $R = R_e$ shows that $\Delta E_g < 0$ and $\Delta E_u > 0$ where the zero of energy corresponds to widely separated H and H^+, and $|\Delta E_u| > |\Delta E_g|$. *Therefore, we conclude that only an H_2^+ molecule described by ψ_g is a stable molecule.* The energy of a molecule described by ψ_u is greater than for the nonbonded state, which makes the molecule unstable with respect to dissociation. Next, consider the denominators in the expressions for ΔE_u and ΔE_g. Because $1 + S_{ab} > 1 - S_{ab}$ and $|\Delta E_u| > |\Delta E_g|$, the u state is raised in energy relative to the nonbonded state more than the g state is lowered.

The preceding explanation was qualitative in nature. How is a quantitative calculation carried out? The Schrödinger equation is numerically solved for H_2^+ at a number of different fixed values of R, giving the function $E(R, \zeta)$. The energy is minimized with respect to ζ at each of the R values in a variational calculation. The equilibrium distance, R_e, which corresponds to the minimum in energy, is of particular interest, and ζ has the value 1.24 for ψ_g and 0.90 for ψ_u at $R_e = 2.00\ a_0$, and $S_{ab} = 0.46$. The result that $\zeta > 1$ for ψ_g shows that the optimal H1s AO to use in constructing ψ_g is contracted relative to a free H atom. This means that the electron in H_2^+ in the ψ_g state is pulled in closer to each of the nuclei than it would be in a free hydrogen atom. The opposite is true for the ψ_u state.

The $E(R)$ curves that result from a minimization of the energy with respect to ζ are shown schematically in Figure 12.4. The most important conclusions that can be drawn from this figure are that the ψ_g state describes a stable H_2^+ molecule, because the energy has a well-defined minimum at $R = R_e$ and that ψ_u does not describe a bound state of H and H^+ because $E_u(R) > 0$ for all R.

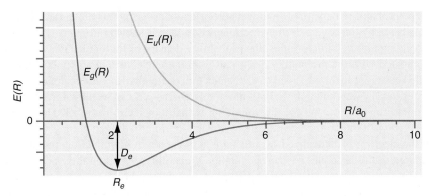

FIGURE 12.4

Schematic energy functions $E(R)$ are shown for the g and u states in the approximate solution discussed. The zero of energy corresponds to infinitely separated H and H$^+$ species.

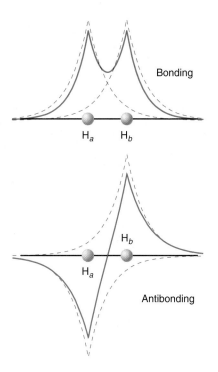

The calculated binding energy D_e in this simple model is 2.36 eV, which is reasonably close to the exact value of 2.70 eV and the exact and calculated R_e values are both 2.00 a_0. The fact that these values are quite close to the exact values validates the assumption that the exact molecular wave function is quite similar to ψ_g. The ψ_g and ψ_u wave functions are referred to as **bonding** and **antibonding molecular orbitals**, respectively, to emphasize their relationship to the chemical bond.

What have we learned so far about the origin of the chemical bond? It is tempting to attribute the binding to H_{ab} or K and, within the formalism that we have used, this is correct. However, other formalisms for solving the Schrödinger equation for the H$_2^+$ molecule do not give rise to the H_{ab} integral. We should, therefore, look for an explanation of chemical binding that is independent of the formalism used. For this reason, we seek the origin of the chemical bond in the differences between ψ_g and ψ_u.

12.4 A CLOSER LOOK AT THE H$_2^+$ MOLECULAR WAVE FUNCTIONS ψ_g AND ψ_u

The values of ψ_g and ψ_u along the molecular axis are shown in Figure 12.5 together with the atomic orbitals from which they are derived. Note that the two wave functions are quite different. The bonding orbital has no nodes, and the amplitude of ψ_g is quite high between the nuclei. The antibonding wave function has a node midway between the nuclei, and ψ_u has its maximum positive and negative amplitudes at the nuclei. Note that the increase in the number of nodes in the wave function with energy is similar to the other quantum mechanical systems that have been studied to this point. Both wave functions are correctly normalized in three dimensions.

Figure 12.6 shows contour plots of ψ_g and ψ_u evaluated in the $z = 0$ plane. If we compare Figures 12.5 and 12.6, we can see that the node midway between the H atoms in ψ_u corresponds to a nodal plane.

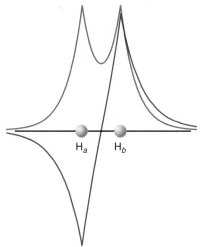

FIGURE 12.5

Molecular wave functions ψ_g and ψ_u (solid lines), evaluated along the internuclear axis are shown in the top two panels. The unmodified ($\zeta = 1$) H1s orbitals from which they were generated are shown as dashed lines. The bottom panel shows a direct comparison of ψ_g and ψ_u.

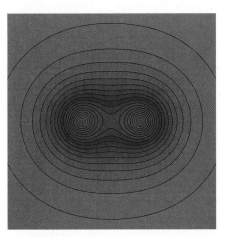 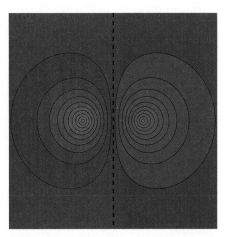

FIGURE 12.6

Contour plots of ψ_g (left) and ψ_u (right). The minimum amplitude is shown as blue, and the maximum amplitude is shown as red for each plot. The dashed line indicates the position of the nodal plane in ψ_u.

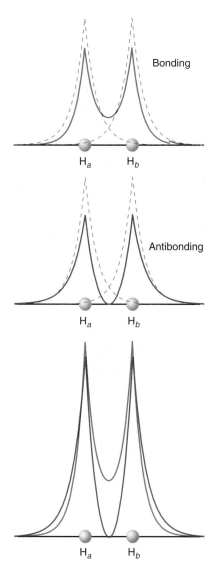

FIGURE 12.7

The upper two panels show the probability densities ψ_g^2 and ψ_u^2 along the internuclear axis for the bonding and antibonding wave functions. The dashed lines show $\frac{1}{2}\psi_{H1s_a}^2$ and $\frac{1}{2}\psi_{H1s_b}^2$, which are the probability densities for unmodified ($\zeta = 1$) $H1s$ orbitals on each nucleus. The lowest panel shows a direct comparison of ψ_g^2 and ψ_u^2. Both molecular wave functions are correctly normalized in three dimensions.

The probability density of finding an electron at various points along the molecular axis is given by the square of the wave function, which is shown in Figure 12.7. For the antibonding and bonding orbitals, the probability density of finding the electron in H_2^+ is compared with the probability density of finding the electron in a hypothetical non-bonded case. For the nonbonded case, the electron is equally likely to be found on each nucleus in $H1s$ AOs and $\zeta = 1$. Two important conclusions can be drawn from this figure. First, for both ψ_g and ψ_u, the volume in which the electron can be found is large compared with the volume accessible to an electron in a hydrogen atom. This tells us that the electron is delocalized over the whole molecule in both the bonding and antibonding orbitals. Second, we can see that the probability of finding the electron in the region between the nuclei is quite different for ψ_g and ψ_u. For the antibonding orbital, the probability is zero midway between the two nuclei, but for the bonding orbital, it is quite high. This difference is what makes the g state a bonding state and the u state an antibonding state.

This pronounced difference between ψ_g^2 and ψ_u^2 is explored further in Figure 12.8. The *difference* between the probability density for these orbitals and the hypothetical nonbonding state is shown in this figure. This difference tells us how the electron density would change if we could suddenly switch on the interaction at the equilibrium geometry. We see that for the antibonding state, electron density would move from the region between the two nuclei to the outer regions of the molecule. For the bonding state, electron density would move both to the region between the nuclei and closer to each nucleus. The origin of the density increase between the nuclei for the bonding orbital is the interference term $2\psi_{H1s_a}\psi_{H1s_b}$ in $(\psi_{H1s_a} + \psi_{H1s_b})^2$. The origin of the density increase near each nucleus is the increase in ζ from 1.00 to 1.24 in going from the free atom to the H_2^+ molecule.

In looking at Figure 12.8, note that for ψ_g the probability density is increased relative to the nonbonding case in the region between the nuclei, and decreased by the same amount outside of this region. The opposite is true for ψ_u. This is true because the integrated probability density over all space is one because the molecular wave functions are normalized. Although it may not be apparent in Figures 12.5 to 12.8, the wave functions satisfy this requirement. Only small changes in the probability density outside of the region between the nuclei are needed to balance larger changes in this region, because the integration volume outside of the region between the nuclei is much larger. The data shown as a line plot in Figure 12.8 are shown as a contour plot in Figure 12.9. Red and blue correspond to the most positive and least positive values for $\Delta\psi_g^2$ and $\Delta\psi_u^2$, respectively. The outermost contour for $\Delta\psi_g^2$ in Figure 12.9 corresponds to a negative value, and it is seen that the corresponding area is large. The product of the small negative charge in $\Delta\psi_g^2$ with the large volume corresponding to the contour area is equal in magnitude and opposite in sign to the increase in $\Delta\psi_g^2$ in the bonding region.

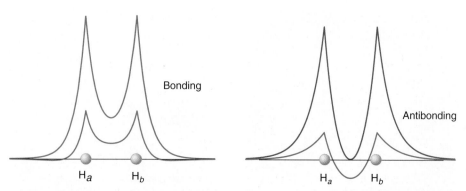

FIGURE 12.8

The red curve shows ψ_g^2 (left panel) and the dark blue curve shows ψ_u^2 (right panel). The light blue curves show the differences $\Delta\psi_g^2 = \psi_g^2 - 1/2(\psi_{H1s_a})^2 - 1/2(\psi_{H1s_b})^2$ (left panel) and $\Delta\psi_u^2 = \psi_u^2 - 1/2(\psi_{H1s_a})^2 - 1/2(\psi_{H1s_b})^2$ (right panel). These differences are a measure of the change in electron density near the nuclei due to bond formation. A charge buildup occurs for the bonding orbital and a charge depletion occurs for the antibonding orbital in the region between the nuclei.

The comparison of the electron charge densities associated with ψ_g and ψ_u helps us to understand the important ingredients in chemical bond formation. For both states, the electronic charge is subject to **delocalization** over the whole molecule. However, charge is also localized in the molecular orbitals, and this **localization** is different in the bonding and antibonding states. In the bonding state, the electronic charge redistribution relative to the nonbonded state leads to a charge buildup both near the nuclei and between the nuclei. In the antibonding state, the electronic charge redistribution leads to a charge buildup outside of the region between the nuclei. We conclude that electronic charge buildup between the nuclei is an essential ingredient of a chemical bond.

We now look at how this charge redistribution affects the kinetic and potential energy of the H_2^+ molecule. A more detailed account is given by N. C. Baird [*J. Chemical Education*, 63 (1986), 660]. The **virial theorem** is very helpful in this context. The virial theorem applies to atoms or molecules described either by exact wave functions or by approximate wave functions if these wave functions have been optimized with respect to all possible parameters. This theorem says that for a Coulomb potential, the average kinetic and potential energies are related by

$$\langle E_{potential} \rangle = -2 \langle E_{kinetic} \rangle \qquad \textbf{(12.14)}$$

Because $E_{total} = E_{potential} + E_{kinetic}$, it follows that

$$\langle E_{total} \rangle = -\langle E_{kinetic} \rangle = \frac{1}{2} \langle E_{potential} \rangle \qquad \textbf{(12.15)}$$

Because this equation applies both to the nonbonded case and to the H_2^+ molecule at its equilibrium geometry, the change in total, kinetic, and potential energies associated with bond formation is given by

$$\langle \Delta E_{total} \rangle = -\langle \Delta E_{kinetic} \rangle = \frac{1}{2} \langle \Delta E_{potential} \rangle \qquad \textbf{(12.16)}$$

For the molecule to be stable, $\langle \Delta E_{total} \rangle < 0$ and, therefore, $\langle \Delta E_{kinetic} \rangle > 0$ and $\langle \Delta E_{potential} \rangle < 0$. Bond formation must lead to an increase in the kinetic energy and a decrease in the potential energy. How does this result relate to the competing effects of charge localization and delocalization that we saw for ψ_g and ψ_u?

Imagine that we could break down the change in the electron charge distribution as the bond is formed into two separate steps. First, we bring the proton and H atom to a distance R_e and let them interact, keeping the effective nuclear charge at the value $\zeta = 1$. In this step, the kinetic energy of the electron decreases, and it can be shown that the potential energy changes little. Therefore, the total energy will decrease. Why is the kinetic energy lower? The explanation follows directly from our analysis of the particle in the one-dimensional box: as the box length increases, the kinetic energy decreases. Similarly, as the electron is delocalized over the whole space of the molecule, the kinetic energy decreases. By looking only at this first step, we see that electron delocalization alone will lead to bond formation. However, the total energy of the molecule can be reduced further at the fixed internuclear distance R_e by optimizing ζ. At the optimal value of $\zeta = 1.24$, some of the electron charge is withdrawn from the region between the nuclei and redistributed around the two nuclei. Because the size of the "box" around each atom is decreased, the kinetic energy of the molecule is increased. This increase is sufficiently large that $\langle \Delta E_{kinetic} \rangle > 0$ for the overall two-step process.

However, increasing ζ from 1.0 to 1.24 decreases the potential energy of the molecule because of the increased Coulombic interaction between the electron and the two protons. The result is that $\langle \Delta E_{potential} \rangle$ is lowered more than $\langle \Delta E_{kinetic} \rangle$ is raised. Therefore, the total energy of the molecule decreases further in this second step. Although the changes in $\langle \Delta E_{potential} \rangle$ and $\langle \Delta E_{kinetic} \rangle$ are both quite large, $\langle \Delta E_{total} \rangle$ changes very little as ζ increases from 1.0 to 1.24. Although $\langle \Delta E_{kinetic} \rangle > 0$ for bond formation, the dominant driving force for bond formation is electron delocalization which is associated with $\langle \Delta E_{kinetic} \rangle < 0$.

At this point, we summarize what we have learned about the chemical bond. We have carried out an approximate solution of the Schrödinger equation for the simplest molecule imaginable and have developed a way to generate molecular wave functions, starting with

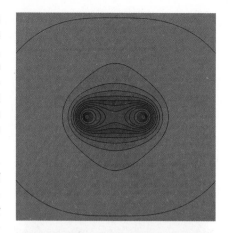

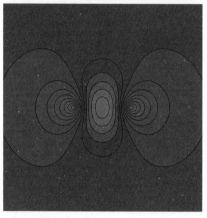

FIGURE 12.9

Contour plots of $\Delta \psi_g^2$ (top) and $\Delta \psi_u^2$ (bottom). The red areas in the top image correspond to positive values for $\Delta \psi_g^2$, and the gray area corresponds to negative values for $\Delta \psi_g^2$. The blue area in the bottom image corresponds to negative values for $\Delta \psi_u^2$, and the red areas just outside of the bonding region correspond to positive values for $\Delta \psi_u^2$. The color in the corners of each contour plot corresponds to $\Delta \psi^2 = 0$.

atomic orbitals. We conclude that both charge delocalization and localization play a role in chemical bond formation. Delocalization promotes bond formation because the kinetic energy is lowered as the electron occupies a larger region in the molecule than it would in the atom. However, localization through the contraction of atomic orbitals and the accumulation of electron density between the atoms in the state described by ψ_g lowers the total energy even further. Although electron delocalization is sufficient to stabilize the molecule, the contraction of the atomic orbitals leads to an additional increase in the bond energy. Both localization and delocalization play a role in bond formation, and it is this complex interplay between opposites that leads to a strong chemical bond.

12.5 COMBINING ATOMIC ORBITALS TO FORM MOLECULAR ORBITALS

In the preceding discussion for H_2^+, we found that combining the two localized atomic orbitals ϕ_{H1s_a} and ϕ_{H1s_b} gave rise to two delocalized molecular wave functions, ψ_g and ψ_u, which are also called **molecular orbitals (MOs)**. For H_2^+, both atomic orbitals (AOs) have the same energy, and for this case we found that $|c_a| = |c_b|$ or $c_a = \pm c_b$. In other words, the weights of the AOs in the MOs are equal. In this section, we examine how the MOs and their associated energies depend on the energies of the two AOs used to form MOs and introduce the molecular orbital energy diagram.

To simplify the mathematics, we consider a diatomic molecule and assume that a MO is generated by combining only one AO on each atom. The AOs are the basis functions for the MO. Such a small basis set is inadequate for quantitative calculations, and in the computational problems at the end of the chapter, you will use much larger basis sets. We anticipate that just as for H_2^+, we will obtain two MOs, with different energies.

$$\psi_b = c_{1b}\phi_1 + c_{2b}\phi_2$$
$$\psi_a = c_{1a}\phi_1 + c_{2a}\phi_2 \tag{12.17}$$

where ϕ_1 and ϕ_2 are AOs on the different atoms. For example, for HF, ϕ_1 would be the H1s AO and ϕ_2 would be the F2p_z AO. How do we determine the energies corresponding to ψ_b and ψ_a and the values of the coefficients in Equation (12.17)?

To do so, we write $\psi_1 = c_1\phi_1 + c_2\phi_2$ and minimize the MO energy with respect to the values of the AO coefficients. The expectation value for ε is given by

$$
\langle \varepsilon \rangle = \frac{\displaystyle\int \psi_1^* \hat{H} \psi_1 \, d\tau}{\displaystyle\int \psi_1^* \psi_1 \, d\tau}
$$

$$
= \frac{\displaystyle\int (c_1\phi_1 + c_2\phi_2)^* \hat{H}(c_1\phi_1 + c_2\phi_2) \, d\tau}{\displaystyle\int (c_1\phi_1 + c_2\phi_2)^* (c_1\phi_1 + c_2\phi_2) \, d\tau}
$$

$$
= \frac{(c_1)^2 \displaystyle\int \phi_1^* \hat{H} \phi_1 \, d\tau + (c_2)^2 \displaystyle\int \phi_2^* \hat{H} \phi_2 \, d\tau + 2c_1c_2 \displaystyle\int \phi_1^* \hat{H} \phi_2 \, d\tau}{(c_1)^2 \displaystyle\int \phi_1^* \phi_1 \, d\tau + (c_2)^2 \displaystyle\int \phi_2^* \phi_2 \, d\tau + 2c_1c_2 \displaystyle\int \phi_1^* \phi_2 \, d\tau} \tag{12.18}
$$

Because the AOs are normalized, the first two integrals in the denominator of the last line of Equation (12.18) have the value one. Using the notation introduced in Section 12.4 for integrals, Equation (12.18) can be written as follows:

$$
\langle \varepsilon \rangle = \frac{(c_1)^2 H_{11} + (c_2)^2 H_{22} + 2c_1c_2 H_{12}}{(c_1)^2 + (c_2)^2 + 2c_1c_2 S_{12}} \tag{12.19}
$$

To minimize ε with respect to the coefficients, ε is first differentiated with respect to c_1 and c_2. We then set the two resulting expressions equal to zero, and solve for c_1 and c_2.

By multiplying both sides of the equation by the denominator before differentiating, the following two equations are obtained:

$$(2c_1 + 2c_2 S_{12})\varepsilon + \frac{\partial \varepsilon}{\partial c_1}((c_1)^2 + (c_2)^2 + 2c_1 c_2 S_{12}) = 2c_1 H_{11} + 2c_2 H_{12}$$

$$(2c_2 + 2c_1 S_{12})\varepsilon + \frac{\partial \varepsilon}{\partial c_2}((c_1)^2 + (c_2)^2 + 2c_1 c_2 S_{12}) = 2c_2 H_{22} + 2c_2 H_{22} \quad \textbf{(12.20)}$$

Setting $\partial \varepsilon / \partial c_1$ and $\partial \varepsilon / \partial c_2 = 0$, and rearranging these two equations results in the following two linear equations for c_1 and c_2, which are called the **secular equations:**

$$c_1(H_{11} - \varepsilon) + c_2(H_{12} - \varepsilon S_{12}) = 0$$

$$c_1(H_{12} - \varepsilon S_{12}) + c_2(H_{22} - \varepsilon) = 0 \quad \textbf{(12.21)}$$

As shown in the Math Supplement (Appendix A), these equations have a solution other than $c_1 = c_2 = 0$ only if the **secular determinant** satisfies the condition

$$\begin{vmatrix} H_{11} - \varepsilon & H_{12} - \varepsilon S_{12} \\ H_{12} - \varepsilon S_{12} & H_{22} - \varepsilon \end{vmatrix} = 0 \quad \textbf{(12.22)}$$

The secular determinant is a 2×2 determinant because the basis set consists of only one AO on each atom.

Expanding the determinant generates a quadratic equation for the MO energy ε. The two solutions are

$$\varepsilon = \frac{1}{2 - 2S_{12}^2}[H_{11} + H_{22} - 2S_{12}H_{12}] \pm \frac{1}{2 - 2S_{12}^2}$$

$$\times \left[\sqrt{(H_{11}^2 + 4H_{12}^2 + H_{22}^2 - 4S_{12}H_{12}H_{22} - 2H_{11}(H_{22} + 2S_{12}H_{12} - 2S_{12}^2 H_{22}))} \right]$$

$$\textbf{(12.23)}$$

For a homonuclear diatomic molecule, $H_{11} = H_{22}$, and Equation (12.12) simplifies to

$$\varepsilon_b = \frac{H_{11} + H_{12}}{1 + S_{12}} \quad \text{and} \quad \varepsilon_a = \frac{H_{11} - H_{12}}{1 - S_{12}} \quad \textbf{(12.24)}$$

where ε_b corresponds to the lower energy MO and refers to the bonding MO. We recognize these solutions from our discussion of the H_2^+ molecule. Substituting ε_b in Equations (12.21), we find that $c_1 = c_2$, whereas if ε_a is substituted in the same equations, we obtain $c_1 = -c_2$, corresponding to the lower and higher energy MO, respectively. These results are also familiar, although they were derived from symmetry considerations in Section 12.2.

EXAMPLE PROBLEM 12.2

Show that substituting $\varepsilon_b = \dfrac{H_{11} + H_{12}}{1 + S_{12}}$ in Equations (12.21) gives the result $c_1 = c_2$.

Solution

$$c_1\left(H_{11} - \frac{H_{11} + H_{12}}{1 + S_{12}}\right) + c_2\left(H_{12} - \frac{H_{11} + H_{12}}{1 + S_{12}}S_{12}\right) = 0$$

$$c_1([1 + S_{12}]H_{11} - [H_{11} + H_{12}]) + c_2([1 + S_{12}]H_{12} - [H_{11} + H_{12}]S_{12}) = 0$$

$$c_1(H_{11} + S_{12}H_{11} - H_{11} - H_{12}) + c_2(H_{12} + H_{12}S_{12} - H_{11}S_{12} - H_{12}S_{12}) = 0$$

$$c_1(H_{11}S_{12} - H_{12}) - c_2(H_{11}S_{12} - H_{12}) = 0$$

$$c_1 = c_2$$

Convince yourself that substitution in the second of the two equations (12.21) gives the same result.

We next consider the case where the AO energies are not equal. To be specific, let ϕ_1 be a hydrogen $1s$ orbital and let ϕ_2 be a fluorine $2p_z$ orbital in the molecule HF. The two MOs have the form

$$\psi_b = c_{1b}\phi_{H1s} + c_{2b}\phi_{F2p_z} \quad \text{and} \quad \psi_a = c_{1a}\phi_{H1s} + c_{2a}\phi_{F2p_z} \qquad \textbf{(12.25)}$$

and normalization requires that

$$(c_{1b})^2 + (c_{2b})^2 + 2c_{1b}c_{2b}S_{12} = 1 \quad \text{and}$$
$$(c_{1a})^2 + (c_{2a})^2 + 2c_{1a}c_{2a}S_{12} = 1 \qquad \textbf{(12.26)}$$

To calculate ε_a, ε_b, c_{1a}, c_{2a}, c_{1b}, and c_{2b}, we need numerical values for H_{11}, H_{22}, H_{12}, and S_{12}. To a good approximation, H_{11} and H_{22} correspond to the first ionization energies of H and F, respectively, and fitting experimental data gives the approximate empirical relation $H_{12} = -1.75 S_{12}\sqrt{H_{11}H_{22}}$. We assume that $S_{12} = 0.30$, so that $H_{11} = -13.6$ eV, $H_{22} = -18.6$ eV, and $H_{12} = -8.35$ eV. We are looking for trends rather than striving for accuracy, so these approximate values are sufficiently good for our purposes. Substituting these values in Equations (12.23) gives the MO energy levels shown below. Example Problem 12.3 shows how to obtain the corresponding values of the coefficients.

$$\varepsilon_b = -19.6 \text{ eV} \qquad \psi_b = 0.34\phi_{H1s} + 0.84\phi_{F2p_z}$$
$$\varepsilon_a = -10.3 \text{ eV} \qquad \psi_a = 0.99\phi_{H1s} - 0.63\phi_{F2p_z} \qquad \textbf{(12.27)}$$

Note that the magnitudes of the coefficients c_1 and c_2 are not equal. The coefficient of the lower energy AO has the larger magnitude in the bonding MO and the smaller magnitude in the antibonding MO. We present the MO energy results for H_2 and HF in the form of a **molecular orbital energy diagram** in Figure 12.10. By convention, the AOs for the bonded atoms are shown on the left and right sides of this diagram. The MOs generated by combining the AOs are shown in the middle. The MOs are depicted by drawing symbols at each atom whose size is proportional to the magnitude of the AO. The sign of the coefficient is indicated by the color of the symbol. Note that the wave function with blue and red interchanged is not distinguishable from the original wave function.

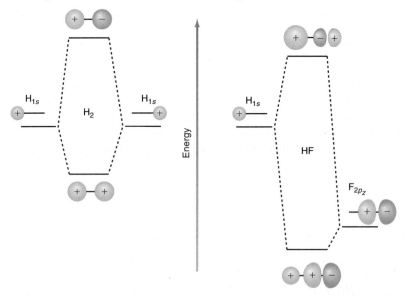

FIGURE 12.10

Molecular orbital energy diagram for a qualitative description of bonding in H_2 and HF. The atomic orbitals are shown to the left and right, and the molecular orbitals are shown in the middle. Dashed lines connect the MO with the AOs from which it was constructed. Shaded circles have a diameter proportional to the coefficients c_{ij}. Red and blue shading signifies positive and negative signs of the AO coefficients, respectively.

EXAMPLE PROBLEM 12.3

Calculate c_{1a} and c_{2a} for the antibonding HFMO for which $\varepsilon_2 = -10.3$ eV. Calculate c_{1b} and c_{2b} for the HF bonding MO for which $\varepsilon_1 = -19.6$ eV. Assume that $S_{12} = 0.30$.

Solution

We first obtain the result $H_{12} = -1.75 S_{12}\sqrt{H_{11}H_{22}} = -8.35$ eV. We calculate c_{1a}/c_{2a} and c_{1b}/c_{2b} by substituting the values for ε_a and ε_b in the first equation in Equations (12.21). Both equations give the same result:

$c_{1a}(H_{11} - \varepsilon) + c_{2a}(H_{12} - \varepsilon_a S_{12}) = 0.$

For $\varepsilon_a = -10.3\ eV$, $c_{1a}(-13.6 + 10.3) + c_{2a}(-8.35 + 0.3 \times 10.3) = 0$

$$\frac{c_{1a}}{c_{2a}} = -1.58$$

Using this result in the normalization equation $c_{1a}^2 + c_{2a}^2 + 2c_{1a}c_{2a}S_{12} = 1$

$c_{1a} = 0.99$, $c_{2a} = -0.63$, and $\psi_a = 0.99\phi_{H1s} - 0.63\phi_{F2p_z}$

For $\varepsilon_b = -19.6$ eV, $c_{1b}(-13.6 + 19.6) + c_{2b}(-8.35 + 0.3 \times 19.6) = 0$

$$\frac{c_{1b}}{c_{2b}} = 0.41$$

Using this result in the normalization equation $c_{1b}^2 + c_{2b}^2 + 2c_{1b}c_{2b}S_{12} = 1$

$c_{1b} = 0.34$, $c_{2b} = 0.84$, and $\psi_b = 0.34\phi_{H1s} + 0.84\phi_{F2p_z}$

Note that in the bonding MO, the coefficients of the AOs have the same sign. In the antibonding MO, they have the opposite sign.

Because of the relative signs of the AO coefficients, the lower energy MO has no nodes in the bonding region and is called the bonding MO. The higher energy MO has a node in the bonding region and is called the antibonding MO. The relative values for the AO coefficients for HF are depicted by the size of the AO. We generalize these results to other diatomic molecules in the bullets below.

- Two interacting AOs give rise to two MOs. Each of these MOs has a different energy than the AO energies from which the MOs originated. A necessary condition for this splitting of energy levels to occur is that the AOs have a nonzero overlap in the molecule and that H_{12} is not zero.

- The energy of the MO that has the form $\psi_b = |c_{1b}|\phi_1 + |c_{2b}|\phi_2$, with **in-phase AOs**, is lower than that of the lower energy AO by $\Delta\varepsilon_+$. This MO is called the bonding orbital. The energy of the MO that has the form $\psi_a = |c_{1a}|\phi_1 - |c_{2a}|\phi_2$, with **out-of-phase AOs**, is higher than the higher energy AO by $\Delta\varepsilon_-$. This MO is called the antibonding orbital.

- The energy splitting $\Delta\varepsilon_+ + \Delta\varepsilon_-$ increases with the overlap S_{ab} and with H_{12}. The inequality $|\Delta\varepsilon_+| < |\Delta\varepsilon_-|$ is always satisfied.

- The relative contribution of two AOs in a MO is measured by the relative magnitude of their coefficients which can be calculated from the secular equations. If two interacting AOs have the same energy, their coefficients in the MO have the same magnitude. They have the same sign in the bonding orbital and opposite signs in the antibonding orbital. If the AO energies are not equal, the magnitude of the coefficient of the lower energy AO is *greater* in the bonding orbital and *smaller* in the antibonding orbital.

The relative magnitude of the coefficients of the AOs gives information about the charge distribution in the molecule, within the framework of the following simple model. Consider an electron in the HF bonding MO described by $\psi_b = 0.34\phi_{H1s} + 0.84\phi_{F2p_z}$. Because of the association made in the first postulate between $|\psi|^2$ and probability, the individual terms in $\int \psi_b^* \psi_b\, d\tau = (c_{1b})^2 + (c_{2b})^2 + 2c_{1b}c_{2b}S_{12} = 1$ can be interpreted in the following way. We associate $(c_{1b})^2 = 0.12$ with the probability of finding the electron around the H atom, $(c_{2b})^2 = 0.71$ with the probability of finding the electron around the F atom, and $2c_{1b}c_{2b}S_{12} = 0.17$ with the probability of finding the electron shared by the F and H atoms. We divide the shared probability equally between the atoms. This gives the probabilities of $(c_{1b})^2 + c_{1b}c_{2b}S_{12} = 0.21$ and $(c_{2b})^2 + c_{1b}c_{2b}S_{12} = 0.79$ for finding the electron on the H and F atoms, respectively. This result is reasonable given the known electronegativities of F and H. Note that, although this method of assigning charge due to Robert Mulliken is reasonable, there is no unique way to distribute the electron charge in an MO among atoms because

TABLE 12.1	AO COEFFICIENTS AND MO ENERGIES FOR DIFFERENT VALUES OF H_{22}		
H_{22} (eV)	c_{1b}	c_{2b}	ε_b (eV)
−18.6	0.345	0.840	−19.9
−23.6	0.193	0.925	−24.1
−33.6	0.055	0.982	−33.7
−43.6	0.0099	1.00	−43.6

the charge on an atom is not a quantum mechanical observable. For a pictorial explanation of this assertion, see Figure 15.23. Note, however, that the charge transfer is in the opposite direction for the antibonding MO. We find that the shared probability has a positive sign for a bonding orbital and a negative sign for an antibonding orbital. This is a useful criterion for distinguishing between bonding and antibonding MOs.

The results for HF show that the bonding MO has a greater amplitude on F which has the lower energy AO. In other words, the bonding MO is more localized on F than on H. We generalize this result to a molecule HX where the AO energy of X lies significantly lower than that of H by calculating ε_b, c_{1b}, and c_{2b} for different AO energies of X. The results are shown in Table 12.1 where $H_{11} = -13.6$ eV and $S_{12} = 0.30$.

Note that as the X AO energy becomes more negative, the X AO coefficient $\rightarrow 1$ and the H AO coefficient $\rightarrow 0$ in the bonding MO. It is also seen that the MO energy approaches the lower AO energy as the X AO energy becomes more negative. This result shows that an MO formed from AOs that differ substantially in energy is largely localized on the atom with the lower AO energy rather than being delocalized over the entire molecule. *Molecular orbitals which we initially assumed to be delocalized can turn out to be localized on a single atom if the AO energies differ sufficiently.*

12.6 MOLECULAR ORBITALS FOR HOMONUCLEAR DIATOMIC MOLECULES

In this section, we develop a qualitative picture of the shape and spatial extent of molecular orbitals for diatomic molecules. Following the same path used in going from the H atom to many-electron atoms, we construct MOs for many-electron molecules on the basis of the excited states of the H_2^+ molecule. These MOs are useful in describing bonding in first and second row homonuclear diatomic molecules. Heteronuclear diatomic molecules are discussed in Section 12.8.

All MOs for homonuclear diatomics can be divided into two groups with regard to each of two **symmetry operations**. The first of these is rotation about the molecular axis which is taken to be along the z axis. If this rotation leaves the MO unchanged, it has no nodes that contain this axis, and the MO has σ **symmetry**. Combining s AOs always gives rise to σ MOs for diatomic molecules. If the MO has one nodal plane containing the molecular axis, the MO has π **symmetry**. Combining p_x or p_y AOs always gives rise to π MOs if the AOs have a common nodal plane. The second operation is inversion through the center of the molecule. Placing the origin at the center of the molecule, inversion corresponds to $\psi(x, y, z) \rightarrow \psi(-x, -y, -z)$. If this operation leaves the MO unchanged, the MO has g symmetry. If $\psi(x, y, z) \rightarrow -\psi(-x, -y, -z)$, the MO has u symmetry. All MOs are constructed using $n = 1$ and $n = 2$. MOs have either σ or π and either g or u symmetry. Molecular orbitals for H_2^+ of g and u symmetry are shown in Figure 12.11. Note that $1\sigma_g$ and $1\pi_u$ are bonding MOs, whereas $1\sigma_u^*$ and $1\pi_g^*$ are antibonding MOs showing that u and g can not be uniquely associated with bonding and antibonding.

Why did we choose the $2p_z$ orbital on F to combine with the H1s orbital in HF? Only atomic orbitals of the same symmetry can combine with one another to form a molecular orbital. For this example, we consider only s and p electrons. Figure 12.12 shows that a net nonzero overlap between two atomic orbitals occurs only if both AOs are either cylindrically symmetric with respect to the molecular axis (σ MOs) or if both have a common nodal plane that coincides with the molecular axis (π MOs).

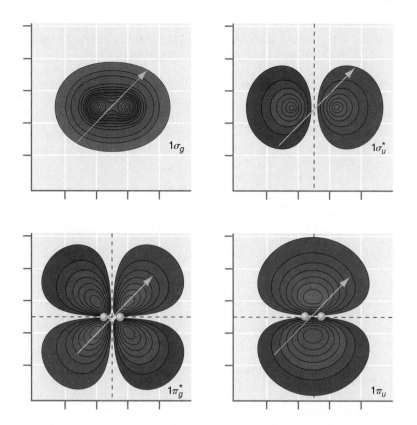

FIGURE 12.11

Contour plots of several bonding and anti-bonding orbitals of H_2^+. Red and blue contours correspond to the most positive and least positive amplitudes, respectively. The yellow arrows show the transformation $(x, y, z) \rightarrow (-x, -y, -z)$ for each orbital. If the amplitude of the wave function changes sign under this transformation, it has u symmetry. If it is unchanged, it has g symmetry.

Two different notations are commonly used to describe MOs in homonuclear diatomic molecules. In the first, the MOs are classified according to symmetry and increasing energy. For instance, a $2\sigma_g$ orbital has the same symmetry, but a higher energy than the $1\sigma_g$ orbital. In the second notation, the integer indicating the relative energy is omitted, and the AOs from which the MOs are generated are listed instead. For instance, the $\sigma_g(2s)$ MO has a higher energy than the $\sigma_g(1s)$ MO. The superscript * is used to designate antibonding orbitals. Two types of MOs can be generated by combining $2p$ AOs. If the axis of the $2p$ orbital lies on the intermolecular axis (by convention the z axis), a σ MO is generated. This specific molecular orbital is called a $3\sigma_u$ or $\sigma_u(2p_z)$ orbital. Adding $2p_x$ (or $2p_y$) orbitals on each atom gives a π molecular orbital because the MO has a nodal plane containing the molecular axis. These MOs are degenerate in energy and are called $1\pi_u$ or $\pi_u(2p_x)$ and $\pi_u(2p_y)$ MOs.

In principle, we should take linear combinations of all the basis functions of the same symmetry (either σ or π) when constructing MOs. However, as shown in Table 12.1, little mixing occurs between AOs of the same symmetry if they have greatly different orbital energies. For example, the mixing between $1s$ and $2s$ AOs for the second row homonuclear diatomics can be neglected at our level of discussion. However, for these same molecules, the $2s$ and $2p_z$ AOs both have σ symmetry and will mix if their energies are not greatly different. Because the energy difference between the $2s$ and $2p_z$ atomic orbitals increases in the sequence Li $\rightarrow$ F, **s–p mixing** decreases for the second row diatomics in the order Li_2, B_2, ..., O_2, F_2. It is useful to think of MO formation in these molecules as a two-step process. We first create separate MOs from the $2s$ and $2p$ AOs, and subsequently combine the MOs of the same symmetry to create new MOs that include $s-p$ mixing.

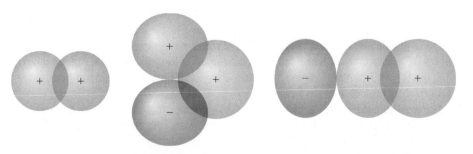

FIGURE 12.12

The overlap between two $1s$ orbitals $(\sigma + \sigma)$, a $1s$ and a $2p_x$ or $2p_y$ $(\sigma + \pi)$, and a $1s$ and a $2p_z$ $(\sigma + \sigma)$ are depicted from left to right. Note that the two shaded areas in the middle panel have opposite signs, so the net overlap of these two atomic orbitals of different symmetry is zero.

TABLE 12.2 MOLECULAR ORBITALS USED TO DESCRIBE CHEMICAL BONDING IN HOMONUCLEAR DIATOMIC MOLECULES

MO Designation	Alternate	Character	Atomic Orbitals
$1\sigma_g$	$\sigma_g(1s)$	Bonding	$1s$
$1\sigma_u^*$	$\sigma_u^*(1s)$	Antibonding	$1s$
$2\sigma_g$	$\sigma_g(2s)$	Bonding	$2s\ (2p_z)$
$2\sigma_u^*$	$\sigma_u^*(2s)$	Antibonding	$2s\ (2p_z)$
$3\sigma_g$	$\sigma_g(2p_z)$	Bonding	$2p_z\ (2s)$
$3\sigma_u^*$	$\sigma_u^*(2p_z)$	Antibonding	$2p_z\ (2s)$
$1\pi_u$	$\pi_u(2p_x, 2p_y)$	Bonding	$2p_x, 2p_y$
$1\pi_g^*$	$\pi_g^*(2p_x, 2p_y)$	Antibonding	$2p_x, 2p_y$

Are the contributions from the s and p AOs equally important in MOs that exhibit $s-p$ mixing? The answer is no because, as discussed in Section 12.3, the AO closest in energy to the resulting MO has the largest coefficient c_{ij} in $\psi_j(1) = \Sigma_i c_{ij}\phi_i(1)$. Therefore, the $2s$ AO is the major contributor to the 2σ MO because the MO energy is closer to the $2s$ than to the $2p$ orbital energy. Applying the same reasoning, the $2p_z$ atomic orbital is the major contributor to the 3σ MO. The MOs used to describe chemical bonding in first and second row homonuclear diatomic molecules are shown in Table 12.2. The AO that is the major contributor to the MO is shown in the last column, and the minor contribution is shown in parentheses. For the sequence of molecules $H_2 \rightarrow N_2$, the MO energy calculated using the Hartree–Fock method increases in the sequence $1\sigma_g < 1\sigma_u^* < 2\sigma_g < 2\sigma_u^* < 1\pi_u < 3\sigma_g < 1\pi_g^* < 3\sigma_u^*$. Moving across the periodic table to O_2 and F_2, the relative order of the $1\pi_u$ and $3\sigma_g$ MOs changes. Note that the first four MO energies follow the AO sequence, but that the σ and π MOs generated from $2p$ AOs have different energies.

It is useful to have an understanding of the spatial extent of these MOs. Figure 12.13 shows contour plots of the first few H_2^+ MOs, including only the major AO in each case (no $s-p$ mixing). The orbital exponent has not been optimized and $\zeta = 1$ for all AOs. Inclusion of the minor AO for the $2\sigma_g, 2\sigma_u^*, 3\sigma_g$, and $3\sigma_u^*$ MOs alters the plots in Figure 12.13 at a minor rather than a major level.

We next discuss the most important features of these plots. As might be expected, the $1\sigma_g$ orbital has no nodes, whereas the $2\sigma_g$ orbital has a nodal surface and the $3\sigma_g$ orbital has two nodal surfaces. All σ_u^* orbitals have a nodal plane perpendicular to the internuclear axis. The π orbitals have a nodal plane containing the internuclear axis. The amplitude for all the antibonding σ MOs is zero midway between the atoms on the molecular axis. This means that the probability density for finding electrons in this region will be small. The antibonding $1\sigma_u^*$ and $3\sigma_u^*$ orbitals have a nodal plane, and the $2\sigma_u^*$ orbital has both a nodal plane and a nodal surface. The $1\pi_u$ orbital has no nodal plane other than on the intermolecular axis, whereas the $1\pi_g^*$ orbital has one nodal plane in the bonding region.

Note that the MOs made up of AOs with $n = 1$ do not extend as far away from the nuclei as the MOs made up of AOs with $n = 2$. In other words, electrons that occupy valence AOs are more likely to overlap with their counterparts on neighboring atoms than are electrons in core AOs. This fact is important in understanding which electrons participate in making bonds in molecules, as well as in understanding reactions between molecules. The MOs shown in Figure 12.13 are specific to H_2^+ and have been calculated using $R = 2.00\ a_0$ and $\zeta = 1$. The detailed shape of these MOs varies from molecule to molecule and depends primarily on the effective nuclear charge, ζ, and the bond length. We can get a qualitative idea of what the MOs look like for other molecules by using the H_2^+ MOs with the effective nuclear charge obtained from Hartree–Fock calculations for the molecule of interest.

For example, the bond length for F_2 is ~35% greater than that for $\zeta = 8.65$ and 5.1 for the $1s$ and $2p$ orbitals, respectively. Because $\zeta > 1$, the amplitude of the fluorine

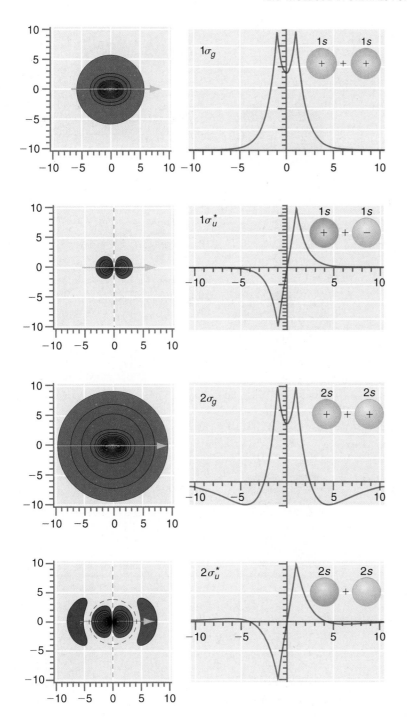

FIGURE 12.13

MOs based on the ground and excited states for H_2^+ generated from $1s$, $2s$, and $2p$ atomic orbitals. Contour plots are shown on the left and line scans along the path indicated by the yellow arrow are shown on the right. Red and blue contours correspond to the most positive and least positive amplitudes, respectively. Dashed lines and curves indicate nodal surfaces. Lengths are in units of a_0, and $R_e = 2.00\ a_0$.

AOs falls off much more rapidly with the distance from the nucleus than is the case for the H_2^+ molecule. Figure 12.14 shows $1\sigma_g$, $3\sigma_u^*$, and $1\pi_u$ MOs for these ζ values generated using the H_2^+ AOs. Note how much more compact the AOs and MOs are compared with $\zeta = 1$. The overlap between the $1s$ orbitals used to generate the lowest energy MO in F_2 is very small. For this reason, electrons in this MO do not make a major contribution to the chemical bond in F_2. Note also that the $3\sigma_u^*$ orbital for F_2 exhibits three nodal surfaces between the atoms rather than one node shown in Figure 12.13 for H_2^+ with $\zeta = 1$. Unlike the $1\pi_u$ MO for H_2^+, the F_2 $1\pi_u$ MO shows distinct contributions from each atom, because the amplitude of the $2p$ AOs falls off rapidly along the internuclear axis. However, apart from these differences, the general features shown in Figure 12.13 are common to the MOs of all first and second row homonuclear diatomics.

FIGURE 12.13

(continued)

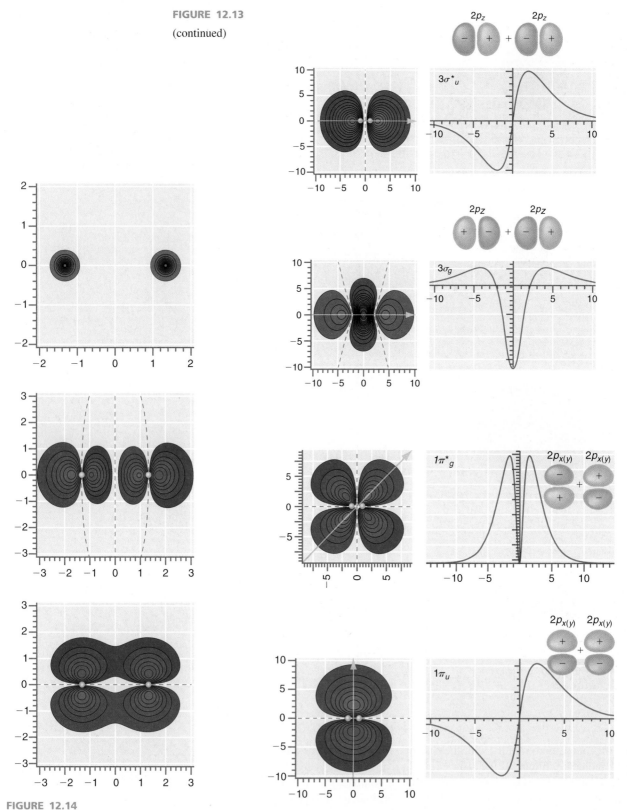

FIGURE 12.14

Contour plots for the $1\sigma_g$ (top), $3\sigma_u^*$ (center), and $1\pi_u$ (bottom) H_2^+ MOs with ζ values appropriate to F_2. Red and blue contours correspond to the most positive and least positive amplitudes, respectively. Dashed lines indicate nodal surfaces. Light circles indicate position of nuclei. Lengths are in units of a_0, and $R_e = 2.66\ a_0$.

12.7 THE ELECTRONIC STRUCTURE OF MANY-ELECTRON MOLECULES

To this point, our discussion has been qualitative in nature. The interaction of two AOs has been shown to give two MOs and the shape and a framework of molecular orbitals, based on the H_2^+ orbitals, has been introduced that can be used for many-electron diatomic molecules. To calculate aspects of diatomic molecules such as the MO energies, the bond length, and the dipole moment, the Schrödinger equation must be solved numerically. As for many-electron atoms, the starting point for quantitative molecular calculations is the Hartree–Fock model. As the formulation of the model is more complex for molecules than for atoms, we refer the interested reader to sources such as I. N. Levine, *Quantum Chemistry*. As was discussed for many-electron atoms in Chapter 10, the crucial input for a calculation is the expansion of the one-electron molecular orbitals $\psi_j(1)$ in a basis set of the N functions $\phi_i(1)$, and a variety of basis sets is available in commercially available computational chemistry software.

$$\psi_j(1) = \sum_{i=1}^{N} c_{ij}\phi_i(1) \qquad (12.28)$$

Although calculations using the Hartree–Fock model generally give sufficiently accurate values for bond lengths in diatomic molecules and bond angles in polyatomic molecules, accurate energy level calculations require electron correlation to be taken into account as discussed in Chapter 15.

Once the MO energy levels have been calculated, a **molecular configuration** is obtained by putting two electrons in each MO, in order of increasing orbital energy, until all electrons have been accommodated. If the degeneracy of an energy level is greater than one, Hund's first rule is followed and the electrons are placed in the MOs in such a way that the total number of unpaired electrons is maximized.

We first discuss the molecular configurations for H_2 and He_2. The MO energy diagrams in Figure 12.15 show the number and spin of the electrons rather than the magnitude and sign of the AO coefficients as was the case in Figure 12.10. What can you say about the magnitude and sign of the AO coefficients for each of the four MOs in Figure 12.15?

The interaction of $1s$ orbitals on each atom gives rise to a bonding and an antibonding MO as shown schematically in Figure 12.15. Each MO can hold two electrons of opposite spin. The configurations for H_2 and He_2 are $(1\sigma_g)^2$ and $(1\sigma_g)^2(1\sigma_u^*)^2$, respectively. You should consider two cautionary remarks about the interpretation of molecular orbital energy diagrams. First, just as for the many-electron atom, the total energy of a molecule is not the sum of the MO energies. Therefore, it is not always valid to draw conclusions about the stability or bond strength of a molecule solely on the basis of the orbital energy diagram. Secondly, the words *bonding* and *antibonding* give information about the relative signs of the AO coefficients in the MO, but they do not convey whether the electron is bound to the molecule. The total energy for any stable molecule is lowered by adding electrons to any orbital for which the energy is less than zero. For example, O_2^- is a stable species compared to O_2 and an electron at infinity, even though the additional electron is placed in an antibonding MO.

For H_2, both electrons are in the $1\sigma_g$ MO, which is lower in energy than the $1s$ AOs. Calculations show that the $1\sigma_u^*$ MO energy is greater than zero. In this case, the total energy is lowered by putting electrons in the $1\sigma_g$ orbital and rises if electrons are additionally put into the $1\sigma_u^*$ as would be the case for H_2^-. In the MO model, He_2 has two electrons in each of the $1\sigma_g$ and $1\sigma_u^*$ orbitals. Because the energy of the $1\sigma_u^*$ orbital is greater than zero, He_2 is not a stable molecule in this model. In fact, He_2 is stable only below ~5 K as a result of a very weak van der Waals interaction, rather than chemical bond formation.

The preceding examples used a single $1s$ orbital on each atom to form molecular orbitals. We now discuss the molecules F_2 and N_2, for which both s and p AOs can contribute to the MOs. Combining n AOs generates n MOs, so combining the $1s$, $2s$, $2p_x$, $2p_y$, and $2p_z$ AOs on N and F generates 10 MOs for F_2 and N_2. Although MOs with contributions from the $1s$ and $2s$ AOs are in principle possible, mixing does not occur for either molecule because the AOs have very different energies. Mixing between $2s$, $2p_x$ and $2p_y$, or $2p_x$ and $2p_y$ AOs does not occur becaue the net overlap is zero. We next consider mixing between the $2s$ and $2p_z$ AOs. For F_2, $s-p$ mixing can be neglected, because the $2s$ AO lies 21.6 eV below the $2p$ AO. The F_2 MOs, in order of increasing energy, are $1\sigma_g < 1\sigma_u^* < 2\sigma_g < 2\sigma_u^* < 3\sigma_g < 1\pi_u = 1\pi_u < 1\pi_g^* = 1\pi_g^*$ and the configuration for F_2 is $(1\sigma_g)^2(1\sigma_u^*)^2(2\sigma_g)^2(2\sigma_u^*)^2(3\sigma_g)^2(1\pi_u)^2(1\pi_u)^2(1\pi_g^*)^2(1\pi_g^*)^2$. For this molecule, the 2σ MOs are quite well described by a single $2s$ AO on each atom,

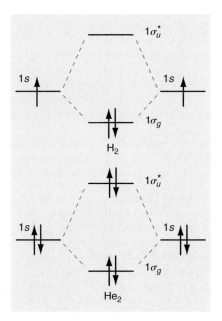

FIGURE 12.15

Atomic and molecular orbital energies and occupation for H_2 and He_2. Upward- and downward-pointing arrows indicate α and β spins. The energy splitting between the MO levels is not to scale.

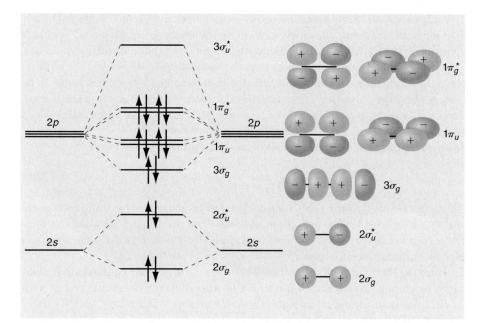

FIGURE 12.16

Schematic MO energy diagram for the valence electrons in F_2. The degenerate p and π orbitals are shown slightly offset in energy. The dominant atomic orbital contributions to the MOs are shown as solid lines. Minor contributions due to $s-p$ mixing have been neglected. The MOs are schematically depicted to the right of the figure. The $1\sigma_g$ and $1\sigma_u^*$ MOs are not shown.

and the 3σ MOs are quite well described by a single $2p_z$ AO on each atom. Because the $2p_x$ and $2p_y$ AOs have a net zero overlap with each other, each of the doubly degenerate $1\pi_u$ and $1\pi_g^*$ molecular orbitals originates from a single AO on each atom. Figure 12.16 shows a molecular orbital energy diagram for F_2. Note that the $3\sigma_g$ and $3\sigma_u^*$ MOs have a greater energy separation than the $1\pi_u$ and $1\pi_g^*$ MOs. This is the case because the overlap of the $2p_z$ AOs is greater than the overlap of the $2p_x$ or $2p_y$ AOs.

For N_2, the $2s$ AO lies below the $2p$ AO by only 12.4 eV, and in comparison to F_2, $s-p$ mixing is not negligible. The MOs, in order of increasing energy, are $1\sigma_g < 1\sigma_u^* < 2\sigma_g < 2\sigma_u^* < 1\pi_u = 1\pi_u < 3\sigma_g < 1\pi_g^* = 1\pi_g^*$ and the configuration is $(1\sigma_g)^2(1\sigma_u^*)^2(2\sigma_g)^2(2\sigma_u^*)^2(1\pi_u)^2(1\pi_u)^2(3\sigma_g)^2$. Because of $s-p$ mixing, the 2σ and 3σ MOs have significant contributions from both $2s$ and $2p_z$ AOs with the result that that the $3\sigma_g$ MO is higher in energy than the $1\pi_u$ MO. A MO energy diagram for N_2 is shown in Figure 12.17. The shape of the 2σ and 3σ N_2 MOs schematically indicates $s-p$ mixing. The $2\sigma_g$ MO has more bonding character, because the probability of finding the electron between the atoms is higher than it was without $s-p$ mixing. Applying the same reasoning, the $2\sigma_u^*$ MO has become less antibonding and the $3\sigma_g$ MO

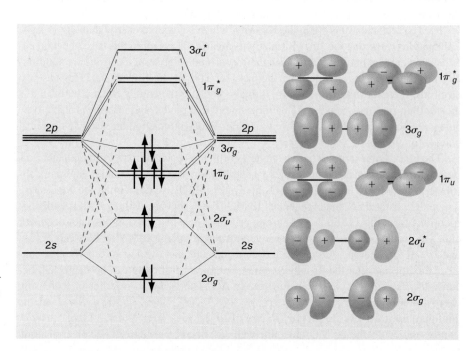

FIGURE 12.17

Schematic MO energy diagram for the valence electrons in N_2. The degenerate p and π orbitals are shown slightly offset in energy. The dominant AO contributions to the MOs are shown as solid lines. Lesser contributions arising from $s-p$ mixing are shown as dashed lines. The MOs are schematically depicted to the right of the figure. The $1\sigma_g$ and $1\sigma_u^*$ MOs are not shown.

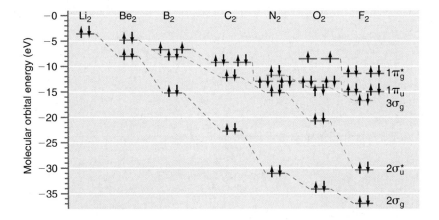

FIGURE 12.18

Molecular orbital energy levels for occupied MOs of the second row diatomics. The $1\sigma_g$ and $1\sigma_u^*$ orbitals lie at much lower values of energy and are not shown.

[From calculations by E. R. Davidson, unpublished]

has become less bonding for N_2 in comparison with F_2. We can see from the overlap in the AOs that the triple bond in N_2 arises from electron occupation of the $3\sigma_g$ and the pair of $1\pi_u$ MOs.

On the basis of this discussion of H_2, He_2, N_2, and F_2, the MO formalism is extended to all first and second row homonuclear diatomic molecules. After the relative energies of the molecular orbitals are established from numerical calculations, the MOs are filled in the sequence of increasing energy, and the number of unpaired electrons for each molecule can be predicted. The results for the second row are shown in Figure 12.18. Using Hund's first rule, we see that both B_2 and O_2 are predicted to have two unpaired electrons; therefore, these molecules should have a net magnetic moment (they are paramagnetic), whereas all other homonuclear diatomics should have a zero net magnetic moment (they are diamagnetic). These predictions are in good agreement with experimental measurements, which provides strong support for the validity of the MO model.

Figure 12.18 shows that the energy of the molecular orbitals tends to decrease with increasing atomic number in this series. This is a result of the increase in ζ in going across the periodic table. The larger effective nuclear charge and the smaller atomic size leads to a lower AO energy, which in turn leads to a lower MO energy. However, the $3\sigma_g$ orbital energy falls more rapidly across this series than the $1\pi_u$ orbital. This occurs because the degree of $s-p$ mixing falls when going from Li_2 to F_2. As a result, an inversion occurs in the order of molecular orbital energies between the $1\pi_u$ and $3\sigma_g$ orbitals for O_2 and F_2 relative to the other molecules in this series.

12.8 BOND ORDER, BOND ENERGY, AND BOND LENGTH

Molecular orbital theory has shown its predictive power by providing an explanation of the observed net magnetic moment in B_2 and O_2 and the absence of a net magnetic moment in the other second row diatomic molecules. We now show that the theory can also provide an understanding of trends in the binding energy and the vibrational force constant for these molecules. Figure 12.19 shows data for these observables for the series $H_2 \rightarrow Ne_2$. As the number of electrons in the diatomic molecule increases, the bond energy has a pronounced maximum for N_2 and a smaller maximum for H_2. The vibrational force constant k shows the same trend. The bond length increases as the bond energy and force constant decrease in the series $Be_2 \rightarrow N_2$, but it exhibits a more complicated trend for the lighter molecules. All of these data can be qualitatively understood using molecular orbital theory.

Consider the MO energy diagrams for H_2 and He_2. For simplicity, we assume that the total energy of the molecules can be approximated by the sum of the orbital energies. Because the bonding orbital is lower in energy than the atomic orbitals from which it was created, putting electrons into a bonding orbital leads to an energy lowering with respect to the atoms. This makes the molecule more stable than the separated atoms, which is

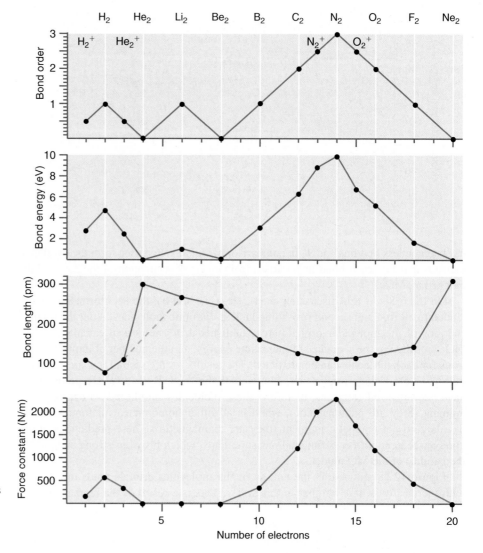

FIGURE 12.19

Bond energy, bond length, and vibrational force constant of the first 10 diatomic molecules as a function of the number of electrons in the molecule. The upper panel shows the calculated bond order for these molecules. The dashed line indicates the dependence of the bond length on the number of electrons if the He₂ data point is omitted.

characteristic of a chemical bond. Similarly, putting two electrons into each of the bonding and antibonding orbitals leads to a total energy that is greater than that of the separated molecules. Therefore, the molecule is unstable with respect to dissociation into two atoms. This result suggests that stable bond formation requires more electrons to be in bonding than in antibonding orbitals. We introduce the concept of **bond order**, which is defined as

$$\text{Bond order} = 1/2[(\text{total bonding electrons}) - (\text{total antibonding electrons})]$$

We expect the bond energy to be very small for a bond order of zero, and to increase with increasing bond order. As shown in Figure 12.19, the bond order shows the same trend as the bond energies. The bond order also tracks the vibrational force constant very well. Again, we can explain the data by associating a stiffer bond with a higher bond order. This agreement is a good example of how a model becomes validated and useful when it provides an understanding for different sets of experimental data.

The relationship between the bond length and the number of electrons in the molecule is influenced both by the bond order and by the variation of the atomic radius with the effective nuclear charge. For a given atomic radius, the bond length is expected to vary inversely with the bond order. This trend is approximately followed for the series Be₂ → N₂ in which the atomic radii are not constant, but decrease steadily. The bond length increases in going from He₂⁺ to Li₂ because the valence electron in Li is in the 2s rather than the 1s AO. The correlation between bond order and bond length also breaks down for He₂ because the atoms are not really chemically bonded. On balance, the trends shown in Figures 12.18 and 12.19 provide significant support for the concepts underlying molecular orbital theory.

EXAMPLE PROBLEM 12.4

Arrange the following in terms of increasing bond energy and bond length on the basis of their bond order: N_2^+, N_2, N_2^-, and N_2^{2-}.

Solution

The ground-state configurations for these species are

$$N_2^+: (1\sigma_g)^2(1\sigma_u^*)^2(2\sigma_g)^2(2\sigma_u^*)^2(1\pi_u)^2(1\pi_u)^2(3\sigma_g)^1$$

$$N_2: (1\sigma_g)^2(1\sigma_u^*)^2(2\sigma_g)^2(2\sigma_u^*)^2(1\pi_u)^2(1\pi_u)^2(3\sigma_g)^2$$

$$N_2^-: (1\sigma_g)^2(1\sigma_u^*)^2(2\sigma_g)^2(2\sigma_u^*)^2(1\pi_u)^2(1\pi_u)^2(3\sigma_g)^2(1\pi_g^*)^1$$

$$N_2^{2-}: (1\sigma_g)^2(1\sigma_u^*)^2(2\sigma_g)^2(2\sigma_u^*)^2(1\pi_u)^2(1\pi_u)^2(3\sigma_g)^2(1\pi_g^*)^1(1\pi_g^*)^1$$

In this series, the bond order is 2.5, 3, 2.5, and 2. Therefore, the bond energy is predicted to follow the order $N_2 > N_2^+$, $N_2^- > N_2^{2-}$ using the bond order alone. However, because of the extra electron in the antibonding $1\pi_g^*$ MO, the bond energy in N_2^- will be less than that in N_2^+. Because bond lengths decrease as the bond strength increases, the bond length will follow the opposite order.

Looking back at what we have learned about homonuclear diatomic molecules, several important concepts stand out. Combining atomic orbitals on each atom to form molecular orbitals provides a way to generate molecular configurations for molecules. Although including many AOs on each atom (that is, using a larger basis set) is necessary to calculate accurate MO energies, important trends can be predicted using a minimal basis set of one or two AOs per atom. The symmetry of atomic orbitals is important in predicting whether they contribute to a given molecular orbital. Molecular orbitals originating from s and p atomic orbitals are either of the σ or π type and, for a homonuclear diatomic molecule, have either u or g symmetry. The concept of bond order allows us to understand why He_2, Be_2, and Ne_2 are not stable and why the bond in N_2 is so strong.

12.9 HETERONUCLEAR DIATOMIC MOLECULES

New issues arise when we consider chemical bond formation in heteronuclear diatomic molecules. As discussed in Section 12.5 for the case of two interacting AOs of different energies, the coefficient of the lower energy AO is greater than that of the higher energy AO in the bonding MO. The opposite is true for the antibonding MO. Because the two atoms are dissimilar, the u and g symmetries do not apply because inversion interchanges the nuclei. However, the MOs will still have either σ or π symmetry. Therefore, the MOs on a heteronuclear diatomic molecule are numbered differently than for the molecules Li_2-N_2:

Homonuclear	$1\sigma_g$	$1\sigma_u^*$	$2\sigma_g$	$2\sigma_u^*$	$1\pi_u$	$3\sigma_g$	$1\pi_g^*$	$3\sigma_u^*$	....
Heteronuclear	1σ	2σ	3σ	4σ	1π	5σ	2π	6σ	....

The symbol * is usually added to the MOs for the heteronuclear molecule to indicate an antibonding MO, as shown for HF in Figure 12.20.

To illustrate the differences between homonuclear and heteronuclear diatomic molecules, we consider HF and construct MOs using the $1s$ AO on H and the $2s$ and $2p$ AOs on F. The molecular orbital energy diagram for HF is shown in Figure 12.20. The AOs on the two atoms that give rise to the MOs are shown on the right side of the diagram, with the size of the orbital proportional to its coefficient in the MO. Numerical calculations show that the $2s$ electrons are almost completely localized on the F atom. The 1π electrons are completely localized on the F atom because the $2p_x$ and $2p_y$ orbitals on F have a zero net overlap with the $1s$ orbital on H. Electrons in MOs localized on a single atom

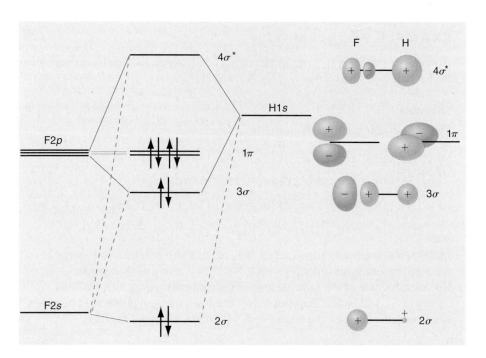

FIGURE 12.20

Schematic energy diagram showing the relationship between the atomic and molecular orbital energy levels for the valence electrons in HF. The degenerate p and π orbitals are shown slightly offset in energy. The dominant atomic orbital contributions to the MOs are shown as solid lines. Lesser contributions are shown as dashed lines. The MOs are depicted to the right of the figure. We assign the $1s$ electrons on F to the 1σ MO, which is actually localized on the F atom.

are referred to as nonbonding electrons. The mixing of $2s$ and $2p$ AOs in the 3σ and $4\sigma^*$ MOs changes the electron distribution in the HF molecule somewhat when compared with a homonuclear diatomic molecule. The 3σ MO has less bonding character and the $4\sigma^*$ MO has less antibonding character. Note that the total bond order is approximately one because the 3σ MO is largely localized on the F atom, the 3σ MO is not totally bonding, and the 1π MOs are completely localized on the F atom. The MO energy diagram depicts the MOs in terms of their constituent AOs. MOs 2 through 4 obtained in calculations using a 28-member basis set are shown in Figure 12.21.

As discussed in Section 12.5 the squares of the coefficients of the H1s and F2p_z AOs are closely related to the probabilities of finding the electron on H and F, respectively. Hartree–Fock calculations using a 28-member basis set give a charge of $+0.48$ and -0.48 on the H and F atoms, respectively. The greater the difference between the coefficients, the larger the molecule's dipole moment. This can also be concluded from our discussion of the results presented in Figure 12.10. These calculated charges give rise to a dipole moment of 2.03 debye (1 debye $= 3.34 \times 10^{-30}$ C m), which is in good agreement with the experimental value of 1.91 debye. As expected, in the bonding orbital the electron density is much greater on the more electronegative fluorine than on the hydrogen. However, in the antibonding $4\sigma^*$ orbital, this polarity is reversed. As you will see in the end-of-chapter problems, the estimated dipole moment is smaller in the excited state than in the ground state.

FIGURE 12.21

The 2σ, 3σ, and 1π MOs for HF are shown from left to right.

12.10 THE MOLECULAR ELECTROSTATIC POTENTIAL

As discussed in Section 12.5, the charge on an atom in a molecule is not a quantum mechanical observable and, consequently, atomic charges cannot be assigned uniquely. However, we know that the electron charge is not uniformly distributed in a polar molecule. For example, the region around the oxygen atom in H_2O has a net negative charge, whereas the region around the hydrogen atoms has a net positive charge. How can this nonuniform charge distribution be discussed? To do so, we introduce the **molecular electrostatic potential**, which is the electrical potential felt by a test charge at various points in the molecule.

The molecular electrostatic potential is calculated by considering the contribution of the valence electrons and the atomic nuclei separately. Consider the nuclei first. For a point charge of magnitude q, the electrostatic potential, $\phi(r)$, at a distance r from the charge, is given by

$$\phi(r) = \frac{q}{4\pi\varepsilon_0 r} \tag{12.29}$$

Therefore, the contribution to the molecular electrostatic potential from the atomic nuclei is given by

$$\phi_{nuclei}(x_1, y_1, z_1) = \sum_i \frac{q_i}{4\pi\varepsilon_0 r_i} \tag{12.30}$$

where q_i is the atomic number of nucleus i, and r_i is the distance of nucleus i from the observation point with the coordinates (x_1, y_1, z_1). The sum extends over all atoms in the molecule.

The electrons in the molecule can be considered as a continuous charge distribution with a density at a point with the coordinates (x, y, z) that is related to the n-electron wave function by

$$\rho(x, y, z) = -e \int \cdots \int (\psi(x, y, z; x_1, y_1, z_1; \ldots; x_n, y_n, z_n))^2$$
$$\times\, dx_1\, dy_1\, dz_1 \ldots dx_n\, dy_n\, dz_n \tag{12.31}$$

The integration is over the position variables of all n electrons. Combining the contributions of the nuclei and the electrons, the molecular electrostatic potential is given by

$$\phi(x_1, y_1, z_1) = \sum_i \frac{q_i}{4\pi\varepsilon_0 r_i} - e \iiint \frac{\rho(x, y, z)}{4\pi\varepsilon_0 r_e}\, dx\, dy\, dz \tag{12.32}$$

where r_e is the distance of an infinitesimal volume element of electron charge from the observation point with the coordinates (x_1, y_1, z_1).

The molecular electrostatic potential must be calculated numerically using the Hartree–Fock method or other methods discussed in Chapter 15. To visualize the polarity in a molecule, it is convenient to display a contour of constant electron density around the molecule and then display the values of the molecular electrostatic potential on the density contour using a color scale, as shown for HF in Figure 12.22. Negative values of the electrostatic potential, shown in red, are found near atoms to which electron charge transfer occurs. For HF, this is the region around the fluorine atom. Positive values of the molecular electrostatic potential, shown in blue, are found around atoms from which electron transfer occurs, as for the hydrogen atom in HF.

The calculated molecular electrostatic potential function identifies regions of a molecule that are either electron rich or depleted in electrons. We can use this function to predict regions of a molecule that are susceptible to nucleophilic or electrophilic attack as in enzyme–substrate reactions. The molecular electrostatic potential is particularly useful because it can also be used to obtain a set of atomic charges that is more reliable than the Mulliken model discussed in Section 12.5. This is done by initially choosing a set of atomic charges and calculating an approximate molecular electrostatic potential around a molecule using the set of charges in Equation (12.30). These atomic charges are varied systematically, subject to the constraint that the total charge is zero for a neutral molecule, until optimal agreement is obtained between the approximate and the accurate molecular electrostatic potential calculated from Equation (12.32). The atomic charges obtained in computational chemistry software such as *Spartan* are calculated in this way as discussed in Chapter 15.

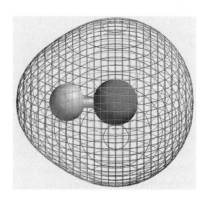

FIGURE 12.22

The grid shows a surface of constant electron density for the HF molecule. The fluorine atom is shown in green. The color shading on the grid indicates the value of the molecular electrostatic potential. Red and blue correspond to negative and positive values, respectively.

Vocabulary

π symmetry
σ symmetry
antibonding molecular orbital
antisymmetric wave function
atomic orbital (AO)
bond energy
bond order
bonding molecular orbital
Born–Oppenheimer approximation
delocalization

g symmetry
in-phase AO
localization
molecular configuration
molecular electrostatic potential
molecular orbital (MO)
molecular orbital energy diagram
molecular wave function
out-of-phase AO
overlap integral

secular determinant
secular equations
$s-p$ mixing
symmetric wave function
symmetry operation
u symmetry
variational parameter
virial theorem

Questions on Concepts

Q12.1 The following images show contours of constant electron density for H_2 calculated using the methods described in Chapter 15. The values of electron density are (a) 0.10, (b) 0.15, (c) 0.20, (d) 0.25, and (e) 0.30 electron/a_0^3.

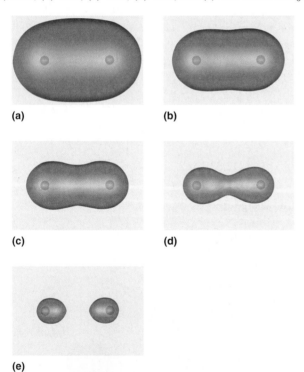

(a) (b)

(c) (d)

(e)

a. Explain why the apparent size of the H_2 molecule as approximated by the volume inside the contour varies in the sequence a–e.
b. Notice the neck that forms between the two hydrogen atoms in contours c and d. What does neck formation tell you about the relative density in the bonding region and in the region near the nuclei?
c. Explain the shape of the contours in image e by comparing this image with Figures 12.8 and 12.9.
d. Estimate the electron density in the bonding region midway between the H atoms by estimating the value of the electron density at which the neck disappears.

Q12.2 Consider the molecular electrostatic potential map for the NH_3 molecule shown here. Is the hydrogen atom (shown as a white sphere) an electron acceptor or an electron donor in this molecule?

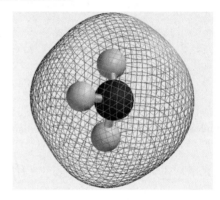

Q12.3 Give examples of AOs for which the overlap reaches its maximum value only as the internuclear separation approaches zero in a diatomic molecule. Also give examples of AOs for which the overlap goes through a maximum value and then decreases as the internuclear separation approaches zero.

Q12.4 Why is it reasonable to approximate H_{11} and H_{22} by the appropriate ionization energy of the corresponding neutral atom?

Q12.5 Identify the molecular orbitals for F_2 in the images shown here in terms of the two designations discussed in Section 12.7. The molecular axis is the z axis, and the y axis is tilted slightly out of the plane of the image.

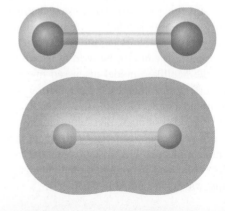

Q12.6 The molecular electrostatic potential maps for LiH and HF are shown here. Does the apparent size of the hydrogen atom (shown as a white sphere) tell you whether it is an electron acceptor or an electron donor in these molecules?

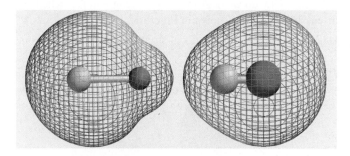

Q12.7 For H_2^+, explain why H_{aa} is the total energy of an undisturbed hydrogen atom separated from a bare proton by the distance R.

Q12.8 Distinguish between the following concepts used to describe chemical bond formation: basis set, minimal basis set, atomic orbital, molecular orbital, and molecular wave function.

Q12.9 Consider the molecular electrostatic potential map for the BH_3 molecule shown here. Is the hydrogen atom (shown as a white sphere) an electron acceptor or an electron donor in this molecule?

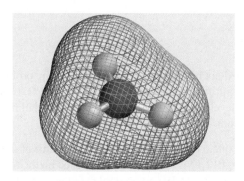

Q12.10 Using Figures 12.8 and 12.9, explain why $\Delta\psi_g^2 < 0$ and $\Delta\psi_u^2 > 0$ outside of the bonding region of H_2^+.

Q12.11 Consider the molecular electrostatic potential map for the BeH_2 molecule shown here. Is the hydrogen atom (shown as a white sphere) an electron acceptor or an electron donor in this molecule?

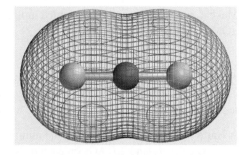

Q12.12 Why are MOs on heteronuclear diatomics not labeled with g and u subscripts?

Q12.13 See Question Q12.5.

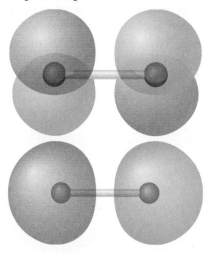

Q12.14 What is the justification for saying that, in expanding MOs in terms of AOs, the equality $\psi_j(1) = \Sigma_i c_{ij}\phi_i(1)$ can in principle be satisfied?

Q12.15 Why are the magnitudes of the coefficients c_a and c_b in the H_2^+ wave functions ψ_g and ψ_u equal?

Q12.16 Explain why $s-p$ mixing is more important in Li_2 than in F_2.

Q12.17 Justify the Born–Oppenheimer approximation based on vibrational frequencies and the timescale for electron motion.

Q12.18 Why can you conclude that the energy of the antibonding MO in H_2^+ is raised more than the energy of the bonding MO is lowered?

Q12.19 Does the total energy of a molecule rise or fall when an electron is put in an antibonding orbital?

Q12.20 Consider the molecular electrostatic potential map for the LiH molecule shown here. Is the hydrogen atom (shown as a white sphere) an electron acceptor or an electron donor in this molecule?

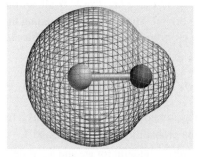

Q12.21 Consider the molecular electrostatic potential map for the H_2O molecule shown here. Is the hydrogen atom (shown as a white sphere) an electron acceptor or an electron donor in this molecule?

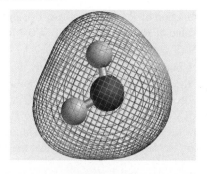

Q12.22 For the case of two H1s AOs, the value of the overlap integral S_{ab} is never exactly zero even at very large separation of the H atoms. Explain this statement.

Q12.23 If there is a node in ψ_u, is the electron in this wave function really delocalized? How does it get from one side of the node to the other?

Q12.24 See Question Q12.5

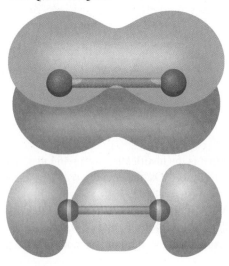

Q12.25 By considering each term in

$$K = \int \phi_{H1s_b}^* \left(\frac{e^2}{4\pi\varepsilon_0 r_b} \right) \phi_{H1s_a} d\tau$$

and

$$J = \int \phi_{H1s_a}^* \left(\frac{e^2}{4\pi\varepsilon_0 r_b} \right) \phi_{H1s_a} d\tau$$

explain why the values of J and K are positive for H_2^+.

Q12.26 Why do we neglect the bond length in He_2 when discussing the trends shown in Figure 12.19?

Q12.27 Explain why the nodal structures of the $1\sigma_g$ MOs in H_2 and F_2 differ.

Q12.28 See Question Q12.5.

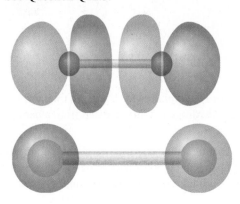

Q12.29 In discussing Figure 12.10, the following statement is made: Note that the wave function with blue and red interchanged is not distinguishable from the original wave function. Justify this statement.

Problems

Problem numbers in **red** indicate that the solution to the problem is given in the *Student's Solutions Manual*.

P12.1 Using ζ as a variational parameter in the normalized function $\psi_{H1s} = 1/\sqrt{\pi}(\zeta/a_0)^{3/2}e^{-\zeta r/a_0}$ allows one to vary the size of the orbital. Show this by calculating the probability of finding the electron inside a sphere of radius $2a_0$ for different values of ζ using the standard integral

$$\int x^2 e^{-ax} dx = -e^{-ax}\left(\frac{2}{a^3} + 2\frac{x}{a^2} + \frac{x^2}{a} \right)$$

a. Obtain an expression for the probability as a function of ζ.
b. Evaluate the probability for $\zeta = 1, 2,$ and 3.

P12.2 The overlap integral for ψ_g and ψ_u as defined in Section 12.3 is given by

$$S_{ab} = e^{-\zeta R/a_0}\left(1 + \zeta\frac{R}{a_0} + \frac{1}{3}\zeta^2\frac{R^2}{a_0^2} \right)$$

Plot S_{ab} as a function of R/a_0 for $\zeta = 0.8, 1.0,$ and 1.2. Estimate the value of R/a_0 for which $S_{ab} = 0.4$ for each of these values of ζ.

P12.3 Sketch out a molecular orbital energy diagram for CO and place the electrons in the levels appropriate for the ground state. The AO ionization energies are O2s: 32.3 eV; O2p: 15.8 eV; C2s: 19.4 eV; and C2p: 10.9 eV. The MO energies follow the sequence (from lowest to highest) $1\sigma, 2\sigma, 3\sigma, 4\sigma, 1\pi, 5\sigma, 2\pi, 6\sigma$. Assume that the 1s

orbital need not be considered and define the 1σ orbital as originating primarily from the 2s AOs on C and O. Connect each MO level with the level of the major contributing AO on each atom.

P12.4 Explain the difference in the appearance of the MOs in Problem P12.12 with those for HF. Based on the MO energies, do you expect LiH^+ to be stable? Do you expect LiH^- to be stable?

P12.5 Calculate the bond order in each of the following species. Predict which of the two species in the following pairs has the higher vibrational frequency:

a. Li_2 or Li_2^+ c. O_2 or O_2^+
b. C_2 or C_2^+ d. F_2 or F_2^-

P12.6 Make a sketch of the highest occupied molecular orbital (HOMO) for the following species:

a. N_2^+ b. Li_2^+ c. O_2^- d. H_2^- e. C_2^+

P12.7 The ionization energy of CO is greater than that of NO. Explain this difference based on the electron configuration of these two molecules.

P12.8 A Hartree–Fock calculation using the minimal basis set of the 1s, 2s, $2p_x$, $2p_y$, and $2p_z$ AOs on each of N and O generated the energy eigenvalues and AO coefficients listed in the following table:

MO	$\varepsilon(eV)$	c_{N1s}	c_{N2s}	c_{N2p_z}	c_{N2p_x}	c_{N2p_y}	c_{O1s}	c_{O2s}	c_{O2p_z}	c_{O2p_x}	c_{O2p_y}
3	−41.1	−0.13	+0.39	+0.18	0	0	−0.20	+0.70	+0.18	0	0
4	−24.2	−0.20	0.81	−0.06	0	0	0.16	−0.71	−0.30	0	0
5	−18.5	0	0	0	0	0.70	0	0	0	0	0.59
6	−15.2	+0.09	−0.46	+0.60	0	0	+0.05	−0.25	−0.60	0	0
7	−15.0	0	0	0	0.49	0	0	0	0	0.78	0
8	−9.25	0	0	0	0	0.83	0	0	0	0	−0.74

a. Designate the MOs in the table as σ or π symmetry and as bonding or antibonding. Assign the MOs to the following images, in which the O atom is red. The molecular axis is the z axis.

b. Explain why MOs 5 and 7 do not have the same energy. Why is the energy of MO 5 lower than that of MO 7?

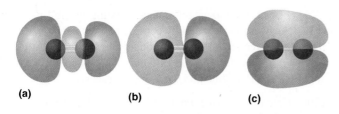

(a) (b) (c)

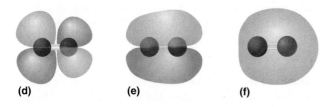

(d) (e) (f)

P12.9 Calculate the value for the coefficients of the AOs in Example Problem 12.3 for $S_{12} = 0.6$. How are they different from the values calculated in that problem for $S_{12} = 0.3$? Can you offer an explanation for the changes?

P12.10 Arrange the following in terms of decreasing bond energy and bond length: O_2^+, O_2, O_2^-, and O_2^{2-}.

P12.11 Predict the bond order in the following species:

a. N_2^+ d. H_2^-

b. Li_2^+ e. C_2^+

c. O_2^-

P12.12 Images of molecular orbitals for LiH calculated using the minimal basis set are shown here. In these images, the smaller atom is H. The H $1s$ AO has a lower energy than the Li $2s$ AO. The energy of the MOs is (left to right) −63.9 eV, −7.92 eV, and +2.14 eV. Make a molecular orbital diagram for this molecule, associate the MOs with the images, and designate the MOs in the images below as filled or empty. Which MO is the HOMO? Which MO is the LUMO? Do you expect the dipole moment in this molecule to have the negative end on H or Li?

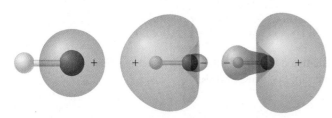

P12.13 What is the electron configuration corresponding to O_2, O_2^-, and O_2^+? What do you expect the relative order of bond strength to be for these species? Which, if any, have unpaired electrons?

P12.14 Calculate the dipole moment of HF for the bonding MO in Equation (12.27). Use the method outlined in Section 12.3 and the results of Example Problem 12.3 to calculate the charge on each atom. The bond length in HF is 91.7 pm. The experimentally determined dipole moment of ground-state HF is 1.91 debye, where 1 debye = 3.33×10^{-30} C m. Compare your result with this value. Does the simple theory give a reliable prediction of the dipole moment?

P12.15 Evaluate the energy for the two MOs generated by combining two H$1s$ AOs. Use Equation (12.23) and carry out the calculation for $S_{12} = 0.1, 0.2,$ and 0.6 to mimic the effect of decreasing the atomic separation in the molecule. Use the parameters $H_{11} = H_{22} = -13.6$ eV and $H_{12} = -1.75 S_{12} \sqrt{H_{11}H_{22}}$. Explain the trend that you observe in the results.

P12.16 Show that calculating E_u in the manner described by Equation (12.7) gives the result $E_u = (H_{aa} - H_{ab})/(1 - S_{ab})$.

P12.17 A surface displaying a contour of the total charge density in LiH is shown here. The molecular orientation is the same as in P12.12. What is the relationship between this surface and the MOs displayed in Problem P12.12? Why does this surface closely resemble one of the MOs?

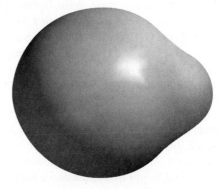

P12.18 Sketch the molecular orbital energy diagram for the radical OH based on what you know about the corresponding diagram for HF. How will the diagrams differ? Characterize the HOMO and LUMO as antibonding, bonding, or nonbonding.

P12.19 The bond dissociation energies of the species NO, CF^-, and CF^+ follow the trend $CF^+ > NO > CF^-$. Explain this trend using MO theory.

P12.20 The expressions $(c_{11})^2 + c_{11}c_{21}S_{12}$ and $(c_{12})^2 + c_{11}c_{21}S_{12}$ for the probability of finding an electron on the H and F atoms in HF, respectively, were derived in

Section 12.5. Use your results from Problem P12.22 and these expressions to calculate the probability of finding an electron in the bonding orbital on the F atom for $S_{12} = 0.1, 0.2$, and 0.6. Explain the trend shown by these results.

P12.21 Follow the procedure outlined in Section 12.3 to determine c_u in Equation (12.4).

P12.22 Evaluate the energy for the two MOs generated by combining a H1s and a F2p AO. Use Equation (12.23) and carry out the calculation for $S_{12} = 0.1, 0.2$, and 0.6 to mimic

the effect of increasing the atomic separation in the molecule. Use the parameters $H_{11} = -13.6$ eV, $H_{22} = -18.6$ eV, and $H_{12} = -1.75 S_{12}\sqrt{H_{11}H_{22}}$. Explain the trend that you observe in the results.

P12.23 Calculate the bond order in each of the following species. Which of the species in part a−d do you expect to have the shorter bond length?

a. Li_2 or Li_2^+ c. O_2 or O_2^+

b. C_2 or C_2^+ d. F_2 or F_2^-

Computational Problems

More detailed instructions on carrying out these calculations using Spartan Physical Chemistry are found on the book website at *www.chemplace.com*.

C12.1 According to Hund's rules, the ground state of O_2 should be a triplet because the last two electrons are placed in a doubly degenerate set of π MOs. Calculate the energy of the singlet and triplet states of O_2 using the B3LYP method and the 6-31G* basis set. Does the singlet or triplet have the lower energy? Both states will be populated if the energy difference $\Delta E \sim kT$. For which temperature is this the case?

C12.2 If the ground state of oxygen is a diradical, you might think that O_2 would dimerize to form square planar O_4 to achieve a molecule in which all electrons are paired. Optimize the geometry and calculate the energies of triplet O_2 and singlet O_4 using the B3LYP method and the 6-31G* basis set. Do you predict O_4 to be more or less stable than 2 O_2 molecules? Use a nonplanar shape in building your O_4 molecule. Is the geometry optimized molecule planar or nonplanar?

C12.3 O_6 might be more stable than O_4 because the bond angle is larger, leading to less steric strain. Optimize the geometry and compare the energy of O_6 with 1.5 times the energy of O_4 using the B3LYP method and the 6-31G* basis set. Is O_6 more stable than O_4? Use a nonplanar shape in building your O_6 molecule. Is the geometry optimized molecule planar or nonplanar?

C12.4 In a LiF crystal, both the Li and F are singly ionized species. Optimize the geometry and calculate the charge on Li and F in a single LiF molecule using the B3LYP method and the 6-31G* basis set. Are the atoms singly ionized? Compare the value of the bond length with the distance between Li^+ and F^- ions in the crystalline solid.

C12.5 Does LiF dissociate into neutral atoms or into Li^+ and F^-? Answer this question by comparing the energy difference between reactants and products for the reactions $LiF(g) \rightarrow Li(g) + F(g)$ and $LiF(g) \rightarrow Li^+(g) + F^-(g)$ using the B3LYP method and the 6-31G* basis set.

C12.6 Calculate Hartree–Fock MO energy values for HF using the MP2 method and the 6-31G* basis set. Make a molecular energy diagram to scale omitting the lowest energy MO. Why can you neglect this MO? Characterize the other MOs as bonding, antibonding, or nonbonding.

C12.7

a. Based on the molecular orbital energy diagram in Problem C12.6, would you expect triplet neutral HF in which an electron is promoted from the 1 π to the $4\sigma^*$ MO to be more or less stable than singlet HF?

b. Calculate the equilibrium bond length and total energy for singlet and triplet HF using the MP2 method and the 6-31G* basis set. Using the frequency as a criterion, are both stable molecules? Compare the bond lengths and vibrational frequencies.

c. Calculate the bond energy of singlet and triplet HF by comparing the total energies of the molecules with the total energy of F and H. Are your results consistent with the bond lengths and vibrational frequencies obtained in part b)?

C12.8 Computational chemistry allows you to carry out calculations for hypothetical molecules that do not exist in order to see trends in molecular properties. Calculate the charge on the atoms in singlet HF and in triplet HF for which the bond length is fixed at 10% greater than the bond length for singlet HF. Are the trends that you see consistent with those predicted by Figure 12.4? Explain your answer.

Web-Based Simulations, Animations, and Problems

W12.1 Two atomic orbitals are combined to form two molecular orbitals. The energy levels of the molecular orbitals and the coefficients of the atomic orbitals in each MO are

calculated by varying the relative energy of the AOs and the overlap, S_{12}, using sliders.

Molecular Structure and Energy Levels for Polyatomic Molecules

For diatomic molecules, the only structural element is the bond length, whereas in polyatomic molecules, both bond lengths and bond angles determine the energy of the molecule. In this chapter, we discuss both localized and delocalized bonding models that enable the structures of small molecules to be predicted. We also discuss the energy levels for two particular classes of polyatomic molecules that require a delocalized description of bonding: conjugated and aromatic molecules, and solids.

13.1 LEWIS STRUCTURES AND THE VSEPR MODEL

In Chapter 12, we discussed chemical bonding and the electronic structure of diatomic molecules. Molecules with more than two atoms introduce a new aspect to our discussion of chemical bonding, namely, bond angles. In this chapter, the discussion of bonding is expanded to include the structure of small molecules. This will allow us to answer questions such as "Why is the bond angle 104.5° in H_2O and 92.2° in H_2S?" One way to answer this question is to say that the angles 104.5° and 92.2° in H_2O and in H_2S minimize the total energy of these molecules. This statement is correct, but it does not really answer the question. As will be shown in Chapter 15, numerical quantum mechanical calculations of bond angles are in very good agreement with experimentally determined values. This result confirms that the approximations made in the calculation are valid and gives confidence that bond angles in molecules for which there are no data can be calculated. However, numerical calculations alone do not provide an understanding about *why* a bond angle of 104.5° minimizes the energy for H_2O, whereas a bond angle of 92.2° minimizes the energy for H_2S.

Qualitative theoretical models that for example address how the bond angle in a class of molecules such as H_2X with X equal to O, S, Se, or Te depends on X are useful for this purpose. Because they can be applied to a class of molecules, they have predictive power. Gaining a qualitative understanding of why small molecules have a particular structure is the primary goal of this chapter.

Molecular structure is addressed from two different vantage points. The significant divide between these points of view is their description of the electrons in a molecule as being localized, as in the valence bond (VB) model, or delocalized, as in the molecular orbital (MO) model. As discussed in Chapter 12, MO theory is based on electron orbitals that are delocalized over the entire molecule. By contrast, a **Lewis structure** represents molecular fluorine as :F̈—F̈:, which is a description in terms of localized bonds and lone pairs. These two viewpoints seem to be irreconcilable at first glance. However, as shown in Section 13.6, each point of view can be reformulated in the language of the other.

We first discuss molecular structure using **localized bonding models**. We do so because there is a long tradition in chemistry of describing chemical bonds in terms of the interaction between neighboring atoms. A great deal was known about the thermochemical properties, stoichiometry, and structure of molecules before the advent of quantum mechanics. For instance, scientists knew that a set of two atom bond enthalpies could be extracted from experimental measurements. Using these bond enthalpies, the enthalpy of formation of molecules can be calculated with reasonable accuracy as the sum of the enthalpies of all the individual bonds in the molecule. Similarly, the bond length between two specific atoms, O—H, is found to be nearly the same in many different compounds. As discussed in Chapter 8, the characteristic vibrational frequency of a group such as —OH is largely independent of the composition of the rest of the molecule. Results such as these give strong support for the idea that a molecule can be described by a set of coupled, but nearly independent chemical bonds between adjacent atoms. The molecule can be assembled by linking these chemical bonds.

Figure 13.1 shows how a structural formula is used to describe ethanol. This structural formula provides a pictorial statement of a localized bonding model. However, a picture like this raises a number of questions. How can the line connecting two bonded atoms be described using the language of quantum mechanics? Localized bonding models imply that electrons are constrained within certain boundaries. However, we know from studying the particle in the box, the hydrogen atom, and many-electron atoms that localizing electrons results in a high energy cost. We also know that it is not possible to distinguish one electron from another, so does it make sense to assign some electrons in F_2 to lone pairs and others to the bond? Quantum mechanics seems better suited to a delocalized picture, with orbitals extending over the whole molecule, than to a localized picture. Yet, the preceding discussion provides ample evidence for a local model of chemical bonding. Reconciling the localized and delocalized models of chemical bonding and molecular structure is a major theme of this chapter.

A useful place to start a discussion of localized bonding is with Lewis structures. Lewis structures emphasize the pairing of electrons as the basis for chemical bond formation. Bonds are shown as connecting lines and electrons not involved in the bonds are indicated by dots. Lewis structures for a few representative small molecules are shown here:

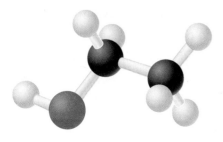

FIGURE 13.1

Ethanol depicted in the form of a ball-and-stick model.

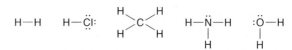

Lewis structures are useful in understanding the stoichiometry of a molecule and in emphasizing the importance of nonbonding electron pairs, also called **lone pairs**. Lewis structures are less useful in predicting the geometrical structure of molecules.

However, the **valence shell electron pair repulsion (VSEPR) model** provides a qualitative theoretical understanding of molecular structures using the Lewis concepts of localized bonds and lone pairs. The basic assumptions of the VSEPR model can be summarized in the following statements about a central atom bonded to several atomic ligands and/or lone pairs:

• The ligands and lone pairs around a central atom act as if they repel one another. They adopt an arrangement in three dimensions that maximizes their angular separation.

- A lone pair occupies more angular space than a ligand.
- The amount of angular space occupied by a ligand increases with its electronegativity and decreases as the electronegativity of the central atom increases.
- A multiply bonded ligand occupies more angular space than a singly bonded ligand.

As Figure 13.2 shows, the structure of a large number of molecules can be understood using the VSEPR model. For example, the decrease in the bond angle for the molecules CH_4, NH_3, and H_2O can be explained on the basis of the greater angular space occupied by lone pairs than by ligands. The tendency of lone pairs to maximize their angular separation also explains why XeF_2 is linear, SO_2 is bent, and IF_4^- is planar. However, in some cases the model is inapplicable or does not predict the correct structure. For instance, a radical, such as CH_3, that has an unpaired electron is planar and, therefore, does not fit into the VSEPR model. Alkaline earth dihalides such as CaF_2 and $SrCl_2$ are angular rather than linear as would be predicted by the model. SeF_6^{2-} and $TeCl_6^{2-}$ are octahedral even though they each have a lone pair in addition to the six ligands. This result indicates that lone pairs do not always exert an influence on molecular shape. In addition, lone pairs do not play as strong a role as the model suggests in transition metal complexes.

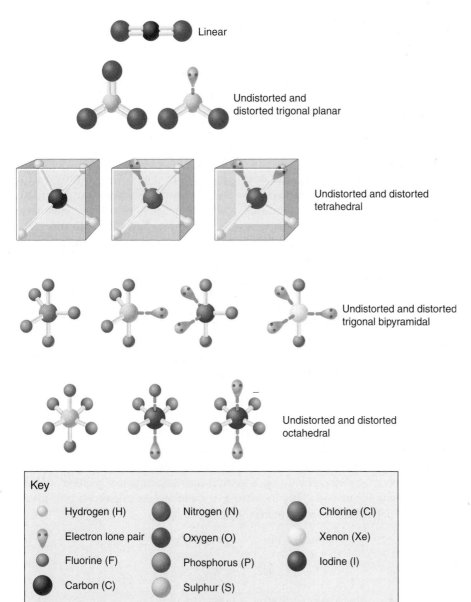

Linear

Undistorted and distorted trigonal planar

Undistorted and distorted tetrahedral

Undistorted and distorted trigonal bipyramidal

Undistorted and distorted octahedral

Key

Hydrogen (H) Nitrogen (N) Chlorine (Cl)

Electron lone pair Oxygen (O) Xenon (Xe)

Fluorine (F) Phosphorus (P) Iodine (I)

Carbon (C) Sulphur (S)

FIGURE 13.2

Examples of correctly predicted molecular shapes using the VSEPR model.

EXAMPLE PROBLEM 13.1

Using the VSEPR model, predict the shape of NO_3^- and OCl_2.

Solution

The following Lewis structure shows one of the three resonance structures of the nitrate ion:

$$\left[\begin{array}{c} :\!\ddot{O}: \\ \| \\ :\!\ddot{O}\!-\!\overset{}{N}\!-\!\ddot{O}: \end{array} \right]^{-}$$

Because the central nitrogen atom has no lone pairs and the three oxygens are equivalent, the nitrate ion should be planar with a 120° bond angle. This is the observed structure.

The Lewis structure for OCl_2 is

$$:\!\ddot{C}l \overset{\ddot{O}}{\diagdown}\diagup \ddot{C}l:$$

The central oxygen atom is surrounded by two ligands and two lone pairs. The ligands and lone pairs are described by a distorted tetrahedral arrangement, leading to a bent molecule. The bond angle should be near the tetrahedral angle of 109.5°. The observed bond angle is 111°.

13.2 DESCRIBING LOCALIZED BONDS USING HYBRIDIZATION FOR METHANE, ETHENE, AND ETHYNE

As discussed earlier, the VSEPR model is useful in predicting the shape of a wide variety of molecules. However, the rules used in its application do not specifically use the vocabulary of quantum mechanics. However, VB theory does use the concept of localized orbitals to explain molecular structure. In the VB model, AOs on the same atom are combined to generate a set of directed orbitals in a process called **hybridization**. The combined orbitals are referred to as **hybrid orbitals**. In keeping with a local picture of bonding, the hybrid orbitals need to contribute independently to the electron density and to the energy of the molecule to the maximum extent possible, because this allows the assembly of the molecule out of separate and largely independent parts. This requires the set of hybrid orbitals to be orthogonal.

How is hybridization used to describe molecular structure? Consider the sequence of molecules methane, ethene, and ethyne. From previous chemistry courses, you know that carbon in these molecules is characterized by the sp^3, sp^2, and sp **hybridizations**, respectively. What is the functional form associated with these different hybridizations? We construct the hybrid orbitals for ethene to illustrate the procedure.

To model the three σ bonds in ethene, the carbon AOs are hybridized to the configuration $1s^2 2p_y^1(\psi_a)^1(\psi_b)^1(\psi_c)^1$ rather than to the configuration $1s^2 2s^2 2p^2$, which is appropriate for an isolated carbon atom. The orbitals ψ_a, ψ_b, and ψ_c are the wave functions that are used in a valence bond model for the three σ bonds in ethene. We next formulate ψ_a, ψ_b, and ψ_c in terms of the $2s$, $2p_x$, and $2p_z$ AOs on carbon.

The three sp^2-hybrid orbitals ψ_a, ψ_b, and ψ_c must satisfy the geometry shown schematically in Figure 13.3. They lie in the $x-z$ plane and are oriented at 120° to one another. The appropriate linear combination of carbon AOs is

$$\psi_a = c_1 \phi_{2p_z} + c_2 \phi_{2s} + c_3 \phi_{2p_x}$$

$$\psi_b = c_4 \phi_{2p_z} + c_5 \phi_{2s} + c_6 \phi_{2p_x}$$

$$\psi_c = c_7 \phi_{2p_z} + c_8 \phi_{2s} + c_9 \phi_{2p_x} \tag{13.1}$$

How can c_1 through c_9 be determined? A few aspects of the chosen geometry simplify the task of determining the coefficients. Because the 2s orbital is spherically symmetrical, it will contribute equally to each of the hybrid orbitals. Therefore, $c_2 = c_5 = c_8$. These three coefficients must satisfy the equation $\sum_i (c_{2si})^2 = 1$, where the subscript 2s refers to the 2s AO. This equation states that all of the 2s contributions to the hybrid orbitals must be accounted for. We choose $c_2 < 0$ in the preceding equations to make the 2s orbital,

$$\psi_{200}(r) = \frac{1}{\sqrt{32\pi}}\left(\frac{1}{a_0}\right)^{3/2}\left(2 - \frac{r}{a_0}\right)e^{-r/2a_0}$$

have a positive amplitude in the bonding region. (For graphs of the 2s AO amplitude versus r, see Figures 9.5 and 9.6.) Therefore, we conclude that

$$c_2 = c_5 = c_8 = -\frac{1}{\sqrt{3}}$$

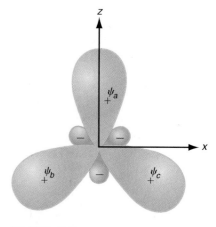

FIGURE 13.3

Geometry of the sp^2-hybrid orbitals used in Equation (13.1). In this and in most of the figures in this chapter, we use a "slimmed down" picture of hybrid orbitals to separate individual orbitals. A more correct form for $s-p$ hybrid orbitals is shown in Figure 13.5.

From the orientation of the orbitals seen in Figure 13.3, $c_3 = 0$ because ψ_a is oriented on the z axis. Because the hybrid orbital points along the positive z axis, $c_1 > 0$. We can also conclude that $c_4 = c_7$, that both are negative, and that $-c_6 = c_9$ with $c_9 > 0$. Based on these considerations, Equation (13.1) simplifies to

$$\psi_a = c_1\phi_{2p_z} - \frac{1}{\sqrt{3}}\phi_{2s}$$

$$\psi_b = c_4\phi_{2p_z} - \frac{1}{\sqrt{3}}\phi_{2s} - c_6\phi_{2p_x}$$

$$\psi_c = c_4\phi_{2p_z} - \frac{1}{\sqrt{3}}\phi_{2s} + c_6\phi_{2p_x} \qquad \textbf{(13.2)}$$

As shown in Example Problem 13.2, the remaining unknown coefficients can be determined by normalizing and orthogonalizing ψ_a, ψ_b, and ψ_c.

EXAMPLE PROBLEM 13.2

Determine the three unknown coefficients in Equation (13.2) by normalizing and orthogonalizing the hybrid orbitals.

Solution

We first normalize ψ_a. Terms such as $\int \phi_{2p_x}^* \phi_{2p_z} d\tau$ and $\int \phi_{2s}^* \phi_{2p_x} d\tau$ do not appear in the following equations because all of the AOs are orthogonal to one another. Evaluation of the integrals is simplified because the individual AOs are normalized.

$$\int \psi_a^* \psi_a \, d\tau = (c_1)^2 \int \phi_{2p_z}^* \phi_{2p_z} \, d\tau + \left(-\frac{1}{\sqrt{3}}\right)^2 \int \phi_{2s}^* \phi_{2s} \, d\tau = 1$$

$$= (c_1)^2 + \frac{1}{3} = 1$$

which tells us that $c_1 = \sqrt{2/3}$. Orthogonalizing ψ_a and ψ_b, we obtain

$$\int \psi_a^* \psi_b \, d\tau = c_4\sqrt{\frac{2}{3}} \int \phi_{2p_z}^* \phi_{2p_z} \, d\tau + \left(-\frac{1}{\sqrt{3}}\right)^2 \int \phi_{2s}^* \phi_{2s} \, d\tau = 0$$

$$= c_4\sqrt{\frac{2}{3}} + \frac{1}{3} = 0 \quad \text{and}$$

$$c_4 = -\sqrt{\frac{1}{6}}$$

Normalizing ψ_b, we obtain

$$\int \psi_b^* \psi_b \, d\tau = \left(-\frac{1}{\sqrt{6}}\right)^2 \int \phi_{2p_z}^* \phi_{2p_z} \, d\tau$$

$$+ \left(-\frac{1}{\sqrt{3}}\right)^2 \int \phi_{2s}^* \phi_{2s} \, d\tau + (-c_6)^2 \int \phi_{2p_x}^* \phi_{2p_x} \, d\tau$$

$$= (c_6)^2 + \frac{1}{3} + \frac{1}{6} = 1 \quad \text{and}$$

$$c_6 = +\frac{1}{\sqrt{2}}$$

We have chosen the positive root so that the coefficient of ϕ_{2px} in ψ_b is negative. Using these results, the normalized and orthogonal set of hybrid orbitals is

$$\psi_a = \sqrt{\frac{2}{3}} \phi_{2p_z} - \frac{1}{\sqrt{3}} \phi_{2s}$$

$$\psi_b = -\frac{1}{\sqrt{6}} \phi_{2p_z} - \frac{1}{\sqrt{3}} \phi_{2s} - \frac{1}{\sqrt{2}} \phi_{2p_x}$$

$$\psi_c = -\frac{1}{\sqrt{6}} \phi_{2p_z} - \frac{1}{\sqrt{3}} \phi_{2s} + \frac{1}{\sqrt{2}} \phi_{2p_x}$$

Convince yourself that ψ_c is normalized and orthogonal to ψ_a and ψ_b.

How can the 2s and 2p character of the hybrids be quantified? Because the sum of the squares of the coefficients for each hybrid orbital equals one, the p and s character of the hybrid orbital can be calculated. The fraction of 2p character in ψ_b is $1/6 + 1/2 = 2/3$. The fraction of 2s character is $1/3$. Because the ratio of the 2p to 2s character is 2:1, one refers to sp^2 hybridization.

How do we know that these hybrid orbitals are oriented with respect to one another as shown in Figure 13.3? Because ψ_a has no component of the $2p_x$ orbital, it must lie on the z axis, corresponding to a value of zero for the polar angle θ. To demonstrate that the ψ_b orbital is oriented as shown, we find its maximum value with respect to the variable θ, which is measured from the z axis.

EXAMPLE PROBLEM 13.3

Demonstrate that the hybrid orbital ψ_b has the orientation shown in Figure 13.3.

Solution

To carry out this calculation, we have to explicitly include the θ dependence of the $2p_x$ and $2p_z$ orbitals from Chapter 9. In doing so, we set the azimuthal angle ϕ, discussed in Section 20.3, equal to zero:

$$\frac{d\psi_b}{d\theta} = \left[\frac{1}{\sqrt{32\pi}}\left(\frac{\zeta}{a_0}\right)^{3/2} e^{-\zeta r/2a_0}\right]$$

$$\times \frac{d}{d\theta}\left(-\frac{1}{\sqrt{6}}\frac{\zeta r}{a_0}\cos\theta - \frac{1}{\sqrt{3}}\left[2 - \frac{\zeta r}{a_0}\right] - \frac{1}{\sqrt{2}}\frac{\zeta r}{a_0}\sin\theta\right) = 0$$

which simplifies to

$$\frac{1}{\sqrt{6}}\sin\theta - \frac{1}{\sqrt{2}}\cos\theta = 0 \text{ or } \tan\theta = \sqrt{3}$$

This value for $\tan\theta$ is satisfied by $\theta = 60°$ and $240°$. Applying the condition that $d^2\psi_b/d\theta^2 < 0$ for the maximum, we conclude that $\theta = 240°$ corresponds to the maximum and $\theta = 60°$ corresponds to the minimum. Similarly, it can be shown that ψ_c has its maximum value at $120°$ and a minimum at $300°$.

Example Problem 13.3 shows that sp^2 hybridization generates three equivalent hybrid orbitals that are separated by an angle of 120°. By following the procedure outlined earlier, it can be shown that the set of orthonormal sp-hybrid orbitals that are oriented 180° apart is

$$\psi_a = \frac{1}{\sqrt{2}}(-\phi_{2s} + \phi_{2p_z})$$

$$\psi_b = \frac{1}{\sqrt{2}}(-\phi_{2s} - \phi_{2p_z}) \qquad \textbf{(13.3)}$$

and that the set of tetrahedrally oriented orthonormal hybrid orbitals for sp^3 hybridization that are oriented 109.4° apart is

$$\psi_a = \frac{1}{2}(-\phi_{2s} + \phi_{2p_x} + \phi_{2p_y} + \phi_{2p_z})$$

$$\psi_b = \frac{1}{2}(-\phi_{2s} - \phi_{2p_x} - \phi_{2p_y} + \phi_{2p_z})$$

$$\psi_c = \frac{1}{2}(-\phi_{2s} + \phi_{2p_x} - \phi_{2p_y} - \phi_{2p_z})$$

$$\psi_d = \frac{1}{2}(-\phi_{2s} - \phi_{2p_x} + \phi_{2p_y} - \phi_{2p_z}) \qquad \textbf{(13.4)}$$

By combining s and p orbitals, at most four hybrid orbitals can be generated. To describe bonding around a central atom with coordination numbers greater than four, d orbitals need to be included in forming the hybrids. Although hybrid orbitals with d character are not discussed here, the principles used in constructing them are the same as those outlined earlier.

The properties of C—C single bonds depend on the hybridization of the carbon atoms, as shown in Table 13.1. The most important conclusion that can be drawn from this table for the discussion in the next section is that increasing the s character in $s-p$ hybrids increases the bond angle. Note also that the C—C single bond length becomes shorter as the s character of the hybridization increases, and that the C—C single bond energy increases as the s character of the hybridization increases.

TABLE 13.1 C—C BOND TYPES

Carbon—Carbon Single Bond Types	σ Bond Hybridization	s-to-p Ratio	Angle between Equivalent σ Bonds (°)	Carbon—Carbon Single Bond Length (pm)
$\geq$C—C$\leq$	sp^3	1:3	109.4	154
$>$C—C$\leq$	sp^2	1:2	120	146
$\equiv$C—C$\equiv$	sp	1:1	180	138

13.3 CONSTRUCTING HYBRID ORBITALS FOR NONEQUIVALENT LIGANDS

In the preceding section, the construction of hybrid orbitals for equivalent ligands was considered. However, in general, molecules contain nonequivalent ligands as well as nonbonding electron lone pairs. How can hybrid orbitals be constructed for such molecules if the bond angles are not known? By considering the experimentally determined structures of a wide variety of molecules, Henry Bent formulated the following guidelines:

• Central atoms that obey the octet rule can be classified into three structural types. Central atoms that are surrounded by a combination of four single bonds or electron pairs are to a first approximation described by a tetrahedral geometry and sp^3

hybridization. Central atoms that form one double bond and a combination of two single bonds or electron pairs are to a first approximation described by a trigonal geometry and sp^2 hybridization. Central atoms that form two double bonds or one triple bond and either a single bond or an electron pair are to a first approximation described by a linear geometry and sp hybridization.

• The presence of different ligands is taken into account by assigning a different hybridization to all nonequivalent ligands and lone pairs. The individual hybridization is determined by the electronegativity of each ligand. A nonbonding electron pair can be considered to be electropositive or, equivalently, to have a small electronegativity. Bent's rule states that atomic s character concentrates in hybrid orbitals directed toward electropositive ligands and that p character concentrates in hybrid orbitals directed toward electronegative ligands.

We now apply these guidelines to H_2O. The oxygen atom in H_2O is to a first approximation described by a tetrahedral geometry and sp^3 hybridization. However, because the H atoms are more electronegative than the electron pairs, the p character of the hybrid orbitals directed toward the hydrogen atoms will be greater than that of sp^3 hybridization. Because Table 13.1 shows that increasing the p character decreases the bond angle, Bent's rule says that the H—O—H bond angle will be less than 109.4°. Note that the effect of Bent's rule is the same as the effect of the VSEPR rules listed in Section 13.1. However, the hybridization model provides a basis for the rules.

Although useful in predicting bond angles, Bent's rule is not quantitative. To make it predictive, a method is needed to assign a hybridization to a specific combination of two atoms that is independent of the other atoms in the molecule. Several authors have developed methods that meet this need, for example, D. M. Root *et al.* in *J. American Chemical Society* 115 (1993), 4201–4209.

EXAMPLE PROBLEM 13.4

a. Use Bent's rule to decide if the X—C—X bond angle in H_2CO is larger or smaller than in F_2CO.

$$\begin{array}{c} X \\ X \end{array}\!\!\!>\!C\!=\!O$$

b. Use Bent's rule to estimate whether the H—C—H bond angle in FCH_3 and $ClCH_3$ differ from 109.5°.

Solution

a. To first order, the carbon atom exhibits sp^2 hybridization. Because F is more electronegative than H, the hybridization of the C—F ligand contains more p character than does the C—H ligand. Therefore, the F—C—F bond angle will be smaller than the H—C—H bond angle.

b. For both FCH_3 and $ClCH_3$, H is more electropositive than the halogen atom so that the C—H bonds have greater s character than the C-halogen bond. This makes the H—C—H bond angle greater than 109.5° in both molecules.

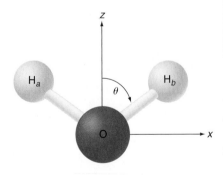

FIGURE 13.4

Coordinate system used to generate the hybrid orbitals on the oxygen atom that are suitable for describing the structure of H_2O.

To test the predictions of Bent's rule, the hybrid bonding and lone pair orbitals for water are constructed, using the known bond angle, and their individual hybridizations are determined. The configuration of the water molecule can be written in the form $1s^2_{oxygen}(\psi_{OH})^2(\psi_{OH})^2(\psi_{lone\ pair})^2(\psi_{lone\ pair})^2$. Each ψ_{OH} and $\psi_{lone\ pair}$ describes localized hybrid orbitals. From the known geometry, the two bonding orbitals are oriented at 104.5° with respect to one another, as shown in Figure 13.4. In constructing the hybrids, the lone pair and bond orbitals are required to maximize their angular separation. Starting with this input, how do we construct ψ_{OH} and $\psi_{lone\ pair}$ starting with the atomic orbitals on hydrogen and oxygen? To describe the H_2O molecule, a pair of orthogonal equivalent s–p hybrids, called ψ_a and ψ_b, is constructed on the oxygen atom. The calculation is initially carried out for an arbitrary angle.

The hybrid orbitals are described by

$$\psi_a = N\left[\cos\theta\,\phi_{2p_z} + \sin\theta\,\phi_{2p_x} - \alpha\phi_{2s}\right]$$

$$\psi_b = N\left[\cos\theta\,\phi_{2p_z} - \sin\theta\,\phi_{2p_x} - \alpha\phi_{2s}\right] \qquad \textbf{(13.5)}$$

where N is a normalization constant and α is the relative amplitude of the 2s and 2p orbitals.

To derive Equation (13.5), visualize ϕ_{2p_x} and ϕ_{2p_z} as vectors along the x and z directions. Because the 2s orbital has one radial node, the 2s orbital coefficient in Equation (13.5) is negative, which generates a positive amplitude at the position of the H atom. The two hybrid orbitals are orthogonal only if

$$\int \psi_a^* \psi_b \, d\tau = N^2 \int \left[\cos\theta\,\phi_{2p_z} + \sin\theta\,\phi_{2p_x} - \alpha\phi_{2s}\right]$$

$$\times \left[\cos\theta\,\phi_{2p_z} - \sin\theta\,\phi_{2p_x} - \alpha\phi_{2s}\right] d\tau$$

$$= N^2\left[\cos^2\theta\int\phi_{2p_z}^*\phi_{2p_z}\,d\tau - \sin^2\theta\int\phi_{2p_x}^*\phi_{2p_x}\,d\tau + \alpha^2\int\phi_{2s}^*\phi_{2s}\,d\tau\right] = 0 \quad \textbf{(13.6)}$$

Terms such as $\int\phi_{2p_x}^*\phi_{2p_z}\,d\tau$ and $\int\phi_{2s}^*\phi_{2p_x}\,d\tau$ do not appear in this equation because all of the atomic orbitals are orthogonal to one another. Because each of the AOs is normalized, Equation (13.6) reduces to

$$N^2[\cos^2\theta - \sin^2\theta + \alpha^2] = N^2[\cos 2\theta + \alpha^2] = 0 \quad \text{or}$$

$$\cos 2\theta = -\alpha^2 \qquad \textbf{(13.7)}$$

In simplifying this equation, we have used the identity $\cos^2 x - \sin^2 y = \cos(x+y)\cos(x-y)$. Because $\alpha^2 > 0$, $\cos 2\theta < 0$ and the bond angle $180° \geq 2\theta \geq 90°$. What has this calculation shown? We have demonstrated that it is possible to create two hybrid orbitals separated by a bonding angle in this angular range simply by varying the relative contributions of the 2s and 2p orbitals to the hybrid. Discussion of the energy cost of hybridization is deferred until later in this section.

The hybrid orbitals in Equation (13.5) are not specific to a particular molecule other than that the two atoms that bond to the central oxygen atom are identical. We now calculate the value of α that generates the correct bond angle in H_2O. Calculating α by substituting the known value $\theta = 52.25°$ in Equation (13.7), we find that the unnormalized hybrid orbitals that describe bonding in water are

$$\psi_a = N[0.61\,\phi_{2p_z} + 0.79\,\phi_{2p_x} - 0.50\,\phi_{2s}]$$

$$\psi_b = N[0.61\,\phi_{2p_z} - 0.79\,\phi_{2p_x} - 0.50\,\phi_{2s}] \qquad \textbf{(13.8)}$$

EXAMPLE PROBLEM 13.5

Normalize the hybrid orbitals given in Equation (13.8).

Solution

$$\int \psi_a^* \psi_a \, d\tau = N^2(0.61)^2 \int \phi_{2p_z}^*\phi_{2p_z}\,d\tau$$

$$+ N^2(0.79)^2 \int \phi_{2p_x}^*\phi_{2p_x}\,d\tau + N^2(0.50)^2 \int \phi_{2s}^*\phi_{2s}\,d\tau = 1$$

Other terms do not contribute because the atomic orbitals are orthogonal to one another.

$$\int \psi_a^* \psi_a \, d\tau = N^2(0.61)^2 + N^2(0.79)^2 + N^2(0.50)^2 = 1.25\,N^2 = 1$$

$$N = 0.89$$

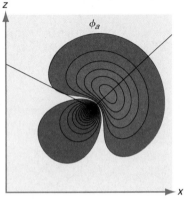

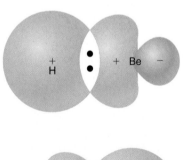

FIGURE 13.5

Directed hybrid bonding orbitals for H_2O. The black lines show the bond angle and orbital orientation. Red and blue contours correspond to the most positive and least positive values of the amplitude.

Using the result of Example Problem 13.5, the normalized hybrid orbitals can be written as follows:

$$\psi_a = 0.55\,\phi_{2p_z} + 0.71\,\phi_{2p_x} - 0.45\,\phi_{2s}$$
$$\psi_b = 0.55\,\phi_{2p_z} - 0.71\,\phi_{2p_x} - 0.45\,\phi_{2s} \qquad \textbf{(13.9)}$$

Because the sum of the squares of the coefficients for each hybrid orbital equals one, we can calculate their p and s character. The fraction of $2p$ character is $(0.55)^2 + (0.71)^2 = 0.80$. The fraction of $2s$ character is $(-0.45)^2 = 0.20$. Therefore, the hybridization of the bonding hybrid orbitals is described as sp^4. This differs from the first approximation sp^3, as predicted by Bent's rule.

The two hybrid orbitals are shown in Figure 13.5. Note that each of the directed hybrid orbitals lies along one of the bonding directions and has little amplitude along the other bonding direction. These hybrid orbitals could be viewed as the basis for the line connecting bonded atoms in the Lewis structure for water. Figure 13.5 shows a realistic representation of the hybrid orbitals that you should compare with the "slimmed down" version of Figure 13.3. The calculation of the hybrid orbitals representing the lone pairs is left as an end-of-chapter problem.

This calculation for H_2O illustrates how to construct bonding hybrid orbitals with a desired relative orientation. To this point, the energetics of this process have not been discussed. In many-electron atoms, the $2p$ orbital energy is greater than that for the $2s$ orbital. How can these orbitals be mixed in all possible proportions without putting energy into the atom? To create the set of occupied hybrid orbitals on an isolated ground-state oxygen atom would indeed require energy; however, the subsequent formation of bonds to the central atom lowers the energy, leading to an overall decrease in the energy of the molecule relative to the isolated atoms after bond formation. In the language of the hybridization model, the energy cost of promoting the electrons from the $1s^2 2s^2 2p^4$ configuration to the $1s^2 \psi_c^2 \psi_d^2 \psi_a^1 \psi_b^1$ configuration is more than offset by the energy gained in forming two O—H bonds.

Keep in mind that the individual steps in the formation of the H_2O molecule such as promotion of the O atom, followed by the creation of O—H bonds, are only an aid in describing the formation of H_2O, rather than a series of actual events. Ultimately, H_2O has a bond angle of $104.5°$ because this arrangement of the nuclei and the associated electrons minimizes the total energy of the molecule. The language of the hybridization model should not be taken too literally because neither the promotion process, nor orbitals, are observables, and reality cannot be assigned to quantities that cannot be measured. The reality of orbitals is discussed in Section 13.6 in more detail.

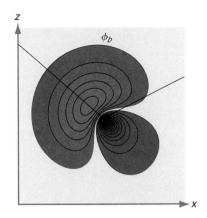

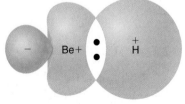

FIGURE 13.6

Bonding in BeH_2 using two sp-hybrid orbitals on Be. The two Be—H hybrid bonding orbitals are shown separately.

13.4 USING HYBRIDIZATION TO DESCRIBE CHEMICAL BONDING

By using the hybridization model to create local bonding orbitals, the concepts inherent in Lewis structures can be given a quantum mechanical basis. As an example, consider BeH_2, which is not observed as an isolated molecule because it forms a solid through polymerization of BeH_2 units stabilized by hydrogen bonds. We consider only a single BeH_2 unit. Be has the configuration $1s^2 2s^2 2p^0$, and because it has no unpaired electrons, it is not obvious how bonding to the H atoms can be explained in the Lewis model. Within the framework outlined earlier, the $2s$ and $2p$ orbitals are hybridized to create bonding hybrids on the Be atom. Because the bond angle is known to be $180°$, two equivalent and orthogonal sp-hybrid orbitals are constructed as given by Equation (13.3). This allows Be to be described as $1s^2 (\psi_a)^1 (\psi_b)^1$. In this configuration, Be has two unpaired electrons and, therefore, the hybridized atom is divalent. The orbitals are depicted schematically in Figure 13.6. To make a connection to Lewis structures, the bonding electron pair is placed in the overlap region between the Be and H orbitals as indicated by the dots. In reality, the bonding electron density is distributed over the entire region in which the orbitals have a nonzero amplitude. We return to BeH_2 in Section 13.6 where we compare localized and delocalized bonding models for this molecule.

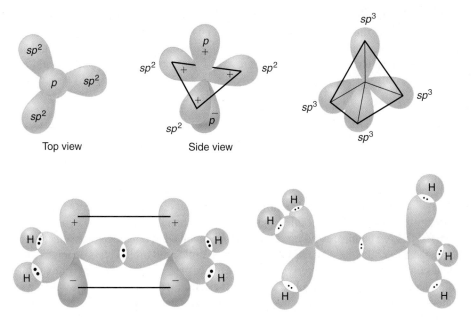

FIGURE 13.7

The top panel shows the arrangement of the hybrid orbitals for sp^2 and sp^3 carbon. The bottom panel shows a schematic depiction of bonding in ethene (left) and ethane (right) using hybrid bonding orbitals.

The chemically most important use of the hybridization model is in describing bonding in molecules containing carbon. Figure 13.7 depicts valence bond hybridization in ethene and ethane. For ethene, each carbon atom is promoted to the $1s^2 2p_y^1 (\psi_a)^1 (\psi_b)^1 (\psi_c)^1$ configuration before forming four C—H σ bonds, a C—C σ bond, and a C—C π bond. The maximal overlap between the p orbitals to create a π bond in ethene occurs when all atoms lie in the same plane. The double bond is made from one σ and one π bond. For ethane, each carbon atom is promoted to the $1s^2 (\psi_a)^1 (\psi_b)^1 (\psi_c)^1 (\psi_d)^1$ configuration before forming six C—H σ bonds and a C—C σ bond.

In closing this discussion of hybridization, it may be useful to emphasize the advantages of the model and to point out some of its shortcomings. The main advantage of hybridization is that it is an easily understandable model with considerable predictive power. It retains the main features of Lewis structures in describing local orthogonal bonds between adjacent atoms in terms of electron pairing, and it justifies Lewis structures in the language of quantum mechanics. Hybridization also provides a theoretical basis for the VSEPR rules through the construction of localized orbitals for bonding electrons and lone pairs.

Hybridization also offers more than a useful framework for understanding bond angles in molecules. Because the $2s$ AO is lower in energy than the $2p$ level in many-electron atoms, the electronegativity of a hybridized atom increases with increasing s character. Therefore, the hybridization model predicts that a sp-hybridized carbon atom is more electronegative than an sp^3-hybridized carbon atom. Evidence for this effect can be seen in the observation that the positive end of the dipole moment in N≡C—Cl is on the Cl atom. We conclude that the carbon atom in the cyanide group is more electronegative than a chlorine atom. Because increased s character leads to shorter bond lengths, and because shorter bonds generally have a greater bond strength, the hybridization model provides a correlation of s character and bond strength. These examples illustrate that the hybridization model has greater predictive power than the VSEPR model.

The hybridization model also has a few shortcomings. For known bond angles, the hybridization can be calculated as was done for ethane and H_2O. However, semiempirical prescriptions must be used to estimate the s and p character of a hybrid orbital for a molecule in the absence of structural information. It is also more straightforward to construct an appropriate hybridization for symmetric molecules such as methane than for molecules with electron lone pairs and several different ligands bonded to the central atom. Additionally, the depiction of bonding hybrids that is usually used (as in Figure 13.7) seems to imply that the electron density is highly concentrated along the bonding directions. This is not true as can be seen by looking at the realistic representation of hybrid orbitals in Figure 13.5. Finally, the conceptual formalism used in creating hybrid orbitals assumes much more detail than can be verified by experiments. The individual steps of promoting electrons to unoccupied orbitals and then hybridizing them may be useful as a rationalization of the observed geometry, but these steps should not be taken literally.

13.5 PREDICTING MOLECULAR STRUCTURE USING QUALITATIVE MOLECULAR ORBITAL THEORY

We now consider a **delocalized bonding model** of the chemical bond. MO theory approaches the structure of molecules quite differently than the VSEPR and hybridization local models of chemical bonding. The electrons involved in bonding are assumed to be delocalized over the molecule. Each one-electron molecular orbital σ_j is expressed as a linear combination of atomic orbitals such as $\sigma_j(k) = \sum_i c_{ij}\phi_i(k)$, which refers to the jth molecular orbital for electron k. The many-electron wave function ψ is written as a Slater determinant in which the individual entries are the $\sigma_j(k)$.

In quantitative molecular orbital theory, which will be discussed in the next chapter, structure emerges naturally as a result of solving the Schrödinger equation and determining the atomic positions for which the energy has its minimum value. Although this concept can be formulated in a few words, carrying out this procedure is a complex exercise in numerical computing. In this section, our focus is on a more qualitative approach that conveys the spirit of molecular orbital theory, but which can be written down without extensive mathematics.

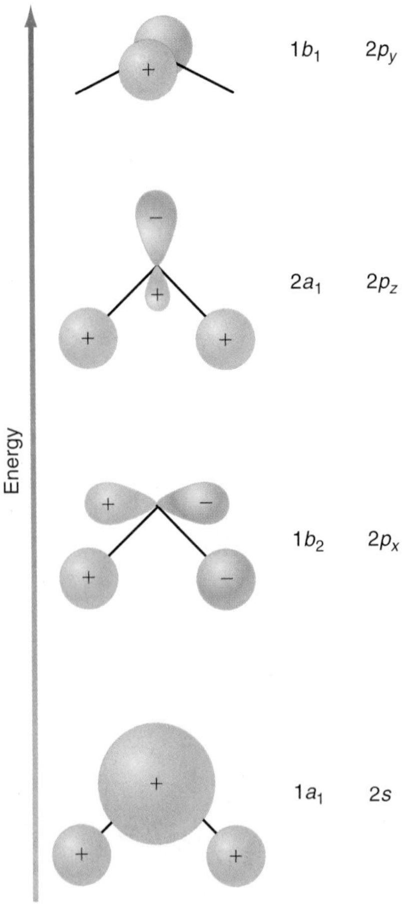
Energy
1b1
2py
2a1
2pz
1b2
2px
1a1
2s

FIGURE 13.8

The valence MOs occupied in the ground state of water are shown in order of increasing orbital energy. The MOs are depicted in terms of the AOs from which they are constructed. The second column gives the MO symmetry, and the third column lists the dominant AO orbital on the oxygen atom.

To illustrate this approach, we use qualitative MO theory to understand the bond angle in two triatomic molecules of the type H_2A, where A is one of the atoms in the sequence Be $\rightarrow$ O, and show that a qualitative picture of the optimal bond angle can be obtained by determining how the energy of the individual occupied molecular orbitals varies with the bond angle. In doing so, we assume that the total energy of the molecule is proportional to the sum of the orbital energies. This assumption can be justified, although we do not do so here. Experimentally, we know that BeH_2 is a linear molecule, and H_2O has a bond angle of 104.5°. How can this difference be explained using MO theory?

We begin by making educated guesses about the nature of the occupied MOs for these molecules. The basis set used here consists of the $1s$ orbitals on H and the $1s$, $2s$, $2p_x$, $2p_y$, and $2p_z$ orbitals on atom A. Seven MOs can be generated using these seven AOs. Water has 10 electrons, and four MOs accommodate the eight valence electrons. We omit the two lowest energy levels generated from the $1s$ oxygen AOs from this discussion because the corresponding electrons are localized on the oxygen atom. Therefore, we begin numbering the MOs with those generated from the $2s$ electrons on oxygen.

Recall that the orbital energy increases with the number of nodes for the particle in the box, the harmonic oscillator, and the H atom. We also know that the lower the AO energies, the lower the MO energy will be. The occupied MOs for water are shown in Figure 13.8, and each MO is depicted in terms of the AOs from which it is constructed. The relative MO orbital energies are discussed later. The MOs are labeled according to their symmetry with respect to a set of rotation and reflection operations that leaves the water molecule unchanged. We will discuss the importance of molecular symmetry in constructing MOs from AOs at some length in Chapter 16. However, in the present context it is sufficient to think of these designations simply as labels. Because the $2s$ AOs are lower in energy than the $2p$ AOs, the MO with no nodes designated $1a_1$ in Figure 13.8 is expected to have the lowest energy of all possible valence MOs. The next higher MOs involve $2p$ orbitals on the O atom.

The three $2p$ orbitals are differently oriented with respect to the plane containing the H atoms. As a result, the MOs that they generate are quite different in energy. Assume that the H_2A molecule lies in the x–z plane with the z axis bisecting the H—A—H angle as shown in Figure 13.4. The $1b_2$ MO, generated using the $2p_x$ AO, and the $2a_1$ MO, generated using the $2p_z$ AO, each have no nodes in the O—H region and, therefore,

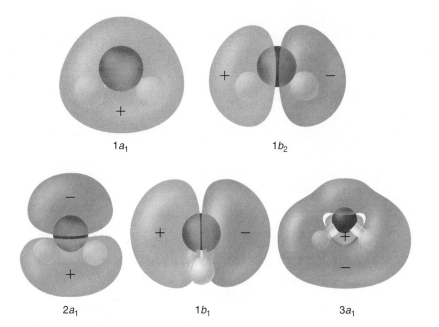

$1a_1$ $1b_2$

$2a_1$ $1b_1$ $3a_1$

FIGURE 13.9

The first five valence MOs for H_2O are depicted. The $1b_1$ and $3a_1$ MOs are the HOMO and LUMO, respectively. Note that the $1b_1$ MO is the AO corresponding to the nonbonding $2p_y$ electrons on oxygen. Note that the plane of the molecule has been rotated for the $1b_1$ MO to better display the nodal structure.

have binding character. However, because each has one node, both MOs have a higher energy than the $1a_1$ MO. It turns out that the MO generated using the $2p_x$ AO has a lower orbital energy than that generated using the $2p_z$ AO. Note that some $s-p$ mixing has been incorporated in the $2a_1$ and $3a_1$ MO generated from the $2s$ and $2p_z$ AO. Having discussed the MOs formed from the $2p_x$ and $2p_z$ AOs, we turn to the $2p_y$ AO. The $2p_y$ orbital has no net overlap with the H atoms and gives rise to the $1b_1$ nonbonding MO that is localized on the O atom. Because this MO is not stabilized through interaction with the H $1s$ AOs, it has the highest energy of all the MOs considered. Antibonding MOs that are higher in energy can be generated by combining out-of-phase AOs. Numerically calculated molecular orbitals are depicted in Figure 13.9.

The preceding discussion about the relative energy of the MOs is sufficient to allow us to draw the MO energy diagram shown in Figure 13.10. The MO energy levels in this figure are drawn for a particular bond angle near 105°, but the energy levels vary with 2θ, as shown in Figure 13.11, in what is known as a **Walsh correlation diagram**. You should make sure you understand the trends shown in this figure because the variation of each MO energy with angle is ultimately responsible for BeH_2 being linear and H_2O being bent. We next discuss the variation of the MO energy with 2θ.

How does the $1a_1$ energy vary with 2θ? The overlap between the s orbitals on A and H is independent of 2θ, but as this angle decreases from 180°, the overlap between the H atoms increases. This stabilizes the molecule and, therefore, the $1a_1$ energy decreases.

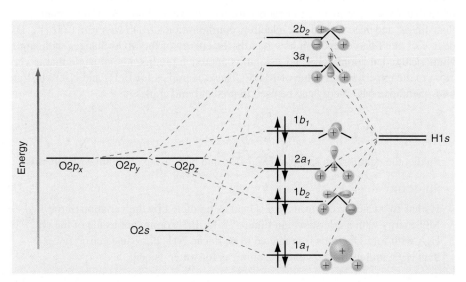

FIGURE 13.10

Molecular orbital energy-level diagram for H_2O at its equilibrium geometry.

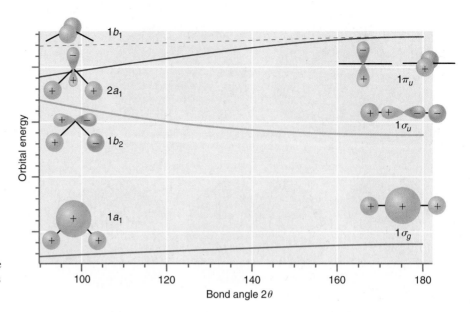

FIGURE 13.11

Schematic variation of the MO energies for water with bond angle. The symbols used on the left to describe the MOs are based on symmetry considerations and are valid for $2\theta < 180°$. This nomenclature is described in Chapter 17.

By contrast, the overlap between the $2p_y$ orbital and the H $1s$ orbitals is a maximum at 180° and therefore the $1b_2$ energy increases as 2θ decreases. Which of these MO energies changes more rapidly with the bond angle? Because the effect of the H—H overlap on the $1a_1$ energy is a secondary effect, the $1b_2$ energy falls more rapidly with increasing 2θ than the $1a_1$ energy increases.

We now consider the $2a_1$ and $1b_1$ energies. The $2p_y$ and $2p_z$ orbitals are nonbonding and degenerate for a linear H_2A molecule. However, as 2θ decreases from 180°, the $2p_z$ orbital has a net overlap with the H $1s$ AOs and has increasingly more bonding character. Therefore, the $2a_1$ MO energy decreases as 2θ decreases from 180°. The $1b_1$ MO remains nonbonding as 2θ decreases from 180°, but electron repulsion effects lead to a slight decrease of the MO energy. These variations in the MO energies are depicted in Figure 13.11.

Using the MO energy diagram of Figure 13.10, let's consider the molecules BeH_2 and H_2O. BeH_2 has four valence electrons that are placed in the two lowest lying MOs, $1a_1$ and $1b_2$. Because the $1b_2$ orbital energy decreases with increasing 2θ more than the $2a_1$ orbital energy increases, the total energy of the molecule is minimized if $2\theta = 180°$. This qualitative argument predicts that BeH_2, as well as any other four-valence electron H_2A molecule, is linear and has the valence electron configuration $(1\sigma_g)^2(1\sigma_u)^2$. Note that the description of H_2A in terms of σ MOs with g or u symmetry applies only to a linear molecule, whereas a description in terms of $1a_1$ and $1b_2$ applies to all bent molecules.

We now consider H_2O, which has eight valence electrons. In this case, the lowest four MOs are doubly occupied. At what angle is the total energy of the molecule minimized? For water, the decrease in the energy of the $1a_1$ and $2a_1$ MOs as 2θ decreases more than offsets the increase in energy for the $1b_2$ MO. Therefore, H_2O is bent rather than linear and has the valence electron configuration $(1a_1)^2(1b_2)^2(2a_1)^2(1b_1)^2$. The degree of bending depends on how rapidly the energy of the MOs changes with angle. Numerical calculations for water using this approach predict a bond angle that is very close to the experimental value of 104.5°. These examples for BeH_2 and H_2O illustrate how qualitative MO theory can be used to predict bond angles.

EXAMPLE PROBLEM 13.6

Predict the equilibrium shape of H_3^+, LiH_2, and NH_2 using qualitative MO theory.

Solution

H_3^+ has two valence electrons and is bent as predicted by the variation of the $1a_1$ MO energy with angle shown in Figure 13.11. LiH_2 or any molecule of the type H_2A with four electrons is predicted to be linear. NH_2 has one electron fewer than H_2O, and using the same reasoning as for water, is bent.

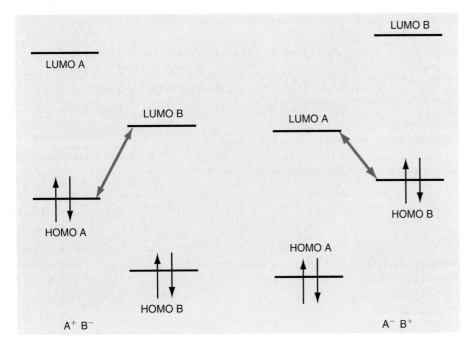

FIGURE 13.12

Interaction between two species A and B. The difference in energies between the HOMO and LUMO orbitals on A and B will determine the direction of charge transfer.

Qualitative **molecular orbital theory** can be used to gain insight into chemical reactions. Recall from Chapter 12 that higher energy MOs extend further out from the center of the molecule than do low energy MOs. Therefore, the higher energy MOs will be the most involved MOs in chemical reactions between molecules. In MO theory, the **highest occupied molecular orbital (HOMO)** and **lowest unoccupied molecular orbital (LUMO)** play an important role in chemical reactions and are called **frontier orbitals**. Consider a reaction involving electron transfer between molecules A and B. The energies of the HOMO and LUMO orbitals are specific to each molecule and could be aligned in either of two ways as is indicated in Figure 13.12.

The degree to which MOs mix is proportional to $S^2/\Delta E$, in which S is the overlap, and ΔE is the difference between the MO energies on species A and B. Assuming that S is not very different for the orbitals being considered, the mixing is dominated by ΔE. For the relative energies of the orbitals on the left side of Figure 13.12, the energy separation between HOMO A and LUMO B is much smaller than that between LUMO A and HOMO B. This means that HOMO A and LUMO B mix more readily than LUMO A and HOMO B, which results in a finite probability of finding additional electron density on B. This is equivalent to a partial charge transfer from A to B. For the opposite relative energies of the orbitals, shown on the right side of Figure 13.12, the same reasoning predicts that electron transfer will occur from B to A rather than from A to B.

13.6 HOW DIFFERENT ARE LOCALIZED AND DELOCALIZED BONDING MODELS?

Molecular orbital theory and hybridization-based valence bond theory have been developed using delocalized and localized bonding, respectively. These models approach the chemical bond from very different starting points. However, it is instructive to compare the molecular wave functions generated by these models using BeH_2 as an example. We have already discussed BeH_2 using hybridization in Section 13.4 and now formulate the many-electron wave function using the MO model. To minimize the size of the determinant in Equation (13.11), we assume that the Be $1s$ electrons are not delocalized over the molecule. With this assumption, BeH_2 has the configuration $(1s_{Be})^2(1\sigma_g)^2(1\sigma_u)^2$. On the basis of the symmetry requirements posed on the MOs by the linear geometry (see Chapter 17), the two lowest energy MOs are

$$\sigma_g = c_1(\phi_{H1sA} + \phi_{H1sB}) + c_2\phi_{Be2s}$$
$$\sigma_u = c_3(\phi_{H1sA} - \phi_{H1sB}) - c_4\phi_{Be2p_z} \qquad \textbf{(13.10)}$$

The many-electron determinantal wave function that satisfies the Pauli requirement is

$$\psi(1,2,3,4) = \frac{1}{4!}\begin{vmatrix} \sigma_g(1)\alpha(1) & \sigma_g(1)\beta(1) & \sigma_u(1)\alpha(1) & \sigma_u(1)\beta(1) \\ \sigma_g(2)\alpha(2) & \sigma_g(2)\beta(2) & \sigma_u(2)\alpha(2) & \sigma_u(2)\beta(2) \\ \sigma_g(3)\alpha(3) & \sigma_g(3)\beta(3) & \sigma_u(3)\alpha(3) & \sigma_u(3)\beta(3) \\ \sigma_g(4)\alpha(4) & \sigma_g(4)\beta(4) & \sigma_u(4)\alpha(4) & \sigma_u(4)\beta(4) \end{vmatrix} \quad \textbf{(13.11)}$$

Each entry in the determinant is an MO multiplied by a spin function.

We now use a property of a determinant that you will prove in the end-of-chapter problems for a 2×2 determinant, namely,

$$\begin{vmatrix} a & c \\ b & d \end{vmatrix} = \begin{vmatrix} a & \gamma a + c \\ b & \gamma b + d \end{vmatrix} \quad \textbf{(13.12)}$$

This equation says that one can add a column of the determinant multiplied by an arbitrary constant γ to another column *without changing the value of the determinant*. For reasons that will become apparent shortly, we replace the MOs σ_g and σ_u with the new hybrid MOs $\sigma' = \sigma_g + (c_1/c_3)\sigma_u$ and $\sigma'' = \sigma_g - (c_1/c_3)\sigma_u$. These hybrid MOs are related to the AOs by

$$\sigma' = 2c_1\phi_{H1sA} + \left(c_2\phi_{Be2s} - \frac{c_1c_4}{c_3}\phi_{Be2p_z}\right)$$

$$\sigma'' = 2c_1\phi_{H1sB} + \left(c_2\phi_{Be2s} + \frac{c_1c_4}{c_3}\phi_{Be2p_z}\right) \quad \textbf{(13.13)}$$

Transforming from σ_g and σ_u to σ' and σ'' requires two steps like the one in Equation (13.12). Note that with this transformation, ϕ_{H1sB} no longer appears in σ', and ϕ_{H1sA} no longer appears in σ''. Because of the property of determinants cited earlier, neither $\psi(1,2,3,4)$—*nor any molecular observable*—will be affected by this change in the MOs. Therefore, the configurations $(1s_{Be})^2 (1\sigma_g)^2(1\sigma_u)^2$ and $(1s_{Be})^2$ $[1\sigma_g + (c_1/c_3)1\sigma_u]^2[1\sigma_g - (c_1/c_3)1\sigma_u]^2$ are completely equivalent, and no experiment can distinguish between them.

Why have we made this particular change? Equation (13.13) and Figure 13.13 show that the new MOs σ' and σ'' are localized bonding MOs, one combining the $1s$ orbital on H_A with an s–p hybrid AO on Be, and the other combining the $1s$ orbital on H_B with an s–p hybrid AO on Be. In other words, the two delocalized MOs σ_g and σ_u have

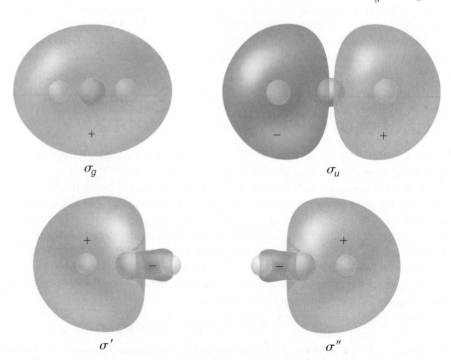

FIGURE 13.13

Schematic representation of the delocalized MOs σ_g and σ_u and the localized bonding orbitals σ' and σ''.

been transformed into two localized hybrid bonding orbitals *without changing the molecular wave function* $\psi(1,2,3,4)$. This result can be generalized to the statement that for any closed-shell molecular configuration, the set of delocalized MOs can be transformed into a set of localized orbitals predominantly involving two neighboring atoms. Such a transformation is not possible for open-shell molecules or the conjugated and aromatic molecules discussed in the next section.

As the BeH_2 example shows, the distinction between localized and delocalized orbitals is not as clear-cut as it seemed to be at the beginning of this chapter. In understanding this result, you should distinguish between observables and elements of a model that are not amenable to measurement. Although the many-electron wave function $\psi(1, \ldots, n)$ cannot be determined, the probability density $|\psi(1, \ldots, n)|^2$ can be measured using techniques such as X-ray and electron diffraction. Therefore, the magnitude of the many-electron molecular wave function is an observable. By contrast, the individual one-electron MOs and AOs that are used to construct the Slater determinant for a many-electron wave function are not observables and are not amenable to direct measurement. They are a part of the model that has been constructed to explain the properties of many-electron atoms and molecules. However, ultimately only $\psi(1, \ldots, n)$, and not the basis functions in which $\psi(1, \ldots, n)$ is expanded is meaningful. Because the two very different sets of basis functions, σ_g and σ_u and σ' and σ'', can be used equally well to construct $\psi(1,2,3,4)$, it is not meaningful to view either as uniquely corresponding to reality.

Having said this, working with σ' and σ'' has some disadvantages, because they are not eigenfunctions of the total energy operator, $\hat{H}$. This means that we cannot assign orbital energies to these functions or draw energy-level diagrams such as those shown in Figure 13.10. By contrast, the delocalized MOs that provide solutions to the molecular Hartree–Fock equations have a well-defined orbital energy, which allows MO energy diagrams to be constructed. Because of these advantages, delocalized MOs rather than localized orbitals are generally used to describe the electron configuration of molecules.

13.7 QUALITATIVE MOLECULAR ORBITAL THEORY FOR CONJUGATED AND AROMATIC MOLECULES: THE HÜCKEL MODEL

The molecules in the preceding sections can be discussed using either a localized or a delocalized model of chemical bonding. This is not the case for conjugated and aromatic molecules where a delocalized model must be used. **Conjugated molecules** such as 1,3-butadiene have a planar carbon backbone with alternating single and double bonds. Butadiene has single and double bond lengths of 147 and 134 pm, respectively. The single bonds are shorter than the single bond length in ethane (154 pm), which suggests that a delocalized π network is formed. Such a delocalized network can be modeled in terms of the coupling between sp^2-hybridized carbon atoms in a σ bonded carbon backbone. The lowering of the total energy that can be attributed to the formation of the π network is responsible for the reduced reactivity of conjugated molecules compared to molecules with isolated double bonds.

Aromatic molecules are a special class of conjugated molecules. They are based on ring structures that are particularly stable in chemical reactions. The presence of "closed circuits" of mobile electrons is required for a molecule to be aromatic. Because such currents imply electron delocalization, bonding in aromatic molecules cannot be explained by electron pairing in localized bonds. For example, benzene has six C—C bonds of equal length, 139 pm, a value between the single and double bonds lengths in 1,3-butadiene. This suggests that the six π electrons are distributed over all six carbon atoms. Therefore, a delocalized model is required to discuss aromatic molecules.

Erich Hückel formulated a useful application of qualitative MO theory to calculate the energy levels of the delocalized π electrons in conjugated and aromatic molecules. Despite its simplicity, the **Hückel model** correctly predicts the stabilization that arises from delocalization and predicts which of many possible cyclic polyenes will be aromatic. In the Hückel model, the π network of MOs can be treated separately from the σ network of the carbon backbone. The Hückel model uses hybridization and the localized valence bond model to describe the σ bonded skeleton and MO theory to describe the delocalized π electrons.

In the Hückel theory, the p atomic orbitals that combine to form π MOs are treated separately from the sp^2 σ bonded carbon backbone. For the four-carbon π network in butadiene, the π MO can be written in the form

$$\psi_\pi = c_1\phi_{2pz1} + c_2\phi_{2pz2} + c_3\phi_{2pz3} + c_4\phi_{2pz4} \tag{13.14}$$

As was done in Section 12.2, the variational method is used to calculate the coefficients that give the lowest energy for the two MOs that result from combining two AOs. We obtain the following secular equations:

$$c_1(H_{11} - \varepsilon S_{11}) + c_2(H_{12} - \varepsilon S_{12}) + c_3(H_{13} - \varepsilon S_{13}) + c_4(H_{14} - \varepsilon S_{14}) = 0$$

$$c_1(H_{21} - \varepsilon S_{21}) + c_2(H_{22} - \varepsilon S_{22}) + c_3(H_{23} - \varepsilon S_{23}) + c_4(H_{24} - \varepsilon S_{24}) = 0$$

$$c_1(H_{31} - \varepsilon S_{31}) + c_2(H_{32} - \varepsilon S_{32}) + c_3(H_{33} - \varepsilon S_{33}) + c_4(H_{34} - \varepsilon S_{34}) = 0$$

$$c_1(H_{41} - \varepsilon S_{41}) + c_2(H_{42} - \varepsilon S_{42}) + c_3(H_{43} - \varepsilon S_{43}) + c_4(H_{44} - \varepsilon S_{44}) = 0$$

$$\tag{13.15}$$

Similar to the discussion in Chapter 12, integrals of the type H_{aa} are called Coulomb integrals, integrals of the type H_{ab} are called resonance integrals, and integrals of the type S_{ab} are called overlap integrals. Rather than evaluate these integrals, in the Hückel model thermodynamic and spectroscopic data obtained from different conjugated molecules are used to obtain their values. Because it relies on both theoretical and experimental input, the Hückel model is a **semiempirical theory**.

In the Hückel model, the Coulomb and resonance integrals are assumed to be the same for all conjugated hydrocarbons and are given the symbols α and β, respectively, where α is the negative of the ionization energy of the $2p$ orbital, and β, which is negative, is usually left as an adjustable parameter. Do not confuse this notation with spin up and spin down. The **secular determinant** that is used to obtain the MO energies and the coefficients of the AOs for 1,3-butadiene is

$$\begin{vmatrix} H_{11} - \varepsilon S_{11} & H_{12} - \varepsilon S_{12} & H_{13} - \varepsilon S_{13} & H_{14} - \varepsilon S_{14} \\ H_{21} - \varepsilon S_{21} & H_{22} - \varepsilon S_{22} & H_{23} - \varepsilon S_{23} & H_{24} - \varepsilon S_{24} \\ H_{31} - \varepsilon S_{31} & H_{32} - \varepsilon S_{32} & H_{33} - \varepsilon S_{33} & H_{34} - \varepsilon S_{34} \\ H_{41} - \varepsilon S_{41} & H_{42} - \varepsilon S_{42} & H_{43} - \varepsilon S_{43} & H_{44} - \varepsilon S_{44} \end{vmatrix} \tag{13.16}$$

Several simplifying assumptions are made in the Hückel model to make it easier to solve secular determinants. The first is $S_{ii} = 1$ and $S_{ij} = 0$ unless $i = j$. This is a rather drastic simplification, because if the overlap between adjacent atoms were zero, no bond formation would occur. It is also assumed that $H_{ij} = \beta$ if $i = j \pm 1$, $H_{ij} = \alpha$ if $i = j$, and $H_{ij} = 0$ otherwise. Setting $H_{ij} = 0$ for nonadjacent carbon atoms amounts to saying that the primary interaction is between the neighboring $2p_z$ orbitals. The result of the simplifying assumptions is that all elements of the determinant that are more than one position removed from the diagonal are zero.

With these assumptions, the secular determinant for butadiene is

$$\begin{vmatrix} \alpha - \varepsilon & \beta & 0 & 0 \\ \beta & \alpha - \varepsilon & \beta & 0 \\ 0 & \beta & \alpha - \varepsilon & \beta \\ 0 & 0 & \beta & \alpha - \varepsilon \end{vmatrix} = 0 \tag{13.17}$$

As shown in Example Problem 13.9, this determinant has the solutions for the π orbital energies shown in Figure 13.14.

Consider several of the results shown in Figure 13.14. First, the coefficients of the AOs in the different MOs are not the same. Secondly, a pattern of nodes is seen that is identical to the other quantum mechanical systems that have been solved. The ground state has no nodes, and successively higher MOs have an increasing number of nodes. Recall that nodes correspond to regions in which the probability of finding the electron is low. Because the nodes appear between the carbon atoms, they add an antibonding character to the MO, which increases the MO energy.

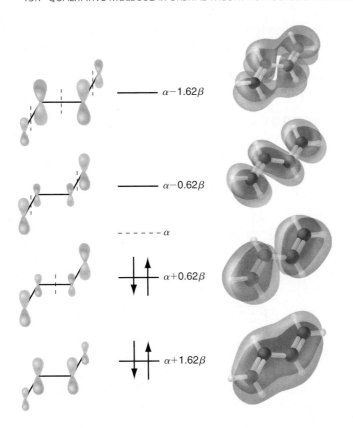

$\alpha - 1.62\beta$

$\alpha - 0.62\beta$

$- - - - - \alpha$

$\alpha + 0.62\beta$

$\alpha + 1.62\beta$

FIGURE 13.14

Energy levels and molecular orbitals for butadiene in the Hückel approximation. The sizes of the $2p_z$ AOs in the left column are proportional to their coefficients in the MO. Calculated MOs (see text) are shown in the right column. Red and blue lobes refer to positive and negative amplitudes, respectively. The dashed lines indicate nodal planes.

EXAMPLE PROBLEM 13.7

Solve the secular determinant for butadiene to obtain the MO energies.

Solution

The 4×4 secular determinant can be expanded to yield (see the Math Supplement, Appendix A) the following equation:

$$\begin{vmatrix} \alpha - \varepsilon & \beta & 0 & 0 \\ \beta & \alpha - \varepsilon & \beta & 0 \\ 0 & \beta & \alpha - \varepsilon & \beta \\ 0 & 0 & \beta & \alpha - \varepsilon \end{vmatrix}$$

$$= (\alpha - \varepsilon) \begin{vmatrix} \alpha - \varepsilon & \beta & 0 \\ \beta & \alpha - \varepsilon & \beta \\ 0 & \beta & \alpha - \varepsilon \end{vmatrix} - \beta \begin{vmatrix} \beta & \beta & 0 \\ 0 & \alpha - \varepsilon & \beta \\ 0 & \beta & \alpha - \varepsilon \end{vmatrix}$$

$$= (\alpha - \varepsilon)^2 \begin{vmatrix} \alpha - \varepsilon & \beta \\ \beta & \alpha - \varepsilon \end{vmatrix} - \beta(\alpha - \varepsilon) \begin{vmatrix} \beta & \beta \\ 0 & \alpha - \varepsilon \end{vmatrix} - \beta^2 \begin{vmatrix} \alpha - \varepsilon & \beta \\ \beta & \alpha - \varepsilon \end{vmatrix} + \beta^2 \begin{vmatrix} 0 & \beta \\ 0 & \alpha - \varepsilon \end{vmatrix}$$

$$= (\alpha - \varepsilon)^4 - (\alpha - \varepsilon)^2\beta^2 - (\alpha - \varepsilon)^2\beta^2 - (\alpha - \varepsilon)^2\beta^2 + \beta^4$$

$$= (\alpha - \varepsilon)^4 - 3(\alpha - \varepsilon)^2\beta^2 + \beta^4 = \frac{(\alpha - \varepsilon)^4}{\beta^4} - \frac{3(\alpha - \varepsilon)^2}{\beta^2} + 1 = 0$$

This equation can be written in the form of a quadratic equation:

$$\frac{(\alpha - \varepsilon)^2}{\beta^2} = \frac{3 \pm \sqrt{5}}{2}$$

which has the four solutions $\varepsilon = \alpha \pm 1.62\beta$ and $\varepsilon = \alpha \pm 0.62\beta$.

The effort required to obtain the energy levels and the AO coefficients for a monocyclic polyene can be greatly simplified by using the following geometrical construction: inscribe a regular polygon with the shape of the polyene in a circle of radius 2β

with one vertex of the polygon pointing directly downward. Draw a horizontal line at each point for which the polygon is tangent to the circle. These lines correspond to the energy levels, with the center of the circle corresponding to the energy α. This method is illustrated in Example Problem 13.8.

EXAMPLE PROBLEM 13.8

Use the inscribed polygon method to calculate the Hückel MO energy levels for benzene.

Solution

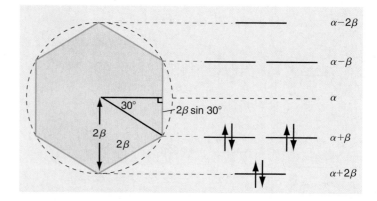

The geometrical construction shows that the energy levels are $\alpha + 2\beta$, $\alpha + \beta$, $\alpha - \beta$, and $\alpha - 2\beta$. The sum of the orbital energies for the six π electrons is $6\alpha + 8\beta$.

The benzene MOs and their energies are shown in Figure 13.15. Note that the energy levels for $\alpha + \beta$ and $\alpha - \beta$ are doubly degenerate. As was the case for butadiene,

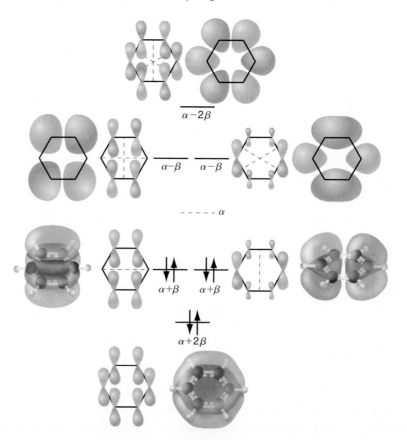

FIGURE 13.15

Energy levels and molecular orbitals for benzene in the Hückel approximation. The sizes of the $2p_z$ AOs are proportional to their coefficients in the MO. Calculated MOs (see text) are shown in a 3D perspective for the filled MOs and as an on-top view for the unfilled MOs. Red and blue lobes refer to positive and negative amplitudes. Thin dashed lines indicate nodal planes.

the lowest MO has no nodes and the energy of the MO increases with the number of nodal planes. The average orbital energy of a π electron in benzene is

$$\frac{1}{6}\Big[2(\alpha + 2\beta) + 4(\alpha + \beta)\Big] = \alpha + 1.33\beta$$

The energy levels for the smaller monocyclic polyenes $(CH)_m$, where m is the number of π bonded carbons, exhibit the pattern shown in Figure 13.16. The energy value α separates the bonding and antibonding MOs.

This figure provides a justification for the following **Hückel rules** for a monocyclic conjugated system with N π electrons:

- If $N = 4n + 2$, where n is an integer 0, 1, 2, ..., the molecule is stabilized through the π delocalization network.

- If $N = 4n + 1$ or $4n + 3$, the molecule is a free radical.

- If $N = 4n$, the molecule has two unpaired electrons and is very reactive.

The justification for these rules can be understood from Figure 13.16. For each cyclic polyene, the lowest energy level is nondegenerate and has the energy $\alpha + 2\beta$. All other levels are doubly degenerate, with the exception of the highest level if m is even. The maximum stabilization is attained if $N = 4n + 2$, because all π electrons are paired and in bonding orbitals for which $\varepsilon < \alpha$. For $n = 1$, six π electrons correspond to the maximal stabilization. Benzene, for which $m = 6$, is an example of this case. Next consider benzene with one fewer or one more π electron. Because $N = 5$ or 7, both species are radicals because the highest occupied energy level is not filled, and both are less stable than benzene. What happens to a system of maximum stabilization if two electrons are removed? Because all energy levels, except the lowest (and if m is even, the highest), are doubly degenerate, each of the degenerate levels has an occupancy of one for $N = 4n$, and the molecule is a diradical.

These rules can be used to make useful predictions. For example, C_3H_3, which is formed from cyclopropene by the removal of one H atom, should be more stable as $C_3H_3^+ (N = 2)$ than as neutral $C_3H_3 (N = 3)$ or $C_3H_3^- (N = 4)$. Undistorted cyclobutadiene $(N = 4)$ with four π electrons will be a diradical and, therefore, very reactive. The maximum stabilization for C_5H_5 is for $N = 6$, as is seen in Figure 13.16. Therefore, $C_5H_5^-$ is predicted to be more stable than C_5H_5 or $C_5H_5^+$. These predictions have been verified by experiment and show that the Hückel model has considerable predictive power, despite its significant approximations.

At present, the Hückel model is primarily useful for explaining the $4n + 2$ rule. Computational chemistry software available in student editions can rapidly obtain MOs and their corresponding energy levels on personal computers, making the determination of α and β from experimental data obsolete. The calculated MOs and energy levels shown in Figures 13.14 and 13.15 were obtained in this way using the B3LYP method of density functional theory and the 6-31G* basis set (see Chapter 15).

We now discuss the **resonance stabilization energy** that arises in aromatic compounds through the presence of closed circuits of mobile electrons. No unique method is available for calculating this stabilization energy. However, a reasonable way to determine this energy is to compare the π network energy of the cyclic polyene with that of a linear polyene that consists of alternating double and single bonds with the same number and arrangement of hydrogen atoms. In some cases, this may be a hypothetical molecule whose π network energy can be calculated using the method outlined earlier. As has been shown by L. Schaad and B. Hess [*J. Chemical Education* 51 (1974), 640–643], meaningful results for the resonance stabilization energy can be obtained only if the reference molecule is similar in all aspects except one: it is a linear rather than a cyclic polyene. For benzene, the reference molecule has the total π energy $6\alpha + 7.608\beta$. From Figure 13.16, note that the corresponding value for benzene is $6\alpha + 8\beta$. Therefore, the resonance stabilization energy per π electron in benzene is $(8.000\beta - 7.608\beta)/6 = 0.065\beta$. By considering suitable reference compounds, these authors have calculated the resonance stabilization energy for a large number of compounds, some of which are shown in Figure 13.17.

Figure 13.17 indicates that benzene and benzocyclobutadiene have the greatest resonance delocalization energy on this basis. Molecules with negative values for the

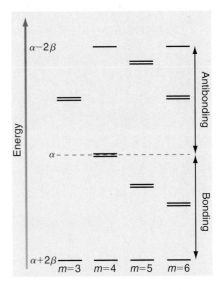

FIGURE 13.16

The energy of the π MOs is shown for cyclic polyenes described by the formula $(CH)_m$ with $m = 3$ to 6 π bonded carbons. The doubly degenerate pairs are shown slightly separated in energy for clarity.

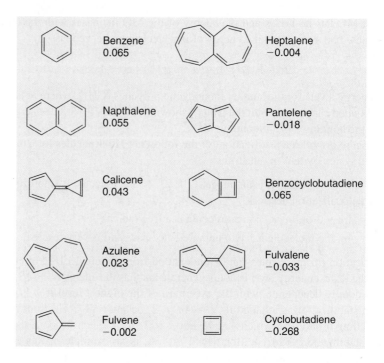

FIGURE 13.17

The resonance delocalization energy per π electron of a number of cyclic polyenes is shown in units of β.

resonance stabilization energy are predicted to be more stable as linear polyenes than as cyclic polyenes and are referred to as **antiaromatic molecules**. Note that these calculations only give information on the π network energy, and that the total energy of the molecule is assumed to be proportional to the sum of the occupied π orbital energies. We have also ignored the possible effect of strain energy that arises if the C—C—C bonding angles are significantly different from 120°, which is optimal for sp^2 hybridization. For example, cyclobutadiene, for which the bond angle is 90°, has an appreciable strain energy associated with the σ bonded backbone, which destabilizes the molecule relative to a linear polyene. A ranking of the degree of aromaticity based on experimental data such as thermochemistry, reactivity, and chemical shifts using nuclear magnetic resonance spectroscopy (see Chapter 17) is in good agreement with the predictions of the Hückel model if the appropriate reference molecule is used to calculate the resonance stabilization energy.

Although the examples used here to illustrate aromaticity are planar compounds, this is not a requirement for aromaticity. Sandwich compounds such as ferrocene as well as the fullerenes also show aromatic behavior. For these molecules, the closed circuits of mobile electrons extend over all three dimensions.

These calculations for conjugated and aromatic molecules show the power of the Hückel model in obtaining useful results with minimal computational effort and without evaluating any integrals or even using numerical values for α and β. Fewer simplifying assumptions are made in the extended Hückel model, which treats the σ and π electrons similarly. It is a very useful semiempirical method that is widely used in quantum chemistry.

13.8 FROM MOLECULES TO SOLIDS

The Hückel model is also useful for understanding the energy levels in a solid, which can be thought of as a giant molecule. In discussing the application of the particle in the box model to solids in Chapter 5, we learned that a solid has an energy spectrum that has both continuous and discrete aspects. Within a range of energies called a band, the energy spectrum is continuous. However, the energy bands are separated by **band gaps** in which no quantum states are allowed. The Hückel model (Figure 13.18) is useful in developing an understanding of how this energy spectrum is generated.

Consider a one-dimensional chain of atoms in which π bonds are formed. Combining $N \, 2p_x$ atomic orbitals creates the same number of π MOs as was seen for ethene, butadiene, and benzene. Hückel theory predicts that the difference in energy between the lowest and highest energy MO depends on the length of the conjugated chain, but

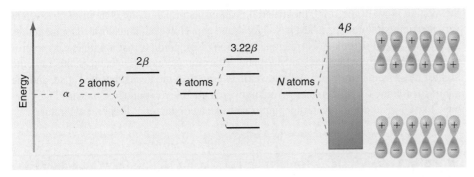

FIGURE 13.18

MOs generated in an atom chain using the Hückel model. As N becomes very large, the energy spectrum becomes continuous. The energy range of the MOs is shown in units of β.

approaches the value 4β as the chain becomes infinitely long. All N energy levels still must lie in the range between $\alpha + 2\beta$ and $\alpha - 2\beta$. Therefore, as $N \to \infty$, the spacing between adjacent levels becomes vanishingly small, and the energy spectrum becomes continuous, generating a band.

The wave function of the long one-dimensional chain is schematically indicated in Figure 13.18. At the bottom of the band, all AOs are in phase (fully bonding), but at the top of the band the AOs on adjacent atoms are out of phase (fully antibonding). At energies near the middle of the band, the nodal spacing is intermediate between N and one atomic spacing, making the state partially bonding. The energy versus distance curves from Figure 12.4 can be applied to this case. This has been done in Figure 13.19. For a two-atom solid (diatomic molecule), the wave function is either fully bonding or fully antibonding. For a long chain, all possible wave functions between fully bonding and fully antibonding are possible. Therefore, the entire energy range shown as shaded in Figure 13.19 is allowed.

We consider a specific example that demonstrates the contrasts among a conductor, a semiconductor, and an insulator. In solids, separate bands are generated from different AOs, such as the $3s$ and $3p$ AOs on Mg. Magnesium, with the $[\text{Ne}]3s^2$ atomic configuration, has two $3s$ valence electrons that go into a band generated from the overlap of the $3s$ electrons on neighboring Mg atoms. Because N Mg atoms generate N MOs, each of which can be doubly occupied, the $2N$ Mg valence electrons completely fill the $3s$-generated band (lower band in Figure 13.19). If there were a gap between this and the next highest band (upper band in Figure 13.19), which is generated from the $3p$ electrons, Mg would be an insulator. However, in this case, the $3s$ and $3p$ bands overlap, and the result is that the unoccupied states in the overlapping bands are only infinitesimally higher in energy than the highest filled state. For this reason, Mg is a conductor.

If there is a gap between a completely filled band and the empty band of next higher energy, the solid is either an insulator or a semiconductor. The distinction between a semiconductor and an insulator is the width of the energy gap. If $E_{gap} \gg kT$ at temperatures below

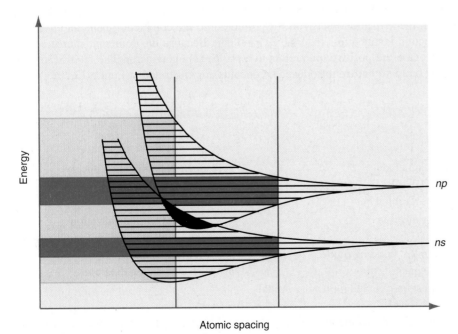

FIGURE 13.19

Bands generated from two different AOs are shown. The width in energy of the band depends on the atomic spacing. For the equilibrium spacing indicated by the red bar, the two bands overlap and all energy values between the top and bottom of the yellow shaded area are allowed. This is not true for significantly larger or shorter atomic spacings, and the solid would exhibit a band gap at the spacing indicated by the blue bar. In this case the two narrow bands indicated by the green areas do not overlap.

the melting point of the solid, the material is an insulator. Diamond is an insulator even at very high temperatures because it has a large band gap. However, if at elevated temperatures $E_{gap} \sim kT$, the Boltzmann distribution [Equation (2.2)] predicts that it will be easy to promote an electron from the filled valence band to the empty conduction band. In this case, the highest filled state is infinitesimally lower in energy than the lowest unfilled state, and the solid is a conductor. Silicon and germanium are called semiconductors because they behave like an insulator at low temperatures and like a conductor at higher temperatures.

13.9 MAKING SEMICONDUCTORS CONDUCTIVE AT ROOM TEMPERATURE

In its pure state, silicon is conductive to an appreciable extent only at temperatures greater than 900 K because it has a band gap of 1.1 eV. Yet computers and other devices that are based on silicon technology function at room temperature. For this to happen, these devices must transmit electrical currents at 300. K. What enables silicon to be conductive at such low temperatures? The key to changing the properties of Si is the introduction of other atoms that occupy Si sites in the silicon crystal structure. The introduction of these foreign atoms in the Si crystal lattice is called **doping**.

Silicon is normally doped using atoms such as boron or phosphorus, which have one fewer or one more valence electron than silicon, respectively. Typically, the dopant concentration is on the order of a part per million relative to the Si concentration. How does this make Si conductive at lower temperatures? The Coulomb potential associated with the phosphorus atoms overlaps with the Coulomb potentials of the neighboring Si atoms, and the valence electrons of the P atom become delocalized throughout the crystal and form a separate band as discussed in Section 5.4. Because P has one more valence electron than Si, this band is only partially filled. As indicated in Figure 13.20, this band is located 0.04 eV below the bottom of the empty conduction band. These atoms can be ionized to populate the empty Si conduction band. Importantly, it is the 0.04 eV rather than the Si band gap of 1.1 eV that must be comparable to kT in order to ionize the dopant, producing delocalized electrons in the conduction band. Therefore, phosphorus-doped silicon is conductive at 300 K, where $kT = 0.04$ eV. Because the dominant charge carriers are negative, one refers to an n-type semiconductor.

If boron is introduced as a dopant, the Si crystal site has one valence electron fewer than the neighboring sites and acts like a positive charge, which is referred to as a hole. The hole is delocalized throughout the lattice and acts like a mobile positive charge because electrons from adjacent Si atoms can fill it leaving the B atom with an extra negative charge while the hole jumps from Si to Si atom. In this case, the dopant band is 0.045 eV above the top of the filled valence band. As for the n-type semiconductor, the activation energy to lift promote into the conduction band and to induce conduction is much less than the Si band gap. Because the dominant charge carriers in this case are positive, one refers to a p-type semiconductor. The modifications to the Si **band structure** introduced by dopants are illustrated in Figure 13.20.

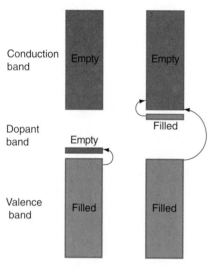

FIGURE 13.20

Modification of the silicon band structure generated by the introduction of dopants. The excitation that leads to conduction is from the dopant band to the valence band (p-type), as shown on the left, or conduction band (n-type), as shown on the right, rather than across the Si band gap as indicated by the right-most curved arrow. Occupied and unoccupied bands are indicated by a darker and lighter shading, respectively. Energy increases vertically in the figure.

Vocabulary

antiaromatic molecules
aromatic molecules
band gap
band structure
conjugated molecules
delocalized bonding model
doping
frontier orbitals
highest occupied molecular orbital
(HOMO)

Hückel model
Hückel rules
hybrid orbital
hybridization
Lewis structure
localized bonding model
lone pair
lowest unoccupied molecular orbital
(LUMO)
molecular orbital theory

resonance stabilization energy
secular determinant
semiempirical theory
sp, sp^2, and sp^3 hybridization
valence shell electron pair repulsion
(VSEPR) model
Walsh correlation diagram

Questions on Concepts

Q13.1 For what condition is it difficult to say whether a material is a semiconductor or an insulator?

Q13.2 Under what condition on the coefficients in an MO will the MO represent a localized bond?

Q13.3 What is the in-plane amplitude of the wave functions describing the π network in the conjugated molecules shown in Figures 13.13 to 13.15?

Q13.4 What experimental evidence can you cite in support of the hypothesis that the electronegativity of a hybridized atom increases with increasing s character?

Q13.5 Explain why all possible wave functions between the fully bonding and the fully antibonding are possible for the bands shown in Figure 13.19.

Q13.6 On the basis of what you know about the indistinguishability of electrons and the difference between the wave functions for bonding electrons and lone pairs, discuss the validity and usefulness of the Lewis structure for the fluorine molecule ($:\ddot{F}—\ddot{F}:$).

Q13.7 What evidence can you find in Table 13.1 that C—C sp bonds are stronger than sp^3 bonds?

Q13.8 How is it possible that a semiconductor would become metallic if the nearest neighbor spacing could be changed sufficiently?

Q13.9 What is the rationale for setting $H_{ij} = 0$ for nonadjacent atoms in the Hückel model?

Q13.10 A certain cyclic polyene is known to be nonplanar. Are the MO energy levels of this molecule well described by the Hückel model? Justify your answer.

Q13.11 Why are localized and delocalized models equally valid for describing bonding in closed-shell molecules? Why can't experiments distinguish between these models?

Q13.12 The hybridization model assumes that atomic orbitals are recombined to prepare directed orbitals that have the bond angles appropriate for a given molecule. What aspects of the model can be tested by experiments, and what aspects are conjectures that are not amenable to experimental verification?

Q13.13 Why can't localized orbitals be represented in an MO energy diagram?

Q13.14 In using the sum of the occupied MO energies to predict the bond angle in H_2A molecules, the total energy of the molecule is assumed to be proportional to the sum of the occupied MO energies. This assumption can be justified. Do you expect this sum to be greater than or smaller than the total energy? Justify your answer.

Q13.15 In explaining molecular structure, the MO model uses the change in MO energy with bond angle. Explain why the decrease in energy of the $1a_1$ and $2a_1$ MOs as 2θ decreases more than offsets the increase in energy for the $1b_2$ MO for water.

Problems

P13.1 Show that the determinantal property

$$\begin{vmatrix} a & c \\ b & d \end{vmatrix} = \begin{vmatrix} a & \gamma a + c \\ b & \gamma b + d \end{vmatrix}$$

used in the discussion of localized and delocalized orbitals in Section 13.6 is correct.

P13.2 Predict whether LiH_2^+ and NH_2^- should be linear or bent based on the Walsh correlation diagram in Figure 13.11. Explain your answers.

P13.3 Use the framework described in Section 13.3 to construct normalized hybrid bonding orbitals on the central oxygen in O_3 that are derived from $2s$ and $2p$ atomic orbitals. The bond angle in ozone is 116.8°.

P13.4 Are the localized bonding orbitals in Equation (13.13) defined by

$$\sigma' = 2c_1\phi_{H1sA} + \left(c_2\phi_{Be2s} - \frac{c_1 c_4}{c_3}\phi_{Be2p_z}\right) \text{ and}$$

$$\sigma'' = 2c_1\phi_{H1sB} + \left(c_2\phi_{Be2s} + \frac{c_1 c_4}{c_3}\phi_{Be2p_z}\right)$$

orthogonal? Answer this question by evaluating the integral $\int (\sigma')^* \sigma'' \, d\tau$.

P13.5 Use the method described in Example Problem 13.3 to show that the sp-hybrid orbitals $\psi_a = 1/\sqrt{2}(-\phi_{2s} + \phi_{2p_z})$ and $\psi_b = 1/\sqrt{2}(-\phi_{2s} - \phi_{2p_z})$ are oriented 180° apart.

P13.6 Use the formula $\cos 2\theta = -\alpha^2$ and the method in Section 13.2 to derive the formula $\psi_a = 1/\sqrt{2}(-\phi_{2s} + \phi_{2p_z})$ and $\psi_b = 1/\sqrt{2}(-\phi_{2s} - \phi_{2p_z})$ for two sp hybrid orbitals directed 180° apart. Show that these hybrid orbitals are orthogonal.

P13.7 Use the geometrical construction shown in Example Problem 13.8 to derive the π electron MO levels for cyclobutadiene. What is the total π energy of the molecule? How many unpaired electrons will the molecule have?

P13.8 Show that two of the set of four equivalent orbitals appropriate for sp^3 hybridization,

$$\psi_a = \frac{1}{2}(-\phi_{2s} + \phi_{2p_x} + \phi_{2p_y} + \phi_{2p_z}) \quad \text{and}$$

$$\psi_b = \frac{1}{2}(-\phi_{2s} - \phi_{2p_x} - \phi_{2p_y} + \phi_{2p_z})$$

are orthogonal.

P13.9   Show that the water hybrid bonding orbitals given by $\psi_a = 0.55\,\phi_{2p_z} + 0.71\,\phi_{2p_x} - 0.45\,\phi_{2s}$ and $\psi_b = 0.55\,\phi_{2p_z} - 0.71\,\phi_{2p_x} - 0.45\,\phi_{2s}$ are orthogonal.

P13.10   Predict which of the bent molecules, BH_2 or NH_2, should have the larger bond angle on the basis of the Walsh correlation diagram in Figure 13.11. Explain your answer.

P13.11   Determine the AO coefficients for the lowest energy Hückel π MO for butadiene.

P13.12   Derive two additional mutually orthogonal hybrid orbitals for the lone pairs on oxygen in H_2O, each of which is orthogonal to ψ_a and ψ_b by following the steps below.

a. Starting with the following formulas for the lone pair orbitals

$$\psi_c = d_1\phi_{2p_z} + d_2\phi_{2p_y} + d_3\phi_{2s} + d_4\phi_{2p_x}$$
$$\psi_d = d_5\phi_{2p_z} + d_6\phi_{2p_y} + d_7\phi_{2s} + d_8\phi_{2p_x}$$

use symmetry conditions to determine d_2 and d_4, and to determine the ratio of d_3 to d_7 and of d_4 to d_8.

b. Use the condition that the sum of the square of the coefficients over all the hybrid orbitals and lone pair orbital is one to determine the unknown coefficients.

P13.13   Use the geometrical construction shown in Example Problem 13.8 to derive the π electron MO levels for the cyclopentadienyl radical. What is the total π energy of the molecule? How many unpaired electrons will the molecule have?

P13.14   Use the Boltzmann distribution to answer parts (a) and (b):

a. Calculate the ratio of electrons at the bottom of the conduction band to those at the top of the valence band for pure Si at 300. K. The Si band gap is 1.1 eV.

b. Calculate the ratio of electrons at the bottom of the conduction band to those at the top of the dopant band for P-doped Si at 300. K. The top of the dopant band lies 0.040 eV below the bottom of the Si conduction band.

Assume for these calculations that the ratio of the degeneracies is unity. What can you conclude about the room temperature conductivity of these two materials on the basis of your calculations?

P13.15   Use the VSEPR method to predict the structures of the following:

a. PCl_5 b. SO_2 c. XeF_2 d. XeF_5

P13.16   The allyl cation $CH_2{=}CH{-}CH_2^+$ has a delocalized π network that can be described by the Hückel method. Derive the MO energy levels of this species and place the electrons in the levels appropriate for the ground state. Using the butadiene MOs as an example, sketch what you would expect the MOs to look like. Classify the MOs as bonding, antibonding, or nonbonding.

P13.17   Write down and solve the secular determinant for the π system of ethylene in the Hückel model. Determine the coefficients for the $2p_z$ AOs on each of the carbons and make a sketch of the MOs. Characterize the MOs as bonding and antibonding.

P13.18   Use the geometrical construction shown in Example Problem 13.8 to derive the energy levels of the cycloheptatrienyl cation. What is the total π energy of the molecule? How many unpaired electrons will the molecule have? Would you expect this species, the neutral species, or the anion to be aromatic? Justify your answer.

P13.19   Use the framework described in Section 13.3 to derive the normalized hybrid lone pair orbital on the central oxygen in O_3 that is derived from 2s and 2p atomic orbitals. The bond angle in ozone is 116.8°.

P13.20   Using your results from Problem P13.12,

a. Calculate the s and p character of the water lone pair hybrid orbitals.

b. Show that the lone pair orbitals are orthogonal to each other and to the hybrid bonding orbitals.

P13.21   Use the VSEPR method to predict the structures of the following:

a. PF_3 b. CO_2 c. BrF_5 d. SO_3^{2-}

P13.22   Predict whether the ground state or the first excited state of CH_2 should have the larger bond angle on the basis of the Walsh correlation diagram shown in Figure 13.11. Explain your answer.

Computational Problems

More detailed instructions on carrying out these calculations using Spartan Physical Chemistry are found on the book website at *www.chemplace.com*.

C13.1   Calculate the bond angles in NH_3 and in NF_3 using the density functional method with the B3LYP functional and the 6-31G* basis set. Compare your result with literature values. Do your results agree with the predictions of the VSEPR model and Bent's rule?

C13.2   Calculate the bond angles in H_2O and in H_2S using the density functional method with the B3LYP functional and the 6-31G* basis set. Compare your result with literature values.

Do your results agree with the predictions of the VSEPR model and Bent's rule?

C13.3   Calculate the bond angle in ClO_2 using the density functional method with the B3LYP functional and the 6-31G* basis set. Compare your result with literature values. Does your result agree with the predictions of the VSEPR model?

C13.4 SiF_4 has four ligands and one lone pair on the central S atom. Which of the following structures do you expect to be the equilibrium form based on a calculation using the density functional method with the B3LYP functional and the 6-31G* basis set? In (a) the structure is a trigonal bipyramid, (b) is a see-saw structure, and (c) is a square planar structure.

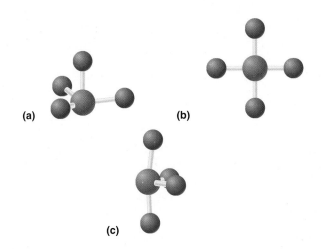

(a)
(b)
(c)

C13.5   Calculate the bond angles in singlet BeH_2, doublet NH_2, and singlet BH_2 using the Hartree–Fock method and the 6-31G* basis set. Explain your results using the Walsh diagram of Figure 13.11.

C13.6   Calculate the bond angle in singlet LiH_2^+ using the Hartree–Fock method and the 6-31G* basis set. Can you explain your results using the Walsh diagram of Figure 13.11? (Hint: Determine the calculated bond lengths in the LiH_2^+ molecule.)

C13.7   Calculate the bond angle in singlet and triplet CH_2 and doublet CH_2^+ using the Hartree–Fock method and the 6-31G* basis set. Can you explain your results using the Walsh diagram of Figure 13.11?

C13.8   Calculate the bond angle in singlet NH_2^+, doublet NH_2, and singlet NH_2^- using the Hartree–Fock method and the 6-31G* basis set. Can you explain your results using the Walsh diagram of Figure 13.11?

C13.9   How essential is coplanarity to conjugation? Answer this question by calculating the total energy of butadiene using the Hartree–Fock method and the 6-31G* basis set for dihedral angles of 0, 45, and 90 degrees.

C13.10   Calculate the equilibrium structures for singlet and triplet formaldehyde using the density functional method with the B3LYP functional and the 6-31G* basis set. Choose (a) planar and (b) pyramidal starting geometries. Calculate vibrational frequencies for both starting geometries. Are any of the frequencies imaginary? Explain your results.

C13.11   Calculate the equilibrium structure for Cl_2O using the density functional method with the B3LYP functional and the 6-31G* basis set. Obtain an infrared spectrum and activate the normal modes. What are the frequencies corresponding to the symmetric stretch, the asymmetric stretch, and the bending modes?

C13.12   Calculate the equilibrium structures for PF_3 using the density functional method with the B3LYP functional and the 6-31G* basis set. Obtain an infrared spectrum and activate the normal modes. What are the frequencies corresponding to the symmetric stretch, the symmetric deformation, the degenerate stretch, and the degenerate deformation modes?

C13.13   Calculate the equilibrium structures for C_2H_2 using the density functional method with the B3LYP functional and the 6-31G* basis set. Obtain an infrared spectrum and activate the normal modes. What are the frequencies corresponding to the symmetric C—H stretch, the antisymmetric C—H stretch, the C—C stretch, and the two bending modes?

C13.14   Calculate the structure of $N\equiv C-Cl$ using the density functional method with the B3LYP functional and the 6-31G* basis set. Which is more electronegative, the Cl or the cyanide group? What result of the calculation did you use to answer this question?

14

Electronic Spectroscopy

Absorption of visible or ultraviolet light can lead to transitions between the ground state and excited electronic states of atoms and molecules. Vibrational transitions that occur together with electronic transitions are governed by the Franck-Condon factors rather than the $\Delta n = \pm 1$ dipole selection rule. The excited state can relax to the ground state through a combination of fluorescence, internal conversion, intersystem crossing, and phosphorescence. Fluorescence is very useful in analytical chemistry and can detect as little as 2×10^{-13} mol/L of a strongly fluorescing species. Ultraviolet photoemission can be used to obtain information about the orbital energies of molecules. Linear and circular dichroism spectroscopy can be used to determine the secondary and tertiary structure of biomolecules in solution.

14.1 THE ENERGY OF ELECTRONIC TRANSITIONS

In Chapter 8 spectroscopy and the basic concepts relevant to transitions between energy levels of a molecule were introduced. Recall that the energy spacing between rotational levels is much less than the spacing between vibrational levels. Extending this comparison to electronic states, it is found that $\Delta E_{electronic} \gg \Delta E_{vibrational} \gg \Delta E_{rotational}$. Whereas rotational and vibrational transitions are induced by microwave and infrared radiation, electronic transitions are induced by visible and ultraviolet (UV) radiation. Just as an absorption spectrum in the infrared exhibits both rotational and vibrational transitions, an absorption spectrum in the visible and UV range exhibits a number of electronic transitions, and a specific electronic transition will contain vibrational and rotational fine structure.

Electronic excitations are responsible for giving color to the objects we observe, because the human eye is sensitive to light only in the limited range of wavelengths in which some electronic transitions occur. Either the reflected or the transmitted light is observed, depending on whether the object is opaque or transparent. Transmitted and reflected light complement the absorbed light. For example, a leaf is green because chlorophyll absorbs in the blue (450-nm) and red (650-nm) regions of the visible light spectrum. Electronic excitations can be detected (at a limited resolution) without the aid of a spectrometer because the human eye is a very sensitive detector of radiation. At a wavelength of 500 nm, the human eye can detect one part in 10^6 of the intensity of sunlight on a bright day. This corresponds to as few as 500 photons per second incident on an area of 1 mm^2.

Because the electronic spectroscopy of a molecule is directly linked to its energy levels, which are in turn determined by its structure and chemical composition, UV-visible spectroscopy provides a very useful qualitative tool for identifying molecules. In addition, for a given molecule, electronic spectroscopy can be used to determine energy levels in molecules. However, the UV and visible photons that initiate an electronic excitation perturb a molecule far more than rotational or vibrational excitation. For example, the bond length in electronically excited states of O_2 is as much as 30% longer than that in the ground state. Whereas in its ground state, formaldehyde is a planar molecule, it is pyramidal in its lowest two excited states. As might be expected from such changes in geometry, the chemical reactivity of excited-state species can be quite different from the reactivity of the ground-state molecule.

14.2 MOLECULAR TERM SYMBOLS

We begin our discussion of electronic excitations by introducing **molecular term** symbols, which describe the electronic states of molecules in the same way that atomic term symbols describe atomic electronic states. The following discussion is restricted to diatomic molecules. A quantitative discussion of electronic spectroscopy requires a knowledge of molecular term symbols. However, electronic spectroscopy can be discussed at a qualitative level without discussing molecular term symbols. To do so, move directly to Section 14.4.

The component of **L** and **S** along the molecular axis (M_L and M_S), which is chosen to be the z axis, and S are the only good quantum numbers by which to specify individual states in diatomic molecules. Therefore, term symbols for molecules are defined using these quantities. As for atoms, only unfilled subshells need to be considered to obtain molecular term symbols. As discussed in Chapter 13, in the first and second row diatomic molecules, the molecular orbitals (MOs) are either of the σ or π type. Just as for atoms, the quantum numbers m_{li} and m_{si} can be added to generate M_L and M_S for the molecule because they are scalars rather than vectors. The addition process is described by the equations

$$M_L = \sum_{i=1}^{n} m_{li} \quad \text{and} \quad M_S = \sum_{i=1}^{n} m_{si} \tag{14.1}$$

in which m_{li} and m_{si} are the z components of orbital and spin angular momentum for the ith electron in its molecular orbital. The molecular orbitals have either σ symmetry, in which the orbital is unchanged by rotation around the molecular axis, or π symmetry, in which the MO has a nodal plane passing through the molecular axis. For a σ orbital, $m_l = 0$, and for a π orbital, $m_l = \pm 1$. Note that the $m_l = 0$ value does not occur for a π MO because this value corresponds to the $2p_z$ AO, which forms a σ MO. M_S is calculated from the individual spin angular momentum vector components $m_{si} = \pm 1/2$ in the same way for molecules as for atoms (see Chapter 10). The allowed values of the quantum numbers S and L can be calculated from $-L \leq M_L \leq L$ and $-S \leq M_S \leq S$ to generate a molecular term symbol of the form $^{2S+1}\Lambda$, where $\Lambda = |M_L|$. For molecules, the following symbols are used for different Λ values to avoid confusion with atomic terms:

$$
\begin{array}{c|cccc}
\Lambda & 0 & 1 & 2 & 3 \\
\text{Symbol} & \Sigma & \Pi & \Delta & \Phi
\end{array}
\qquad \textbf{(14.2)}
$$

A g or u right subscript is added to the molecular term symbol for homonuclear diatomics as illustrated in Example Problem 14.1. Because heteronuclear diatomics do not possess an inversion center, they do not have g or u symmetry. This formalism will become clearer with a few examples.

EXAMPLE PROBLEM 14.1

What is the molecular term symbol for the H_2 molecule in its ground state? In its first two excited states?

Solution

In the ground state, the H_2 molecule is described by the $(1\sigma_g)^2$ configuration. For both electrons, $m_l = 0$. Therefore, $\Lambda = 0$, and we are dealing with a Σ term. Because of the Pauli principle, one electron has $m_s = +1/2$ and the other has $m_s = -1/2$. Therefore, $M_S = 0$ and it follows that $S = 0$. It remains to be determined whether the MO has g or u symmetry. Each term in the antisymmetrized MO is of the form $\sigma_g \times \sigma_g$. Recall that the products of two even or odd functions is even, and the product of an odd and an even function is odd. Therefore, the product of two g (or two u) functions is a g function, and the ground state of the H_2 molecule is $^1\Sigma_g$.

In the first excited state, the configuration is $(1\sigma_g)^1(1\sigma_u^*)^1$, and because the electrons are in separate MOs, this configuration leads to both **singlet states** and **triplet states**. Again, because $m_l = 0$ for both electrons, we are dealing with a Σ term. Because the two electrons are in different MOs, $m_s = \pm 1/2$ for each electron, giving m_s values of -1, 0 (twice), and $+1$. This is consistent with $S = 1$ and $S = 0$. Because the product of a u and a g function is a u function, both singlet and triplet states are u functions. Therefore, the first two excited states are described by the terms $^3\Sigma_u$ and $^1\Sigma_u$. Using Hund's first rule, we conclude that the triplet state is lower in energy than the singlet state.

In a more complete description, an additional subscript $+$ or $-$ is added to Σ terms only, depending on whether the antisymmetrized molecular wave function changes sign $(-)$ or remains unchanged $(+)$ in a reflection through any plane containing the molecular axis. The assignment of $+$ or $-$ to the terms is an advanced topic that is discussed in Supplemental Section 14.14. For our purposes, the following guidelines are sufficient for considering the ground state of second row homonuclear diatomic molecules:

- If all MOs are filled, $+$ applies.

- If all partially filled MOs have σ symmetry, $+$ applies.

- For partially filled MOs of π symmetry (for example, B_2 and O_2), if Σ terms arise, the triplet state is associated with $-$, and the singlet state is associated with $+$.

These guidelines do not apply to excited states. We conclude that the term corresponding to the π^2 ground-state configuration of O_2 is designated by $^3\Sigma_g^-$. The other terms that arise from the ground-state configuration are discussed in Example Problem 14.2.

EXAMPLE PROBLEM 14.2

Determine the possible molecular terms for O_2, which has the following configuration:

$$(1\sigma_g)^2(1\sigma_u^*)^2(2\sigma_g)^2(2\sigma_u^*)^2(3\sigma_g)^2(1\pi_u)^2(1\pi_u)^2(1\pi_g^*)^1(1\pi_g^*)^1$$

Solution

Only the last two electrons contribute to nonzero net values of M_L and M_S, because the other subshells are filled. The various possibilities for combining the orbital and spin angular momenta of these two electrons in a way consistent with the Pauli principle are given in the following table. The Λ values are determined as discussed for atomic terms in Chapter 10. Because $M_L \leq L$, the first two entries in the table belong to a Δ term. Because $M_S = 0$ for both entries, it is a $^1\Delta$ term. Of the remaining four entries, two have $|M_S| = 1$, corresponding to a triplet term. One of the two other entries with $M_S = 0$ must also belong to this term. Because $M_L = 0$ for all four entries, it is a $^3\Sigma$ term. The remaining entry corresponds to a $^1\Sigma$ term.

m_{l1}	m_{l2}	$M_L = m_{L1} + m_{L2}$	m_{s1}	m_{s2}	$M_S = m_{s1} + m_{s2}$	Term
1	1	2	+1/2	−1/2	0	$\left.\begin{array}{c} \\ \end{array}\right\}{}^1\Delta$
−1	−1	−2	+1/2	−1/2	0	
1	−1	0	+1/2	+1/2	1	$\left.\begin{array}{c} \\ \end{array}\right\}{}^3\Sigma$
1	−1	0	−1/2	−1/2	−1	
1	−1	0	+1/2	−1/2	0	$\left.\begin{array}{c} \\ \end{array}\right\}{}^1\Sigma, {}^3\Sigma$
1	−1	0	−1/2	+1/2	0	

The next task is the assignment of the g or u label to these molecular terms. Because both of the electrons are in an MO of g symmetry, the overall symmetry of the term will be g in all cases.

The $+$ and $-$ symbols are assigned in Supplemental Section 14.14. We show there that the singlet term is $^1\Sigma_g^+$ and the triplet term is $^3\Sigma_g^-$. By Hund's first rule, the $^3\Sigma_g^-$ term is lowest in energy and is the ground state. Experimentally, the $^1\Delta_g$ and $^1\Sigma_g^+$ terms are found to lie 0.98 and 1.62 eV higher in energy, respectively, than the ground state.

In terms of arrows indicating the spin orientations, the allowed combinations of m_l and m_s in the table can be represented in a shorthand notation by

$$(1\sigma_g)^2(1\sigma_u^*)^2(2\sigma_g)^2(2\sigma_u^*)^2(3\sigma_g)^2(1\pi_u)^2(1\pi_u)^2(1\pi_g^*)^1(1\pi_g^*)^1$$

$$(\uparrow\downarrow)\ (\uparrow\downarrow)\ (\uparrow\downarrow)\ (\uparrow\downarrow)\ (\uparrow\downarrow)\ (\uparrow\downarrow)\ (\uparrow\downarrow)\ (\uparrow)\ (\downarrow)\quad {}^1\Sigma_g^+, {}^1\Delta_g$$

$$\left\{ \begin{array}{l} (1\sigma_g)^2(1\sigma_u^*)^2(2\sigma_g)^2(2\sigma_u^*)^2(3\sigma_g)^2(1\pi_u)^2(1\pi_u)^2(1\pi_g^*)^1(1\pi_g^*)^1 \\ (\uparrow\downarrow)\ (\uparrow\downarrow)\ (\uparrow\downarrow)\ (\uparrow\downarrow)\ (\uparrow\downarrow)\ (\uparrow\downarrow)\ (\uparrow\downarrow)\ (\uparrow)\ (\uparrow) \\ \left[\begin{array}{l} (\uparrow\downarrow)\ (\uparrow\downarrow)\ (\uparrow\downarrow)\ (\uparrow\downarrow)\ (\uparrow\downarrow)\ (\uparrow\downarrow)\ (\uparrow\downarrow)\ (\uparrow)\ (\downarrow) \\ \qquad\qquad\qquad\qquad + \\ (\uparrow\downarrow)\ (\uparrow\downarrow)\ (\uparrow\downarrow)\ (\uparrow\downarrow)\ (\uparrow\downarrow)\ (\uparrow\downarrow)\ (\uparrow\downarrow)\ (\downarrow)\ (\uparrow) \end{array}\right] \\ (\uparrow\downarrow)\ (\uparrow\downarrow)\ (\uparrow\downarrow)\ (\uparrow\downarrow)\ (\uparrow\downarrow)\ (\uparrow\downarrow)\ (\uparrow\downarrow)\ (\downarrow)\ (\downarrow) \end{array}\right\} {}^3\Sigma_g^-$$

Note that this notation with arrows pointing up and down to indicate α and β spins is inadequate because it is not possible to represent the different values of m_{l1} and m_{l2}.

On the basis of this discussion, the ground-state terms for the first row homonuclear diatomic molecules are listed in Table 14.1. The procedure for heteronuclear diatomics is similar, but differs in that the numbering of the MOs is different, and the g and u symmetries do not apply.

TABLE 14.1	TERMS FOR GROUND-STATE SECOND ROW DIATOMICS	
Molecule	**Electron Configuration**	**Ground-State Term**
H_2^+	$(1\sigma_g)^1$	$^2\Sigma_g^+$
H_2	$(1\sigma_g)^2$	$^1\Sigma_g^+$
He_2^+	$(1\sigma_g)^2(1\sigma_u^*)^1$	$^2\Sigma_u^+$
Li_2	$(1\sigma_g)^2(1\sigma_u^*)^2(2\sigma_g)^2$	$^1\Sigma_g^+$
B_2	$(1\sigma_g)^2(1\sigma_u^*)^2(2\sigma_g)^2(2\sigma_u^*)^2(1\pi_u)^1(1\pi_u)^1$	$^3\Sigma_g^-$
C_2	$(1\sigma_g)^2(1\sigma_u^*)^2(2\sigma_g)^2(2\sigma_u^*)^2(1\pi_u)^2(1\pi_u)^2$	$^1\Sigma_g^+$
N_2^+	$(1\sigma_g)^2\,(1\sigma_u^*)^2\,(2\sigma_g)^2\,(2\sigma_u^*)^2\,(1\pi_u)^2\,(1\pi_u)^2\,(3\sigma_g)^1$	$^2\Sigma_g^+$
N_2	$(1\sigma_g)^2(1\sigma_u^*)^2(2\sigma_g)^2(2\sigma_u^*)^2(1\pi_u)^2(1\pi_u)^2(3\sigma_g)^2$	$^1\Sigma_g^+$
O_2^+	$(1\sigma_g)^2(1\sigma_u^*)^2(2\sigma_g)^2(2\sigma_u^*)^2(3\sigma_g)^2(1\pi_u)^2(1\pi_u)^2(1\pi_g^*)^1$	$^2\Pi_g$
O_2	$(1\sigma_g)^2(1\sigma_u^*)^2(2\sigma_g)^2(2\sigma_u^*)^2(3\sigma_g)^2(1\pi_u)^2(1\pi_u)^2(1\pi_g^*)^1(1\pi_g^*)^1$	$^3\Sigma_g^-$
F_2	$(1\sigma_g)^2(1\sigma_u^*)^2(2\sigma_g)^2(2\sigma_u^*)^2(3\sigma_g)^2(1\pi_u)^2(1\pi_u)^2(1\pi_g^*)^2(1\pi_g^*)^2$	$^1\Sigma_g^+$

14.3 TRANSITIONS BETWEEN ELECTRONIC STATES OF DIATOMIC MOLECULES

Diatomic molecules have the most easily interpretable electronic spectra, because the spacing between the various rotational-vibrational-electronic states is sufficiently large to allow individual states to be resolved. Potential energy curves for the five lowest lying bound states of O_2 are shown in Figure 14.1. Vibrational energy levels are indicated schematically in the figure, but rotational levels are not shown. Note that the lowest four states all dissociate to give two ground-state ^3P oxygen atoms, whereas the highest energy state shown dissociates to give one ^3P and one ^1D oxygen atom. The letter X before $^3\Sigma_g^-$ indicates that the term symbol refers to the ground state. Electronic states of higher energy are designated by $A, B, C, \ldots$, if they have the same multiplicity, $2S + 1$, as the ground state, and $a, b, c, \ldots$, if they have a different multiplicity.

The bond length of excited-state molecules is generally greater and the binding energy generally less than that for the ground state. This is the case because the excited states generally have a greater antibonding character than the ground states, and the decrease in bond order leads to a smaller bond energy, a larger bond length, and a lower vibrational frequency for the excited-state species. You will address the fact that the bond lengths for the first two excited states are similar to that for the ground state in the end-of-chapter questions.

Although a symbol such as $^3\Sigma_g^-$ completely describes the quantum state for a ground-state O_2 molecule, it is also useful to associate a **molecular configuration** with the state. Starting with a configuration makes it easier to visualize a transition in terms of promoting an electron from an occupied to an unoccupied level. To what configurations do the excited states shown in Figure 14.1 correspond? The $X^3\Sigma_g^-$, $a^1\Delta_g$, and $b^1\Sigma_g^+$ states all belong to the ground-state configuration $(1\sigma_g)^2(1\sigma_u^*)^2(2\sigma_g)^2(2\sigma_u^*)^2(3\sigma_g)^2(1\pi_u)^2(1\pi_u)^2(1\pi_g^*)^1(1\pi_g^*)^1$, but are associated with different M_L and M_S values as was shown in Example Problem 14.2. The $A^3\Sigma_u^+$ and $B^3\Sigma_u^-$ states are associated with the $(1\sigma_g)^2(1\sigma_u^*)^2(2\sigma_g)^2(2\sigma_u^*)^2(3\sigma_g)^2(1\pi_u)^1(1\pi_u)^2(1\pi_g^*)^1(1\pi_g^*)^2$ configuration. Keep in mind that although a molecular term can be associated with a configuration, in general, several molecular terms are generated from the same configuration.

Spectroscopy involves transitions between molecular states. What selection rules govern transitions between different electronic states? The selection rules for molecular electronic transitions are most well defined for lower molecular weight diatomic molecules in which spin-orbit coupling is not important. This is the case if the atomic number of the atoms, Z, is less than 40. For these molecules, the selection rules are

$$\Delta\Lambda = 0, \pm 1, \text{ and } \Delta S = 0 \qquad (14.3)$$

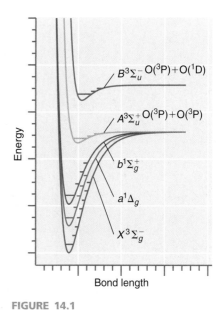

FIGURE 14.1

Potential energy curves for the ground state of O_2 and for the four lowest excited states. The spectroscopic designation of the states is explained in the text. Horizontal lines indicate vibrational levels for each state.

Recall that Λ is the component of the total orbital angular momentum $\mathbf{L}$ along the molecular axis. The value $\Delta\Lambda = 0$ applies for a $\Sigma \leftrightarrow \Sigma$ transition, and $\Delta\Lambda = \pm1$ applies to $\Sigma \leftrightarrow \Pi$ transitions. Further selection rules are associated with the $+/-$ and g/u parities. For homonuclear diatomics, $u \leftrightarrow g$ transitions are allowed, but $u \leftrightarrow u$ and $g \leftrightarrow g$ transitions are forbidden. The transitions $\Sigma^- \leftrightarrow \Sigma^-$ and $\Sigma^+ \leftrightarrow \Sigma^+$ are allowed, but $\Sigma^+ \leftrightarrow \Sigma^-$ transitions are forbidden. All of these selection rules can be derived by calculating the transition dipole element defined in Section 8.5.

With these selection rules in mind, we consider the possible transitions among the states shown in Figure 14.1 for O_2. The $X^3\Sigma_g^- \rightarrow a^1\Delta_g$ and $X^3\Sigma_g^- \rightarrow b^1\Sigma_g^+$ transitions are forbidden because of the $\Delta S = 0$ selection rule and because $g \leftrightarrow g$ transitions are forbidden. The $X^3\Sigma_g^- \rightarrow A^3\Sigma_u^+$ transition is forbidden because $\Sigma^+ \leftrightarrow \Sigma^-$ transitions are forbidden. Therefore, the lowest allowed transition originating from the ground state is $X^3\Sigma_g^- \rightarrow B^3\Sigma_u^-$. Absorption from the ground state into various vibrational levels of the $B^3\Sigma_u^-$ excited state occurs in a band between 175- and 200-nm wavelengths. An interesting consequence of these selection rules is that if transitions from the ground state to the first two excited states were allowed, O_2 would absorb light in the visible part of the spectrum, and the Earth's atmosphere would not be transparent.

If sufficient energy is taken up by the molecule, dissociation can occur through the pathway

$$O_2 + h\nu \rightarrow 2O\cdot \qquad (14.4)$$

The maximum wavelength consistent with this reaction is 242 nm. This reaction is an example of a **photodissociation** reaction. This particular reaction is of great importance in the stratosphere because it is the only significant pathway for forming the atomic oxygen needed for ozone production through the reaction

$$O\cdot + O_2 + M \rightarrow O_3 + M^* \qquad (14.5)$$

where M designates a gas-phase spectator species that takes up energy released in the O_3 formation reaction. Because O_3 absorbs UV radiation strongly over the 220- to 350-nm range, it plays a vital role in filtering out UV radiation from the sunlight incident on the planet.

14.4 THE VIBRATIONAL FINE STRUCTURE OF ELECTRONIC TRANSITIONS IN DIATOMIC MOLECULES

Each of the molecular bound states shown in Figure 14.1 has well-defined vibrational and rotational energy levels. As discussed in Chapter 8, changes in the vibrational state can occur together with a change in the rotational state. Similarly, the vibrational and rotational quantum numbers can change during electronic excitation. We next discuss the vibrational excitation and de-excitation associated with electronic transitions, but do not discuss the associated rotational transitions. We will see that the $\Delta n = \pm1$ selection rule for vibrational transitions within a given electronic state does not hold for transitions between two electronic states.

What determines Δn in a vibrational transition between electronic states? This question can be answered by looking more closely at the **Born–Oppenheimer approximation**, which was introduced in Chapter 12. This approximation can be expressed mathematically by stating that the total wave function for the molecule can be factored into two parts. The part that depends only on the position of the nuclei, $(\mathbf{R}_1, \ldots, \mathbf{R}_m)$ is associated with vibration of the molecule. The second part depends only on the position of the electrons, $(\mathbf{r}_1, \ldots, \mathbf{r}_n)$ at a fixed position of all the nuclei. This part describes electron "motion" in the molecule:

$$\psi(\mathbf{r}_1, \ldots, \mathbf{r}_n, \mathbf{R}_1, \ldots, \mathbf{R}_m) = \psi^{electronic}(\mathbf{r}_1, \ldots, \mathbf{r}_n, \mathbf{R}_1^{fixed}, \ldots, \mathbf{R}_m^{fixed})$$
$$\times \phi^{vibrational}(\mathbf{R}_1, \ldots, \mathbf{R}_m) \qquad (14.6)$$

As discussed in Section 8.5, the spectral line corresponding to an electronic transition (initial → final) has a measurable intensity only if the value of the transition dipole moment is different from zero:

$$\mu^{fi} = \int \psi_f^*(\mathbf{r}_1, \ldots, \mathbf{r}_n, \mathbf{R}_1, \ldots, \mathbf{R}_m) \hat{\mu} \psi_i(\mathbf{r}_1, \ldots, \mathbf{r}_n, \mathbf{R}_1, \ldots, \mathbf{R}_m) d\tau \neq 0 \qquad (14.7)$$

The superscripts and subscripts f and i refer to the final and initial states in the transition. In Equation (14.7), the dipole moment operator $\hat{\mu}$ is given by

$$\hat{\mu} = -e \sum_{j=1}^{n} \mathbf{r}_j \qquad (14.8)$$

where the summation is over the positions of the electrons.

Because the total wave function can be written as a product of electronic and vibrational parts, Equation (14.7) becomes

$$\mu^{fi} = \int \left(\phi_f^{vibrational}(\mathbf{R}_1, \ldots, \mathbf{R}_m) \right)^* \phi_i^{vibrational}(\mathbf{R}_1, \ldots, \mathbf{R}_m) \, d\tau$$

$$\times \int \left(\psi_f^{electronic}(\mathbf{r}_1, \ldots, \mathbf{r}_n, \mathbf{R}_1^{fixed}, \ldots, \mathbf{R}_m^{fixed}) \right)^* \hat{\mu} \, \psi_i^{electronic}(\mathbf{r}_1, \ldots, \mathbf{r}_n, \mathbf{R}_1^{fixed}, \ldots, \mathbf{R}_m^{fixed}) \, d\tau$$

$$= S \int \psi_f^*(\mathbf{r}_1, \ldots, \mathbf{r}_n, \mathbf{R}_1^{fixed}, \ldots, \mathbf{R}_m^{fixed}) \hat{\mu} \, \psi_i(\mathbf{r}_1, \ldots, \mathbf{r}_n, \mathbf{R}_1^{fixed}, \ldots, \mathbf{R}_m^{fixed}) \, d\tau \qquad (14.9)$$

Note that the first of the two product integrals in Equation (14.9) represents the overlap between the vibrational wave functions in the ground and excited states. The magnitude of the square of this integral for a given transition is known as the **Franck-Condon factor** and is a measure of the expected intensity of an electronic transition. The Franck-Condon factor replaces the selection rule $\Delta n = \pm 1$ obtained for pure vibrational transitions derived in Section 8.4 as a criterion for the intensity of a transition:

$$S^2 = \left| \int (\phi_f^{vibrational})^* \phi_i^{vibrational} \, d\tau \right|^2 \qquad (14.10)$$

The **Franck-Condon principle** states that transitions between electronic states correspond to vertical lines on an energy versus internuclear distance diagram. The basis of this principle is that electronic transitions occur on a timescale that is very short compared to the vibrational period of a molecule. Therefore, the atomic positions are fixed during the transition. As Equation (14.10) shows, the probability of a vibrational-electronic transition is governed by the overlap between the final and initial vibrational wave functions at fixed values of the internuclear distances. Is it necessary to consider all vibrational levels in the ground state as an initial state for an electronic transition? As discussed in Chapter 8, nearly all of the molecules in the ground state have the vibrational quantum number $n = 0$, for which the maximum amplitude of the wave function is at the equilibrium bond length. As shown in Figure 14.2, vertical transitions predominantly occur from this ground vibrational state to several vibrational states in the upper electronic state.

How does the Franck-Condon principle determine the n values in the excited state that give the most intense spectral lines? The most intense spectral transitions are to vibrational states in the upper electronic state that have the largest overlap with the ground vibrational state in the lower electronic state. As Figure 7.4 shows, the vibrational wave functions have their largest amplitude near the R value at which the energy level meets the potential curve, because this corresponds to the classical turning point. For the example shown in Figure 14.2, the overlap $\left| \int (\phi_f^{vibrational})^* \phi_i^{vibrational} \, d\tau \right|$ is greatest between the $n = 0$ vibrational state of the ground electronic state and the $n = 4$ vibrational state of the excited electronic state. Although this transition has the maximum overlap and generates the most intense spectral line, other states close in energy to the most probable state will also give rise to spectral lines. Their intensity is lower because the vibrational wave functions of the ground and excited states have a smaller overlap.

The fact that a number of vibrational transitions are observed in an electronic transition is very useful in obtaining detailed information about both the ground electronic state potential energy surface and that of the electronic state to which the transition occurs. For example, vibrational transitions are observed in the electronic spectra of O_2 and N_2, although neither of these molecules absorbs energy in the infrared. Because multiple vibrational peaks are often observed in electronic spectra, the bond strength of the molecule in the excited states can be determined by fitting the observed frequencies of the transitions to a model potential such as the Morse potential discussed in

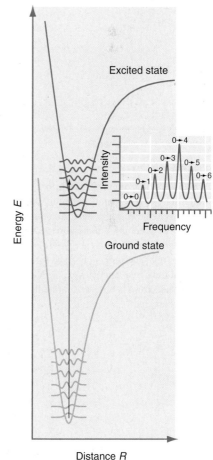

FIGURE 14.2

The relation between energy and bond length is shown for two electronic states. Only the lowest vibrational energy levels and the corresponding wave functions are shown. The vertical line shows the most probable transition predicted by the Franck-Condon principle. The inset shows the relative intensities of different vibrational lines in an absorption spectrum for the potential energy curves shown.

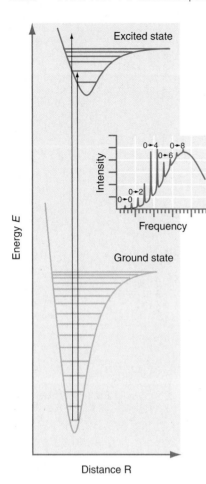

FIGURE 14.3

For absorption from the ground vibrational state of the ground electronic state to the excited electronic state, a continuous energy spectrum will be observed for sufficiently high photon energy. A discrete energy spectrum is observed for an incident light frequency $\nu < E/h$. A continuous spectrum is observed for higher frequencies.

Section 8.3. Because the excited state can also correspond to a photodissociation product, electronic spectroscopy can be used to determine the vibrational force constant and bond energy of highly reactive species such as the CN radical.

For the example shown in Figure 14.2, the molecule will exhibit a discrete energy spectrum in the visible or UV region of the spectrum. However, for some conditions the electronic absorption spectrum for a diatomic molecule is continuous. A continuous spectrum is observed if the photon energy is sufficiently high that excitation occurs to an unbound region of an excited state. This is illustrated in Figure 14.3. In this case, a discrete energy spectrum is observed for low photon energy and a **continuous energy spectrum** is observed for incident light frequencies $\nu > E/h$, where E corresponds to the energy of the transition to the highest bound state in the excited state potential. A purely continuous energy spectrum at all energies is observed if the excited state is a nonbinding state, such as that corresponding to the first excited state for H_2^+.

The preceding discussion briefly summarizes the most important aspects of the electronic spectroscopy of diatomic molecules. In general, the vibrational energy levels for these molecules are sufficiently far apart that individual transitions can be resolved. We next consider polyatomic molecules, for which this is not usually the case.

14.5 UV-VISIBLE LIGHT ABSORPTION IN POLYATOMIC MOLECULES

Many rotational and vibrational transitions are possible if an electronic transition occurs in polyatomic molecules. Large molecules have large moments of inertia, and as Equation (8.15) shows, this leads to closely spaced rotational energy levels. A large molecule may have ~1000 rotational levels in an interval of 1 cm^{-1}. For this reason, individual spectral lines overlap so that broad bands are often observed in UV-visible absorption spectroscopy. This is schematically indicated in Figure 14.4. An electronic transition in an atom gives a sharp line. An electronic transition in a diatomic molecule has additional structure resulting from vibrational and rotational transitions that can often be resolved into individual peaks. However, the many rotational and vibrational transitions possible in a polyatomic molecule generally overlap, giving rise to a broad, nearly featureless band. This overlap makes it difficult to extract information on the initial and final states involved in an electronic transition in polyatomic molecules. In addition, there are no good angular momentum quantum numbers for triatomic and larger molecules. Therefore, the main selection rule that applies is $\Delta S = 0$, together with selection rules based on the symmetry of the initial and final states.

The number of transitions observed can be reduced dramatically by obtaining spectra at low temperatures. Low-temperature spectra for individual molecules can be

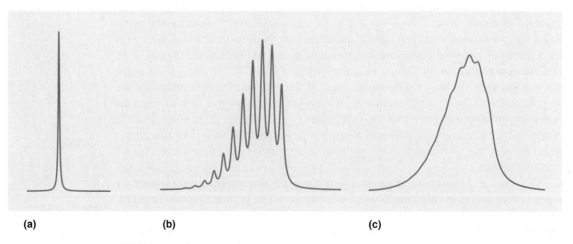

(a)   **(b)**   **(c)**

FIGURE 14.4

The intensity of absorption in a small part of the UV-visible range of the electromagnetic spectrum is shown schematically for **(a)** an atom, **(b)** a diatomic molecule, and **(c)** a polyatomic molecule.

obtained either by embedding the molecule of interest in a solid rare gas matrix at cryogenic temperatures or by expanding gaseous He containing the molecules of interest in dilute concentration through a nozzle into a vacuum. The He gas as well as the molecules of interest are cooled to very low temperatures in the expansion. An example of the elimination of spectral congestion through such a gas expansion is shown in Figure 14.5. The temperature of 9 K is reached by simply expanding the 300 K gas mixture into a vacuum using a molecular beam apparatus.

The concept of chromophores is particularly useful for discussing the electronic spectroscopy of polyatomic molecules. As discussed in Chapter 8, characteristic vibrational frequencies are associated with two neighboring atoms in larger molecules. Similarly, the absorption of UV and visible light in larger molecules can be understood by visualizing the molecule as a system of coupled atoms, such as $-C=C-$ or $-O-H$, that are called **chromophores**. A chromophore is a chemical entity embedded within a molecule that absorbs radiation at nearly the same wavelength in different molecules. Common chromophores in electronic spectroscopy are $C=C$, $C=O$, $C\equiv N$, or $C=S$ groups. Each chromophore has one or several characteristic absorption frequencies in the UV; and the UV absorption spectrum of the molecule, to a first approximation, can be thought of as arising from the sum of the absorption spectra of its chromophores. The wavelengths and absorption strengths associated with specific chromophores are discussed in Section 14.6.

As discussed in Chapter 13, the electronic structure of molecules can be viewed in either a localized or delocalized framework. In viewing the transitions involved in electronic spectroscopy in this way, it is useful to work from a localized bonding model. However, the electrons in radicals and those in delocalized π bonds in conjugated and aromatic molecules need to be described in a delocalized rather than a localized binding model.

What transitions are most likely to be observed in electronic spectroscopy? The fundamental excitation is the promotion of an electron from its HOMO to an excited-state MO. Consider the electronic ground-state configuration of formaldehyde, H_2CO, and those of its lowest lying electronically excited states. In a localized bonding model, the $2s$ and $2p$ electrons combine to form sp^2-hybrid orbitals on the carbon atom as shown in Figure 14.6.

We write the ground-state configuration in the localized orbital notation $(1s_O)^2(1s_C)^2$ $(2s_O)^2(\sigma_{CH})^2(\sigma'_{CH})^2(\sigma_{CO})^2(\pi_{CO})^2(n_O)^2(\pi^*_{CO})^0$ to emphasize that the $1s$ and $2s$ electrons on oxygen and the $1s$ electrons on carbon remain localized on the atoms and are not involved in the bonding. There is also an electron lone pair in a nonbonding MO, designated by n_O, localized on the oxygen atom. Bonding orbitals are primarily localized on adjacent $C-H$ or $C-O$ atoms as indicated in the configuration. The $C-H$ bonds and one of the $C-O$ bonds are σ bonds, and the remaining $C-O$ bond is a π bond.

What changes in the occupation of the MO energy levels can be associated with the electronic transitions observed for formaldehyde? To answer this question, it is useful to generalize the results obtained for MO formation in diatomic molecules to the CO chromophore in formaldehyde. In a simplified picture of this molecule, we expect that the σ_{CO} orbital formed primarily from the $2p_z$ orbital on O and one of the sp^2-hybrid orbitals on C has the lowest energy, and that the antibonding combination of the same orbitals has the highest energy. The π orbital formed from the $2p$ levels on each atom has the next lowest energy, and the antibonding π^* combination has the next highest energy. The lone pair electrons that occupy the $2p$ orbital on O have an energy intermediate between the π and π^* levels. The very approximate molecular orbital energy diagram shown in Figure 14.7 is sufficient to discuss the transitions that formaldehyde undergoes in the UV-visible region.

From the MO energy diagram, we conclude that the nonbonding orbital on O derived from the $2p$ AO is the HOMO, and the empty π^* orbital is the LUMO. The lowest excited state is reached by promoting an electron from the n_O to the π^*_{CO} orbital and is called an **$n \rightarrow \pi^*$ transition**. The resulting state is associated with the configuration $(1s_O)^2(1s_C)^2(2s_O)^2(\sigma_{CH})^2(\sigma'_{CH})^2(\sigma_{CO})^2(\pi_{CO})^2(n_O)^1(\pi^*_{CO})^1$. The next excited state is reached by promoting an electron from the π_{CO} to the π^*_{CO} MO and is called a **$\pi \rightarrow \pi^*$ transition**. The resulting state is associated with the configuration $(1s_O)^2$ $(1s_C)^2(2s_O)^2(\sigma_{CH})^2(\sigma'_{CH})^2(\sigma_{CO})^2(\pi_{CO})^1(n_O)^2(\pi^*_{CO})^1$.

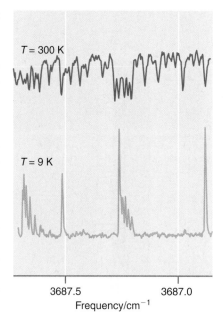

3687.5   3687.0
Frequency/cm^{-1}

FIGURE 14.5

A small portion of the electronic absorption spectrum of methanol is shown at 300 and 9 K using expansion of a dilute mixture of methanol in He through a nozzle into a vacuum. At 300 K, the molecule absorbs almost everywhere in the frequency range. At 9 K, very few rotational and vibrational states are populated, and individual spectral features corresponding to rotational fine structure are observed.

[Reproduced by permission of Elsevier Science from the OH Stretching Fundamental of Methanol. P. Carrick, R. F. Curl, M. Dawes, E. Koester, K. K. Murray, M. Petri, and M. L. Richnow, *Journal of Molecular Structure 223*, 171–184 Fig 4 (1990) by Elsevier Science Ltd.]

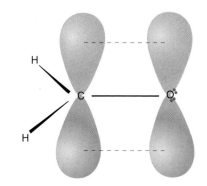

FIGURE 14.6

Valence bond picture of the formaldehyde molecule. The solid lines indicate σ bonds and the dashed lines indicate a π bond. The nonequivalent lone pairs on oxygen are also shown.

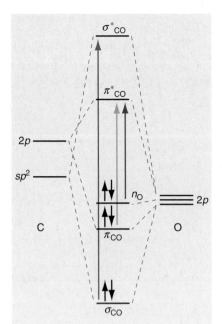

FIGURE 14.7

A simplified MO energy diagram is shown for the C—O bonding interaction in formaldehyde. The most important allowed transitions between these levels are shown. Only one of the sp^2 orbitals on carbon is shown because the other two hybrid orbitals form σ_{CH} bonds.

However, as was the case for atoms, these configurations do not completely describe the quantum states because the alignment of the spins in the unfilled orbitals is not specified by the configuration. Because each of the excited-state configurations just listed has two half-filled MOs, both singlet and triplet states arise from each configuration. The relative energy of these states is indicated in Figure 14.8. Just as for diatomic molecules, for the same configuration, triplet states lie lower in energy than singlet states. The difference in energy between the singlet and triplet states is specific to a molecule, but typically lies between 2 and 10 eV.

The energy difference between the initial and final states determines the frequency of the spectral line. Although large variations can occur among different molecules for a given type of transition, generally the energy increases in the sequence $n \rightarrow \pi^*$, $\pi \rightarrow \pi^*$, and $\sigma \rightarrow \sigma^*$. The $\pi \rightarrow \pi^*$ transitions require multiple bonds, and occur in alkenes, alkynes, and aromatic compounds. The $n \rightarrow \pi^*$ transitions require both a nonbonding electron pair and multiple bonds and occur in molecules containing carbonyls, thiocarbonyls, nitro, azo, and imine groups and in unsaturated halocarbons. The **$\sigma \rightarrow \sigma^*$ transitions** are seen in many molecules, particularly in alkanes, in which none of the other transitions is possible.

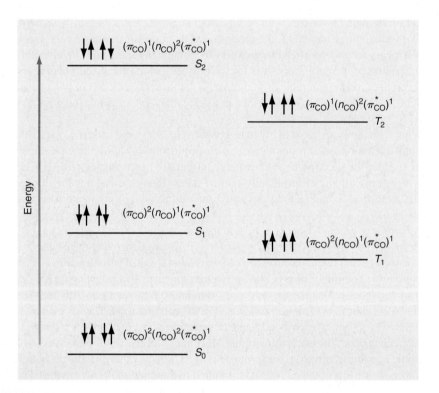

FIGURE 14.8

The ground state of formaldehyde is a singlet and is designated S_0. Successively higher energy singlet and triplet states are designated S_1, S_2, T_1, and T_2. The electron configurations and the alignment of the unpaired spins for the states involved in the most important transitions are also shown. The energy separation between the singlet and triplet states has been exaggerated in this figure.

14.6 TRANSITIONS AMONG THE GROUND AND EXCITED STATES

We next generalize the preceding discussion for formaldehyde to an arbitrary molecule. What transitions can take place among ground and excited states? Consider the energy levels for such a molecule shown schematically in Figure 14.9. The ground state is in general a singlet state and the excited states can be either a singlet or triplet state. We include only one excited singlet and triplet state in addition to the ground state and consider the possible transitions among these states. The restriction is justified because

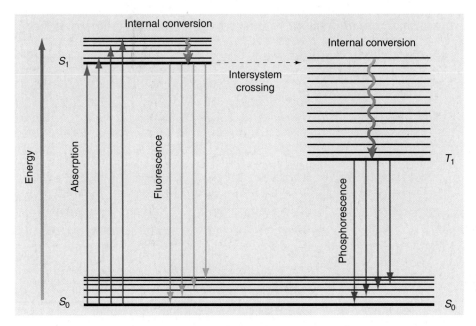

FIGURE 14.9

Possible transitions among the ground and excited electronic states are indicated. The spacing between vibrational levels is exaggerated in this diagram. Rotational levels have been omitted for reasons of clarity.

an initial excitation to higher lying states will rapidly decay to the lowest lying state of the same multiplicity through a process called internal conversion, which is discussed later. The diagram also includes vibrational levels associated with each of the electronic levels. Rotational levels are omitted to simplify the diagram. The fundamental rule governing transitions is that all transitions must conserve energy and angular momentum. For transitions within a molecule, this condition can be satisfied by transferring energy between electronic, vibrational, and rotational states. Alternatively, energy can be conserved by transferring energy between a molecule and its surroundings.

Three types of transitions are indicated in Figure 14.9. **Radiative transitions**, in which a photon is absorbed or emitted, are indicated by solid vertical lines. **Nonradiative transitions**, in which energy is transferred between different degrees of freedom of a molecule or to the surroundings, are indicated by wavy vertical lines. The dashed line indicates nonradiative transitions between singlet and triplet states, which are forbidden by the dipole selection rule. The pathway by which a molecule in an excited state decays to the ground state depends on the rates of a number of competing processes. In the next two sections, these processes are discussed individually.

14.7 SINGLET–SINGLET TRANSITIONS: ABSORPTION AND FLUORESCENCE

As discussed in Section 14.5, an absorption band in an electronic spectrum can be associated with a specific chromophore. Whereas in atomic spectroscopy the selection rule $\Delta S = 0$ is strictly obeyed, in molecular spectroscopy one finds instead that spectral lines for transitions corresponding to $\Delta S = 0$ are much stronger than those for which this condition is not fulfilled. It is useful to quantify what is meant by strong and weak absorption. If I_0 is the incident light intensity at the frequency of interest and I_t is the intensity of transmitted light, the dependence of I_t/I_0 on the concentration c and the path length l is described by **Beer's law**:

$$log\left(\frac{I_t}{I_0}\right) = -\varepsilon l c \qquad (14.11)$$

The **molar extinction coefficient** ε is a measure of the strength of the transition. It is independent of the path length and concentration and is characteristic of the chromophore. The **integral absorption coefficient**, $A = \int \varepsilon(\nu)\, d\nu$, in which the integration over the spectral line includes associated vibrational and rotational transitions, is a measure of the probability that an incident photon will be absorbed in a specific electronic transition.

TABLE 14.2   **CHARACTERISTIC PARAMETERS FOR COMMON CHROMOPHORES**

Chromophore	Transition	λ_{max} (nm)	ε_{max} (dm^3 mol^{-1} cm^{-1})
N=O	$n \rightarrow \pi^*$	660	200
N=N	$n \rightarrow \pi^*$	350	100
C=O	$n \rightarrow \pi^*$	280	20
NO$_2$	$n \rightarrow \pi^*$	270	20
C$_6$H$_6$ (benzene)	$\pi \rightarrow \pi^*$	260	200
C=N	$n \rightarrow \pi^*$	240	150
C=C—C=O	$\pi \rightarrow \pi^*$	220	2×10^5
C=C—C=C	$\pi \rightarrow \pi^*$	220	2×10^5
S=O	$n \rightarrow \pi^*$	210	1.5×10^3
C=C	$\pi \rightarrow \pi^*$	180	1×10^3
C—C	$\sigma \rightarrow \sigma^*$	<170	1×10^3
C—H	$\sigma \rightarrow \sigma^*$	<170	1×10^3

The terms A and ε depend on the frequency, and ε measured at the maximum of the spectral line, ε_{max}, has been tabulated for many chromophores. Some characteristic values for spin-allowed transitions are given in Table 14.2.

In Table 14.2, note the large enhancement of ε_{max} that occurs for conjugated bonds. As a general rule, ε_{max} lies between 10 and 5×10^4 dm^3 mol^{-1} cm^{-1} for spin-allowed transitions ($\Delta S = 0$), and between 1×10^{-4} and 1 dm^3 mol^{-1}cm^{-1} for singlet–triplet transitions ($\Delta S = 1$). Therefore, the attenuation of light passing through the sample resulting from singlet–triplet transitions will be smaller by a factor of ~10^4 to 10^7 than the attenuation from singlet–singlet transitions. This illustrates that in an absorption experiment, transitions for which $\Delta S = 1$ are not totally forbidden if spin-orbit coupling is not negligible, but are typically too weak to be of much importance. However, as discussed in Section 14.8, singlet–triplet transitions are important for phosphorescence.

The excited-state molecule can return to the ground state through radiative or non-radiative transitions involving collisions with other molecules. What determines which of these two pathways will be followed? An isolated excited-state molecule (for instance, in interstellar space) cannot exchange energy with other molecules through collisions and, therefore, nonradiative transitions (other than isoenergetic internal electronic-to-vibrational energy transfer) will not occur. However, excited-state molecules in a crystal, in solution, or in a gas undergo frequent collisions with other molecules in which they lose energy and return to the lowest vibrational state of S_1. This process generally occurs much faster than a radiative transition directly from a vibrationally excited state in S_1 to a vibrational state in S_0. An example of a nonradiative transition induced by collisions is **internal conversion**, which is the decay from a higher vibrational state to the ground vibrational state of the same multiplicity indicated in Figure 14.9. Once in the lowest vibrational state of S_1, either of two events can occur. The molecule can undergo a radiative transition to a vibrational state in S_0 in a process called **fluorescence**, or it can make a nonradiative transition to an excited vibrational state of T_1 in a process called **intersystem crossing**. Intersystem crossing violates the $\Delta S = 0$ selection rule and, therefore, occurs at a very low rate in comparison with the other processes depicted in Figure 14.9.

Because internal conversion to the ground vibrational state of S_1 is generally fast in comparison with fluorescence to S_0, the vibrationally excited-state molecule will relax to the ground vibrational state of S_1 before undergoing fluorescence. As a result of the relaxation, the fluorescence spectrum is shifted to lower energies relative to the absorption spectrum, as shown in Figure 14.10. When comparing absorption and fluorescence spectra, it is often seen that for potentials that are symmetric about the minimum (for example, a harmonic potential), the band of lines corresponding to absorption and fluorescence are mirror images of one another. This relationship is shown in Figure 14.10.

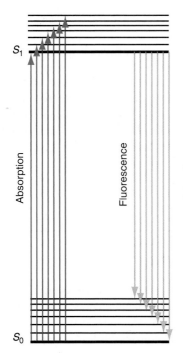

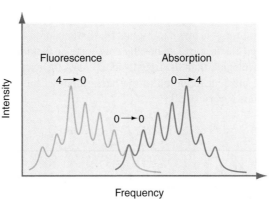

FIGURE 14.10

Illustration of the absorption and fluorescence bands expected if internal conversion is fast relative to fluorescence. The relative intensities of individual transitions within the absorption and fluorescence bands are determined by the Franck-Condon principle.

14.8 INTERSYSTEM CROSSING AND PHOSPHORESCENCE

Although intersystem crossing between singlet and triplet electronic states is forbidden by the $\Delta S = 0$ selection rule, the probability of this happening is high for many molecules. The probability of intersystem crossing transitions is enhanced by two factors: a very similar molecular geometry in the excited singlet and triplet states, and a strong spin-orbit coupling, which allows the spin flip associated with a singlet–triplet transition to occur. The processes involved in phosphorescence are illustrated in a simplified fashion in Figure 14.11 for a diatomic molecule.

Imagine that a molecule is excited from S_0 to S_1. This is a dipole-allowed transition, so it has a high probability of occurring. Through collisions with other molecules, the excited-state molecule loses vibrational energy and decays to the lowest vibrational state of S_1. As shown in Figure 14.11, the potential energy curves can overlap such that an excited vibrational state in S_1 can have approximately the same energy as an excited vibrational state in T_1. In this case, the molecule has the same geometry and energy in both singlet and triplet states. In Figure 14.11, this occurs for $n = 4$ in the state S_1. If the spin-orbit coupling is strong enough to initiate a spin flip, the molecule can cross over to the triplet state without a change in geometry or energy. Through internal conversion, it will rapidly relax to the lowest vibrational state of T_1. At this point, it can no longer make a transition back to S_1 because the ground vibrational state of T_1 is lower than any state in S_1.

However, from the ground vibrational state of T_1, the molecule can decay radiatively to the ground state in the dipole transition forbidden process called **phosphorescence**. Is this a high probability event? As discussed earlier, the lifetime of the ground vibrational state of T_1 can be very long compared to a vibrational period. Therefore, nonradiative processes involving collisions between molecules or with the walls of the reaction vessel can compete effectively with phosphorescence. Because it is a forbidden transition and because of the competition from nonradiative processes, the probability for a $T_1 \rightarrow S_0$ phosphorescence transition is generally much lower than for fluorescence. It usually lies in the range of 10^{-2} to 10^{-5}. The distinction between allowed and forbidden transitions is experimentally accessible through the lifetime of the excited state. Fluorescence is an allowed transition, and the excited-state lifetime is short, typically less than 10^{-7} s.

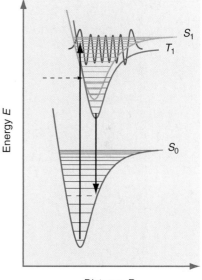

FIGURE 14.11

Process giving rise to phosphorescence illustrated for a diatomic molecule. Absorption from S_0 leads to population of excited vibrational states in S_1. The molecule has a finite probability of making a transition to an excited vibrational state of T_1 if it has the same geometry in both states and if there are vibrational levels of the same energy in both states. The dashed arrow indicates the coincidence of vibrational energy levels in T_1 and S_1. For reasons of clarity, only the lowest vibrational levels in T_1 are shown. The initial excitation to S_1 occurs to a vibration state of maximum overlap with the ground state of S_0 as indicated by the blue vibrational wave function.

By contrast, phosphorescence is a forbidden transition and, therefore, the excited-state lifetime is long, typically longer than 10^{-3} s.

Fluorescence can be induced using broadband radiation or highly monochromatic laser light. Fluorescence spectroscopy is well suited for detecting very small concentrations of a chemical species if the wavelength of the emission lies in the visible-UV part of the electromagnetic spectrum where there is little background noise near room temperature. As shown in Figure 14.10, relaxation to lower vibrational levels within the excited electronic state has the consequence that the fluorescence signal occurs at a longer wavelength than the light used to create the excited state. Therefore, the contribution of the incident radiation to the background at the wavelength used to detect the fluorescence is very small.

14.9 FLUORESCENCE SPECTROSCOPY AND ANALYTICAL CHEMISTRY

We now describe a particularly powerful application of fluorescence spectroscopy, namely, the sequencing of the human genome. The goal of the human genome project was to determine the sequence of the four bases, A, C, T, and G, in DNA that encode all the genetic information necessary for propagating the human species. A sequencing technique based on laser-induced fluorescence spectroscopy that has been successfully used in this effort can be divided into three parts.

In the first part, a section of DNA is cut into small lengths of 1000 to 2000 base pairs using mechanical shearing. Each of these pieces is replicated to create many copies, and these replicated pieces are put into a solution with a mixture of the four bases, A, C, T, and G. A reaction is set in motion that leads to the strands growing in length through replication. A small fraction of each of the A, C, T, and G bases in solution that are incorporated into the pieces of DNA has been modified in two ways. The modified base terminates the replication process. It also contains a dye chosen to fluoresce strongly at a known wavelength. The initial segments continue to grow if they incorporate unmodified bases, and no longer grow if they incorporate one of the modified bases. As a result of these competing processes involving the incorporation of modified or unmodified bases, a large number of partial replicas of the whole DNA are created, each of which is terminated in the base that has a fluorescent tag built into it. The ensemble of these partial replicas contains all possible lengths of the original DNA segment that terminate in the particular base chosen. If the lengths of these segments can be measured, then the positions of the particular base in the DNA segment can be determined.

The lengths of the partial replicas are measured using capillary electrophoresis coupled with detection using laser-induced fluorescence spectroscopy. In this method, a solution containing the partial replicas is passed through a glass capillary filled with a gel. An electrical field along the capillary causes the negatively charged DNA partial replicas to travel down the column with a speed that depends inversely on their length. Because of the different migration speeds, a separation in length occurs as the partial replicas pass through the capillary. At the end of the capillary, the partial replicas emerge from the capillary into a buffer solution that flows past the capillary, forming a sheath. The flow pattern of the buffer solution is carefully controlled to achieve a focusing of the emerging stream containing the partial replicas to a diameter somewhat smaller than the inner diameter of the capillary. A schematic diagram of such a sheath flow cuvette electrophoresis apparatus is shown in Figure 14.12. An array of capillaries is used rather than a single capillary in order to obtain the multiplexing advantage of carrying out several experiments in parallel.

The final part of the sequencing procedure is to measure the time that each of the partial replicas spent in transit through the capillary, which determines its length, and to identify the terminating base. The latter task is accomplished by means of laser-induced fluorescence spectroscopy. A narrow beam of visible laser light is passed through all the capillaries in series. Because of the very dilute solutions involved, the attenuation of the laser beam by each successive capillary is very small. The fluorescent light emitted

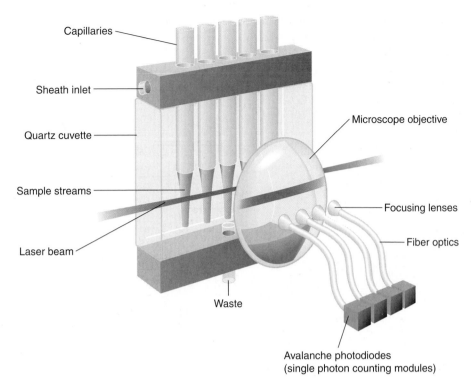

FIGURE 14.12

Schematic diagram of the application of fluorescence spectroscopy in the sequencing of the human genome. [After J. Zhang *et al.*, *Nucleic Acids Research* 27 (1999), 36e.]

from each of the capillaries is directed to light-sensing photodiodes by means of a microscope objective and individual focusing lenses. A rotating filter wheel between the microscope objective and the focusing lenses allows a discrimination to be made among the four different fluorescent dyes with which the bases were tagged. The sensitivity of the system shown in Figure 14.12 is 130 ± 30 molecules in the volume illuminated by the laser. This corresponds to a concentration of 2×10^{-13} mol/L! This extremely high sensitivity is a result of coupling the sensitive fluorescence technique to a sample cell designed with a very small sampling volume. Matching the laser beam diameter to the sample size and reducing the size of the cuvette result in a significant reduction in background noise. Commercial versions of this approach utilizing 96 parallel capillaries played a major part in the first phase of the sequencing of the human genome.

14.10 ULTRAVIOLET PHOTOELECTRON SPECTROSCOPY

Spectroscopy in general, and electronic spectroscopy in particular, gives information on the energy difference between the initial and final states rather than the energy levels involved in the transition. However, the energy of both occupied and unoccupied molecular orbitals is of particular interest to chemists. Information at this level of detail cannot be obtained directly from a UV absorption spectrum, because only a difference between energy levels is measured. However, information about the orbitals involved in the transition can be extracted from an experimentally obtained spectrum using a model. For example, the molecular orbital model described in Chapter 12 can be used to calculate the orbital energy levels for a molecule. With these results, an association can be made between energy-level differences calculated from observed spectral peaks and orbital energy levels obtained from the model.

Of all the possible forms of electronic spectroscopy, **UV photoelectron spectroscopy** comes closest to the goal of directly identifying the orbital energy level from which an electronic transition originates. What is the principle of this spectroscopy? As in the photoelectric effect discussed in Chapter 1, an incident photon of sufficiently high energy ejects an electron from one of the filled valence orbitals of the molecule, creating

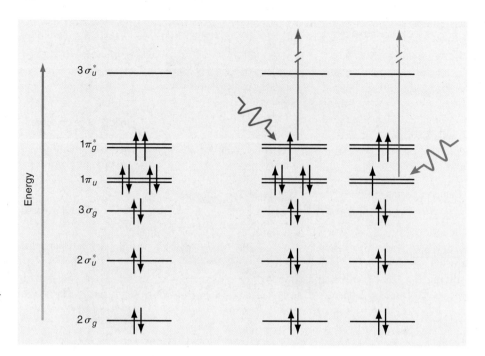

FIGURE 14.13

The ground-state molecular orbital diagram of O_2 is shown on the left. An incident UV photon can eject an electron from one of the occupied MOs, generating an O_2^+ ion and an unbound electron whose kinetic energy can be measured as shown for two different MOs in the center and right of the figure . Electrons ejected from different MOs will differ in their kinetic energy.

a positive ion as shown in Figure 14.13, using O_2 as an example. The kinetic energy of the ejected electron is related to the total energy required to form the positive ion via **photoionization**, by

$$E_{kinetic} = h\nu - \left[E_f + \left(n_f + \frac{1}{2} \right) h\nu_{vibration} \right] \qquad \textbf{(14.12)}$$

where E_f is the energy of the cation, which is formed by the removal of the electron, in its ground state. Equation (14.12) takes vibrational excitation of the cation into account, which by conservation of energy leads to a lower kinetic energy for the photoejected electron. Because in general either the initial or final state is a radical, a delocalized MO model must be used to describe UV photoelectron spectroscopy.

Under the assumptions to be discussed next, the measured value of $E_{kinetic}$ can be used to obtain the energy of the orbital, $\varepsilon_{orbital}$, from which the electron originated. The energy of the cation, E_f, which can be determined directly from a photoelectron spectrum, is equal to $\varepsilon_{orbital}$ if the following assumptions are valid:

- The nuclear positions are unchanged in the transition (Born–Oppenheimer approximation).
- The orbitals for the atom and ion are the same **(frozen orbital approximation)**. This assumes that the electron distribution is unchanged in the ion, even though the ion has one fewer electron.
- The total electron correlation energy in the molecule and ion are the same.

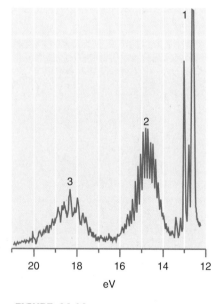

FIGURE 14.14

UV photoelectron spectrum of gas-phase H_2O. Three groups of peaks are seen. The structure within each group results from vibrational excitation of the cation formed in the photoionization process.

[From C. R. Brundle and D. W. Turner, *Proc. Phys.* Soc. A307 (1968), 27.]

The association of E_f with $\varepsilon_{orbital}$ for the neutral molecule under these assumptions is known as **Koopmans' theorem**. In comparing spectra obtained for a large number of molecules with high-level numerical calculations, the measured and calculated orbital energies are often found to differ by approximately 1 to 3 eV. The difference results primarily from the last two assumptions not being entirely satisfied.

This discussion suggests that a photoelectron spectrum consists of a series of peaks, each of which can be associated with a particular molecular orbital of the molecule. Figure 14.14 shows a photoelectron spectrum obtained for gas-phase water molecules for a photon energy $h\nu = 21.8$ eV, corresponding to a strong UV emission peak from a helium discharge lamp. Each of the three groups of peaks can be associated with a particular molecular orbital of H_2O, and the approximately equally spaced peaks within a group correspond to vibrational excitations of the cation formed in the photoionization process.

An analysis of this spectrum offers a good opportunity to compare and contrast localized and delocalized models of chemical bonding in molecules. It turns out that the assignment of peaks in a molecular photoelectron spectrum to individual localized orbitals is not valid. This is the case because the molecular wave function must exhibit the symmetry of the molecule. This important topic will be discussed in some detail in Chapter 17. Using the photoelectron spectrum of H_2O as an example, we show that the correct assignment of peaks in photoelectron spectra is to delocalized linear combinations of the localized orbitals, rather than to individual localized orbitals.

In a localized bonding model, water has two lone pairs and two O—H bonding orbitals. Because the lone pairs and the bonding orbitals are identical except for their orientation, one might expect to observe one group of peaks associated with the lone pair and one group of peaks associated with the bonding orbital in the photoelectron spectrum. In fact, four rather than two groups are observed if the photon energy is significantly higher than that used to obtain the data shown in Figure 14.14. In the localized bonding model, this discrepancy can be understood in terms of the coupling between the lone pairs and between the bonding orbitals. The coupling leads to **symmetric combinations** and **antisymmetric combinations**, just as was observed for vibrational spectroscopy in Section 8.5. In the molecular orbital model, the result can be understood by solving the Hartree–Fock equations, which generates four distinct MOs. We refer to these MOs as symmetric (S) or antisymmetric (A) and as having lone pair (or nonbonding) character (n) or sigma character (σ).

We now return to the photoelectron spectrum of Figure 14.14. The group of peaks below 13eV can be attributed to ε_{nA}. The corresponding MO wave function is the $1b_1$ orbital of Figure 13.8, which can be associated with the antisymmetric combination of the lone pairs. The group of peaks between 14 and 16 eV can be attributed to ε_{nS}. The corresponding wave function is the $2a_1$ orbital, which can be associated with the symmetric combination of the lone pairs. The group between 17 and 20 eV can be attributed to $\varepsilon_{\sigma A}$. The corresponding wave function is the $1b_2$ orbital, which can be associated with the antisymmetric combination of the bonding orbitals. The group attributed to $\varepsilon_{\sigma S}$ lies at higher ionization energies than were accessible in the experiment and, therefore, is not observed. The corresponding wave function is the $1a_1$ orbital, which can be associated with the symmetric combination of the bonding orbitals. The nomenclature used for these MOs, which was introduced in Chapter 13, will be explained in Chapter 16.

The preceding analysis leaves us with the following question: *Why* do **equivalent bonds** or lone pairs give rise to several different orbital energies? A nonmathematical explanation follows. Although the localized bonding orbitals are equivalent and orthonormal, the electron distribution in one O—H bond is not independent of the electron distribution in the other O—H bond because of Coulombic interactions between the two bonding regions. Therefore, an electronic excitation in one local bonding orbital changes the potential energy felt by the electrons in the region of the other local bonding orbital. This interaction leads to a coupling between the two localized bonds. By forming symmetric and antisymmetric combinations of the local orbitals, the coupling is removed. However, the local character of the bonding orbitals has also been removed. Therefore, the decoupled molecular wave functions cannot be identified with a state that is localized in only one of the two O—H regions. Only the decoupled wave functions, and not the localized orbitals, are consistent with the symmetry of the molecule.

In the case of water, the two equivalent localized O—H bonds give rise to two distinct orbital energies. However, in highly symmetric molecules, the number of distinct orbital energies can be less than the number of equivalent localized bonds. For instance, the three equivalent localized N—H bonds in NH_3 give rise to two distinct orbital energies, and the four equivalent localized C—H bonds in CH_4 give rise to two distinct orbital energies. The reason for these differences will become apparent after molecular symmetry is discussed in Chapter 16.

14.11 SINGLE MOLECULE SPECTROSCOPY

Spectroscopic measurements as described above are generally carried out in a sample cell in which a very large number of the molecules of interest, called an ensemble, are present. In general, the local environment of the molecules in an ensemble is not identical, which

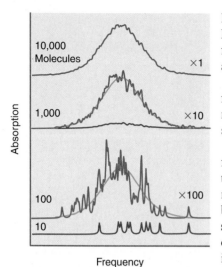

FIGURE 14.15

The absorption spectrum of an individual molecule is narrow, but the peak occurs over a range of frequencies for different molecules as shown in the lowest curve for 10 molecules in the sampling volume. As the number of molecules in the sampling volume is increased, the observed peak shows inhomogeneous broadening and is characteristic of the ensemble rather than of an individual molecule.

leads to inhomogeneous broadening of an absorption line as discussed in Section 8.9. Figure 14.15 shows how a broad absorption band arises if the corresponding narrow bands for individual molecules in the ensemble are slightly shifted in frequency because of variations in the immediate environment of a molecule.

Clearly, more information is obtained from the spectra of the individual molecules than from the inhomogeneously broadened band. The "true" absorption band for an individual molecule is observed only if the number of molecules in the volume being sampled is very small, for example, the bottom spectrum in Figure 14.15.

Single molecule spectroscopy is particularly useful in understanding the structure-function relationship for biomolecules. The **conformation** of a biomolecule refers to the arrangement of its constituent atoms in space and can be discussed in terms of primary, secondary, and tertiary structure. The **primary structure** is determined by the backbone of the molecule, for example, peptide bonds in a polypeptide. The term **secondary structure** refers to the local conformation of a part of the polypeptide. Two common secondary structures of polypeptides are the α-helix and the β-sheet as shown in Figure 14.16. **Tertiary structure** refers to the overall shape of the molecule; globular proteins are folded into a spherical shape, whereas fibrous proteins have polypeptide chains that arrange into parallel strands or sheets.

Keep in mind that the conformation of a biomolecule in solution is not static. Collisions with solvent and other solute molecules continuously change the energy and the conformation of a dissolved biomolecule with time. What are the consequences of such conformational changes for an enzyme? Because the activity is intimately linked to structure, conformational changes lead to fluctuations in activity, making an individual enzyme molecule alternately active and inactive as a function of time. Spectroscopic

FIGURE 14.16

The **(a)** α-helix and **(b)** β-sheet are two important forms in which proteins are found in aqueous solution. In both structures, hydrogen bonds form between imino (—NH—) groups and carbonyl groups.

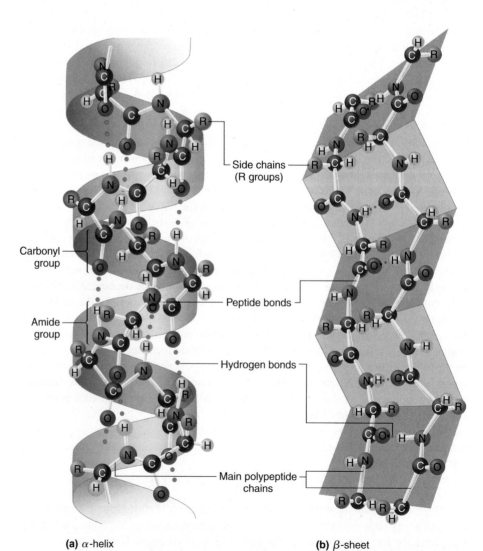

(a) α-helix   **(b)** β-sheet

measurements carried out on an ensemble of enzyme molecules give an average over all possible conformations, and hence over all possible activities for the enzyme. Such measurements are of limited utility in understanding how structure and chemical activity are related. As we will show in the next section, single molecule spectroscopy can go beyond the ensemble limit and gives information on the possible conformations of biomolecules and on the timescales on which transitions to different conformations take place.

To carry out single molecule spectroscopy, the number of molecules in the sampling volume must be reduced to approximately one. How is a spectrum of individual biomolecules in solution obtained? Only molecules in the volume that is both illuminated by the light source and imaged by the detector contribute to the measured spectrum. For this number to be approximately one, a laser focused to a small diameter is used to excite the molecules of interest, and a confocal microscope is used to collect the photons emitted in fluorescence from a small portion of the much larger cylindrical volume of the solution illuminated by the laser. In a confocal microscope, the sampling volume is at one focal point and the detector is behind a pinhole aperture located at the other focal point of the microscope imaging optics. Because photons that originate outside of a volume of approximately $1 \ \mu m \times 1 \ \mu m \times 1 \ \mu m$ centered at the focal point are not imaged on the pinhole aperture, they cannot reach the detector. Therefore, the confocal arrangement rather than the solution volume illuminated with the laser determines the sampling volume. To ensure that no more than a few molecules are likely to be found in the sampling volume, the concentration of the biomolecule must be less than $\sim 1 \times 10^{-6}$ M. Because the number of photons emitted in single molecule spectroscopy is small, it is important to ensure that spurious photons which could originate from scattered laser light or from Raman scattering outside of sampling volume do not reach the detector. Fluorescence spectroscopy is well suited for single molecule studies, because as discussed in Section 14.8, internal conversion ensures that the emitted photons have a lower frequency than the laser used to excite the molecule. Therefore, optical filters can be used to ensure that scattered laser light does not reach the detector. If the molecules being investigated are immobilized, they can also be imaged using the same experimental techniques. Figure 14.17 shows an image of individual Rhodamine B dye molecules tethered to a glass slide. The apparent size of the molecules is determined by the wavelength of the light used, and not by the actual molecular size. For visible light the apparent molecular size is ~ 300 nm, so that both small and very large molecules have the same apparent size.

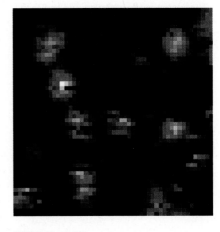

FIGURE 14.17

Microscope image of single Rhodamine B dye molecules on glass obtained using a confocal scanning microscope. The bright spots in the image correspond to fluorescence from single molecules. The image dimension is 5 μm by 5 μm.

14.12 FLUORESCENT RESONANCE ENERGY TRANSFER (FRET)

FRET is a form of single molecule spectroscopy that has proved to be very useful in studying biochemical systems. An electronically excited molecule can lose energy by either radiative or nonradiative events as discussed in Section 14.9. We refer to the molecule that loses energy as the **donor**, and the molecule that accepts the energy as the **acceptor**. If the emission spectrum of the donor overlaps the absorption spectrum of the acceptor as shown in Figure 14.18, then we refer to **resonance energy transfer**, meaning that the photon energy for fluorescence in the donor is equal to the photon energy for absorption in the acceptor as shown in Figure 14.19. Under resonance conditions, the energy transfer from the donor to the acceptor can occur with a high efficiency.

The probability for resonant energy transfer is strongly dependent on the distance between the two molecules. It was shown by Theodor Förster that the rate at which resonant energy transfer occurs decreases as the sixth power of the donor-acceptor distance.

$$k_{ret} = \frac{1}{\tau_D^{\circ}} \left(\frac{R_0}{r} \right)^6 \qquad \textbf{(14.13)}$$

In Equation (14.13), τ_D° is the lifetime of the donor in its excited state and R_0, the critical Förster radius, is the distance at which the resonance transfer rate and the rate for

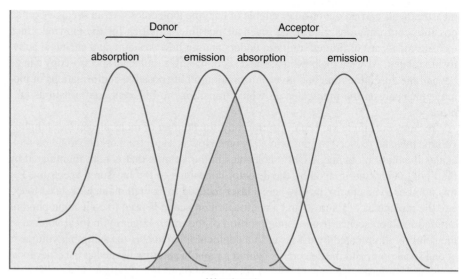

FIGURE 14.18

The energy levels of the ground and excited state donor and acceptor are shown. Resonant energy transfer only occurs if the donor and acceptor match up in energy.

spontaneous decay of the excited state donor are equal. Both these quantities can be determined experimentally. The sensitive dependence of the resonance energy transfer on the donor-acceptance distance makes it possible to use FRET as a **spectroscopic ruler** to measure donor-acceptor distances in the 10–100 nm range.

Figure 14.20 illustrates how FRET can be used to determine the conformation of a biomolecule. Schuler *et al.* attached dyes acting as donor and acceptor molecules to the ends of polyproline peptides of defined length containing between 6 and 40 proline residues. The donor absorbed a photon from a laser, and the efficiency with which the photon was transferred to the acceptor was measured. As Equation (14.13) shows, the efficiency falls off as the sixth power of the donor-acceptor distance. The results for a large number of measurements are shown in Figure 14.20b. A range of values for the efficiency is seen for each polypeptide. This is the case because the peptide is not a rigid rod, and a single molecule can have a variety of possible conformations, each of which has a different donor-acceptor distance. The width of the distribution in efficiency increases with the peptide length because more twists and turns can occur in a longer strand.

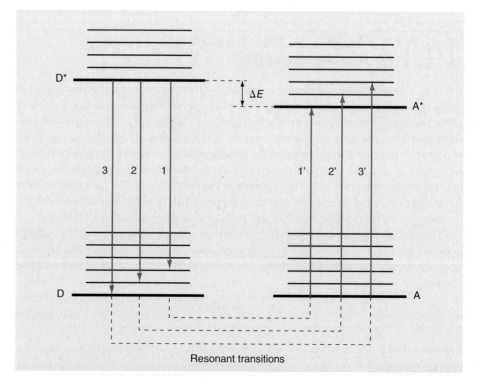

FIGURE 14.19

The emission spectrum of the excited donor occurs at a longer wavelength than the absorption as discussed in Section 14.7. If the emission spectrum of the donor overlaps the absorption spectrum of the acceptor, resonant energy transfer between the donor and acceptor can occur. Note the shift in the wavelength of the light emitted by the acceptor and the light absorbed by the donor. This shift allows the use of optical filters to detect acceptor emission in the presence of scattered light from the laser used to excite the donor.

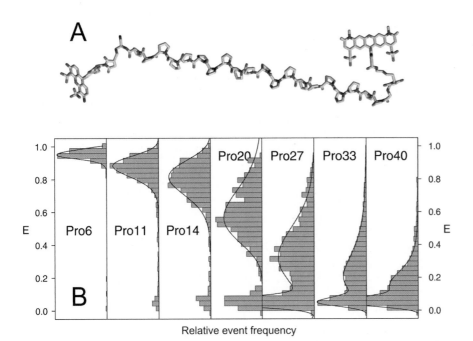

FIGURE 14.20

(A) Donor (left) and acceptor (right) dyes are attached to a polyproline peptide which becomes increasingly flexible as its length is increased. **(B)** The efficiency, E, of resonant energy transfer from the donor to acceptor is shown for peptides of different lengths. The length of the bars represents the relative event frequency for a large number of measurements on individual molecules. Note that the width of the distribution in E increases with the length of the peptide. The peak at zero efficiency is an experimental artifact due to inactive acceptors.

[From Schuler *et al.*, *Proceedings of the National Academy of Sciences* 102 (2005), 2754.]

An interesting application of single molecule FRET is in probing the conformational flexibility of single-stranded DNA in solution. The conformational flexibility of single-stranded DNA plays an important role in many DNA processes such as replication, repair, and transcription. Such a strand can be viewed as a flexible rod, approximately 2 nm in diameter. To put the dimensions of a strand in perspective, if it were a rubber tube of 1 cm in diameter, its length would be nearly 1 kilometer. A single molecule of DNA can be as long as ~1 cm in length, yet must fit in the nucleus of a cell, which is typically ~1 μm in diameter. To do so, the conformation of a DNA strand might take the form of a tangled mess such as a very long piece of spaghetti coiled upon itself as shown in Figure 14.21.

Such a complex conformation is best described by a statistical model, one of which is called the worm-like chain model.

In the **worm-like chain model**, the strand takes the form of a flexible rod that is continuously and randomly curved in all possible directions. However, there is an energetic cost of bending the strand, which is available to the strand through the energy transferred in collisions with other species in solution. This energy depends linearly on the absolute temperature. The energy required to bend the rod depends on the radius of curvature; a very gentle bend with a large radius of curvature requires much less energy than a sharp hairpin turn. In the limit of zero kelvin, the collisional energy transfer approaches zero, and the strand takes the form of a rigid rod. As the temperature increases, fluctuations in the radius of curvature increasingly occur along the rod. This behavior is described in the worm-like chain model by the **persistence length**, which is the length you can travel along the rod in a straight line before the rod bends in a different direction. As the temperature increases from zero kelvin to room temperature, a worm-like chain changes in conformation from a rigid rod of infinite persistence length to the tangled mess depicted in Figure 14.21, which has a very small persistence length.

How well does the worm-like chain model describe single-stranded DNA? This was tested by M. C. Murphy *et al.*, who attached flexible single-stranded DNAs to a rigid tether, which was immobilized by bonding the biotin at the end of the tether to a streptavidin-coated quartz surface as shown in Figure 14.22.

A donor fluorophore was attached to the free end of the flexible strand and an acceptor was attached to the rigid end. The length of the strand was varied between 10 and 70 nucleotides corresponding to distances from ~60 nm to ~420 nm between the donor and acceptor. After measuring τ_D° and R_0, the efficiency of resonant energy transfer from the donor to the acceptor was measured as a function of the strand length in NaCl solution whose concentration ranged for 2.5×10^{-3} M to 2 M. The results are shown in

FIGURE 14.21

The conformation of a long rod-like molecule in solution can be highly tangled.

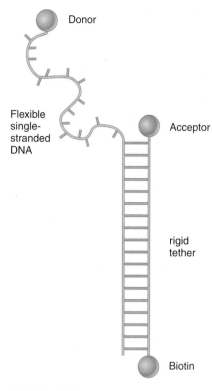

Donor

Flexible single-stranded DNA

Acceptor

rigid tether

Biotin

FIGURE 14.22

A donor and acceptor are attached to opposite ends of flexible single strands of DNA. The strands are attached to a silica substrate. The rigid tether has the function of moving the acceptor away from the quartz surface into the solution.

[From M. C. Murphy *et al.*, *Biophysical Journal* 86 (2004), 1430.]

Figure 14.23, and compared with calculations in which the persistence length was used as a parameter.

It is seen that the worm-like chain model represents the data well and that the persistence length decreases from 3 nm at low NaCl concentration to 1.5 nm at the highest concentration. The decrease in persistence length as the NaCl concentration increases can be attributed to a reduction in the repulsive interaction between the charged phosphate groups on the DNA through screening of the charge by the ionic solution (see Section 10.4). In this case, FRET measurements have provided a validation of the worm-like chain model for the conformation of biomolecules.

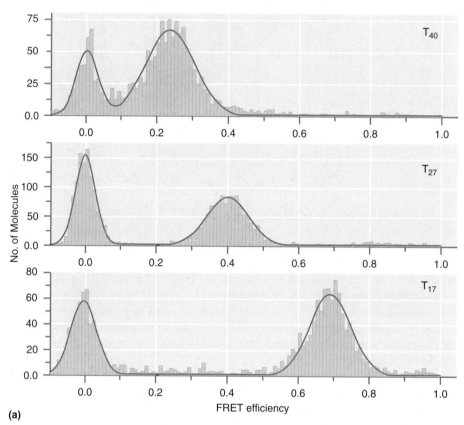

(a)

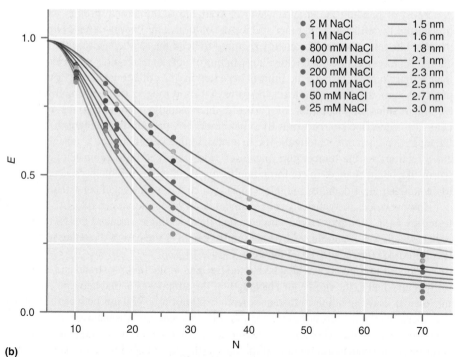

(b)

FIGURE 14.23

(a) The FRET efficiency is shown for Poly dT ssDNA of length 40, 27, and 17 nucleotides (top to bottom panel). The peak at zero efficiency is an experimental artifact due to inactive acceptors. **(b)** The FRET efficiency is shown as a function of N, the number of nucleotides, for various concentrations of NaCl. The various curves are calculated curves using the persistence length as a parameter. The best fit curves and the corresponding persistence length are shown for each salt concentration.

[From M. C. Murphy *et al.*, *Biophysical Journal* 86 (2004), 2530.]

Similar studies have been carried out using electron transfer reactions rather than resonant energy transfer between the donor and acceptor in order to probe the time scale of the conformational fluctuations of a single protein molecule. Yang *et al.* [*Science* 302 (2003), 262] found that conformational fluctuations occur over a wide range of timescales ranging from hundreds of microseconds to seconds. This result suggests that there are many different pathways that lead from one conformer to another and provides valuable data to researchers who model protein folding and other aspects of the conformational dynamics of biomolecules.

14.13 LINEAR AND CIRCULAR DICHROISM

Because the structure of a molecule is closely linked to its reactivity, it is a goal of chemists to understand the structure of a molecule of interest. This is a major challenge in the case of biomolecules because the larger the molecule, the more challenging it is to determine the structure. However, there are techniques available to determine aspects of the molecular structure of biomolecules, although they do not give the positions of all atoms in the molecule. Linear and circular dichroism are particularly useful in giving information on the secondary structure of biomolecules.

As discussed in Section 8.1, light is a transverse electromagnetic wave that interacts with molecules through a coupling of the electric field, **E**, of the light to the permanent or transient dipole moment, $\boldsymbol{\mu}$, of the molecule. Both **E** and $\boldsymbol{\mu}$ are vectors, and in classical physics the strength of the interaction is proportional to the scalar product $\mathbf{E} \cdot \boldsymbol{\mu}$. In quantum mechanics, the strength of the interaction is proportional to $\mathbf{E} \cdot \boldsymbol{\mu}^{fi}$, where the **transition dipole moment** is defined by

$$\boldsymbol{\mu}^{fi} = \int \psi_f^*(\tau)\, \boldsymbol{\mu}(\tau) \psi_i(\tau)\, d\tau \qquad (14.14)$$

In Equation (14.14), τ is a shorthand symbol for the spatial coordinates x, y, z or r, θ, ϕ and ψ_i and ψ_f refer to the initial and final states in the transition in which a photon is absorbed or emitted. The spatial orientation of $\boldsymbol{\mu}^{fi}$ is determined by evaluating an integral such as Equation (14.14), which goes beyond the level of this text. We show the orientation of $\boldsymbol{\mu}^{fi}$ for the amide group, which is the building block for the backbone of a polypeptide for a given $\pi \rightarrow \pi^*$ transition in Figure 14.24.

Many biomolecules have a long rod-like shape and can be oriented by embedding them in a film and then stretching the film. For such a sample, the molecule and therefore $\boldsymbol{\mu}^{fi}$ has a well-defined orientation in space. The electric field E can also be oriented in a plane with any desired orientation using a polarization filter, in which **linearly polarized light** is generated, as shown in Figure 14.25.

If the plane of polarization is varied with respect to the molecular orientation, the measured absorbance, A, will vary. It has a maximum value if **E** and $\boldsymbol{\mu}^{fi}$ are parallel, and is zero if **E** and $\boldsymbol{\mu}^{fi}$ are perpendicular. In **linear dichroism spectroscopy**, the variation of the absorbance with the orientation of plane-polarized light is measured. It is useful because it allows the direction of $\boldsymbol{\mu}^{fi}$ to be determined for an oriented molecule whose secondary structure is not known. One measures the absorbance with **E** parallel and perpendicular to the molecular axis. The difference $A_\parallel - A_\perp$ relative to the absorbance for randomly polarized light is the quantity of interest.

We illustrate the application of linear dichroism spectroscopy in determining the secondary structure of a polypeptide in the following discussion. The amide groups shown in Figure 14.25 interact with one another because of their close spacing, and the interaction can give rise to a splitting of the transition into two separate peaks.

The orientation of $\boldsymbol{\mu}^{fi}$ for each component depends on the polypeptide secondary structure. For the case of an α-helix, a transition near 208 nm with $\boldsymbol{\mu}^{fi}$ parallel to the helix axis and a transition near 190 nm with $\boldsymbol{\mu}^{fi}$ perpendicular to the helix axis is predicted from

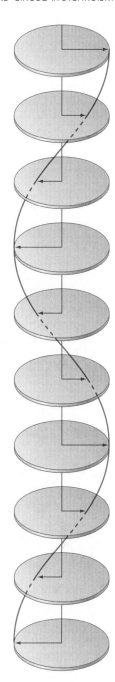

FIGURE 14.24

The arrows in successive images indicate the direction of the electric field vector as a function of time or distance. For linearly polarized light, the amplitude to the electric field vector changes periodically, but is confined to the plane of polarization.

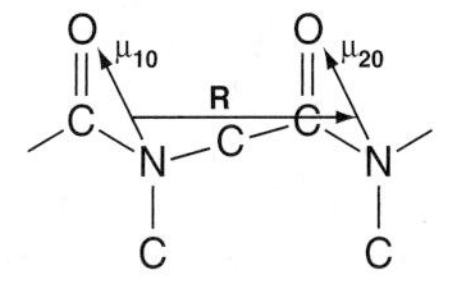

FIGURE 14.25

The amide bonds in a polypeptide chain are shown. The transition dipole moment is shown for a $\pi \rightarrow \pi^*$ transition.

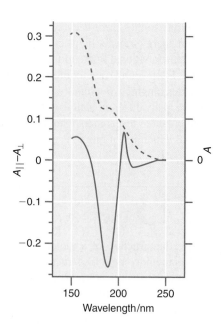

FIGURE 14.26

The normal isotropic absorbance, A (dashed line), and $A_{\parallel} - A_{\perp}$ (solid line) are shown as a function of the wavelength for an oriented film of poly(γ-ethyl-L-glutamate) in which it has the conformation of an α-helix. [From J. Brahms *et al.*, *Proceedings of the National Academy of Sciences USA* 60 (1968), 1130.]

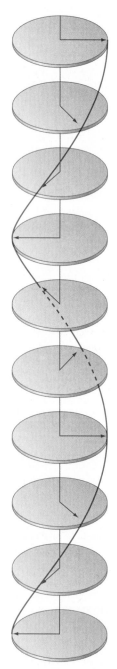

FIGURE 14.27

The arrows in successive images indicate the direction of the electric field vector as a function of time or distance. For circularly polarized light, the amplitude to the electric field vector is constant, but its plane of polarization undergoes a periodic variation.

theory. Figure 14.26 shows the absorbance for randomly polarized light and $A_{\parallel} - A_{\perp}$ for a polypeptide. The data show that $A_{\parallel} < A_{\perp}$ for the transition near 190 nm and that $A_{\parallel} > A_{\perp}$ near 200 nm. This shows that the secondary structure of this polypeptide is an α-helix. Note that the absorption for randomly polarized light gives no structural information.

Because molecules must be oriented in space for linear dichroism spectroscopy, it cannot be used for biomolecules in a static solution. For solutions, circular dichroism spectroscopy is widely used to obtain secondary structural information. In this spectroscopy, circularly polarized light, which is depicted in Figure 14.27, is passed through the solution.

Biomolecules are optically active, meaning that they do not possess a center of inversion. For an optically active molecule, the absorption for circularly polarized light in which the direction of rotation is clockwise (R) differs from that in which the direction of rotation is counterclockwise (L). This difference in A can be expressed as a difference in the extinction coefficient ε.

$$\Delta A(\lambda) = A_L(\lambda) - A_R(\lambda) = [\varepsilon_L(\lambda) - \varepsilon_R(\lambda)]lc = \Delta\varepsilon lc \qquad \textbf{(14.15)}$$

In Equation (14.15), l is the path length in the sample cell, and c is the concentration. In practice, the difference between $A_L(\lambda)$ and $A_R(\lambda)$ is usually expressed as the molar residual ellipticity, which is the shift in the phase angle θ between the components of the circularly polarized light in the form

$$\theta = 2.303 \times (A_L - A_R) \times 180/(4\pi) \text{ degrees} \qquad \textbf{(14.16)}$$

Circular dichroism can only be observed if ε is nonzero, and is usually observed in the visible part of the light spectrum.

As in the case of linear dichroism, $\Delta A(\lambda)$ for a given transition is largely determined by the secondary structure, and is much less sensitive to other aspects of the conformation. A derivation of how $\Delta A(\lambda)$ depends on the secondary structure is beyond the level of this text, but it can be shown that common secondary structures such as the α-helix, the β-sheet, a single turn, and a random coil have a distinctly different $\varepsilon(\lambda)$ dependence as shown in Figure 14.28. In this range of wavelengths, the absorption corresponds to $\pi \rightarrow \pi^*$ transitions of the amide group.

The differences between the $\Delta\varepsilon(\lambda)$ curves are sufficient that the observed $\Delta\varepsilon(\lambda)$ curve obtained for a protein of unknown secondary structure in solution can be expressed in the form

$$\Delta\varepsilon_{observed}(\lambda) = \sum_i F_i \Delta\varepsilon_i(\lambda) \qquad \textbf{(14.17)}$$

where $\Delta\varepsilon_i(\lambda)$ is the curve corresponding to one of the secondary structures in Figure 14.28, and F_i is the fraction of the peptide chromophores in that particular secondary structure. A best fit of the data to Equation (14.17) using available software allows a determination of the F_i to be made.

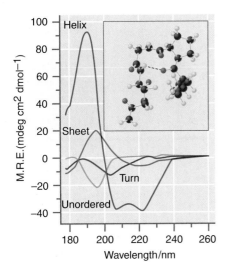

FIGURE 14.28

The mean residual ellipticity θ is shown as a function of wavelength for biomolecules having different secondary structures. Because the curves are distinctly different, circular dichroism spectra can be used to determine the secondary structure for optically active molecules. The inset shows the hydrogen bonding between different amide groups that generates different secondary structures.

[From J. T. Pelton, *Science* 291 (2001), 2175.]

Figure 14.29 shows the results of an application of circular dichroism in determining the secondary structure of α-synuclein bound to unilamellar phospholipid vesicles, which were used as a model for cell membranes. α-Synuclein is a small soluble protein of 140–143 amino acids that is found in high concentration in presynaptic nerve terminals. A mutation in this protein has been linked to Parkinson's disease and it is believed to be a precursor in the formation of extracellular plaques in Alzheimer's disease.

As can be seen by comparing the spectra in Figure 14.29 with those of Figure 14.28, the conformation of α-synuclein in solution is that of a random coil. However, upon binding to unilamellar phospholipid vesicles, the circular dichroism spectrum is dramatically changed and is characteristic of an α-helix. These results show that the binding of α-synuclein requires a conformational change. This conformational change can be understood from the known sequence of amino acids in the protein. By forming an α-helix, the polar and nonpolar groups in the protein are shifted to opposite sides of the helix. This allows the polar groups to associate with the acidic phospholipids, leading to a stronger binding than would be the case for a random coil.

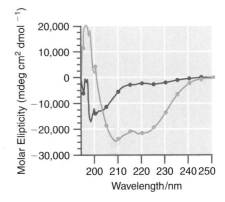

FIGURE 14.29

The molar ellipticity is shown as a function of the wavelength for α-synuclein in solution (filled squares) and for α-synuclein bound to unilamellar phospholipid vesicles (filled circles).

[From W. S. Davidson *et al.*, *Journal of Biological Chemistry* 273 (1998), 9443.]

14.14 ASSIGNING + AND − TO Σ TERMS OF DIATOMIC MOLECULES

In this section we go beyond the conclusions stated in Section 14.2 and illustrate how the + and − symmetry designations are applied to Σ terms. A more complete discussion can be found in *Quantum Chemistry*, fifth edition, by I. Levine, or in *Atoms and Molecules* by M. Karplus and R. N. Porter.

Recall that only unfilled MOs need to be considered in generating term symbols from a molecular configuration. The + and − designations refer to the change in sign of the molecular wave function on reflection in a plane that contains the molecular axis. If there is no change in sign, the + designation applies; if the wave function does change sign, the − designation applies. In the simplest case, all MOs are filled or the unpaired electrons are all in σ MOs. For such states, the + sign applies because there is no change in the sign of the wave function as a result of the reflection operation, as can be seen in Figure 14.30.

We next discuss molecular terms that do not fit into these categories, using O_2 as an example. The configuration for ground-state O_2 is $(1\sigma_g)^2(1\sigma_u^*)^2(2\sigma_g)^2(2\sigma_u^*)^2(3\sigma_g)^2$ $(1\pi_u)^2(1\pi_u)^2(1\pi_g^*)^1(1\pi_g^*)^1$, where we associate the partially filled MOs with the out-of-phase combinations of the $2p_x$ and $2p_y$ AOs as shown in Figure 14.31. Because filled MOs can be ignored, O_2 has a π^2 configuration, with one electron on each of the two degenerate π MOs. Recall that, in general, a configuration gives rise to several quantum states. Because the two electrons are in different $1\pi_g^*$ MOs, all six combinations of ± 1 for m_l and $\pm 1/2$ for m_s which do not assume that electrons are distinguishable

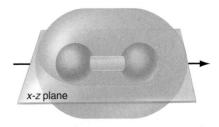

FIGURE 14.30

Reflection of a σ MO in a plane passing through the molecular axis, leaving the wave function unchanged.

x-z plane

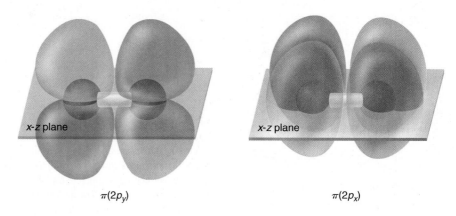

x-z plane

x-z plane

$\pi(2p_y)$

$\pi(2p_x)$

FIGURE 14.31

The two degenerate $1\pi_g^*$ wave functions are depicted.

are possible. For example, the Σ terms for which $M_L = m_{l1} + m_{l2} = 0$ occur as singlet and triplet terms. To satisfy the Pauli exclusion principle, the overall wave function (which is a product of spin and spatial parts) must be antisymmetric in the exchange of two electrons.

However, just as the $2p_x$ and $2p_y$ AOs are not eigenfunctions of the operator $\hat{l}_z$, as discussed in Section 7.5, the MOs depicted in Figure 14.31 are not eigenfunctions of the operator $\hat{L}_z$. To discuss the assignment of $+$ and $-$ to molecular terms, we can only use wave functions that are eigenfunctions of $\hat{L}_z$. In the cylindrical coordinates appropriate for a diatomic molecule, $\hat{L}_z = -i\hbar(\partial/\partial\phi)$, where ϕ is the angle of rotation around the molecular axis, and the eigenfunctions of this operator have the form $\psi(\phi) = Ae^{-i\Lambda\phi}$, as shown in Section 7.2. We cannot depict these complex functions, because this requires a six-dimensional space, rather than the three-dimensional space required to depict real functions.

The O_2 molecule has a π^2 configuration, and antisymmetric molecular wave functions can be formed either by combining symmetric spatial functions with antisymmetric spin functions or vice versa. All possible combinations are shown in the following equations. The subscript $+1$ or -1 on the spatial function indicates the value of m_l.

$$\psi_1 = \pi_{+1}\pi_{+1}(\alpha(1)\beta(2) - \beta(1)\alpha(2))$$

$$\psi_2 = \pi_{-1}\pi_{-1}(\alpha(1)\beta(2) - \beta(1)\alpha(2))$$

$$\psi_3 = (\pi_{+1}\pi_{-1} + \pi_{-1}\pi_{+1})(\alpha(1)\beta(2) - \beta(1)\alpha(2))$$

$$\psi_4 = (\pi_{+1}\pi_{-1} - \pi_{-1}\pi_{+1})\alpha(1)\alpha(2)$$

$$\psi_5 = (\pi_{+1}\pi_{-1} - \pi_{-1}\pi_{+1})(\alpha(1)\beta(2) + \beta(1)\alpha(2))$$

$$\psi_6 = (\pi_{+1}\pi_{-1} - \pi_{-1}\pi_{+1})\beta(1)\beta(2)$$

As shown in Section 10.8, the first three wave functions are associated with singlet states, and the last three are associated with triplet states. Because $\Lambda = |M_L|$, ψ_1 and ψ_2 belong to a Δ term, and ψ_3 through ψ_6 belong to Σ terms.

We next determine how these six wave functions are changed on reflection through a plane containing the molecular axis. As shown in Figure 14.32, reflection through such a plane changes the rotation angle $+\phi$ into $-\phi$. As a consequence, each eigenfunction of $\hat{L}_z$, $Ae^{-i\Lambda\phi}$ is transformed into $Ae^{+i\Lambda\phi}$, which is equivalent to changing the sign of M_L. Therefore, $\pi_{+1} \rightarrow \pi_{-1}$ and $\pi_{-1} \rightarrow \pi_{+1}$. Note that reflection does not change the sign of the wave function for ψ_1 through ψ_3 because $(-1) \times (-1) = 1$. Therefore, the plus sign applies and the term corresponding to ψ_3 is $^1\Sigma_g^+$. However, reflection does change the sign of the wave function for ψ_4 through ψ_6 because $(-1) \times (+1) = -1$; therefore, the minus sign applies. Because these three wave functions belong to a triplet term, the term symbol is $^3\Sigma_g^-$. A similar analysis can be carried out for other configurations.

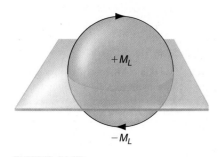

$+M_L$

$-M_L$

FIGURE 14.32

The rotation angle ϕ is transformed into $-\phi$ through reflection in a plane that contains the molecular axis. This is equivalent to changing $+M_L$ into $-M_L$.

Vocabulary

acceptor

antisymmetric combination

Beer's law

Born–Oppenheimer approximation

chromophore

conformation

continuous energy spectrum

donor

equivalent bonds

fluorescence

Franck-Condon factor

Franck-Condon principle

frozen orbital approximation

integral absorption coefficient

internal conversion

intersystem crossing

Koopmans' theorem

linear dichroism spectroscopy

linearly polarized light

molar extinction coefficient

molecular configuration

molecular term

$n \rightarrow \pi^*$ transition

nonradiative transition

$\pi \rightarrow \pi^*$ transition

persistence length

phosphorescence

photodissociation

photoionization

primary structure

radiative transition

resonance energy transfer

$\sigma \rightarrow \sigma^*$ transition

secondary structure

singlet state

spectroscopic ruler

symmetric combination

tertiary structure

transition dipole moment

triplet state

UV photoelectron spectroscopy

worm-like chain model

Questions on Concepts

Q14.1 Predict the number of unpaired electrons and the ground-state term for the following:

a. BO   b. LiO

Q14.2 How can FRET give information about the tertiary structure of a biological molecule in solution?

Q14.3 Photoionization of a diatomic molecule produces a singly charged cation. For the molecules listed here, calculate the bond order of the neutral molecule and the lowest energy cation:

a. H_2 b. O_2 c. F_2 d. NO

Q14.4 What would the intensity versus frequency plot in Figure 14.10 look like if fluorescence were fast with respect to internal conversion?

Q14.5 What aspect of the confocal microscope makes single molecule spectroscopy in solutions possible?

Q14.6 Explain why the fluorescence and absorption groups of peaks in Figure 14.10 are shifted and show mirror symmetry for idealized symmetrical ground-state and excited-state potentials.

Q14.7 The rate of fluorescence is in general higher than that for phosphorescence. Can you explain this fact?

Q14.8 Can linear dichroism spectroscopy be used for molecules in a static solution or in a flowing solution? Explain your answer.

Q14.9 Predict the number of unpaired electrons and the ground-state term for the following:

a. NO   b. CO

Q14.10 How many distinguishable states belong to the following terms:

a. $^1\Sigma_g^+$ b. $^3\Sigma_g^-$ c. $^2\Pi$ d. $^2\Delta$

Q14.11 Explain why the spectator species M in Equation (14.5) is needed to make the reaction proceed.

Q14.12 Because internal conversion is in general very fast, the absorption and fluorescence spectra are shifted in frequency as shown in Figure 14.10. This shift is crucial in making fluorescence spectroscopy capable of detecting very small concentrations. Can you explain why?

Q14.13 Make a sketch, like that in the inset of Figure 14.2 of what you might expect the electronic spectrum to look like for the ground and excited states shown below.

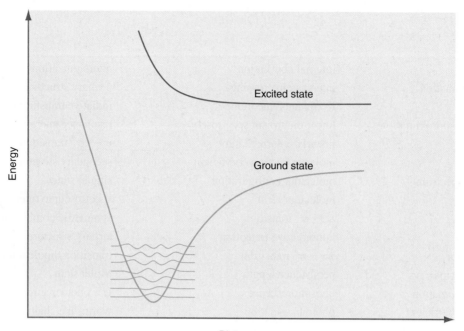

Distance

Q14.14 Why are the spectra of the individual molecules shown in the bottom trace of Figure 14.15 shifted in frequency?

Q14.15 Suppose you obtain the UV photoelectron spectrum shown here for a gas-phase molecule. Each of the groups corresponds to a cation produced by ejecting an electron from a different MO. What can you conclude about the bond length of the cations in the three states formed relative to the ground-state neutral molecule? Use the relative intensities of the individual vibrational peaks in each group to answer this question.

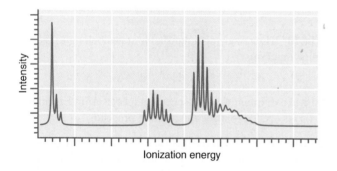

Ionization energy

Q14.16 The ground state of O_2^+ is $X^2\Pi_g$, and the next few excited states, in order of increasing energy, are $a^4\Pi_u$, $A^2\Pi_u$, $b^4\Sigma_g^-$, $^2\Delta_g$, $^2\Sigma_g^-$, and $c^4\Sigma_u^-$. On the basis of selection rules, which of the excited states can be accessed from the ground state by absorption of UV light?

Q14.17 The relative intensities of vibrational peaks in an electronic spectrum are determined by the Franck-Condon factors. How would the potential curve for the excited state in Figure 14.2 need to be shifted along the distance axis for the $n = 0 \rightarrow n' = 0$ transition to have the highest intensity? The term n refers to the vibrational quantum number in the ground state, and n' refers to the vibrational quantum number in the excited state.

Q14.18 a. Calculate the bond order for O_2 in the $X^3\Sigma_g^-$, $a^1\Delta_g$, $b^1\Sigma_g^+$, $A^3\Sigma_u^+$, and $B^3\Sigma_u^-$ states. Arrange these states in order of increasing bond length on the basis of bond order. Do your results agree with the potential energy curves shown in Figure 14.1?

b. For which of the molecules do you expect the $n = 0 \rightarrow n' = 1$ vibrational peak to have a higher intensity than the $n = 0 \rightarrow n' = 0$ vibrational peak? The term n refers to the vibrational quantum number in the ground state, and n' refers to the vibrational quantum number in the excited state.

Q14.19 How can circular dichroism spectroscopy be used to determine the secondary structure of a biomolecule?

Q14.20 What does the word resonance in FRET refer to?

Q14.21 In a simple model used to analyze UV photoelectron spectra, the orbital energies of the neutral molecule and the cation formed by ejection of an electron are assumed to be the same. In fact, some relaxation occurs to compensate for the reduction in the number of electrons by one. Would you expect the orbital energies to increase or decrease in the relaxation? Explain your answer.

15

Computational Chemistry

Warren J. Hehre, CEO, Wavefunction, Inc.

To the memory of Sir John Pople, 1925–2004

The Schrödinger equation can be solved exactly only for atoms or molecules containing one electron. For this reason, numerical methods that allow us to calculate approximate wave functions and values for observables such as energy, equilibrium bond lengths and angles, and dipole moments are at the heart of computational chemistry. The starting point for our discussion is the Hartree–Fock molecular orbital model. Although this model gives good agreement with experiment for some variables such as bond lengths and angles, it is inadequate for calculating many other observables. By extending the model to include electron correlation in a more realistic manner, and by judicious choice of a basis set, more accurate calculations can be made. The configuration interaction, Møller-Plesset, and density functional methods are discussed in this chapter, and the trade-off between computational cost and accuracy is emphasized. The 37 problems provided with this chapter are designed to give the student a working, rather than a theoretical, knowledge of computational chemistry.

15.1 THE PROMISE OF COMPUTATIONAL CHEMISTRY

Calculations on molecules based on quantum mechanics, once a mere novelty, are now poised to complement experiments as a means to uncover and explore new chemistry. The most important reason for this is that the theories underlying the calculations have now evolved to the point at which a variety of important quantities, among them molecular equilibrium geometry and reaction energetics, can be obtained with sufficient accuracy to actually be of use. Also important are the spectacular advances in computer hardware that have been made during the past decade. Taken together, this means that good theories can now be routinely applied to real systems. Finally, current computer software can be easily and productively used with little special training.

In making these quantum mechanics calculations, however, significant obstacles remain. For one, the chemist is confronted with many choices to make and few guidelines on how to make these choices. The fundamental problem is that the mathematical equations that arise from the application of quantum mechanics to chemistry—and that ultimately govern molecular structure and properties—cannot be solved analytically. Approximations need to be made in order to realize equations that can actually be solved. Severe approximations may lead to methods that can be widely applied, but may not yield accurate information. Less severe approximations may lead to methods that are more accurate, but too costly to apply routinely. In short, no one method of calculation is likely to be ideal for all applications, and the ultimate choice of specific methods rests on a balance between accuracy and cost. We equate cost with the computational time required to carry out the calculation.

The purpose of this chapter is to guide the student past the point of merely thinking about quantum mechanics as one of several components of a physical chemistry course and to instead have the student actually use quantum mechanics to address real chemical problems. The chapter starts with the many-electronic Schrödinger equation and then outlines the approximations that need to be made to transform this equation into what is now commonly known as Hartree–Fock (HF) theory. In the spirit of emphasizing the concepts rather than the theoretical framework, mathematical descriptions of the theoretical models discussed appear in boxes. A detailed understanding of this framework, however desirable, is not necessary to apply quantum mechanics to chemistry.

A focus on the limitations of Hartree–Fock theory leads to ways to improve on it and to a range of practical quantum chemical models. A few of these models are examined in detail and their performance and cost discussed. Finally, a series of graphical techniques is presented to portray the results of quantum chemical calculations. Aside from its practical focus, what sets this chapter apart from the remainder of this text is the problems. None of these are of the pencil-and-paper type; instead they require use of a quantum chemical program[1] on a digital computer. For the most part, the problems are open ended (as is an experimental laboratory) meaning that the student is free to explore. Problems that use the quantum chemical models under discussion are referenced throughout the chapter. Working problems as they are presented, before proceeding to the next section, is strongly recommended.

This Icon Indicates That Relevant Computational Problems are Available in the End-of-Chapter Problems.

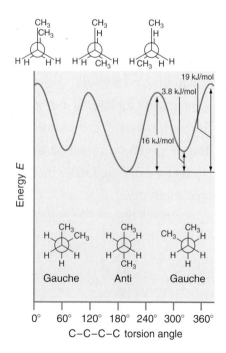

FIGURE 15.1

The energy of *n*-butane is shown as a function of the CCCC torsion angle, which is the reaction coordinate.

15.2 POTENTIAL ENERGY SURFACES

Chemists are familiar with the plot of energy versus the torsion angle involving the central carbon–carbon bond in *n*-butane. Figure 15.1 reveals three energy minima, corresponding to staggered structures, and three energy maxima, corresponding to eclipsed structures. One of the minima is given by a torsion angle of 180° (the so-called *anti* structure), and it is lower in energy and distinct from the other two minima with torsion angles of approximately 60° and 300° (so-called *gauche* structures), which are identical. Similarly, one of the energy maxima corresponding to a torsion angle of 0° is distinct from the other two maxima with torsion angles of approximately 120° and 240°, which are identical.

---

[1]The problems have been designed with the capabilities of the Student Edition of the Spartan molecular modeling program in mind. Other programs that allow equilibrium and transition-state geometry optimization, conformational searching, energy, property, and graphical calculations using Hartree–Fock, and density functional and MP2 models can also be used. The only exceptions are problems that appear early in the chapter before calculation models have been fully introduced. These problems make use of precalculated Spartan files that will be made available to students. See *www.chemplace.com*.

Eclipsed forms of *n*-butane are not stable molecules; instead they correspond only to hypothetical structures between *anti* and *gauche* minima. Thus, any sample of *n*-butane is made up of only two distinct compounds, *anti n*-butane and *gauche n*-butane. The relative abundance of the two compounds as a function of temperature is given by the Boltzmann distribution (see the discussion in Section 2.1).

The important geometrical coordinate in the example of Figure 15.1 can be clearly identified as a torsion involving one particular carbon–carbon bond. More generally, the important coordinate will be some combination of bond distances and angles and will be referred to simply as the **reaction coordinate**. This leads to a general type of plot in which the energy is given as a function of the reaction coordinate. Diagrams like this are commonly referred to as **reaction coordinate diagrams** and provide essential connections between important chemical observables—structure, stability, reactivity, and selectivity—and energy.

15.2.1 Potential Energy Surfaces and Geometry

The positions of the energy minima along the reaction coordinate give the equilibrium structures of the reactants and products as shown in Figure 15.2. Similarly, the position of the energy maximum defines the transition state. For example, where the reaction involves *gauche n*-butane going to the more stable *anti* conformer, the reaction coordinate may be thought of as a simple torsion about the central carbon–carbon bond, and the individual reactant, transition-state, and product structures in terms of this coordinate are depicted in Figure 15.3.

Equilibrium structure (geometry) can be determined from experiments as long as the molecule can be prepared and is sufficiently long lived to be subject to measurement. On the other hand, the geometry of a transition state cannot be established from measurement. This is simply because the transition state does not exist in terms of a sufficiently large population of molecules on which measurements can be performed.

Both equilibrium and transition-state structure can be determined from calculations. The former requires a search for an energy minimum on a potential energy surface, whereas the latter requires a search for an energy maximum along the reaction coordinate (and a minimum along each of the remaining coordinates). To see what is actually involved, the qualitative picture provided earlier must be replaced by a rigorous mathematical treatment. Reactants, products, and transition states are all stationary points on the potential energy diagram. In the one-dimensional case (the reaction coordinate diagram alluded to previously), this means that the first derivative of the potential energy with respect to the reaction coordinate is zero:

$$\frac{dV}{dR} = 0 \qquad \textbf{(15.1)}$$

The same must be true in dealing with a many-dimensional potential energy diagram (a potential energy surface). Here all partial derivatives of the energy with respect to each of the $3N - 6$ (N atoms) independent geometrical coordinates (R_i) are zero:

$$\frac{\partial V}{\partial R_i} = 0 \quad i = 1, 2, \dots, 3N - 6 \qquad \textbf{(15.2)}$$

In the one-dimensional case, reactants and products are energy minima and are characterized by a positive second energy derivative:

$$\frac{d^2V}{dR^2} > 0 \qquad \textbf{(15.3)}$$

The transition state is an energy maximum and is characterized by a negative second energy derivative:

$$\frac{d^2V}{dR^2} < 0 \qquad \textbf{(15.4)}$$

Problem P15.1

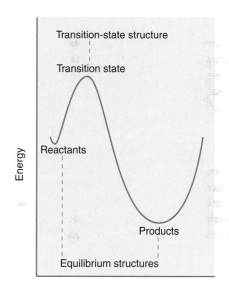

FIGURE 15.2

A reaction coordinate diagram shows the energy as a function of the reaction coordinate. Reactants and products correspond to minima, and the transition state corresponds to a maximum along this path.

FIGURE 15.3

The structure of the reactant, product, and transition state in the "reaction" of *gauche* *n*-butane to *anti n*-butane.

In the many-dimensional case, each independent coordinate, R_i, gives rise to $3N - 6$ second derivatives:

$$\frac{\partial^2 V}{\partial R_i R_1}, \frac{\partial^2 V}{\partial R_i R_2}, \frac{\partial^2 V}{\partial R_i R_3}, \ldots, \frac{\partial^2 V}{\partial R_i R_{3N-6}} \tag{15.5}$$

This leads to a matrix of second derivatives (the so-called Hessian):

$$\begin{bmatrix} \dfrac{\partial^2 V}{\partial R_1^2} & \dfrac{\partial^2 V}{\partial R_1 R_2} & \cdots & \\ \dfrac{\partial^2 V}{\partial R_2 R_1} & \dfrac{\partial^2 V}{\partial R_2^2} & \cdots & \\ \cdots & \cdots & & \\ \cdots & \cdots & & \dfrac{\partial^2 V}{\partial R_{3N-6}^2} \end{bmatrix} \tag{15.6}$$

In this form, it is not possible to say whether any given coordinate corresponds to an energy minimum, an energy maximum, or neither. To see the correspondence, the original set of geometrical coordinates (R_i) is replaced by a new set of coordinates (ξ_i), which leads to a matrix of second derivatives that is diagonal:

$$\begin{bmatrix} \dfrac{\partial^2 V}{\partial \xi_1^2} & 0\ldots & & 0 \\ 0 & \dfrac{\partial^2 V}{\partial \xi_2^2}\ldots & & 0 \\ \cdots & \cdots & & \cdots \\ 0 & 0\ldots & & \dfrac{\partial^2 V}{\partial \xi_{3N-6}^2} \end{bmatrix} \tag{15.7}$$

The ξ_i are unique and referred to as **normal coordinates**. Stationary points for which all second derivatives (in normal coordinates) are positive are energy minima:

$$\frac{\partial^2 V}{\partial \xi_i^2} > 0 = 1, 2, \ldots, 3N - 6 \tag{15.8}$$

These correspond to equilibrium forms (reactants and products). Stationary points for which all but one of the second derivatives are positive are so-called (first-order) saddle points and may correspond to transition states. If they do, the coordinate for which the second derivative is negative is referred to as the reaction coordinate (ξ_p):

$$\frac{\partial^2 V}{\partial \xi_p^2} < 0 \tag{15.9}$$

15.2.2 Potential Energy Surfaces and Vibrational Spectra

The vibrational frequency for a diatomic molecule A-B is given by Equation (15.10) as discussed in Section 7.1:

$$\nu = \frac{1}{2\pi}\sqrt{\frac{k}{\mu}} \tag{15.10}$$

In this equation, k is the force constant, which is in fact the second energy derivative of the potential energy, V, with respect to the bond length, R, at its equilibrium position

$$k = \frac{d^2 V(R)}{dR^2} \tag{15.11}$$

and μ is the reduced mass,

$$\mu = \frac{m_A m_B}{m_A + m_B} \tag{15.12}$$

where m_A and m_B are masses of atoms A and B.

Polyatomic systems are treated in a similar manner. Here, the force constants are the elements in the diagonal representation of the Hessian [Equation (15.7)]. Each vibrational mode is associated with a particular motion of atoms away from their equilibrium positions on the potential energy surface. Low frequencies correspond to motions in shallow regions of the surface, whereas high frequencies correspond to motions in steep regions. Note that one of the elements of the Hessian for a transition state will be a negative number, meaning that the corresponding frequency will be imaginary [the square root of a negative number as in Equation (15.10)]. This normal coordinate refers to motion along the reaction coordinate.

15.2.3 Potential Energy Surfaces and Thermodynamics

The relative stability of reactant and product molecules is indicated on the potential energy surface by their energies. The thermodynamic state functions' internal energy, U, and enthalpy, H, can be obtained from the energy of a molecule calculated by quantum mechanics, as discussed in Section 15.8.4.

The most common case is, as depicted in Figure 15.4, the one in which energy is released in the reaction. This kind of reaction is said to be **exothermic**, and the difference in stabilities of reactant and product is simply the enthalpy difference ΔH. For example, the reaction of *gauche* n-butane to *anti* n-butane is exothermic, and $\Delta H = -3.8$ kJ/mol.

Thermodynamics tells us that if we wait long enough the amount of products in an exothermic reaction will be greater than the amount of reactants. The actual ratio of the number of molecules of products ($n_{products}$) to reactants ($n_{reactants}$) also depends on the temperature and follows from the Boltzmann distribution:

$$\frac{n_{products}}{n_{reactants}} = \exp\left[-\frac{E_{products} - E_{reactants}}{kT}\right] \tag{15.13}$$

where $E_{products}$ and $E_{reactants}$ are the energies per molecule of the products and reactants, respectively, T is the temperature, and k is the Boltzmann constant. The Boltzmann distribution tells us the relative amounts of products and reactants at equilibrium. Even small energy differences between major and minor products lead to large product ratios, as shown in Table 15.1. The product formed in greatest abundance is that with the lowest energy, irrespective of the reaction pathway. In this case, the product is referred to as the **thermodynamic product** and the reaction is said to be **thermodynamically controlled**.

15.2.4 Potential Energy Surfaces and Kinetics

A potential energy surface also reveals information about the rate at which a reaction occurs. The difference in energy between reactants and the transition state as shown in Figure 15.5 determines the kinetics of reaction. The absolute reaction rate depends both on the concentrations of the reactants, $[A]^a$, $[B]^b$, ..., where $a, b, \ldots$ are typically integers or half integers, and a quantity k', called the **rate constant**:

$$\text{Rate} = k'[A]^a[B]^b[C]^c \ldots \tag{15.14}$$

The rate constant is given by the Arrhenius equation and depends on the temperature:

$$k' = A \exp\left[-\frac{(E_{transition\ state} - E_{reactants})}{kT}\right] \tag{15.15}$$

Here, $E_{transition\ state}$ and $E_{reactants}$ are the energies per molecule of the transition state and the reactants, respectively. Note that the rate constant and the overall rate do not depend on the energies of reactants and products, but only on the difference in energies between reactants and the transition state. This difference is commonly referred to as the **activation energy** and is usually given the symbol $\Delta E^{\ddagger}$. Other factors such as

Problems P15.2–P15.3

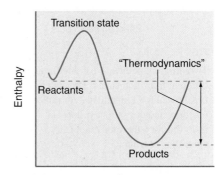

FIGURE 15.4

The energy difference between the reactants and products determines the thermodynamics of a reaction.

TABLE 15.1 THE RATIO OF THE MAJOR TO MINOR PRODUCT IS SHOWN AS A FUNCTION OF THE ENERGY DIFFERENCE BETWEEN THESE PRODUCTS

Energy Difference kJ/mol	Major: Minor (at Room Temperature)
2	~80 : 20
4	~90 : 10
8	~95 : 5
12	~99 : 1

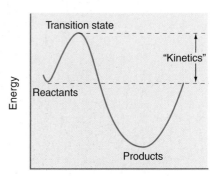

FIGURE 15.5

The energy difference between the reactants and transition state determines the rate of a reaction.

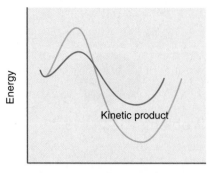

FIGURE 15.6

Two different pathways passing through different transition states. The path in red is followed for a kinetically controlled reaction, and the path in yellow is followed for a thermodynamically controlled reaction.

Problem P15.4

the likelihood of encounters between molecules and the effectiveness of these encounters in promoting reactions are taken into account by way of the **preexponential factor**, A, which is generally assumed to have the same value for reactions involving a single set of reactants going to different products or for reactions involving closely related reactants.

In general, the lower the activation energy, the faster the reaction. In the limit $\Delta E^{\ddagger} = 0$, the reaction rate will be limited entirely by how rapidly the molecules can move. Such limiting reactions are known as **diffusion-controlled reactions**. The product formed in greatest amount in a kinetically controlled reaction is that proceeding via the lowest energy transition state, irrespective of whether or not this is the thermodynamically stable product. For example, a **kinetically controlled** reaction will proceed along the red pathway in Figure 15.6, and the product formed is different than that corresponding to equilibrium in the system. The **kinetic product** ratio shows a dependence on activation energy differences that is analogous to that of Equation (15.13) with $E_{transition\ state}$ in place of $E_{product}$.

15.3 HARTREE–FOCK MOLECULAR ORBITAL THEORY: A DIRECT DESCENDANT OF THE SCHRÖDINGER EQUATION

The Schrödinger equation is deceptive in that, although it is remarkably easy to write down for any collection of nuclei and electrons, it has proven to be insolvable except for the one-electron case (the hydrogen atom). This situation was elaborated as early as 1929 by Dirac, one of the early founders of quantum mechanics:

The underlying physical laws necessary for the mathematical theory of a large part of physics and the whole of chemistry are thus completely known, and the difficulty is only that the exact application of these laws leads to equations much too complicated to be solvable.

P. A. M. Dirac, 1902–1984

To realize a practical quantum mechanical theory, it is necessary to make three approximations to the general multinuclear, multielectron Schrödinger equation:

$$\hat{H}\Psi = E\Psi \tag{15.16}$$

where E is the total energy of the system and Ψ is the n-electron wave function that depends both on the identities and positions of the nuclei and on the total number of electrons. The Hamiltonian $\hat{H}$ provides the recipe for specifying the kinetic and potential energies for each of the particles:

$$\hat{H} = -\frac{\hbar^2}{2m_e}\sum_i^{electrons}\nabla_i^2 - \frac{\hbar^2}{2}\sum_A^{nuclei}\frac{1}{M_A}\nabla_A^2 - \frac{e^2}{4\pi\varepsilon_0}\sum_i^{electrons}\sum_A^{nuclei}\frac{Z_A}{r_{iA}}$$

$$+ \frac{e^2}{4\pi\varepsilon_0}\sum_{i>}^{electrons}\sum_j^{electrons}\frac{1}{r_{ij}} + \frac{e^2}{4\pi\varepsilon_0}\sum_{A>}^{nuclei}\sum_B^{nuclei}\frac{Z_AZ_B}{R_{AB}} \tag{15.17}$$

where Z_A is the nuclear charge, M_A is the mass of nucleus A, m_e is the mass of the electron, R_{AB} is the distance between nuclei A and B, r_{ij} is the distance between electrons i and j, r_{iA} is the distance between electron i and nucleus A, ε_0 is the permittivity of free space, and $\hbar$ is the Planck constant divided by 2π.

The first approximation takes advantage of the fact that nuclei move much more slowly than do electrons. We assume that the nuclei are stationary from the perspective of the electrons (see Section 12.2), which is known as the **Born–Oppenheimer approximation**. This assumption leads to a nuclear kinetic energy term in Equation (15.17), the second term, which is zero, and a nuclear–nuclear Coulombic energy term, the last term, which is constant. What results is the **electronic Schrödinger equation**:

$$\hat{H}^{el}\Psi^{el} = E^{el}\Psi^{el} \tag{15.18}$$

$$\hat{H} = -\frac{\hbar^2}{2m_e}\sum_i^{electrons}\nabla_i^2 - \frac{e^2}{4\pi\varepsilon_0}\sum_i^{electrons}\sum_A^{nuclei}\frac{Z_A}{r_{iA}} + \frac{e^2}{4\pi\varepsilon_0}\sum_j^{electrons}\sum_{>}^{electrons}\frac{1}{r_{ij}} \tag{15.19}$$

The (constant) nuclear–nuclear Coulomb energy, the last term in Equation (15.17) needs to be added to E^{el} to get the total energy. Note that nuclear mass does not appear in the electronic Schrödinger equation. To the extent that the Born–Oppenheimer approximation is valid, this means that isotope effects on molecular properties must have a different origin.

Equation (15.18), like Equation (15.16), is insolvable for the general (many-electron) case and further approximations need to be made. The most obvious thing to do is to assume that electrons move independently of each other, which is what is done in the **Hartree–Fock approximation**. In practice, this can be accomplished by assuming that individual electrons are confined to functions called spin orbitals, χ_i. Each of the N electrons feels the presence of an average field made up of all of the other $(N-1)$ electrons. To ensure that the total (many-electron) wave function Ψ is anti-symmetric upon interchange of electron coordinates, it is written in the form of a single determinant called the **Slater determinant** (see Section 10.3):

$$\Psi = \frac{1}{\sqrt{n!}}\begin{vmatrix} \chi_1(1) & \chi_2(1)\cdots & \chi_n(1) \\ \chi_1(2) & \chi_2(2)\cdots & \chi_n(2) \\ \cdots & \cdots & \cdots \\ \chi_1(n) & \chi_2(n) & \chi_n(n) \end{vmatrix} \tag{15.20}$$

Individual electrons are represented by different rows in the determinant, which means that interchanging the coordinates of two electrons is equivalent to interchanging two rows in the determinant, multiplying its value by -1. Spin orbitals are the product of spatial functions or molecular orbitals, ψ_i, and spin functions, α or β. The fact that there are only two kinds of spin functions (α and β) leads to the conclusion that two electrons at most may occupy a given molecular orbital. Were a third electron to occupy the orbital, two rows in the determinant would be the same, as was shown in Section 10.3. Therefore, the value of the determinant would be zero. Thus, the notion that electrons are paired is a consequence of the Hartree–Fock approximation through the use of a determinant for the wave function. The set of molecular orbitals leading to the lowest energy is obtained by a process referred to as a **self-consistent-field (SCF) procedure**, which was discussed in Section 10.5 for atoms and in Section 13.1 for molecules.

The Hartree–Fock approximation leads to a set of differential equations, the **Hartree–Fock equations**, each involving the coordinates of a single electron. Although they can be solved numerically, it is advantageous to introduce an additional approximation in order to transform the Hartree–Fock equations into a set of algebraic equations. The basis for this approximation is the expectation that the one-electron solutions for many-electron molecules will closely resemble the one-electron wave functions for the hydrogen atom. After all, molecules are made up of atoms, so why shouldn't molecular solutions be made up of atomic solutions? As discussed in Section 13.2, the molecular orbitals ψ_i are expressed as linear combinations of a basis set of prescribed functions known as basis functions, ϕ:

$$\psi_i = \sum_\mu^{basis\ functions} c_{\mu i}\phi_\mu \tag{15.21}$$

In this equation, the coefficients $c_{\mu i}$ are the (unknown) molecular orbital coefficients. Because the ϕ are usually centered at the nuclear positions, they are referred to as atomic orbitals, and Equation (15.21) is called the **linear combination of atomic orbitals (LCAO) approximation**. Note, that in the limit of a complete (infinite) basis set, the LCAO approximation is exact.

MATHEMATICAL FORMULATION OF THE HARTREE–FOCK METHOD

The Hartree–Fock and LCAO approximations, taken together and applied to the electronic Schrödinger equation, lead to a set of matrix equations now known as the **Roothaan–Hall equations:**

$$\mathbf{Fc} = \varepsilon \mathbf{Sc} \tag{15.22}$$

where $\mathbf{c}$ are the unknown molecular orbital coefficients [see Equation (15.21), ε are orbital energies, $\mathbf{S}$ is the overlap matrix, and $\mathbf{F}$ is the Fock matrix, which is analogous to the Hamiltonian in the Schrödinger equation:

$$F_{\mu\nu} = H_{\mu\nu}^{core} + J_{\mu\nu} - K_{\mu\nu} \tag{15.23}$$

where H^{core} is the so-called core Hamiltonian, the elements of which are given by

$$H_{\mu\nu}^{core} = \int \phi_\mu(1) \left[-\frac{\hbar^2}{2m_e}\nabla^2 - \frac{e^2}{4\pi\varepsilon_0} \sum_A^{nuclei} \frac{Z_A}{r} \right] \phi_\nu(1)\, d\tau \tag{15.24}$$

Coulomb and exchange elements are given by:

$$J_{\mu\nu} = \sum_\lambda^{basis\ functions} \sum_\sigma P_{\lambda\sigma}(\mu\nu|\lambda\sigma) \tag{15.25}$$

$$K_{\mu\nu} = \frac{1}{2} \sum_\lambda^{basis\ functions} \sum_\sigma P_{\lambda\sigma}(\mu\lambda|\nu\sigma) \tag{15.26}$$

where $\mathbf{P}$ is called the density matrix, the elements of which involve a product of two molecular orbital coefficients summed over all occupied molecular orbitals (the number of which is simply half the total number of electrons for a closed-shell molecule):

$$P_{\lambda\sigma} = 2 \sum_i^{\substack{occupied\ molecular \\ orbitals}} c_{\lambda i} c_{\sigma i} \tag{15.27}$$

and $(\mu\nu|\lambda\sigma)$ are two-electron integrals, the number of which increases as the fourth power of the number of basis functions. Therefore, the cost of a calculation rises rapidly with the size of the basis set:

$$(\mu\nu|\lambda\sigma) = \iint \phi_\mu(1)\phi_\nu(1)\left[\frac{1}{r_{12}}\right]\phi_\lambda(2)\phi_\sigma(2)\, d\tau_1\, d\tau_2 \tag{15.28}$$

Methods resulting from solution of the Roothaan–Hall equations are called **Hartree–Fock models**. The corresponding energy in the limit of a complete basis set is called the **Hartree–Fock energy**.

15.4 PROPERTIES OF LIMITING HARTREE–FOCK MODELS

As discussed earlier in Section 10.5, total energies obtained from limiting (complete basis set) Hartree–Fock calculations will be too large (positive). This can be understood by recognizing that the Hartree–Fock approximation leads to replacement of instantaneous interactions between individual pairs of electrons with a picture in which each electron interacts with a charge cloud formed by all other electrons. The loss of flexibility causes electrons to get in each other's way to a greater extent than would actually be the case, leading to an overall electron repulsion energy that is too large and, hence, a total energy that is too large. The direction of the error in the total energy is also a direct consequence of the fact that Hartree–Fock models are variational. The limiting Hartree–Fock energy must be larger than (or at best equal to) the energy that would result from solution of the exact Schrödinger equation.

The difference between the limiting Hartree–Fock energy and the exact Schrödinger energy is called the **correlation energy**. The name *correlation* stems from the idea that the motion of one electron necessarily adjusts to or correlates with the motions of all other electrons. Any restriction on the freedom of electrons to move independently will, therefore, reduce their ability to correlate with other electrons.

The magnitude of the correlation energy may be quite large in comparison with typical bond energies or reaction energies. However, a major part of the total correlation energy may be insensitive to molecular structure, and Hartree–Fock models, which provide an incomplete account of correlation, may provide acceptable accounts of the energy change in some types of chemical reactions. It is also often the case that other properties, such as equilibrium geometries and dipole moments, are less influenced by correlation effects than are total energies. The sections that follow explore to what extent these conclusions are valid.

It is important to realize that calculations cannot actually be carried out at the Hartree–Fock limit. Presented here under the guise of limiting Hartree–Fock quantities are the results of calculations performed with a relatively large and flexible basis set, specifically the 6-311+G** basis set. (Basis sets are discussed at length in Section 15.7.) Although such a treatment leads to total energies that are higher than actual limiting Hartree–Fock energies by several tens to several hundreds of kilojoules per mole (depending on the size of the molecule), it is expected that errors in relative energies as well as in geometries, vibrational frequencies, and properties such as dipole moments will be much smaller.

15.4.1 Reaction Energies

The most easily understood problem with limiting Hartree–Fock models is uncovered in comparisons of **homolytic bond dissociation** energies. In such a reaction, a bond is broken leading to two radicals, for example, in methanol:

$$CH_3\!-\!OH \longrightarrow \; \cdot CH_3 + \; \cdot OH$$

As seen from the data in Table 15.2, Hartree–Fock dissociation energies are too small. In fact, limiting Hartree–Fock calculations suggest an essentially zero O—O bond energy in hydrogen peroxide and a negative F—F "bond energy" in the fluorine molecule! Something is seriously wrong. To see what is going on, consider the analogous bond dissociation reaction in hydrogen molecule:

$$H\!-\!H \longrightarrow \; \cdot H + \; \cdot H$$

Each of the hydrogen atoms that make up the product contains only a single electron, and its energy is given exactly by the (limiting) Hartree–Fock model. On the other hand, the reactant contains two electrons and, according to the variation principle, its energy must be too high (too positive). Therefore, the bond dissociation energy must be too low (too negative). To generalize, because the products of a homolytic bond dissociation reaction will contain one fewer electron pair than the reactant, the products would be expected to have lower correlation energy. The correlation energy associated with an electron pair is greater than that for a separated pair of electrons.

TABLE 15.2 *HOMOLYTIC BOND DISSOCIATION ENERGIES (kJ/mol)*

Molecule (bond)	Hartree–Fock Limit	Experiment	Δ
$CH_3\!-\!CH_3 \longrightarrow \; \cdot CH_3 + \; \cdot CH_3$	276	406	−130
$CH_3\!-\!NH_2 \longrightarrow \; \cdot CH_3 + \; \cdot NH_2$	238	389	−151
$CH_3\!-\!OH \longrightarrow \; \cdot CH_3 + \; \cdot OH$	243	410	−167
$CH_3\!-\!F \longrightarrow \; \cdot CH_3 + \; \cdot F$	289	477	−188
$NH_2\!-\!NH_2 \longrightarrow \; \cdot NH_2 + \; \cdot NH_2$	138	289	−151
$HO\!-\!OH \longrightarrow \; \cdot OH + \; \cdot OH$	−8	230	−238
$F\!-\!F \longrightarrow \; \cdot F + \; \cdot F$	−163	184	−347

TABLE 15.3 RELATIVE ENERGIES OF STRUCTURAL ISOMERS (kJ/mol)

Reference Compound	Isomer	Hartree–Fock Limit	Experiment	Δ
Acetonitrile	Methyl isocyanide	88	88	0
Acetaldehyde	Oxirane	134	113	21
Acetic acid	Methyl formate	71	75	−4
Ethanol	Dimethyl ether	46	50	−4
Propyne	Allene	8	4	4
	Cyclopropene	117	92	25
Propene	Cyclopropane	42	29	13
1,3-Butadiene	2-Butyne	29	38	−9
	Cyclobutene	63	46	17
	Bicyclo[1.1.0]butane	138	109	29

The poor results seen for homolytic bond dissociation reactions do not necessarily carry over into other types of reactions as long as the total number of electron pairs is maintained. A good example is found in energy comparisons among structural isomers (see Table 15.3). Although bonding may be quite different in going from one isomer to another, for example, one single and one double bond in propene versus three single bonds in cyclopropane, the total number of bonds is the same in reactants and products:

$$CH_3CH{=}CH_2 \longrightarrow \underset{H_2C-CH_2}{\overset{CH_2}{\triangle}}$$

The errors noted here are an order of magnitude less than those found for homolytic bond dissociation reactions, although in some of the comparisons they are still quite large, in particular, where small-ring compounds are compared with (unsaturated) acyclics, for example, propene with cyclopropane.

The performance of limiting Hartree–Fock models for reactions involving even more subtle changes in bonding is better still. For example, the data in Table 15.4 show that calculated energies of protonation of nitrogen bases relative to the energy of protonation of methylamine as a standard, for example, pyridine relative to methylamine, are typically in reasonable accord with their respective experimental values:

pyridinium + $CH_3NH_2 \longrightarrow$ pyridine + $CH_3NH_3{}^+$

TABLE 15.4 PROTON AFFINITIES OF NITROGEN BASES RELATIVE TO THE PROTON AFFINITY OF METHYLAMINE (kJ/mol)

Base	Hartree–Fock Limit	Experiment	Δ
Ammonia	−50	−38	−12
Aniline	−25	−10	−15
Methylamine	0	0	—
Dimethylamine	29	27	2
Pyridine	29	29	0
Trimethylamine	50	46	4
Diazabicyclooctane	75	60	15
Quinuclidine	92	75	17

15.4.2 Equilibrium Geometries

Systematic discrepancies are also noted in comparisons involving limiting Hartree–Fock and experimental equilibrium geometries. Two comparisons are provided. The first (Table 15.5) involves the geometries of the hydrogen molecule, lithium hydride, methane, ammonia, water, and hydrogen fluoride, whereas the second (Table 15.6) involves AB bond distances in two-heavy-atom hydrides, H_mABH_n. Most evident is the fact that, aside from lithium hydride, all calculated bond distances are shorter than experimental values. In the case of bonds to hydrogen, the magnitude of the error increases with the electronegativity of the heavy atom. In the case of the two-heavy-atom hydrides, the error increases substantially when two electronegative elements are involved in the bond. Thus, although errors in bond distances for methylamine, methanol, and methyl fluoride are fairly small, those for hydrazine, hydrogen peroxide, and fluorine molecule are much larger.

The reason for this trend—limiting Hartree–Fock bond distances being shorter than experimental values—as well as the reason that lithium hydride is an exception will

TABLE 15.5 STRUCTURES OF ONE-HEAVY-ATOM HYDRIDES
(bond distances, Å; bond angles, °)

Molecule	Geometrical Parameter	Hartree–Fock Limit	Experiment	Δ
H_2	r(HH)	0.736	0.742	−0.006
LiH	r(LiH)	1.607	1.596	+0.011
CH_4	r(CH)	1.083	1.092	−0.009
NH_3	r(NH)	1.000	1.012	−0.012
	<(HNH)	107.9	106.7	−1.2
H_2O	r(OH)	0.943	0.958	−0.015
	<(HOH)	106.4	104.5	+1.9
HF	r(FH)	0.900	0.917	−0.017

TABLE 15.6 BOND DISTANCES IN TWO HEAVY METAL HYDRIDES (Å)

Molecule (Bond)	Hartree–Fock Limit	Experiment	Δ
Ethane (H_3C—CH_3)	1.527	1.531	−0.004
Methylamine (H_3C—NH_2)	1.453	1.471	−0.018
Methanol (H_3C—OH)	1.399	1.421	−0.022
Methyl fluoride (H_3C—F)	1.364	1.383	−0.019
Hydrazine (H_2N—NH_2)	1.412	1.449	−0.037
Hydrogen peroxide (HO—OH)	1.388	1.452	−0.064
Fluorine (F—F)	1.330	1.412	−0.082
Ethylene (H_2C=CH_2)	1.315	1.339	−0.024
Formaldimine (H_2C=NH)	1.247	1.273	−0.026
Formaldehyde (H_2C=O)	1.178	1.205	−0.027
Diimide (NH=NH)	1.209	1.252	−0.043
Oxygen (O=O)	1.158	1.208	−0.050
Acetylene (HC≡CH)	1.185	1.203	−0.018
Hydrogen cyanide (HC≡N)	1.124	1.153	−0.029
Nitrogen (N≡N)	1.067	1.098	−0.031

TABLE 15.7 SYMMETRIC STRETCHING FREQUENCIES IN DIATOMIC AND SMALL POLYATOMIC MOLECULES (cm^{-1})

Molecule	Hartree–Fock Limit	Experiment	Δ
Lithium fluoride	927	914	13
Fluorine	1224	923	301
Lithium hydride	1429	1406	23
Carbon monoxide	2431	2170	261
Nitrogen	2734	2360	374
Methane	3149	3137	12
Ammonia	3697	3506	193
Water	4142	3832	310
Hydrogen fluoride	4490	4139	351
Hydrogen	4589	4401	188

become evident when we examine how Hartree–Fock models can be extended to treat electron correlation in Section 15.6.

15.4.3 Vibrational Frequencies

A few comparisons of limiting Hartree–Fock and experimental symmetric stretching frequencies for diatomic and small polyatomic molecules are provided in Table 15.7. (Note that the experimentally measured frequencies have been corrected for anharmonic behavior before being compared with calculated **harmonic frequencies**.) The systematic error in equilibrium bond distances for limiting Hartree–Fock models (calculated distances are shorter than experimental lengths) seems to be paralleled by a systematic error in stretching frequencies (calculated frequencies are larger than experimental frequencies). This is not unreasonable: too short a bond implies too strong a bond, which translates to a frequency that is too large. Note, however, that homolytic bond dissociation energies from limiting Hartree–Fock models are actually smaller (not larger) than experimental values, an observation that might imply that frequencies should be smaller (not larger) than experimental values. The reason for the apparent contradiction is that the Hartree–Fock model does not dissociate to the proper limit of two radicals as a bond is stretched.

15.4.4 Dipole Moments

Electric dipole moments for a few simple molecules from limiting Hartree–Fock calculations are compared with experimental moments in Table 15.8. The calculations reproduce the overall ordering of dipole moments. Although the sample is too small to generalize, the calculated values are consistently larger than the corresponding

TABLE 15.8 ELECTRIC DIPOLE MOMENTS (debyes)

Molecule	Hartree–Fock Limit	Experiment	Δ
Methylamine	1.5	1.31	0.2
Ammonia	1.7	1.47	0.2
Methanol	1.9	1.70	0.2
Hydrogen fluoride	2.0	1.82	0.2
Methyl fluoride	2.2	1.85	0.3
Water	2.2	1.85	0.3

experimental quantities. This might seem to be at odds with the notion that limiting Hartree–Fock bond lengths in these same molecules are smaller than experimental distances (which would imply dipole moments should be smaller than experimental values). We address this issue later in Section 15.8.8.

15.5 THEORETICAL MODELS AND THEORETICAL MODEL CHEMISTRY

As discussed in the preceding sections, limiting Hartree–Fock models do not provide results that are identical to experimental results. This is, of course, a direct consequence of the Hartree–Fock approximation, which replaces instantaneous interactions between individual electrons by interactions between a particular electron and the average field created by all other electrons. Because of this, electrons get in each other's way to a greater extent than they should. This leads to an overestimation of the electron–electron repulsion energy and too high a total energy.

At this point it is instructive to introduce the idea of a **theoretical model**, that is, a detailed recipe starting from the electronic Schrödinger equation and ending with a useful scheme, as well as the notion that any given theoretical model necessarily leads to a set of results, a **theoretical model chemistry**. At the outset, we might anticipate that the less severe the approximations that make up a particular theoretical model, the closer will be its results to experiment. The terms *theoretical model* and *theoretical model chemistry* were introduced by Sir John Pople, who in 1998 received the Nobel Prize in chemistry for his work in bringing quantum chemistry into widespread use.

All possible theoretical models may be viewed in the context of the two-dimensional diagram shown in Figure 15.7. The horizontal axis relates the extent to which the motions of electrons in a many-electron system are independent of each other or, alternately, the degree to which electron correlation is taken into account. At the extreme left are Hartree–Fock models. The vertical axis designates the basis set, which is used to represent the individual molecular orbitals. At the top is a so-called minimal basis set, which involves the fewest possible functions discussed in Section 15.7.1, while at the very bottom is the hypothetical complete basis set. The bottom of the column of Hartree–Fock models (at the far left) is called the Hartree–Fock limit.

Proceeding all the way to the right in Figure 15.7 (electron correlation fully taken into account) and then all the way to the bottom (both complete basis set and electron correlation taken into account) on this diagram is functionally equivalent to solving the Schrödinger equation exactly—something that, as stated earlier, cannot be realized. Note, however, that starting from some position on the diagram, that is, some level of treatment of electron correlation and some basis set, if moving down and to the

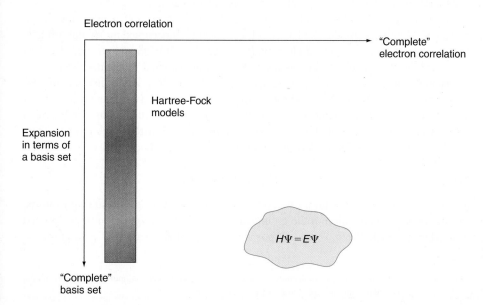

FIGURE 15.7

Different theoretical models can be classified by the degree to which electron correlation is taken into account and by the size of the basis set used.

right produces no significant change in a particular property of interest, then we can reasonably conclude that further motion would also not result in change in this property. In effect, this would signal that the exact solution has been achieved.

To the extent that it is possible, any theoretical model should satisfy a number of conditions. Most importantly, it should yield a unique energy, among other molecular properties, given only the kinds and positions of the nuclei, the total number of electrons, and the number of unpaired electrons. A model should not appeal in any way to chemical intuition. Also important is that, if at all possible or practical, the magnitude of the error of the calculated energy should increase roughly in proportion to molecular size, that is, the model should be **size consistent**. Only then is it reasonable to anticipate that reaction energies can be properly described. Less important, but highly desirable, is that the model energy should represent a bound to the exact energy, that is, the model should be **variational**. Finally, a model needs to be practical, that is, able to be applied not only to very simple or idealized systems, but also to problems that are actually of interest. Were this not an issue, then it would not be necessary to move beyond the Schrödinger equation itself.

Hartree–Fock models, which have previously been discussed, are well defined and yield unique properties. They are both size consistent and variational. Most importantly, Hartree–Fock models are presently applicable to molecules comprising upward of 50 to 100 atoms. We have already seen that limiting Hartree–Fock models also provide excellent descriptions of a number of important chemical observables, most important among them, equilibrium geometry and the energies of some kinds of reactions. We shall see in Section 15.8 that "practical" Hartree–Fock models are also quite successful in similar situations.

15.6 MOVING BEYOND HARTREE–FOCK THEORY

We next discuss improvements to the Hartree–Fock model that have the effect of moving down and to the right in Figure 15.7. Because these improvements increase the cost of a calculation, it is important to ask if they are necessary for a given calculation. This question must be answered by determining the extent to which the value of the observable of interest has the desired accuracy. Sections 15.8.1 through 15.8.11 explicitly address this question for a number of important observables, among them equilibrium geometries, reaction energies, and dipole moments.

Two fundamentally different approaches for moving beyond Hartree–Fock theory have received widespread attention. The first increases the flexibility of the Hartree–Fock wave function (associated with the electronic ground state) by combining it with wave functions corresponding to various excited states. The second introduces an explicit term in the Hamiltonian to account for the interdependence of electron motions.

Solution of the Roothaan-Hall equations results in a set of molecular orbitals, each of which is doubly occupied,[2] and a set of higher energy unoccupied molecular orbitals. The number of occupied molecular orbitals is equal to half of the number of electrons for closed-shell molecules, whereas the number of unoccupied molecular orbitals depends on the choice of basis set. Typically this number is much larger than the number of occupied molecular orbitals, and for the hypothetical case of a complete basis set it is infinite. The unoccupied molecular orbitals play no part in establishing the Hartree–Fock energy nor any ground-state properties obtained from Hartree–Fock models. They are, however, the basis for models that move beyond Hartree–Fock theory.

15.6.1 Configuration Interaction Models

It can be shown that in the limit of a complete basis set, the energy resulting from the optimum linear combination of the ground-state electronic configuration (that obtained from Hartree–Fock theory) and all possible excited-state electronic configurations

---

[2]This is valid for the vast majority of molecules. Radicals have one singly occupied molecular orbital and the oxygen molecule has two singly occupied molecular orbitals.

formed by promotion of one or more electrons from occupied to unoccupied molecular orbitals is the same as would result from solution of the full many-electron Schrödinger equation. An example of such a promotion is shown in Figure 15.8.

This result, referred to as **full configuration interaction**, while interesting, is of no practical value simply because the number of excited-state electronic configurations is infinite. Practical configuration interaction models may be realized first by assuming a finite basis set and then by restricting the number of excited-state electronic configurations included in the mixture. Because of these two restrictions, the final energy is not the same as would result from solution of the exact Schrödinger equation. Operationally, what is required is first to obtain the Hartree–Fock wave function, and then to write a new wave function as a sum, the leading term of which, Ψ_0, is the Hartree–Fock wave function, and remaining terms, Ψ_s, are wave functions derived from the Hartree–Fock wave function by electron promotions:

$$\Psi = a_0\Psi_0 + \sum_{s>0} a_s\Psi_s \quad (15.29)$$

The unknown linear coefficients, a_s, are determined by solving Equation (15.30):

$$\sum_s (H_{st} - E\delta_{st})a_s = 0 \quad (15.30)$$

where the matrix elements are given by

$$H_{st} = \int \dots \int \Psi_s \hat{H} \Psi_t \, d\tau_1 \, d\tau_2 \dots d\tau_n \quad (15.31)$$

The lowest energy wave function obtained from solution of Equation (15.30) corresponds to the energy of the electronic ground state.

One approach for limiting the number of electron promotions is referred to as the **frozen core approximation**. In effect, this eliminates any promotions from molecular orbitals that correspond essentially to (combinations of) inner-shell or core electrons. Although the total contribution to the energy arising from inner-shell promotions is not insignificant, experience suggests that this contribution is nearly identical for the same types of atoms in different molecules. A more substantial approximation is to limit the number of promotions based on the total number of electrons involved, that is, **single-electron promotions**, **double-electron promotions**, and so on. Configuration interaction based on single-electron promotions only, the so-called **CIS method**, leads to no improvement of the (Hartree–Fock) energy or wave function. The simplest procedure to use that actually leads to improvement over Hartree–Fock is the so-called **CID method**, which is restricted to double-electron promotions:

$$\Psi_{CID} = a_0\Psi_0 + \sum_{i<j}^{occ}\overset{molecular\ orbitals}{\sum} \sum_{a<b}^{unocc}\sum a_{ij}^{ab}\Psi_{ij}^{ab} \quad (15.32)$$

A somewhat less restricted recipe, the so-called CISD method, considers both single- and double-electron promotions:

$$\Psi_{CISD} = a_0\Psi_0 + \sum_i^{occ}\sum_a^{unocc} a_i^a\Psi_i^a + \sum_{i<j}^{occ}\sum_{a<b}^{unocc} a_{ij}^{ab}\Psi_{ij}^{ab} \quad (15.33)$$

Solution of Equation (15.30) for either CID or CISD methods is practical for reasonably large systems. Both methods are obviously well defined and they are variational. However, neither method (or any limited configuration interaction method) is size consistent. This can easily be seen by considering the CISD description of a two-electron system, for example, a helium atom as shown in Figure 15.9, using just two basis functions, which leads to one occupied and one unoccupied molecular orbital. In this case the CISD description for the isolated atom is exact (within the confines of the basis set), meaning that all possible electron promotions have been explicitly considered. Similarly, the description of two helium atoms treated independently is exact.

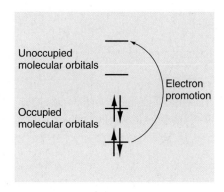

FIGURE 15.8

Electron promotion from occupied to unoccupied molecular orbitals.

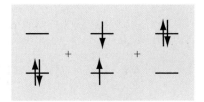

FIGURE 15.9

The CISD description of He.

FIGURE 15.10

The CISD description of He$_2$ restricted to one- and two-electron promotions.

Next consider the corresponding CISD treatment of two helium atoms together but at infinite separation, as shown in Figure 15.10. This description is not exact because three- and four-electron promotions have not been taken into account. Thus, the calculated energies of two helium atoms treated separately and two helium atoms at infinite separation will be different. Size consistency is a very important attribute for any quantum chemical model, and its absence for any practical configuration interaction models makes them much less appealing than they otherwise might be.

15.6.2 Møller-Plesset Models

Practical size-consistent alternatives to configuration interaction models are **Møller-Plesset models**, in particular, the second-order Møller-Plesset model (MP2). Møller-Plesset models are based on the recognition that, while the Hartree–Fock wave function Ψ_0 and ground-state energy E_0 are approximate solutions to the Schrödinger equation, they are exact solutions to an analogous problem involving the Hartree–Fock Hamiltonian, $\hat{H}_0$, in place of the exact Hamiltonian, $\hat{H}$. Assuming that the Hartree–Fock wave function Ψ and energy are, in fact, very close to the exact wave function and ground-state energy E, the exact Hamiltonian can then be written in the following form:

$$\hat{H} = \hat{H}_0 + \lambda \hat{V} \tag{15.34}$$

In Equation (15.34), $\hat{V}$ is a small perturbation and λ is a dimensionless parameter. Expanding the exact wave function and energy in terms of the Hartree–Fock wave function and energy yields

$$E = E^{(0)} + \lambda E^{(1)} + \lambda^2 E^{(2)} + \lambda^3 E^{(3)} + \dots \tag{15.35}$$

$$\Psi = \Psi_0 + \lambda \Psi^{(1)} + \lambda^2 \Psi^{(2)} + \lambda^3 \Psi^{(3)} + \dots \tag{15.36}$$

MATHEMATICAL FORMULATION OF MØLLER-PLESSET MODELS

Substituting the expansions of Equations (15.34) to (15.36) into the Schrödinger equation and gathering terms in λ^n yields

$$\hat{H}_0 \Psi_0 = E^{(0)} \Psi_0 \tag{15.37a}$$

$$\hat{H}_0 \Psi^{(1)} + \hat{V} \Psi_0 = E^{(0)} \Psi^{(1)} + E^{(1)} \Psi_0 \tag{15.37b}$$

$$\hat{H}_0 \Psi^{(2)} + \hat{V} \Psi^{(1)} = E^{(0)} \Psi^{(2)} + E^{(1)} \Psi^{(1)} + E^{(2)} \Psi_0 \tag{15.37c}$$

$$\dots$$

Multiplying each of the Equations (15.37c) by Ψ_0 and integrating over all space yields the following expression for the nth-order (MPn) energy:

$$E^{(0)} = \int \dots \int \Psi_0 \hat{H}_0 \Psi_0 \, d\tau_1 \, d\tau_2 \dots d\tau_n \tag{15.38a}$$

$$E^{(1)} = \int \dots \int \Psi_0 \hat{V} \Psi_0 \, d\tau_1 \, d\tau_2 \dots d\tau_n \tag{15.38b}$$

$$E^{(2)} = \int \dots \int \Psi_0 \hat{V} \Psi^{(1)} d\tau_1 \, d\tau_2 \dots d\tau_n \tag{15.38c}$$

$$\dots$$

In this framework, the Hartree–Fock energy is the sum of the zero- and first-order Møller-Plesset energies:

$$E^{(0)} + E^{(1)} = \int \cdots \int \Psi_0 (\hat{H}_0 + \hat{V}) \Psi_0 \, d\tau_1 \, d\tau_2 \ldots d\tau_n \qquad (15.39)$$

The first correction, $E^{(2)}$ can be written as follows:

$$E^{(2)} = \overset{\substack{molecular\ orbitals \\ occ}}{\sum_{i<j}\sum} \; \overset{unocc}{\sum_{a<b}\sum} \frac{[(ij\|ab)]^2}{(\varepsilon_a + \varepsilon_b - \varepsilon_i - \varepsilon_j)} \qquad (15.40)$$

where ε_i and ε_j are energies of occupied molecular orbitals, and ε_a and ε_b are energies of unoccupied molecular orbitals. The integrals $(ij\|ab)$ over filled (i and j) and empty (a and b) molecular orbitals account for changes in electron–electron interactions as a result of electron promotion,

$$(ia\|jb) = -(ib|ja) \qquad (15.41)$$

in which the integrals $(ij|ab)$ and $(ib|ja)$ involve molecular orbitals rather than basis functions, for example,

$$(ia|jb) = \int \psi_i(1)\psi_a(1)\left[\frac{1}{r_{12}}\right]\psi_j(2)\psi_b(2)\, d\tau_1\, d\tau_2 \qquad (15.42)$$

The two integrals are related by a simple transformation,

$$(ij|ab) = \overset{basis\ functions}{\sum_\mu \sum_\nu} \; \sum_\lambda \sum_\sigma c_{\mu i} c_{\nu j} c_{\lambda a} c_{\sigma b} (\mu\nu|\lambda\sigma) \qquad (15.43)$$

where $(\mu\nu|\lambda\sigma)$ are given by Equation (15.28).

The MP2 model is well defined and leads to unique results. As mentioned previously, MP2 is size consistent, although (unlike configuration interaction models) it is not variational. Therefore, the calculated energy may be lower than the exact value.

15.6.3 Density Functional Models

The second approach for moving beyond the Hartree–Fock model is now commonly known as **density functional theory**. It is based on the availability of an exact solution for an idealized many-electron problem, specifically an electron gas of uniform density. The part of this solution that relates only to the exchange and correlation contributions is extracted and then directly incorporated into an SCF formalism much like Hartree–Fock formalism. Because the new exchange and correlation terms derive from idealized problems, density functional models, unlike configuration interaction and Møller-Plesset models, do not limit to the exact solution of the Schrödinger equation. In a sense, they are empirical in that they incorporate external data (the form of the solution of the idealized problem). What makes density functional models of great interest is their significantly lower computation cost than either configuration interaction or Møller-Plesset models. For his discovery, leading up to the development of practical density functional models, Walter Kohn was awarded the Nobel Prize in chemistry in 1998.

The Hartree–Fock energy may be written as a sum of the kinetic energy, E_T, the electron–nuclear potential energy, E_V, and Coulomb, E_J, and exchange, E_K, components of the electron–electron interaction energy:

$$E^{HF} = E_T + E_V + E_J + E_K \qquad (15.44)$$

The first three of these terms carry over directly to density functional models, whereas the Hartree–Fock exchange energy is replaced by a so-called exchange/correlation energy, E_{XC}, the form of which follows from the solution of the idealized electron gas problem:

$$E^{DFT} = E_T + E_V + E_J + E_{XC} \qquad (15.45)$$

Except for E_T, all components depend on the total electron density, $\rho(\mathbf{r})$:

$$\rho(\mathbf{r}) = 2 \sum_i^{\text{orbitals}} |\psi_i(\mathbf{r})|^2 \tag{15.46}$$

The ψ_i are orbitals, strictly analogous to molecular orbitals in Hartree–Fock theory.

MATHEMATICAL FORMULATION OF DENSITY FUNCTIONAL THEORY

Within a finite basis set (analogous to the LCAO approximation for Hartree–Fock models), the components of the density functional energy, E^{DFT}, can be written as follows:

$$E_T = \sum_\mu^{\text{basis functions}} \sum_\nu \int \phi_\mu(\mathbf{r}) \left[-\frac{\hbar^2 e^2}{2m_e} \nabla^2 \right] \phi_\nu(\mathbf{r}) \, d\mathbf{r} \tag{15.47}$$

$$E_V = \sum_\mu^{\text{basis functions}} \sum_\nu P_{\mu\nu} \sum_A^{\text{nuclei}} \int \phi_\mu(\mathbf{r}) \left[-\frac{Z_A e^2}{4\pi\varepsilon_0 |\mathbf{r} - \mathbf{R}_A|} \right] \phi_\nu(\mathbf{r}) \, d\mathbf{r} \tag{15.48}$$

$$E_J = \frac{1}{2} \sum_\mu^{\text{basis functions}} \sum_\nu \sum_\lambda \sum_\sigma P_{\mu\nu} P_{\lambda\sigma} (\mu\nu|\lambda\sigma) \tag{15.49}$$

$$E_{XC} = \int f(\rho(\mathbf{r}), \nabla\rho(\mathbf{r}) \dots) \, d\mathbf{r} \tag{15.50}$$

where Z is the nuclear charge, $|\mathbf{r} - \mathbf{R}_A|$ is the distance between the nucleus and the electron, $\mathbf{P}$ is the density matrix [Equation (15.27)], and the $(\mu\nu|\lambda\sigma)$ are two-electron integrals [Equation (15.28)]. The $f(\rho(\mathbf{r}), \nabla\rho(\mathbf{r}), \dots)$ is the so-called exchange/correlation functional, which depends on the electron density. In the simplest form of the theory, it is obtained by fitting the density resulting from the idealized electron gas problem to a function. Better models result from also fitting the gradient of the density. Minimizing E^{DFT} with respect to the unknown orbital coefficients yields a set of matrix equations, the Kohn-Sham equations, analogous to the Roothaan–Hall equations [Equation (15.22)]:

$$\mathbf{Fc} = \varepsilon \mathbf{Sc} \tag{15.51}$$

Here the elements of the Fock matrix are given by

$$F_{\mu\nu} = H_{\mu\nu}^{core} + J_{\mu\nu} - F_{\mu\nu}^{XC} \tag{15.52}$$

and are defined analogously to Equations (15.25) and (15.26), respectively, and $\mathbf{F}^{XC}$ is the exchange/correlation part, the form of which depends on the particular exchange/correlation functional used. Note that substitution of the Hartree–Fock exchange, $\mathbf{K}$, for $\mathbf{F}^{XC}$ yields the Roothaan–Hall equations.

Density functional models are well defined and yield unique results. They are neither size consistent nor variational. Note that if the exact exchange/correlation functional had been known for the problem at hand (rather than only for the idealized many-electron gas problem), then the density functional approach would be exact. Although better forms of such functionals are constantly being developed, at present, there is no systematic way to improve the functional to achieve an arbitrary level of accuracy.

15.6.4 Overview of Quantum Chemical Models

An overview of quantum chemical models, starting with the Schrödinger equation, and including Hartree–Fock models, configuration interaction and Møller-Plesset models, and density functional models, is provided in Figure 15.11.

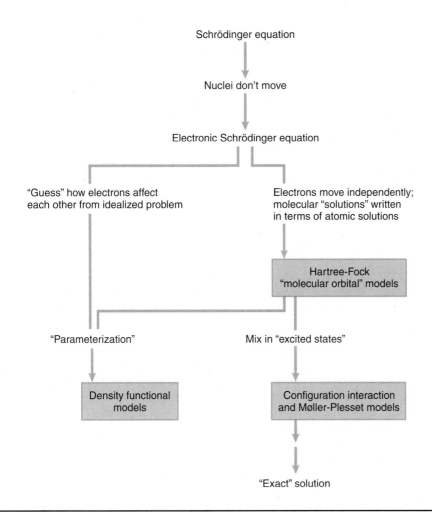

Schrödinger equation

↓

Nuclei don't move

↓

Electronic Schrödinger equation

"Guess" how electrons affect
each other from idealized problem

Electrons move independently;
molecular "solutions" written
in terms of atomic solutions

↓

Hartree-Fock
"molecular orbital" models

"Parameterization"

Mix in "excited states"

↓

Density functional
models

Configuration interaction
and Møller-Plesset models

↓

"Exact" solution

FIGURE 15.11

Schematic diagram showing how quantum chemical models are related to one another.

15.7 GAUSSIAN BASIS SETS

The LCAO approximation requires the use of a basis set made up of a finite number of well-defined functions centered on each atom. The obvious choice for the functions would be those corresponding closely to the exact solution of the hydrogen atom, that is, a polynomial in the Cartesian coordinates multiplying an exponential in r. However, the use of these functions was not cost effective, and early numerical calculations were carried out using nodeless **Slater-type orbitals** (STOs), defined by

$$\phi(r, \theta, \phi) = \frac{(2\zeta/a_0)^{n+1/2}}{[(2n)!]^{1/2}} r^{n-1} e^{-\zeta r/a_0} Y_l^m(\theta, \phi) \qquad \textbf{(15.53)}$$

The symbols n, m, and l denote the usual quantum numbers and ζ is the effective nuclear charge. Use of these so-called Slater functions was entertained seriously in the years immediately following the introduction of the Roothaan–Hall equations, but soon abandoned because they lead to integrals that are difficult if not impossible to evaluate analytically. Further work showed that the cost of calculations can be further reduced if the AOs are expanded in terms of **Gaussian functions**, which have the form

$$g_{ijk}(r) = N x^i y^j z^k e^{-\alpha r^2} \qquad \textbf{(15.54)}$$

In this equation, x, y, and z are the position coordinates measured from the nucleus of an atom; i, j, and k are nonnegative integers, and α is an orbital exponent. An s-type function (zeroth order Gaussian) is generated by setting $i = j = k = 0$; a p-type function (first order Gaussian) is generated if one of i, j, and k is 1 and the remaining two are 0; and a d-type function (second order Gaussian) is generated by all combinations that give $i + j + k = 2$. Note that this recipe leads to six rather than five d-type functions, but appropriate combinations of these six functions give the usual five d-type functions and a sixth function that has s symmetry.

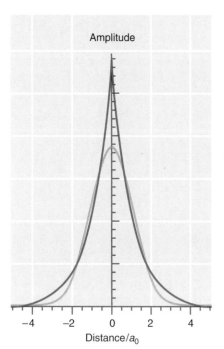

Amplitude

Distance/a_0

FIGURE 15.12

A cut through the hydrogen atom $1s$ AO (red curve) is compared with a single Gaussian function (yellow curve). Note that the AO has a cusp at the nucleus, whereas the Gaussian function has zero slope at the nucleus. Note also that the Gaussian function falls off more rapidly with distance because of the r^2 dependence in the exponent.

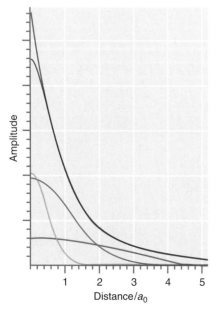

Amplitude

Distance/a_0

FIGURE 15.13

A hydrogen atom $1s$ AO (red curve) can be fit with the three Gaussian functions with different α values shown in green, yellow, and cyan. Both the α values and the coefficients multiplying the Gaussian functions are optimized in the best fit function, shown in blue.

Gaussian functions lead to integrals that are easily evaluated. With the exception of so-called semi-empirical models, which do not actually entail evaluation of large numbers of difficult integrals, all practical quantum chemical models now make use of Gaussian functions.

Given the different radial dependence of STOs and Gaussian functions, it is not obvious at first glance that Gaussian functions are appropriate choices for AOs. Figure 15.12 shows a comparison of the two functional forms. The solution to this problem is to approximate the STO by a linear combination of Gaussian functions having different α values, rather than by a single Gaussian function. For example, a best fit to a $1s$-type STO using three Gaussians is shown in Figure 15.13. We can see that, although the region near the nucleus is not fit well, in the bonding region beyond 0.5 a_0, the fit is very good. The fit near the nucleus can be improved by using more Gaussian functions.

In practice, instead of taking individual Gaussian functions as members of the basis set, a normalized linear combination of Gaussian functions with fixed coefficients is constructed to provide a best fit to an AO. The value of each coefficient is optimized either by seeking minimum atom energies or by comparing calculated and experimental results for "representative" molecules. These linear combinations are called **contracted functions**. The contracted functions become the elements of the basis set. Although the coefficients in the contracted functions are fixed, the coefficients $c_{\mu i}$ in Equation (15.21) are variable and are optimized in the solution of the Schrödinger equation.

15.7.1 Minimal Basis Sets

Although there is no limit to the number of functions that can be placed on an atom, there is a minimum number. The minimum number is the number of functions required to hold all the electrons of the atom while still maintaining its overall spherical nature. This simplest representation or **minimal basis set** involves a single ($1s$) function for hydrogen and helium, a set of five functions ($1s$, $2s$, $2p_x$, $2p_y$, $2p_z$) for lithium to neon, and a set of nine functions ($1s$, $2s$, $2p_x$, $2p_y$, $2p_z$, $3s$, $3p_x$, $3p_y$, $3p_z$) for sodium to argon. Note that although $2p$ functions are not occupied in the lithium or beryllium atoms (and $3p$ functions are not occupied in the sodium or magnesium atoms), they are needed to provide proper descriptions of the bonding in molecular systems. For example, the bonding in a molecule such as lithium fluoride involves electron donation from a lone pair on fluorine to an appropriate (p-type) empty orbital on lithium (back bonding) as shown in Figure 15.14.

Of the minimal basis sets that have been devised, perhaps the most widely used and extensively documented is the **STO-3G basis set**. Here, each of the basis functions is expanded in terms of three Gaussian functions, where the values of the Gaussian exponents and the linear coefficient have been determined by least squares as best fits to Slater-type (exponential) functions.

The STO-3G basis set and all minimal basis sets have two obvious shortcomings: the first is that all basis functions are either themselves spherical or come in sets that, when taken together, describe a sphere. This means that atoms with spherical molecular environments or nearly spherical molecular environments will be better described than atoms with aspherical environments. This suggests that comparisons among different molecules will be biased in favor of those incorporating the most spherical atoms. The second shortcoming follows from the fact that basis functions are atom centered. This restricts their ability to describe electron distributions between nuclei, which are a critical element of chemical bonds. Minimal basis sets such as STO-3G are primarily of historical interest and have largely been replaced in practical calculations by split-valence basis sets and polarization basis sets, which have been formulated to address these two shortcomings. These basis sets are discussed in the following two subsections.

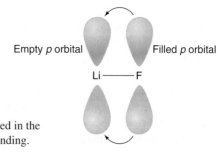

Empty p orbital — Filled p orbital

Li ——— F

FIGURE 15.14

$2p$ orbitals need to be included in the Li basis set to allow back bonding.

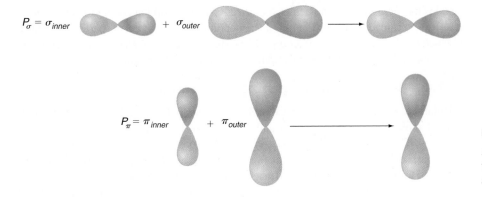

FIGURE 15.15

A split-valence basis set provides a way to allow the electron distribution about an atom to be nonspherical.

15.7.2 Split-Valence Basis Sets

The first shortcoming of a minimal basis set, namely, a bias toward atoms with spherical environments, can be addressed by providing two sets of valence basis functions: an inner set, which is more tightly held and an outer set, which is more loosely held. The iterative process leading to solution of the Roothaan–Hall equations adjusts the balance of the two parts independently for the three Cartesian directions, by adjusting the individual molecular orbital coefficients. For example, the proper linear combination to produce a molecular orbital suitable for σ bonding might involve a large coefficient (σ_{inner}) multiplying the inner basis function (in the σ direction) and a small coefficient (σ_{outer}) multiplying the outer basis function, whereas that to produce a molecular orbital suitable for π bonding might involve a small coefficient (π_{inner}) multiplying the inner basis function and a large coefficient (π_{outer}) multiplying the outer basis function as shown in Figure 15.15. The fact that the three Cartesian directions are treated independently of each other means that the atom (in the molecule) may be nonspherical.

A **split-valence basis set** represents core atomic orbitals by one set of functions and valence atomic orbitals by two sets of functions, $1s$, $2s^i$, $2p_x^i$, $2p_y^i$, $2p_z^i$, $2s^o$, $2p_x^o$, $2p_y^o$, $2p_z^o$ for lithium to neon and $1s$, $2s$, $2p_x$, $2p_y$, $2p_z$, $3s^i$, $3p_x^i$, $3p_y^i$, $3p_z^i$, $3s^o$, $3p_x^o$, $3p_y^o$, $3p_z^o$ for sodium to argon. Note that the valence $2s$ ($3s$) functions are also split into inner (superscript i) and outer (superscript o) components, and that hydrogen atoms are also represented by inner and outer valence ($1s$) functions. Among the simplest split-valence basis sets are 3-21G and 6-31G. Each core atomic orbital in the 3-21G basis set is expanded in terms of three Gaussians, whereas basis functions representing inner and outer components of valence atomic orbitals are expanded in terms of two and one Gaussians, respectively. The 6-31G basis sets are similarly constructed, with core orbitals represented in terms of six Gaussians and valence orbitals split into three and one Gaussian components. Expansion coefficients and Gaussian exponents for 3-21G and 6-31G basis sets have been determined by Hartree–Fock energy minimization on atomic ground states.

15.7.3 Polarization Basis Sets

The second shortcoming of a minimal (or split-valence) basis set, namely, that the basis functions are centered on atoms rather than between atoms, can be addressed by providing d-type functions on main-group elements (where the valence orbitals are of s and p type), and (optionally) p-type functions on hydrogen (where the valence orbital is of s type). This allows displacement of electron distributions away from the nuclear positions, as depicted in Figure 15.16.

The inclusion of **polarization functions** can be thought about either in terms of hybrid orbitals, for example, pd and sp hybrids, or alternatively in terms of a Taylor series expansion of a function (d functions are the first derivatives of p functions and p functions are the first derivatives of s functions). Although the first way of thinking is quite familiar to chemists (Pauling hybrids), the second offers the advantage of knowing what steps might be taken next to effect further improvement, that is, adding second and third derivatives.

Among the simplest polarization basis sets is 6-31G*, constructed from 6-31G by adding a set of d-type polarization functions written in terms of a single Gaussian for each heavy (non-hydrogen) atom. A set of six second-order Gaussians is added in the case of 6-31G*. Gaussian exponents for polarization functions have been chosen to give the lowest

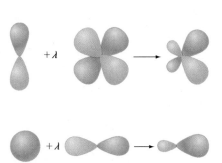

FIGURE 15.16

Polarization functions shift the center of the electron distribution to the bonding region between atoms.

energies for representative molecules. Polarization of the *s* orbitals on hydrogen atoms is necessary for an accurate description of the bonding in many systems (particularly those in which hydrogen is a bridging atom). The 6-31G** basis set is identical to 6-31G*, except that it also provides *p*-type polarization functions for hydrogen.

15.7.4 Basis Sets Incorporating Diffuse Functions

Calculations involving anions, for example, absolute acidity calculations, and calculations of molecules in excited states and of UV absorption spectra often pose special problems. This is because the highest energy electrons for such species may only be loosely associated with specific atoms (or pairs of atoms). In these situations, basis sets may need to be supplemented by **diffuse functions**, such as diffuse *s*- and *p*-type functions, on heavy (non-hydrogen) atoms (designated with a plus sign as in 6-31 + G* and 6-31 + G**). It may also be desirable to provide hydrogens with diffuse *s*-type functions (designated by two plus signs as in 6-31+ +G* and 6-31+ +G**).

15.8 SELECTION OF A THEORETICAL MODEL

By now, it should be apparent to the reader that many different models are available and useful in describing molecular geometry, reaction energies, and other properties. All of these models ultimately stem from the electronic Schrödinger equation, and they differ from each other both in the manner in which they treat electron correlation and in the nature of the atomic basis set. Each distinct combination (a theoretical model) leads to a scheme with its own particular characteristics (a theoretical model chemistry).

Hartree–Fock models may be seen as the parent model in that they treat electron correlation in the simplest possible manner, in effect, replacing instantaneous electron–electron interactions with average interactions. Despite their simplicity, Hartree–Fock models have proven to be remarkably successful in a large number of situations and remain a mainstay of computational chemistry.

As discussed earlier, **correlated models** can be broadly divided into two categories: density functional models, which provide an explicit empirical term in the Hamiltonian to account for electron correlation, and configuration interaction and Møller-Plesset models, which start from the Hartree–Fock description and then optimally mix together wave functions corresponding to the ground and various excited states. Each of these models exhibits its own particular characteristics.

Of course, no single theoretical model is likely to be ideal for all applications. A great deal of effort has gone into defining the limits of different models and judging the degree of success and the pitfalls of each. Most simply, success depends on the ability of a model to consistently reproduce known (experimental) data. This assumes that reliable experimental data are available or, at least, that errors in the data have been quantified. These include data on the geometries and conformations (shapes) of stable molecules, the enthalpies of chemical reactions (thermodynamics), and on such properties as vibrational frequencies (infrared spectra) and dipole moments. Quantum mechanical models may also be applied to high-energy molecules (reactive intermediates) for which reliable experimental data may be difficult to come by, and to reaction transition states, which may not even be directly observed much less characterized. Although no experimental transition-state structures are available with which to compare the results of the calculations, experimental kinetic data may be interpreted to provide information about activation energies. As an alternative, transition-state geometries can instead be compared with the results of high-level quantum chemical calculations.

The success of a quantum chemical model is not an absolute. Different properties and certainly different problems may require different levels of confidence to actually be of value. Neither is success sufficient. A model also needs to be practical for the task at hand. The nature and size of the system needs to be taken into account, as do the available computational resources and the experience and patience of the practitioner. Practical models usually do share one feature in common, in that they are not likely to be the best possible treatments to have been formulated. Compromise is almost always an essential component of model selection.

Oddly enough, the main problem faced by those who wish to apply computation to investigate chemistry is not the lack of suitable models but rather the excess of models. Quite simply, there are too many choices. In this spirit, consideration from this point on will be limited to just four theoretical models: Hartree–Fock models with 3-21G split-valence and 6-31G* polarization basis sets, the B3LYP/6-31G* density functional model, and the MP2/6-31G* model. Although all of these models can be routinely applied to molecules of considerable size, they differ by two orders of magnitude in the amount of computer time they require. Thus, it is quite important to know where the less time-consuming models perform satisfactorily and where the more time-consuming models are needed. Note that although this set of models has been successfully applied to a wide range of chemical problems, some problems may require more accurate and more time-consuming models.

It is difficult to quantify the overall computation time of a calculation, because it depends not only on the specific system and task at hand, but also on the sophistication of the computer program and the experience of the user. For molecules of moderate size (say, 10 atoms other than H), the HF/6-31G*, B3LYP/6-31G*, and MP2/6-31G* models would be expected to exhibit overall computation times in a ratio of roughly 1:1.5:10. The HF/3-21G model will require a third to half of the computation time required by the corresponding HF/6-31G* model, whereas the computation time of Hartree–Fock, B3LYP, and MP2 models with basis sets larger than 6-31G* will increase roughly as the cube (HF and B3LYP) and the fifth power (MP2) of the total number of basis functions. Geometry optimizations and frequency calculations are typically an order of magnitude more time-consuming than energy calculations, and the ratio will increase with increasing complexity (number of independent geometrical variables) of the system. Transition-state geometry optimizations are likely to be even more time-consuming than equilibrium geometry optimizations, due primarily to a poorer initial guess of the geometry.

Only a few calculated properties are examined in this discussion: equilibrium bond distances, reaction energies, conformational energy differences, and dipole moments. Comparisons between the results of the calculations and experimental data are few for each of these, but sufficient to establish meaningful trends.

15.8.1 Equilibrium Bond Distances

A comparison of calculated and experimentally determined carbon–carbon bond distances in hydrocarbons is provided in Table 15.9. Whereas errors in measured bond distances are typically on the order of ± 0.02 Å, experimental data for hydrocarbons and other small molecules presented here are better, and comparisons with the results of calculations to 0.01 Å are meaningful. In terms of mean absolute error, all four models perform admirably. The

TABLE 15.9 BOND DISTANCES IN HYDROCARBONS (Å)

| Bond | Hydrocarbon | Hartree–Fock | | B3LYP | MP2 | |
		3-21G	6-31G*	6-31G*	6-31G*	Experiment
C—C	But-1-yne-3-ene	1.432	1.439	1.424	1.429	1.431
	Propyne	1.466	1.468	1.461	1.463	1.459
	1,3-Butadiene	1.479	1.467	1.458	1.458	1.483
	Propene	1.510	1.503	1.502	1.499	1.501
	Cyclopropane	1.513	1.497	1.509	1.504	1.510
	Propane	1.541	1.528	1.532	1.526	1.526
	Cyclobutane	1.543	1.548	1.553	1.545	1.548
C=C	Cyclopropene	1.282	1.276	1.295	1.303	1.300
	Allene	1.292	1.296	1.307	1.313	1.308
	Propene	1.316	1.318	1.333	1.338	1.318
	Cyclobutene	1.326	1.322	1.341	1.347	1.332
	But-1-yne-ene	1.320	1.322	1.341	1.344	1.341
	1,3-Butadiene	1.320	1.323	1.340	1.344	1.345
	Cyclopentadiene	1.329	1.329	1.349	1.354	1.345
Mean absolute error		0.011	0.011	0.006	0.007	—

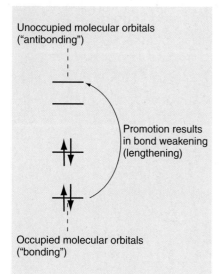

FIGURE 15.17

The promotion of electrons to unfilled orbitals reduces the bond strength and leads to bond lengthening.

B3LYP/6-31G* and MP2/6-31G* models perform better than the two Hartree–Fock models, due in most part to a sizable systematic error in carbon–carbon double bond lengths. With one exception, Hartree–Fock double bond lengths are shorter than experimental distances. This is easily rationalized. Treatment of electron correlation (for example, in the MP2 model) involves the promotion of electrons from occupied molecular orbitals (in the Hartree–Fock wave function) to unoccupied molecular orbitals. Because occupied molecular orbitals are (generally) net bonding in character, and because unoccupied molecular orbitals are (generally) net antibonding in character, any promotions should result in bond weakening (lengthening) as illustrated in Figure 15.17. This in turn suggests that bond lengths from limiting Hartree–Fock models are necessarily shorter than exact values. Apparently, Hartree–Fock models with 3-21G and 6-31G* basis set are close enough to the limit for this behavior to be seen.

Consistent with such an interpretation, B3LYP/6-31G* and MP2/6-31G* double bond lengths do not show a systematic trend and are both smaller and larger than experimental values.

Similar comments can be made regarding CN and CO bond distances (Table 15.10). In terms of mean absolute error, the performance of the B3LYP/6-31G* and MP2/6-31G* models is similar to that previously noted for CC bonds in hydrocarbons, but the two Hartree–Fock models do not fare as well. Note that although bond distances from the HF/6-31G* model are constantly smaller than measured values, in accord with the picture presented for hydrocarbons, HF/3-21G bond lengths do not show such a trend. It appears that the 3-21G basis set is not large enough to closely mirror the Hartree–Fock limit in this instance. Most bond distances obtained from the B3LYP/6-31G* and MP2/6-31G* models are actually slightly larger than experimental distances. (The CN bond length in formamide is the only significant exception.) Bond lengthening from the corresponding (6-31G* basis set) Hartree–Fock model is a direct consequence of treatment of electron correlation.

In summary, all four models provide a plausible account of equilibrium bond lengths. Similar comments also apply to bond angles and more generally to the structures of larger molecules.

15.8.2 Finding Equilibrium Geometries

As detailed at the start of this chapter, an equilibrium structure is a point on a multidimensional potential energy surface for which all first energy derivatives with respect to the individual geometrical coordinates are zero, and for which the diagonal representation of the matrix of second energy derivatives has all positive elements. In simple terms, an equilibrium structure corresponds to the bottom of a well on the overall potential energy surface.

Not all equilibrium structures correspond to (kinetically) stable molecules, meaning that not all equilibrium structures will correspond to detectable (let alone characterizable) molecules. Stability also implies that the well is deep enough to preclude the molecule from being transformed into other molecules. Equilibrium structures that no doubt exist but cannot be detected easily are commonly referred to as **reactive intermediates**.

TABLE 15.10 BOND DISTANCES IN MOLECULES WITH HETEROATOMS (Å)

| Bond | Hydrocarbon | Hartree–Fock | | B3LYP | MP2 | |
		3-21G	6-31G*	6-31G*	6-31G*	Experiment
C—N	Formamide	1.351	1.349	1.362	1.362	1.376
	Methyl isocyanide	1.432	1.421	1.420	1.426	1.424
	Trimethylamine	1.471	1.445	1.456	1.455	1.451
	Aziridine	1.490	1.448	1.473	1.474	1.475
	Nitromethane	1.497	1.481	1.499	1.488	1.489
C—O	Formic acid	1.350	1.323	1.347	1.352	1.343
	Furan	1.377	1.344	1.364	1.367	1.362
	Dimethyl ether	1.435	1.392	1.410	1.416	1.410
	Oxirane	1.470	1.401	1.430	1.439	1.436
Mean absolute error		0.017	0.018	0.005	0.005	—

Geometry optimization does not guarantee that the final geometry will have a lower energy than any other geometry of the same molecular formula. All that it guarantees is that the geometry will correspond to a local minimum, that is, a geometry the energy of which is lower than that of any similar geometry. However, the resulting structure may still not be the lowest energy structure possible for the molecule. Other local minima that are actually lower in energy may exist and be accessible via low-energy rotations about single bonds or puckering of rings. The full collection of local minima are referred to as **conformers**. Finding the lowest energy conformer or global minimum requires repeated geometry optimization starting with different initial geometries as discussed in Section 15.8.6.

Finding an equilibrium structure is not as difficult a chore as it might first appear. For one, chemists know a great deal about what molecules look like and can usually provide an excellent starting structure. Also, optimization to a minimum is an important task in many fields of science and engineering, and very good algorithms exist with which to accomplish it.

Geometry optimization is an iterative process. The energy and energy gradient (first derivatives with respect to all geometrical coordinates) are calculated for the initial geometry, and this information is then used to project a new geometry. This process needs to continue until the lowest energy or optimized geometry is reached. Three criteria must be satisfied before a geometry is accepted as optimized. First, successive geometry changes must not lower the energy by more than a specified (small) value. Second, the energy gradient must closely approach zero. Third, successive iterations must not change any geometrical parameter by more than a specified (small) value.

In principle, geometry optimization carried out in the absence of symmetry must result in a local energy minimum. On the other hand, the imposition of symmetry may result in a geometry that is not an energy minimum. The most conservative tactic is always to optimize geometry in the absence of symmetry. If this is not practical, and if there is any doubt whatsoever that the symmetrical structure actually corresponds to an energy minimum, then it is always possible to verify that the geometry located indeed corresponds to a local minimum by calculating vibrational frequencies for the final (optimized) geometry. These should all be real numbers. The presence of an imaginary frequency indicates that the corresponding coordinate is not an energy minimum.

Problems P15.5–P15.12

15.8.3 Reaction Energies

Reaction energy comparisons are divided into three parts: bond dissociation energies, energies of reactions relating structural isomers, and relative proton affinities. Bond dissociation reactions are the most disruptive, because they lead to a change in the number of electron pairs. Structural isomer comparisons maintain overall electron pair count, but swap bonds of one kind for those of another. Relative proton affinity comparisons are least disruptive in that they maintain the numbers of each kind of formal chemical bond and lead only to subtle changes in the molecular environment.

A comparison of homolytic bond dissociation energies based on calculation and on experimental thermochemical data is provided in Table 15.11. Hartree–Fock models with the 3-21G and 6-31G* basis set turn in a very poor performance, paralleling the

TABLE 15.11 HOMOLYTIC BOND DISSOCIATION ENERGIES (kJ/mol)					
	Hartree–Fock		B3LYP	MP2	
Bond Dissociation Reaction	**3-21G**	**6-31G***	**6-31G***	**6-31G***	**Experiment**
$CH_3-CH_3 \rightarrow \cdot CH_3 + \cdot CH_3$	285	293	406	414	406
$CH_3-NH_2 \rightarrow \cdot CH_3 + \cdot NH_2$	247	243	372	385	389
$CH_3-OH \rightarrow \cdot CH_3 + \cdot OH$	222	247	402	410	410
$CH_3-F \rightarrow \cdot CH_3 + \cdot F$	247	289	473	473	477
$NH_2-NH_2 \rightarrow \cdot NH_2 + \cdot NH_2$	155	142	293	305	305
$HO-OH \rightarrow \cdot OH + \cdot OH$	13	0	226	230	230
$F-F \rightarrow \cdot F + \cdot F$	−121	−138	176	159	159
Mean absolute error	190	186	9	2	—

TABLE 15.12 RELATIVE ENERGIES ISOMER - REFERENCE COMPOUND OF STRUCTURAL ISOMERS (kJ/mol)

		Hartree–Fock		B3LYP	MP2	
Reference Compound	Isomer	3-21G	6-31G*	6-31G*	6-31G*	Experiment
Acetonitrile	Methyl isocyanide	88	100	113	121	88
Acetaldehyde	Oxirane	142	130	117	113	113
Acetic acid	Methyl formate	54	54	50	59	75
Ethanol	Dimethyl ether	25	29	21	38	50
Propyne	Allene	13	8	−13	21	4
	Cyclopropene	167	109	92	96	92
Propene	Cyclopropane	59	33	33	17	29
1,3-butadiene	2-Butyne	17	29	33	17	38
	Cyclobutane	75	54	50	33	46
	Bicyclo [1.1.0] butane	192	126	117	88	109
Mean absolute error		32	13	12	15	—

poor performance of limiting Hartree–Fock models (see discussion in Section 15.4.1). Bond energies are far too small, consistent with the fact that the total correlation energy for the radical products is smaller than that for the reactant due to a decrease in the number of electron pairs. B3LYP/6-31G* and especially MP2/6-31G* models fare much better (results for the latter are well inside the experimental error bars).

"Which of several possible structural isomers is most stable?" and "What are the relative energies of any reasonable alternatives?" are without doubt two of the most commonly asked questions relating to thermochemistry. The ability to pick out the lowest energy isomer and at least rank the energies of higher energy isomers is essential to the success of any model. A few comparisons of this kind are found in Table 15.12.

In terms of mean absolute error, three of the four models provide similar results. The HF/3-21G model is inferior. None of the models is actually up to the standard that would make it a useful reliable replacement for experimental data (<5 kJ/mol). More detailed comparisons provide insight. For example, Hartree–Fock models consistently disfavor small-ring cyclic structures over their unsaturated cyclic isomers. This is also generally the case for the B3LYP/6-31G* model (albeit with much reduced error), but not with the MP2/6-31G* model, which generally shows the opposite behavior.

The final comparison (Table 15.13) is between proton affinities of a variety of nitrogen bases and that of methylamine as a standard, that is,

$$BH^+ + NH_3 \longrightarrow B + NH_4^+$$

TABLE 15.13 PROTON AFFINITIES OF NITROGEN BASES RELATIVE TO THE PROTON AFFINITY OF METHYLAMINE (kJ/mol)

	Hartree–Fock		B3LYP	MP2	
Base	3-21G	6-31G*	6-31G*	6-31G*	Experiment
Ammonia	−42	−46	−42	−42	−38
Aniline	−38	−17	−21	−13	−10
Methylamine	0	0	0	0	0
Dimethylamine	29	29	25	25	27
Pyridine	17	29	25	13	29
Trimethylamine	46	46	38	38	46
Diazabicyclooctane	67	71	59	54	60
Quinuclidine	79	84	75	71	75
Mean absolute error	8	5	4	6	—

This type of comparison is important not only because proton affinity (basicity) is an important property in its own right, but also because it typifies property comparisons among sets of closely related compounds. The experimental data derive from equilibrium measurements in the gas phase and are accurate to ±4 kJ/mol. In terms of mean absolute error, all four models turn in similar and respectable accounts over what is a considerable range (>100 kJ/mol) of experimental proton affinities. The HF/3-21G model is clearly the poorest performer, due primarily to underestimation of the proton affinities of aniline and pyridine.

Problems P15.13–P15.16

15.8.4 Energies, Enthalpies, and Gibbs Energies

Quantum chemical calculations account for reaction thermochemistry by combining the energies of reactant and product molecules at 0 K. Additionally, the residual energy of vibration (the so-called zero point energy discussed in Section 7.1) is ignored. On the other hand, experimental thermochemical comparisons are most commonly based on enthalpies or Gibbs energies of 1 mol of real (vibrating) molecules at some finite temperature (typically 298.15 K). The connection between the various quantities involves the mass, equilibrium geometry, and set of vibrational frequencies for each of the molecules in the reaction. Calculating thermodynamic quantities is straightforward but, because it requires frequencies, consumes significant computation time, and is performed only where necessary.

We start with two familiar thermodynamic relationships:

$$\Delta G = \Delta H - T\Delta S$$

$$\Delta H = \Delta U + \Delta(PV) \approx \Delta U$$

where G is the Gibbs energy, H is the enthalpy, S is the entropy, U is the internal energy, and T, P, and V are the temperature, pressure, and volume, respectively. For most cases, the $\Delta(PV)$ term can be ignored, meaning that the $\Delta U = \Delta H$ at 0 K. Three steps are required to obtain ΔG the first two to relate the quantum mechanical energy at 0 K to the internal energy at 298 K, and the third to calculate the Gibbs energy.

1. *Correction of the internal energy for finite temperature.* The change in internal energy from 0 K to a finite temperature, T, $\Delta U(T)$, is given by

$$\Delta U(T) = \Delta U_{trans}(T) + \Delta U_{rot}(T) + \Delta U_{vib}(T)$$

$$\Delta U_{trans}(T) = \frac{3}{2}RT$$

$$\Delta U_{rot}(T) = \frac{3}{2}RT \ (RT \text{ for a linear molecule})$$

$$\Delta U_{vib}(T) = U_{vib}(T) - U_{vib}(0\,K) = N_A \sum_i^{vibrational\ frequencies} \frac{h\nu_i}{e^{h\nu_i/kT} - 1}$$

The ν_i are vibrational frequencies, N_A is Avogadro's number, and R, k, and h are the gas constant, the Boltzmann constant, and the Planck constant, respectively.

2. *Correction for zero point vibrational energy.* The zero point vibrational energy, $U_{vib}(0)$, of n moles of a molecule at 0 K is given by

$$U_{vib}(0) = nN_A E_{zero\ point} = \frac{1}{2}nN_A \sum_i^{vibrational\ frequencies} h\nu_i$$

where N_A is Avogadro's number. This calculation also requires knowledge of the vibrational frequencies.

3. *Entropy.* The absolute entropy, S, of n moles of a molecule may be written as a sum of terms:

$$S = S_{trans} + \alpha S_{rot} + S_{vib} + S_{el} - nR[\ln(nN_A) - 1]$$

$$S_{trans} = nR\left[\frac{3}{2} + \ln\left(\left(\frac{nRT}{P}\right)\left(\frac{2\pi mkT}{h^2}\right)^{3/2}\right)\right]$$

$$S_{rot} = nR\left[\frac{3}{2} + \ln\left(\left(\frac{\sqrt{\pi}}{\sigma}\right)\left(\frac{1}{\beta hcB_A}\right)^{1/2}\left(\frac{1}{\beta hcB_B}\right)^{1/2}\left(\frac{1}{\beta hcB_C}\right)^{1/2}\right)\right]$$

$$S_{vib} = nR\sum_i^{vibrational\ frequencies}\left[\left(\frac{\mu_i}{e^{\mu_i} - 1}\right) + \ln\left(\frac{1}{1 - e^{-\mu_i}}\right)\right]$$

$$S_{el} = nR \ln g_0$$

In these equations, m is the molecular mass, B_i is the rotational constant, σ is the symmetry number, $\mu_i = h\nu_i/kT$, $\beta = 1/kT$, c is the speed of light, and g_0 is the degeneracy of the electronic ground state (normally equal to one).

Note that molecular structure enters into the rotational entropy, and the vibrational frequencies enter into the vibrational entropy. The translational entropy cancels in a (mass) balanced reaction, and the electronic entropy is usually zero because for most molecules $g_0 = 1$. Note also that the expression provided for the vibrational contribution to the entropy goes to infinity as the vibrational frequency goes to zero. This is clearly wrong and has its origin in the use of the linear harmonic oscillator approximation to derive the expression. Unfortunately, low-frequency modes are the major contributors to the vibrational entropy, and caution must be exercised when using the preceding formulas for the case of frequencies below approximately 300 cm^{-1}. In this case, the molecular partition function must be evaluated term by term rather than assuming the classical limit.

Problem P15.17

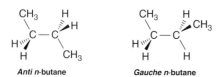

Anti n-butane   *Gauche n-butane*

FIGURE 15.18

Structures of two *n*-butane conformers.

Problems P15.18–P15.20

15.8.5 Conformational Energy Differences

Rotation around single bonds may give rise to rotational isomers (conformers). Because bond rotation is almost always a very low energy process, this means that more than one conformer may be present at equilibrium. For example, *n*-butane exists as a mixture of *anti* and *gauche* conformers, as shown in Figure 15.18. The same reasoning carries over to molecules incorporating flexible rings, where conformer interconversion may be viewed in terms of a process involving restricted rotation about the bonds in the ring.

Knowledge of the conformer of lowest energy and, more generally, the distribution of conformers is important because many molecular properties depend on detailed molecular shape. For example, whereas *gauche n*-butane is a polar molecule (albeit very weakly polar), *anti n*-butane is nonpolar, and the value of the dipole moment for an actual sample of *n*-butane would depend on how much of each species was actually present.

Experimentally, a great deal is known about the conformational preferences of molecules in the solid state (from X-ray crystallography). Far less is known about the conformations of isolated (gas-phase) molecules, although there are sufficient data to allow gross assessment of practical quantum chemical models. Experimental conformational energy differences are somewhat more scarce, but accurate data are available for a few very simple (two-conformer) systems. Comparison of these data with the results of calculations for hydrocarbons is provided in Table 15.14. These are expressed in terms of the energy of the high-energy conformer relative to that of the low-energy conformer.

All models correctly assign the ground-state conformer in all molecules. In terms of mean absolute error, the MP2/6-31G* model provides the best description of conformational energy differences and the HF/6-31G* model the worst description. Hartree–Fock models consistently overestimate differences (the sole exception is for the *trans/gauche* energy difference in 1,3-butadiene from the 3-21G model), in some cases by large amounts (nearly 5 kJ/mol for the *equatorial/axial* energy difference in *tert*-butylcyclohexane from the 3-21G model). Correlated models also typically (but not always) overestimate energy differences, but the magnitudes of the errors are much smaller than those seen for Hartree–Fock models.

TABLE 15.14 CONFORMATIONAL ENERGY IN HYDROCARBONS (kJ/mol)

Hydrocarbon	Low-Energy/ High-Energy Conformer	Hartree–Fock		B3LYP	MP2	
		3-21G	6-31G*	6-31G*	6-31G*	Experiment
n-Butane	*anti/gauche*	3.3	4.2	3.3	2.9	2.80
1-Butene	*skew/cis*	3.3	2.9	1.7	2.1	0.92
1,3-Butadiene	*trans/gauche*	11.3	13.0	15.1	10.9	12.1
Cyclohexane	*chair/twist-boat*	27.2	28.5	26.8	27.6	19.7–25.9
Methylcyclohexane	*equatorial/axial*	7.9	9.6	8.8	7.9	7.32
tert-Butylcyclohexane	*equatorial/axial*	27.2	25.5	22.2	23.4	22.6
cis-1,3-Dimethylcyclohexane	*equatorial/axial*	26.4	27.2	25.1	23.8	23.0
Mean absolute error		1.9	2.3	1.3	0.9	—

15.8.6 Determining Molecular Shape

Many molecules can (and do) exist in more than one shape, arising from different arrangements around single bonds and/or flexible rings. The problem of identifying the lowest energy conformer (or the complete set of conformers) in simple molecules such as *n*-butane and cyclohexane is straightforward, but rapidly becomes difficult as the number of conformational degrees of freedom increases, due simply to the large number of arrangements that need to be examined. For example, a systematic search on a molecule with N single bonds and step size of $360°/M$, would need to examine M^N conformers. For a molecule with three single bonds and a step size of $120°$ ($M = 3$), this leads to 27 conformers; for a molecule with eight single bonds, more than 6500 conformers would need to be considered. It is clear that it will not always be possible to look everywhere, and sampling techniques will need to replace systematic procedures for complex molecules. The most common of these are so-called Monte Carlo methods (which randomly sample different conformations) and molecular dynamics techniques (which follow motion among different conformers in time).

15.8.7 Alternatives to Bond Rotation

Single-bond rotation (including restricted bond rotation in flexible rings) is the most common mechanism for conformer interconversion, but it is by no means the only mechanism. At least two other processes are known: inversion and pseudorotation. Inversion is normally associated with pyramidal nitrogen or phosphorus and involves a planar (or nearly planar) transition state, for example, in ammonia, as shown in Figure 15.19.

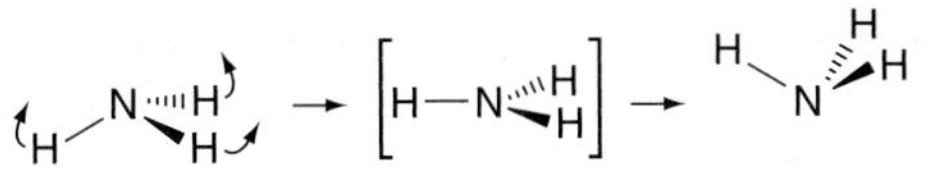

FIGURE 15.19

Inversion of NH_3 leads to its mirror image.

Note that the starting and ending molecules are mirror images. Were the nitrogen to be bonded to three different groups and were the nitrogen lone pair to be counted as a fourth group, inversion would result in a change in chirality at this center. Pseudorotation, which is depicted in Figure 15.20, is normally associated with trigonal bipyramidal

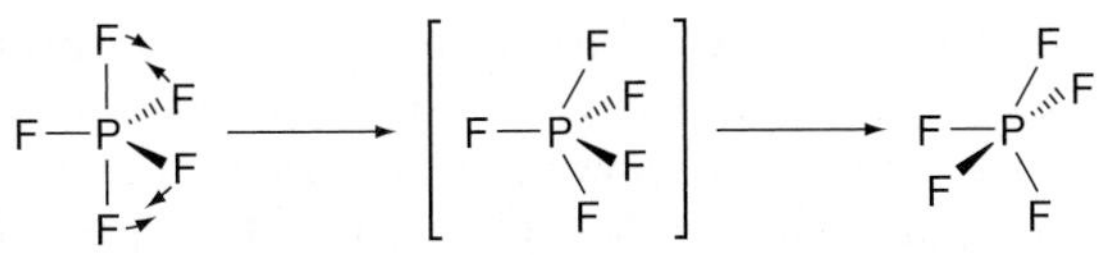

FIGURE 15.20

Pseudorotation leads to exchange of *equatorial* and *axial* positions at a trigonal bipyramidal phosphorus center.

Problem P15.21

phosphorus and involves a square-based-pyramidal transition state. Note that pseudo-rotation interconverts *equatorial* and *axial* positions on phosphorus.

Both inversion of pyramidal nitrogen and pseudorotation around trigonal bipyramidal phosphorus are very low energy processes ($<20-30$ kJ/mol) and generally proceed rapidly at 298 K. On the other hand, inversion of pyramidal phosphorus is more difficult (100 kJ/mol) and is inhibited at 298 K.

15.8.8 Dipole Moments

Calculated dipole moments for a selection of diatomic and small polyatomic molecules are compared with experimental values in Table 15.15. The experimental data cover a wide spectrum of molecules, from carbon monoxide, which is close to being nonpolar, to lithium fluoride, which is close to being fully ionic. All models provide a good overall account of this range. In terms of mean absolute error, the HF/3-21G model fares worst and the MP2/6-31G* model fares best, but the differences are not large.

TABLE 15.15 DIPOLE MOMENTS IN DIATOMIC AND SMALL POLYATOMIC MOLECULES (debyes)

Molecule	Hartree–Fock		B3LYP	MP2	
	3-21G	6-31G*	6-31G*	6-31G*	Experiment
Carbon monoxide	0.4	0.3	0.1	0.2	0.11
Ammonia	1.8	1.9	1.9	2.0	1.47
Hydrogen fluoride	2.2	2.0	1.9	1.9	1.82
Water	2.4	2.2	2.1	2.2	1.85
Methyl fluoride	2.3	2.0	1.7	1.9	1.85
Formaldehyde	2.7	2.7	2.2	2.3	2.34
Hydrogen cyanide	3.0	3.2	2.9	3.0	2.99
Lithium hydride	6.0	6.0	5.6	5.8	5.83
Lithium fluoride	5.8	6.2	5.6	5.9	6.28
Mean absolute error	0.3	0.2	0.2	0.1	—

Note that dipole moments from the two Hartree–Fock models are consistently larger than experimental values, the only exception being for lithium fluoride. This is in accord with the behavior of the limiting Hartree–Fock model (see discussion earlier in Section 15.4.4) and may now easily be rationalized. Recognize that electron promotion from occupied to unoccupied molecular orbitals (either implicit or explicit in all electron correlation models) takes electrons from "where they are" (negative regions) to "where they are not" (positive regions), as illustrated in Figure 15.21. In formaldehyde, for example, the lowest energy promotion is from a nonbonded lone pair localized on oxygen into a π^* orbital principally concentrated on carbon. As a result, electron correlation acts to reduce overall charge separation and to reduce the dipole moment in comparison with the Hartree–Fock value. This is supported by the fact that dipole moments from correlated (B3LYP/6-31G* and MP2/6-31G*) calculations are not consistently larger than experimental values.

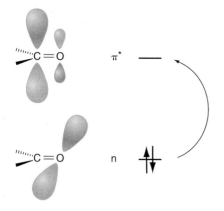

FIGURE 15.21

Accounting for electron correlation involves excitations such as the $n \rightarrow \pi^*$ transition in formaldehyde, which moves charge from oxygen to carbon and reduces the dipole moment in the carbonyl group.

15.8.9 Atomic Charges: Real or Make Believe?

Charges are part of the everyday language of chemistry and, aside from geometries and energies, are certainly the most commonly demanded quantities from quantum chemical calculations. Charge distributions not only assist chemists in assessing overall molecular structure and stability, but they also tell them about the chemistry that molecules can undergo. Consider, for example, the two resonance structures that a chemist would draw for acetate anion, $CH_3CO_2^-$, as shown in Figure 15.22. This figure indicates that the two CO bonds are equivalent and should be intermediate in length between single

FIGURE 15.22

The two Lewis structures of the acetate ion.

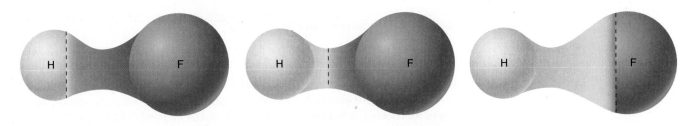

FIGURE 15.23

Three different ways of partitioning the electrons in hydrogen fluoride between hydrogen and fluorine.

and double linkages, and that the negative charge is evenly distributed on the two oxygens. Taken together, these two observations suggest that the acetate ion is delocalized and therefore particularly stable.

Despite their obvious utility, atomic charges are not measurable properties, nor can they be determined uniquely from calculations. Although the total charge on a molecule (the total nuclear charge and the sum of the charges on all of the electrons) is well defined, and although overall charge distribution may be inferred from such observables as the dipole moment, it is not possible to assign discrete atomic charges. To do this would require accounting both for the nuclear charge and for the charge of any electrons uniquely associated with the particular atom. Although it is reasonable to assume that the nuclear contribution to the total charge on an atom is simply the atomic number, it is not at all obvious how to partition the total electron distribution by atoms. Consider, for example, the electron distribution for the heteronuclear diatomic molecule hydrogen fluoride, shown in Figure 15.23. Here, the surrounding contour is a particular electron density surface that, for example, corresponds to a van der Waals surface and encloses a large fraction of the total electron density. In this picture, the surface has been drawn to suggest that more electrons are associated with fluorine than with hydrogen. This is entirely reasonable, given the known polarity of the molecule, that is, $^{\delta+}H—F^{\delta-}$, as evidenced experimentally by the direction of its dipole moment. It is, however, not at all apparent how to divide this surface between the two nuclei. Are any of the divisions shown in Figure 15.23 better than the others? No! Atomic charges are not molecular properties, and it is not possible to provide a unique definition (or even a definition that will satisfy all). We can calculate (and measure using X-ray diffraction) molecular charge distributions, that is, the number of electrons in a particular volume of space, but it is not possible to uniquely partition them among the atomic centers.

Despite the obvious problem with their definition, atomic charges are still useful, and several recipes have been formulated to calculate them. The simplest of these, now referred to as **Mulliken population analysis**, was discussed in Section 13.3.

MATHEMATICAL DESCRIPTION OF THE MULLIKEN POPULATION ANALYSIS

The Mulliken population analysis starts from the definition of the electron density, $\rho(\mathbf{r})$, in the framework of the Hartree–Fock model:

$$\rho(\mathbf{r}) = \sum_{\mu}^{basis\ functions} \sum_{\nu} P_{\mu\nu}\phi_{\mu}(\mathbf{r})\phi_{\nu}(\mathbf{r}) \qquad (15.55)$$

where $P_{\mu\nu}$ is an element of the density matrix [see Equation (15.27)], and the summations are carried out over all atom-centered basis functions, ϕ_{μ}. Summing over basis functions and integrating over all space leads to an expression for the total number of electrons, n:

$$\int \rho(\mathbf{r})\,d\mathbf{r} = \sum_{\mu}^{basis\ functions} \sum_{\nu} P_{\mu\nu} \int \phi_{\mu}(\mathbf{r})\phi_{\nu}(\mathbf{r})\,d\mathbf{r}$$

$$= \sum_{\mu}^{basis\ functions} \sum_{\nu} P_{\mu\nu} S_{\mu\nu} = n \qquad (15.56)$$

where $S_{\mu\nu}$ are elements of the overlap matrix:

$$S_{\mu\nu} = \int \phi_\mu(\mathbf{r})\phi_\nu(\mathbf{r}) \, d\mathbf{r} \tag{15.57}$$

Analogous expressions can be constructed for correlated models. The important point is that it is possible to equate the total number of electrons in a molecule to a sum of products of density matrix and overlap matrix elements as follows:

$$\sum_\mu^{\substack{basis\ functions}} \sum_\nu^{\substack{basis\ functions}} P_{\mu\nu}S_{\mu\nu} = \sum_\mu^{\substack{basis\ functions}} P_{\mu\mu} + 2 \sum_{\mu\neq\nu}^{\substack{basis\ functions}} P_{\mu\nu} S_{\mu\nu} = n \tag{15.58}$$

It is reasonable (but not necessarily correct) to assign any electrons associated with a particular diagonal element, $\mu\mu$, to that atom on which the basis function ϕ_μ is located. It is also reasonable to assign electrons associated with off-diagonal elements, $\mu\nu$, where both ϕ_μ and ϕ_ν reside on the same atom, to that atom. However, it is not apparent how to partition electrons from density matrix elements, $\mu\nu$, where ϕ_μ and ϕ_ν reside on different atoms. Mulliken provided a recipe. Give each atom half of the total, which is very simple but completely arbitrary! According to Mulliken's scheme, the gross electron population, q_μ, for basis function ϕ_μ is given by:

$$q_\mu = P_{\mu\mu} + \sum_\nu^{\substack{basis\ functions}} P_{\mu\nu}S_{\mu\nu} \tag{15.59}$$

Atomic electron populations, q_A, and atomic charges, Q_A, follow, where Z_A is the atomic number of atom A:

$$q_A = \sum_\mu^{\substack{basis\ functions \\ on\ atom\ A}} q_\mu \tag{15.60}$$

$$Q_A = Z_A - q_A \tag{15.61}$$

An entirely different approach to providing atomic charges is to fit the value of some property that has been calculated based on the exact wave function with that obtained from representation of the electronic charge distribution in terms of a collection of atom-centered charges. One choice of property is the electrostatic potential, ε_p. This represents the energy of interaction of a unit positive charge at some point in space, p, with the nuclei and the electrons of a molecule:

$$\varepsilon_p = \sum_A^{\substack{nuclei}} \frac{Z_A e^2}{4\pi\varepsilon_0 R_{Ap}} - \frac{e^2}{4\pi\varepsilon_0} \sum_\mu^{\substack{basis\ functions}} \sum_\nu P_{\mu\nu} \int \frac{\phi_\mu(\mathbf{r})\phi_\nu(\mathbf{r})}{r_p} \, d\mathbf{r} \tag{15.62}$$

Z_A are atomic numbers, $P_{\mu\nu}$ are elements of the density matrix, and R_{Ap} and r_p are distances separating the point charges from the nuclei and electrons, respectively. The first summation is over nuclei and the second pair of summations is over basis functions.

Operationally, electrostatic-fit charges are obtained by first defining a grid of points surrounding the molecule, then calculating the electrostatic potential at each of these grid points, and finally providing a best (least-squares) fit of the potential at the grid points to an approximate electrostatic potential, ε_p^{approx}, based on replacing the nuclei and electron distribution by a set of atom-centered charges, Q_A, subject to overall charge balance:

$$\varepsilon_p^{approx} = \sum_A^{\substack{nuclei}} \frac{e^2 Q_A}{4\pi\varepsilon_0 R_{Ap}} \tag{15.63}$$

Problem P15.22

The lack of uniqueness of the procedure results from selection of the grid points.

15.8.10 Transition-State Geometries and Activation Energies

Quantum chemical calculations need not be limited to the description of the structures and properties of stable molecules, that is, molecules that can actually be observed and characterized experimentally. They may as easily be applied to molecules that are highly reactive (reactive intermediates) and, even more interesting, to transition states, which cannot be observed let alone characterized. However, activation energies (the energy difference between the reactants and the transition state) can be inferred from experimental kinetic data. The complete absence of experimental data on transition-state geometries complicates assessment of the performance of different models. However, it is possible to get around this by assuming that some particular (high-level) model yields reasonable geometries for the transition state, and then to compare the results of the other models with this standard. The MP2/6-311+G** model has been selected as the standard.

The most conspicuous difference between the structure data presented in Table 15.16 and previous comparisons involving equilibrium bond distances is the much larger variation among different models. This should not come as a surprise. Transition states represent a compromise situation in which some bonds are being broken while others are being formed, and the potential energy surface around the transition state would be expected to be flat, meaning that large changes in geometry are expected to lead only to small changes in the energy. In terms of mean absolute deviations from the standard, the MP2/6-31G* model fares best and the two Hartree–Fock models fare worst, but all models give reasonable results. In terms of individual comparisons, the largest deviations among different models correspond to making and breaking single bonds. In such situations, the potential energy surface is expected to be quite flat.

As discussed in Section 15.2.4, an experimental activation energy can be obtained from the temperature dependence of the measured reaction rate by way of the Arrhenius equation, Equation (15.15). This first requires that a rate law be postulated [Equation (15.14)]. Association of the activation energy with the difference in energies between reactants and transition state (as obtained from quantum chemical calculations) requires the further assumption that all reacting molecules pass through the transition state. In effect, this implies that all reactants have the same energy, or that none has energy in excess of that needed to reach the transition state. This is the essence of transition-state theory.

TABLE 15.16 KEY BOND DISTANCES IN TRANSITION STATES FOR ORGANIC REACTIONS (Å)

Reaction/Transition State	Bond Length	Hartree–Fock		B3LYP	MP2	
		3-21G	6-31G*	6-31G*	6-31G*	6-311+G**
	a	1.88	1.92	1.90	1.80	1.80
	b	1.29	1.26	1.29	1.31	1.30
	c	1.37	1.37	1.38	1.38	1.39
	d	2.14	2.27	2.31	2.20	2.22
	e	1.38	1.38	1.38	1.39	1.39
	f	1.39	1.39	1.40	1.41	1.41
	a	1.40	1.40	1.42	1.43	1.43
	b	1.37	1.38	1.39	1.39	1.39
	c	2.11	2.12	2.11	2.02	2.07
	d	1.40	1.40	1.41	1.41	1.41
	e	1.45	1.45	1.48	1.55	1.53
	f	1.35	1.36	1.32	1.25	1.25
	a	1.39	1.38	1.40	1.40	1.40
	b	1.37	1.37	1.38	1.38	1.38
	c	2.12	2.26	2.18	2.08	2.06
	d	1.23	1.22	1.24	1.25	1.24
	e	1.88	1.74	1.78	1.83	1.83
	f	1.40	1.43	1.42	1.41	1.41
Mean absolute deviation from MP2/6-311+G**		0.05	0.05	0.03	0.01	—

TABLE 15.17 ABSOLUTE ACTIVATION ENERGIES FOR ORGANIC REACTIONS (kJ/mol)

Reaction	Hartree–Fock 3-21G	Hartree–Fock 6-31G*	B3LYP 6-31G*	MP2 6-31G*	MP2 6-311+G**	Experiment
$CH_3NC \longrightarrow CH_3CN$	238	192	172	180	172	159
$HCO_2CH_2CH_3 \longrightarrow HCO_2H + C_2H_4$	259	293	222	251	234	167, 184
(structure → structure)	192	238	142	117	109	151
(structure → structure)	176	205	121	109	105	130
(structure + ‖ → structure)	126	167	84	50	38	84
(structure → structure + C_2H_4)	314	356	243	251	230	—
$HCNO + C_2H_2 \longrightarrow$ (structure)	105	146	50	33	38	—
(structure → structure)	230	247	163	159	142	—
(structure → structure)	176	197	151	155	142	—
(structure → structure + CO_2)	247	251	167	184	172	—
(structure + $SO_2 \longrightarrow$ structure + SO_2)	205	205	92	105	92	—
Mean absolute deviation from MP2/6-311+G**	71	100	17	13	—	—

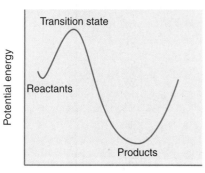

FIGURE 15.24

The potential energy surface for a reaction is typically represented by a one-dimensional representation of the energy as a function of the reaction coordinate.

Absolute activation energies for a small series of organic reactions are provided in Table 15.17. As with transition-state geometries, results from the practical models are compared with those of the standard, MP2/6-311+G**. Overall, the performance of Hartree–Fock models is very poor. In most cases, the activation energies are overestimated by large amounts. This is not surprising in view of previous comparisons involving homolytic bond dissociation energies (see Table 15.11), which were too small. The argument that might be given here is that a transition state is typically more tightly bound than the reactants, meaning that correlation effects will be greater. The B3LYP/6-31G* and MP2/6-31G* perform much better, and lead to errors (relative to the standard) that are comparable to those previously noted for reaction energy comparisons.

15.8.11 Finding a Transition State

The usual picture of a chemical reaction in terms of a one-dimensional potential energy (or reaction coordinate) diagram is shown in Figure 15.24. The vertical axis corresponds to the energy of the system, and the horizontal axis (reaction coordinate) corresponds to the geometry of the system. The starting point on the diagram (reactants) is an energy minimum, as is the ending point (products). Motion along the reaction coordinate is assumed to be continuous and to pass through a single energy maximum called the transition state. As described in Section 15.2.1, a transition state on a real many-dimensional potential energy surface corresponds to a point that is actually an energy minimum in all but one dimension and an energy maximum along the reaction coordinate. The obvious analogy is to the crossing of a mountain range, the goal of which is simply to get from one side of the range to the other side with minimal effort.

Crossing over the top of a "mountain" (pathway A), which corresponds to crossing through an energy maximum on a (two-dimensional) potential energy surface, accomplishes the goal, as shown in Figure 15.25. However, it is not likely to be the chosen pathway. This is because less effort (energy) will be expended by going through a "pass" between two "mountains" (pathway B), a maximum in one dimension but a minimum in the other dimension. This is referred to as a saddle point and corresponds to a transition state.

A single molecule may have many transition states (some corresponding to real chemical reactions and others not), and merely finding *a* transition state does not guarantee that it is *the* transition state, meaning that it is at the top of the lowest energy pathway that smoothly connects reactants and products. Although it is possible to verify the smooth connection of reactants and products, it will generally not be possible to know with complete certainty that what has been identified as the transition state is in fact the lowest energy structure over which the reaction might proceed, or whether in fact the actual reaction proceeds over a transition state that is not the lowest energy structure.

The fact that transition states, like the reactants and products of a chemical reaction, correspond to well-defined structures, means that they can be fully characterized from calculation. However, this is one area where the results of calculation cannot be tested, except with reference to chemical intuition. For example, it is reasonable to expect that the transition state for the unimolecular isomerization of methyl isocyanide to acetonitrile takes the form of a three-membered ring,

$$H_3C-N{\equiv}C \longrightarrow \underset{N=C}{\overset{CH_3}{\diagdown}} \longrightarrow N{\equiv}C-CH_3$$

in accord with the structure actually calculated, which is shown in Figure 15.26.

It is also reasonable to expect that the transition state for pyrolysis of ethyl formate leading to formic acid and ethylene will take the form of a six-membered ring:

$$\underset{H_2CH}{\overset{H_2C}{\mid}}\overset{O}{\underset{O}{\diagup}}\overset{}{\underset{}{\diagdown}}C \longrightarrow \underset{H_2C}{\overset{H_2C}{\mid}}\overset{O}{\underset{H}{\cdots}}\overset{}{\underset{}{}}CH \longrightarrow \underset{H_2C}{\overset{H_2C}{\parallel}} + \underset{HO}{\overset{O}{\diagdown}}CH$$

This expectation agrees with the result of the calculation, as shown in Figure 15.27.

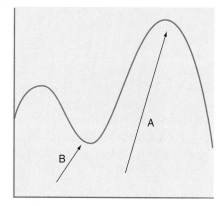

FIGURE 15.25

A reaction in two dimensions is analogous to crossing a mountain range. Pathway A, which goes over the top of a mountain, requires more effort to traverse than pathway B, which goes through a pass between two mountains.

FIGURE 15.26

The calculated transition state for the isomerization of methyl isocyanide to acetonitrile is consistent with a three-membered ring in the reaction scheme shown.

Problems P15.23–P15.25

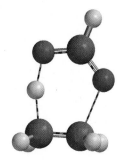

FIGURE 15.27

The calculated transition state in the pyrolysis of ethyl formate is consistent with a six-membered ring in the reaction scheme shown.

15.9 GRAPHICAL MODELS

In addition to numerical quantities (bond lengths and angles, energies, dipole moments, and so on), quantum chemical calculations furnish a wealth of information that is best displayed in the form of images. Among the results of calculations that have proven to be of value are the molecular orbitals themselves, the electron density, and the electrostatic potential. These can all be expressed as three-dimensional functions of the coordinates. One way to display them on a two-dimensional video screen (or on a printed page) is to define a surface of constant value, a so-called isovalue surface or, more simply, isosurface:

$$f(x, y, z) = constant \tag{15.64}$$

The value of the constant may be chosen to reflect a particular physical observable of interest, for example, the "size" of a molecule in the case of display of electron density.

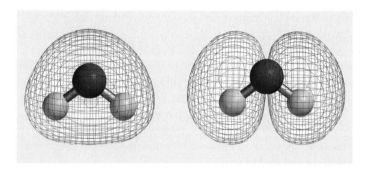

FIGURE 15.28

These molecular orbitals can be identified with the O — H bonds in water.

15.9.1 Molecular Orbitals

As detailed in Section 15.3, molecular orbitals, ψ, are written in terms of linear combinations of basis functions, ϕ, which are centered on the individual nuclei:

$$\psi_i = \sum_{\mu}^{basis\ functions} c_{\mu i}\phi_\mu \tag{15.65}$$

Although it is tempting to associate a molecular orbital with a particular bond, more often than not this is inappropriate. Molecular orbitals will generally be spread out (delocalized) over the entire molecule, whereas bonds are normally associated with a pair of atoms. Also, molecular orbitals, unlike bonds, show the symmetry of the molecule. For example, the equivalence of the two OH bonds in water is revealed by the two molecular orbitals best describing OH bonding as shown in Figure 15.28.

Molecular orbitals, in particular, the highest energy occupied molecular orbital (the **HOMO**) and the lowest energy unoccupied molecular orbital (the **LUMO**), are often quite familiar to chemists. The former holds the highest energy (most available) electrons and should be subject to attack by electrophiles, whereas the latter provides the lowest energy space for additional electrons and should be subject to attack by nucleophiles. For example, the HOMO in acetone is in the plane of the molecule, indicating that attack by an electrophile, for example, a proton, will occur here, as shown in Figure 15.29, whereas the LUMO is out of plane on the carbonyl carbon, consistent with the known nucleophilic chemistry.

15.9.2 Orbital Symmetry Control of Chemical Reactions

Woodward and Hoffmann, building on the earlier ideas of Fukui, first clearly pointed out how the symmetries of the HOMO and LUMO (together referred to as the **frontier molecular orbitals**) could be used to rationalize why some chemical reactions proceed easily whereas others do not. For example, the fact that the HOMO in *cis*-1,3-butadiene is able to interact favorably with the LUMO in ethene suggests that the two molecules

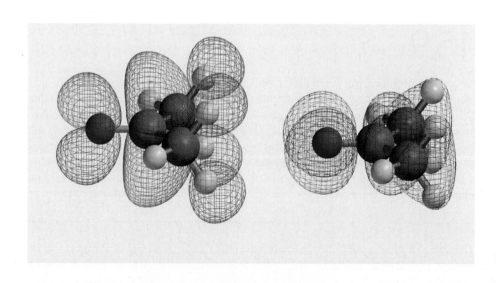

FIGURE 15.29

The HOMO (left) and LUMO (right) for acetone identify regions where electrophilic and nucleophilic attack, respectively, are likely to occur.

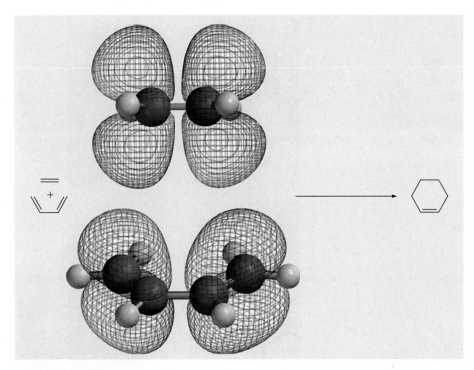

FIGURE 15.30

The HOMO of butadiene (bottom) is able to interact with the LUMO of ethene (top), resulting in cycloaddition, in agreement with experiment.

Problems P15.26–P15.31

should readily combine in a concerted manner to form cyclohexene in a process called Diels-Alder cycloaddition. This process is depicted in Figure 15.30. On the other hand, interaction between the HOMO on one ethene and the LUMO on another ethene is not favorable, as illustrated in Figure 15.31, and concerted addition to form cyclobutane would not be expected. Reactions which are allowed or forbidden because of orbital symmetry have been collected under what is now known as the "Woodward-Hoffmann" rules. For their work, Hoffmann and Fukui shared the Nobel Prize in chemistry in 1981.

15.9.3 Electron Density

The electron density, $\rho(\mathbf{r})$, is a function of the coordinates $\mathbf{r}$, defined such that $\rho(\mathbf{r})d\mathbf{r}$ is the number of electrons inside a small volume $d\mathbf{r}$. This is what is measured in an X-ray diffraction experiment. Electron density $\rho(\mathbf{r})$ is written in terms of a sum of products of basis functions, ϕ_μ:

$$\rho(\mathbf{r}) = \sum_{\mu}^{basis\ functions} \sum_{\nu} P_{\mu\nu}\phi_\mu(\mathbf{r})\phi_\nu(\mathbf{r}) \qquad (15.66)$$

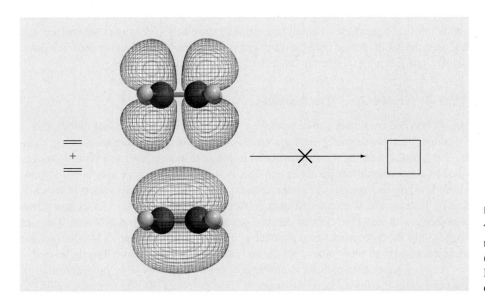

FIGURE 15.31

The HOMO of ethene (bottom) is not able to interact with the corresponding LUMO (top), suggesting that cycloaddition is not likely to occur, in agreement with experiment.

FIGURE 15.32

Electron density surfaces for cyclohexanone corresponding to three different values of the electron density: 0.4 electron/a_0^3 (left), 0.1 electrons/a_0^3 (center), and 0.002 electrons/a_0^3 (right). Conventional skeletal and space-filling models appear underneath the last two electron density surfaces.

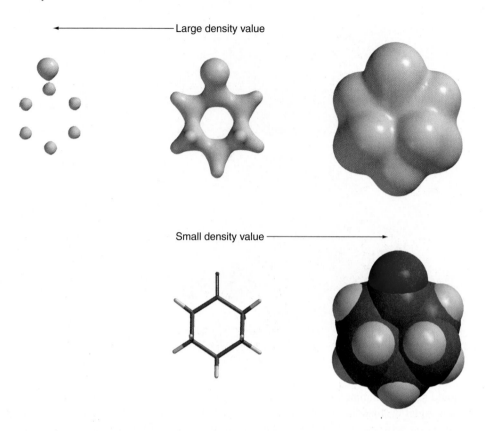

Large density value

Small density value

where $P_{\mu\nu}$ are elements of the density matrix [Equation (15.27)]. The electron density can be portrayed in terms of a surface (an **electron density surface**) with the size and shape of the surface being given by the value of the density, for example, in cyclohexanone in Figure 15.32.

Depending on the value, isodensity surfaces can either serve to locate atoms (left image in Figure 15.32), to delineate chemical bonds (center image), or to indicate overall molecular size and shape (right image). The regions of highest electron density surround the heavy (non-hydrogen) atoms in a molecule. This is the basis of X-ray crystallography, which locates atoms by identifying regions of high electron density. Also interesting are regions of lower electron density. For example, a 0.1 electrons/a_0^3 isodensity surface for cyclohexanone conveys essentially the same information as a conventional skeletal structure model, that is, it depicts the locations of bonds. A surface of 0.002 electrons/a_0^3 provides a good fit to conventional space-filling models and, hence, serves to portray overall molecular size and shape. As is the case with the space-filling model, this definition of molecular size is completely arbitrary (except that it closely matches experimental data on how closely atoms fit together in crystalline solids). A single parameter, namely, the value of the electron density at the surface, has replaced the set of atomic radii used for space-filling models. These latter two electron density surfaces are examined in more detail in the following section.

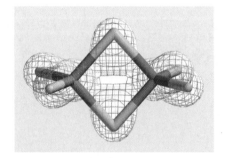

FIGURE 15.33

An electron density surface for diborane shows that there is no boron–boron bond.

15.9.4 Where Are the Bonds in a Molecule?

An electron density surface can be employed to reveal the location of bonds in a molecule. Of course, chemists routinely employ a variety of tactics to depict chemical bonding, ranging from pencil sketches (Lewis structures) to physical models such as Dreiding models. The most important advantage of electron density surfaces is that they can be applied to elucidate bonding and not only to portray bonding in cases where the location of bonds is known. For example, the electron density surface for diborane (Figure 15.33) clearly shows a molecule with very little electron density concentrated between the two borons. This fact suggests that the appropriate Lewis structure of the two shown in Figure 15.34 is the one that lacks a boron–boron bond, rather than the one that shows the two borons directly bonded.

Another important application of electron density surfaces is to the description of the bonding in transition states. An example is the pyrolysis of ethyl formate, leading to

FIGURE 15.34

Two possible Lewis structures of diborane differ in that only one has a boron–boron bond.

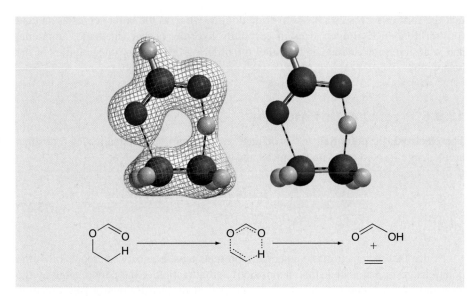

FIGURE 15.35

An electron density structure for the transition state in the pyrolysis of ethyl formate shows a six-membered ring consistent with the conventional Lewis picture shown in the reaction scheme.

Problem P15.32

formic acid and ethylene, which is illustrated in Figure 15.35. The electron density surface offers clear evidence of a **late transition state**, meaning that the CO bond is nearly fully cleaved and the migrating hydrogen is more tightly bound to oxygen (as in the product) than to carbon (as in the reactant).

15.9.5 How Big Is a Molecule?

The size of a molecule can be defined according to the amount of space that it takes up in a liquid or solid. The so-called space-filling or CPK model has been formulated to portray molecular size, based on fitting the experimental data to a set of atomic radii (one for each atom type). Although this simple model is remarkably satisfactory overall, some problematic cases do arise, in particular for atoms that may adopt different oxidation states, for example, Fe^0 in $FeCO_5$ versus Fe^{II} in $FeCl_4^{2-}$.

Because the electrons—not the underlying nuclei—dictate overall molecular size, the electron density provides an alternate measure of how much space molecules actually take up. Unlike space-filling models, electron density surfaces respond to changes in the chemical environment and allow atoms to adjust their sizes in response to different environments. An extreme example concerns the size of hydrogen in main-group hydrides.

As seen in Figure 15.36, electron density surfaces reveal that the hydrogen in lithium hydride is much larger than that in hydrogen fluoride, consistent with the fact

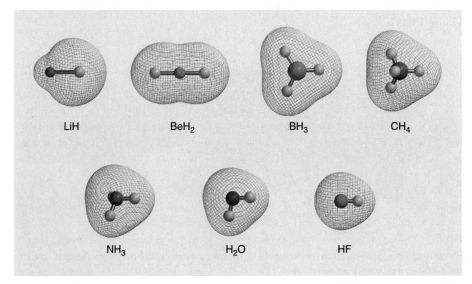

FIGURE 15.36

Electron density surfaces are shown for hydrides of lithium to fluorine.

Problems P15.33–P.15.34

that the former serves as a base (hydride donor), whereas the latter serves as an acid (proton donor). Hydrogen sizes in beryllium hydride, borane, methane, ammonia, and water are intermediate and parallel the ordering of the electronegativities of the heavy atom.

15.9.6 Electrostatic Potential

The **electrostatic potential**, ε_p, is defined as the energy of interaction of a positive point charge located at p with the nuclei and electrons of a molecule:

$$\varepsilon_p = \sum_A^{nuclei} \frac{e^2 Z_A}{4\pi\varepsilon_0 R_{Ap}} - \sum_\mu^{basis\ functions} \sum_\nu P_{\mu\nu} \int \frac{\phi_\mu^*(\mathbf{r})\phi_\nu(\mathbf{r})}{\mathbf{r}_p} d\tau \qquad \textbf{(15.67)}$$

Notice that the electrostatic potential represents a balance between repulsion of the point charge by the nuclei (first summation) and attraction of the point charge by the electrons (second summation). $P_{\mu\nu}$ are elements of the density matrix [see Equation (15.27)] and the ϕ are atomic basis functions.

15.9.7 Visualizing Lone Pairs

The octet rule dictates that each main-group atom in a molecule will be surrounded by eight valence electrons. These electrons can either be tied up in bonds (two electrons for a single bond, four electrons for a double bond, 6 electrons for a triple bond), or can remain with the atom as a nonbonded or lone pair of electrons. Although you cannot actually see bonds, you can see their consequence (the atoms to which bonds are made). On this basis, lone pairs would seem to be completely invisible, because there are no telltale atoms. However, the fact that the electrons in lone pairs should be highly accessible suggests another avenue. Regions of space around a molecule where the potential is negative suggest an excess of electrons. To the extent that lone pairs represent electron-rich environments, they should be revealed by electrostatic potential surfaces. A good example is provided by negative electrostatic potential surfaces for ammonia, water, and hydrogen fluoride, as shown in Figure 15.37.

The electron-rich region in ammonia is in the shape of a lobe pointing in the fourth tetrahedral direction, whereas that in water takes the form of a crescent occupying two tetrahedral sites. At first glance, the electrostatic potential surface for hydrogen fluoride is nearly identical to that in ammonia. Closer inspection reveals that rather than pointing away from the fluorine (as it points away from ammonia), the surface encloses the atom. All in all, these three surfaces are entirely consistent with conventional Lewis structures for the three hydrides shown in Figure 15.38.

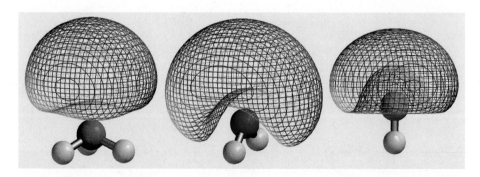

FIGURE 15.37

Electrostatic potential surfaces for ammonia (left), water (center), and hydrogen fluoride (right) are useful in depicting lone pairs.

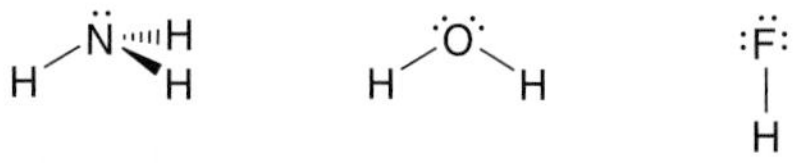

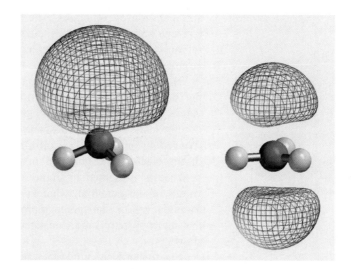

FIGURE 15.38

Lewis structures for ammonia, water, and hydrogen fluoride.

FIGURE 15.39

Electrostatic potential surfaces show that the lone pair is directed to one side of pyramidal ammonia, but is equally distributed on both sides of planar ammonia.

A related comparison between electrostatic potential surfaces for ammonia in both the observed pyramidal and unstable trigonal planar geometries is shown in Figure 15.39. As previously mentioned, the former depicts a lobe pointing in the fourth tetrahedral direction, and the electrostatic potential surface for the planar trigonal arrangement shows two equal out-of-plane lobes. This is, of course, consistent with the fact that pyramidal ammonia has a dipole moment (with the negative end pointing in the direction of the lone pair), whereas planar ammonia does not have a dipole moment.

15.9.8 Electrostatic Potential Maps

Graphical models need not be restricted to portraying a single quantity. Additional information can be presented in terms of a property map on top of an isosurface, where different colors can be used to portray different property values. Most common are maps on electron density surfaces. Here the surface can be used to designate overall molecular size and shape, and the colors to represent the value of some property at various locations on the surface. The most commonly used property map is the **electrostatic potential map**, schematically depicted in Figure 15.40. This gives the value of the electrostatic potential at locations on a particular surface, most commonly a surface of electron density corresponding to overall molecular size.

To see how an electrostatic potential map (and by implication any property map) is constructed, first consider both a density surface and a particular (negative) electrostatic potential surface for benzene, as shown in Figure 15.41. Both of these

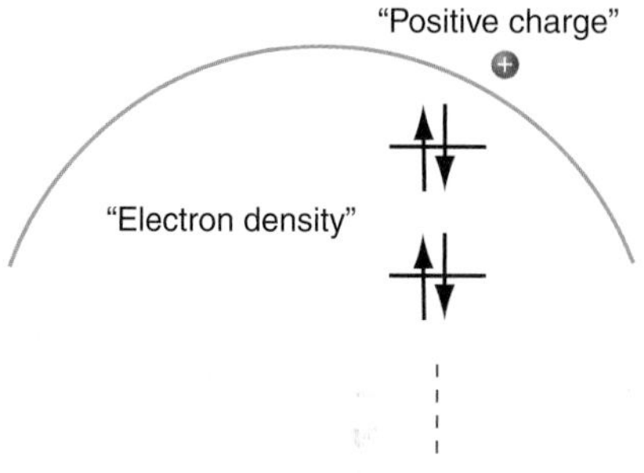

"Positive charge"

"Electron density"

FIGURE 15.40

An electrostatic potential map shows the value of the electrostatic potential at all locations on a surface of electron density (corresponding to overall size and shape).

Problem P15.35

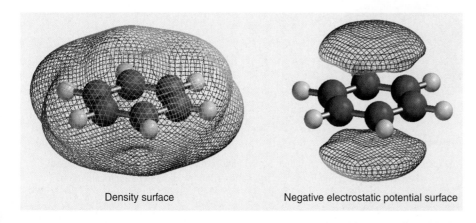

Density surface

Negative electrostatic potential surface

FIGURE 15.41

An electron density surface and a negative electrostatic potential surface are shown for benzene.

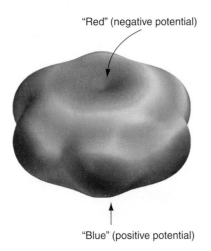

"Red" (negative potential)

"Blue" (positive potential)

FIGURE 15.42

An electrostatic potential map of benzene.

surfaces convey structure. The density surface reveals the size and shape of benzene, and the negative electrostatic potential surface delineates in which regions surrounding benzene a particular (negative) electrostatic potential will be felt.

Next, consider making a map of the value of the electrostatic potential on the density surface (an electrostatic potential map), using colors to designate values of the potential. This leaves the density surface unchanged (insofar as it represents the size and shape of benzene), but replaces the gray-scale image (conveying only structural information) with a color image (conveying the value of the electrostatic potential *in addition to* structure). An electrostatic map for benzene is presented in Figure 15.42. Colors near red represent large negative values of the potential, whereas colors near blue represent large positive values (orange, yellow, and green represent intermediate values of the potential). Note that the π system is red, consistent with the (negative) potential surface previously shown.

Electrostatic potential maps are used for a myriad of purposes other than rapidly conveying which regions of a molecule are likely to be electron rich and which are likely to be electron poor. For example, they can be used to distinguish between molecules in which charge is localized from those where it is delocalized.

Compare the electrostatic potential maps in Figure 15.43 for the planar (top) and perpendicular (bottom) structures of the benzyl cation. The latter reveals a heavy concentration of positive charge (blue color) on the benzylic carbon and perpendicular to the plane of the ring. This is consistent with the notion that only a single Lewis structure can be drawn. On the other hand, planar benzyl cation shows no such buildup of positive charge on the benzylic carbon, but rather delocalization onto *ortho* and *para* ring carbons, consistent with the fact that several Lewis structures can be drawn.

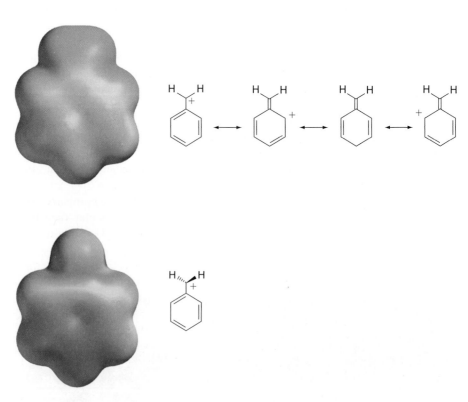

FIGURE 15.43

An electrostatic potential map for planar benzyl cation (top) shows delocalization of positive charge, whereas that for perpendicular benzyl cation (bottom) shows charge localization.

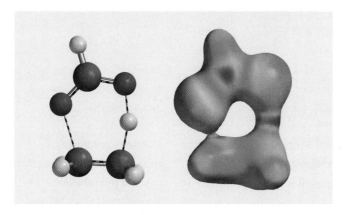

FIGURE 15.44

An electrostatic potential map is useful in depicting the charge distribution in the transition state for the pyrolysis of ethyl formate.

Electrostatic potential maps can also be employed to characterize transition states in chemical reactions. A good example is pyrolysis of ethyl formate (leading to formic acid and ethylene):

Here, the electrostatic potential map shown in Figure 15.44 (based on an electron density surface appropriate to identify bonds) clearly shows that the hydrogen being transferred (from carbon to oxygen) is positively charged, that is, it is an electrophile.

Problems P15.36–P15.37

15.10 CONCLUSION

Quantum chemical calculations are rapidly becoming a viable alternative to experiments as a means to investigate chemistry. Continuing rapid advances in computer hardware and software technology will only further this trend and lead to even wider adoption among mainstream chemists. Calculations are already able to properly account for molecular structure and energetics, among other important quantities. Perhaps most intriguing is the ability of the calculations to deal with highly reactive molecules, which may be difficult to synthesize, and with reaction transition states, which cannot be observed at all. In this regard, calculations open up entirely new avenues for chemical research.

Quantum chemical calculations do have limitations. Most conspicuous of these is the trade-off between accuracy and cost. Practical quantum chemical models do not always yield results that are sufficiently accurate to actually be of value, and models that are capable of yielding accurate results may not yet be practical for the system of interest. Second, a number of quantities important to chemists cannot yet be routinely and reliably obtained from calculations. The most important limitation, however, is that, for the most part, calculations apply strictly to isolated molecules (gas phase), whereas much if not most chemistry is carried out in solution. Practical models that take solvent into account need to be developed and then tested to determine their accuracy and limitations.

The prognosis is very bright. Your generation will have at its disposal a whole range of powerful tools for exploring and understanding chemistry, just like the generations before you were given technologies such as the laser to exploit. It is all part of the natural evolution of the science.

Vocabulary

activation energy
Born–Oppenheimer approximation
CID method
CIS method
conformer
contracted functions
correlated models
correlation energy
density functional theory
diffuse functions
diffusion-controlled reactions
double-electron promotions
electron density surface
electronic Schrödinger equation
electrostatic potential
electrostatic potential map
exothermic
frontier molecular orbitals
frozen core approximation
full configuration interaction

Gaussian functions
harmonic frequencies
Hartree–Fock approximation
Hartree–Fock energy
Hartree–Fock equations
Hartree–Fock model
highest-occupied molecular orbital (HOMO)
homolytic bond dissociation
kinetic product
kinetically controlled
late transition state
linear combination of atomic orbitals (LCAO) approximation
lowest-occupied molecular orbital (LUMO)
minimal basis set
Møller-Plesset models
Mulliken population analysis
normal coordinates

polarization basis set
polarization functions
preexponential factor
rate constant
reaction coordinate
reaction coordinate diagrams
reactive intermediates
Roothaan–Hall equations
self-consistent-field (SCF) procedure
size consistent
single-electron promotions
Slater determinant
Slater-type orbitals
split-valence basis set
STO-3G basis set
theoretical model
theoretical model chemistry
thermodynamic product
thermodynamically controlled
variational

Problems

P15.1 The assumption that the reaction coordinate in going from *gauche* to *anti* n-butane is a simple torsion is an over-simplification, because other geometrical changes no doubt also occur during rotation around the carbon–carbon bond, for example, changes in bond lengths and angles. Examine the energy profile for *n*-butane ("n-butane" on the Spartan download) and plot the change in distance of the central CC bond and CCC bond angle as a function of the torsion angle. Are the bond length and bond angle nearly identical or significantly different (>0.02 Å and >2°) for the two equilibrium forms of *n*-butane? Are the two parameters nearly identical or significantly different between the *anti* form and either or both of the transition states? Explain your results.

P15.2 Ammonia provides a particularly simple example of the dependence of vibrational frequencies on the atomic masses and of the use of vibrational frequencies to distinguish between a stable molecule and a transition state. First examine the vibrational spectrum of pyramidal ammonia ("ammonia" on the Spartan download).

a. How many vibrational frequencies are there? How does this number relate to the number of atoms? Are all frequencies real numbers or are one or more imaginary numbers? Describe the motion associated with each frequency and characterize each as being primarily bond stretching, angle bending, or a combination of the two. Is bond stretching or angle bending easier? Do the stretching motions each involve a single NH bond or do they involve combinations of two or three bonds?

b. Next, consider changes to the vibrational frequencies of ammonia as a result of substituting deuteriums for hydrogens ("perdeuteroammonia" on the Spartan download). Are the frequencies in ND_3 larger, smaller, or unchanged from those in NH_3? Are any changes greater for motions that are primarily bond stretching or motions that are primarily angle bending?

c. Finally, examine the vibrational spectrum of an ammonia molecule that has been constrained to a planar geometry ("planar ammonia" on the Spartan download). Are all the frequencies real numbers? If not, describe the motions associated with any imaginary frequencies and relate them to the corresponding motion(s) in the pyramidal equilibrium form.

P15.3 The presence of the carbonyl group in a molecule is easily confirmed by an intense line in the infrared spectrum around 1700 cm^{-1} that corresponds to the $C=O$ stretching vibration. Locate this line in the calculated infrared spectrum of acetone ("acetone" on the Spartan download) and note its position in the overall spectrum (relative to the positions of the other lines) and the intensity of the absorption.

a. Speculate why this line is a reliable diagnostic for carbonyl functionality.

b. Examine the lowest frequency mode for acetone and then the highest frequency mode. Describe each and relate to the relative ease of difficulty of the associated motion.

P15.4 Chemists recognize that the cyclohexyl radical is likely to be more stable than cyclopentylmethyl radical, because they know that six-membered rings are more stable than five-membered rings and, more importantly, that secondary radicals are more stable than primary radicals. However, much important chemistry is not controlled by what is most stable (thermodynamics) but rather by what forms most readily (kinetics). For example, loss of bromine from 6-bromohexene leading initially to hex-5-enyl radical, results primarily in product from cyclopentylmethyl radical.

The two possible interpretations for the experimental result are that the reaction is thermochemically controlled but that our understanding of radical stability is wrong or that the reaction is kinetically controlled.

a. First, see if you can rule out the first possibility. Examine structures and total energies for cyclohexyl and cyclopentylmethyl radicals ("cyclohexyl and cyclopentylmethyl radicals" on the Spartan download). Which radical, cyclohexyl or cyclopentylmethyl, is more stable (lower in energy)? Is the energy difference large enough such that only the more stable radical is likely to be observed? (Recall that at room temperature an energy difference of 12 kJ/mol corresponds to a product ratio of >99:1.) Do you conclude that ring closure is under thermodynamic control?

b. The next objective is to establish which ring closure, to cyclohexyl radical or to cyclopentylmethyl radical, is easier; that is, which product, cyclohexane or methylcyclopentane, is the kinetic product? Examine structures and total energies for the transition states for the two ring closures ("to cyclohexyl and cyclopentylmethyl radicals" on the Spartan download). Which radical, cyclohexyl or cyclopentylmethyl, is more easily formed?

c. Consider the following relationships between transition-state energy difference, $\Delta E^{\ddagger}$, and the ratio of major to minor (kinetic) products, calculated from the Boltzmann distribution:

$\Delta E^{\ddagger}$ (kJ/mol)	Major : Minor (room temperature)
4	~90:10
8	~95:5
12	~99:1

What is the approximate ratio of products suggested by the calculations? How does this compare with what is observed? Do you conclude that ring closure is under kinetic control?

P15.5 VSEPR (valence state electron pair repulsion) theory was formulated to anticipate the local geometry about an atom in a molecule (see discussion in Section 14.1). All that is required is the number of electron pairs surrounding the atom, broken down into bonded pairs and nonbonded (lone) pairs. For example, the carbon in carbon tetrafluoride is surrounded by four electron pairs, all of them tied up in CF bonds, whereas the sulfur in sulfur tetrafluoride is surrounded by five electron pairs, four of which are tied up in SF bonds with the fifth being a lone pair.

VSEPR theory is based on two simple rules. The first is that electron pairs (either lone pairs or bonds) will seek to avoid each other as much as possible. Thus, two electron pairs will lead to a linear geometry, three pairs to a trigonal planar geometry, four pairs to a tetrahedral geometry, five pairs to a trigonal bipyramidal geometry, and six pairs to an octahedral geometry. Although this knowledge is sufficient to assign a geometry for a molecule such as carbon tetrafluoride (tetrahedral), it is not sufficient to specify the geometry of a molecule such as sulfur tetrafluoride. Does the lone pair assume an *equatorial* position on the trigonal bipyramid leading to a seesaw geometry, or an *axial* position leading to a trigonal pyramidal geometry?

Seesaw **Trigonal pyramidal**

The second rule, that lone pairs take up more space than bonds, clarifies the situation. The seesaw geometry in which the lone pair is 90° to two of the SF bonds and 120° to the other two bonds is preferable to the trigonal pyramidal geometry in which three bonds are 90° to the lone pair.

Although VSEPR theory is easy to apply, its results are strictly qualitative and often of limited value. For example, although the model tells us that sulfur tetrafluoride adopts a seesaw geometry, it does not reveal whether the trigonal pyramidal structure (or any other structure) is an energy minimum, and if it is, what its energy is relative to the seesaw form. Also it has little to say when more than six electron pairs are present. For example, VSEPR theory tells us that xenon hexafluoride is not octahedral, but it does not tell us what geometry the molecule actually assumes. Hartree–Fock molecular orbital calculations provide an alternative.

a. Optimize the structure of SF_4 in a seesaw geometry (C_{2v} symmetry) using the HF/3-21G model and calculate vibrational frequencies (the infrared spectrum). This calculation is necessary to verify that the energy is at a minimum. Next, optimize the geometry of SF_4 in a trigonal pyramidal geometry and calculate its vibrational frequencies. Is the seesaw structure an energy minimum?

What leads you to your conclusion? Is it lower in energy than the corresponding trigonal pyramidal structure in accordance with VSEPR theory? What is the energy difference between the two forms? Is it small enough that both might actually be observed at room temperature? Is the trigonal pyramidal structure an energy minimum?

b. Optimize the geometry of XeF_6 in an octahedral geometry (O_h symmetry) using the HF/3-21G model and calculate vibrational frequencies. Next, optimize XeF_6 in a geometry that is distorted from octahedral (preferably a geometry with C_1 symmetry) and calculate its vibrational frequencies. Is the octahedral form of XeF_6 an energy minimum? What leads you to your conclusion? Does distortion lead to a stable structure of lower energy?

P15.6 Each of the carbons in ethane is surrounded by four atoms in a roughly tetrahedral geometry; each carbon in ethene is surrounded by three atoms in a trigonal planar geometry and each carbon in acetylene by two atoms in a linear geometry. These structures can be rationalized by suggesting that the valence $2s$ and $2p$ orbitals of carbon are able to combine either to produce four equivalent sp^3 hybrids directed toward the four corners of a tetrahedron, or three equivalent sp^2 hybrids directed toward the corners of an equilateral triangle with a p orbital left over, or two equivalent sp hybrids directed along a line with two p orbitals left over. The $2p$ atomic orbitals extend farther from carbon than the $2s$ orbital. Therefore, sp^3 hybrids will extend farther than sp^2 hybrids, which in turn will extend farther than sp hybrids. As a consequence, bonds made with sp^3 hybrids should be longer than those made with sp^2 hybrids, which should in turn be longer than those made with sp hybrids.

a. Obtain equilibrium geometries for ethane, ethene, and acetylene using the HF/6-31G* model. Is the ordering in CH bond lengths what you expect on the basis of the hybridization arguments? Using the CH bond length in ethane as a standard, what is the percent reduction in CH bond lengths in ethene? In acetylene?

b. Obtain equilibrium geometries for cyclopropane, cyclobutane, cyclopentane, and cyclohexane using the HF/6-31G* model. Are the CH bond lengths in each of these molecules consistent with their incorporating sp^3-hybridized carbons? Note any exceptions.

c. Obtain equilibrium geometries for propane, propene, and propyne using the HF/6-31G* model. Is the ordering of bond lengths the same as that observed for the CH bond lengths in ethane, ethene, and acetylene? Are the percent reductions in bond lengths from the standard (propane) similar ($\pm 10\%$) to those seen for ethene and acetylene (relative to ethane)?

P15.7 The bond angle about oxygen in alcohols and ethers is typically quite close to tetrahedral ($109.5°$), but opens up significantly in response to extreme steric crowding, for example, in going from *tert*-butyl alcohol to di-*tert*-butyl ether:

This is entirely consistent with the notion that while lone pairs take up space, they can be "squeezed" to relieve crowding. Another way to relieve unfavorable steric interactions (without changing the position of the lone pairs) is to increase the CO bond distance.

a. Build *tert*-butyl alcohol and di-*tert*-butyl ether and optimize the geometry of each using the HF/6-31G* model. Are the calculated bond angles involving oxygen in accord with the values given earlier, in particular with regard to the observed increase in bond angle? Do you see any lengthening of the CO bond in the ether over that in the alcohol? If not, or if the effect is very small (<0.01 Å), speculate why not.

b. Next, consider the analogous trimethylsilyl compounds Me_3SiOH and $Me_3SiOSiMe_3$. Calculate their equilibrium geometries using the HF/6-31G* model. Point out any similarities and any differences between the calculated structures of these compounds and their *tert*-butyl analogues. In particular, do you see any widening of the bond angle involving oxygen in response to increased steric crowding? Do you see lengthening of the SiO bond in the ether over that of the alcohol? If not, rationalize what you do see.

P15.8 Water contains two acidic hydrogens that can act as hydrogen-bond donors and two lone pairs that can act as hydrogen-bond acceptors:

Given that all are tetrahedrally disposed around oxygen, this suggests two reasonable structures for the hydrogen-bonded dimer of water, $(H_2O)_2$, one with a single hydrogen bond and one with two hydrogen bonds:

Whereas the second seems to make better use of water's attributes, in doing so, it imposes geometrical restrictions on the dimer.

Build the two dimer structures. Take into account that the hydrogen-bond distance (O · · · H) is typically on the order of 2 Å. Optimize the geometry of each using the HF/6-31G* model and, following this, calculate vibrational frequencies.

Which structure, singly or doubly hydrogen bonded, is more stable? Is the other (higher energy) structure also an energy minimum? Explain how you reached your conclusion. If the dimer with the single hydrogen bond is more stable, speculate what this has told you about the geometric requirements of hydrogen bonds. Based on your experience with water dimer, suggest a "structure" for liquid water.

P15.9 For many years, a controversy raged concerning the structures of so-called "electron-deficient" molecules, that is, molecules with insufficient electrons to make normal two-atom, two-electron bonds. Typical is ethyl cation, $C_2H_5^+$, formed from protonation of ethene.

Open **Hydrogen bridged**

Is it best represented as an open Lewis structure with a full positive charge on one of the carbons, or as a hydrogen-bridged structure in which the charge is dispersed onto several atoms? Build both open and hydrogen-bridged structures for ethyl cation. Optimize the geometry of each using the B3LYP/6-31G* model and calculate vibrational frequencies. Which structure is lower in energy, the open or hydrogen-bridged structure? Is the higher energy structure an energy minimum? Explain your answer.

P15.10 One of the most powerful attractions of quantum chemical calculations over experiments is their ability to deal with any molecular system, stable or unstable, real or imaginary. Take as an example the legendary (but imaginary) kryptonite molecule. Its very name gives us a formula, KrO_2^{2-}, and the fact that this species is isoelectronic with the known linear molecule, KrF_2, suggests that it too should be linear.

a. Build KrF_2 as a linear molecule (F—Kr—F), optimize its geometry using the HF/6-31G* model, and calculate vibrational frequencies. Is the calculated KrF bond distance close to the experimental value (1.89 Å)? Does the molecule prefer to be linear or does it want to bend? Explain how you reached this conclusion.

b. Build KrO_2^{2-} as a linear molecule (or as a bent molecule if the preceding analysis has shown that KrF_2 is not linear), optimize its structure using the HF/6-31G* model, and calculate vibrational frequencies. What is the structure of KrO_2^{2-}?

P15.11 Discussion of the VSEPR model in Section 14.1 suggested a number of failures, in particular, in CaF_2 and $SrCl_2$, which (according to the VSEPR) should be linear but which are apparently bent, and in SeF_6^{2-} and $TeCl_6^{2-}$, which should not be octahedral but apparently are. Are these really failures or does the discrepancy lie with the fact that the experimental structures correspond to the solid rather than the gas phase (isolated molecules)?

a. Obtain equilibrium geometries for linear CaF_2 and $SrCl_2$ and also calculate vibrational frequencies (infrared spectra). Use the HF/3-21G model, which has actually proven to be quite successful in describing the structures of main-group inorganic molecules. Are the linear structures for CaF_2 and $SrCl_2$ actually energy minima? Elaborate. If one or both are not, repeat your optimization starting with a bent geometry.

b. Obtain equilibrium geometries for octahedral SeF_6^{2-} and $TeCl_6^{2-}$ and also calculate vibrational frequencies. Use the HF/3-21G model. Are the octahedral structures for SeF_6^{2-} and $TeCl_6^{2-}$ actually energy minima? Elaborate. If one or both are not, repeat your optimization starting with distorted structures (preferably with C_1 symmetry).

P15.12 Benzyne has long been implicated as an intermediate in nucleophilic aromatic substitution, for example,

Although the geometry of benzyne has yet to be conclusively established, the results of a ^{13}C labeling experiment leave little doubt that two (adjacent) positions on the ring are equivalent:

There is a report, albeit controversial, that benzyne has been trapped in a low-temperature matrix and its infrared spectrum recorded. Furthermore, a line in the spectrum at 2085 cm^{-1} has been assigned to the stretching mode of the incorporated triple bond.

Optimize the geometry of benzyne using the HF/6-31G* model and calculate vibrational frequencies. For reference, perform the same calculations on 2-butyne. Locate the $C \equiv C$ stretching frequency in 2-butyne and determine an appropriate scaling factor to bring it into agreement with the corresponding experimental frequency (2240 cm^{-1}). Then, identify the vibration corresponding to the triple-bond stretch in benzyne and apply the same scaling factor to this frequency. Finally, plot the calculated infrared spectra of both benzyne and 2-butyne.

Does your calculated geometry for benzyne incorporate a fully formed triple bond? Compare with the bond in 2-butyne as a standard. Locate the vibrational motion in benzyne corresponding to triple bond stretch. Is the corresponding (scaled) frequency significantly different (>100 cm^{-1}) from the frequency assigned in the experimental investigation? If it is, are you able to locate any frequencies from your calculation that would fit with the assignment of a benzyne mode at 2085 cm^{-1}? Elaborate. Does the calculated infrared spectrum provide further evidence for or against the experimental observation? (*Hint*: Look at the intensity of the triple-bond stretch in 2-butyne.)

P15.13 All chemists know that benzene is unusually stable, that is, it is aromatic. They are also well aware that many other similar molecules are stabilized by aromaticity to some extent and, more often than not, can recognize aromatic molecules as those with delocalized bonding. What most chemists are unable to do, however, is to "put a number" on the aromatic stabilization afforded benzene or to quantify aromatic stabilization among different molecules. This is not to say that methods have not been proposed (for a discussion see Section 14.7), but rather that these methods have rarely been applied to real molecules.

Assigning a value to aromatic stabilization is actually quite straightforward. Consider a hypothetical reaction in which a molecule of hydrogen is added to benzene to yield 1,3-cyclohexadiene. Next, consider analogous hydrogenation reactions of 1,3-cyclohexadiene (leading to cyclohexene) and of cyclohexene (leading to cyclohexane):

benzene   1,3-cyclohexadiene   cyclohexene   cyclohexane

+ H₂ +25 kJ/mol   + H₂ −109 kJ/mol   + H₂ −117 kJ/mol

Addition of H_2 to benzene trades an H—H bond and a C—C π bond for two C—H bonds, but in so doing destroys the aromaticity, whereas H_2 addition to either 1,3-cyclohexadiene or cyclohexene trades the same bonds but does not result in any loss of aromaticity (there is nothing to lose). Therefore, the difference in the heats of hydrogenation (134 kJ/mol referenced to 1,3-cyclohexadiene and 142 kJ/mol referenced to cyclohexene) is a measure of the aromaticity of benzene.

Reliable quantitative comparisons require accurate experimental data (heats of formation). These will generally be available only for very simple molecules and will almost never be available for novel interesting compounds. As a case in point, consider to what extent, if any, the 10 π-electron molecule 1,6-methanocyclodeca-1,3,5,7,9-pentaene ("bridged naphthalene") is stabilized by aromaticity. Evidence provided by the X-ray crystal structure suggests a fully delocalized π system. The 10 carbons that make up the base are very nearly coplanar and all CC bonds are intermediate in length between normal single and double linkages, just as they are in naphthalene:

1.38
1.42
1.40

1,6-Methanocyclodeca-1,3,5,7,9-pentaene

1.37
1.41
1.42

Naphthalene

Calculations provide a viable alternative to experiment for thermochemical data. Although absolute hydrogenation energies may be difficult to describe with currently practical models, hydrogenation energies relative to a closely related standard compound are much easier to accurately describe. In this case, the natural standard is benzene.

a. Optimize the geometries of benzene, 1,3-cyclohexadiene, naphthalene, and 1,2-dihydronaphthalene using the HF/6-31G* model. Evaluate the energy of the following reaction, relating the energy of hydrogenation of naphthalene to that of benzene (as a standard):

naphthalene   1,3-cyclohexadiene   1,2-dihydronaphthalene   benzene

On the basis of relative hydrogenation energies, would you say that naphthalene is stabilized (by aromaticity) to about the same extent as is benzene or to a lesser or greater extent? Try to explain your result.

b. Optimize the geometries of 1,6-methanocyclodeca-1,3,5,7,9-pentaene and its hydrogenation product using the HF/6-31G* model. Evaluate the energy of hydrogenation relative to that of naphthalene. On the basis of relative hydrogenation energies, would you say that the bridged naphthalene is stabilized to about the same extent as is naphthalene or to a lesser or greater extent? Try to explain your result.

P15.14 Singlet and triplet carbenes exhibit different properties and show markedly different chemistry. For example, a singlet carbene will add to a *cis*-disubstituted alkene to produce only *cis*-disubstituted cyclopropane products (and to a *trans*-disubstituted alkene to produce only *trans*-disubstituted cyclopropane products), whereas a triplet carbene will add to produce a mixture of *cis* and *trans* products.

The origin of the difference lies in the fact that triplet carbenes are biradicals (or diradicals) and exhibit chemistry similar to that exhibited by radicals, whereas singlet carbenes incorporate both a nucleophilic site (a low-energy unfilled molecular orbital) and an electrophilic site (a high-energy filled molecular orbital); for example, for singlet and triplet methylene:

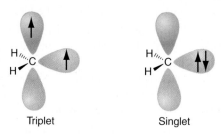

Triplet   Singlet

It should be possible to take advantage of what we know about stabilizing radical centers versus stabilizing empty orbitals and use that knowledge to design carbenes that will either be singlets or triplets. Additionally, it should be possible to say with confidence that a specific carbene of interest will either be a singlet or a triplet and, thus, to anticipate its chemistry.

The first step is to pick a model and then to establish the error in the calculated singlet–triplet energy separation in methylene where the triplet is known experimentally to be approximately 42 kJ/mol lower in energy than the singlet. This can then be applied as a correction for calculated singlet–triplet separations in other systems.

a. Optimize the structures of both the singlet and triplet states of methylene using both Hartree–Fock and B3LYP density functional models with the 6-31G* basis set. Which state (singlet or triplet) is found to be of lower energy according to the HF/6-31G* calculations? Is the singlet or the triplet unduly favored at this level of calculation? Rationalize your result. (*Hint*: Triplet methylene contains one fewer electron pair than singlet methylene.) What energy correction needs to be applied to calculated singlet–triplet energy separations? Which state (singlet or triplet) is found to be of lower energy according to the B3LYP/6-31G* calculations? What energy correction needs to be applied to calculated energy separations?

b. Proceed with either the HF/6-31G* or B3LYP/6-31G* model, depending on which leads to better agreement for the singlet–triplet energy separation in methylene. Optimize singlet and triplet states for cyanomethylene, methoxymethylene, and cyclopentadienylidene:

Cyanomethylene   Methoxymethylene   Cyclopentadienylidene

Apply the correction obtained in the previous step to estimate the singlet–triplet energy separation in each. For each of the three carbenes, assign the ground state as singlet or triplet. Relative to hydrogen (in methylene), has the cyano substituent in cyanomethylene and the methoxy substituent in methoxymethylene led to favoring of the singlet or the triplet? Rationalize your result by first characterizing cyano and methoxy substituents as π donors or π acceptors, and then speculating about how a donor or acceptor would stabilize or destabilize singlet and triplet methylene. Has incorporation into a cyclopentadienyl ring led to increased preference for a singlet or triplet ground state (relative to the preference in methylene)? Rationalize your result. (*Hint*: Count the number of π electrons associated with the rings in both singlet and triplet states.)

P15.15 Electron-donating groups on benzene promote electrophilic aromatic substitution and lead preferentially to so-called *ortho* and *para* products over *meta* products, whereas electron-withdrawing groups retard substitution and lead preferentially to *meta* products (over *ortho* and *para* products), for example, for electrophilic alkylation:

We can expect the first step in the substitution to be addition of the electrophile, leading to a positively charged adduct:

So-called benzenium ions have been characterized spectroscopically and X-ray crystal structures for several are known.

Will the stabilities of benzenium ion intermediates anticipate product distribution?

a. Optimize the geometries of benzene, aniline, and nitrobenzene using the HF/3-21G model. You will need their energies to ascertain the relative reactivities of the three substituted benzenes. Also, optimize the geometry of benzenium ion using the HF/3-21G model. A good guess is a planar six-membered ring comprising five sp^2 carbons and an sp^3 carbon with bond distances between sp^2 carbons intermediate in length between single and double bonds. It should have C_{2v} symmetry. In terms of ring bond distances, how does your calculated structure compare with the experimental X-ray geometry of heptamethylbenzenium ion?

b. Optimize the geometries of methyl cation adducts of benzene, aniline (*meta* and *para* isomers only), and nitrobenzene (*meta* and *para* isomers only) using the HF/3-21G model. Use the calculated structure of parent benzenium ion as a template. Which isomer, *meta* or *para*, of the aniline adduct is more stable? Which isomer of the nitrobenzene adduct is more stable? Considering only the lower energy isomer for each system, order the binding energies of methyl cation adducts of benzene, aniline, and nitrobenzene, that is: E (substituted benzene methyl cation adduct) – E (substituted benzene) – E (methyl cation). You will need to calculate the energy of methyl cation using the HF/3-21G model. Which aromatic should be most reactive? Which should be least reactive? Taken as a whole, do your results provide support for the involvement of benzenium ion adducts in electrophilic aromatic substitution? Explain.

P15.16 Aromatics such as benzene typically undergo substitution when reacted with an electrophile such as Br_2, whereas alkenes such as cyclohexene most commonly undergo addition:

Addition   Substitution

What is the reason for the change in preferred reaction in moving from the alkene to the arene? Use the Hartree–Fock 6-31G* model to obtain equilibrium geometries and energies for reactants and products of both addition and substitution reactions of both cyclohexene and benzene (four reactions in total). Assume *trans* addition products (1,2-dibromocyclohexane and 5,6-dibromo-1,3-cyclohexadiene). Is your result consistent with what is actually observed? Are all four reactions exothermic? If one or more are not exothermic, provide a rationale as to why.

P15.17 Evaluate the difference between change in energy at 0 K in the absence of zero point vibration and both change in enthalpy and in free energy for real molecules at 298 K. Consider both a unimolecular isomerization that does not lead to a net change in the number of molecules and a thermal decomposition reaction that leads to an increase in the number of molecules.

a. Calculate ΔU, $\Delta H(298)$, and $\Delta G(298)$ for the following isomerization reaction:

$$CH_3N \equiv C \longrightarrow CH_3C \equiv N$$

a. Obtain equilibrium geometries for both methyl isocyanide and acetonitrile using the B3LYP/6-31G* density functional model. Do the calculated values for ΔU and ΔH (298) differ significantly (by more than 10%)? If so, is the difference due primarily to the temperature correction or to the inclusion of zero point energy (or to a combination of both)? Is the calculated value for ΔG (298) significantly different from that of ΔH (298)?

b. Repeat your analysis (again using the B3LYP/6-31G* model) for the following pyrolysis reaction:

$$HCO_2CH_2CH_3 \longrightarrow HCO_2H + H_2C \equiv CH_2$$

Do these two reactions provide a similar or a different picture as to the importance of relating experimental thermochemical data to calculated ΔG values rather than ΔU values? If different, explain your result.

P15.18 Hydrazine would be expected to adopt a conformation in which the NH bonds stagger. There are two likely candidates, one with the lone pairs on nitrogen *anti* to each other and the other with the lone pairs *gauche*:

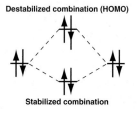

Anti hydrazine   *Gauche* hydrazine

On the basis of the same arguments made in VSEPR theory (electron pairs take up more space than bonds) you might expect that *anti* hydrazine would be the preferred structure.

a. Obtain energies for the *anti* and *gauche* conformers of hydrazine using the HF/6-31G* model. Which is the more stable conformer? Is your result in line with what you expect from VSEPR theory?

You can rationalize your result by recognizing that when electron pairs interact they form combinations, one of which is stabilized (relative to the original electron pairs) and one of which is destabilized. The extent of destabilization is greater than that of stabilization, meaning that overall interaction of two electron pairs is unfavorable energetically:

Destabilized combination (HOMO)

Stabilized combination

b. Measure the energy of the highest occupied molecular orbital (the HOMO) for each of the two hydrazine conformers. This corresponds to the higher energy (destabilized) combination of electron pairs. Which hydrazine conformer (*anti* or *gauche*) has the higher HOMO energy? Is this also the higher energy conformer? If so, is the difference in HOMO energies comparable to the difference in total energies between the conformers?

P15.19 Diels-Alder cycloaddition of 1,3-butadiene with acrylonitrile requires that the diene be in a *cis* (or *cis*-like) conformation:

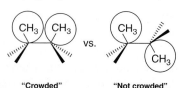

In fact, 1,3-butadiene exists primarily in a *trans* conformation, the *cis* conformer being approximately 9 kJ/mol less stable and separated from the *trans* conformer by a low-energy barrier. At room temperature, only about 5% of butadiene molecules will be in a *cis* conformation. Clearly, rotation into a *cis* conformation is required before reaction can proceed.

Conduct a search for a substituted 1,3-butadiene that actually prefers to exist in a *cis* (or *cis*-like) conformation as opposed to a *trans* conformation. The only restriction you need to be aware of is that the diene needs to be electron rich in order to be reactive. Restrict your search to alkyl and alkoxy substituents as well as halogen. Use the HF/3-21G model. Report your successes and provide rationales.

P15.20 The energy of rotation about a single bond is a periodic function of the torsion angle, ϕ, and is, therefore, appropriately described in terms of a truncated Fourier series, the simplest acceptable form of which is given by

$$V(\phi) = \frac{1}{2}V_1(1 - \cos \phi) + \frac{1}{2}V_2(1 - \cos 2\phi)$$
$$+ \frac{1}{2}V_3(1 - \cos 3\phi)$$
$$= V_1(\phi) + V_2(\phi) + V_3(\phi)$$

Here, V_1 is the onefold component (periodic in 360°), V_2 is the twofold component (periodic in 180°), and V_3 is the threefold component (periodic in 120°).

A Fourier series is an example of an orthogonal polynomial, meaning that the individual terms which it comprises are independent of each other. It should be possible, therefore, to dissect a complex rotational energy profile into a series of N-fold components and to interpret each of these components independent of all others. The one-fold component is quite easy to rationalize. For example, the onefold term for rotation about the central bond in *n*-butane no doubt reflects the crowding of methyl groups, whereas the onefold term in 1,2-difluoroethane probably reflects differences in electrostatic interactions as represented by bond dipoles:

Bond dipoles add   **Bond dipoles cancel**

The threefold component represents the difference in energy between eclipsed and staggered arrangements about a single bond. However, the twofold component is perhaps the most interesting of the three and is what concerns us here. It relates to the difference in energy between planar and perpendicular arrangements.

Optimize the geometry of dimethyl peroxide (CH_3OOCH_3) subject to the COOC dihedral angle being held at 0°, 20°, 40°, ..., 180° (10 optimizations in total). Use the B3LYP/6-31G* density functional model. Construct a plot of energy versus dihedral angle and fit this to a three-term Fourier series. Does the Fourier series provide a good fit to your data? If so, what is the dominant term? Rationalize it. What is the second most important term? Rationalize your result.

P15.21 Pyramidal inversion in the cyclic amine aziridine is significantly more difficult than inversion in an acyclic amine, for example, requiring 80 kJ/mol versus 23 kJ/mol in dimethylamine according to HF/6-31G* calculations. One plausible explanation is that the transition state for inversion needs to incorporate a planar trigonal nitrogen center, which is obviously more difficult to achieve in aziridine, where one bond angle is constrained to a value of around 60°, than it is in dimethylamine. Such an interpretation suggests that the barriers to inversion in the corresponding four- and five-membered ring amines (azetidine and pyrrolidine) should also be larger than normal and that the inversion barrier in the six-membered ring amine (piperidine) should be quite close to that for the acyclic.

Aziridine   Azetidine   Pyrrolidine   Piperidine   Dimethylamine

Optimize the geometries of aziridine, azetidine, pyrrolidine, and piperidine using the HF/6-31G* model. Starting from these optimized structures, provide guesses at the respective inversion transition states by replacing the tetrahedral nitrogen center with a trigonal center. Obtain transition states using the same Hartree–Fock model and calculate inversion barriers. Calculate vibrational frequencies to verify that you have actually located the appropriate inversion transition states.

Do the calculated inversion barriers follow the order suggested in the preceding figure? If not, which molecule(s) appear to be anomalous? Rationalize your observations by considering other changes in geometry from the amine to the transition state.

P15.22 Molecules such as dimethylsulfoxide and dimethylsulfone can either be represented as *hypervalent*, that is, with more than the normal complement of eight valence electrons

around sulfur, or as *zwitterions*, in which sulfur bears a positive charge:

Atomic charges obtained from quantum chemical calculations can help to decide which representation is more appropriate.

a. Obtain equilibrium geometries for dimethylsulfide, $(CH_3)_2S$, and dimethylsulfoxide using the HF/3-21G model and obtain charges at sulfur based on fits to the electrostatic potential. Is the charge on sulfur in dimethylsulfoxide about the same as that on sulfur in dimethylsulfide (normal sulfur), or has it increased by one unit, or is it somewhere between? Would you conclude that dimethylsulfoxide is best represented as a hypervalent molecule, as a zwitterion, or something between? See if you can support your conclusion with other evidence (geometries, dipole moments, and so on).

b. Repeat your analysis for dimethylsulfone. Compare your results for the charge at sulfur to those for dimethylsulfide and dimethylsulfoxide.

P15.23 Hydroxymethylene has never actually been observed, although it is believed to be an intermediate both in the photofragmentation of formaldehyde to hydrogen and carbon monoxide,

$$H_2CO \xrightarrow{h\nu} [H\ddot{C}OH] \longrightarrow H_2 + CO$$

and in the photodimerization of formaldehyde in an argon matrix:

$$H_2CO \xrightarrow{h\nu} [H\ddot{C}OH] \xrightarrow{H_2CO} HOCH_2CHO$$

Does hydroxymethylene actually exist? To have a chance "at life," it must be separated from both its rearrangement product (formaldehyde) and from its dissociation product (hydrogen and carbon monoxide) by a sizable energy barrier (>80 kJ/mol). Of course, it must also actually be a minimum on the potential energy surface.

a. First calculate the energy difference between formaldehyde and hydroxymethylene and compare your result to the indirect experimental estimate of 230 kJ/mol. Try two different models, B3LYP/6-31G* and MP2/6-31G*. Following calculation of the equilibrium geometry for hydroxymethylene, obtain vibrational frequencies. Is hydroxymethylene an energy minimum? How do you know? Is the energy difference inferred from experiment reasonably well reproduced with one or both of the two models?

b. Proceed with the model that gives the better energy difference and try to locate transition states both for isomerization of hydroxymethylene to formaldehyde and for dissociation to hydrogen and carbon monoxide. Be certain to calculate vibrational frequencies for the two transition states. On the basis of transition states you have located, would you expect that both isomerization and dissociation reactions are available to hydroxymethylene? Explain. Do both suggest that hydroxymethylene is in a deep enough energy well to actually be observed?

P15.24 The three vibrational frequencies in H_2O (1595, 3657, and 3756 cm^{-1}) are all much larger than the corresponding frequencies in D_2O (1178, 1571, and 2788 cm^{-1}). This follows from the fact that vibrational frequency is given by the square root of a (mass-independent) quantity, which relates to the curvature of the energy surface at the minima, divided by a quantity that depends on the masses of the atoms involved in the motion.

As discussed in Section 15.8.4, vibrational frequencies enter into both terms required to relate the energy obtained from a quantum chemical calculation (stationary nuclei at 0 K) to the enthalpy obtained experimentally (vibrating nuclei at finite temperature), as well as the entropy required to relate enthalpies to free energies. For the present purpose, focus is entirely on the so-called zero point energy term, that is, the energy required to account for the latent vibrational energy of a molecule at 0 K.

The zero point energy is given simply as the sum over individual vibrational energies (frequencies). Thus, the zero point energy for a molecule in which isotopic substitution has resulted in an increase in mass will be reduced from that in the unsubstituted molecule:

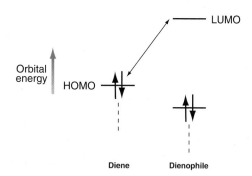

A direct consequence of this is that enthalpies of bond dissociation for isotopically substituted molecules (light to heavy) are smaller than those for unsubstituted molecules.

a. Perform B3LYP/6-31G* calculations on HCl and on its dissociation products, chlorine atom and hydrogen atom. Following geometry optimization on HCl, calculate the vibrational frequency for both HCl and DCl and evaluate the zero point energy for each. In terms of a percentage of the total bond dissociation energy, what is the change noted in going from HCl to DCl?

d_1-Methylene chloride can react with chlorine atoms in either of two ways: by hydrogen abstraction (producing HCl) or by deuterium abstraction (producing DCl):

Which pathway is favored on the basis of thermodynamics and which is favored on the basis of kinetics?

b. Obtain the equilibrium geometry for dichloromethyl radical using the B3LYP/6-31G* model. Also obtain vibrational frequencies for both the unsubstituted and the deuterium-substituted radical and calculate zero point energies for the two abstraction pathways (you already have zero point energies for HCl and DCl). Which pathway is favored on the basis of thermodynamics? What would you expect the (thermodynamic) product ratio to be at room temperature?

c. Obtain the transition state for hydrogen abstraction from methylene chloride using the B3LYP/6-31G* model. A reasonable guess is shown here:

Calculate vibrational frequencies for the two possible structures with one deuterium and evaluate the zero point energies for these two structures. (For the purpose of zero point energy calculation, ignore the imaginary frequency corresponding to the reaction coordinate.) Which pathway is favored on the basis of kinetics? Is it the same or different from the thermodynamic pathway? What would you expect the (kinetic) product ratio to be at room temperature?

P15.25 Diels-Alder reactions commonly involve electron-rich dienes and electron-deficient dienophiles:

Y = R, OR
X = CN, CHO, CO$_2$H

The rate of these reactions generally increases with the π-donor ability of the diene substituent, Y, and with the π-acceptor ability of the dienophile substituent, X. The usual interpretation is that electron donors will push up the energy of the HOMO on the diene and that electron acceptors will push down the energy of the LUMO on the dienophile:

The resulting decrease in the HOMO–LUMO gap leads to a stronger interaction between diene and dienophile and to a decrease in the activation barrier.

a. Obtain equilibrium geometries for acrylonitrile, 1,1-dicyanoethylene, cis- and trans-1,2-dicyanoethylene, tricyanoethylene, and tetracyanoethylene using the HF/3-21G model.

| 0 | 4.64 | 1.94 |
| Acrylonitrile | 1, 1-Dicyanoethene | cis-1,2-Dicyanoethene |

| 1.89 | 5.66 | 7.61 |
| trans-1, 2-Dicyanoethene | Tricyanoethene | Tetracyanoethene |

Plot the LUMO energy for each dienophile versus the log of the observed relative rate for its addition to cyclopentadiene (listed below the structures in the preceding figure). Is there a reasonable correlation between LUMO energy and relative rate?

b. Obtain transition-state geometries for Diels-Alder cycloadditions of acrylonitrile and cyclopentadiene and tetracyanoethylene and cyclopentadiene using the HF/3-21G model. Also obtain a geometry for cyclopentadiene. Calculate activation energies for the two reactions.

How does the calculated difference in activation energies compare with the experimental difference (based on a value of 7.61 for the difference in the log of the rates and assuming 298 K)?

P15.26 It is well known that cyanide acts as a "carbon" and not a "nitrogen" nucleophile in S_N2 reactions, for example,

$$:N\equiv C:\overset{\frown}{\ }CH_3\overset{\frown}{-}I \longrightarrow :N\equiv C-CH_3 + I^-$$

How can this behavior be rationalized with the notion that nitrogen is in fact more electronegative than carbon and, therefore, would be expected to hold any excess electrons?

a. Optimize the geometry of cyanide using the HF/3-21G model and examine the HOMO. Describe the shape of the HOMO of cyanide. Is it more concentrated on carbon or nitrogen? Does it support the picture of cyanide acting as a carbon nucleophile? If so, explain why your result is not at odds with the relative electronegativities of carbon and nitrogen.

Why does iodide leave following nucleophilic attack by cyanide on methyl iodide?

b. Optimize the geometry of methyl iodide using the HF/3-21G model and examine the LUMO. Describe the shape of the LUMO of methyl iodide. Does it anticipate the loss of iodide following attack by cyanide? Explain.

P15.27 At first glance, the structure of diborane would seem unusual. Why shouldn't the molecule assume the same geometry as ethane, which after all has the same number of heavy atoms and the same number of hydrogens?

Diborane   Ethane

The important difference between the two molecules is that diborane has two fewer electrons than ethane and is not able to make the same number of bonds. In fact, it is ethene which has the same number of electrons, to which diborane is structurally related.

Obtain equilibrium geometries for both diborane and ethene using the HF/6-31G* model and display the six valence molecular orbitals for each. Associate each valence orbital in ethene with its counterpart in diborane. Focus on similarities in the structure of the orbitals and not on their position in the lists of orbitals. To which orbital in diborane does the π orbital in ethene (the HOMO) best relate? How would you describe this orbital in diborane? Is it B—B bonding, B—H bonding, or both?

P15.28 Molecular orbitals are most commonly delocalized throughout the molecule and exhibit distinct bonding or antibonding character. Loss of an electron from a specific molecular orbital from excitation by light or by ionization would, therefore, be expected to lead to distinct changes in bonding and changes in molecular geometry.

a. Obtain equilibrium geometries for ethene, formaldimine, and formaldehyde using the HF/6-31G* model and display the highest occupied and lowest unoccupied molecular orbitals (HOMO and LUMO, respectively) for each. What would happen to the geometry around carbon (remain planar versus pyramidalize), to the C=X bond length, and (for formaldimine) to the C=NH bond angle if an electron were to be removed from the HOMO of ethene, formaldimine, and formaldehyde?

b. Obtain equilibrium geometries for radical cations of ethene, formaldimine, and formaldehyde using the HF/6-31G* model. Are the calculated geometries of these species, in which an electron has been removed from the corresponding neutral molecule, in line with your predictions based on the shape and nodal structure of the HOMO?

Unoccupied molecular orbitals are also delocalized and also show distinct bonding or antibonding character. Normally, this is of no consequence. However, were these orbitals to become occupied (from excitation or from capture of an electron), then changes in molecular geometry would also be expected. What would happen to the geometry around carbon, to the C=X bond length, and (for formaldimine) to the C=NH bond angle, if an electron were to be added to the LUMO of ethene, formaldimine, and formaldehyde?

c. Obtain equilibrium geometries for the radical anions of ethene, formaldimine, and formaldehyde using the HF/6-31G* model. Are the calculated geometries of these species, in which an electron has been added to the corresponding neutral molecule, in line with your predictions based on the shape and nodal structure of the LUMO?

The first excited state of formaldehyde (the so-called $n \rightarrow \pi^*$ state) can be thought of as arising from the promotion of one electron from the HOMO (in the ground state of formaldehyde) to the LUMO. The experimental equilibrium geometry of the molecule shows lengthening of the CO bond and a pyramidal carbon (ground-state values are shown in parentheses):

1.32Å (1.21Å)
154° (180°)

d. Rationalize this experimental result on the basis of what you know about the HOMO and LUMO in formaldehyde and your experience with calculations on the radical cation and radical anion of formaldehyde.

P15.29 BeH$_2$ is linear, whereas CH$_2$ with two additional electrons and H$_2$O with four additional electrons are both bent to a similar degree. Could these changes in geometry have been anticipated by examining the shapes of the bonding molecular orbitals?

a. Perform a series of geometry optimizations on BeH$_2$ with the bond angle constrained at 90°, 100°, 110°, . . . , 180° (10 optimizations in total). Use the HF/6-31G* model. Plot the total energy, along with the HOMO and LUMO energies versus bond angle. Also, display the HOMO and LUMO for one of your structures of intermediate bond angle.

   Does the energy of the HOMO of BeH$_2$ increase (more positive) or decrease in going from a bent to a linear structure, or does it remain constant, or is the energy at a minimum or maximum somewhere between? Would this result have been anticipated by examining the shape and nodal structure of the HOMO?

   Does the energy of the LUMO of BeH$_2$ increase or decrease with increase in bond angle, or does it remain constant, or is the energy at a minimum or maximum somewhere between? Rationalize your result by reference to the shape and nodal structure of the LUMO. What do you anticipate would happen to the geometry of BeH$_2$ as electrons are added to the LUMO? Take a guess at the structure of BH$_2^-$ (one electron added to the LUMO) and singlet CH$_2$ (two electrons added to the LUMO).

b. Optimize the geometries of (singlet) BH$_2^-$ and singlet CH$_2$ using the HF/6-31G* model. Are the results of the quantum chemical calculations in line with your qualitative arguments?

c. Perform a series of geometry optimizations on singlet CH$_2$ with the bond angle constrained to 90°, 100°, 110°, . . . , 180°. Plot the total energy as a function of the angle as well as the HOMO and LUMO energies.

   Display the LUMO for some intermediate structure. Does the plot of HOMO energy versus angle in CH$_2$ mirror the plot of LUMO energy versus angle in BeH$_2$? Rationalize your answer. Does the energy of the LUMO in CH$_2$ increase, decrease, or remain constant with increase in bond angle (or is it at a minimum or maximum somewhere between)? Is the change in LUMO energy smaller, larger, or about the same as the change in the energy of the HOMO over the same range of bond angles? Rationalize these two observations by reference to the shape and nodal structure of the LUMO. What do you anticipate would happen to the geometry of CH$_2$ as electrons are added to the LUMO? Take a guess at the structure of NH$_2^-$ (one electron added to the LUMO) and H$_2$O (two electrons added to the LUMO).

d. Optimize the geometries of NH$_2^-$ and H$_2$O using the HF/6-31G* model. Are the results of the quantum chemical calculations in line with your qualitative arguments?

P15.30 Olefins assume planar (or nearly planar) geometries wherever possible. This ensures maximum overlap between p orbitals and maximum π-bond strength. Any distortion away from planarity should reduce orbital overlap and bond strength. In principle, π-bond strength can be determined experimentally, by measuring the activation energy required for *cis-trans* isomerization, for example, in *cis*-1,2-dideuteroethylene:

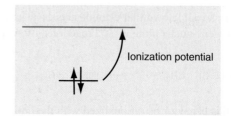

Another measure of π-bond strength, at least π-bond strength relative to a standard, is the energy required to remove an electron from the π orbital, or the ionization potential:

Non-planar olefins might be expected to result from incorporation to a *trans* double bond into a small ring. Small-ring cycloalkenes prefer *cis* double bonds, and the smallest *trans* cycloalkene to actually have been isolated is cyclooctene. It is known experimentally to be approximately 39 kJ/mol less stable than *cis*-cyclooctene. Is this a measure of reduction in π bond strength?

   Optimize the geometries of both *cis*- and *trans*-cyclooctene using the HF/3-21G model. (You should first examine the possible conformers available to each of the molecules.) Finally, calculate and display the HOMO for each molecule.

   Is the double bond in *trans*-cyclooctene significantly distorted from its ideal planar geometry? If so, would you characterize the distortion as puckering of the double bond carbons or as twisting around the bond, or both? Does the HOMO in *trans*-cyclooctene show evidence of distortion? Elaborate. Is the energy of the HOMO in *trans*-cyclooctene significantly higher (less negative) than that in *cis*-cyclooctene? How does the energy difference compare to the experimentally measured difference in ionization potentials between the two isomers (0.29 eV)? How does the difference in HOMO energies (ionization potentials) relate to the calculated (measured) difference in isomer energies?

P15.31 Singlet carbenes add to alkenes to yield cyclopropanes. Stereochemistry is maintained, meaning that *cis*- and *trans*-substituted alkenes give *cis*- and *trans*-substituted cyclopropanes, respectively; for example:

This implies that the two σ bonds are formed more or less simultaneously, without the intervention of an intermediate that would allow *cis-trans* isomerization.

Locate the transition state for addition of singlet difluorocarbene and ethene using the HF/3-21G model and, following this, calculate vibrational frequencies. When completed, verify that you have in fact found a transition state and that it appears to be on the way to the correct product.

What is the orientation of the carbene relative to ethene in your transition state? Is it the same orientation as adopted in the product (1,1-difluorocyclopropane)? If not, what is the reason for the difference? (*Hint*: Consider that the π electrons on ethylene need to go into a low-lying unoccupied molecular orbital on the carbene. Build difluorocarbene and optimize its geometry using the HF/3-21G model and display the LUMO.)

P15.32 Further information about the mechanism of the ethyl formate pyrolysis reaction can be obtained by replacing the static picture with a movie, that is, an animation along the reaction coordinate. Bring up "ethyl formate pyrolysis" (on the Spartan download) and examine the change in electron density as the reaction proceeds. Do hydrogen migration and CO bond cleavage appear to occur in concert or is one leading the other?

P15.33 Do related molecules with the same number of electrons occupy the same amount of space, or are other factors (beyond electron count) of importance when dictating overall size requirements? Obtain equilibrium geometries for methyl anion, ammonia, and hydronium cation using the HF/6-31G* model and compare electron density surfaces corresponding to enclosure of 99% of the total electron density. Do the three molecules take up the same amount of space? If not, why not?

P15.34 Lithium provides a very simple example of the effect of oxidation state on overall size. Perform HF/6-31G* calculations on lithium cation, lithium atom, and lithium anion, and compare the three electron density surfaces corresponding to enclosure of 99% of the total electron density. Which is smallest? Which is largest? How does the size of lithium relate to the number of electrons? Which surface most closely resembles a conventional space-filling model? What, if anything does this tell you about the kinds of molecules that were used to establish the space-filling radius for lithium?

P15.35 A surface for which the electrostatic potential is negative delineates regions in a molecule that are subject to electrophilic attack. It can help you to rationalize the widely different chemistry of molecules that are structurally similar.

Optimize the geometries of benzene and pyridine using the HF/3-21G model and examine electrostatic potential surfaces corresponding to −100 kJ/mol. Describe the potential surface for each molecule. Use it to rationalize the following experimental observations: (1) Benzene and its derivatives undergo electrophilic aromatic substitution far more readily than do pyridine and its derivatives;

(2) protonation of perdeuterobenzene (C_6D_6) leads to loss of deuterium, whereas protonation of perdeuteropyridine (C_5D_5N) does not lead to loss of deuterium; and (3) benzene typically forms π-type complexes with transition models, whereas pyridine typically forms σ-type complexes.

P15.36 Hydrocarbons are generally considered to be nonpolar or weakly polar at best, characterized by dipole moments that are typically only a few tenths of a debye. For comparison, dipole moments for molecules of comparable size with heteroatoms are commonly several debyes. One recognizable exception is azulene, which has a dipole moment of 0.8 debye:

Azulene   **Naphthalene**

Optimize the geometry of azulene using the HF/6-31G* model and calculate an electrostatic potential map. For reference, perform the same calculations on naphthalene, a nonpolar isomer of azulene. Display the two electrostatic potential maps side by side and on the same (color) scale. According to its electrostatic potential map, is one ring in azulene more negative (relative to naphthalene as a standard) and one ring more positive? If so, which is which? Is this result consistent with the direction of the dipole moment in azulene? Rationalize your result. (*Hint*: Count the number of π electrons.)

P15.37 Chemists know that nitric and sulfuric acids are strong acids and that acetic acid is a weak acid. They would also agree that ethanol is at best a very weak acid. Acid strength is given directly by the energetics of deprotonation (heterolytic bond dissociation); for example, for acetic acid:

$$CH_3CO_2H \longrightarrow CH_3CO_2^- + H^+$$

As written, this is a highly *endothermic* process, because not only is a bond broken but two charged molecules are created from the neutral acid. It occurs readily in solution only because the solvent acts to disperse charge.

Acid strength can be calculated simply as the difference in energy between the acid and its conjugate base (the energy of the proton is 0). In fact, acid strength comparisons among closely related systems, for example, carboxylic acids, are quite well described with practical quantum chemical models. This is consistent with the ability of the same models to correctly account for relative base strengths (see discussion in Section 15.8.3).

Another possible measure of acid strength is the degree of positive charge on the acidic hydrogen as measured by the electrostatic potential. It is reasonable to expect that the more positive the potential in the vicinity of the hydrogen, the more easily it will dissociate and the stronger the acid. This kind of measure, were it to prove successful, offers an advantage over the calculation of reaction energy, in that only the acid (and not the conjugate base) needs to be considered.

a. Obtain equilibrium geometries for nitric acid, sulfuric acid, acetic acid, and ethanol using the HF/3-21G model, and compare electrostatic potential maps. Be certain to choose the same (color) scale for the four acids. For which acid is the electrostatic potential in the vicinity of (the acidic) hydrogen most positive? For which is it least positive? Do electrostatic potential maps provide a qualitatively correct account of the relative acid strength of these four compounds?

b. Obtain equilibrium geometries for several of the carboxylic acids found in the following table using the HF/3-21G model and display an electrostatic potential map for each.

Acid	pK_a	Acid	pK_a
Cl_3CCO_2H	0.7	HCO_2H	3.75
HO_2CCO_2H	1.23	trans-$ClCH$=$CHCO_2H$	3.79
Cl_2CHCO_2H	1.48	$C_6H_5CO_2H$	4.19
$NCCH_2CO_2H$	2.45	p-ClC_6H_4CH=$CHCO_2H$	4.41
$ClCH_2CO_2H$	2.85	trans-CH_3CH=$CHCO_2H$	4.70
trans-HO_2CCH=$CHCO_2H$	3.10	CH_3CO_2H	4.75
p-$HO_2CC_6H_4CO_2H$	3.51	$(CH_3)_3CCO_2H$	5.03

"Measure" the most positive value of the electrostatic potential associated with the acidic hydrogen in each of these compounds and plot this against experimental pK_a (given in the preceding table). Is there a reasonable correlation between acid strengths and electrostatic potential at hydrogen in this closely related series of acids?

Molecular Symmetry

The combination of group theory and quantum mechanics provides a powerful tool for understanding the consequences of molecular symmetry. In this chapter, after a brief description of the most important aspects of group theory, several applications are discussed. They include using molecular symmetry to decide which atomic orbitals contribute to molecular orbitals, understanding the origin of spectroscopic selection rules, identifying the normal modes of vibration for a molecule, and determining if a particular molecular vibration is infrared active and/or Raman active.

16.1 SYMMETRY ELEMENTS, SYMMETRY OPERATIONS, AND POINT GROUPS

An individual molecule has an inherent symmetry based on the spatial arrangement of its atoms. For example, after a rotation of benzene by 60° about an axis that is perpendicular to the plane of the molecule and that passes through the center of the molecule, the molecule cannot be distinguished from the original configuration. Solid benzene in a crystalline form has additional symmetries that arise from the way in which individual benzene molecules are arranged in the crystal structure. These symmetry elements are essential in discussing diffraction of X-rays. However, in this chapter, the focus is on the symmetry of an individual molecule.

Why is molecular symmetry useful to chemists? The symmetry of a molecule determines a number of its important properties. For example, CF_4 has no dipole moment, but H_2O has a dipole moment because of the symmetry of these molecules. All molecules have vibrational modes. However, the number of vibrational modes that are infrared and Raman active and the degeneracy of a given vibrational frequency depend on the molecular symmetry. Symmetry also determines the selection rules for transitions

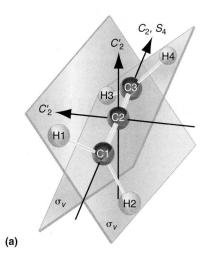

(a)

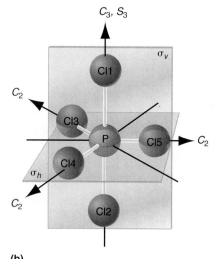

(b)

between states of the molecule in all forms of spectroscopy, and symmetry determines which atomic orbitals contribute to a given molecular orbital.

The focus in this chapter is on applying the predictive power of group theory to problems of interest in quantum chemistry, rather than on formally developing the mathematical framework. Therefore, results from group theory that are needed for specific applications are introduced without their derivations. These results are highlighted in shaded text boxes. Readers who wish to see these results derived or discussed in more detail are referred to standard texts such as *Symmetry and Structure* by S. F. A. Kettle, *Molecular Symmetry and Group Theory* by R. L. Carter, and *Chemical Applications of Group Theory*, by F. A. Cotton. In Sections 16.1 through 16.5, the essentials of group theory that are needed to address problems of chemical interest are discussed. With a working knowledge of reducible and irreducible representations and character tables, several applications of group theory to chemistry are presented in the rest of the chapter. In Section 16.6, group theory is used to construct molecular orbitals (MOs) that incorporate the symmetry of the molecule under consideration from atomic orbitals (AOs). In Section 16.7, we discuss the normal modes for the vibration of molecules, and in Section 16.8, we show that symmetry determines whether a given vibrational mode of a molecule is infrared or Raman active. We will also show that symmetry determines the number of normal modes that have the same vibrational frequency.

We begin our discussion of molecular symmetry by discussing symmetry elements and symmetry operations. **Symmetry elements** are geometric entities such as axes, planes, or points with respect to which operations can be carried out. **Symmetry operations** are actions with respect to the symmetry elements that leave the molecule in a configuration that cannot be distinguished from the original configuration. There are only five different types of symmetry elements for an isolated molecule, although a molecule may require several elements of each type—n-fold rotation axes, n-fold rotation-reflection axes, or mirror planes—to fully define its symmetry. These elements and operations are listed in Table 16.1. Operators are indicated by a caret above the symbol.

Whereas other symmetry elements generate a single operation, C_n and S_n axes generate n operations. We choose the direction of rotation to be counterclockwise. However, if carried through consistently, either direction can be used. Examples of these symmetry elements are illustrated in Figure 16.1 for allene and PCl_5. Consider first the following symmetry elements for allene:

- A rotation of $360°/2 = 180°$ about the C_2 **rotation axes** passing through the carbon atoms leaves the molecule in a position that is indistinguishable from its initial position.
- A rotation of $360°/4 = 90°$ about the twofold axis discussed in the previous point, followed by a reflection through a plane perpendicular to the axis that passes through the central carbon atom also leaves the allene molecule unchanged. The combined operation is called an S_4 fourfold **rotation-reflection axis**. This axis and the C_2 rotation axis of the previous bullet are collinear.
- Two further C_2 rotation axes exist in this molecule. Both pass through the central carbon atom (C2). Consider the two planes shown in Figure 16.1a. One contains H1 and H2, and the other contains H3 and H4. The two C_2 axes bisect the angle between the two planes and, therefore, are perpendicular to one another.

TABLE 16.1 SYMMETRY ELEMENTS AND THEIR CORRESPONDING OPERATIONS

Symmetry Elements		Symmetry Operations	
E	Identity	$\hat{E}$	leave molecule unchanged
C_n	n-Fold rotation axis	$\hat{C}_n, \hat{C}_n^2, \ldots, \hat{C}_n^n$	rotate about axis by $360°/n$ 1, 2, ..., n times (indicated by superscript)
σ	Mirror plane	$\hat{\sigma}$	reflect through the mirror plane
i	Inversion center	$\hat{i}$	$(x, y, z) \rightarrow (-x, -y, -z)$
S_n	n-Fold rotation-reflection axis	$\hat{S}_n$	rotate about axis by $360°/n$, and reflect through a plane perpendicular to the axis.

- The molecule contains two mirror planes, as shown. Because they contain the main twofold axis, which is referred to as the vertical axis, they are designated with σ_v.

These symmetry elements for allene are shown in Figure 16.1. Consider next the PCl_5 molecule, which has the following symmetry elements:

- A threefold rotation axis C_3 that passes through Cl1, Cl2, and the central P atom.
- A **mirror plane**, σ_h, that passes through the centers of the three equatorial Cl atoms. Reflection through this plane leaves the equatorial Cl atoms in their original location and exchanges the axial Cl atoms.
- Three C_2 axes that pass through the central P atom and one of the equatorial Cl atoms.
- Three mirror planes, σ_v, that contain Cl1, Cl2, and P as well as one of Cl3, Cl4, or Cl5.

One of these planes is shown in Figure 16.1b. As we will see in Section 16.2, allene and PCl_5 can each be assigned to a group on the basis of symmetry elements of the molecule.

What is the relationship between symmetry elements, the symmetry operators, and the group? A set of symmetry elements forms a **group** if the following statements are true about their corresponding operators:

- The successive application of two operators is equivalent to one of the operations of the group. This guarantees that the group is closed.
- An **identity operator**, $\hat{E}$, exists that commutes with any other operator and leaves the molecule unchanged. Although this operator seems trivial, it plays an important role as we will see later. The identity operator has the property that $\hat{A}\hat{E} = \hat{E}\hat{A} = \hat{A}$ where $\hat{A}$ is an arbitrary element of the group.
- The group contains an **inverse operator** for each element in the group. If $\hat{B}^{-1}$ is the inverse operator of $\hat{B}$, then $\hat{B}\hat{B}^{-1} = \hat{B}^{-1}\hat{B} = \hat{E}$. If $\hat{A} = \hat{B}^{-1}$, then $\hat{A}^{-1} = \hat{B}$. In addition, $\hat{E} = \hat{E}^{-1}$.
- The operators are **associative**, meaning that $\hat{A}(\hat{B}\hat{C}) = (\hat{A}\hat{B})\hat{C}$.

The groups of interest in this chapter are called **point groups** because the set of symmetry elements intersects in a point or set of points. To utilize the power of group theory in chemistry, molecules are assigned to point groups on the basis of the symmetry elements characteristic of the particular molecule. Each point group has its own set of symmetry elements and corresponding operations. We work with several of these groups in more detail in the following sections.

16.2 ASSIGNING MOLECULES TO POINT GROUPS

How is the point group to which a molecule belongs determined? The assignment is made using the logic diagram of Figure 16.2. To illustrate the use of this logic diagram, we assign NF_3, CO_2, and $Au(Cl_4)^-$ to specific point groups. In doing so, it is useful to first identify the major symmetry elements. After a tentative assignment of a point group is made based on these symmetry elements, it is necessary to verify that the other symmetry elements of that group are also present in the molecule. We start at the top of the diagram and follow the branching points.

NF_3 is a pyramidal molecule that has a threefold axis (C_3) passing through the N atom and a point in the plane of the F atoms that is equidistant from all three F atoms. NF_3 has no other rotation axes. The molecule has three mirror planes in which the C_3 axis, the N atom, and one F atom lie. These planes are perpendicular to the line connecting the other two fluorine atoms. Because the C_3 axis lies in the mirror plane, we conclude that NF_3 belongs to the C_{3v} group. The pathway through the logic diagram of Figure 16.2 is shown as a red line.

Carbon dioxide is a linear molecule with an **inversion center**. These symmetry characteristics uniquely specify CO_2 as belonging to the $D_{\infty h}$ group. The ∞ appears rather than the subscript n because any rotation about the molecular axis leaves the molecule unchanged.

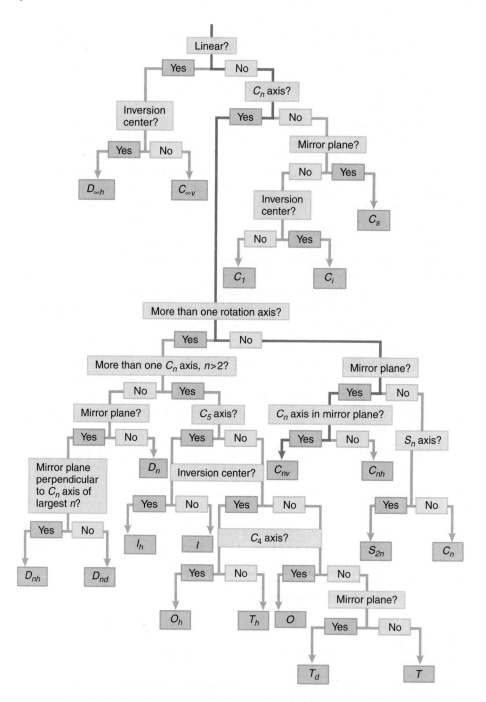

FIGURE 16.2

Logic diagram indicating how molecules are assigned to point groups. The red line indicates how NF_3 is assigned to the C_{3v} point group.

$Au(Cl_4)^-$ is a square planar complex with a C_4 axis. It has C_2 axes perpendicular to the C_4 axis, but no other C_n axis with $n > 2$. It has mirror planes, one of which is perpendicular to the C_4 axis. Therefore, this complex belongs to the D_{4h} group. Trace the paths through the logic diagram for these molecules to see if you would have made the same assignments.

These examples illustrate how a given molecule can be assigned to a point group, but have only utilized a few of the symmetry operations of a given group. A number of point groups applicable to small molecules are listed in Table 16.2. All symmetry elements of the group are listed. Note that several groups have different categories or **classes** of symmetry elements such as C_n and σ, which are indicated by single and double primes. Classes are defined in Section 16.3.

The preceding discussion of the symmetry elements of a group has been of a general nature. In the following section, we discuss the symmetry elements of the C_{2v} group, to which water belongs, in greater detail.

TABLE 16.2   SELECTED POINT GROUPS AND THEIR ELEMENTS

Point Group	Symmetry Elements	Example Molecule
C_s	E, σ	BFClBr (planar)
C_2	E, C_2	H_2O_2
C_{2v}	E, C_2, σ, σ'	H_2O
C_{3v}	$E, C_3, C_3^2, 3\sigma$	NF_3
$C_{\infty v}$	$E, C_\infty, \infty\sigma$	HCl
C_{2h}	E, C_2, σ, i	$trans$-$C_2H_2F_2$
D_{2h}	$E, C_2, C_2', C_2'', \sigma, \sigma', \sigma'', i$	C_2F_4
D_{3h}	$E, C_3, C_3^2, 3C_2, S_3, S_3^2, \sigma, 3\sigma'$	SO_3
D_{4h}	$E, C_4, C_4^3, C_2, 2C_2', 2C_2'', i, S_4, S_4^3, \sigma, 2\sigma', 2\sigma''$	XeF_4
D_{6h}	$E, C_6, C_6^5, C_3, C_3^2, C_2, 3C_2', 3C_2'', i, S_3, S_3^2, S_6, S_6^5, \sigma, 3\sigma', 3\sigma''$	C_6H_6 (benzene)
$D_{\infty h}$	$E, C_\infty, S_\infty, \infty C_2, \infty\sigma, \sigma', i$	H_2, CO_2
T_d	$E, 4C_3, 4C_3^2, 3C_2, 3S_4, 3S_4^3, 6\sigma$	CH_4
O_h	$E, 4C_3, 4C_3^2, 6C_2, 3C_4, 3C_2, i, 3S_4, 3S_4^3, 4S_6, 4S_6^5, 3\sigma, 6\sigma'$	SF_6

16.3 THE H₂O MOLECULE AND THE C₂ᵥ POINT GROUP

To gain practice in working with the concepts introduced in the preceding section, we next consider a specific molecule, express the symmetry operators mathematically, and show that the symmetry elements form a group. We do so by representing the operators as matrices and showing that the requirements that the elements of any group must meet for this particular group.

Figure 16.3 shows all the symmetry elements for the water molecule. By convention, the rotation axis of highest symmetry (principal rotation axis), C_2, is oriented

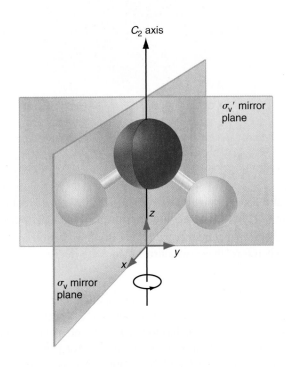

FIGURE 16.3

The water molecule is shown together with its symmetry elements. Convince yourself that the two mirror planes are in different classes.

along the z axis. The C_2 axis passes through the O atom. The molecule has two mirror planes oriented at 90° to one another, and their line of intersection is the C_2 axis. Because the mirror planes contain the principal rotation axis, the symmetry planes are referred to as vertical planes and designated by the subscript v. Mirror planes perpendicular to the principal rotation axis are referred to as horizontal and are designated by the subscript h. The molecule lies in the plane designated σ'_v, and the second mirror plane, designated σ_v, bisects the H—O—H bond angle. As shown in Example Problem 16.1, these two mirror planes belong to different classes and, therefore, have different symbols.

Elements that belong to the same class can be transformed into one another by other symmetry operations of the group. For example, the operators $\hat{C}_n, \hat{C}_n^2, \ldots, \hat{C}_n^n$, belong to the same class.

EXAMPLE PROBLEM 16.1

a. Are the three mirror planes for the NF$_3$ molecule in the same or in different classes?

b. Are the two mirror planes for H$_2$O in the same or in different classes?

Solution

a. NF$_3$ belongs to the C_{3v} group, which contains the rotation operators $\hat{C}_3, \hat{C}_3^2 = (\hat{C}_3)^{-1}$, and $\hat{C}_3^3 = \hat{E}$ and the vertical mirror planes $\hat{\sigma}_v(1)$, $\hat{\sigma}_v(2)$, and $\hat{\sigma}_v(3)$. These operations and elements are illustrated by this figure:

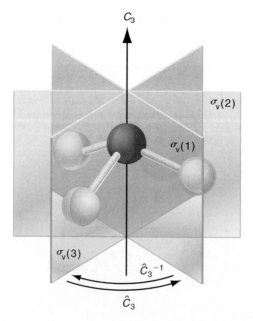

We see that $\hat{C}_3$ converts $\sigma_v(1)$ to $\sigma_v(3)$, and $\hat{C}_3^2 = (\hat{C}_3)^{-1}$ converts $\sigma_v(1)$ to $\sigma_v(2)$. Therefore, all three mirror planes belong to the same class.

b. Figure 16.3 shows that neither the $\hat{C}_2$ nor the $\hat{E}$ operation converts σ_v to σ'_v. Therefore, these two mirror planes are in different classes.

Using the logic diagram of Figure 16.2, we conclude that H$_2$O belongs to the C_{2v} group. This point group is given the shorthand notation C_{2v} because it has a C_2 axis and vertical mirror planes. The C_{2v} group has four symmetry elements: the identity element, a C_2 rotation axis, and two mutually perpendicular mirror planes. The corresponding operators are the identity operator $\hat{E}$ and the operators $\hat{C}_2$, $\hat{\sigma}$, and $\hat{\sigma}'$.

To understand how these operators act, we must introduce mathematical representations of the operators and then carry out the operations. To do so, the operators of the C_{2v} group are represented by 3×3 matrices, which act on a vector in three-dimensional space. See the Math Supplement (Appendix A) for an introduction to working with matrices.

Consider the effect of the symmetry operators on an arbitrary vector $\mathbf{r} = (x_1, y_1, z_1)$, originating at the intersection of the mirror planes and the C_2 axis. The vector $\mathbf{r}$ is converted to the vector (x_2, y_2, z_2) through the particular symmetry operation. We begin with a counterclockwise rotation by the angle θ about the z axis. As Example Problem 16.2 shows, the transformation of the components of the vector is described by Equation (16.1):

$$\begin{pmatrix} x_2 \\ y_2 \\ z_2 \end{pmatrix} = \begin{pmatrix} \cos\theta & -\sin\theta & 0 \\ \sin\theta & \cos\theta & 0 \\ 0 & 0 & 1 \end{pmatrix} \begin{pmatrix} x_1 \\ y_1 \\ z_1 \end{pmatrix} \qquad \textbf{(16.1)}$$

EXAMPLE PROBLEM 16.2

Show that a rotation about the z axis can be represented by the matrix

$$\begin{pmatrix} \cos\theta & -\sin\theta & 0 \\ \sin\theta & \cos\theta & 0 \\ 0 & 0 & 1 \end{pmatrix}$$

Show that for a rotation of 180° this matrix takes the form

$$\begin{pmatrix} -1 & 0 & 0 \\ 0 & -1 & 0 \\ 0 & 0 & 1 \end{pmatrix}$$

Solution

The z coordinate is unchanged in a rotation about the z axis, so we need only consider the vectors $\mathbf{r}_1 = (x_1, y_1)$ and $\mathbf{r}_2 = (x_2, y_2)$ in the x–y plane. These equations can be derived from the figure that follows:

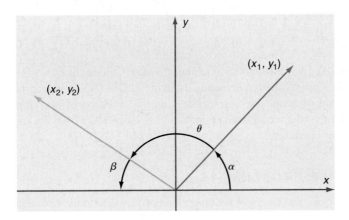

$$\theta = 180° - \alpha - \beta$$
$$x_1 = r\cos\alpha, \quad y_1 = r\sin\alpha$$
$$x_2 = -r\cos\beta, \quad y_2 = r\sin\beta$$

Using the identities $\cos(\phi \pm \delta) = \cos\phi\cos\delta \mp \sin\phi\sin\delta$ and $\sin(\phi \pm \delta) = \sin\phi\cos\delta \pm \cos\phi\sin\delta$, x_2 and y_2 can be expressed in terms of θ and α.

$$x_2 = -r\cos\beta = -r\cos(180° - \alpha - \theta)$$
$$= r\sin 180° \sin(-\theta - \alpha) - r\cos 180° \cos(-\theta - \alpha)$$
$$= r\cos(-\theta - \alpha) = r\cos(\theta + \alpha) = r\cos\theta\cos\alpha - r\sin\theta\sin\alpha$$
$$= x_1\cos\theta - y_1\sin\theta$$

Using the same procedure, it can be shown that $y_2 = x_1 \sin \theta + y_1 \cos \theta$.

The coordinate z is unchanged in the rotation, so that $z_2 = z_1$. The three equations

$$x_2 = x_1 \cos \theta - y_1 \sin \theta$$
$$y_2 = x_1 \sin \theta + y_1 \cos \theta \quad \text{and}$$
$$z_2 = z_1$$

can be expressed in the matrix form

$$\begin{pmatrix} x_2 \\ y_2 \\ z_2 \end{pmatrix} = \begin{pmatrix} \cos \theta & -\sin \theta & 0 \\ \sin \theta & \cos \theta & 0 \\ 0 & 0 & 1 \end{pmatrix} \begin{pmatrix} x_1 \\ y_1 \\ z_1 \end{pmatrix}$$

Because $\cos(180°) = -1$ and $\sin(180°) = 0$, the matrix for 180° rotation around the z axis takes the form

$$\begin{pmatrix} -1 & 0 & 0 \\ 0 & -1 & 0 \\ 0 & 0 & 1 \end{pmatrix}$$

The effect of the four operators, $\hat{E}$, $\hat{C}_2$, $\hat{\sigma}_v$, and $\hat{\sigma}'_v$ on $\mathbf{r}$ can also be deduced from Figure 16.4. Convince yourself, using Example Problem 16.1 and Figure 16.4, that the symmetry operators of the C_{2v} group have the following effect on the vector (x, y, z):

$$\hat{E}\begin{pmatrix} x \\ y \\ z \end{pmatrix} \Rightarrow \begin{pmatrix} x \\ y \\ z \end{pmatrix}, \ \hat{C}_2\begin{pmatrix} x \\ y \\ z \end{pmatrix} \Rightarrow \begin{pmatrix} -x \\ -y \\ z \end{pmatrix}, \ \hat{\sigma}_v\begin{pmatrix} x \\ y \\ z \end{pmatrix} \Rightarrow \begin{pmatrix} x \\ -y \\ z \end{pmatrix}, \ \hat{\sigma}'_v\begin{pmatrix} x \\ y \\ z \end{pmatrix} \Rightarrow \begin{pmatrix} -x \\ y \\ z \end{pmatrix} \quad \textbf{(16.2)}$$

Given these results, the operators $\hat{E}$, $\hat{C}_2$, $\hat{\sigma}_v$, and $\hat{\sigma}'_v$ can be described by the following 3×3 matrices:

$$\hat{E}: \begin{pmatrix} 1 & 0 & 0 \\ 0 & 1 & 0 \\ 0 & 0 & 1 \end{pmatrix} \quad \hat{C}_2: \begin{pmatrix} -1 & 0 & 0 \\ 0 & -1 & 0 \\ 0 & 0 & 1 \end{pmatrix} \quad \hat{\sigma}_v: \begin{pmatrix} 1 & 0 & 0 \\ 0 & -1 & 0 \\ 0 & 0 & 1 \end{pmatrix} \quad \hat{\sigma}'_v: \begin{pmatrix} -1 & 0 & 0 \\ 0 & 1 & 0 \\ 0 & 0 & 1 \end{pmatrix} \quad \textbf{(16.3)}$$

Equation (16.3) gives a formulation of the symmetry operators as 3×3 matrices. Do these operators satisfy the requirements listed in Section 16.1 for the corresponding elements to form a group? We begin answering this question by showing in Example Problem 16.3 that the successive application of two operators is equivalent to applying one of the four operators.

EXAMPLE PROBLEM 16.3

Evaluate $\hat{C}_2\,\hat{\sigma}_v$ and $\hat{C}_2\,\hat{C}_2$. What operation is equivalent to the two sequential operations?

Solution

$$\hat{C}_2\hat{\sigma}_v = \begin{pmatrix} -1 & 0 & 0 \\ 0 & -1 & 0 \\ 0 & 0 & 1 \end{pmatrix}\begin{pmatrix} 1 & 0 & 0 \\ 0 & -1 & 0 \\ 0 & 0 & 1 \end{pmatrix} = \begin{pmatrix} -1 & 0 & 0 \\ 0 & 1 & 0 \\ 0 & 0 & 1 \end{pmatrix} = \hat{\sigma}'_v$$

$$\hat{C}_2\hat{C}_2 = \begin{pmatrix} -1 & 0 & 0 \\ 0 & -1 & 0 \\ 0 & 0 & 1 \end{pmatrix}\begin{pmatrix} -1 & 0 & 0 \\ 0 & -1 & 0 \\ 0 & 0 & 1 \end{pmatrix} = \begin{pmatrix} 1 & 0 & 0 \\ 0 & 1 & 0 \\ 0 & 0 & 1 \end{pmatrix} = \hat{E}$$

We see that the product of the two operators is another operator of the group.

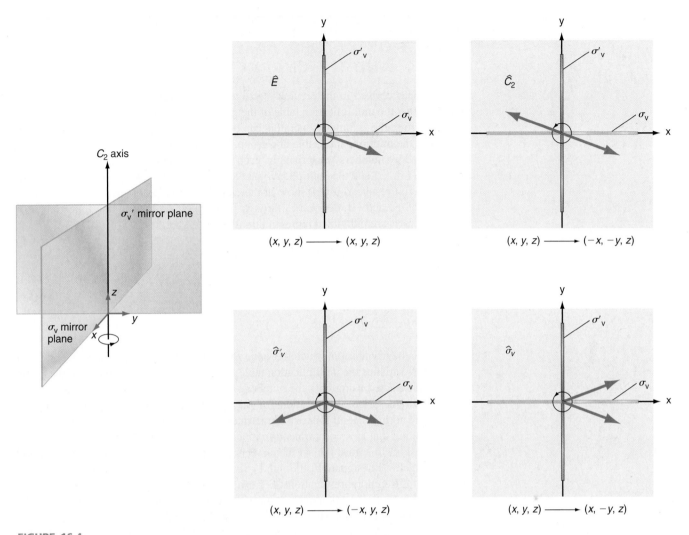

FIGURE 16.4

Schematic of the effect of the four symmetry operations of the C_{2v} group on an arbitrary vector (x, y, z). The symmetry elements are shown on the left. Because z is unchanged through any of the operations, it is sufficient to determine the changes in the $x-y$ coordinates through the symmetry operations. This is shown on the right side of the figure viewed along the C_2 axis. The wide lines along the x and y axis represent the σ_v and σ_v' mirror planes, respectively. The red vector is transformed into the green vector in each case.

By repeating the procedure from Example Problem 16.3 with all possible combinations of operators, Table 16.3 can be generated. This table shows that, as required, the result of any two successive operations is another of these four symmetry operations. The table also shows that $\hat{C}_2\hat{C}_2 = \hat{\sigma}_v\hat{\sigma}_v = \hat{\sigma}_v'\hat{\sigma}_v' = \hat{E}$. Each operator has an inverse operator in the group, and in this particular case, each operator is its own inverse operator. The operations are also associative, which can be shown by evaluating an arbitrary combination of three operators such as $\hat{\sigma}_v(\hat{C}_2\hat{\sigma}_v') - (\hat{\sigma}_v\hat{C}_2)\hat{\sigma}_v'$. If the operators are associative, this expression will equal zero. Using the multiplication table to evaluate the products in parentheses in the following equation, the result is

$$\hat{\sigma}_v(\hat{C}_2\hat{\sigma}_v') - (\hat{\sigma}_v\hat{C}_2)\hat{\sigma}_v' = \hat{\sigma}_v\hat{\sigma}_v - \hat{\sigma}_v'\hat{\sigma}_v' = \hat{E} - \hat{E} = 0 \qquad (16.4)$$

You can convince yourself that any other combination of three operators will give the same result. We have now shown that the four symmetry elements characteristic of the water molecule satisfy the requirements of a group.

In this section, it was useful to express the operators of the C_{2v} group as 3×3 matrices in order to generate the group multiplication table. It turns out that these operators can be expressed in many different ways. This important topic is discussed in the following section.

TABLE 16.3 MULTIPLICATION TABLE FOR OPERATORS OF THE C₂ᵥ GROUP

Second Operation	First Operation			
	$\hat{E}$	$\hat{C}_2$	$\hat{\sigma}_v$	$\hat{\sigma}_v'$
$\hat{E}$	$\hat{E}$	$\hat{C}_2$	$\hat{\sigma}_v$	$\hat{\sigma}_v'$
$\hat{C}_2$	$\hat{C}_2$	$\hat{E}$	$\hat{\sigma}_v'$	$\hat{\sigma}_v$
$\hat{\sigma}_v$	$\hat{\sigma}_v$	$\hat{\sigma}_v'$	$\hat{E}$	$\hat{C}_2$
$\hat{\sigma}_v'$	$\hat{\sigma}_v'$	$\hat{\sigma}_v$	$\hat{C}_2$	$\hat{E}$

16.4 REPRESENTATIONS OF SYMMETRY OPERATORS, BASES FOR REPRESENTATIONS, AND THE CHARACTER TABLE

The matrices derived in the previous section are called **representations** of that group, meaning that the multiplication table of the group can be reproduced with the matrices. For this group the symmetry operators can be represented by numbers, and these numbers obey the multiplication table of a group. How can the operators of the C_{2v} group be represented by numbers? Surprisingly, each operation can be represented by either the number $+1$ or -1 and the multiplication table is still satisfied. As shown later, this is far from a trivial result. You will show in the end-of-chapter problems that the following four sets of $+1$ and -1, denoted Γ_1 through Γ_4, each satisfy the C_{2v} multiplication table and, therefore, are individual representations of the C_{2v} group:

Representation	E	C_2	σ_v	σ_v'
Γ_1	1	1	1	1
Γ_2	1	1	-1	-1
Γ_3	1	-1	1	-1
Γ_4	1	-1	-1	1

Other than the trivial set in which the value zero is assigned to all operators, no other set of numbers satisfies the multiplication table. The fact that a representation of the group can be constructed using only the numbers $+1$ and -1 means that 1×1 matrices are sufficient to describe all operations of the C_{2v} group. This conclusion can also be reached by noting that all four 3×3 matrices derived in the previous section are diagonal, meaning that x, y, and z transform independently in Equation (16.2).

It is useful to regard the set of numbers for an individual representation as a row vector, which we designate Γ_1 through Γ_4 for the C_{2v} group. Each group has an infinite number of different representations. For example, had we considered a Cartesian coordinate system at the position of each atom in water, we could have used 9×9 matrices to describe the operators. However, a much smaller number of representations, called **irreducible representations**, play a fundamental role in group theory. The irreducible representations are the matrices of smallest dimension that obey the multiplication table of the group. We cite the following theorem from group theory:

A group has as many irreducible representations as it has classes of symmetry elements.

Irreducible representations play a central role in discussing molecular symmetry. We explore irreducible representations in greater depth in the next section.

Because the C_{2v} group has four classes of symmetry elements, only four different irreducible representations of this group are possible. This is an important result that we will return to in Section 16.5. The usefulness of these representations in quantum chemistry can be seen by considering the effect of symmetry operations on the oxygen AOs in H_2O. Consider the three different $2p$ atomic orbitals on the oxygen atoms shown in Figure 16.5.

How are the three oxygen $2p$ orbitals transformed under the symmetry operations of the C_{2v} group? Numbers are assigned to the transformation of the $2p$ orbitals in the following way. If the sign of each lobe is unchanged by the operation, $+1$ is assigned to the transformation. If the sign of each lobe is changed, -1 is assigned to the transformation. These are the only possible outcomes for the symmetry operations of the C_{2v} group. We consider the $2p_z$ AO first. Figure 16.5 shows that the sign of each lobe remains the same after each operation. Therefore, we assign $+1$ to each operation. For the $2p_x$ AO, the $\hat{C}_2$ rotation and the $\hat{\sigma}_v'$ reflection change the sign of each lobe, but the sign of each lobe is unchanged after the $\hat{E}$ and $\hat{\sigma}_v$ operations. Therefore, we assign $+1$ to the $\hat{E}$ and $\hat{\sigma}_v$ operators and -1 to the $\hat{C}_2$ and $\hat{\sigma}_v'$ operators. Similarly, for the $2p_y$ AO, we assign $+1$ to the $\hat{E}$ and $\hat{\sigma}_v'$ operators and -1 to the $\hat{C}_2$ and $\hat{\sigma}_v$ operators. Note that if you

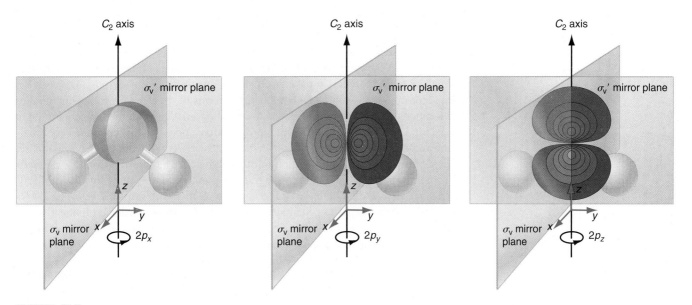

FIGURE 16.5

The three p orbitals on the oxygen atoms transform differently under the symmetry operations of the C_{2v} group.

arrange the numbers $+1$ and -1 obtained separately for the $2p_z$, $2p_x$, and $2p_y$ orbitals in the order $\hat{E}$, $\hat{C}_2$, $\hat{\sigma}_v$, and $\hat{\sigma}'_v$, the sequences that we have just derived are identical to the first, third, and fourth representations for the C_{2v} group.

Because each of the $2p_z$, $2p_x$, and $2p_y$ AOs can be associated with a different representation, each of these AOs **forms a basis** for one of the representations. Had we considered an unoccupied $3d_{xy}$ AO on the oxygen, we would have found that it forms a basis for the second representation. In the nomenclature used in group theory, one says that an AO, or any other function, **belongs to a particular representation** if it forms a basis for that representation. To this point, we have shown that 3×3 matrices, an appropriate set of the numbers $+1$ and -1, and the AOs of oxygen all form a basis for the C_{2v} point group.

The information on the possible representations discussed can be assembled in a form known as a **character table**. Each point group has its unique character table. The character table for the C_{2v} group is as follows:

	E	C_2	σ_v	σ'_v			
A_1	1	1	1	1	z	x^2, y^2, z^2	$2p_z(O)$
A_2	1	1	-1	-1	R_z	xy	$3d_{xy}(O)$
B_1	1	-1	1	-1	x, R_y	xz	$2p_x(O)$
B_2	1	-1	-1	1	y, R_x	yz	$2p_y(O)$

Much of the information in this character table was derived in order to make the origin of the individual entries clear. However, this task is not necessary, because character tables for point groups in this standard format are widely accessible and are listed in Appendix C.

The character table is the single most important result of group theory for chemists. Therefore, the structure and individual entries in the character table are now discussed in detail. The leftmost column in a character table shows the symbol for each irreducible representation. By convention, a representation that is symmetric $(+1)$ with respect to rotation about the principal axis, C_2 in this case, is given the symbol A. A representation that is antisymmetric (-1) with respect to rotation about the principal axis is given the symbol B. The subscript 1 (2) is used for representations that are symmetric (antisymmetric) with respect to a C_2 axis perpendicular to the principal axis. If such an axis is not an element of the group, the symmetry with respect to a vertical mirror plane, $\hat{\sigma}_v$ in this case, is used. The representation in which all entries are $+1$ is called the **totally symmetric representation**. Every group has a totally symmetric representation.

The next section of the table (columns 2 through 5) has an entry for each operation of the group in each representation. These entries are called **characters**. The right section of the table (columns 6 through 8) shows several of the many possible bases for each representation. Column 6 shows bases in terms of the three Cartesian coordinates and rotations about the three axes. Column 8 shows the AOs on the oxygen atom that can be used as bases for the different representations; note that this column is not usually shown in character tables. It is shown here because we will work further with this set of **basis functions**. The information in this column can be inferred from the previous two columns as the p_x, p_y, and p_z AOs transform as x, y, and z, respectively. Similarly, the d_{z^2}, d_{xy}, d_{yz}, and $d_{x^2-y^2}$, and d_{xy} AOs transform as their subscript indices. The s AOs are a basis for A_1 because of their spherical symmetry. Next consider columns 6 and 7 in this section, which have entries based on the x, y, and z coordinates and rotations about the axes designated R_x, R_y, and R_z. We show later that the R_z rotation and the different coordinate combinations are bases for the indicated representations.

How can it be shown that the indicated functions are bases for the four irreducible representations? Equation (16.2) shows that the effect of any of the C_{2v} operators on the components x, y, and z of an arbitrary three-dimensional vector are $x \rightarrow \pm x$, $y \rightarrow \pm y$, and $z \rightarrow z$. Because z does not change sign under any of the operators, all characters for the representation have the value $+1$. Therefore, z is a basis for the A_1 representation. Similarly, because x^2, y^2, and z^2 do not change under any of the operations, these functions are also bases for the A_1 representation. Equation (16.2) shows that the product $xy \rightarrow xy$ for $\hat{E}$ and $\hat{C}_2$ and $xy \rightarrow -xy$ for $\hat{\sigma}_v$ and $\hat{\sigma}'_v$. Therefore, the product xy is a basis for the A_2 representation. Because z does not change sign under any operation, xz and yz transform as x and y. Therefore, Equation (16.2) shows that the functions x and xz are bases for the B_1 representation, and y and yz are bases for the B_2 representation.

Example Problem 16.1 demonstrated that in the operation R_z (C_2 in this case), $x \rightarrow -x$, $y \rightarrow -y$, and $z \rightarrow z$. Therefore, the product xy is unchanged because $xy \rightarrow (-x)(-y) = xy$. This shows that both R_z and xy are bases for the A_2 representation. We will not prove that R_x and R_y are bases for the B_1 and B_2 representations, but the procedure to do so is the same as for the other representations. As we saw in Section 16.3, the rotation operators are three-dimensional matrices. Therefore, in contrast to the coordinate bases, the rotation operators are bases for **reducible representations**, because their dimension is greater than one.

As shown earlier, all irreducible representations of the C_{2v} group are one dimensional. However, it is useful to consider reducible representations for this group such as R_x, R_y, and R_z, all of which are three dimensional, to visualize how individual operators act on an arbitrary vector. Some of the groups discussed in this chapter also have irreducible representations whose dimensionality is two or three. Therefore, before we begin to work on problems of chemical interest using character tables, it is necessary to discuss the dimensionality of irreducible representations.

16.5 THE DIMENSION OF A REPRESENTATION

The bases for the different representations of the C_{2v} group include either x or y or z, but not a linear combination of two coordinates such as $x + y$. This is the case because under any transformation $(x, y, z) \rightarrow (x', y', z')$, x' is only a function of x as opposed to being a function of x and y or x and z or x, y, and z. Similar statements can be made for y' and z'. As a consequence, all of the matrices that describe the operators for the C_{2v} group have a diagonal form, as shown in Equation (16.3).

The matrix generated by two successive operations of diagonal matrices, which is denoted by $\hat{R}''' = \hat{R}'\hat{R}''$, is also a diagonal matrix whose elements are given by

$$\hat{R}'''_{ii} = \hat{R}'_{ii}\,\hat{R}''_{ii} \tag{16.5}$$

The **dimension of a representation** is defined as the size of the matrix used to represent the symmetry operations. As discussed earlier, the matrices of Equation (16.5) form a three-dimensional representation of the C_{2v} group. However, because all of the matrices are diagonal, the 3×3 matrix operations can be reduced to three 1×1 matrix operations, which consist of the numbers $+1$ and -1. Therefore, the three-dimensional reducible representation of Equation (16.5) can be reduced to three one-dimensional representations.

Point groups can also have **two-dimensional** and **three-dimensional irreducible representations**. If x' and/or $y' = f(x, y)$ for a representation, then the basis will be (x, y) and the dimension of that irreducible representation is two. At least one of the matrices representing the operators will have the form

$$\begin{pmatrix} a & b & 0 \\ c & d & 0 \\ 0 & 0 & e \end{pmatrix}$$

in which entries a through e are in general nonzero. If x' and/or y' and/or $z' = f(x, y, z)$ the dimension of the representation is three and at least one of the operators will have the form

$$\begin{pmatrix} a & b & c \\ d & e & f \\ g & h & j \end{pmatrix}$$

in which entries a through j are in general nonzero.

How does one know how many irreducible representations a group has and what their dimension is? The following result of group theory is used to answer this question:

The dimension of the different irreducible representations, d_j, and the **order of the group**, h, defined as the number of symmetry elements in the group, are related by the equation

$$\sum_{j=1}^{N} d_j^2 = h \qquad (16.6)$$

This sum is over the irreducible representations of the group.

Because every point group contains the one-dimensional totally symmetric representation, at least one of the $d_j = 1$ We apply this formula to the C_{2v} representations. This group has four elements, and all belong to different classes. Therefore, there are four different representations. The only set of nonzero integers that satisfies the equation

$$d_1^2 + d_2^2 + d_3^2 + d_4^2 = 4 \qquad (16.7)$$

is $d_1 = d_2 = d_3 = d_4 = 1$. We conclude that all of the irreducible representations of the C_{2v} group are one dimensional. Because a 1×1 matrix cannot be reduced to one of lower dimensionality, all one-dimensional representations are irreducible.

For the C_{2v} group, the number of irreducible representations is equal to the number of elements and classes. More generally, the number of irreducible representations is equal to the number of classes for any group. Recall that all operators generated from a single symmetry element and successive applications of other operators of the group belong to the same class. For example, consider NF_3, which belongs to the C_{3v} group. As shown in Example Problem 16.1, the C_3 and C_3^2 rotations of the C_{3v} group belong to the same class. The three σ_v mirror planes also belong to the same class because the second and third planes are generated from the first by applying $\hat{C}_3$ and $\hat{C}_3^2$. Therefore, the C_{3v} group has six elements, but only three classes.

We next show that the C_{3v} point group has one representation that is not one dimensional. Using the result from Example Problem 16.2, the matrix that describes a 120° rotation is

$$\hat{C}_3 = \begin{pmatrix} \cos\theta & -\sin\theta & 0 \\ \sin\theta & \cos\theta & 0 \\ 0 & 0 & 1 \end{pmatrix} = \begin{pmatrix} -1/2 & -\sqrt{3}/2 & 0 \\ \sqrt{3}/2 & -1/2 & 0 \\ 0 & 0 & 1 \end{pmatrix} \qquad (16.8)$$

The other operators in this group have a diagonal form. The $\hat{C}_3$ operator does not have a diagonal form, and $\hat{C}_3$ acting on the vector (x, y, z) mixes x and y. However, z' depends only on z and not on x or y. Therefore, it is possible to reduce the 3×3 matrix operator for $\hat{C}_3$ into separate irreducible 2×2 and 1×1 matrix operators. We conclude that the C_{3v} point group contains a two-dimensional irreducible representation. Example Problem 16.4 shows how to determine the number and dimension of the remaining irreducible representations for the C_{3v} group.

EXAMPLE PROBLEM 16.4

The C_{3v} group has the elements $\hat{E}$, $\hat{C}_3$, and $\hat{C}_3^2$ and three σ_v mirror planes. How many different irreducible representations does this group have, and what is the dimensionality of each irreducible representation?

Solution

The order of the group is the number of elements, so $h = 6$. The number of representations is the number of classes. As discussed earlier, $\hat{C}_3$ and $\hat{C}_3^2$ belong to one class, and the same is true of the three σ_v reflections. Although the group has six elements, it has only three classes. Therefore, the group has three irreducible representations. The equation $l_1^2 + l_2^2 + l_3^2 = 6$ is solved to find the dimension of the representations, and one of the values must be 1. The only possible solution is $l_1 = l_2 = 1$ and $l_3 = 2$. We see that the C_{3v} group contains one two-dimensional representation and two one-dimensional representations.

To gain practice in working with irreducible representations of more than one dimension, the matrices for the individual operations that describe the two-dimensional representation in the C_{3v} group are derived next. Example Problem 16.2 shows how to set up the matrices for rotation operators. Figure 16.6 shows how the $x-y$ coordinate system is transformed by a mirror plane, σ.

The values x' and y' are related to x and y by

$$x' = -x\cos 2\theta - y\sin 2\theta$$
$$y' = -x\sin 2\theta + y\cos 2\theta \qquad \text{(16.9)}$$

Equation (16.9) is used to evaluate the 2×2 matrices for the mirror planes $\hat{\sigma}$, $\hat{\sigma}'$, and $\hat{\sigma}''$ at 0, $\pi/3$, and $2\pi/3$, and Equation (16.1) is used to evaluate the 2×2 matrices for $\hat{C}_3$ and $\hat{C}_3^2$. The resulting operators for the two-dimensional representation of the C_{3v} group are shown in Equation (16.10). Remember that $\hat{\sigma}$, $\hat{\sigma}'$, and $\hat{\sigma}''$ all belong to the one class, as do $\hat{C}_3$ and $\hat{C}_3^2$.

$$\hat{E} = \begin{pmatrix} 1 & 0 \\ 0 & 1 \end{pmatrix}$$

$$\hat{\sigma} = \begin{pmatrix} -\cos 0 & -\sin 0 \\ -\sin 0 & \cos 0 \end{pmatrix} = \begin{pmatrix} -1 & 0 \\ 0 & 1 \end{pmatrix}$$

$$\hat{\sigma}' = \begin{pmatrix} -\cos(2\pi/3) & -\sin(2\pi/3) \\ -\sin(2\pi/3) & \cos(2\pi/3) \end{pmatrix} = \begin{pmatrix} 1/2 & \sqrt{3}/2 \\ \sqrt{3}/2 & -1/2 \end{pmatrix}$$

$$\hat{\sigma}'' = \begin{pmatrix} -\cos(4\pi/3) & -\sin(4\pi/3) \\ -\sin(4\pi/3) & \cos(4\pi/3) \end{pmatrix} = \begin{pmatrix} 1/2 & -\sqrt{3}/2 \\ -\sqrt{3}/2 & -1/2 \end{pmatrix}$$

$$\hat{C}_3 = \begin{pmatrix} \cos(2\pi/3) & -\sin(2\pi/3) \\ \sin(2\pi/3) & \cos(2\pi/3) \end{pmatrix} = \begin{pmatrix} -1/2 & -\sqrt{3}/2 \\ \sqrt{3}/2 & -1/2 \end{pmatrix}$$

$$\hat{C}_3^2 = \begin{pmatrix} \cos(4\pi/3) & -\sin(4\pi/3) \\ \sin(4\pi/3) & \cos(4\pi/3) \end{pmatrix} = \begin{pmatrix} -1/2 & \sqrt{3}/2 \\ -\sqrt{3}/2 & -1/2 \end{pmatrix} \qquad \text{(16.10)}$$

How is the character table for the C_{3v} group constructed? In particular, how are characters assigned to the two-dimensional representation, which is generally called E? (Don't

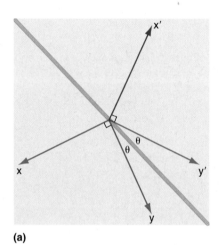

(a)

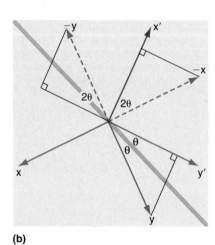

(b)

FIGURE 16.6

Schematic depiction of the transformation of $x-y$ coordinates effected by reflection through a mirror plane, σ, containing the z axis. **(a)** The $x-y$ coordinate system for which the y axis is rotated by θ relative to the mirror plane is reflected through the mirror plane (yellow line). This operation generates the $x'-y'$ coordinate system. **(b)** The geometry used to derive Equation 16.9 is shown.

confuse this symbol for a two-dimensional representation with the operator $\hat{E}$.) The following theorem of group theory is used:

> The character for an operator in a representation of dimension higher than one is given by the sum of the diagonal elements of the matrix.

Using this rule, we see that the character of $\hat{\sigma}, \hat{\sigma}'$, and $\hat{\sigma}''$ is 0, and the character of $\hat{C}_3$ and $\hat{C}_3^2$ is -1. As expected, the character of all elements in a class is the same. Recall also that every group has a totally symmetric representation in which all characters are $+1$.

Because the C_{3v} group contains three classes, it must have three irreducible representations. We enter the information that we obtained earlier for A_1 and E in the following partially completed character table. All of the symmetry operators of a class are grouped together in a character table. For example, in the following listing, the elements C_3 and C_3^2 are listed as $2C_3$ to make the notation compact.

	E	$2C_3$	$3\sigma_v$
A_1	1	1	1
?	a	b	c
E	2	-1	0

How can the values for the characters a, b, and c be obtained? We use another result from group theory:

> If the set of characters associated with a representation of the group is viewed as a vector, $\Gamma_i = \chi_i(\hat{R}_j)$, with one component for each element of the group, the following condition holds:
>
> $$\Gamma_i\Gamma_k = \sum_{j=1}^{h}\chi_i(\hat{R}_j)\,\chi_k(\hat{R}_j) = h\delta_{ik}, \text{ where } \delta_{ik} = 0 \text{ if } i \neq k \text{ and } 1 \text{ if } i = k \quad \textbf{(16.11)}$$
>
> or, equivalently, $\Gamma_i\Gamma_k = \chi_i(\hat{R}_j) \cdot \chi_k(\hat{R}_j) = h\delta_{ik}$. The sum is over all elements of the group.

EXAMPLE PROBLEM 16.5

Determine the unknown coefficients a, b, and c for the preceding partially completed character table and assign the appropriate symbol to the irreducible representation.

Solution

From Example Problem 16.4, we know that the unknown representation is one dimensional. From Equation (16.11), we know that the $\chi_i(\hat{R}_j)$ for different values of the index i are orthogonal. Therefore,

$$\vec{\chi}_? \cdot \vec{\chi}_{A_1} = a + b + b + c + c + c = a + 2b + 3c = 0$$
$$\vec{\chi}_? \cdot \vec{\chi}_E = 2a - b - b = 2a - 2b = 0$$

We could also have taken the sum over classes and multiplied each term by the number of elements in the class, because all elements in a class have the same character. We also know that $a = 1$ because it is the character of the identity operator. Solving the equations gives the results of $b = 1$ and $c = -1$. Because the character of C_3 is $+1$, and the character of σ_v is -1, the unknown representation is designated A_2. Table 16.4 shows the completed C_{3v} character table.

Note that the two-dimensional basis functions occur in pairs. You will be asked to verify that z and R_z are bases for the A_1 and A_2 representations, respectively, in the end-of-chapter problems.

TABLE 16.4 THE C₃ᵥ CHARACTER TABLE

	E	$2C_3$	$3\sigma_v$		
A_1	1	1	1	z	$x^2 + y^2, z^2$
A_2	1	1	-1	R_z	
E	2	-1	0	$(x, y), (R_x, R_y)$	$(x^2 - y^2, xy), (xz, yz)$

16.6 USING THE C₂ᵥ REPRESENTATIONS TO CONSTRUCT MOLECULAR ORBITALS FOR H₂O

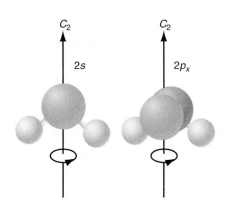

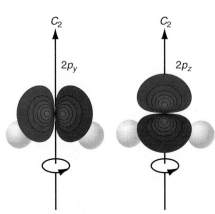

FIGURE 16.7

Depiction of the oxygen atomic orbitals that will be considered as contributors to the MO formed using $\phi_+ = \phi_{H1sA} + \phi_{H1sB}$.

A number of aspects of group theory have been discussed in the preceding sections. In particular, the structure of character tables, which are the most important result of group theory for chemists, has been explained. We now illustrate the usefulness of character tables to solve a problem of chemical interest, namely, the construction of MOs that incorporate the symmetry of a molecule. Why is this necessary?

To answer this question, consider the relationship among the total energy operator, the molecular wave functions ψ_j, and the symmetry of the molecule. A molecule that has undergone one of its symmetry operations, $\hat{A}$, is indistinguishable from the original molecule. Therefore, $\hat{H}$ must also be unchanged under this and any other symmetry operation of the group, because the total energy of the molecule is the same in any of its equivalent positions. If this is the case, then $\hat{H}$ belongs to the totally symmetric representation.

Because the order of applying $\hat{H}$ and $\hat{A}$ to the molecule is immaterial, it follows that $\hat{H}$ and $\hat{A}$ commute. Therefore, as discussed in Chapter 6, eigenfunctions of $\hat{H}$ can be found that are simultaneously eigenfunctions of $\hat{A}$ and of all other operators of the group. These **symmetry-adapted MOs** are of central importance in quantum chemistry. In this section, we illustrate how to generate symmetry-adapted MOs from AOs. Not all AOs contribute to a particular symmetry-adapted MO. Invoking the symmetry of a molecule results in a set of MOs consisting of fewer AOs than would have been obtained had the molecular symmetry been neglected.

Consider a specific example. Which of the AOs on oxygen contribute to the symmetry-adapted MOs on water? We begin by asking which of the four oxygen valence AOs can be combined with the hydrogen AOs to form symmetry-adapted MOs. All possible combinations are shown in Figure 16.7. In order to take the symmetry of the water molecule into account, the hydrogen AOs will appear as in-phase or out of phase combinations. Consider first the in-phase combination, $\phi_+ = \phi_{H1sA} + \phi_{H1sB}$.

The overlap integral S_{+j} between the orbital ϕ_+ and an oxygen AO ϕ_j is defined by

$$S_{+j} = \int \phi_+^* \phi_j \, d\tau \tag{16.12}$$

Only the oxygen AOs that have a nonzero overlap with the hydrogen AOs are useful in forming chemical bonds. Because S_{+j} is just a number, it cannot change upon applying any of the operators of the C_{2v} group to the integral. In other words, S_{+j} belongs to the A_1 representation. The same must be true of the integrand and, therefore, the integrand must also belong to the A_1 representation. If ϕ_+ belongs to one representation and ϕ_j belongs to another, what can be said about the symmetry of the direct product $\phi_+ \cdot \phi_j$? A result of group theory is used to answer this question:

The character for an operator $\hat{R}$ ($\hat{E}$, $\hat{C}_2$, $\hat{\sigma}_v$, or $\hat{\sigma}'_v$ for the C_{2v} group) of the direct product of two representations is given by

$$\chi_{product}(\hat{R}) = \chi_i(\hat{R})\chi_j(\hat{R}) \tag{16.13}$$

For example, if ϕ_+ belongs to A_2, and ϕ_j belongs to B_2, $\Gamma_{product}$ can be calculated from the $\chi_{product}$ terms as follows:

$$\Gamma_{product} = \chi_{A_2} \cdot \chi_{B_2} = [1 \times 1 \quad 1 \times (-1) \quad (-1) \times (-1) \quad (-1) \times 1]$$
$$= (1 \quad -1 \quad 1 \quad -1) \tag{16.14}$$

Looking at the C_{2v} character table, we can see that the direct product $A_2 \cdot B_2$ belongs to B_1.

How is this result useful in deciding which of the oxygen AOs contribute to the symmetry-adapted water MOs? Because the integrand must belong to the A_1 representation, each character of the representation of $\phi_+^* \phi_j$ must be equal to one. We conclude that

$$\sum_{k=1}^{h} \chi_+(\hat{R}_k) \chi_j(\hat{R}_k) = h \tag{16.15}$$

However, according to Equation (16.11), this equation is never satisfied if the two representations to which the orbitals belong denoted $+$ and j are different. We conclude that *the overlap integral between two combinations of AOs is nonzero only if the combinations belong to the same representation.*

Using this result, which of the oxygen AOs in Figure 16.7 form symmetry-adapted MOs with the combination $\phi_+ = \phi_{H1sA} + \phi_{H1sB}$? The orbital ϕ_+ is unchanged by any of the symmetry operators, so it must belong to the A_1 representation. The $2s$ AO on oxygen is spherically symmetrical, so that it transforms as $x^2 + y^2 + z^2$. As the C_{2v} character table shows, the $2s$ AO belongs to the A_1 representation, as does the $2p_z$ orbital. By contrast, the $2p_x$ and $2p_y$ AOs on oxygen belong to the B_1 and B_2 representations, respectively. Therefore, only the oxygen $2s$ AO and $2p_z$ AOs belong to the same irreducible representation as ϕ_+, and only these AOs will contribute to MOs involving ϕ_+. The combinations of the $2s$ AO and $2p_z$ oxygen AOs with $\phi_+ = \phi_{H1sA} + \phi_{H1sB}$ result in the $1a_1$, the $2a_1$, and the $3a_1$, MOs for water. We now have an explanation for the nomenclature introduced in Figure 16.7 for the symmetry-adapted water MOs. The a_1 refers to the particular irreducible representation of the C_{2v} group, and the integer 1, 2, ..., refers to the lowest, next lowest, ..., energy MO belonging to the A_1 representation.

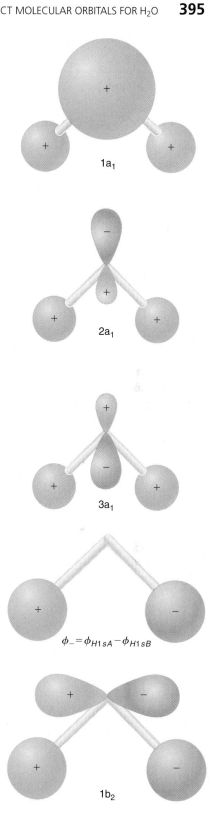

$1a_1$

$2a_1$

$3a_1$

$\phi_- = \phi_{H1sA} - \phi_{H1sB}$

$1b_2$

$2b_2$

EXAMPLE PROBLEM 16.6

Which of the oxygen AOs shown in Figure 16.7 will participate in forming symmetry-adapted water MOs with the antisymmetric combination of hydrogen AOs defined by $\phi_- = \phi_{H1sA} - \phi_{H1sB}$?

Solution

The antisymmetric combination of the H AOs is given by $\phi_- = \phi_{H1sA} - \phi_{H1sB}$, shown in the margin. By considering the C_{2v} operations shown in Figure 16.4, convince yourself that the characters for the different operations are $\hat{E}: +1$, $\hat{C}_2: -1$, $\hat{\sigma}_v: -1$, and $\hat{\sigma}_v': +1$. Therefore, ϕ_- belongs to the B_2 representation. Of the valence oxygen AOs, only the $2p_y$ orbital belongs to the B_2 representation. Therefore, the only symmetry adapted MOs formed from ϕ_- and the $2s$ and $2p$ orbitals that have a nonzero overlap among the AOs are the MOs denoted $1b_2$ and $2b_2$, which are shown in the margin and in Figure 13.7. This nomenclature indicates that they are the lowest and next lowest energy MOs of B_2 symmetry.

We now make an important generalization of the result just obtained. The same symmetry considerations used for the overlap integral apply in evaluating integrals of the type $H_{ab} = \int \psi_a^* \hat{H} \psi_b \, d\tau$. As shown in Chapters 12 and 13, such integrals appear whenever the total energy is calculated. The value of H_{ab} is zero unless $\psi_a^* \hat{H} \psi_b$ belongs to the A_1 representation. Because $\hat{H}$ belongs to the A_1 representation, H_{ab} will be zero unless ψ_a and ψ_b belong to the same representation (not necessarily the A_1 representation). Only then will the integrand $\psi_a^* \hat{H} \psi_b$ contain the A_1 representation. This important result is of great help in evaluating entries in a secular determinant such as those encountered in Chapter 12.

In the preceding discussion, symmetry-adapted MOs for H_2O were generated from AOs that belong to the different irreducible representations of the C_{2v} group in an ad hoc manner. In Supplemental Section 16.9, we discuss a powerful method, called the projection operator method, that allows the symmetry-adapted MOs to be constructed for arbitrary molecules.

16.7 THE SYMMETRIES OF THE NORMAL MODES OF VIBRATION OF MOLECULES

The vibrational motions of individual atoms in a molecule might appear to be chaotic and independent of one another. However, the selection rules for infrared vibrational and Raman spectroscopy are characteristic of the normal modes of a molecule, which can be described in the following way. In a normal mode vibration, each atom is displaced from its equilibrium position by a vector that can but need not lie along the bond direction (for example in a bending mode). The directions and magnitudes of the displacements are not the same for all atoms. The following can be said of the motion of the atoms in the normal modes:

- During a vibrational period, the center of mass of the molecule remains fixed and all atoms in the molecule undergo in-phase periodic motion about their equilibrium positions.
- All atoms in a molecule reach their minimum and maximum amplitudes at the same time.
- These collective motions are called **normal modes**, and the frequencies are called the **normal mode frequencies**.
- The frequencies measured in vibrational spectroscopy are the normal mode frequencies.
- All normal modes are independent in the harmonic approximation, meaning that excitation of one normal mode does not transfer vibrational energy into another normal mode.
- Any seemingly random motion of the atoms in a molecule can be expressed as a linear combination of the normal modes of that molecule.

How many normal modes does a molecule have? An isolated atom has three translational degrees of freedom; therefore, a molecule consisting of n atoms has $3n$ degrees of freedom. Three of these are translations of the molecule and are not of interest here. A nonlinear molecule with n atoms has three degrees of rotational freedom, and the remaining $3n - 6$ internal degrees of freedom correspond to normal modes of vibration. Because a linear molecule has only two degrees of rotational freedom, it has $3n - 5$ normal modes of vibration. For a diatomic molecule, there is only one vibrational mode, and the motion of the atoms is directed along the bond. In the harmonic approximation,

$$V(x) = \frac{1}{2}kx^2 = \frac{1}{2}\left(\frac{d^2V}{dx^2}\right)x^2$$

where x is the displacement from the equilibrium position in the center of mass coordinates. For a molecule with N vibrational degrees of freedom, the potential energy is given by

$$V(q_1, q_2, \ldots, q_N) = \frac{1}{2}\sum_{i=1}^{N}\sum_{j=1}^{N}\frac{\partial^2 V}{\partial q_i \partial q_j}\, q_i q_j \qquad \textbf{(16.16)}$$

where the q_i designates the individual normal mode displacements. Classical mechanics allows us to find a new set of vibrational coordinates $Q_j(q_1, q_2, \ldots, q_N)$ that simplify Equation (16.16) to the form

$$V(Q_1, Q_2, \ldots, Q_N) = \frac{1}{2}\sum_{i=1}^{N}\left(\frac{\partial^2 V}{\partial Q_i^2}\right)Q_i^2 \qquad \textbf{(16.17)}$$

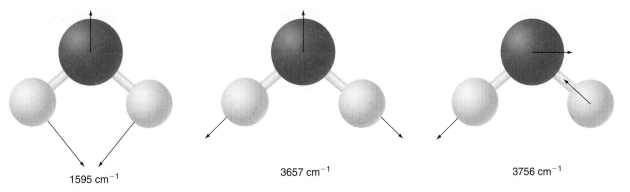

1595 cm^{-1} 3657 cm^{-1} 3756 cm^{-1}

FIGURE 16.8

The normal modes of H_2O are depicted, with the vectors indicating atomic displacements (not to scale). From left to right, the modes correspond to a bond bending, an O—H symmetric stretch, and an O—H asymmetric stretch. The experimentally observed frequencies are indicated.

The $Q_j(q_1, q_2, \ldots, q_N)$ are known as the **normal coordinates** of the molecule. This transformation has significant advantages in describing vibrational motion. Because there are no cross terms of the type $Q_i Q_j$ in the potential energy, the vibrational modes are independent in the harmonic approximation, meaning that

$$\psi_{vibrational}(Q_1, Q_2, \ldots, Q_N) = \psi_1(Q_1)\psi_2(Q_2)\ldots\psi_N(Q_N) \quad \text{and}$$

$$E_{vibrational} = \sum_{i=1}^{N}\left(n_j + \frac{1}{2}\right)h\nu_j \qquad \textbf{(16.18)}$$

Because of the transformation to normal coordinates, each of the normal modes contributes independently to the energy, and the vibrational motions of different normal modes are not coupled, consistent with the properties in the first paragraph of this section. Finding the normal modes is a nontrivial but straightforward exercise that can be done most easily using numerical methods. The calculated normal modes of H_2O are shown in Figure 16.8. The arrows show the displacement of each atom at a given time. After half the vibrational period, each arrow has the same magnitude, but the direction is opposite to that shown in the figure. We do not carry out a normal mode calculation, but focus instead on the symmetry properties of the normal modes.

Just as the $2p$ atomic orbitals on the oxygen atom belong to individual representations of the C_{2v} group, the normal modes of a molecule belong to individual representations. The next task is to identify the symmetry of the three different normal modes of H_2O. To do so, a coordinate system is set up at each atom and a matrix representation formed that is based on the nine x, y, and z coordinates of the atoms in the molecule. Figure 16.9 illustrates the geometry under consideration.

Consider the C_2 operation. By visualizing the motion of the coordinate systems on the three atoms, convince yourself that the individual coordinates are transformed as follows under this operation:

$$\begin{pmatrix} x_1 \\ y_1 \\ z_1 \\ x_2 \\ y_2 \\ z_2 \\ x_3 \\ y_3 \\ z_3 \end{pmatrix} \Longrightarrow \begin{pmatrix} -x_3 \\ -y_3 \\ z_3 \\ -x_2 \\ -y_2 \\ z_2 \\ -x_1 \\ -y_1 \\ z_1 \end{pmatrix} \qquad \textbf{(16.19)}$$

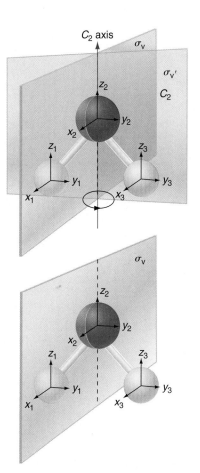

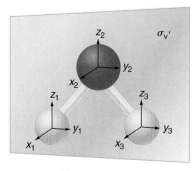

FIGURE 16.9

Transformations of coordinate systems on the atom in water. Each is considered to be a separate entity under symmetry operations of the C_{2v} group.

You can also convince yourself that the 9×9 matrix that describes this transformation is

$$
\hat{C}_2 = \begin{pmatrix}
0 & 0 & 0 & 0 & 0 & 0 & -1 & 0 & 0 \\
0 & 0 & 0 & 0 & 0 & 0 & 0 & -1 & 0 \\
0 & 0 & 0 & 0 & 0 & 0 & 0 & 0 & 1 \\
0 & 0 & 0 & -1 & 0 & 0 & 0 & 0 & 0 \\
0 & 0 & 0 & 0 & -1 & 0 & 0 & 0 & 0 \\
0 & 0 & 0 & 0 & 0 & 1 & 0 & 0 & 0 \\
-1 & 0 & 0 & 0 & 0 & 0 & 0 & 0 & 0 \\
0 & -1 & 0 & 0 & 0 & 0 & 0 & 0 & 0 \\
0 & 0 & 1 & 0 & 0 & 0 & 0 & 0 & 0
\end{pmatrix}
\qquad (16.20)
$$

Note that this matrix has a simple structure. It consists of identical 3×3 subunits shown in boxes. More importantly, the diagonal elements of the subunits lie along the diagonal of the 9×9 matrix only if the atom is not shifted to another position through the transformation. Because H atoms 1 and 3 exchange places under the C_2 operation, they do not contribute to the character of the operator, which is the sum of the diagonal elements of the matrix. This result leads to the following guidelines for calculating the character of each element of the group in the 9×9 matrix representation:

- If the atom remains in the same position under the transformation, and the sign of x, y, or z is not changed, the value $+1$ is associated with each unchanged coordinate.

- If the sign of x, y, or z is changed, the value -1 is associated with each changed coordinate.

- If the coordinate system is exchanged with the position of another coordinate system, the value zero is associated with each of the three coordinates.

- Recall that only the diagonal elements contribute to the character. Therefore, only atoms that are not shifted by an operation contribute to the character.

This procedure is applied to the water molecule. Because nothing changes under the E operation, the character of $\hat{E}$ is 9. Under the rotation of $180°$, the two H atoms are interchanged, so that none of the six coordinates contributes to the character of the $\hat{C}_2$ operator. On the oxygen atom, $x \rightarrow -x$, $y \rightarrow -y$, and $z \rightarrow z$. Therefore, the character of $\hat{C}_2$ is -1. For the $\hat{\sigma}_v$ operation, the H atoms are again interchanged so that they do not contribute to the character of $\hat{\sigma}_v$. On the oxygen atom, $x \rightarrow x$, $y \rightarrow -y$, and $z \rightarrow z$. Therefore, the character of $\hat{\sigma}_v$ is $+1$. For the $\hat{\sigma}'_v$ operation, on the H atoms, $x \rightarrow -x$, $y \rightarrow y$, and $z \rightarrow z$, so that the two H atoms contribute 2 to the $\hat{\sigma}'_v$ character. On the oxygen atom, $x \rightarrow -x$, $y \rightarrow y$, and $z \rightarrow z$ so that the O atom contributes $+1$ to the $\hat{\sigma}'_v$ character. Therefore, the total character of $\hat{\sigma}'_v$ is $+3$. These considerations show that the reducible representation formed using the coordinate systems on the three atoms as a basis is

E	C_2	σ_v	σ'_v
9	-1	1	3

$\qquad (16.21)$

This is a reducible representation because it is a nine-dimensional representation, whereas all irreducible representations of the C_{2v} group are one dimensional. To use this result to characterize the symmetry of the normal modes of water, it is necessary to decompose this reducible representation into the irreducible representations that it contains, as follows:

The general method for decomposing a reducible representation into its irreducible representations utilizes the vector properties of the representations introduced in Section 16.3. Take the scalar product between the reducible representation $\Gamma_{reducible}(\hat{R}_j)$ and each of the irreducible representations $\Gamma_i(\hat{R}_j)$ in turn, and divide by the order of the group. The result of this procedure is a positive integer n_i that is the number of times each representation appears in the irreducible representation. This statement is expressed by the equation

$$n_i = \frac{1}{h}\Gamma_i\Gamma_{reducible} = \frac{1}{h}\boldsymbol{\chi}_i(\hat{R}_j) \cdot \boldsymbol{\chi}_{reducible}(\hat{R}_j) = \frac{1}{h}\sum_{j=1}^{h} \chi_i(\hat{R}_j)\,\chi_{reducible}(\hat{R}_j),$$

for $i = 1, 2, \ldots, N$

$$(16.22)$$

We calculate the contribution of the individual irreducible representations to this reducible presentation using Equations (16.22):

$$n_{A_1} = \frac{1}{h}\sum_{j=1}^{h}\chi_{A_1}(\hat{R}_j)\,\chi_{reducible}(\hat{R}_j) = \frac{1\times 9 - 1\times 1 + 1\times 1 + 1\times 3}{4} = 3$$

$$n_{A_2} = \frac{1}{h}\sum_{j=1}^{h}\chi_{A_2}(\hat{R}_j)\,\chi_{reducible}(\hat{R}_j) = \frac{1\times 9 - 1\times 1 + (-1)\times 1 + (-1)\times 3}{4} = 1$$

$$n_{B_1} = \frac{1}{h}\sum_{j=1}^{h}\chi_{B_1}(\hat{R}_j)\,\chi_{reducible}(\hat{R}_j) = \frac{1\times 9 - (-1)\times 1 + 1\times 1 + (-1)\times 3}{4} = 2$$

$$n_{B_2} = \frac{1}{h}\sum_{j=1}^{h}\chi_{B_2}(\hat{R}_j)\,\chi_{reducible}(\hat{R}_j) = \frac{1\times 9 - (-1)\times 1 + (-1)\times 1 + 1\times 3}{4} = 3$$

$$(16.23)$$

This calculation shows that $\Gamma_{reducible} = 3A_1 + A_2 + 2B_1 + 3B_2$. However, not all of these representations describe vibrational normal modes. The translation of the molecules along the x, y, and z axes as well as their rotation about the same axes must be separated out to obtain the representations of the vibrational normal modes. This can be done by subtracting the representations belonging to x, y, and z as well as to R_x, R_y, and R_z. Representations for these degrees of freedom can be determined from the C_{2v} character table. Eliminating them gives the representations of the three vibrational modes as

$$\Gamma_{reducible} = 3A_1 + A_2 + 2B_1 + 3B_2 - (B_1 + B_2 + A_1) - (B_2 + B_1 + A_2)$$
$$= 2A_1 + B_2$$

$$(16.24)$$

This calculation has shown that the symmetry of the H_2O molecule dictates the symmetry of the normal modes. Of the three normal modes, one belongs to B_2 and two belong to A_1. The normal mode calculations outlined here give the modes shown in Figure 16.8.

How can these modes be assigned to different irreducible representations of the C_{2v} group? The arrows on each atom in Figure 16.8 show the direction and magnitude of the displacement at a given time. Because all atoms undergo in-phase periodic motion about their initial positions simultaneously, all displacement vectors are reversed after half a period. If the set of displacement vectors is to be a basis for a representation, they must transform as the characters of the particular representation. Consider first the 1595-cm^{-1} normal mode. The direction and magnitude of each vector is unaffected by each of the operations E, C_2, σ_v, and σ'_v. Therefore, this mode must belong to the A_1 representation. The same is true of the 3657-cm^{-1} normal mode. By contrast, the displacement vector on the O atom is reversed upon carrying out the C_2 operation for the 3756-cm^{-1} normal mode. Because the H atoms are interchanged, their displacement vectors do not contribute to the character of the C_2 operation, which is -1. Therefore, this mode must belong to either the B_1 or B_2 representations. Which of these is appropriate can be decided by examining the effect of the σ'_v operation on the individual displacement vectors. Because the vectors lie in the mirror plane, they are unchanged in the reflection, corresponding to a character of $+1$. Therefore, the 3756-cm^{-1} normal mode belongs to the B_2 representation.

The water molecule is small enough that the procedure described can be carried out without a great deal of effort. For larger molecules, the effort is significantly greater, but the normal modes and the irreducible representations to which they belong can be calculated using widely available quantum chemistry software. Many of these programs allow an animation of the vibration to be displayed, which is helpful in assigning the dominant motion to a stretch or a bend. Normal mode animations for several molecules are explored in the Web-based problems of Chapter 8 and in the computational problems for Chapter 13.

16.8 SELECTION RULES AND INFRARED VERSUS RAMAN ACTIVITY

We next show that the selection rule for infrared absorption spectroscopy, $\Delta n = +1$, can be derived using group theory. More importantly, we show that for allowed transitions, $\Gamma_{reducible}$ as calculated in the previous section must contain the A_1 representation. As discussed in Section 8.3, for most molecules, only the $n = 0$ vibrational state is populated to a significant extent at 300 K. The molecule can be excited to a state with $n_j > 0$ through the absorption of infrared energy if the dipole matrix element satisfies the condition given by

$$\mu_{Q_j}^{m \leftarrow 0} = \left(\frac{\partial \mu}{\partial Q_j}\right) \int \psi_m^*(Q_j)\hat{\mu}(Q_j)\psi_0(Q_j)dQ_j \neq 0,$$

$$\text{where } j = \text{one of } 1, 2, \ldots, 3N - 6 \tag{16.25}$$

We have modified Equation (8.6), which is applicable to a diatomic molecule, to the more general case of a polyatomic molecule and expressed the position variable in terms of the normal coordinate. To simplify the mathematics, the electric field is oriented along the normal coordinate. From Chapter 7, ψ_0, ψ_1, and ψ_2 are given by

$$\psi_0(Q_j) = \left(\frac{\alpha_j}{\pi}\right)^{1/4} e^{-\frac{1}{2}\alpha_j Q_j^2}$$

$$\psi_1(Q_j) = \left(\frac{4\alpha_j^3}{\pi}\right)^{1/4} Q_j e^{-\frac{1}{2}\alpha_j Q_j^2}$$

$$\psi_2(Q_j) = \left(\frac{\alpha_j}{4\pi}\right)^{1/4} (2\alpha_j Q_j^2 - 1)e^{-\frac{1}{2}\alpha_j Q_j^2} \tag{16.26}$$

and the dipole moment operator is given by

$$\hat{\mu}(Q_j) = \mu_e + \left[\left(\frac{\partial \mu}{\partial Q_j}\right)Q_j + \ldots\right]$$

where μ_e is the static dipole moment. Higher terms are neglected in the harmonic approximation.

For what final states ψ_f will Equation (16.25) for the transition dipole moment be satisfied? Section 16.6 demonstrated that for the integral to be nonzero, the integrand $\psi_m^*(Q_j)\mu(Q_j)\psi_0(Q_j)$ must belong to the A_1 representation. The C_{2v} character table shows that Q_j^2 is a basis for this representation, so the integrand must be a function of Q_j^2 only. We know that ψ_0 is an even function of Q_j, $\psi_0(Q_j) = \psi_0(-Q_j)$, and that μ is an odd function of Q_j, $\mu(Q_j) = -\mu(-Q_j)$. Under what condition will the integrand be an even function of Q_j? It will be an even function only if ψ_m^* is an odd function of Q_j, $\psi_m^*(Q_j) = f(Q_j)$. Because of this restriction, $n = 0 \rightarrow n = 1$ is an allowed transition, but $n = 0 \rightarrow n = 2$ is not allowed in the dipole approximation. This is the same conclusion that was reached in Section 8.4, using a different line of reasoning.

The preceding discussion addressed the selection rule, but did not address the symmetry requirements for the normal modes that satisfy Equation (16.25). Because $\hat{\mu}(Q_j) = f(Q_j)$ and transforms as x, y, or z and $\psi_0(Q_j) = f(Q_j^2)$ transform as x^2, y^2, or z^2, $\psi_m(Q_j)$ must transform as x, y, or z, in order for $\psi_m^*(Q_j)\mu(Q_j)\psi_0(Q_j)$ to transform as x^2, y^2, or z^2. This gives us the requirement that a normal mode is infrared active; it must have x, y, or z as a basis. For H_2O, this means that the normal modes must belong to A_1, B_1, or B_2. Because, as shown in Equation (16.24), the three normal modes belong to A_1 and B_2, we conclude that all are infrared active. As discussed in more advanced texts, normal modes of a molecule are Raman active if the bases of the representation to which the normal mode belongs are the x^2, y^2, z^2, xy, yz, or xz functions. By looking at the C_{2v} character table, we can see that all three normal modes of water are Raman active. It is not generally the case that all normal modes are both infrared and Raman active for a molecule.

Based on this discussion, recall the infrared absorption spectrum for CH_4 shown in Figure 8.9. Although CH_4 has $3n - 6 = 9$ normal modes, only two peaks are observed. Methane belongs to the T_d point group, and an analysis equivalent to that which led to Equation (16.24) shows that

$$\Gamma_{reducible} = A_1 + E + 2T_2 \qquad (16.27)$$

The dimensions of the representations are one for A_1, two for E, and three for T_2. Therefore, all nine normal modes are accounted for. An examination of the character table for the T_d group shows that only the T_2 representations have x, y, or z as a basis. Therefore, only six of the nine normal modes of methane are infrared active. Why are only two peaks observed in the spectrum? The following result of group theory is used:

> All normal modes that belong to a particular representation have the same frequency.

Therefore, each of the T_2 representations has three degenerate vibrational frequencies. For this reason, only two vibrational frequencies are observed in the infrared absorption spectrum of CH_4 shown in Figure 8.9. However, each frequency corresponds to three distinct but degenerate normal modes.

SUPPLEMENTAL

16.9 USING THE PROJECTION OPERATOR METHOD TO GENERATE MOs THAT ARE BASES FOR IRREDUCIBLE REPRESENTATIONS

In Section 16.6, symmetry-adapted MOs for H_2O were generated from AOs that belong to the different irreducible representations of the C_{2v} group in an ad hoc manner. We next discuss a powerful method, called the **projection operator method**, that allows the same end to be achieved for arbitrary molecules. The method is applied to ethene, which belongs to the D_{2h} point group.

The symmetry elements for ethene and the D_{2h} character table are shown in Table 16.5 and Figure 16.10. Aside from the identity element, the group contains three C_2 axes and three mirror planes, all of which form separate classes, as well as an inversion center. Irreducible representations in groups with an inversion center have the subscript g or u denoting that they are symmetric $(+1)$ or antisymmetric (-1) with respect to the inversion center.

TABLE 16.5   THE CHARACTER TABLE FOR THE D_{2h} POINT GROUP

	E	$C_2(z)$	$C_2(y)$	$C_2(x)$	i	$\sigma(xy)$	$\sigma(xz)$	$\sigma(yz)$		
A_g	1	1	1	1	1	1	1	1		x^2, y^2, z^2
B_{1g}	1	1	−1	−1	1	1	−1	−1	R_z	xy
B_{2g}	1	−1	1	−1	1	−1	1	−1	R_y	xz
B_{3g}	1	−1	−1	1	1	−1	−1	1	R_x	yz
A_u	1	1	1	1	−1	−1	−1	−1		
B_{1u}	1	1	−1	−1	−1	−1	1	1	z	
B_{2u}	1	−1	1	−1	−1	1	−1	1	y	
B_{3u}	1	−1	−1	1	−1	1	1	−1	x	

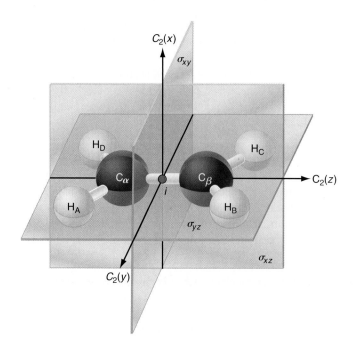

FIGURE 16.10

The symmetry elements of the D_{2h} group are shown using ethene as an example. The symbol at the intersection of the C_2 axes indicates the inversion center. The molecule lies in the y–z plane.

Next consider how the individual H atoms are affected by the symmetry operations of the group. Convince yourself that the result of applying a symmetry operation to the molecule shifts the atom H_A as listed here:

$\hat{E}$	$\hat{C}_2(z)$	$\hat{C}_2(y)$	$\hat{C}_2(x)$
$H_A \to H_A$	$H_A \to H_D$	$H_A \to H_B$	$H_A \to H_C$
$\hat{i}$	$\hat{\sigma}(xy)$	$\hat{\sigma}(xz)$	$\hat{\sigma}(yz)$
$H_A \to H_C$	$H_A \to H_B$	$H_A \to H_D$	$H_A \to H_A$

(16.28)

Next consider the atom C_α and follow the same procedure used earlier for the $2s$, $2p_x$, $2p_y$, and $2p_z$ AOs. Convince yourself that the results shown in Table 16.6 are correct. These results are used to generate symmetry-adapted MOs for ethene using the method described here:

The following procedure, based on the projection operator method, can be used to generate a symmetry-adapted MO from AOs that forms a basis for a given representation. The recipe consists of the following steps:

- Choose an AO on an atom and determine into which AO it is transformed by each symmetry operator of the group.
- Multiply the AO of the transformed species by the character of the operator in the representation of interest for each symmetry operator.
- The resulting linear combination of these AOs forms a MO that is a basis for that representation.

The use of the projection operator method is illustrated in the following two example problems.

TABLE 16.6 EFFECT OF THE SYMMETRY OPERATIONS ON THE CARBON ATOM ORBITALS

	$\hat{E}$	$\hat{C}_2(z)$	$\hat{C}_2(y)$	$\hat{C}_2(x)$	$\hat{i}$	$\hat{\sigma}(xy)$	$\hat{\sigma}(xz)$	$\hat{\sigma}(yz)$
$2s$	$C_\alpha \to C_\alpha$	$C_\alpha \to C_\alpha$	$C_\alpha \to C_\beta$	$C_\alpha \to C_\beta$	$C_\alpha \to C_\beta$	$C_\alpha \to C_\beta$	$C_\alpha \to C_\alpha$	$C_\alpha \to C_\alpha$
$2p_x$	$C_\alpha \to C_\alpha$	$C_\alpha \to -C_\alpha$	$C_\alpha \to -C_\beta$	$C_\alpha \to C_\beta$	$C_\alpha \to -C_\beta$	$C_\alpha \to C_\beta$	$C_\alpha \to C_\alpha$	$C_\alpha \to -C_\alpha$
$2p_y$	$C_\alpha \to C_\alpha$	$C_\alpha \to -C_\alpha$	$C_\alpha \to C_\beta$	$C_\alpha \to -C_\beta$	$C_\alpha \to -C_\beta$	$C_\alpha \to C_\beta$	$C_\alpha \to -C_\alpha$	$C_\alpha \to C_\alpha$
$2p_z$	$C_\alpha \to C_\alpha$	$C_\alpha \to C_\alpha$	$C_\alpha \to -C_\beta$	$C_\alpha \to -C_\beta$	$C_\alpha \to -C_\beta$	$C_\alpha \to -C_\beta$	$C_\alpha \to C_\alpha$	$C_\alpha \to C_\alpha$

EXAMPLE PROBLEM 16.7

Form a linear combination of the H atomic orbitals in ethene that is a basis for the B_{1u} representation. Show that there is no combination of the H atomic orbitals in ethene that is a basis for the B_{3u} representation.

Solution

We take ϕ_{H_A} as the initial orbital, and multiply the AO into which ϕ_{H_A} is transformed by the character of the B_{1u} representation for each operator and sum these terms. The result is

$$\psi_{B_{1u}}^{H} = 1 \times \phi_{H_A} + 1 \times \phi_{H_D} - 1 \times \phi_{H_B} - 1 \times \phi_{H_C} - 1 \times \phi_{H_C}$$
$$- 1 \times \phi_{H_B} + 1 \times \phi_{H_D} + 1 \times \phi_{H_A}$$
$$= \phi_{H_A} + \phi_{H_D} - \phi_{H_B} - \phi_{H_C} - \phi_{H_C} - \phi_{H_B} + \phi_{H_D} + \phi_{H_A}$$
$$= 2(\phi_{H_A} - \phi_{H_B} - \phi_{H_C} + \phi_{H_D})$$

This molecular wave function has not yet been normalized. Pictorially, this combination looks like this:

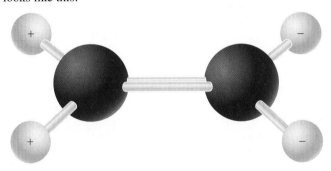

Follow the same procedure to generate the linear combination for the B_{3u} representation:

$$\psi_{B_{3u}}^{H} = \phi_{H_A} - \phi_{H_D} - \phi_{H_B} + \phi_{H_C} - \phi_{H_C} + \phi_{H_B} + \phi_{H_D} - \phi_{H_A} = 0$$

This result shows that there is no linear combination of the H AOs that is a basis for B_{3u}.

EXAMPLE PROBLEM 16.8

Use the same procedure as in Example Problem 16.7 to form a linear combination of the C atomic orbitals in ethene that is a basis for the B_{1u} representation.

Solution

Follow the same procedure outlined in Example Problem 16.7 and apply it to each of the carbon valence AOs:

$2s$: $1 \times \phi_{C_\alpha} + 1 \times \phi_{C_\alpha} - 1 \times \phi_{C_\beta} - 1 \times \phi_{C_\beta} - 1 \times \phi_{C_\beta} - 1 \times \phi_{C_\beta}$
$\quad\quad + 1 \times \phi_{C_\alpha} + 1 \times \phi_{C_\alpha} = 4\phi_{C_\alpha} - 4\phi_{C_\beta}$

$2p_x$: $1 \times \phi_{C_\alpha} - 1 \times \phi_{C_\alpha} + 1 \times \phi_{C_\beta} - 1 \times \phi_{C_\beta} + 1 \times \phi_{C_\beta} - 1 \times \phi_{C_\beta}$
$\quad\quad + 1 \times \phi_{C_\alpha} - 1 \times \phi_{C_\alpha} = 0$

$2p_y$: $1 \times \phi_{C_\alpha} - 1 \times \phi_{C_\alpha} - 1 \times \phi_{C_\beta} + 1 \times \phi_{C_\beta} + 1 \times \phi_{C_\beta} - 1 \times \phi_{C_\beta}$
$\quad\quad - 1 \times \phi_{C_\alpha} + 1 \times \phi_{C_\alpha} = 0$

$2p_z$: $1 \times \phi_{C_\alpha} + 1 \times \phi_{C_\alpha} + 1 \times \phi_{C_\beta} + 1 \times \phi_{C_\beta} + 1 \times \phi_{C_\beta} + 1 \times \phi_{C_\beta}$
$\quad\quad + 1 \times \phi_{C_\alpha} + 1 \times \phi_{C_\alpha} = 4\phi_{C_\alpha} + 4\phi_{C_\beta}$

This result shows that the appropriate linear combination of carbon AOs to construct the symmetry-adapted MO that is a basis for the B_{1u} representation is

$$\psi_{B_{1u}}^{C} = c_1(\phi_{C_{2s\alpha}} - \phi_{C_{2s\beta}}) + c_2(\phi_{C_{2pz\alpha}} + \phi_{C_{2pz\beta}})$$

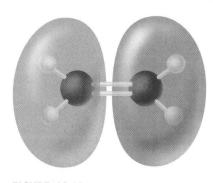

FIGURE 16.11

The ethene MO that is a basis of the B_{1u} representation.

Combining the results of the last two example problems, we find that the symmetry-adapted MO that includes AOs on all atoms and is also a representation of the B_{1u} representation is

$$\psi_{B1u} = c_1(\phi_{C_{2s\alpha}} - \phi_{C_{2s\beta}}) + c_2(\phi_{C_{2pz\alpha}} + \phi_{C_{2pz\beta}}) + c_3(\phi_{H_A} - \phi_{H_B} - \phi_{H_C} + \phi_{H_D})$$

An image of this molecular orbital is shown in Figure 16.11.

The values of the AO coefficients in the MOs cannot be obtained from symmetry considerations, but we can determine which coefficients are zero, equal in magnitude, and equal or opposite in sign. For the case of interest, c_1 through c_3 must be determined in a variational calculation in which the total energy of the molecule is minimized. Note that without taking symmetry into consideration, 12 coefficients would have been required to specify the wave function (one AO on each H, and four AOs on each C). We see that forming the symmetry-adapted MO significantly reduces the number of coefficients required in the calculation from 12 to just 3. This example shows the simplification of the molecular wave function that is obtained by forming symmetry-adapted MOs.

Vocabulary

associative operation

basis function

belongs to a particular representation

character

character table

class

dimension of a representation

forms a basis

group

identity operator

inverse operator

inversion center

irreducible representation

mirror plane

normal coordinate

normal mode

normal mode frequency

order of a group

point group

projection operator method

reducible representation

representation

rotation axis

rotation-reflection axis

symmetry elements

symmetry operations

symmetry-adapted MO

three-dimensional irreducible representation

totally symmetric representation

two-dimensional irreducible representation

Questions on Concepts

Q16.1 Can a molecule with an inversion center have a dipole moment? Give an example of a molecule with this symmetry element and explain your reasoning.

Q16.2 Which of the three normal modes of H_2O in Figure 16.8 is best described as a bending mode? Does the bond angle remain unchanged in any of the modes? Which requires less energy, bond bending or bond stretching?

Q16.3 Why does the list of elements for the D_{6h} group in Table 16.2 not list the elements C_6^2, C_6^3, and C_6^4?

Q16.4 Why does the list of elements for the D_{6h} group in Table 16.2 not list the elements S_6^2, S_6^3, and S_6^4?

Q16.5 How are quantum mechanical calculations in the LCAO-MO model simplified through the construction of symmetry-adapted MOs?

Q16.6 Some symmetry operations can be carried out physically using a ball-and-stick model of a molecule and others can only be imagined. Give two examples of each category.

Q16.7 Why does the C_{3v} group have a two-dimensional irreducible representation? Answer this question by referring to the form of the matrices that represent the operations of the group.

Q16.8 Can NH_3 have molecular orbitals that are triply degenerate in energy?

Q16.9 Can a molecule with an inversion center be superimposed on its mirror image and therefore be chiral? Give an example of a molecule with this symmetry element and explain your reasoning.

Q16.10 Why are all one-dimensional representations irreducible?

Q16.11 What is the difference between a symmetry element and a symmetry operation?

Q16.12 Can a molecule with D_{2h} symmetry have a dipole moment? Give an example of a molecule with this symmetry and explain your reasoning.

Q16.13 Can a molecule with C_{3h} symmetry have a dipole moment? Give an example of a molecule with this symmetry and explain your reasoning.

Q16.14 Explain why only two peaks are observed in the infrared spectrum of methane although six of the nine normal modes are infrared active.

Q16.15 Explain why the overlap integral between two combinations of AOs is nonzero only if the combinations belong to the same representation.

Problems

Problem numbers in **red** indicate that the solution to the problem is given in the *Student's Solutions Manual*.

P16.1 Show that a molecule with an inversion center implies the presence of an S_2 element.

P16.2 Use the 3×3 matrices for the C_{2v} group in Equation (16.2) to verify the group multiplication table for the following successive operations:

a. $\hat{\sigma}_v \hat{\sigma}_v'$ b. $\hat{\sigma}_v \hat{C}_2$ c. $\hat{C}_2 \hat{C}_2$

P16.3 Use the logic diagram of Figure 16.2 to determine the point group for the planar molecule *trans*−HBrC=CBrH. Indicate your decision-making process as was done in the text for NH_3.

P16.4 The D_3 group has the following classes: E, $2C_3$, and $3C_2$. How many irreducible representations does this group have and what is the dimensionality of each?

P16.5 Benzene, C_6H_6, belongs to the D_{6h} group. The reducible representation for the vibrational modes is

$$\Gamma_{reducible} = 2A_{1g} + A_{2g} + A_{2u} + 2B_{1u} + 2B_{2g}$$
$$+ 2B_{2u} + E_{1g} + 3E_{1u} + 4E_{2g} + 2E_{2u}$$

a. How many vibrational modes does benzene have?

b. How many of these modes are infrared active and to which representation do they belong?

c. Which of the infrared active modes are degenerate in energy and what is the degeneracy for each?

d. How many of these modes are Raman active and to which representation do they belong?

e. Which of the Raman active modes are degenerate in energy and what is the degeneracy for each?

f. Which of the infrared modes are also Raman active?

P16.6 NH_3 belongs to the C_{3v} group. The reducible representation for the vibrational modes is $\Gamma_{reducible} = 2A_1 + 2E$

a. How many vibrational modes does NH_3 have?

b. How many of these modes are infrared active and to which representation do they belong?

c. Are any of the infrared active modes degenerate in energy?

d. How many of these modes are Raman active and to which representation do they belong?

e. Are any of the Raman active modes degenerate in energy?

f. How many modes are both infrared and Raman active?

P16.7 XeF_4 belongs to the D_{4h} point group with the following symmetry elements: E, C_4, C_4^2, C_2, C_2', C_2'', i, S_4, S_4^2, σ, $2\sigma'$, and $2\sigma''$. Make a drawing similar to Figure 16.1 showing these elements.

P16.8 Methane belongs to the T_d group. The reducible representation for the vibrational modes is $\Gamma_{reducible} = A_1 + E + 2T_2$.

a. Show that the A_1 and T_2 representations are orthogonal to each other and to the other representations in the table.

b. What is the symmetry of each of the vibrational modes that gives rise to Raman activity? Are any of the Raman active modes degenerate in energy?

P16.9 Use the 3×3 matrices for the C_{2v} group in Equation (16.2) to verify the associative property for the following successive operations:

a. $\hat{\sigma}_v(\hat{\sigma}_v' \hat{C}_2) = (\hat{\sigma}_v \hat{\sigma}_v')\hat{C}_2$

b. $(\hat{\sigma}_v \hat{E})\hat{C}_2 = \hat{\sigma}_v(\hat{E}\hat{C}_2)$

P16.10 Use the logic diagram of Figure 16.2 to determine the point group for allene. Indicate your decision-making process as was done in the text for NH_3.

P16.11 To determine the symmetry of the normal modes of methane, an analysis of the transformation of individual coordinate systems on the five atoms is carried out, as shown in Figure 16.9 for H_2O. After the rotational and translational representations are removed, the following reducible representation $\chi_{reducible}$ is obtained for the vibrational modes:

E	$8C_3$	$3C_2$	$6C_4$	$6\sigma_d$
9	0	1	−1	3

Using the character table for the T_d group, verify that $\Gamma_{reducible} = A_1 + E + 2T_2$.

P16.12 Use the logic diagram of Figure 16.2 to determine the point group for the planar molecule *cis*−HBrC=CCIH. Indicate your decision-making process as was done in the text for NH_3.

P16.13 Decompose the following reducible representation into irreducible representations of the C_{2v} group:

E	C_2	σ_v	σ_v'
4	0	0	0

P16.14 Show that z is a basis for the A_1 representation and that R_z is a basis for the A_2 representation of the C_{3v} group.

P16.15 Use the logic diagram of Figure 16.2 to determine the point group for PCl_5. Indicate your decision-making process as was done in the text for NH_3.

P16.16 Use the method illustrated in Example Problem 16.2 to generate a 3×3 matrix for the following:

a. $\hat{C}_6$ operator

b. $\hat{S}_4$ operator

c. $\hat{i}$ operator

P16.17 Consider the function $f(x, y) = xy$ integrated over a square region in the $x-y$ plane centered at the origin.

a. Draw contours of constant f values (positive and negative) in the plane and decide whether the integral can have a nonzero value.

b. Use the information that the square has D_{4h} symmetry and determine which representation the integrand belongs to. Decide whether the integral can have a nonzero value from this information.

P16.18 Use the logic diagram of Figure 16.2 to determine the point group for CH_3Cl. Indicate your decision-making process as was done in the text for NH_3.

P16.19 CH_4 belongs to the T_d point group with the following symmetry elements: E, $4C_3$, $4C_3^2$, $3C_2$, $3S_4$, $3S_4^3$, and 6σ. Make drawings similar to Figure 16.1 showing these elements.

P16.20 Show that the presence of a C_2 axis and a mirror plane perpendicular to the rotation axis imply the presence of a center of inversion.

P16.21 Decompose the following reducible representation into irreducible representations of the C_{3v} group:

E	$2C_3$	$3\sigma_v$
5	2	-1

P16.22 Assume that a central atom in a molecule has ligands with C_{4v} symmetry. Decide by evaluating the appropriate transition dipole element if the transition $p_x \rightarrow p_z$ is allowed with the electric field in the z direction.

P16.23 Show that a molecule with a C_n axis cannot have a dipole moment perpendicular to the axis.

P16.24 The C_{4v} group has the following classes: E, $2C_4$, C_2, $2\sigma_v$ and $2\sigma_d$. How many irreducible representations does this group have and what is the dimensionality of each? σ_d refers to a dihedral mirror plane. For example in the molecule BrF_5, the σ_v mirror planes each contain two of the equatorial F atoms, whereas the dihedral mirror planes do not contain the equatorial F atoms.

P16.25 Use the 2×2 matrices of Equation (16.10) to derive the multiplication table for the C_{3v} group.

Nuclear Magnetic Resonance Spectroscopy

Although the nuclear magnetic moment interacts only weakly with an external magnetic field, this interaction provides a very sensitive probe of the local electron distribution in a molecule. A nuclear magnetic resonance (NMR) spectrum can distinguish between inequivalent nuclei such as ^1H at different sites in a molecule. Individual spins, such as nearby ^1H nuclei, can couple to generate a multiplet splitting of NMR peaks. This splitting can be used to determine the structure of small organic molecules. NMR can also be used as a nondestructive imaging technique that is widely used in medicine and in the study of materials. Pulsed NMR and 2D Fourier transform techniques provide a powerful combination to determine the structure of large molecules of biological interest.

17.1 INTRINSIC NUCLEAR ANGULAR MOMENTUM AND MAGNETIC MOMENT

Recall that the electron has an intrinsic magnetic moment. Some, but not all, nuclei also have an intrinsic magnetic moment. Because the **nuclear magnetic moment** of the proton is about 2000 times weaker than that of the electron magnetic moment, it has an insignificant effect on the one-electron energy levels in the hydrogen atom. The nuclear magnetic moment does not generate chemical effects in that the reactivity of a molecule containing ^{12}C with zero nuclear spin is no different than the reactivity of a molecule containing ^{13}C with nuclear spin 1/2. However, the nuclear magnetic moment gives rise to an important spectroscopy. As shown later, a nucleus with a nonzero nuclear spin is an extremely sensitive probe of the local electron distribution within a molecule.

Because of this sensitivity, nuclear magnetic resonance spectroscopy is arguably the single most important spectroscopic technique used by chemists today. NMR spectroscopy can be used to determine the structure of complex biomolecules, to map out the electron distribution in molecules, to study the kinetics of chemical transformations, and to nondestructively image internal organs in the human body. What is the basis for this spectroscopy?

Whereas electrons only have the spin quantum number 1/2, nuclear spins can take on integral multiples of 1/2. For example, ^{12}C and ^{16}O have spin 0, ^1H and ^{19}F have spin 1/2, and ^2H and ^{14}N have spin 1. The nuclear magnetic moment $\boldsymbol{\mu}$ and the nuclear angular momentum $\mathbf{I}$ are proportional to one another according to

$$\boldsymbol{\mu} = g_N \frac{e\hbar}{2m_{proton}}\mathbf{I} = g_N\beta_N\mathbf{I} = \gamma\hbar\mathbf{I} \qquad (17.1)$$

In the SI system of units, $\boldsymbol{\mu}$ has the units of ampere $(\text{meter})^2 = \text{joule }(\text{tesla})^{-1}$, and $\mathbf{I}$ has the units of joule second. In these equations, the quantity $\beta_N = e\hbar/2m_{proton}$, which has the value $5.0507866 \times 10^{-27}$ J T^{-1}, is called the **nuclear magneton** and $\gamma = g_N\beta_N/\hbar$ is called the **magnetogyric ratio**. Just as for the orbital angular momentum (see Chapter 7), the z component of the intrinsic nuclear angular momentum can take on the values $m_z\hbar$ with $-I \leq m_z \leq I$, where $|\mathbf{I}| = \hbar\sqrt{I(I+1)}$. Because m_{proton} is greater than m_e by about a factor of 2000, the nuclear magnetic moment is much smaller than the electron magnetic moment for the same value of I. The **nuclear g factor** g_N, which is a dimensionless number, is characteristic of a particular nucleus. Values of these quantities for the nuclei most commonly used in NMR spectroscopy are shown in Table 17.1. Because the abundantly occurring nuclei ^{12}C and ^{16}O have no nuclear magnetic moment, they do not have a signature in an NMR experiment. In the rest of this chapter, we focus our attention on ^1H. However, this formalism can be applied to other spin-active nuclei in a straightforward manner.

As we learned when considering the electron spin, the quantum mechanical operators for orbital and spin angular momentum have the same commutation relations. The same relations also apply to nuclear spin. Therefore, we can immediately conclude that we can only know the magnitude of the nuclear angular momentum and one of its components simultaneously. The other two components remain unknown. As for electron spin, the nuclear angular momentum is quantized in units of $\hbar/2$. For ^1H, which has the spin quantum number 1/2, the operator $\hat{I}^2$ has two eigenfunctions that are usually called α and β. They correspond to $I_z = +(1/2)\hbar$ and $I_z = -(1/2)\hbar$, respectively. These functions α and β satisfy the relations

$$\hat{I}^2\alpha = \frac{1}{2}\left(\frac{1}{2}+1\right)\hbar^2\alpha; \qquad \hat{I}_z\alpha = +\frac{1}{2}\hbar\alpha$$

$$\hat{I}^2\beta = \frac{1}{2}\left(\frac{1}{2}+1\right)\hbar^2\beta; \qquad \hat{I}_z\beta = -\frac{1}{2}\hbar\beta \qquad (17.2)$$

Note that by convention the same nomenclature is used for the electron spin and nuclear spin eigenfunctions. Although this presents a possibility for confusion, it emphasizes the fact that both sets of eigenfunctions have the same relationship to their angular momentum operators.

TABLE 17.1 PARAMETERS FOR SPIN-ACTIVE NUCLEI

Nucleus	Isotopic Abundance (%)	Spin	Nuclear g-Factor g_N	Magnetogyric Ratio $\gamma/10^7$ (rad T^{-1}s^{-1})
^1H	99.985	1/2	5.5854	26.75
^{13}C	1.108	1/2	1.4042	6.73
^{31}P	100	1/2	2.2610	10.84
^2H	0.015	1	0.8574	4.11
^{14}N	99.63	1	0.4036	1.93

17.2 THE ENERGY OF NUCLEI OF NONZERO NUCLEAR SPIN IN A MAGNETIC FIELD

Classically, a magnetic moment or dipole can have any orientation in a magnetic field $\mathbf{B}_0$, and its energy in a particular orientation relative to the field (which we choose to lie along the z direction) is given by

$$E = -\boldsymbol{\mu} \cdot \mathbf{B}_0 = -\gamma B_0 m_z \hbar \qquad (17.3)$$

However, we know that I_z for an atom of nuclear spin $1/2$ like ^{1}H can only have two values. Additionally, the magnetic moment of a single spin cannot be oriented parallel to the quantization axis because the components of the angular momentum operator do not commute (see Section 7.4). Therefore, the only allowed energy values for spin $1/2$ are

$$E = -\left(\pm\frac{1}{2}\right)g_N\beta_N B_0 = -\left(\pm\frac{1}{2}\right)\gamma B_0 \qquad (17.4)$$

Although only two discrete energy levels are possible, the energy of these levels is a continuous function of the magnetic field, as shown in Figure 17.1.

Equation (17.4) shows that the two orientations of the magnetic moment have different potential energies. Additionally, a magnetic moment that is not parallel to the magnetic field experiences a force. For a classical magnetic moment, the torque is given by

$$\boldsymbol{\Gamma} = \boldsymbol{\mu} \times \mathbf{B}_0 \qquad (17.5)$$

The torque is perpendicular to the plane containing $\mathbf{B}_0$ and $\boldsymbol{\mu}$ and, therefore, leads to a movement of $\boldsymbol{\mu}$ on the surface of a cone about the magnetic field direction. This motion is called **precession** and is analogous to the motion of a spinning top in a gravitational field. The precession of individual spins is shown in Figure 17.2. In NMR spectroscopy, one deals with a finite volume that contains many individual spins. Therefore, it is useful to define the **macroscopic magnetic moment M**, which is the vector sum of the individual magnetic moments, $\mathbf{M} = \sum_i \boldsymbol{\mu}_i$. Whereas classical mechanics is not appropriate for describing individual nuclear magnetic moments, it is useful for describing the behavior of $\mathbf{M}$. In Figure 17.2, all the individual $\boldsymbol{\mu}_i$ yellow cones have the same magnitude for the z component, but their transverse components are randomly oriented in the $x-y$ plane. Therefore, in a macroscopic sample containing on the order of Avogadro's number of nuclear spins, the transverse component of $\mathbf{M}$ is zero. We conclude that $\mathbf{M}$ lies on the z axis, which corresponds to the field direction.

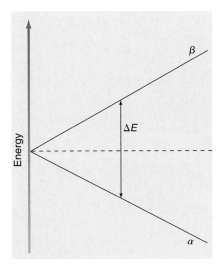

FIGURE 17.1

Energy of a nuclear spin of quantum number $1/2$ as a function of the magnetic field.

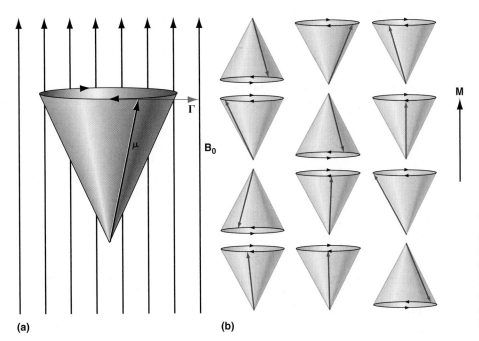

(a)

(b)

FIGURE 17.2

(a) Precession of an individual nuclear spin about the magnetic field direction for an $\boldsymbol{\alpha}$ spin. **(b)** The magnetization vector **M** resulting from summing the individual spin magnetic moments (yellow cones) is oriented parallel to the magnetic field. It has no transverse component.

The frequency with which an individual magnetic moment precesses about the magnetic field direction is given by

$$\nu = \frac{1}{2\pi}\gamma B_0 \quad \text{or} \quad \omega = 2\pi\nu = \gamma B_0 \tag{17.6}$$

and is called the **Larmor frequency**. The Larmor frequency increases linearly with the magnetic field and has characteristic values for different nuclei. For instance, ^1H has a resonance frequency of 500 MHz at a field of approximately 12 T.

In NMR spectroscopy, as in any spectroscopy, a transition must be induced between two different energy levels so that the absorption or emission of the electromagnetic energy that occurs can be detected. As we saw earlier, a spin 1/2 system has only two levels, and their separation increases linearly with B_0. As shown in Example Problem 17.1, the energy separation of these two levels is very small compared to kT. This makes the energy absorption difficult to detect, because the levels are nearly equally populated. Therefore, a major focus within the technology supporting NMR spectroscopy has been the development of very high magnetic fields to increase the energy separation of these two levels. Currently, by means of superconducting magnets, fields of up to approximately 21.1 T can be generated. This is a factor of nearly 10^6 higher than the Earth's magnetic field.

EXAMPLE PROBLEM 17.1

  a. Calculate the two possible energies of the ^1H nuclear spin in a uniform magnetic field of 5.50 T.
  b. Calculate the energy ΔE absorbed in making a transition from the α to the β state. If a transition is made between these levels by the absorption of electromagnetic radiation, what region of the spectrum is used?
  c. Calculate the relative populations of these two states in equilibrium at 300. K.

Solution

  a. The two energies are given by

$$E = \pm\frac{1}{2}g_N\beta_N B_0$$

$$= \pm\frac{1}{2} \times 5.5854 \times 5.051 \times 10^{-27}\,\text{J/T} \times 5.50\,\text{T}$$

$$= \pm 7.76 \times 10^{-26}\,\text{J}$$

  b. The energy difference is given by

$$\Delta E = 2(7.76 \times 10^{-26}\,\text{J})$$

$$= 1.55 \times 10^{-25}\,\text{J}$$

$$\nu = \frac{\Delta E}{h} = \frac{1.55 \times 10^{-25}}{6.626 \times 10^{-34}} = 2.34 \times 10^8\,\text{s}^{-1}$$

This is in the range of frequencies called radio frequencies.

  c. The relative populations of the two states are given by

$$\frac{n_\beta}{n_\alpha} = \exp\left(-\frac{E_\beta - E_\alpha}{kT}\right) = \exp\left(\frac{-2 \times 7.76 \times 10^{-26}\,\text{J}}{1.381 \times 10^{-23}\,\text{J K}^{-1} \times 300.\,\text{K}}\right) = 0.999963$$

$$\frac{n_\alpha - n_\beta}{\frac{1}{2}(n_\beta + n_\alpha)} \approx \frac{(1 - 0.999963)\,n_\alpha}{n_\alpha} = 3.7 \times 10^{-5}$$

From this result, we see that the populations of the two states are the same to within a few parts per million. Note that observing the appropriate rules for significant figures, we would obtain a ratio 1.00.

The solution to part (c) of Example Problem 17.1 shows that because $E_\beta - E_\alpha \ll kT$, $n_\alpha \approx n_\beta$. This result has important consequences for implementing NMR spectroscopy. As we learned in Chapter 8, if a system with only two energy levels is exposed to radiation of frequency $\nu = (E_\beta - E_\alpha)/h$, and if $n_\alpha \approx n_\beta$, the rate of upward transitions is nearly equal to the rate of downward transitions. Therefore, only a very small fraction of the nuclear spins contributes to the NMR signal. More generally, the energy absorbed is proportional to the product of $E_\beta - E_\alpha$ and $n_\alpha - n_\beta$. Both of these quantities increase as the magnetic field B_0 increases, and this is a major reason for carrying out NMR experiments at high magnetic fields.

The energy level diagram of Figure 17.1 indicates that under the condition

$$\nu_0 = \frac{E_\beta - E_\alpha}{h} = \frac{g_N \beta_N B_0}{2\pi} = \frac{\gamma B_0}{2\pi} \qquad \textbf{(17.7)}$$

energy can be absorbed by a sample containing atoms with a nonzero nuclear spin.

How can transitions be induced? As Figure 17.2 shows, the net magnetization induced by the static field $\mathbf{B}_0$ is parallel to the field. Inducing transitions is equivalent to rotating $\mathbf{M}$ away from the direction of $\mathbf{B}_0$. The torque acting on $\mathbf{M}$ is $\mathbf{\Gamma} = \mathbf{M} \times \mathbf{B}_{rf}$, where $\mathbf{B}_{rf}$ is the radio-frequency field inducing the transitions. To obtain the maximum effect, the time-dependent electromagnetic field $\mathbf{B}_{rf}$ should lie in a plane perpendicular to the static field $\mathbf{B}_0$. Equation (17.7) can be satisfied either by tuning the monochromatic radio-frequency input to the resonance value at a constant magnetic field, or vice versa. As we discuss in Supplemental Sections 17.12 through 17.14, modern NMR spectroscopy uses neither of these methods; instead it utilizes radio-frequency pulse techniques.

If no more information than that outlined in the preceding paragraphs could be obtained with this technique, NMR spectroscopy would simply be an expensive tool for quantitatively analyzing the elemental composition of compounds. However, two important aspects of this technique make it very useful for obtaining additional chemical information at the molecular level. The first of these is that the magnetic field in Equation (17.7) is not the applied external field, but rather the local field. As we will see, the local field is influenced by the electron distribution on the atom of interest as well as by the electron distribution on nearby atoms. This difference between the external and induced magnetic fields is the origin of the **chemical shift**. The H atoms in methane and chloroform have a different Larmor frequency because of this chemical shift. The origin of the chemical shift is discussed in Sections 17.3 through 17.6. The second important aspect is that individual magnetic dipoles interact with one another. This leads to a splitting of the energy levels of a two-spin system and the appearance of multiplet spectra in NMR. As discussed in Sections 17.7 and 17.8, the multiplet structure of a NMR resonance absorption gives direct structural information about the molecule.

17.3 THE CHEMICAL SHIFT FOR AN ISOLATED ATOM

When an atom is placed in a magnetic field, a circulation current is induced in the electron charge around the nucleus that generates a secondary magnetic field. The direction of the induced magnetic field at the position of the nucleus of interest opposes the external field; this phenomenon is referred to as a **diamagnetic response**. The origin of this response is shown in Figure 17.3. At distances from the center of the distribution that are large compared to an atomic diameter, the field is the same as that of a magnetic dipole. The z component of the induced magnetic field is given by

$$B_z = \frac{\mu_0}{4\pi} \frac{|\boldsymbol{\mu}|}{r^3} (3\cos^2\theta - 1) \qquad \textbf{(17.8)}$$

In Equation (17.8), μ_0 is the vacuum permeability, $\boldsymbol{\mu}$ is the induced magnetic moment, and θ and r define the coordinates of the observation point relative to the center of the charge distribution. Note that the induced field falls off rapidly with distance.

FIGURE 17.3

The shaded spherical volume represents a negatively charged classical continuous charge distribution. When placed in a magnetic field, the distribution will circulate as indicated by the horizontal orbit, viewed from the perspective of classical electromagnetic theory. The motion will induce a magnetic field at the center of the distribution that opposes the external field. This classical picture is not strictly applicable at the atomic level, but the outcome is the same as a rigorous quantum mechanical treatment.

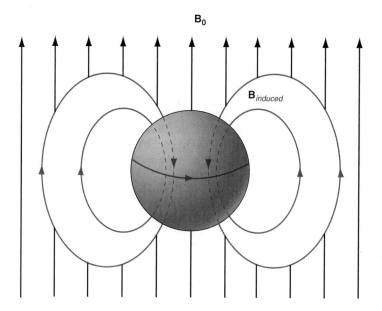

Depending on whether $3\cos^2\theta > 1$ or $3\cos^2\theta < 1$, B_z will add to or subtract, from the external applied magnetic field at the point r, θ. As we will see later, the angular dependence of B_z is important in averaging out the induced magnetic field of freely tumbling molecules in a solution.

For a diamagnetic response, the induced field at the nucleus of interest is opposite in direction and is linearly proportional to the external field in magnitude. Therefore, we can write $\mathbf{B}_{induced} = -\sigma \mathbf{B}_0$, which defines the **shielding constant** σ. The total field at the nucleus is given by the sum of the external and induced fields,

$$\mathbf{B}_{total} = (1 - \sigma)\mathbf{B}_0 \qquad (17.9)$$

and the resonance frequency taking the shielding into account is given by

$$\nu_0 = \frac{\gamma B_0 (1 - \sigma)}{2\pi} \quad \text{or} \quad \omega_0 = \gamma B_0 (1 - \sigma) \qquad (17.10)$$

Because $\sigma > 0$ for a diamagnetic response, the resonance frequency of a nucleus in an atom is lower than would be expected for the bare nucleus. The frequency shift is given by

$$\Delta\nu = \frac{\gamma B_0}{2\pi} - \frac{\gamma B_0 (1 - \sigma)}{2\pi} = \frac{\sigma \gamma B_0}{2\pi} \qquad (17.11)$$

which shows that the electron density around the nucleus reduces the resonance frequency of the nuclear spin. This effect is the basis for the chemical shift in NMR. The shielding constant σ increases with the electron density around the nucleus and, therefore, with the atomic number. Although ^{1}H is the most utilized nuclear probe in NMR, it has the smallest shielding constant of all atoms because it has only one electron orbiting around the nucleus. By comparison, σ for ^{13}C and ^{31}P are a factor of 15 and 54 greater, respectively.

17.4 THE CHEMICAL SHIFT FOR AN ATOM EMBEDDED IN A MOLECULE

We now consider the effect of neighboring atoms in a molecule on the chemical shift of a ^{1}H atom. As we have seen, the frequency shift for an atom depends linearly on the shielding constant σ. Because σ depends on the electron density around the nuclear spin of interest, it will change as neighboring atoms or groups either withdraw or increase electron density from the hydrogen atom of interest. This leads to a shift in frequency that makes NMR a sensitive probe of the chemical environment around a

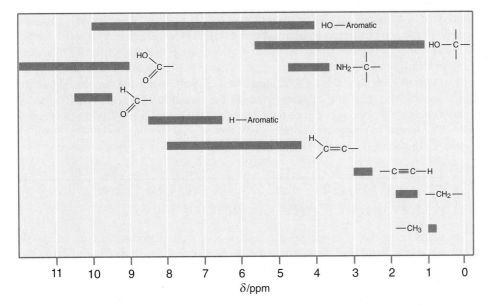

FIGURE 17.4

Chemical shifts δ as defined by Equation (17.12) for ^{1}H in different classes of chemical compounds. Extensive compilations of chemical shifts are available in the chemical literature.

nucleus with nonzero nuclear spin. For ^{1}H, σ is typically in the range of 10^{-5} to 10^{-6}, so that the change in the resonance frequency due to the chemical shift is quite small. It is convenient to define a dimensionless quantity δ to characterize this frequency shift, with δ defined relative to a reference compound by

$$\delta = 10^6 \frac{(\nu - \nu_{ref})}{\nu_{ref}} = 10^6 \frac{\gamma B_0 (\sigma_{ref} - \sigma)}{\gamma B_0 (1 - \sigma_{ref})} \approx 10^6 (\sigma_{ref} - \sigma) \qquad \textbf{(17.12)}$$

For ^{1}H NMR, tetramethylsilane, $(CH_3)_4Si$, is usually used as a reference compound. Defining the chemical shift in this way has the advantage that δ is independent of the frequency, so that all measurements using spectrometers with different magnetic fields will give the same value of δ.

Figure 17.4 illustrates the observed ranges of δ for hydrogen atoms in different types of chemical compounds. The figure shows that the chemical shift for the OH group in alcohols is quite different than the chemical shift for H atoms in a methyl group. It also shows that the range in observed chemical shifts for a class of compounds can be quite large, as is seen for the aromatic alcohols. How can these chemical shifts be understood?

Although a quantitative understanding of these shifts requires the consideration of many factors, two factors are responsible for the major part of the chemical shift: the electronegativity of the neighboring group and the induced magnetic field of the neighboring group at the position of the nucleus of interest. We discuss each of these effects in the next two sections.

17.5 ELECTRONEGATIVITY OF NEIGHBORING GROUPS AND CHEMICAL SHIFTS

Rather than consider individual atoms near the nuclear spin of interest, we consider groups of atoms such as $-OH$ or $-CH_2-$. If a neighboring group is more electronegative than hydrogen, it will withdraw electron density from the region around the ^{1}H nucleus. Therefore, the nucleus is less shielded, and the NMR resonance frequency appears at a larger value of δ. For example, the chemical shift for ^{1}H in the methyl halides follows the sequence $CH_3I < CH_3Br < CH_3Cl < CH_3F$. The range of this effect is limited to about three or four bond lengths as can be shown by considering the chemical shifts in 1-chlorobutane. In this molecule, δ for the ^{1}H on the CH_2 group closest to the Cl is almost 3 ppm larger than the ^{1}H on the terminal CH_3 group, which has nearly the same δ as in propane.

As Figure 17.4 shows, the chemical shifts for different classes of molecules are strongly correlated with their electron-withdrawing ability. Carboxyl groups are very

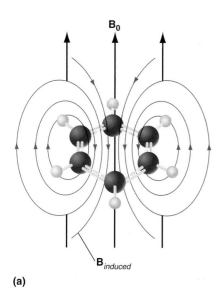

(a)

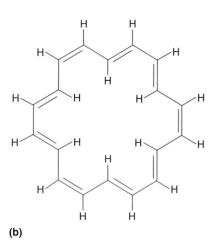

(b)

FIGURE 17.5

(a) The induced magnetic field generated by a circulating ring current in benzene. Note that in the plane of the molecule, the induced field is in the same direction as the external field outside the ring, and in the opposite direction inside the ring.
(b) 18-Annulene provides a confirmation of this model, because δ has the opposite sign for interior and exterior ^{1}H.

effective in withdrawing charge from around the ^{1}H nucleus; therefore, the chemical shift is large and positive. Aldehydes, alcohols, and amines are somewhat less effective in withdrawing electron charge. Aromatic rings are somewhat more effective than double and triple bonds in withdrawing charge. A methyl group attached to an electron-rich atom such as Li or Al will have a negative chemical shift, indicating that the ^{1}H nucleus is more shielded than in $(CH_3)_4Si$. However, the spread in chemical shifts for any of these classes of compounds can be quite large, and the ranges for different classes overlap. The spread and overlap arise because of the induced magnetic field of neighboring groups. This topic is discussed in the next section.

17.6 MAGNETIC FIELDS OF NEIGHBORING GROUPS AND CHEMICAL SHIFTS

The magnetic field at a ^{1}H nucleus is a superposition of the external field, the local field induced by the diamagnetic response of the electrons around the ^{1}H nucleus, and the local induced magnetic fields from neighboring atoms or groups. The value of σ is small for the H atom because the H atom has only one electron; therefore, the magnetic field at a ^{1}H nucleus is often dominated by the local induced magnetic fields from neighboring atoms or groups. We focus on neighboring groups rather than on individual neighboring atoms, because groups can have a high diamagnetic or paramagnetic response. The stronger the magnetic field induced by a diamagnetic or paramagnetic response in a group, the greater the effect it will have at the neighboring ^{1}H nucleus under consideration. It is helpful to think of the neighboring group as a magnetic dipole $\boldsymbol{\mu}$ whose strength and direction are determined by the magnitude and sign of its shielding constant σ. Groups containing delocalized electrons such as aromatic groups, carbonyl, and other groups containing multiple bonds give rise to large values of $\boldsymbol{\mu}$. Aromatic rings generate a large $\boldsymbol{\mu}$ value because the delocalized electrons can give rise to a ring current, as illustrated in Figure 17.5a, just as a current is induced in a macroscopic wire loop by a time-dependent magnetic field. This model of the ring current predicts that the chemical shift of the interior and exterior ^{1}H atoms attached to an aromatic system should be in the opposite direction. In fact, δ for an exterior ^{1}H of 18-annulene is +9.3 and that for an interior ^{1}H is −3.0.

An aromatic ring has a strong **magnetic anisotropy**. A sizable ring current is induced when the magnetic field is perpendicular to the plane of the ring, but the current is negligible when the magnetic field lies in the plane of the ring. This is true of many neighboring groups; the magnitude of $\boldsymbol{\mu}$ depends on the orientation of the group relative to the field.

Although we have considered individual molecules, the NMR signal of a solution sample is generated by the large number of molecules contained in the sampling volume. Therefore, the observed σ is an average over all possible orientations of the molecule, $\sigma_{average} = \langle \sigma_{individual} \rangle$, where $\sigma_{individual}$ applies to a particular orientation. In a gas or a solution, the molecules in the sample have all possible orientations with respect to the magnetic field. To determine how this random orientation affects the spectrum, we must ask if the shielding or deshielding of a ^{1}H by a neighboring group depends on the orientation of the molecule relative to the field.

Consider the induced magnetic field of a neighboring group at the ^{1}H of interest. Because the direction of the induced magnetic field is linked to the external field, rather than to the molecular axis, it retains its orientation relative to the field as the molecule tumbles in a gas or a solution. For the case of an isotropic neighboring group, $\langle B_z \rangle$ is obtained by averaging the induced magnetic field given by Equation (17.8) over all possible angles. Because the induced magnetic field for an isotropic neighboring group is independent of θ $|\boldsymbol{\mu}|$ can be taken out of the integral, leading to

$$\langle B_z \rangle = \frac{\mu_0}{4\pi} \frac{|\boldsymbol{\mu}|}{r^3} \int_0^{2\pi} d\phi \int_0^{\pi} (3\cos^2\theta - 1)\sin\theta \, d\theta = \frac{\mu_0}{2} \frac{|\boldsymbol{\mu}|}{r^3} \left[-\cos^3\theta + \cos\theta \right]_0^{\pi} = 0 \quad \textbf{(17.13)}$$

Equation (17.13) shows that as a result of the tumbling, $\langle B_z \rangle = 0$ unless the neighboring group has a magnetic anisotropy. In this case, $\boldsymbol{\mu}$ depends on θ and ϕ, and $\mu(\theta, \phi)$ must remain inside the integral.

If the neighboring groups are magnetically isotropic and the molecule is tumbling freely, the NMR spectrum of a sample in solution is greatly simplified because $\sigma_{average} = 0$. For a solid in which tumbling cannot occur, there is another way to eliminate dipolar interactions from neighboring groups. This is done by orienting the sample in the static magnetic field, choosing the angle $\theta = 54.74°$ at which $\langle B_z \rangle$ goes to zero. Solid-state NMR spectra and the technique of magic angle spinning are discussed in Section 17.10.

This and the previous section have provided a brief introduction to the origin of the chemical shift in NMR spectroscopy. For ^1H, the range of observed values for δ among different chemical compounds is about 10 ppm. For nuclei in atoms that can exhibit both paramagnetic and diamagnetic behavior, δ can vary over a much wider range. For example, δ for ^{19}F can vary by 1000 ppm for different chemical compounds. Vast libraries of ^1H NMR spectra for different compounds have been assembled and provide chemists with a valuable tool for identifying chemical compounds on the basis of chemical shifts in their NMR spectra.

17.7 MULTIPLET SPLITTING OF NMR PEAKS ARISES THROUGH SPIN–SPIN COUPLING

What might one expect the spectrum of a molecule of ethanol, with three different types of hydrogens, to look like? A good guess is that each group of chemically equivalent protons resonates in a separate frequency range, one corresponding to the methyl group, another to the methylene group, and the third to the OH group. The OH proton is most strongly deshielded (largest δ) because it is directly bound to the electronegative oxygen atom. It is found near 5 ppm. Because the methylene group is closer to the electronegative OH group, the protons are more deshielded and appear at larger values of δ (near 3.5 ppm) than the methyl protons, which are found near 1 ppm. Furthermore, we expect that the areas of the peaks have the ratio $CH_3 : CH_2 : OH = 3 : 2 : 1$ because the NMR signal is proportional to the number of spins. A simulated NMR spectrum for ethanol is shown in Figure 17.6.

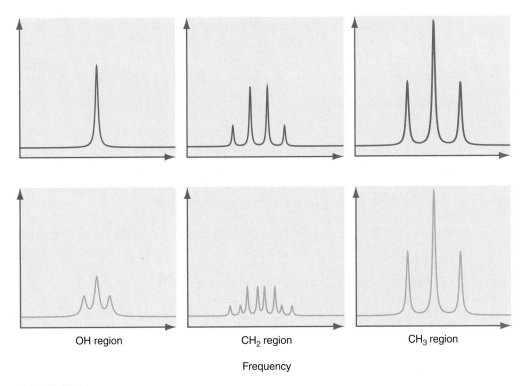

OH region CH₂ region CH₃ region

Frequency

FIGURE 17.6

Simulated NMR spectrum showing the intensity (vertical axis) as a function of frequency for ethanol. The top panel shows the multiplet structure at room temperature. The lower panel shows the multiplet structure observed at lower temperature in acid-free water. The different portions of the spectrum are not to scale, but have the relative areas discussed in the text.

A very important feature that has not been discussed yet is shown in Figure 17.6: the individual peaks are split into **multiplets**. At low temperature and in the absence of acidic protons, the OH proton resonance is a triplet, whereas the CH_3 proton resonance is a triplet and the CH_2 resonance is an octet. At higher temperature, a change in the NMR spectrum is observed. The OH proton resonance is a singlet, the CH_3 proton resonance is a triplet, and the CH_2 resonance is a quartet. How can this splitting be understood? The higher temperature spectrum is the result of rapid transfer of the OH proton between ethanol and water, as discussed later in Section 17.9. For now, we turn our attention to the origin of **multiplet splitting**.

Multiplets arise as a result of spin–spin interactions among different nuclei. We first consider the case of two distinguishable noninteracting spins such as the ^{1}H nuclei of the CH_3 and CH_2 groups in ethanol. We give these spins the labels 1 and 2 and subsequently introduce the interaction. The spin energy operator for the noninteracting spins is

$$\hat{H} = -\gamma B_0(1 - \sigma_1)\hat{I}_{z_1} - \gamma B_0(1 - \sigma_2)\hat{I}_{z_2} \qquad \textbf{(17.14)}$$

and the eigenfunctions of this operator are products of the eigenfunctions of the individual operators $\hat{I}_{z_1}$ and $\hat{I}_{z_2}$:

$$\psi_1 = \alpha(1)\alpha(2)$$
$$\psi_2 = \beta(1)\alpha(2)$$
$$\psi_3 = \alpha(1)\beta(2)$$
$$\psi_4 = \beta(1)\beta(2) \qquad \textbf{(17.15)}$$

We solve the Schrödinger equation for the corresponding eigenvalues, which are as follows (see Example Problem 17.2):

$$E_1 = -\hbar\gamma B_0\left(1 - \frac{\sigma_1 + \sigma_2}{2}\right)$$

$$E_2 = -\frac{\hbar\gamma B_0}{2}(\sigma_1 - \sigma_2)$$

$$E_3 = \frac{\hbar\gamma B_0}{2}(\sigma_1 - \sigma_2)$$

$$E_4 = \hbar\gamma B_0\left(1 - \frac{\sigma_1 + \sigma_2}{2}\right) \qquad \textbf{(17.16)}$$

We have assumed that $\sigma_1 > \sigma_2$.

EXAMPLE PROBLEM 17.2

Show that the total nuclear energy eigenvalue for the wave function $\psi_2 = \beta(1)\alpha(2)$ is

$$E_2 = -\frac{\hbar\gamma B_0}{2}(\sigma_1 - \sigma_2)$$

Solution

$$\hat{H}\psi_2 = [-\gamma B_0(1 - \sigma_1)\hat{I}_{z_1} - \gamma B_0(1 - \sigma_2)\hat{I}_{z_2}]\beta(1)\alpha(2)$$

$$= [-\gamma B_0(1 - \sigma_1)\hat{I}_{z_1}]\beta(1)\alpha(2) + [-\gamma B_0(1 - \sigma_2)\hat{I}_{z_2}]\beta(1)\alpha(2)$$

$$= \frac{\hbar}{2}\gamma B_0(1 - \sigma_1)\beta(1)\alpha(2) - \frac{\hbar}{2}\gamma B_0(1 - \sigma_2)\beta(1)\alpha(2)$$

$$= -\frac{\hbar\gamma B_0}{2}(\sigma_1 - \sigma_2)\beta(1)\alpha(2)$$

$$= E_2\psi_2$$

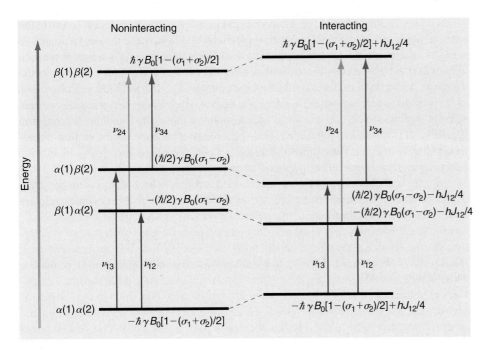

FIGURE 17.7

The energy levels for two noninteracting spins and the allowed transitions between these levels are shown on the left. The same information is shown on the right for interacting spins. The splitting between levels 2 and 3 and the energy shifts of all four levels for interacting spins are greatly magnified to emphasize the spin-spin interactions.

You will calculate the energy eigenvalues for the other eigenfunctions in Equation (17.15) in the end-of-chapter problems. All four energy eigenvalues are plotted in the energy diagram of Figure 17.7. We initially focus on the left half of this figure, which shows the energy levels for noninteracting spins. The selection rule for NMR spectroscopy is that only one of the spins can change in a transition. The four allowed transitions are indicated in the figure. For the noninteracting spin case, $E_2 - E_1 = E_4 - E_3$ and $E_3 - E_1 = E_4 - E_2$. Therefore, the NMR spectrum contains only two peaks corresponding to the frequencies

$$\nu_{12} = \nu_{34} = \frac{E_2 - E_1}{h} = \frac{\gamma B_0 (1 - \sigma_1)}{2\pi}$$

$$\nu_{13} = \nu_{24} = \frac{E_4 - E_2}{h} = \frac{\gamma B_0 (1 - \sigma_2)}{2\pi} \qquad \textbf{(17.17)}$$

You will calculate the allowed frequencies in the end-of-chapter problems. This result shows that the splitting of a single peak into multiplets is not observed for noninteracting spins.

We next consider the case of interacting spins. Because each of the nuclear spins acts like a small bar magnet, they interact with one another through **spin–spin coupling**. There are two different types of spin–spin coupling: through-space vectorial dipole–dipole coupling, which is important in the NMR of solids (see Section 17.10), and through-bond, or scalar, dipole–dipole coupling, which is considered next.

The spin energy operator that takes scalar dipole–dipole coupling into account is

$$\hat{H} = -\gamma B_0 (1 - \sigma_1)\hat{I}_{z_1} - \gamma B_0 (1 - \sigma_2)\hat{I}_{z_2} + \frac{hJ_{12}}{\hbar^2}\hat{I}_1 \cdot \hat{I}_2 \qquad \textbf{(17.18)}$$

In this equation, J_{12}, is called the **coupling constant** and is a measure of the strength of the interaction between the individual magnetic moments. The factor $h/\hbar^2$ in the last term of Equation (17.18) is included to make the units of J_{12} be s^{-1}. What is the origin of this through-bond coupling interaction? Two possibilities are considered: vectorial dipole–dipole coupling and the interaction between nuclear and electron spins.

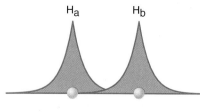

Unpolarized orbital

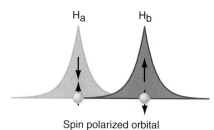

Spin polarized orbital

FIGURE 17.8

Schematic illustration of how spin polarized orbitals couple nuclear spins even though they are highly shielded from one another through the electron density. The upper and lower arrows in the lower part of the figure indicate the electron and nuclear spin, respectively.

Because the directions of the induced magnetic moments $\boldsymbol{\mu}_1$ and $\boldsymbol{\mu}_2$ are linked to the external field, they retain their orientation parallel to the field as the molecule tumbles in a gas or a solution. (Again, we are using a classical picture, and a more rigorous—although less transparent—discussion would refer to the macroscopic magnetization vector **M**, rather than to the individual magnetic moments.) An individual nucleus such as ^{1}H is magnetically isotropic. Therefore, the vectorial dipole–dipole interaction between spins is averaged to zero in a macroscopic sample by molecular tumbling, as shown in Equation (17.13) and does not contribute to the through-bond coupling interaction. Therefore, the spin–spin coupling must be transmitted between nuclei through an interaction between the nuclear and electron spins as shown in Figure 17.8.

An antiparallel orientation of the nuclear and electron spins is favored energetically over a parallel orientation. Therefore, the electrons around a nucleus with β spin are more likely to be of α than β spin. In a molecular orbital connecting two nuclei of nonzero spin, the β electrons around atom H$_a$ are pushed toward atom H$_b$ because of the electron sharing resulting from the chemical bond. Nucleus H$_b$ is slightly lower in energy if it has α rather than β spin, because this generates an antiparallel arrangement of nuclear and electron spins on the atom. This effect is referred to as **spin polarization**. A well-shielded nuclear spin senses the spin orientation of its neighbors through the interaction between the nuclear spin and the electrons. Because this is a very weak interaction and other factors favor molecular orbitals without spin polarization, the degree of spin polarization is very small. However, this very weak interaction is sufficient to account for the parts per million changes in the frequency of NMR transitions.

At this point, we discuss the spin energy operator for interacting spins and use an approximation method to determine the spin energy eigenvalues of this operator. The eigenfunctions are linear combinations of the eigenfunctions for noninteracting spins and need not concern us further. The approximation method is called first-order perturbation theory. It is applicable when we know how to solve the Schrödinger equation for a problem that is very similar to the one of interest. In this case, the problem we know how to solve is for noninteracting spins. If the change in the energy levels brought about by an additional interaction term in the spin energy operator, $\hat{H}_{interaction}$, is small, then we state without proof that the first-order correction to the energy for the case of two interacting spins is given by

$$\Delta E_j = \iint \psi_j^* \hat{H}_{interaction} \psi_j \, d\tau_1 \, d\tau_2 = \frac{4\pi^2}{h} J_{12} \iint \psi_j^* \hat{I}_1 \cdot \hat{I}_2 \psi_j \, d\tau_1 \, d\tau_2 \quad \textbf{(17.19)}$$

In this equation, the wave functions are those for the problem in the absence of $\hat{H}_{interaction}$, and the integration is over the two spin variables.

To evaluate this integral, we write $\hat{I}_1 \cdot \hat{I}_2 = \hat{I}_{1x}\hat{I}_{2x} + \hat{I}_{1y}\hat{I}_{2y} + \hat{I}_{1z}\hat{I}_{2z}$ and must solve equations of the type

$$\Delta E_j = \frac{4\pi^2}{h} J_{12} \iint \alpha^*(1)\alpha^*(1)\hat{I}_{1x}\hat{I}_{2x}\alpha(1)\alpha(2) \, d\tau_1 \, d\tau_2$$

We know that α and β are eigenfunctions of $\hat{I}_z$ and that they are not eigenfunctions of $\hat{I}_x$ and $\hat{I}_y$. The following relations, which are not proved, are used to solve the necessary integrals as shown in Example Problem 17.3:

$$\hat{I}_x\alpha = \frac{\hbar}{2}\beta; \qquad \hat{I}_y\alpha = \frac{i\hbar}{2}\beta; \qquad \hat{I}_z\alpha = \frac{\hbar}{2}\alpha$$

$$\hat{I}_x\beta = \frac{\hbar}{2}\alpha; \qquad \hat{I}_y\beta = -\frac{i\hbar}{2}\alpha; \qquad \hat{I}_z\beta = -\frac{\hbar}{2}\beta \quad \textbf{(17.20)}$$

EXAMPLE PROBLEM 17.3

Show that the energy correction to $\psi_2 = \beta(1)\alpha(2)$ is $\Delta E_2 = -(hJ_{12}/4)$

Solution

We evaluate

$$\Delta E_2 = \frac{4\pi^2}{h} J_{12} \iint \alpha^*(1)\beta^*(2)[\hat{I}_{1x}\hat{I}_{2x} + \hat{I}_{1y}\hat{I}_{2y} + \hat{I}_{1z}\hat{I}_{2z}]\alpha(1)\beta(2)\, d\tau_1\, d\tau_2$$

$$= \frac{4\pi^2}{h} J_{12} \left[\begin{array}{l} \iint \alpha^*(1)\beta^*(2)[\hat{I}_{1x}\hat{I}_{2x}]\alpha(1)\beta(2)\, d\tau_1\, d\tau_2 \\[4pt] + \iint \alpha^*(1)\beta^*(2)[\hat{I}_{1y}\hat{I}_{2y}]\alpha(1)\beta(2)\, d\tau_1\, d\tau_2 \\[4pt] + \iint \alpha^*(1)\beta^*(2)[\hat{I}_{1z}\hat{I}_{2z}]\alpha(1)\beta(2)\, d\tau_1\, d\tau_2 \end{array} \right]$$

$$= \frac{4\pi^2}{h} J_{12} \left[\begin{array}{l} \iint \alpha^*(1)\beta^*(2)\left[\dfrac{\hbar^2}{4}\right]\beta(1)\alpha(2)\, d\tau_1\, d\tau_2 \\[8pt] + \iint \alpha^*(1)\beta^*(2)\left[-\dfrac{i^2\hbar^2}{4}\right]\beta(1)\alpha(2)\, d\tau_1\, d\tau_2 \\[8pt] + \iint \alpha^*(1)\beta^*(2)\left[-\dfrac{\hbar^2}{4}\right]\alpha(1)\beta(2)\, d\tau_1\, d\tau_2 \end{array} \right]$$

Because of the orthogonality of the spin functions, the first two integrals are zero and

$$\Delta E_2 = \frac{4\pi^2}{h}\left(-\frac{\hbar^2}{4}\right) J_{12} \iint \alpha^*(1)\beta^*(2)\alpha(1)\beta(2)\, d\tau_1\, d\tau_2 = \frac{4\pi^2}{h} J_{12}\left(-\frac{\hbar^2}{4}\right) = -\frac{hJ_{12}}{4}$$

Note that because J_{12} has the units of s^{-1}, hJ has the unit joule.

You will use the procedure of Example Problem 17.3 in the end-of-chapter problems to show that the spin energy eigenvalue for a given state is changed relative to the case of noninteracting spins by the amount

$$\Delta E = m_1 m_2 h J_{12} \text{ with } m_1 \text{ and } m_2 = +\frac{1}{2} \text{ for } \alpha \text{ and } -\frac{1}{2} \text{ for } \beta \qquad \textbf{(17.21)}$$

A given energy level is shifted to a higher energy if both spins are of the same orientation, and to a lower energy if the orientations are different. As you will see in the end-of-chapter problems, the frequencies of the allowed transitions including the spin–spin coupling are

$$\nu_{12} = \frac{\gamma B_0(1 - \sigma_1)}{2\pi} - \frac{J_{12}}{2}$$

$$\nu_{34} = \frac{\gamma B_0(1 - \sigma_1)}{2\pi} + \frac{J_{12}}{2}$$

$$\nu_{13} = \frac{\gamma B_0(1 - \sigma_2)}{2\pi} - \frac{J_{12}}{2}$$

$$\nu_{24} = \frac{\gamma B_0(1 - \sigma_2)}{2\pi} + \frac{J_{12}}{2} \qquad \textbf{(17.22)}$$

The energy levels and transitions corresponding to these frequencies are shown on the right side of Figure 17.7. This calculation shows that spin–spin interactions result in the appearance of multiplet splitting in NMR spectra. Each of the two peaks that appeared in the spectrum in the absence of spin–spin interactions is now split into a doublet in

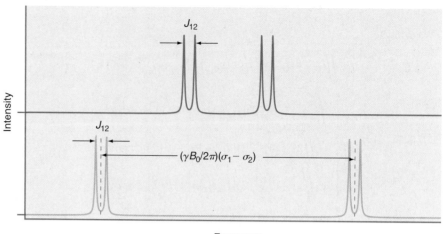

FIGURE 17.9

Splitting of a system of two interacting spins into doublets for two values of B_0. The spacing within the doublet is independent of the magnetic field strength, but the spacing of the doublets increases linearly with B_0.

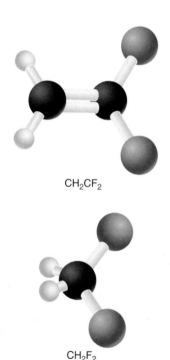

CH$_2$CF$_2$

CH$_2$F$_2$

FIGURE 17.10

The H atoms in CH$_2$CF$_2$ are chemically equivalent, but magnetically inequivalent. The H atoms in CH$_2$F$_2$ are chemically and magnetically equivalent.

which the two components are separated by J_{12} (Figure 17.9). Note that, whereas the separation in frequency of the doublets increases with the magnetic field strength, the splitting within each doublet is unaffected by the magnetic field.

Not all NMR peaks are split into multiplets. To understand this result, it is important to distinguish between **chemically** as opposed to **magnetically equivalent nuclei**. Consider the two molecules shown in Figure 17.10. In both cases, the two H atoms and the two F atoms are chemically equivalent. The nuclei of chemically equivalent atoms are also magnetically equivalent *only* if the interactions that they experience with other nuclei of nonzero spin are identical. Because the two F nuclei in CH$_2$F$_2$ are equidistant from each H atom, the two H—F couplings are identical and the ^{1}H are magnetically equivalent. However, the two H—F couplings in CH$_2$CF$_2$ are different because the spacing between the H and F nuclei is different. Therefore, the ^{1}H nuclei in this molecule are magnetically inequivalent. Multiplet splitting only arises through the interaction of magnetically inequivalent nuclei and is observed in CH$_2$CF$_2$, but not in CH$_2$F$_2$ or the reference compound (CH$_3$)$_4$Si. Because the derivation of this result is somewhat lengthy, it is omitted in this chapter.

17.8 MULTIPLET SPLITTING WHEN MORE THAN TWO SPINS INTERACT

For simplicity, we have considered only the case of two coupled spins in the previous sections. However, many organic molecules have more than two inequivalent protons that are close enough to one another to generate multiplet splittings. In this section, several different coupling schemes are considered. The frequencies for transitions in such a system involving the nuclear spin A can be written as

$$\nu_A = \frac{\gamma_A B(1 - \sigma_A)}{2\pi} - \sum_{X \neq A} J_{AX} m_X m_A \qquad (17.23)$$

where the summation is over all other spin-active nuclei. The strength of the interaction that leads to peak splitting is weak because J_{AX} falls off rapidly with distance. Therefore, the neighboring spins must be rather close in order to generate peak splitting. Experiments have shown that generally only those atoms within three or four bond lengths of the nucleus of interest have a sufficiently strong interaction to generate peak split-

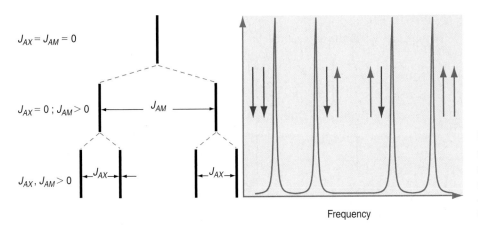

$J_{AX} = J_{AM} = 0$

$J_{AX} = 0 \; ; J_{AM} > 0$ $\leftarrow \qquad J_{AM} \qquad \rightarrow$

$J_{AX}, J_{AM} > 0$ $\leftarrow J_{AX} \rightarrow \leftarrow$ $\leftarrow J_{AX} \rightarrow$

Frequency

FIGURE 17.11

Coupling scheme and expected NMR spectrum for spin A coupled to spins M and X with different coupling constants J_{AX} and J_{AM}. The vertical axis shows the spectrum intensity.

ting. In strongly coupled systems such as those with conjugated bonds, the coupling can still be strong when the spins are farther apart.

To illustrate the effect of spin–spin interaction in generating multiplet splittings, we consider the coupling of three distinct spin $1/2$ nuclei that we label A, M, and X. The two coupling constants are J_{AM} and J_{AX} with $J_{AM} > J_{AX}$. The effect of these couplings can be determined by turning on the couplings individually as indicated in Figure 17.11. The result is that each of the lines in the doublet that arises from turning on the interaction J_{AM} is again split into a second doublet when the interaction J_{AX} is turned on as shown in Figure 17.11.

A special case occurs when A and M are identical so that $J_{AM} = J_{AX}$. The middle two lines for the AMX case now lie at the same frequency, giving rise to the AX_2 pattern shown in Figure 17.12. Because the two lines lie at the same frequency, the resulting spectrum is a triplet with the intensity ratio $1:2:1$. Such a spectrum is observed for the methylene protons in the molecule $CHCl_2 - CH_2 - CHCl_2$.

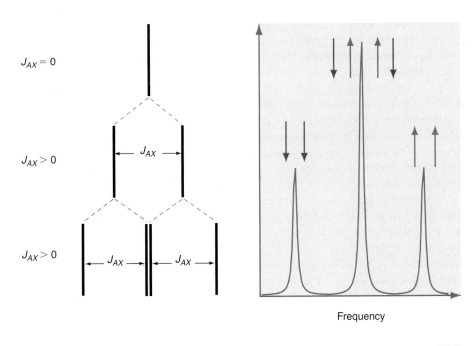

$J_{AX} = 0$

$J_{AX} > 0$ $\leftarrow \quad J_{AX} \quad \rightarrow$

$J_{AX} > 0$ $\leftarrow J_{AX} \rightarrow \leftarrow J_{AX} \rightarrow$

Frequency

FIGURE 17.12

Coupling scheme and expected NMR spectrum for spin A coupled to two spins X. In this case, there is only one coupling constant J_{AX}. The closely spaced pair of lines in the lower part of the left figure actually coincide. They have been shown separated to make their origin clear. The vertical axis shows the spectrum intensity.

EXAMPLE PROBLEM 17.4

Using the same reasoning as that applied to the AX_2 case, predict the NMR spectrum for an AX_3 spin system. Such a spectrum is observed for the methylene protons in the molecule $CH_3 - CH_2 - CCl_3$ where the coupling is to the methyl group hydrogens.

Solution

Turning on each of the interactions in sequence results in the following diagram:

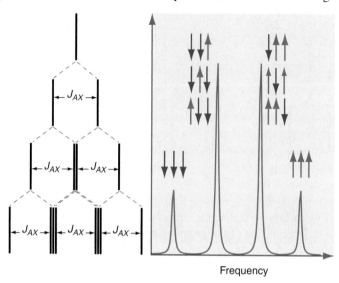

The end result is a quartet with the intensity ratios $1:3:3:1$. These results can be generalized to the rule that if a ^1H nucleus has n equivalent ^1H neighbors, its NMR spectral line will be split into $n+1$ peaks. The relative intensity of these peaks is given by the coefficients in the expansion of $(1+x)^n$, the binomial expression. The closely spaced pair of lines in the left figure actually coincide. They have been shown separated to make their origin clear.

Given the results of the last two sections, we are (almost) at the point of being able to understand the fine structure in the NMR spectrum of ethanol shown in Figure 17.6. As discussed in Section 17.7, the resonance near 5 ppm can be attributed to the OH proton, the resonance near 3.5 ppm can be attributed to the CH_2 protons, and the resonance near 1 ppm can be attributed to the CH_3 protons. This is consistent with the integrated intensities of the peaks, which from low to high δ are in the ratio $1:2:3$. We now consider the multiplet splitting. Invoking the guideline that spins located more than three bonds away do not generate peak splitting, we conclude that the CH_3 resonance is a triplet because it is split by the two CH_2 protons. The OH proton is too distant to generate a further splitting of the CH_3 group. We conclude that the CH_2 resonance is an octet (two pairs of quartets) because it is split by the three equivalent CH_3 protons and the OH proton. We predict that the OH resonance is a triplet because it is split by the two equivalent CH_2 protons. In fact, this is exactly what is observed for the NMR spectrum of ethanol at low temperatures. This example shows the power of NMR spectroscopy in obtaining structural information at the molecular level.

For ethanol at room temperature, these predictions are correct for the CH_3 group, but not for the other groups. The CH_2 hydrogen resonance is a quartet and the OH proton resonance is a singlet. This tells us that there is something that we have overlooked regarding the OH group. What has been overlooked is the rapid exchange of the OH proton with water, a topic that is discussed in the next section.

17.9 PEAK WIDTHS IN NMR SPECTROSCOPY

The ability of any spectroscopic technique to deliver useful information is limited by the width of the peaks in frequency. If two different NMR active nuclei in a sample have characteristic frequencies that are significantly closer than the width of the peaks, it is difficult to distinguish them. For samples in solutions, NMR spectra can exhibit peak widths of as little as 0.1 Hz, whereas for solid samples, peak widths of 10 kHz are not atypical. What are the reasons for such a large variation of peak widths in NMR spectra?

To answer this question, the change in the **magnetization vector M** with time must be considered. The vector **M** has two components: M_z, which is parallel to the static field $\mathbf{B}_0$, and $M_{x\text{-}y}$ or $\mathbf{M}_\parallel$, which is perpendicular to the field. Assume that the system has been perturbed so that **M** is not parallel to $\mathbf{B}_0$. How does the system of spins return to equilibrium? Note first that M_z decays at a different rate than $M_{x\text{-}y}$. It is not surprising that these two processes have different rates. To relax M_z, energy must be transferred to the surroundings, which is usually referred to as the lattice. The characteristic time associated with this process is called the longitudinal or **spin-lattice relaxation time T_1**. The relaxation of $M_{x\text{-}y}$ occurs through a randomization or **dephasing** of the spins and does not involve energy transfer to the surroundings because this component of the magnetization vector is perpendicular to $\mathbf{B}_0$. The characteristic time associated with this process is called the transverse or **spin–spin relaxation time T_2**. Because M_z will return to its initial value only after $\mathbf{M}_\parallel \rightarrow 0$, we conclude that $T_1 \geq T_2$.

The relaxation time T_1 determines the rate at which the energy absorbed from the radio-frequency field is dissipated to the surroundings. If T_1 is not sufficiently small, energy is not lost quickly enough to the surroundings, and the population of the excited state becomes as large as that of the ground state. If the populations of the ground and excited states are equal, the net absorption at the transition frequency is zero, and we say that the transition is a **saturated transition**. In obtaining NMR spectra, the radio-frequency power is kept low in order to avoid saturation.

How is the rate of relaxation of **M** related to the NMR linewidth? In discussing this issue, it is useful to view the experiment from two vantage points, time and frequency. As shown later in Section 17.12, the NMR signal is proportional to $M_{x\text{-}y}$, which decreases with increasing time with the functional form e^{-t/T_2} in the time domain. In a measurement of the peak width as a function of the frequency, we look at the same process in the frequency domain, because the signal in the frequency domain is the Fourier transform of the time-domain signal. Because of this relationship between the two domains, T_2 determines the spectral linewidth. The linewidth can be estimated with the Heisenberg uncertainty principle. The lifetime of the excited state, Δt, and the width in frequency of the spectral line corresponding to the transition to the ground state, $\Delta \nu$, are inversely related by

$$\frac{\Delta E\, \Delta t}{h} \approx 1 \quad \text{or} \quad \Delta \nu \approx \frac{1}{\Delta t} \qquad \textbf{(17.24)}$$

In the NMR experiment, T_2 is equivalent to Δt and, therefore, it determines the width of the spectral line, $\Delta \nu$. For this reason, narrow spectral features correspond to large values of T_2. In solution, T_2 can be orders of magnitude greater than for an ordered or disordered solid of the same substance. Therefore, NMR spectra in solution, in which through-space vectorial dipole–dipole coupling is averaged to zero through the tumbling of molecules resulting in large T_2 values, consist of narrow lines. By contrast, solid-state spectra exhibit broad lines because T_2 is small. The vectorial dipole–dipole coupling is not averaged to zero in this case because the molecules are fixed at their lattice sites.

The lifetime of the excited state in NMR spectroscopy can be significantly changed relative to the preceding discussion if the spins are strongly coupled to their surroundings. For example, this occurs if a proton on a tumbling molecule in solution undergoes a chemical exchange between two different sites. Consider the proton exchange reaction for ethanol:

$$CH_3CH_2OH + H_3O^+ \rightleftharpoons CH_3CH_2OH_2^+ + H_2O \qquad \textbf{(17.25)}$$

The exchange decreases the lifetime of the excited state, or T_2, leading to a broadening of the NMR peak. It turns out that the peaks become significantly broader only if the site exchange time is in the range of 10^{-4} to 10 s. This effect is referred to as **motional broadening**.

For a significantly faster exchange, only a single sharp peak is observed, and this effect is referred to as **motional narrowing**. Because the exchange occurs in times faster than 10^{-4} s, motional narrowing is observed for ethanol at room temperature. For this reason, the portion of the ethanol NMR spectrum shown in Figure 17.6 corresponding to the OH proton is a singlet rather than a triplet. However, at low temperatures and

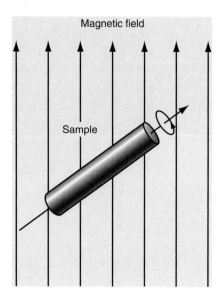

FIGURE 17.13

In magic angle spinning, the sample is rapidly spun about its axis, which is tilted 54.74° with respect to the static magnetic field.

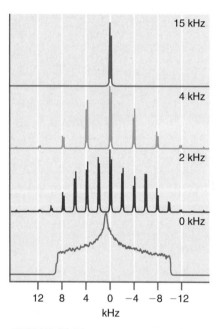

FIGURE 17.14

The ^{13}C NMR spectrum of a powder in which the unit cell contains a molecule with two inequivalent —C=O groups. The light blue spectrum shows the broad and nearly featureless solid-state spectrum. The 15-kHz spectrum (red) shows only two sharp peaks that can be attributed to the two chemically inequivalent —C=O groups. The remaining spectra are taken for different spinning frequencies. The spinning sidebands seen at 2 (dark blue) and 4 kHz (yellow) are experimental artifacts that arise if the spinning frequency is not sufficiently high.

[Published by permission of Gary Drobny, University of Washington.]

under acid-free conditions, the exchange rate can be sufficiently reduced so that the exchange can be ignored. In this case, the OH ^1H signal is a triplet. We now understand why the 300 K CH_2 hydrogen resonance in ethanol is a quartet rather than an octet and why the OH hydrogen resonance is a singlet rather than a triplet.

17.10 SOLID-STATE NMR

Whereas NMR spectra with well-separated narrow peaks are generally observed in solution, this is not the case for solids because direct dipole–dipole coupling between spins is not averaged to zero in solids as it is through molecular tumbling in solution. As we saw in Section 17.3, the magnetic field of a neighboring dipole can increase or decrease the external field $\mathbf{B_0}$ at the position of a spin, leading to a shift in the resonance frequency. The frequency shift resulting from direct coupling between two dipoles i and j is

$$\Delta\nu_{d-d} \propto \frac{3\mu_i\mu_j}{hr_{ij}^3}(3\cos^2\theta_{ij} - 1) \tag{17.26}$$

In this equation, r_{ij} is the distance between the dipoles and θ_{ij} is the angle between the magnetic field direction and the vector connecting the dipoles. Why did we not consider direct dipole–dipole coupling in discussing NMR spectra of solutions? Because molecules in a solution are rapidly tumbling, the time-averaged value of $\cos^2\theta_{ij}$, rather than the instantaneous value, determines $\Delta\nu_{d-d}$. As shown in Section 17.6, $\langle\cos^2\theta_{ij}\rangle = 1/3$ and, therefore, $\Delta\nu_{d-d} = 0$ for rapidly tumbling molecules in solution. By contrast, in solids the relative orientation of all the spin-active nuclei is frozen because of the crystal structure. For this reason, $\Delta\nu_{d-d}$ can be as large as several hundred kilohertz. This leads to very broad spectral features in the NMR spectra of solids. Given this situation, why carry out NMR experiments on solids?

The question can be answered in several ways. First, many materials are only available as solids, so that the option of obtaining solution spectra is not available. Second, useful information about the molecular anisotropy of the chemical shift can be obtained from solid-state NMR spectra. Finally, the technique of magic angle spinning can be used to transform broad solid-state spectra into spectra with linewidths comparable to those obtained in solution, as discussed next.

In general, a sample used in solid-state NMR experiments consists of many individual solid particles that are randomly oriented with respect to one another rather than a single crystal. Now imagine that a molecule in the unit cell is rotating about an axis rather than tumbling freely. Although not derived here, the time average of $3\cos^2\theta_{ij} - 1$ in this case is given by

$$\langle 3\cos^2\theta_{ij} - 1\rangle = (3\cos^2\theta' - 1)\left(\frac{3\cos^2\gamma_{ij} - 1}{2}\right) \tag{17.27}$$

In this equation, θ' is the angle that the sample rotation axis makes with B_0, and γ_{ij} is the angle between the vector $\mathbf{r}_{ij}$ that connects the magnetic dipoles i and j and the rotation axis. If the whole solid sample is rotated rapidly, then all pairs of coupled dipoles in the entire sample have the same value of θ' even though they have different values of γ_{ij}. If we choose to make $\theta' = 54.74°$, then $\langle 3\cos^2\theta' - 1\rangle = 0$, $\Delta\nu_{d-d} = 0$, and the broadening introduced by direct dipole coupling vanishes. Because this choice of θ' has such a dramatic effect, it is referred to as the **magic angle**, and the associated technique is referred to as **magic angle spinning** (Figure 17.13). An example of how a broad solid-state NMR spectrum can be transformed into a sharp spectrum through magic angle spinning is shown in Figure 17.14.

17.11 NMR IMAGING

One of the most important applications of NMR spectroscopy is its use in imaging the interior of solids. In the health sciences, **NMR imaging** has proved to be the most powerful and least invasive technique for obtaining information on soft tissue such as internal organs in humans. How is the spatial resolution needed for imaging obtained using NMR? For imaging, a **magnetic field gradient** is superimposed onto the constant magnetic field normally used in NMR. In this way, the resonance frequency of a given spin depends not only on the identity of the spin (that is, ^1H or ^{13}C), but also on the local magnetic field, which is determined by the location of the spin relative to the poles of the magnet. Figure 17.15 illustrates how the addition of a field gradient to the constant magnetic field allows the spatial mapping of spins to be carried out. Imagine a sphere and a cube containing ^1H_2O immersed in a background that contains no spin-active nuclei. In the absence of the field gradient, all spins in the structures resonate at the same frequency, giving rise to a single NMR peak. However, with the field gradient present, each volume element of the structure along the gradient has a different resonance frequency. The intensity of the NMR peak at each frequency is proportional to the total number of spins in the volume. A plot of the NMR peak intensity versus field strength gives a projection of the volume of the structures along the gradient direction. If a number of scans corresponding to different directions of the gradient are obtained, the three-dimensional structure of the specimen can be reconstructed, provided that the scans cover a range of at least 180°.

The particular usefulness of NMR for imaging biological samples relies on the different properties that can be used to create contrast in an image. In X-ray radiography, the image contrast is determined by the differences in electron density in various parts of the structure. Because carbon has a lower atomic number than oxygen, it does not scatter X-rays as strongly as oxygen. Therefore, fatty tissue appears lighter in a transmission image than tissues with a high density of water. However, this difference in scattering power is small and often gives insufficient contrast. To obtain a higher contrast, material that strongly scatters X-rays is injected or ingested. For NMR spectroscopy, several different properties can be utilized to provide image contrast without adding foreign substances.

The properties include the relaxation times T_1 and T_2, as well as chemical shifts and flow rates. The relaxation time offers the most useful contrast mechanism. The relaxation times T_1 and T_2 for water can vary in biological tissues from 0.1 s to several seconds. The more strongly bound the water is to a biological membrane, the greater the change in its relaxation time relative to freely tumbling water molecules. For example, the brain can be imaged with high contrast because the relaxation times of ^1H in gray matter, white matter, and spinal fluid are quite different. Data acquisition methods have been developed to enhance the signal amplitude for a particular range of relaxation times, enabling the contrast to be optimized for the problem of interest. Figure 17.16 shows an NMR image of a human brain.

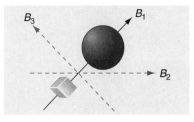

(a)

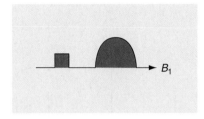

(b)

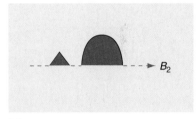

(c)

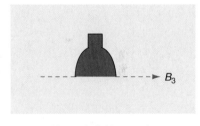

———— Magnetic field strength ————→

(d)

FIGURE 17.15

(**a**) Two structures are shown along with the three gradient directions indicated along which NMR spectra will be taken. In each case, spins within a thin volume element slice along the gradient resonate at the same frequency. This leads to a spectrum that is a projection of the volume onto the gradient axis. Image reconstruction techniques originally developed for X-rays can be used to determine the three-dimensional structure.

(**b–d**) NMR spectra that would be observed along the B_1, B_2, and B_3 directions indicated in part (a).

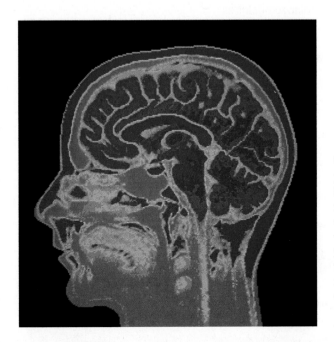

FIGURE 17.16

NMR image of a human brain. The section shown is from a noninvasive scan of the patient's head. The contrast has its origin in the dependence of the relaxation time on the strength of binding of the water molecule to different biological tissues.
[© M. Kulyk/Photo Researchers, Inc.]

Chemical shift imaging can be used to localize metabolic processes and to follow signal transmission in the brain through chemical changes that occur at nerve synapses. One variation of flow imaging is based on the fact that it takes times that are several multiples of T_1 for the local magnetization to achieve its equilibrium value. If, for instance, blood flows into the region under investigation on shorter timescales, it will not have the full magnetization of the spins that have been exposed to the field for much longer times. In such a case, the ^1H_2O in the blood resonates at a different frequency than the surrounding ^1H spins.

NMR imaging also has many applications in materials science, for example, in the measurement of the chemical cross-link density in polymers, the appearance of heterogeneities in elastomers such as rubber through vulcanization or aging, and the diffusion of solvents into polymers. Voids and defects in ceramics and the porosity of ceramics can be detected by nondestructive NMR imaging.

S U P P L E M E N T A L

17.12 THE NMR EXPERIMENT IN THE LABORATORY AND ROTATING FRAMES

As discussed at the beginning of this chapter, NMR peaks can be observed by varying either the magnetic field strength or the frequency of the applied ac field. However, modern NMR spectrometers utilize Fourier transform techniques because they greatly enhance the rate at which information can be acquired. In this and the next section, we describe the principles underlying Fourier transform NMR experiments.

A schematic diagram of the main components of an NMR experiment is shown in Figure 17.17. A sample is placed in a strong static magnetic field $\mathbf{B_0}$ that is directed along the z axis. A coil wound around the sample generates a much weaker oscillating radio-frequency (rf) magnetic field $\mathbf{B_1}$ of frequency ω that is directed along the y axis. A third detector coil used to detect the signal (not shown) is also wound around the sample. The sample under consideration has a single characteristic frequency, $\omega = \omega_0$. Additional frequencies that arise from chemical shifts are considered later. Why are two separate magnetic fields needed for the experiment? The static magnetic field $\mathbf{B_0}$ gives rise to the two energy levels shown as a function of the magnetic field strength in Figure 17.1. It does not induce transitions between the two states. However, the rf field $\mathbf{B_1}$ induces transitions between the two levels if the resonance condition $\omega = \omega_0$ is met.

To see how $\mathbf{B_1}$ induces a transition, we consider an alternative way of representing this rf field. The linearly polarized field $\mathbf{B_1}$ is mathematically equivalent to the superposition of two circularly polarized fields rotating in opposite directions. This can be seen by writing the two circularly polarized fields as

$$\mathbf{B}_1^{cc} = B_1(\mathbf{x}\cos\omega t + \mathbf{y}\sin\omega t)$$

$$\mathbf{B}_1^{c} = B_1(\mathbf{x}\cos\omega t - \mathbf{y}\sin\omega t) \tag{17.28}$$

In these equations, $\mathbf{x}$ and $\mathbf{y}$ are unit vectors along the x and y directions, and the superscripts c and cc refer to clockwise and counterclockwise rotation, respectively, as shown in Figure 17.18. The sum of these fields has zero amplitude in the y direction and an oscillatory amplitude in the x direction. This is analogous to the superposition

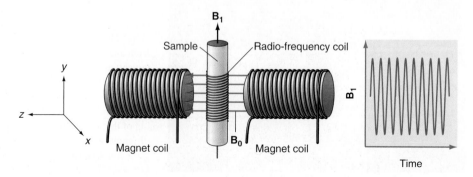

FIGURE 17.17

Schematic of the NMR experiment showing the static field and the rf field coil.

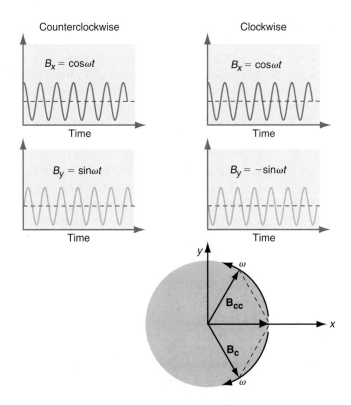

FIGURE 17.18

The superposition of two circularly polarized magnetic fields rotating in opposite directions leads to a linearly polarized magnetic field.

of two traveling waves to create a standing wave, a topic that was considered in Section 2.2. Of the two rotating components, only the counterclockwise component that is rotating in the same direction as the magnetic dipole will induce transitions; therefore, a linearly polarized magnetic field $\mathbf{B}_1$ has the same effect as a circularly polarized field that is rotating counterclockwise in the $x-y$ plane. For this reason, we can associate the part of the linearly polarized field that is effective for NMR spectroscopy with $\mathbf{B}_1^{cc} = B_1(\mathbf{x} \cos \omega t + \mathbf{y} \sin \omega t)$.

At this point, we discuss the precession of $\mathbf{M}$ about the total magnetic field. We consider the precession in the frame of reference rotating about the external magnetic field axis at the frequency ω of the rf field. The resultant magnetic field that is experienced by the nuclear spins is the vector sum of $\mathbf{B}_0$ and $\mathbf{B}_1$ and is depicted in Figure 17.19.

An observer in the laboratory frame sees a static field in the z direction, a circularly polarized field rotating at the frequency ω in the $x-y$ plane, and a resultant field that precesses around the z axis at the frequency ω. The resultant field is the vector sum of the static and rf fields. The total nuclear magnetic moment precesses around the resultant field, and this precession about a vector, which is itself precessing about the z axis, is difficult to visualize. The geometry becomes simpler if we view the motion of the magnetic moment from a frame of reference that is rotating about the z axis at the frequency ω. We choose the zero of time such that $\mathbf{B}_1$ lies along the x axis. According to classical mechanics, in the **rotating frame**, the rf field and the static field are stationary, and the

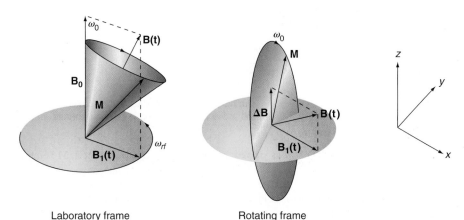

Laboratory frame   Rotating frame

FIGURE 17.19

The NMR experiment as viewed from the laboratory and the rotating frame of reference.

magnetic moment precesses about the resultant field $\Delta\mathbf{B}$, where $\Delta\mathbf{B} = \mathbf{B} - \mathbf{B}_1$, with a frequency $\omega_0 - \omega$. What can we say about the magnitude of the static field, $\Delta\mathbf{B}$, along the z axis in the rotating frame? We know that the torque acting on the magnetization vector is given by $\boldsymbol{\Gamma} = \mathbf{M} \times \mathbf{B}$ and that the magnetic moment has not changed. In order for the precession frequency to decrease from ω to $\omega_0 - \omega$, the apparent static field in the rotating frame must be

$$\Delta\mathbf{B} = \mathbf{B}_0 - \frac{\omega}{\gamma} = \frac{1}{\gamma}(\boldsymbol{\omega}_0 - \boldsymbol{\omega}) \qquad (17.29)$$

As $\boldsymbol{\omega}_{rf}$ approaches the resonance condition $\omega_0 = \gamma\mathbf{B}_0$, $\Delta\mathbf{B}$ approaches zero and $\mathbf{B} = \mathbf{B}_1$. In the rotating frame at resonance, the half-angle of the precession cone increases to 90°, and $\mathbf{M}$ now precesses in the y–z plane at the resonance frequency $\boldsymbol{\omega}$. The usefulness of viewing the NMR experiment in the rotating frame is that it allows the NMR pulse sequences described in the next section to be visualized easily.

SUPPLEMENTAL

17.13 FOURIER TRANSFORM NMR SPECTROSCOPY

NMR spectra can be obtained by scanning the static magnetic field or the frequency of the rf magnetic field. In these methods, data are only obtained at one particular frequency at any one instant of time. Because a sample typically contains different molecules with multiple resonance frequencies ω_0, obtaining data in this way is slow. If instead the rf signal is applied in the form of short pulses in a controlled sequence, information about a wide spectrum of resonance frequencies can be obtained simultaneously. In the following, we illustrate how this method, called **Fourier transform NMR spectroscopy**, is implemented. The procedure is depicted in Figure 17.20.

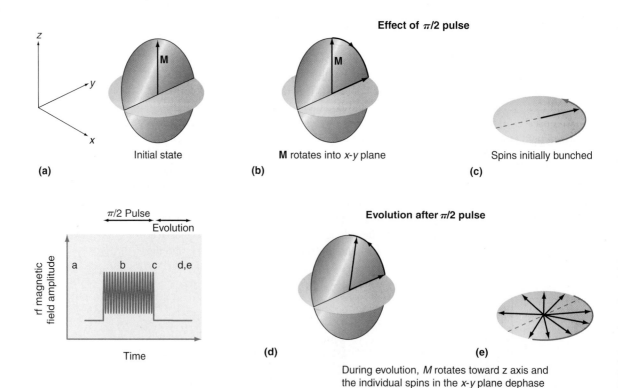

FIGURE 17.20

RF pulse timing and the effect on $\mathbf{M}$ as viewed from the rotating frame. At time a, $\mathbf{M}$ points along the z axis. As the $\pi/2$ pulse is applied, $\mathbf{M}$ precesses in the x–y plane and points along the y axis as shown for time c. After the pulse is turned off, $\mathbf{M}$ relaxes to its initial orientation along the z axis. The z component increases with the relaxation time T_1. Simultaneously, the x–y component of $\mathbf{M}$ decays with the relaxation time T_2 as the individual spins dephase.

At resonance in the rotating frame, the magnetic moment $\mathbf{M}$ is stationary and is aligned along the z axis before the rf pulse is applied. As soon as the pulse is applied, $\mathbf{M}$ begins to precess in the $y-z$ plane. The angle through which $\mathbf{M}$ precesses is given by

$$\alpha = \gamma B_1 t_p = \omega t_p \qquad (17.30)$$

in which t_p is the length of time that the rf field B_1 is on. The pulse length can be chosen so that $\mathbf{M}$ rotates 90°, after which time it lies in the $x-y$ plane. This is called a **$\pi/2$ pulse**. In the $x-y$ plane, the individual spins precess at slightly different frequencies because of their differing local fields which may be caused by different chemical shifts or field inhomogeneities. Immediately after the $\pi/2$ pulse, the spins are bunched together in the $x-y$ plane and $\mathbf{M}$ is aligned along the y axis. However, this is not the lowest energy configuration of the system because $\mathbf{M}$ is perpendicular rather than parallel to the static field. With increasing time, the magnetic moment returns to its equilibrium orientation parallel to the z axis by undergoing spin-lattice relaxation with the characteristic relaxation time T_1.

What happens to the component of $\mathbf{M}$ in the $x-y$ plane? The vector sum of the individual spin magnetic moments in the $x-y$ plane is the transverse magnetic moment component. Because the individual spins precess at different frequencies in the $x-y$ plane, they will fan out, leading to a dephasing of the spins. This process occurs with the spin–spin relaxation time T_2. As the spins dephase, the magnitude of the transverse component decays to its equilibrium value of zero, as shown in Figure 17.21. Three major mechanisms lead to dephasing: unavoidable inhomogeneities in $\mathbf{B_0}$, chemical shifts, and **transverse relaxation** due to spin–spin interactions.

How is the NMR spectrum generated using the Fourier transform technique? This process is indicated in Figure 17.21. The variation of $\mathbf{M}$ with time traces a spiral in which M_z increases and the **transverse magnetization** $M_{x\text{-}y}$ decreases with time. Because the detector coil has its axis along the y axis, it is not sensitive to changes in M_z. However, changes in $M_{x\text{-}y}$ induce a time-dependent voltage in the coil and, for that reason, the evolution of $M_{x\text{-}y}$ with time is shown separately in Figure 17.21. Because $M_{x\text{-}y}$ is a periodic function with the angular frequency ω, the induced voltage in the detector coil is alternately positive and negative. Because of the damping from spin–spin relaxation, its amplitude decays with time as e^{-t/T_2}. The process by which $M_{x\text{-}y}$ decays to its equilibrium value after the rf pulse is turned off is called **free induction decay**. This experiment provides a way to measure T_2.

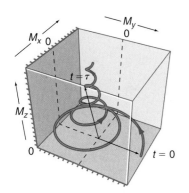
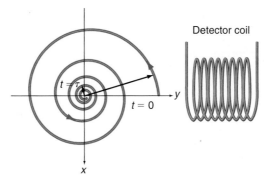

Evolution of $\mathbf{M}$ in three dimensions   Evolution of $\mathbf{M}$ in x-y plane

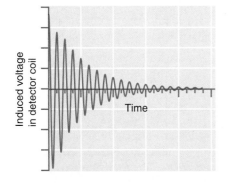

FIGURE 17.21

Evolution of the magnetization vector $\mathbf{M}$ in three dimensions and $M_{x\text{-}y}$ as a function of time. The variation of $M_{x\text{-}y}$ with time leads to exponentially decaying induced rf voltage in the detector coil.

Recall that because of the chemical shifts and unavoidable heterogeneities in the static magnetic field, not all spins have the same resonant frequency. Each group of spins with the same chemical shift gives rise to a different magnetization vector **M** that induces an ac voltage in the detector coil with a frequency equal to its characteristic precession frequency. Because all of these frequencies are contained in the signal, it contains the spectral information in a form that is not easily interpreted. However, by taking the **Fourier transform** of the detector coil signal,

$$I(\omega) = \int_0^\infty I(t)[\cos \omega t + i \sin \omega t]\, dt \qquad (17.31)$$

which is readily accomplished on a laboratory computer, the spectrum can be obtained as a function of frequency rather than time. Examples of the relationship between the free induction decay curves and the spectrum obtained through Fourier transformation are shown in Figure 17.22 for one, two, and three different frequencies.

What is the advantage of the Fourier transform technique over scanning either the magnetic field strength or the rf field strength to obtain an NMR spectrum? By using the Fourier transform technique, the whole spectral range is accessed at all times in which the data are collected. By contrast, in the scanning techniques, the individual frequencies are accessed serially. In any experiment as insensitive as NMR spectroscopy, it is difficult to extract useful signal from a background of noise. Therefore, any method in which more data are collected in a given time is to be preferred. Two arguments may be useful in gaining an understanding of how the method works. The first of these is an analogy with a mechanical resonator. If a bell is struck with a hammer, it will ring with its characteristic frequencies no matter what kind of hammer is used and how it is hit. Similarly, a solution containing precessing spins also has its collection of resonant frequencies and the "right hammer," in this case an rf pulse, excites the spins *at their resonant frequencies* regardless of which additional frequencies are contained in the pulse. The second argument is mathematical in nature. In analogy to the discussion in Section 2.7, many frequency components are required to describe a time-dependent function that changes rapidly over a small time interval. To write the $\pi/2$ pulse of Figure 17.20 as a sum of sine and cosine terms, $f(t) = d_0 + \sum_{n=1}^{m}(c_n \sin n\omega t + d_n \cos n\omega t)$ requires many terms. In this sense, the rf pulse consists of many individual frequencies.

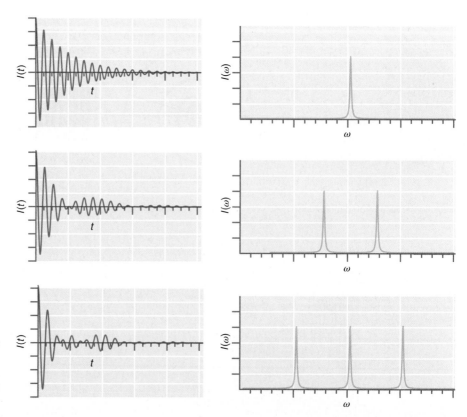

FIGURE 17.22

Free induction decay curves on the left for one, two, and three equal amplitude frequency components. The NMR spectrum on the right is the Fourier transform of the free induction decay curves.

Therefore, the pulse experiment is equivalent to carrying out many parallel experiments with rf magnetic fields of different frequency.

Fourier transform NMR provides the opportunity to manipulate the evolution of **M** by the application of successive rf pulses with varying length, intensity, frequency, and phase. Such a succession of pulses is called a **pulse sequence**. Pulse sequences are designed to manipulate the evolution of spins and reveal interactions between them or to selectively detect certain relaxation pathways. Pulse sequences are the foundation of modern NMR and constitute the basis of multidimensional NMR. The usefulness of these techniques can be understood by describing the spin–echo experiment.

To obtain the frequency spectrum from the free induction decay curve, T_2 must be known. The **spin–echo technique** uses a pulse sequence of particular importance to measure the transverse relaxation time, T_2. The experiment is schematically outlined in Figure 17.23. After an initial $\pi/2$ pulse from a coil along the x axis, the spins begin to fan out in the $x-y$ plane as a result of unavoidable inhomogeneities in $\mathbf{B_0}$ and because of the presence of chemical shifts. The decay of the signal that results from that part of the dephasing which originates from chemical shifts and field inhomogeneities can be eliminated in the following way. Rather than considering the resultant transverse magnetization component M_{x-y}, we consider two spins A and B. Spins A and B correspond to a Larmor frequency slightly higher and slightly lower than the frequency of the rf field, respectively. There is no coupling between the spins in this example; the spin–echo experiment for coupled spins is discussed in the next section. In a frame that is rotating at the average Larmor frequency, one of these spins will move clockwise and the other counterclockwise, as shown in Figure 17.23.

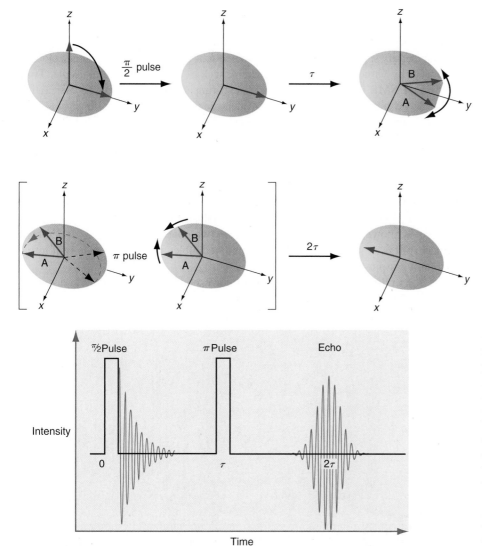

FIGURE 17.23

Schematic representation of the spin–echo experiment. The $\pi/2$ pulse applied along the x axis rotates **M** into the $x-y$ plane. After an evolution time τ in which free induction decay occurs, a π pulse is applied along the x axis. The effect of the π pulse on spin A is shown by the red arc. As a result of the π pulse, the fanning out process resulting from field inhomogeneities and chemical shifts is reversed. An echo will be observed in the detector coil at time 2τ and successive echoes will be observed at further integral multiples of τ. The amplitude of the successive echoes decreases with time because of transverse spin relaxation. The images in the square brackets show the effect of the π pulse on spins A and B. The left image shows the transformation of the spins, and the right image shows the resulting direction of precession.

After a time τ, a π pulse is applied, again along the x axis. This pulse causes the transformation $M_x \rightarrow M_x$ and $M_y \rightarrow -M_y$. As a result, the spins are flipped with respect to the x axis. The direction of precession of spins A and B is unchanged, but after the π pulse the angle between spins A and B now decreases with time. Therefore, the trend toward dephasing is reversed, and the spins will be in phase again after a second time interval τ equal to the initial evolution time. An echo of the original free induction decay signal is observed at 2τ. The amplitude of the echo is smaller than the original signal by the factor e^{-t/T_2} because of the dephasing resulting from transverse relaxation. Therefore, by measuring the amplitude of the echo, T_2 can be determined. Note that only the dephasing that occurs because of the field homogeneities and chemical shifts can be reversed using this technique. The spin–echo experiment is the most accurate method of determining the transverse relaxation time T_2.

SUPPLEMENTAL

17.14 TWO-DIMENSIONAL NMR

A ^{1}H NMR spectrum for a given molecule in solution contains a wealth of information. For large molecules, the density of spectral peaks can be very high as shown in Figure 17.24. Because of the high density, it is difficult to assign individual peaks to a particular ^{1}H in the molecule. One of the major uses of NMR is to determine the structure of molecules in their natural state in solution. To identify the molecule, it is necessary to know which peaks belong to equivalent ^{1}H that are split into a multiplet through coupling to other spins. Similarly, it would be useful to identify those peaks corresponding to ^{1}H that are coupled by through-bond interactions as opposed to through-space interactions. This type of information can be used to identify the structure of the molecule because through-bond interactions only occur over a distance of three to four bond lengths, whereas through-space interactions can identify spins that are more than three to four bond lengths apart, but are close to one another by virtue of a secondary structure, such as a folding of the molecule. **Two-dimensional NMR (2D-NMR)** allows such experiments to be carried out by separating the overlapped spectra of chemically inequivalent spins in multiple dimensions.

What is meant by 2D-NMR? We answer this question by describing how 2D-NMR is used to extract information from the five-peak one-dimensional spectrum shown in Figure 17.25. On the basis of the information contained in this NMR spectrum alone, there is no way to distinguish between peaks that arise from a chemical shift alone and peaks that arise from a chemical shift plus spin–spin coupling. The goal of the following 2D example is to outline how such a separation among the five peaks can be accomplished. For pedagogical reasons, we apply the analysis to a case for which the origin of each peak is known. This spectrum results from two nonequivalent ^{1}H, separated by a chemical shift δ, in which one peak is split into a doublet and the second is split into a triplet through spin–spin coupling.

The key to the separation between peaks corresponding to coupled and uncoupled ^{1}H is the use of an appropriate pulse sequence, which for this case is shown in

FIGURE 17.24

One-dimensional ^{1}H NMR spectrum of a small protein (molecular weight: ~17 kDa) in aqueous solution. The large number of overlapping broad peaks precludes a structural determination on the basis of the spectrum.

[*Published by permission of Rachel Klevit, University of Washington.*]

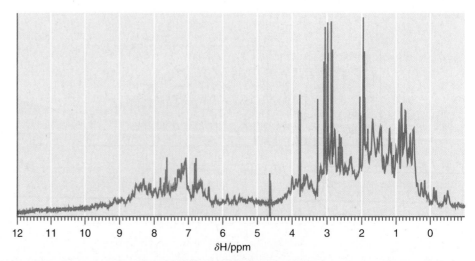

$\delta H/ppm$

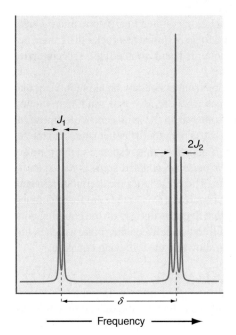

FIGURE 17.25

Illustration of a conventional one-dimensional NMR spectrum consisting of a doublet and a triplet separated by a chemical shift δ.

FIGURE 17.26

An initial $\pi/2$ pulse is applied along the x axis to initiate the experiment. After time t_1, a π pulse is applied, again along the x axis. After a second time interval t_1, the detector is turned on at the time indicated by the dashed line, and the signal is measured by the detector coil along the y axis as a function of the time t_2.

Figure 17.26. We recognize this pulse sequence as that used in the spin–echo experiment, and the effect of this pulse sequence on uncoupled chemically shifted ^{1}H nuclei was discussed in the previous section. In that case, we learned that the spins are refocused into an echo if the first and second time intervals are of equal length. However, we now consider the case of coupled spins. What is changed in the outcome of this experiment through the coupling? This question is answered in Figure 17.27.

As for the spin–echo experiment without coupling, we consider two spins, one higher and one lower in frequency than $\boldsymbol{\nu_0}$. In this case, the frequency difference between the spins is a result of the coupling, as opposed to a chemical shift. From Equation (17.21), the two frequencies are given by

$$\nu_B = \nu_0 - \frac{J}{2} \quad \text{and} \quad \nu_A = \nu_0 + \frac{J}{2} \tag{17.32}$$

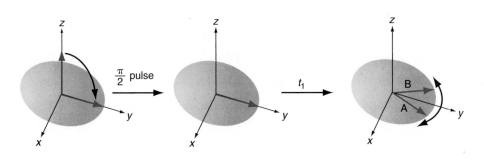

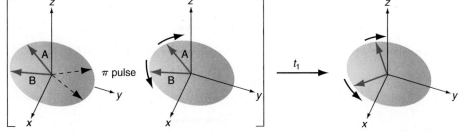

FIGURE 17.27

Illustration of the effect of the pulse sequence in Figure 17.26 on two coupled ^{1}H nuclei. Note the different effect of the π pulse illustrated in the square brackets compared with Figure 17.23. Spin A (B) is rotated as in Figure 17.27, but also converted to spin B (A).

where ν_A originates from the ^1H spin of interest being coupled to a β spin, and ν_B originates from the ^1H spin of interest being coupled to an α spin. The crucial difference between the effect of the pulse sequence on uncoupled and coupled spins occurs as a result of the π pulse.

The effect of the π pulse on coupled spins can be understood by breaking it down into two steps. Initially, the pulse causes the transformation $M_x \rightarrow M_x$ and $M_y \rightarrow -M_y$, which would make the spins rotate toward each other as in the spin–echo experiment for uncoupled spins (see Figure 17.23). However, because both the ^1H under study and the ^1H to which it is coupled make the transitions $\alpha \rightarrow \beta$ and $\beta \rightarrow \alpha$ in response to the π pulse, $\nu_B \rightarrow \nu_A$ and $\nu_A \rightarrow \nu_B$. The total effect of the π pulse on coupled spins is that spins A and B rotate away rather than toward one another. Therefore, after the second time interval t_1, the coupled spins are not refocused on the negative y axis as they are for uncoupled spins. Instead, they are refocused at a later time that depends linearly on the coupling constant J_{12}. This time is determined by the phase difference between the spins, which is linearly proportional to the coupling constant J_{12} as shown in the following equation:

$$\phi = 2\pi J_{12} t_1 \qquad \textbf{(17.33)}$$

What is the effect of this pulse sequence on the uncoupled spins in the sample? All uncoupled chemically shifted ^1H will give a pronounced echo after the second time interval t_1 shown in Figure 17.23, regardless of the value of δ. Therefore, coupled and uncoupled spins behave quite differently in response to the pulse sequence of Figure 17.26.

How can the values for J_{12} and δ contained in the spectrum of Figure 17.25 be separately determined? First, a series of experiments is carried out for different values of t_1. The evolution of M_{x-y} in this time interval depends on both J and δ. The free induction decay curves, $A(t_1, t_2)$, are obtained as a function of t_2 for each value of t_1. Note that the chemically shifted spins are refocused though the spin echo at the zero of t_2. Therefore, the value of $A(t_1, t_2 = 0)$ depends only on J, and not on δ. For all times $t_2 > 0$, the evolution of M_{x-y} once again depends on both J and δ. The set of $A(t_1, t_2)$ are shown in Figure 17.28 in the time interval denoted t_2. Next, each of these signals $A(t_1, t_2)$ is Fourier transformed with respect to t_2 to give $C(t_1, \omega_2)$. The sign of $C(t_1, \omega_2)$ is determined by $A(t_1, t_2 = 0)$, and can be either positive or negative. Each of the $C(t_1, \omega_2)$ for

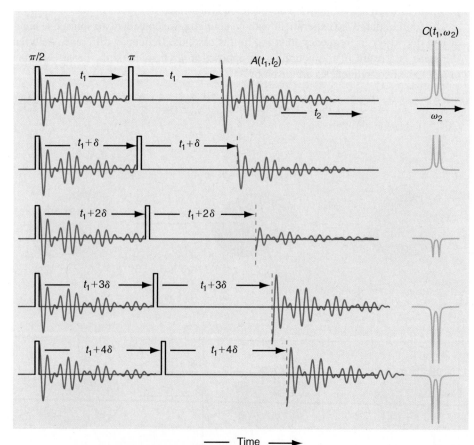

FIGURE 17.28

A series of NMR experiments corresponding to the pulse sequence of Figure 17.26 is shown for different values of t_1 for two coupled spins with a single coupling constant J. The Fourier transformed signal $C(t_1, \omega_2)$ obtained from a Fourier transformation of $A(t_1, t_2)$ with respect to t_2 is shown on the far right side of the figure.

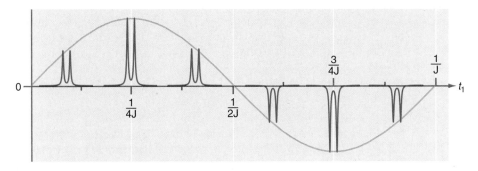

FIGURE 17.29

The function $C(t_1, \omega_2)$ is shown for different values of t_1. The periodicity in time evident in the figure can be expressed as a frequency by carrying out a Fourier transformation with respect to t_1.

given values of t_1 corresponds to the doublet shown in the rightmost column of Figure 17.28. Although $C(t_1, \omega_2)$, in general, exhibits a number of peaks, only two are shown for illustrative purposes. The dependence of $C(t_1, \omega_2)$ on t_1 is shown in Figure 17.29.

A periodic variation of $C(t_1, \omega_2)$ is observed, with the period T given by $T = 1/J$ as shown in Figure 17.29. We conclude that $C(t_1, \omega_2)$ is an amplitude-modulated periodic function whose period is determined by the coupling constant J. The periodicity in time can be converted to a frequency by a further Fourier transformation of $C(t_1, \omega_2)$, this time with respect to the time t_1, to give the function $G(\omega_1, \omega_2)$. Because the experiment has two characteristic frequencies, they can be used to define the two dimensions of the 2D technique. Function $G(\omega_1, \omega_2)$ is closely related to the desired 2D-NMR spectrum. As shown earlier, Fourier transformation with respect to the frequency ω_1 allows us to extract information from the data set on both δ and J_{12}, whereas the frequency ω_2 depends only on δ. Therefore, the information on δ and J_{12} can be independently obtained.

The J_{12} dependence can be separated from $C(t_1, \omega_2)$ to obtain a function $F(\omega_1, \omega_2)$ in which ω_2 depends only on δ and ω_1 depends only on J_{12}. Function $F(\omega_1, \omega_2)$ is referred to as the 2D J-δ spectrum and is shown as a contour plot in Figure 17.30. This function has two maxima along the ω_2 axis corresponding to the two multiplets of the 1D spectrum in Figure 17.25 that are separated by δ. At each of the δ values, further peaks will be observed along the ω_1 axis, with one peak for each member of the multiplet. The measured separation allows the value of J_{12} to be determined. As can be seen from the figure, the pulse sequence of Figure 17.26 allows a clear separation to be made between peaks in the 1D spectrum arising from a chemical shift and those arising from spin–spin coupling. It is clear that the information content of a 2D-NMR spectrum is much higher than that of a 1D spectrum.

The power of 2D-NMR is further illustrated for structural studies with another example. For this example, a pulse sequence is used that reveals the through-bond coupling of two ^1H. This particular 2D technique is called COSY (an acronym for *CO*rrelated *S*pectroscop*Y*). We illustrate the information that can be obtained from a COSY experiment for the molecule 1-bromobutane. The 1D-NMR spectrum of this

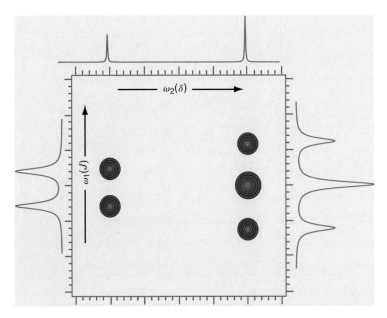

FIGURE 17.30

The two-dimensional function $F(\omega_1, \omega_2)$, corrected mathematically to separate δ and J on the ω_2 axis, is displayed as a contour plot. The horizontal scan above the contour plot shows the δ contribution to the 1D-NMR curve, and the two vertical plots on either side of the contour plot show the spin-spin coupling contribution to the 1D-NMR curve of Figure 17.25.

[*Published by permission of Tom Pratum, University of Washington.*]

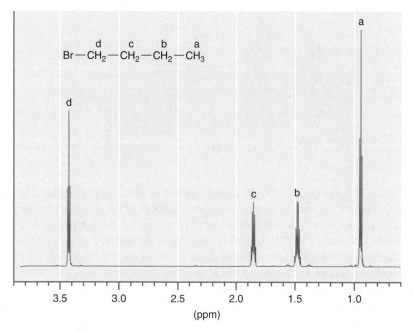

FIGURE 17.31

1D-NMR spectrum of 1-bromobutane. The multiplet splitting is not clearly seen because of the large range of δ used in the plot. The assignment of the individual peaks to equivalent ^{1}H spins in the molecule is indicated.

[*Published by permission of Tom Pratum, University of Washington.*]

molecule is shown in Figure 17.31. It consists of four peaks with multiplet splittings. On the basis of the discussion in Section 17.5, the peak assignments can be made readily by considering the effect of the electronegative Br atom on the different carbon atoms. The ^{1}H in the CH_2 group attached to the Br (d) generate a triplet, those in the adjacent CH_2 group (c) generate a five-peak multiplet, those in the CH_2 group (b) generate a six-peak multiplet, and those in the terminal CH_3 group generate a triplet. The integrated peak areas are in the ratio a:b:c:d = 3:2:2:2. These are the results expected on the basis of the discussion in Sections 17.7 and 17.8.

We now show that 2D-NMR can be used to find out which of the ^{1}H spins are coupled to one another. By applying the COSY pulse sequence, the 2D-NMR spectrum can be obtained as a function of ω_1 and ω_2. The results are shown in Figure 17.32 in the form of a contour plot. The 1D spectrum shown in this figure corresponds to the diagonal in the 2D spectrum representation. Four peaks are seen corresponding to different δ

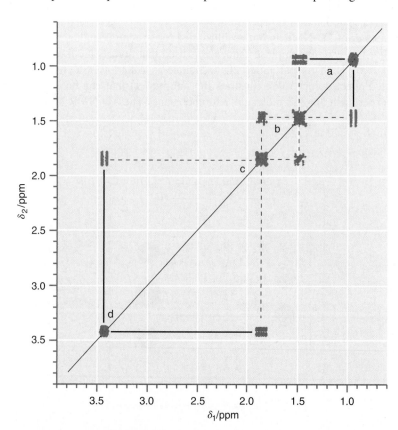

FIGURE 17.32

2D-NMR data for 1-bromobutane in the form of a contour plot. Dashed lines indicate coupling of a set of spins to two other groups. Solid lines indicate coupling of a set of spins to only one other group.

[*Published by permission of Tom Pratum, University of Washington.*]

values. We also see off-diagonal peaks at positions that are symmetrical with respect to the diagonal. These peaks identify spins that are coupled. The strength of the coupling can be determined from the intensity of each peak. We can determine which spins are coupled by moving vertically and horizontally from the off-diagonal peaks until the diagonal is reached. It is seen that spins (d) couple only with spins (c), spins (c) couple with both spins (d) and (b), spins (b) couple with both spins (c) and (a), and spins (a) couple only with spins (b). Therefore, the 2D COSY experiment allows us to determine which spins couple with one another. Note that these results are exactly what we would expect for the structural model shown in Figure 17.31 if coupling is ineffective for spins that are separated by more than three bond lengths.

Again, for pedagogical purposes we have chosen to analyze a simple spin system. This particular 2D-NMR experiment actually gives no more information than could have been deduced from the observed multiplet splitting. However, for large molecules with a molecular weight of several thousand Daltons and many inequivalent ^1H, COSY spectra give detailed information on the through-bond coupling of chemically inequivalent ^1H.

A classically based discussion of the effect of a pulse sequence on a sample containing spin-active nuclei analogous to the spin–echo experiment is not adequate to describe the COSY experiment. The higher level quantum mechanical description of this experiment is discussed in advanced texts. An analogous technique, called NOESY, gives information on the through-space coupling of inequivalent ^1H. These two techniques are just a small subset of the many powerful techniques available to NMR spectroscopists. Because of this diversity of experiments achievable through different pulse sequences, 2D-NMR is a powerful technique for the structural determination of biomolecules.

Vocabulary

chemical shift	magnetic field gradient	pulse sequence
chemical shift imaging	magnetically equivalent nuclei	rotating frame
chemically equivalent nuclei	magnetization vector	saturated transition
coupling constant	magnetogyric ratio	shielding constant
dephasing	motional broadening	spin–echo technique
diamagnetic response	motional narrowing	spin-lattice relaxation time T_1
Fourier transform	multiplet	spin polarization
Fourier transform NMR spectroscopy	multiplet splitting	spin-spin coupling
free induction decay	nuclear g factor	spin-spin relaxation time T_2
Larmor frequency	nuclear magnetic moment	transverse magnetization
macroscopic magnetic moment	nuclear magneton	transverse relaxation
magic angle	NMR imaging	two-dimensional NMR (2D-NMR)
magic angle spinning	$\pi/2$ pulse	
magnetic anisotropy	precession	

Questions on Concepts

Q17.1 Why can the signal loss resulting from spin dephasing caused by magnetic field inhomogeneities and chemical shift be recovered in the spin–echo experiment?

Q17.2 Why do neighboring groups lead to a net induced magnetic field at a given spin in a molecule in the solid state, but not for the same molecule in solution?

Q17.3 Why is it useful to define the chemical shift relative to a reference compound as follows?

$$\delta = 10^6 \frac{(\nu - \nu_{ref})}{\nu_{ref}}$$

Q17.4 What is the advantage of a 2-D NMR experiment over a 1-D NMR experiment?

Q17.5 Why do magnetic field inhomogeneities of only a few parts per million pose difficulties in NMR experiments?

Q17.6 Why does NMR lead to a higher contrast in the medical imaging of soft tissues than X-ray techniques?

Q17.7 Why is the multiplet splitting for coupled spins independent of the static magnetic field?

Q17.8 Why does the H atom on the OH group not lead to a multiplet splitting of the methyl hydrogens of ethanol?

Q17.9 Why are the multiplet splittings in Figure 17.15 not dependent on the static magnetic field?

Q17.10 Redraw Figure 17.2 for β spins. What is the direction of precession for the spins and for the macroscopic magnetic moment?

Q17.11 Why is the measurement time in NMR experiments reduced by using Fourier transform techniques?

Q17.12 Order the molecules CH_3I, CH_3Br, CH_3Cl, and CH_3F in terms of increasing chemical shift for ^1H. Explain your answer.

Q17.13 Explain why $T_1 \geq T_2$.

Q17.14 Explain why two magnetic fields, a static field and a radio-frequency field, are needed to carry out NMR experiments. Why must the two field directions be perpendicular?

Q17.15 Explain the difference in the mechanism that gives rise to through-space dipole–dipole coupling and through-bond coupling.

Problems

P17.1 Predict the number of chemically shifted ^1H peaks and the multiplet splitting of each peak that you would observe for diethyl ether. Justify your answer.

P17.2 Using your results from the previous problems, show that there are four possible transitions between the energy levels of two interacting spins and that the frequencies are given by

$$\nu_{12} = \frac{\gamma B(1 - \sigma_1)}{2\pi} - \frac{J_{12}}{2}$$

$$\nu_{34} = \frac{\gamma B(1 - \sigma_1)}{2\pi} + \frac{J_{12}}{2}$$

$$\nu_{13} = \frac{\gamma B(1 - \sigma_2)}{2\pi} - \frac{J_{12}}{2}$$

$$\nu_{24} = \frac{\gamma B(1 - \sigma_2)}{2\pi} + \frac{J_{12}}{2}$$

P17.3 For a fixed frequency of the radio frequency field, ^1H, ^{13}C, and ^{31}P will be in resonance at different values of the static magnetic field. Calculate the value of $\mathbf{B_0}$ for these nuclei to be in resonance if the radio frequency field has a frequency of 250 MHz.

P17.4 Using the matrix representation of the operators and spin eigenfunctions of Problem P17.7, show that the relationships listed in Equation (17.20) are obeyed.

P17.5 Predict the number of chemically shifted ^1H peaks and the multiplet splitting of each peak that you would observe for bromoethane. Justify your answer.

P17.6 A 250 MHz ^1H spectrum of a compound shows two peaks. The frequency of one peak is 510 Hz higher than that of the reference compound (tetramethylsilane) and the second peak is at a frequency 170 Hz lower than that of the reference compound. What chemical shift should be assigned to these two peaks?

P17.7 The nuclear spin operators can be represented as 2×2 matrices in the form and α and β can be represented as column vectors in the form

$$\alpha = \begin{pmatrix} 1 \\ 0 \end{pmatrix} \text{ and } \beta = \begin{pmatrix} 0 \\ 1 \end{pmatrix}$$

Given that

$$\hat{I}_x = \frac{\hbar}{2}\begin{pmatrix} 0 & 1 \\ 1 & 0 \end{pmatrix}, \quad \hat{I}_y = \frac{\hbar}{2}\begin{pmatrix} 0 & -i \\ i & 0 \end{pmatrix}, \quad \hat{I}_z = \frac{\hbar}{2}\begin{pmatrix} 1 & 0 \\ 0 & -1 \end{pmatrix}$$

and

$$\hat{I}^2 = \left(\frac{\hbar}{2}\right)^2\begin{pmatrix} 3 & 0 \\ 0 & 3 \end{pmatrix}$$

show that

$$\hat{I}^2\alpha = \frac{1}{2}\left(\frac{1}{2} + 1\right)\hbar^2\alpha, \ \hat{I}_z\alpha = +\frac{1}{2}\hbar\alpha, \ \hat{I}^2\beta = \frac{1}{2}\left(\frac{1}{2} + 1\right)\hbar^2\beta,$$

and $\hat{I}_z\beta = -\frac{1}{2}\hbar\beta$

P17.8 Predict the number of chemically shifted ^1H peaks and the multiplet splitting of each peak that you would observe for 1,1,1,2-tetrachloroethane. Justify your answer.

P17.9 Predict the number of chemically shifted ^1H peaks and the multiplet splitting of each peak that you would observe for 1,1,2,2-tetrachloroethane. Justify your answer.

P17.10 Predict the number of chemically shifted ^1H peaks and the multiplet splitting of each peak that you would observe for nitroethane. Justify your answer.

P17.11 Predict the number of chemically shifted ^1H peaks and the multiplet splitting of each peak that you would observe for nitromethane. Justify your answer.

P17.12 Predict the number of chemically shifted ^1H peaks and the multiplet splitting of each peak that you would observe for 1,1,2-trichloroethane. Justify your answer.

P17.13 Calculate the spin energy eigenvalues for the wave functions $\psi_1 = \alpha(1)\alpha(2)$, $\psi_3 = \alpha(1)\beta(2)$, and $\psi_4 = \beta(1)\beta(2)$ [Equation (17.15)] for noninteracting spins.

P17.14 Predict the number of chemically shifted ^1H peaks and the multiplet splitting of each peak that you would observe for 1-chloropropane. Justify your answer.

P17.15 Consider the first-order correction to the energy of interacting spins illustrated in Example Problem 17.3 for ψ_2. Calculate the energy correction to the wave functions $\psi_1 = \alpha(1)\alpha(2)$, $\psi_2 = \beta(1)\alpha(2)$, and $\psi_4 = \beta(1)\beta(2)$. Show that your results are consistent with $\Delta E = m_1 m_2 h J_{12}$ with m_1 and $m_2 = +1/2$ for α and $-1/2$ for β.

Appendix A Math Supplement

A.1 WORKING WITH COMPLEX NUMBERS AND COMPLEX FUNCTIONS

Imaginary numbers can be written in the form

$$z = a + ib \tag{A.1}$$

where a and b are real numbers and $i = \sqrt{-1}$. It is useful to represent complex numbers in the complex plane shown in Figure A.1. The vertical and horizontal axes correspond to the imaginary and real parts of z, respectively.

In the representation shown in Figure A.1, a complex number corresponds to a point in the complex plane. Note the similarity to the polar coordinate system. Because of this analogy, a complex number can be represented either as the pair (a, b), or by the radius vector r and the angle θ. From Figure A.1, it can be seen that

$$r = \sqrt{a^2 + b^2} \quad \text{and} \quad \theta = \cos^{-1}\frac{a}{r} = \sin^{-1}\frac{b}{r} = \tan^{-1}\frac{b}{a} \tag{A.2}$$

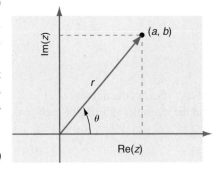

FIGURE A.1

Using the relations between a, b, and r as well as the Euler relation $e^{i\theta} = \cos\theta + i\sin\theta$, a complex number can be represented in either of two equivalent ways:

$$a + ib = r\cos\theta + r\sin\theta = re^{i\theta} = \sqrt{a^2 + b^2}\exp[i\tan^{-1}(b/a)] \tag{A.3}$$

If a complex number is represented in one way, it can easily be converted to the other way. For example, we express the complex number $6 - 7i$ in the form $re^{i\theta}$. The magnitude of the radius vector r is given by $\sqrt{6^2 + 7^2} = \sqrt{85}$. The phase is given by $\tan\theta = (-7/6)$ or $\theta = \tan^{-1}(-7/6)$. Therefore, we can write $6 - 7i$ as $\sqrt{85}\exp[i\tan^{-1}(-7/6)]$.

In a second example, we convert the complex number $2e^{i\pi/2}$, which is in the $re^{i\theta}$ notation, to the $a + ib$ notation. Using the relation $e^{i\alpha} = \exp(i\alpha) = \cos\alpha + i\sin\alpha$, we can write $2e^{i\pi/2}$ as

$$2\left(\cos\frac{\pi}{2} + i\sin\frac{\pi}{2}\right) = 2(0 + i) = 2i$$

The complex conjugate of a complex number z is designated by z^* and is obtained by changing the sign of i, wherever it appears in the complex number. For example, if $z = (3 - \sqrt{5}i)e^{i\sqrt{2}\phi}$, then $z^* = (3 + \sqrt{5}i)e^{-i\sqrt{2}\phi}$. The magnitude of a complex number is defined by $\sqrt{zz^*}$ and is always a real number. This is the case for the previous example:

$$zz^* = (3 - \sqrt{5}i)e^{i\sqrt{2}\phi}(3 + \sqrt{5}i)e^{-i\sqrt{2}\phi}$$
$$= (3 - \sqrt{5}i)(3 + \sqrt{5}i)e^{i\sqrt{2}\phi - i\sqrt{2}\phi} = 14 \tag{A.4}$$

Note also that $zz^* = a^2 + b^2$.

Complex numbers can be added, multiplied, and divided just like real numbers. A few examples follow:

$$(3 + \sqrt{2}i) + (1 - \sqrt{3}i) = [4 + (\sqrt{2} - \sqrt{3})i]$$
$$(3 + \sqrt{2}i)(1 - \sqrt{3}i) = 3 - 3\sqrt{3}i + \sqrt{2}i - \sqrt{6}i^2$$
$$= (3 + \sqrt{6}) + (\sqrt{2} - 3\sqrt{3})i$$
$$\frac{(3 + \sqrt{2}i)}{(1 - \sqrt{3}i)} = \frac{(3 + \sqrt{2}i)(1 + \sqrt{3}i)}{(1 - \sqrt{3}i)(1 + \sqrt{3}i)}$$
$$= \frac{3 + 3\sqrt{3}i + \sqrt{2}i + \sqrt{6}i^2}{4}$$

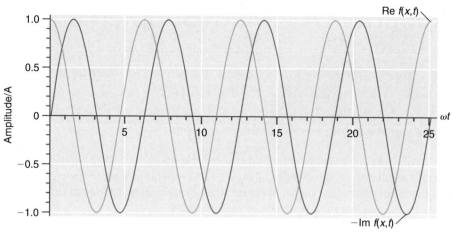

FIGURE A.2

$$= \frac{(3 - \sqrt{6}) + (3\sqrt{3} + \sqrt{2})i}{4}$$

Functions can depend on a complex variable. It is convenient to represent a plane traveling wave usually written in the form

$$\psi(x, t) = A \sin(kx - \omega t) \tag{A.5}$$

in the complex form

$$Ae^{i(kx - \omega t)} = A \cos(kx - \omega t) - iA \sin(kx - \omega t) \tag{A.6}$$

Note that

$$\psi(x, t) = -\operatorname{Im} Ae^{i(kx - \omega t)} \tag{A.7}$$

The reason for working with the complex form rather than the real form of a function is that calculations such as differentiation and integration can be carried out more easily. Waves in classical physics have real amplitudes, because their amplitudes are linked directly to observables. For example, the amplitude of a sound wave is the local pressure that arises from the expansion or compression of the medium through which the wave passes. However, in quantum mechanics, observables are related to $|\psi(x, t)|^2$ rather than $\psi(x, t)$. Because $|\psi(x, t)|^2$ is always real, $\psi(x, t)$ can be complex, and the observables associated with the wave function are still real.

For the complex function $f(x, t) = Ae^{i(kx - \omega t)}$, $zz^* = \psi(x, t)\psi^*(x, t) = Ae^{i(kx - \omega t)}A^* e^{-i(kx - \omega t)} = AA^*$, so that the magnitude of the function is a constant and does not depend on t or x. As Figure A.2 shows, the real and imaginary parts of $Ae^{i(kx - \omega t)}$ depend differently on the variables x and t; they are phase shifted by $\pi/2$. The figure shows the amplitudes of the real and imaginary parts as a function of ωt for $x = 0$.

A.2 DIFFERENTIAL CALCULUS

A.2.1 The First Derivative of a Function

The derivative of a function has as its physical interpretation the slope of the function evaluated at the position of interest. For example, the slope of the function $y = x^2$ at the point $x = 1.5$ is indicated by the line tangent to the curve shown in Figure A.3.

Mathematically, the first derivative of a function $f(x)$ is denoted $f'(x)$ or $df(x)/dx$. It is defined by

$$\frac{df(x)}{dx} = \lim_{h \to 0} \frac{f(x + h) - f(x)}{h} \tag{A.8}$$

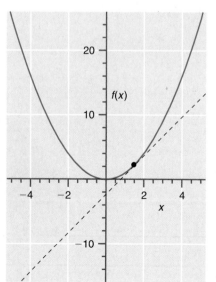

FIGURE A.3

For the function of interest,

$$\frac{df(x)}{dx} = \lim_{h\to 0} \frac{(x+h)^2 - (x)^2}{h}$$

$$= \lim_{h\to 0} \frac{2hx + h^2}{h} = \lim_{h\to 0} 2x + h = 2x \qquad \text{(A.9)}$$

In order for $df(x)/dx$ to be defined over an interval in x, $f(x)$ must be continuous over the interval.

Based on this example, $df(x)/dx$ can be calculated if $f(x)$ is known. Several useful rules for differentiating commonly encountered functions are listed next:

$$\frac{d(ax^n)}{dx} = anx^{n-1}, \quad \text{where } a \text{ is a constant and } n > 0 \qquad \text{(A.10)}$$

For example, $d(\sqrt{3}x^{4/3})/dx = (4/3)\sqrt{3}x^{1/3}$

$$\frac{d(ae^{bx})}{dx} = abe^{bx}, \quad \text{where } a \text{ and } b \text{ are constants} \qquad \text{(A.11)}$$

For example, $d(5e^{3\sqrt{2}x})/dx = 15\sqrt{2}e^{3\sqrt{2}x}$

$$\frac{d(ae^{bx})}{dx} = abe^{bx}, \quad \text{where } a \text{ and } b \text{ are constants}$$

$$\frac{d(a \sin x)}{dx} = a \cos x, \quad \text{where } a \text{ is a constant}$$

$$\frac{d(a \cos x)}{dx} = -a \sin x, \quad \text{where } a \text{ is a constant} \qquad \text{(A.12)}$$

Two useful rules in evaluating the derivative of a function that is itself the sum or product of two functions are as follows:

$$\frac{d[f(x) + g(x)]}{dx} = \frac{df(x)}{dx} + \frac{dg(x)}{dx} \qquad \text{(A.13)}$$

For example,

$$\frac{d(x^3 + \sin x)}{dx} = \frac{dx^3}{dx} + \frac{d\sin x}{dx} = 3x^2 + \cos x$$

$$\frac{d[f(x)g(x)]}{dx} = g(x)\frac{df(x)}{dx} + f(x)\frac{dg(x)}{dx} \qquad \text{(A.14)}$$

For example,

$$\frac{d[\sin(x)\cos(x)]}{dx} = \cos(x)\frac{d\sin(x)}{dx} + \sin(x)\frac{d\cos(x)}{dx}$$

$$= \cos^2 x - \sin^2 x$$

A.2.2 The Reciprocal Rule and the Quotient Rule

How is the first derivative calculated if the function to be differentiated does not have a simple form such as those listed in the preceding section? In many cases, the derivative can be found by using the product and quotient rules stated here:

$$\frac{d\left(\dfrac{1}{f(x)}\right)}{dx} = -\frac{1}{[f(x)]^2}\frac{df(x)}{dx} \qquad \text{(A.15)}$$

For example,

$$\frac{d\left(\dfrac{1}{\sin x}\right)}{dx} = -\frac{1}{\sin^2 x}\frac{d\sin x}{dx} = \frac{-\cos x}{\sin^2 x}$$

$$\frac{d\left[\dfrac{f(x)}{g(x)}\right]}{dx} = \frac{g(x)\dfrac{df(x)}{dx} - f(x)\dfrac{dg(x)}{dx}}{[g(x)]^2} \tag{A.16}$$

For example,

$$\frac{d\left(\dfrac{x^2}{\sin x}\right)}{dx} = \frac{2x\sin x - x^2\cos x}{\sin^2 x}$$

A.2.3 The Chain Rule

In this section, we deal with the differentiation of more complicated functions. Suppose that $y = f(u)$ and $u = g(x)$. From the previous section, we know how to calculate $df(u)/du$. How do we calculate $df(u)/dx$? The answer to this question is stated as the chain rule:

$$\frac{df(u)}{dx} = \frac{df(u)}{du}\frac{du}{dx} \tag{A.17}$$

Several examples illustrating the chain rule follow:

$$\frac{d\sin(3x)}{dx} = \frac{d\sin(3x)}{d(3x)}\frac{d(3x)}{dx} = 3\cos(3x)$$

$$\frac{d\ln(x^2)}{dx} = \frac{d\ln(x^2)}{d(x^2)}\frac{d(x^2)}{dx} = \frac{2x}{x^2} = \frac{2}{x}$$

$$\frac{d\left(x+\dfrac{1}{x}\right)^{-4}}{dx} = \frac{d\left(x+\dfrac{1}{x}\right)^{-4}}{d\left(x+\dfrac{1}{x}\right)}\frac{d\left(x+\dfrac{1}{x}\right)}{dx} = -4\left(x+\frac{1}{x}\right)^{-5}\left(1-\frac{1}{x^2}\right)$$

$$\frac{d\exp(ax^2)}{dx} = \frac{d\exp(ax^2)}{d(ax^2)}\frac{d(ax^2)}{dx} = 2ax\exp(ax^2), \quad \text{where } a \text{ is a constant}$$

A.2.4 Higher Order Derivatives: Maxima, Minima, and Inflection Points

A function $f(x)$ can have higher order derivatives in addition to the first derivative. The second derivative of a function is the slope of a graph of the slope of the function versus the variable. Mathematically,

$$\frac{d^2 f(x)}{dx^2} = \frac{d}{dx}\left(\frac{df(x)}{dx}\right) \tag{A.18}$$

For example,

$$\frac{d^2\exp(ax^2)}{dx^2} = \frac{d}{dx}\left[\frac{d\exp(ax^2)}{dx}\right] = \frac{d[2ax\exp(ax^2)]}{dx}$$

$$= 2a\exp(ax^2) + 4a^2x^2\exp(ax^2), \quad \text{where } a \text{ is a constant}$$

The second derivative is useful in identifying where a function has its minimum or maximum value within a range of the variable, as shown next.

Because the first derivative is zero at a local maximum or minimum, $df(x)/dx = 0$ at the values x_{max} and x_{min}. Consider the function $f(x) = x^3 - 5x$ shown in Figure A.4 over the range $-2.5 \leq x \leq 2.5$.

By taking the derivative of this function and setting it equal to zero, we find the minima and maxima of this function in the range

$$\frac{d(x^3 - 5x)}{dx} = 3x^2 - 5 = 0, \quad \text{which has the solutions } x = \pm\sqrt{\frac{5}{3}} = 1.291$$

The maxima and minima can also be determined by graphing the derivative and finding the zero crossings as shown in Figure A.5.

Graphing the function clearly shows that the function has one maximum and one minimum in the range specified. What criterion can be used to distinguish between these extrema if the function is not graphed? The sign of the second derivative, evaluated at the point for which the first derivative is zero, can be used to distinguish between a maximum and a minimum:

$$\frac{d^2 f(x)}{dx^2} = \frac{d}{dx}\left[\frac{df(x)}{dx}\right] < 0 \quad \text{for a maximum}$$

$$\frac{d^2 f(x)}{dx^2} = \frac{d}{dx}\left[\frac{df(x)}{dx}\right] > 0 \quad \text{for a minimum} \qquad \text{(A.19)}$$

We return to the function graphed earlier and calculate the second derivative:

$$\frac{d^2(x^3 - 5x)}{dx^2} = \frac{d}{dx}\left[\frac{d(x^3 - 5x)}{dx}\right] = \frac{d(3x^2 - 5)}{dx} = 6x$$

By evaluating

$$\frac{d^2 f(x)}{dx^2} \text{ at } x = \pm\sqrt{\frac{5}{3}} = \pm 1.291$$

we see that $x = 1.291$ corresponds to the minimum, and $x = -1.291$ corresponds to the maximum.

If a function has an inflection point in the interval of interest, then

$$\frac{df(x)}{dx} = 0 \quad \text{and} \quad \frac{d^2 f(x)}{dx^2} = 0 \qquad \text{(A.20)}$$

An example for an inflection point is $x = 0$ for $f(x) = x^3$. A graph of this function in the interval $-2 \leq x \leq 2$ is shown in Figure A.6. As you can verify,

$$\frac{dx^3}{dx} = 3x^2 = 0 \text{ at } x = 0 \quad \text{and} \quad \frac{d^2(x^3)}{dx^2} = 6x = 0 \text{ at } x = 0$$

A.2.5 Maximizing a Function Subject to a Constraint

A frequently encountered problem is that of maximizing a function relative to a constraint. We first outline how to carry out a constrained maximization, and subsequently apply the method to maximizing the volume of a cylinder while minimizing its area. The theoretical framework for solving this problem originated with the French mathematician Lagrange, and the method is known as Lagrange's method of undetermined multipliers. We wish to maximize the function $f(x, y)$ subject to the constraint that $\phi(x, y) - C = 0$, where C is a constant. For example, you may want to maximize the area, A, of a rectangle while minimizing its circumference, C. In this case, $f(x, y) = A(x, y) = xy$ and $\phi(x, y) = C(x, y) = 2(x + y)$, where x and y are the length and width of the rectangle. The total differentials of these functions are given by Equation (A.21):

$$df = \left(\frac{\partial f}{\partial x}\right)_y dx + \left(\frac{\partial f}{\partial y}\right)_x dy = 0 \quad \text{and} \quad d\phi = \left(\frac{\partial \phi}{\partial x}\right)_y dx + \left(\frac{\partial \phi}{\partial y}\right)_x dy = 0$$

$$\text{(A.21)}$$

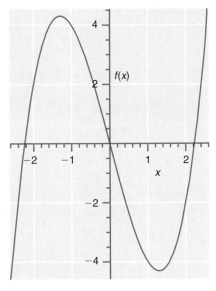

FIGURE A.4

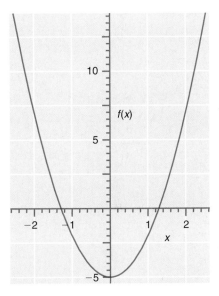

FIGURE A.5

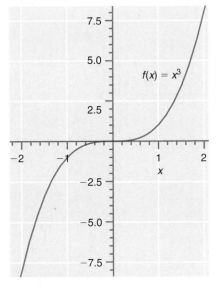

FIGURE A.6

If x and y were independent variables (there is no constraining relationship), the maximization problem would be identical to those dealt with earlier. However, because $d\phi = 0$ also needs to be satisfied, x and y are not independent variables. In this case, Lagrange found that the appropriate function to minimize is $f - \lambda\phi$, where λ is an undetermined multiplier. He showed that each of the expressions in the square brackets in the differential given by Equation (A.22) can be maximized independently. A separate multiplier is required for each constraint:

$$df = \left[\left(\frac{\partial f}{\partial x}\right)_y - \lambda\left(\frac{\partial \phi}{\partial x}\right)_y\right]dx + \left[\left(\frac{\partial f}{\partial y}\right)_x - \lambda\left(\frac{\partial \phi}{\partial y}\right)_x\right]dy \qquad \textbf{(A.22)}$$

We next use this method to maximize the volume, V, of a cylindrical can subject to the constraint that its exterior area, A, be minimized. The functions f and ϕ are given by

$$V = f(r,h) = \pi r^2 h \quad \text{and} \quad A = \phi(r,h) = 2\pi r^2 + 2\pi rh \qquad \textbf{(A.23)}$$

Calculating the partial derivatives and using Equation (A.22), we have

$$\left(\frac{\partial f(r,h)}{r}\right)_h = 2\pi rh \quad \left(\frac{\partial f(r,h)}{h}\right)_r = \pi r^2$$

$$\left(\frac{\partial \phi(r,h)}{r}\right)_h = 4\pi r + 2\pi h \quad \left(\frac{\partial \phi(r,h)}{h}\right)_r = 2\pi r$$

$$(2\pi rh - \lambda[4\pi r + 2\pi h])dr = 0 \quad \text{and} \quad (\pi r^2 - \lambda 2\pi r)dh = 0 \qquad \textbf{(A.24)}$$

Eliminating λ from these two equations gives

$$\frac{2\pi rh}{4\pi r + 2\pi h} = \frac{\pi r^2}{2\pi r} \qquad \textbf{(A.25)}$$

Solving for h in terms of r gives the result $h = 2r$. Note that there is no need to determine the value of multiplier λ. Perhaps you have noticed that beverage cans do not follow this relationship between r and h. Can you think of factors other than minimizing the amount of metal used in the can that might be important in this case?

A.3 SERIES EXPANSIONS OF FUNCTIONS

A.3.1 Convergent Infinite Series

Physical chemists often express functions of interest in the form of an infinite series. For this application, the series must converge. Consider the series

$$a_0 + a_1 x + a_2 x^2 + a_3 x^3 + \cdots a_n x^n + \cdots \qquad \textbf{(A.26)}$$

How can we determine if such a series converges? A useful convergence criterion is the ratio test. If the absolute ratio of successive terms (designated u_{n-1} and u_n) is less than 1 as $n \to \infty$, the series converges. We consider the series of Equation (A.26) with (a) $a_n = n!$ and (b) $a_n = 1/n!$, and apply the ratio test as shown in Equations (A.27a and b).

$$(a)\ \lim_{n\to\infty}\left|\frac{u_n}{u_{n-1}}\right| = \left|\frac{n!x^n}{(n-1)!x^{n-1}}\right| = \lim_{n\to\infty}|nx| > 1 \text{ unless } x = 0 \qquad \textbf{(A.27a)}$$

$$(b)\ \lim_{n\to\infty}\left|\frac{u_n}{u_{n-1}}\right| = \left|\frac{x^n/n!}{x^{n-1}/(n-1)!}\right| = \lim_{n\to\infty}\left|\frac{x}{n}\right| < 1 \text{ for all } x \qquad \textbf{(A.27b)}$$

We see that the infinite series converges if $a_n = 1/n!$ but diverges if $a_n = n!$.

The power series is a particularly important form of a series that is frequently used to fit experimental data to a functional form. It has the form

$$a_0 + a_1 x + a_2 x^2 + a_3 x^3 + a_1 x + a_4 x^4 + \cdots = \sum_{n=0}^{\infty} a_n x^n \qquad \textbf{(A.28)}$$

Fitting a data set to a series with a large number of terms is impractical, and to be useful, the series should contain as few terms as possible to satisfy the desired accuracy. For example, the function $\sin x$ can be fit to a power series over the interval $0 \leq x \leq 1.5$ by the following truncated power series

$$\sin x \approx -1.20835 \times 10^{-3} + 1.02102x - 0.0607398x^2 - 0.11779x^3$$

$$\sin x \approx -8.86688 \times 10^{-5} + 0.996755x + 0.0175769x^2$$
$$-0.200644x^3 - 0.027618x^4 \tag{A.29}$$

The coefficients in Equation (A.29) have been determined using a least squares fitting routine. The first series includes terms in x up to x^3, and is accurate to within 2% over the interval. The second series includes terms up to x^4, and is accurate to within 0.1% over the interval. Including more terms will increase the accuracy further.

A special case of a power series is the geometric series, in which successive terms are related by a constant factor. An example of a geometric series and its sum is given in Equation (A.30). Using the ratio criterion of Equation (A.27), convince yourself that this series converges for $|x| < 1$.

$$a(1 + x + x^2 + x^3 + \cdots) = \frac{a}{1 - x}, \quad \text{for } |x| < 1 \tag{A.30}$$

A.3.2 Representing Functions in the Form of Infinite Series

Assume that you have a function in the form $f(x)$ and wish to express it as a power series in x of the form

$$f(x) = a_0 + a_1 x + a_2 x^2 + a_3 x^3 + \cdots \tag{A.31}$$

To do so, we need a way to find the set of coefficients $(a_0, a_1, a_2, a_3, \cdots)$. How can this be done?

If the functional form $f(x)$ is known, the function can be expanded about a point of interest using the Taylor-Mclaurin expansion. In the vicinity of $x = a$, the function can be expanded in the series

$$f(x) = f(a) + \left(\frac{df(x)}{dx}\right)_{x=a}(x-a) + \frac{1}{2!}\left(\frac{d^2 f(x)}{dx^2}\right)_{x=a}(x-a)^2$$

$$+ \frac{1}{3!}\left(\frac{d^3 f(x)}{dx^3}\right)_{x=a}(x-a)^3 + \cdots \tag{A.32}$$

For example, consider the expansion of $f(x) = e^x$ about $x = 0$. Because $(d^n e^x/dx^n)_{x=0} = 1$ for all values of n, the Taylor-Mclaurin expansion for e^x about $x = 0$ is

$$f(x) = 1 + x + \frac{1}{2!}x^2 + \frac{1}{3!}x^3 + \cdots \tag{A.33}$$

Similarly, the Taylor-Mclaurin expansion for $\ln(1 + x)$ is found by evaluating the derivatives in turn:

$$\frac{d \ln(1 + x)}{dx} = \frac{1}{1 + x}$$

$$\frac{d^2 \ln(1 + x)}{dx^2} = \frac{d}{dx}\frac{1}{(1 + x)} = -\frac{1}{(1 + x)^2}$$

$$\frac{d^3 \ln(1 + x)}{dx^3} = -\frac{d}{dx}\frac{1}{(1 + x)^2} = \frac{2}{(1 + x)^3}$$

$$\frac{d^4 \ln(1 + x)}{dx^4} = \frac{d}{dx}\frac{2}{(1 + x)^3} = \frac{-6}{(1 + x)^4}$$

Each of these derivatives must be evaluated at $x = 0$.

Using these results, the Taylor-Mclaurin expansion for $\ln(1 + x)$ about $x = 0$ is

$$f(x) = x - \frac{x^2}{2!} + \frac{2x^3}{3!} - \frac{6x^4}{4!} + \cdots = x - \frac{x^2}{2} + \frac{x^3}{3} - \frac{x^4}{4} + \cdots \tag{A.34}$$

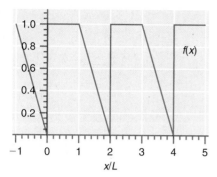

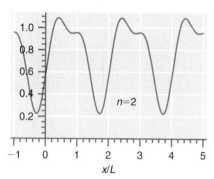

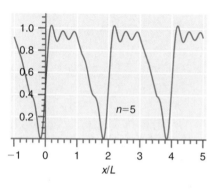

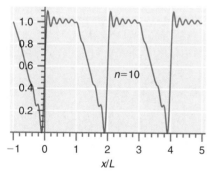

FIGURE A.7

The number of terms that must be included to adequately represent the function depends on the value of x. For $-1 \ll x \ll 1$, the series converges rapidly and, to a very good approximation, we can truncate the Taylor-Mclaurin series after the first one or two terms involving the variable. For the two functions just considered, it is reasonable to write $e^x \approx 1 + x$ and $\ln(1 \pm x) \approx \pm x$ if $-1 \ll x \ll 1$.

A second widely used series is the Fourier sine and cosine series. This series can be used to expand functions that are periodic over an interval $-L \le x \le L$ by the series

$$f(x) = \frac{1}{2}b_0 + \sum_n b_n \cos \frac{n\pi x}{L} + \sum_n a_n \sin \frac{n\pi x}{L} \quad \text{(A.35)}$$

A Fourier series is an infinite series, and the coefficients a_n and b_n can be calculated using the equations

$$a_n = \frac{1}{L} \int_{-L}^{+L} f(x) \sin \frac{n\pi x}{L} dx \quad \text{and} \quad b_n = \frac{1}{L} \int_{-L}^{+L} f(x) \cos \frac{n\pi x}{L} dx \quad \text{(A.36)}$$

The usefulness of the Fourier series is that a function can often be approximated by a few terms, depending on the accuracy desired.

For functions that are either even or odd with respect to the variable x, only either the sine or the cosine terms will appear in the series. For even functions, $f(-x) = f(x)$, and for odd functions, $f(-x) = -f(x)$. Because $\sin(-x) = -\sin(x)$ and $\cos(-x) = \cos(x)$, all coefficients a_n are zero for an even function, and all coefficients b_n are zero for an odd function. Note that Equations (A.29) are not odd functions of x because the function was only fit over the interval $0 \le x \le 1.5$.

Whereas the coefficients for the Taylor-Mclaurin series can be readily calculated, those for the Fourier series require more effort. To avoid mathematical detail here, the Fourier coefficients a_n and b_n are not explicitly calculated for a model function. The coefficients can be easily calculated using a program such as *Mathematica*. Our focus here is to show that periodic functions can be approximated to a reasonable degree by using the first few terms in a Fourier series, rather than to carry out the calculations.

To demonstrate the usefulness of expanding a function in a Fourier series, consider the function

$$f(x) = 1 \quad \text{for } 0 \le x \le L$$
$$f(x) = -x \quad \text{for } -L \le x \le 0 \quad \text{(A.37)}$$

which is periodic in the interval $-L \le x \le L$, in a Fourier series. This function is a demanding function to expand in a Fourier series because the function is discontinuous at $x = 0$ and the slope is discontinuous at $x = 0$ and $x = 1$. The function and the approximate functions obtained by truncating the series at $n = 2$, $n = 5$, and $n = 10$ are shown in Figure A.7. The agreement between the truncated series and the function is reasonably good for $n = 10$. The oscillations seen near $x/L = 0$ are due to the discontinuity in the function. More terms in the series are required to obtain a good fit, if the function changes rapidly in a small interval.

$A.4$ INTEGRAL CALCULUS

A.4.1 Definite and Indefinite Integrals

In many areas of physical chemistry, the property of interest is the integral of a function over an interval in the variable of interest. For example, the total probability of finding a particle within an interval $0 \le x \le a$ is the integral of the probability density $P(x)$ over the interval

$$P_{total} = \int_0^a P(x)\, dx \qquad (A.38)$$

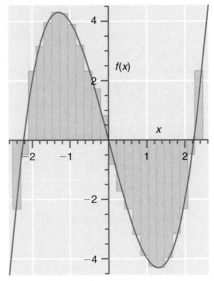

FIGURE A.8

Geometrically, the integral of a function over an integral is the area under the curve describing the function. For example, the integral $\int_{-2.3}^{2.3}(x^3 - 5x)\, dx$ is the sum of the areas of the individual rectangles in Figure A.8 in the limit within which the width of the rectangles approaches zero. If the rectangles lie below the zero line, the incremental area is negative; if the rectangles lie above the zero line, the incremental area is positive. In this case the total area is zero because the total negative area equals the total positive area. This is the case because $f(x)$ is an odd function of x.

The integral can also be understood as an antiderivative. From this point of view, the integral symbol is defined by the relation

$$f(x) = \int \frac{df(x)}{dx}\, dx \qquad (A.39)$$

and the function that appears under the integral sign is called the integrand. Interpreting the integral in terms of area, we evaluate a definite integral, and the interval over which the integration occurs is specified. The interval is not specified for an indefinite integral.

The geometrical interpretation is often useful in obtaining an integral from experimental data when the functional form of the integrand is not known. For our purposes, the interpretation of the integral as an antiderivative is more useful. The value of the indefinite integral $\int (x^3 - 5x)\, dx$ is that function which, when differentiated, gives the integrand. Using the rules for differentiation discussed earlier, you can verify that

$$\int (x^3 - 5x)\, dx = \frac{x^4}{4} - \frac{5x^2}{2} + C \qquad (A.40)$$

Note the constant that appears in the evaluation of every indefinite integral. By differentiating the function obtained upon integration, you should convince yourself that any constant will lead to the same integrand. In contrast, a definite integral has no constant of integration. If we evaluate the definite integral

$$\int_{-2.3}^{2.3}(x^3 - 5x)\, dx = \left(\frac{x^4}{4} - \frac{5x^2}{2} + C\right)_{x=2.3} - \left(\frac{x^4}{4} - \frac{5x^2}{2} + C\right)_{x=-2.3} \qquad (A.41)$$

we see that the constant of integration cancels. Because the function obtained upon integration is an even function of x, $\int_{-2.3}^{2.3}(x^3 - 5x)\, dx = 0$, just as we saw in the geometric interpretation of the integral.

It is useful for the student of physical chemistry to commit the integrals listed next to memory, because they are encountered frequently. These integrals are directly related to the derivatives discussed in Section A.2:

$$\int df(x) = f(x) + C$$

$$\int x^n\, dx = \frac{x^{n+1}}{n+1} + C$$

$$\int \frac{dx}{x} = \ln x + C$$

$$\int e^{ax} = \frac{e^{ax}}{a} + C, \quad \text{where } a \text{ is a constant}$$

$$\int \sin x\, dx = -\cos x + C$$

$$\int \cos x\, dx = \sin x + C$$

However, the primary tool for the physical chemist in evaluating integrals is a good set of integral tables. The integrals that are most frequently used in elementary quantum mechanics are listed here; the first group lists indefinite integrals:

$$\int (\sin ax)\, dx = -\frac{1}{a}\cos ax + C$$

$$\int (\cos ax)\, dx = \frac{1}{a}\sin ax + C$$

$$\int (\sin^2 ax)\, dx = \frac{1}{2}x - \frac{1}{4a}\sin 2ax + C$$

$$\int (\cos^2 ax)\, dx = \frac{1}{2}x + \frac{1}{4a}\sin 2ax + C$$

$$\int (x^2 \sin^2 ax)\, dx = \frac{1}{6}x^3 - \left(\frac{1}{4a}x^2 - \frac{1}{8a^3}\right)\sin 2ax - \frac{1}{4a^2}x\cos 2ax + C$$

$$\int (x^2 \cos^2 ax)\, dx = \frac{1}{6}x^3 + \left(\frac{1}{4a}x^2 - \frac{1}{8a^3}\right)\sin 2ax + \frac{1}{4a^2}x\cos 2ax + C$$

$$\int x^m e^{ax}\, dx = \frac{x^m e^{ax}}{a} - \frac{m}{a}\int x^{m-1} e^{ax}\, dx + C$$

$$\int \frac{e^{ax}}{x^m}\, dx = -\frac{1}{m-1}\frac{e^{ax}}{x^{m-1}} + \frac{a}{m-1}\int \frac{e^{ax}}{x^{m-1}}\, dx + C$$

The following group lists definite integrals.

$$\int_0^a \sin\left(\frac{n\pi x}{a}\right) \times \sin\left(\frac{m\pi x}{a}\right) dx = \int_0^a \cos\left(\frac{n\pi x}{a}\right) \times \cos\left(\frac{m\pi x}{a}\right) dx = \frac{a}{2}\delta_{mn}$$

$$\int_0^a \left[\sin\left(\frac{n\pi x}{a}\right)\right] \times \left[\cos\left(\frac{n\pi x}{a}\right)\right] dx = 0$$

$$\int_0^\pi \sin^2 mx\, dx = \int_0^\pi \cos^2 mx\, dx = \frac{\pi}{2}$$

$$\int_0^\infty \frac{\sin x}{\sqrt{x}}\, dx = \int_0^\infty \frac{\cos x}{\sqrt{x}}\, dx = \sqrt{\frac{\pi}{2}}$$

$$\int_0^\infty x^n e^{-ax}\, dx = \frac{n!}{a^{n+1}} \quad (a > 0, n \text{ positive integer})$$

$$\int_0^\infty x^{2n} e^{-ax^2}\, dx = \frac{1\cdot 3\cdot 5\cdots(2n-1)}{2^{n+1}a^n}\sqrt{\frac{\pi}{a}} \quad (a > 0, n \text{ positive integer})$$

$$\int_0^\infty x^{2n+1} e^{-ax^2}\, dx = \frac{n!}{2\, a^{n+1}} \quad (a > 0, n \text{ positive integer})$$

$$\int_0^\infty e^{-ax^2}\, dx = \left(\frac{\pi}{4a}\right)^{1/2}$$

In the first integral above, $\delta_{mn} = 1$ if $m = n$, and 0 if $m \neq n$.

A.4.2 Multiple Integrals and Spherical Coordinates

In the previous section, integration with respect to a single variable was discussed. Often, however, integration occurs over two or three variables. For example, the wave functions for the particle in a two-dimensional box are given by

$$\psi_{n_x n_y}(x, y) = N \sin \frac{n_x \pi x}{a} \sin \frac{n_y \pi y}{b} \qquad \text{(A.42)}$$

In normalizing a wave function, the integral of $|\psi_{n_x n_y}(x, y)|^2$ is required to equal one over the range $0 \le x \le a$ and $0 \le y \le b$. This requires solving the double integral

$$\int_0^b dy \int_0^a \left(N \sin \frac{n_x \pi x}{a} \sin \frac{n_y \pi y}{b} \right)^2 dx = 1 \qquad \text{(A.43)}$$

to determine the normalization constant N. We sequentially integrate over the variables x and y or vice versa using the list of indefinite integrals from the previous section.

$$\int_0^b dy \int_0^a \left(N \sin \frac{n_x \pi x}{a} \sin \frac{n_y \pi y}{b} \right)^2 dx$$

$$= \left[\frac{1}{2}x - \frac{a}{4n\pi} \sin \frac{2n_x \pi x}{a} \right]_{x=0}^{x=a} \times N^2 \int_0^b \left(\sin \frac{n_y \pi y}{b} \right)^2 dy$$

$$1 = \left[\frac{1}{2}a - \frac{a}{4n\pi}(\sin 2n_x \pi - 0) \right] \times N^2 \int_0^b \left(\sin \frac{n_y \pi y}{b} \right)^2 dy$$

$$1 = N^2 \left[\frac{1}{2}a - \frac{a}{4n\pi}(\sin 2n_x \pi - 0) \right] \times \left[\frac{1}{2}b - \frac{a}{4n\pi}(\sin 2n_y \pi - 0) \right] = \frac{N^2 ab}{4}$$

$$N = \frac{2}{\sqrt{ab}}$$

Convince yourself that the normalization constant for the wave functions of the three-dimensional particle in the box

$$\psi_{n_x n_y n_z}(x, y, z) = N \sin \frac{n_x \pi x}{a} \sin \frac{n_y \pi y}{b} \sin \frac{n_z \pi z}{c} \qquad \text{(A.44)}$$

has the value $N = 2\sqrt{2}/\sqrt{abc}$.

Up to this point, we have considered functions of a single variable. This restricts us to dealing with a single spatial dimension. The extension to three independent variables becomes important in describing three-dimensional systems. The three-dimensional system of most importance to us is the atom. Closed-shell atoms are spherically symmetric, so we might expect atomic wave functions to be best described by spherical coordinates. Therefore, you should become familiar with integrations in this coordinate system. In transforming from spherical coordinates r, θ, and ϕ to Cartesian coordinates x, y, and z, the following relationships are used:

$$x = r \sin \theta \cos \phi$$

$$y = r \sin \theta \sin \phi \qquad \text{(A.45)}$$

$$z = r \cos \theta$$

These relationships are depicted in Figure A.9. For small increments in the variables r, θ, and ϕ, the volume element depicted in this figure is a rectangular solid of volume

$$dV = (r \sin \theta \, d\phi)(dr)(r \, d\theta) = r^2 \sin \theta \, dr \, d\theta \, d\phi \qquad \text{(A.46)}$$

Note in particular that the volume element in spherical coordinates is not $dr \, d\theta \, d\phi$ in analogy with the volume element $dx \, dy \, dz$ in Cartesian coordinates.

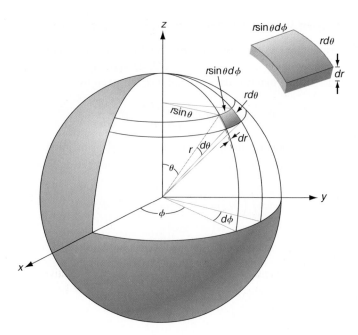

FIGURE A.9

In transforming from Cartesian coordinates x, y, and z to the spherical coordinates r, θ, and ϕ, these relationships are used:

$$r = \sqrt{x^2 + y^2 + z^2} \quad \theta = \cos^{-1} \frac{z}{\sqrt{x^2 + y^2 + z^2}} \text{ and } \phi = \tan^{-1} \frac{y}{x} \quad \textbf{(A.47)}$$

What is the appropriate range of variables to integrate over all space in spherical coordinates? If we imagine the radius vector scanning over the range $0 \leq \theta \leq \pi$; $0 \leq \phi \leq 2\pi$, the whole angular space is scanned. If we combine this range of θ and ϕ with $0 \leq r \leq \infty$, all of the three-dimensional space is scanned. Note that $r = \sqrt{x^2 + y^2 + z^2}$ is always positive.

To illustrate the process of integration in spherical coordinates, we normalize the function $e^{-r} \cos \theta$ over the interval $0 \leq r \leq \infty$; $0 \leq \theta \leq \pi$; $0 \leq \phi \leq 2\pi$:

$$N^2 \int_0^{2\pi} d\phi \int_0^\pi \sin \theta \, d\theta \int_0^\infty (e^{-r} \cos \theta)^2 r^2 \, dr = N^2 \int_0^{2\pi} d\phi \int_0^\pi \cos^2 \theta \sin \theta \, d\theta \int_0^\infty r^2 e^{-2r} \, dr = 1$$

It is most convenient to integrate first over ϕ, giving

$$2\pi N^2 \int_0^\pi \cos^2 \theta \sin \theta \, d\theta \int_0^\infty r^2 e^{-2r} \, dr = 1$$

We next integrate over θ, giving

$$2\pi N^2 \left[\frac{-\cos^3 \pi + \cos^3 0}{3} \right] \times \int_0^\infty r^2 e^{-2r} \, dr = \frac{4\pi N^2}{3} \int_0^\infty r^2 e^{-2r} \, dr = 1$$

We finally integrate over r using the standard integral

$$\int_0^\infty x^n e^{-ax} \, dx = \frac{n!}{a^{n+1}} \ (a > 0, n \text{ positive integer})$$

The result is

$$\frac{4\pi N^2}{3} \int_0^\infty r^2 e^{-2r} dr = \frac{4\pi N^2}{3} \frac{2!}{8} = 1 \quad \text{or} \quad N = \sqrt{\frac{3}{\pi}}$$

We conclude that the normalized wave function is $\sqrt{3/\pi} \, e^{-r} \cos \theta$.

A.5 VECTORS

The use of vectors occurs frequently in physical chemistry. Consider circular motion of a particle at constant speed in two dimensions, as depicted in Figure A.10. The particle is moving in a counterclockwise direction on the ring-like orbit. At any instant in time, its position, velocity, and acceleration can be measured. The two aspects to these measurements are the magnitude and the direction of each of these observables. Whereas a scalar quantity such as speed has only a magnitude, a vector has both a magnitude and a direction.

For the particular case under consideration, the position vectors $\mathbf{r}_1$ and $\mathbf{r}_2$ extend outward from the origin and terminate at the position of the particle. The velocities $\mathbf{v}_1$ and $\mathbf{v}_2$ are related to the position vector as $\mathbf{v} = \lim_{\Delta t \to 0} [\mathbf{r}(t + \Delta t) - \mathbf{r}(t)]/\Delta t$. Therefore, the velocity vector is perpendicular to the position vector. The acceleration vector is defined by $\mathbf{a} = \lim_{\Delta t \to 0} [\mathbf{v}(t + \Delta t) - \mathbf{v}(t)]/\Delta t$. As we see in part (b) of Figure A.10, $\mathbf{a}$ is perpendicular to $\mathbf{v}$, and is antiparallel to $\mathbf{r}$. As this example of a relatively simple motion shows, vectors are needed to describe the situation properly by keeping track of both the magnitude and direction of each of the observables of interest. For this reason, it is important to be able to work with vectors.

In three-dimensional Cartesian coordinates, any vector can be written in the form

$$\mathbf{r} = x_1\mathbf{i} + y_1\mathbf{j} + z_1\mathbf{k} \tag{A.48}$$

where $\mathbf{i}$, $\mathbf{j}$, and $\mathbf{k}$ are the mutually perpendicular vectors of unit length along the x, y, and z axes, respectively, and x_1, y_1, and z_1 are numbers. The length of a vector is defined by the equation

$$|\mathbf{r}| = \sqrt{x_1^2 + y_1^2 + z_1^2} \tag{A.49}$$

This vector is depicted in the three-dimensional coordinate system shown in Figure A.11.

By definition, the angle θ is measured from the z axis, and the angle ϕ is measured in the $x-y$ plane from the x axis. The angles θ and ϕ are related to x_1, y_1, and z_1 by

$$\theta = \cos^{-1}\frac{z_1}{\sqrt{x_1^2 + y_1^2 + z_1^2}} \quad \text{and} \quad \phi = \tan^{-1}\frac{y_1}{x_1} \tag{A.50}$$

We next consider the addition and subtraction of two vectors. Two vectors $\mathbf{a} = x\mathbf{i} + y\mathbf{j} + z\mathbf{k}$ and $\mathbf{b} = x'\mathbf{i} + y'\mathbf{j} + z'\mathbf{k}$ can be added or subtracted according to the equations

$$\mathbf{a} \pm \mathbf{b} = (x \pm x')\mathbf{i} + (y \pm y')\mathbf{j} + (z \pm z')\mathbf{k} \tag{A.51}$$

The addition and subtraction of vectors can also be depicted graphically, as done in Figure A.12.

The multiplication of two vectors can occur in either of two forms. Scalar multiplication of $\mathbf{a}$ and $\mathbf{b}$, also called the dot product of $\mathbf{a}$ and $\mathbf{b}$, is defined by

$$\mathbf{a} \cdot \mathbf{b} = |\mathbf{a}||\mathbf{b}| \cos \alpha \tag{A.52}$$

where α is the angle between the vectors. For $\mathbf{a} = 3\mathbf{i} + 1\mathbf{j} - 2\mathbf{k}$ and $\mathbf{b} = 2\mathbf{i} + -1\mathbf{j} + 4\mathbf{k}$, the vectors in the previous equation can be expanded in terms of their unit vectors:

$$\mathbf{a} \cdot \mathbf{b} = (3\mathbf{i} + 1\mathbf{j} - 2\mathbf{k}) \cdot (2\mathbf{i} + -1\mathbf{j} + 4\mathbf{k})$$
$$= 3\mathbf{i} \cdot 2\mathbf{i} + 3\mathbf{i} \cdot (-1\mathbf{j}) + 3\mathbf{i} \cdot 4\mathbf{k} + 1\mathbf{j} \cdot 2\mathbf{i} + 1\mathbf{j} \cdot (-1\mathbf{j})$$
$$+ 1\mathbf{j} \cdot 4\mathbf{k} - 2\mathbf{k} \cdot 2\mathbf{i} - 2\mathbf{k} \cdot (-1\mathbf{j}) - 2\mathbf{k} \cdot 4\mathbf{k}$$

However, because $\mathbf{i}$, $\mathbf{j}$, and $\mathbf{k}$ are mutually perpendicular vectors of unit length, $\mathbf{i} \cdot \mathbf{i} = \mathbf{j} \cdot \mathbf{j} = \mathbf{k} \cdot \mathbf{k} = 1$ and $\mathbf{i} \cdot \mathbf{j} = \mathbf{i} \cdot \mathbf{k} = \mathbf{j} \cdot \mathbf{k} = 0$. Therefore, $\mathbf{a} \cdot \mathbf{b} = 3\mathbf{i} \cdot 2\mathbf{i} + 1\mathbf{j} \cdot (-1\mathbf{j}) - 2\mathbf{k} \cdot 4\mathbf{k} = -3$.

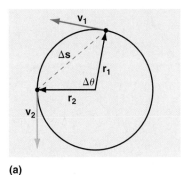

(a)

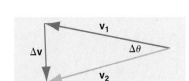

(b)

FIGURE A.10

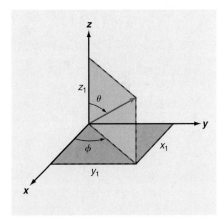

FIGURE A.11

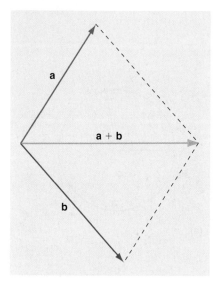

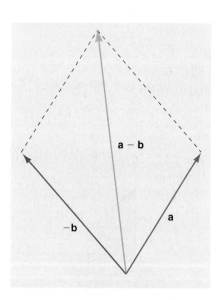

FIGURE A.12

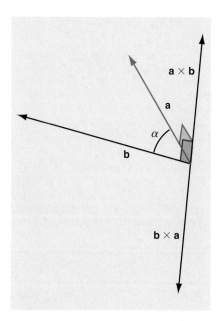

FIGURE A.13

The other form in which vectors are multiplied is the vector product, also called the cross product. The vector multiplication of two vectors results in a vector, whereas the scalar multiplication of two vectors results in a scalar. The cross product is defined by the equation

$$\mathbf{a} \times \mathbf{b} = \mathbf{c}|\mathbf{a}||\mathbf{b}| \sin \alpha \qquad \text{(A.53)}$$

Note that $\mathbf{a} \times \mathbf{b} = -\mathbf{b} \times \mathbf{a}$ as shown in Figure A.13. By contrast, $\mathbf{a} \cdot \mathbf{b} = \mathbf{b} \cdot \mathbf{a}$.

In Equation (A.53), $\mathbf{c}$ is a vector of unit length that is perpendicular to the plane containing $\mathbf{a}$ and $\mathbf{b}$ and has a positive direction found by using the right-hand rule (see Chapter 7) and α is the angle between $\mathbf{a}$ and $\mathbf{b}$.

The cross product between two three-dimensional vectors $\mathbf{a}$ and $\mathbf{b}$ is given by

$$\begin{aligned}
\mathbf{a} \times \mathbf{b} &= (a_x\mathbf{i} + a_y\mathbf{j} + a_z\mathbf{k}) \times (b_x\mathbf{i} + b_y\mathbf{j} + b_z\mathbf{k}) \\
&= a_x\mathbf{i} \times b_x\mathbf{i} + a_x\mathbf{i} \times b_y\mathbf{j} + a_x\mathbf{i} \times b_z\mathbf{k} + a_y\mathbf{j} \times b_x\mathbf{i} + a_y\mathbf{j} \times b_y\mathbf{j} \\
&\quad + a_y\mathbf{j} \times b_z\mathbf{k} + a_z\mathbf{k} \times b_x\mathbf{i} + a_z\mathbf{k} \times b_y\mathbf{j} + a_z\mathbf{k} \times b_z\mathbf{k}
\end{aligned} \qquad \text{(A.54)}$$

However, using the definition of the cross product in Equation (A.53),

$$\mathbf{i} \times \mathbf{i} = \mathbf{j} \times \mathbf{j} = \mathbf{k} \times \mathbf{k} = 0, \quad \mathbf{i} \times \mathbf{j} = \mathbf{k}, \quad \mathbf{i} \times \mathbf{k} = -\mathbf{j}$$
$$\mathbf{j} \times \mathbf{i} = -\mathbf{k}, \quad \mathbf{j} \times \mathbf{k} = \mathbf{i}, \quad \mathbf{k} \times \mathbf{i} = \mathbf{j}, \quad \mathbf{k} \times \mathbf{j} = -\mathbf{i}$$

Therefore, Equation (A.54) simplifies to

$$\mathbf{a} \times \mathbf{b} = (a_yb_z - a_zb_y)\mathbf{i} + (a_zb_x - a_xb_z)\mathbf{j} + (a_xb_y - a_yb_x)\mathbf{k} \qquad \text{(A.55)}$$

As we will see in Section A.7, there is a simple way to calculate cross products using determinants.

The angular momentum $\mathbf{l} = \mathbf{r} \times \mathbf{p}$ is of particular interest in quantum chemistry, because s, p, and d electrons are distinguished by their orbital angular momentum. For the example of the particle rotating on a ring depicted at the beginning of this section, the angular momentum vector is pointing upward in a direction perpendicular to the plane of the page. In analogy to Equation (A.55),

$$\mathbf{l} = \mathbf{r} \times \mathbf{p} = (yp_z - zp_y)\mathbf{i} + (zp_x - xp_z)\mathbf{j} + (xp_y - yp_x)\mathbf{k}$$

A.6 PARTIAL DERIVATIVES

In this section, we discuss the differential calculus of functions that depend on several independent variables. Consider the volume of a cylinder of radius r and height h, for which

$$V = f(r, h) = \pi r^2 h \qquad \text{(A.56)}$$

where V can be written as a function of the two variables r and h. The change in V with a change in r or h is given by the partial derivatives

$$\left(\frac{\partial V}{\partial r}\right)_h = \lim_{\Delta r \to 0} \frac{V(r + \Delta r, h) - V(r, h)}{\Delta r} = 2\pi rh$$

$$\left(\frac{\partial V}{\partial h}\right)_r = \lim_{\Delta h \to 0} \frac{V(r, h + \Delta h) - V(r, h)}{\Delta h} = \pi r^2 \qquad \text{(A.57)}$$

The subscript h in $(\partial V/\partial r)_h$ reminds us that h is being held constant in the differentiation. The partial derivatives in Equation (A.57) allow us to determine how a function changes when one of the variables changes. How does V change if the values of both variables change? In this case, V changes to $V + dV$ where

$$dV = \left(\frac{\partial V}{\partial r}\right)_h dr + \left(\frac{\partial V}{\partial h}\right)_r dh \qquad \text{(A.58)}$$

These partial derivatives are useful in calculating the error in the function that results from errors in measurements of the individual variables. For example, the relative error in the volume of the cylinder is given by

$$\frac{dV}{V} = \frac{1}{V}\left[\left(\frac{\partial V}{\partial r}\right)_h dr + \left(\frac{\partial V}{\partial h}\right)_r dh\right] = \frac{1}{\pi r^2 h}[2\pi rh\, dr + \pi r^2 dh] = \frac{2dr}{r} + \frac{dh}{h}$$

This equation shows that a given relative error in r generates twice the relative error in V as a relative error in h of the same size.

We can also take second or higher derivatives with respect to either variable. The mixed second partial derivatives are of particular interest. The mixed partial derivatives of V are given by

$$\left(\frac{\partial}{\partial h}\left(\frac{\partial V}{\partial r}\right)_h\right)_r = \left(\partial\left(\frac{\partial[\pi r^2 h]}{\partial r}\right)_h \bigg/ \partial h\right)_r = \left(\frac{\partial[2\pi rh]}{\partial h}\right)_r = 2\pi r$$

$$\left(\frac{\partial}{\partial r}\left(\frac{\partial V}{\partial h}\right)_r\right)_h = \left(\partial\left(\frac{\partial[\pi r^2 h]}{\partial h}\right)_r \bigg/ \partial r\right)_h = \left(\frac{\partial[\pi r^2]}{\partial r}\right)_h = 2\pi r \qquad \textbf{(A.59)}$$

For the specific case of V, the order in which the function is differentiated does not affect the outcome. Such a function is called a state function. Therefore, for any state function f of the variables x and y,

$$\left(\frac{\partial}{\partial y}\left(\frac{\partial f(x,y)}{\partial x}\right)_y\right)_x = \left(\frac{\partial}{\partial x}\left(\frac{\partial f(x,y)}{\partial y}\right)_x\right)_y \qquad \textbf{(A.60)}$$

Because Equation (A.60) is satisfied by all state functions, f, it can be used to determine if a function f is a state function.

We demonstrate how to calculate the partial derivatives

$$\left(\frac{\partial f}{\partial x}\right)_y, \left(\frac{\partial f}{\partial y}\right)_x, \left(\frac{\partial^2 f}{\partial x^2}\right)_y, \left(\frac{\partial^2 f}{\partial y^2}\right)_x, \left(\partial\left(\frac{\partial f}{\partial x}\right)_y \bigg/ \partial y\right)_x, \text{and} \left(\partial\left(\frac{\partial f}{\partial y}\right)_x \bigg/ \partial x\right)_y$$

for the function $f(x,y) = ye^{ax} + xy\cos x + y\ln xy$, where a is a real constant:

$$\left(\frac{\partial f}{\partial x}\right)_y = aye^{ax} + \frac{y}{x} + y\cos x - xy\sin x,$$

$$\left(\frac{\partial f}{\partial y}\right)_x = 1 + e^{ax} + x\cos x + \ln xy$$

$$\left(\frac{\partial^2 f}{\partial x^2}\right)_y = a^2 ye^{ax} - \frac{y}{x^2} - 2y\sin x - xy\cos x, \quad \left(\frac{\partial^2 f}{\partial y^2}\right)_x = \frac{1}{y}$$

$$\left(\partial\left(\frac{\partial f}{\partial x}\right)_y \bigg/ \partial y\right)_x = ae^{ax} + \frac{1}{x} + \cos x - x\sin x,$$

$$\left(\partial\left(\frac{\partial f}{\partial y}\right)_x \bigg/ \partial x\right)_y = ae^{ax} + \frac{1}{x} + \cos x - x\sin x$$

Because we have shown that

$$\left(\partial\left(\frac{\partial f}{\partial x}\right)_y \bigg/ \partial y\right)_x = \left(\partial\left(\frac{\partial f}{\partial y}\right)_x \bigg/ \partial x\right)_y$$

$f(x, y)$ is a state function of the variables x and y.

Whereas the partial derivatives tell us how the function changes if the value of one of the variables is changed, the total differential tells us how the function changes when all of the variables are changed simultaneously. The total differential of the function $f(x, y)$ is defined by

$$df = \left(\frac{\partial f}{\partial x}\right)_y dx + \left(\frac{\partial f}{\partial y}\right)_x dy \qquad \textbf{(A.61)}$$

The total differential of the function used earlier is calculated as follows:

$$df = \left(aye^{ax} + \frac{y}{x} + y\cos x - xy\sin x \right) dx + (1 + e^{ax} + x\cos x + \log xy)\, dy$$

Two other important results from multivariate differential calculus are used frequently. For a function $z = f(x,y)$, which can be rearranged to $x = g(y,z)$ or $y = h(x,z)$,

$$\left(\frac{\partial x}{\partial y} \right)_z = \frac{1}{\left(\dfrac{\partial y}{\partial x} \right)_z} \tag{A.62}$$

The other important result that is used frequently is the cyclic rule:

$$\left(\frac{\partial x}{\partial y} \right)_z \left(\frac{\partial y}{\partial z} \right)_x \left(\frac{\partial z}{\partial x} \right)_y = -1 \tag{A.63}$$

Consider an additional example of calculating partial derivatives for a function encountered in quantum mechanics. The Schrödinger equation for the hydrogen atom takes the form

$$-\frac{\hbar^2}{2\mu}\left[\frac{1}{r^2}\frac{\partial}{\partial r}\left(r^2 \frac{\partial\psi(r,\theta,\phi)}{\partial r} \right) + \frac{1}{r^2\sin\theta}\frac{\partial}{\partial\theta}\left(\sin\theta\frac{\partial\psi(r,\theta,\phi)}{\partial\theta} \right) + \frac{1}{r^2\sin\theta}\frac{\partial^2\psi(r,\theta,\phi)}{\partial\phi^2} \right]$$

$$-\frac{e^2}{4\pi\varepsilon_0 r}\psi(r,\theta,\phi) = E\psi(r,\theta,\phi)$$

Note that each of the first three terms on the left side of the equation involves partial differentiation with respect to one of the variables r, θ, and ϕ in turn. Two of the solutions to this differential equation are $(r/a_0)e^{-r/2a_0}\sin\theta e^{\pm i\phi}$. Each of these terms is evaluated separately to demonstrate how partial derivatives are taken in quantum mechanics. Although this is a more complex exercise than those presented earlier, it provides good practice in partial differentiation. For the first term, the partial derivative is taken with respect to r:

$$-\frac{\hbar^2}{2\mu}\frac{1}{\sqrt{64\pi}}\left(\frac{1}{a_0} \right)^{3/2}\left[\frac{1}{r^2}\frac{\partial}{\partial r}\left(r^2\frac{\partial\left(\dfrac{r}{a_0}e^{-r/2a_0}\sin\theta e^{\pm i\phi} \right)}{\partial r} \right) \right]$$

$$= -\frac{\hbar^2}{2\mu}\frac{1}{\sqrt{64\pi}}\left(\frac{1}{a_0} \right)^{3/2}\sin\theta e^{\pm i\phi}\left[\frac{1}{r^2}\frac{\partial}{\partial r}\left(r^2\frac{\partial\left(\dfrac{r}{a_0}e^{-r/2a_0} \right)}{\partial r} \right) \right]$$

$$= -\frac{\hbar^2}{2\mu}\frac{1}{\sqrt{64\pi}}\left(\frac{1}{a_0} \right)^{3/2}\sin\theta e^{\pm i\phi}\left[\frac{1}{r^2}\frac{\partial}{\partial r}\left(r^2\left(\frac{1}{a_0}e^{-r/2a_0} - (r/2a_0^2)e^{-r/2a_0} \right) \right) \right]$$

$$= -\frac{\hbar^2}{2\mu}\frac{1}{\sqrt{64\pi}}\left(\frac{1}{a_0} \right)^{3/2}\sin\theta e^{\pm i\phi}\left[\frac{1}{r^2}\left(\begin{array}{c} -r^2\dfrac{e^{-r/2a_0}}{a_0^2} + r^3\dfrac{e^{-r/2a_0}}{4a_0^3} + 2r\dfrac{e^{-r/2a_0}}{a_0} \\ -2r^2\dfrac{e^{-r/2a_0}}{a_0^2} \end{array} \right) \right]$$

$$= -\frac{\hbar^2}{2\mu}\frac{1}{\sqrt{64\pi}}\left(\frac{1}{a_0} \right)^{3/2}\sin\theta e^{\pm i\phi}e^{-r/2a_0}\frac{(8a_0^2 - 8a_0 r + r^2)}{4a_0^3 r}$$

Partial differentiation with respect to θ is easier, because the terms that depend on r and ϕ are constant:

$$-\frac{\hbar^2}{2\mu}\frac{1}{\sqrt{64\pi}}\left(\frac{1}{a_0} \right)^{3/2}\left[\frac{1}{r^2\sin\theta}\frac{\partial}{\partial\theta}\left(\sin\theta\frac{\partial\left(\dfrac{r}{a_0}e^{-r/2a_0}\sin\theta e^{\pm i\phi} \right)}{\partial\theta} \right) \right]$$

$$= -\frac{\hbar^2}{2\mu}\frac{1}{\sqrt{64\pi}}\left(\frac{1}{a_0} \right)^{3/2}\frac{r}{a_0}e^{-r/2a_0}e^{\pm i\phi}\left[\frac{1}{r^2\sin\theta}\frac{\partial}{\partial\theta}\left(\sin\theta\frac{\partial(\sin\theta)}{\partial\theta} \right) \right]$$

$$= -\frac{\hbar^2}{2\mu}\frac{1}{\sqrt{64\pi}}\left(\frac{1}{a_0}\right)^{3/2}\frac{r}{a_0}e^{-r/2a_0}e^{\pm i\phi}\left[\frac{1}{r^2\sin\theta}\frac{\partial}{\partial\theta}(\sin\theta\cos\theta)\right]$$

$$= -\frac{\hbar^2}{2\mu}\frac{1}{\sqrt{64\pi}}\left(\frac{1}{a_0}\right)^{3/2}\frac{r}{a_0}e^{-r/2a_0}e^{\pm i\theta}\left[\frac{1}{r^2\sin\theta}(\cos^2\theta - \sin^2\theta)\right]$$

Partial differentiation with respect to ϕ is also not difficult, because the terms that depend on r and θ are constant:

$$-\frac{\hbar^2}{2\mu}\frac{1}{\sqrt{64\pi}}\left(\frac{1}{a_0}\right)^{3/2}\left[\frac{1}{r^2\sin\theta}\frac{\partial^2 \frac{r}{a_0}e^{-r/2a_0}\sin\theta\, e^{\pm i\phi}}{\partial\phi^2}\right]$$

$$= -\frac{\hbar^2}{2\mu}\frac{1}{\sqrt{64\pi}}\left(\frac{1}{a_0}\right)^{3/2}\frac{r}{a_0}e^{-r/2a_0}\frac{1}{r^2}\left[\frac{\partial^2 e^{\pm i\phi}}{\partial\phi^2}\right]$$

$$= \frac{\hbar^2}{2\mu}\frac{1}{\sqrt{64\pi}}\left(\frac{1}{a_0}\right)^{3/2}\frac{r}{a_0}e^{-r/2a_0}\frac{1}{r^2}\left[e^{\pm i\phi}\right]$$

A.7 WORKING WITH DETERMINANTS

A determinant of nth order is a square $n \times n$ array of numbers symbolically enclosed by vertical lines. A fifth-order determinant is shown here with the conventional indexing of the elements of the array:

$$\begin{vmatrix} a_{11} & a_{12} & a_{13} & a_{14} & a_{15} \\ a_{21} & a_{22} & a_{23} & a_{24} & a_{25} \\ a_{31} & a_{32} & a_{33} & a_{34} & a_{35} \\ a_{41} & a_{42} & a_{43} & a_{44} & a_{45} \\ a_{51} & a_{52} & a_{53} & a_{54} & a_{55} \end{vmatrix} \qquad \text{(A.64)}$$

A 2×2 determinant has a value that is defined in Equation (A.65). It is obtained by multiplying the elements in the diagonal connected by a line with a negative slope and subtracting from this the product of the elements in the diagonal connected by a line with a positive slope.

$$\begin{vmatrix} a_{11} & a_{12} \\ a_{21} & a_{22} \end{vmatrix} = a_{11}a_{22} - a_{12}a_{21} \qquad \text{(A.65)}$$

The value of a higher order determinant is obtained by expanding the determinant in terms of determinants of lower order. This is done using the method of cofactors. We illustrate the use of method of cofactors by reducing a 3×3 determinant to a sum of 2×2 determinants. Any row or column can be used in the reduction process. We use the first row of the determinant in the reduction. The recipe is spelled out in this equation:

$$\begin{vmatrix} a_{11} & a_{12} & a_{13} \\ a_{21} & a_{22} & a_{23} \\ a_{31} & a_{32} & a_{33} \end{vmatrix} = (-1)^{1+1}a_{11}\begin{vmatrix} a_{22} & a_{23} \\ a_{32} & a_{33} \end{vmatrix} + (-1)^{1+2}a_{12}\begin{vmatrix} a_{21} & a_{23} \\ a_{31} & a_{33} \end{vmatrix}$$

$$+ (-1)^{1+3}a_{13}\begin{vmatrix} a_{21} & a_{22} \\ a_{31} & a_{32} \end{vmatrix}$$

$$= a_{11}\begin{vmatrix} a_{22} & a_{23} \\ a_{32} & a_{33} \end{vmatrix} - a_{12}\begin{vmatrix} a_{21} & a_{23} \\ a_{31} & a_{33} \end{vmatrix} + a_{13}\begin{vmatrix} a_{21} & a_{22} \\ a_{31} & a_{32} \end{vmatrix} \qquad \text{(A.66)}$$

Each term in the sum results from the product of one of the three elements of the first row, $(-1)^{m+n}$, where m and n are the indices of the row and column designating the element, respectively, and the 2×2 determinant obtained by omitting the entire row and column to which the element used in the reduction belongs. The product $(-1)^{m+n}$ and the 2×2

determinant are called the cofactor of the element used in the reduction. For example, the value of the following 3×3 determinant is found using the cofactors of the second row:

$$\begin{vmatrix} 1 & 3 & 4 \\ 2 & -1 & 6 \\ -1 & 7 & 5 \end{vmatrix} = (-1)^{2+1}2\begin{vmatrix} 3 & 4 \\ 7 & 5 \end{vmatrix} + (-1)^{2+2}(-1)\begin{vmatrix} 1 & 4 \\ -1 & 5 \end{vmatrix} + (-1)^{2+3}6\begin{vmatrix} 1 & 3 \\ -1 & 7 \end{vmatrix}$$

$$= -1 \times 2 \times (-13) + 1 \times (-1) \times 9 + (-1) \times 6 \times 10 = -43$$

If the initial determinant is of a higher order than 3, multiple sequential reductions as outlined earlier will reduce it in order by one in each step until a sum of 2×2 determinants is obtained.

The main usefulness for determinants is in solving a system of linear equations. Such a system of equations is obtained in evaluating the energies of a set of molecular orbitals obtained by combining a set of atomic orbitals. Before illustrating this method, we list some important properties of determinants that we will need in solving a set of simultaneous equations.

Property I The value of a determinant is not altered if each row in turn is made into a column or vice versa as long as the original order is kept. By this we mean that the nth row becomes the nth column. This property can be illustrated using 2×2 and 3×3 determinants:

$$\begin{vmatrix} 2 & 1 \\ 3 & -1 \end{vmatrix} = \begin{vmatrix} 2 & 3 \\ 1 & -1 \end{vmatrix} = -5 \quad \text{and} \quad \begin{vmatrix} 1 & 3 & 4 \\ 2 & -1 & 6 \\ -1 & 7 & 5 \end{vmatrix} = \begin{vmatrix} 1 & 2 & -1 \\ 3 & -1 & 7 \\ 4 & 6 & 5 \end{vmatrix} = -43$$

Property II If any two rows or columns are interchanged, the sign of the value of the determinant is changed. For example,

$$\begin{vmatrix} 2 & 1 \\ 3 & -1 \end{vmatrix} = -5, \text{ but } \begin{vmatrix} 1 & 2 \\ -1 & 3 \end{vmatrix} = +5 \quad \text{and}$$

$$\begin{vmatrix} 1 & 3 & 4 \\ 2 & -1 & 6 \\ -1 & 7 & 5 \end{vmatrix} = -43, \text{ but } \begin{vmatrix} 2 & -1 & 6 \\ 1 & 3 & 4 \\ -1 & 7 & 5 \end{vmatrix} = +43$$

Property III If two rows or columns of a determinant are identical, the value of the determinant is zero. For example,

$$\begin{vmatrix} 2 & 1 \\ 2 & 1 \end{vmatrix} = 2 - 2 = 0 \text{ and}$$

$$\begin{vmatrix} 1 & 1 & 4 \\ 2 & 2 & 6 \\ -1 & -1 & 5 \end{vmatrix} = (-1)^{2+1}2\begin{vmatrix} 1 & 4 \\ -1 & 5 \end{vmatrix} + (-1)^{2+2}2\begin{vmatrix} 1 & 4 \\ -1 & 5 \end{vmatrix}$$

$$+ (-1)^{2+3}6\begin{vmatrix} 1 & 1 \\ -1 & -1 \end{vmatrix}$$

$$= -1 \times 2 \times 9 + 1 \times 2 \times 9 + (-1) \times 6 \times 0 = 0$$

Property IV If each element of a row or column is multiplied by a constant, the value of the determinant is multiplied by that constant. For example,

$$\begin{vmatrix} 2 & 1 \\ 3 & -1 \end{vmatrix} = -5 \text{ and } \begin{vmatrix} 8 & 4 \\ 3 & -1 \end{vmatrix} = -20 \quad \text{and}$$

$$\begin{vmatrix} 1 & 2 & -1 \\ 3 & -1 & 7 \\ 4 & 6 & 5 \end{vmatrix} = -43 \text{ and } \begin{vmatrix} 1 & 3\sqrt{2} & 4 \\ 2 & -\sqrt{2} & 6 \\ -1 & 7\sqrt{2} & 5 \end{vmatrix} = -43\sqrt{2}$$

Property V The value of a determinant is unchanged if a row or column multiplied by an arbitrary number is added to another row or column. For example,

$$\begin{vmatrix} 2 & 1 \\ 3 & -1 \end{vmatrix} = \begin{vmatrix} 2+1 & 1 \\ 3-1 & -1 \end{vmatrix} = \begin{vmatrix} 2 & 1 \\ 3 & -1 \end{vmatrix} = -5 \quad \text{and}$$

$$\begin{vmatrix} 1 & 3 & 4 \\ 2 & -1 & 6 \\ -1 & 7 & 5 \end{vmatrix} = \begin{vmatrix} 1 & 3 & 4 \\ 2-1 & -1+7 & 6+5 \\ -1 & 7 & 5 \end{vmatrix} = \begin{vmatrix} 1 & 3 & 4 \\ 2 & -1 & 6 \\ -1 & 7 & 5 \end{vmatrix} = -43$$

How are determinants useful? This question can be answered by illustrating how determinants can be used to solve a set of linear equations:

$$x + y + z = 10$$

$$3x + 4y - z = 12$$

$$-x + 2y + 5z = 26 \tag{A.67}$$

This set of equations is solved by first constructing the 3×3 determinant that is the array of the coefficients of x, y, and z:

$$\mathbf{D}_{coefficients} = \begin{vmatrix} 1 & 1 & 1 \\ 3 & 4 & -1 \\ -1 & 2 & 5 \end{vmatrix} \tag{A.68}$$

Now imagine that we multiply the first column by x. This changes the value of the determinant as stated in Property IV:

$$\begin{vmatrix} 1x & 1 & 1 \\ 3x & 4 & -1 \\ -1x & 2 & 5 \end{vmatrix} = x\mathbf{D}_{coefficients} \tag{A.69}$$

We next add to the first column of $x\mathbf{D}_{coefficients}$ the second column of $\mathbf{D}_{coefficients}$ multiplied by y, and the third column multiplied by z. According to Properties IV and V, the value of the determinant is unchanged. Therefore,

$$\mathbf{D}_{c1} = \begin{vmatrix} 1 & 1 & 1 \\ 3 & 4 & -1 \\ -1 & 2 & 5 \end{vmatrix} = \begin{vmatrix} x+y+z & 1 & 1 \\ 3x+4y-z & 4 & -1 \\ -x+2y+5z & 2 & 5 \end{vmatrix} = \begin{vmatrix} 10 & 1 & 1 \\ 12 & 4 & -1 \\ 26 & 2 & 5 \end{vmatrix} = x\mathbf{D}_{coefficients} \tag{A.70}$$

To obtain the third determinant in the previous equation, the individual equations in Equation (A.67) are used to substitute the constants for the algebraic expression in the preceding determinants. From the previous equation, we conclude that

$$x = \frac{\mathbf{D}_{c1}}{\mathbf{D}_{coefficients}} = \frac{\begin{vmatrix} 10 & 1 & 1 \\ 12 & 4 & -1 \\ 26 & 2 & 5 \end{vmatrix}}{\begin{vmatrix} 1 & 1 & 1 \\ 3 & 4 & -1 \\ -1 & 2 & 5 \end{vmatrix}} = 3$$

To determine y and z, the exact same procedure can be followed, but we substitute instead in columns 2 and 3, respectively. The first step in each case is to multiply all elements of the second (third) row by $y(z)$. If we do so, we obtain the determinants $\mathbf{D}_{c2}$ and $\mathbf{D}_{c3}$:

$$\mathbf{D}_{c2} = \begin{vmatrix} 1 & 10 & 1 \\ 3 & 12 & -1 \\ -1 & 26 & 5 \end{vmatrix} \quad \text{and} \quad \mathbf{D}_{c3} = \begin{vmatrix} 1 & 1 & 10 \\ 3 & 4 & 12 \\ -1 & 2 & 26 \end{vmatrix}$$

and we conclude that

$$y = \frac{\mathbf{D}_{c2}}{\mathbf{D}_{coefficients}} = \frac{\begin{vmatrix} 1 & 10 & 1 \\ 3 & 12 & -1 \\ -1 & 26 & 5 \end{vmatrix}}{\begin{vmatrix} 1 & 1 & 1 \\ 3 & 4 & -1 \\ -1 & 2 & 5 \end{vmatrix}} = 2 \quad \text{and}$$

$$z = \frac{\mathbf{D}_{c3}}{\mathbf{D}_{coefficients}} = \frac{\begin{vmatrix} 1 & 1 & 10 \\ 3 & 4 & 12 \\ -1 & 2 & 26 \end{vmatrix}}{\begin{vmatrix} 1 & 1 & 1 \\ 3 & 4 & -1 \\ -1 & 2 & 5 \end{vmatrix}} = 5$$

This method of solving a set of simultaneous linear equations is known as Cramer's method.

If the constants in the set of equations are all zero, as in Equations A.71a and A.71b,

$$x + y + z = 0$$
$$3x + 4y - z = 0 \qquad \text{(A.71a)}$$
$$-x + 2y + 5z = 0$$

$$3x - y + 2z = 0$$
$$-x + y - z = 0$$
$$(1 + \sqrt{2})x + (1 - \sqrt{2})y + \sqrt{2}z = 0 \qquad \text{(A.71b)}$$

the determinants $\mathbf{D}_{c1}$, $\mathbf{D}_{c2}$, and $\mathbf{D}_{c3}$ all have the value zero. An obvious set of solutions is $x = 0$, $y = 0$ and $z = 0$. For most problems in physics and chemistry, this set of solutions is not physically meaningful and is referred to as the set of trivial solutions. A set of nontrivial solutions only exists if the equation $\mathbf{D}_{coefficients} = 0$ is satisfied. There is no nontrivial solution to the set of Equation (A.71a) because $\mathbf{D}_{coefficients} \neq 0$. There is a set of nontrivial solutions to the set of Equations (A.71b), because $\mathbf{D}_{coefficients} = 0$ in this case.

Determinants offer a convenient way to calculate the cross product of two vectors, as discussed in Section A.5. The following recipe is used:

$$\mathbf{a} \times \mathbf{b} = \begin{vmatrix} \mathbf{i} & \mathbf{j} & \mathbf{k} \\ a_x & a_y & a_z \\ b_x & b_y & b_z \end{vmatrix} = \mathbf{i}\begin{vmatrix} a_y & a_z \\ b_y & b_z \end{vmatrix} - \mathbf{j}\begin{vmatrix} a_x & a_z \\ b_x & b_z \end{vmatrix} + \mathbf{k}\begin{vmatrix} a_x & a_y \\ b_x & b_y \end{vmatrix}$$

$$= (a_y b_z - a_z b_y)\mathbf{i} + (a_z b_x - a_x b_z)\mathbf{j} + (a_x b_y - a_y b_x)\mathbf{k} \quad \text{(A.72)}$$

Note that by referring to Property II, you can show that $\mathbf{b} \times \mathbf{a} = -\mathbf{a} \times \mathbf{b}$.

$A.8$ WORKING WITH MATRICES

Physical chemists find widespread use for matrices. Matrices can be used to represent symmetry operations in the application of group theory to problems concerning molecular symmetry. They can also be used to obtain the energies of molecular orbitals formed through the linear combination of atomic orbitals. We next illustrate the use of matrices for representing the rotation operation that is frequently encountered in molecular symmetry considerations.

Consider the rotation of a three-dimensional vector about the z axis. Because the z component of the vector is unaffected by this operation, we need only consider the effect of the rotation operation on the two-dimensional vector formed by the projection of the three-dimensional vector on the x–y plane. The transformation can be represented by $(x_1, y_1, z_1) \rightarrow (x_2, y_2, z_1)$. The effect of the operation on the x and y components of the vector is shown in Figure A.14.

Next, relationships are derived among (x_1, y_1, z_1), (x_2, y_2, z_1), the magnitude of the radius vector r, and the angles α and β, based on the preceding figure. The magnitude of the radius vector r is

$$r = \sqrt{x_1^2 + y_1^2 + z_1^2} = \sqrt{x_2^2 + y_2^2 + z_1^2} \qquad \text{(A.73)}$$

Although the values of x and y change in the rotation, r is unaffected by this operation. The relationships between x, y, r, α, and β are given by

$$\theta = 180° - \alpha - \beta$$
$$x_1 = r \cos \alpha, \qquad y_1 = r \sin \alpha$$
$$x_2 = -r \cos \beta, \qquad y_1 = r \sin \beta \qquad \text{(A.74)}$$

In the following discussion, these identities are used:

$$\cos(\alpha \pm \beta) = \cos \alpha \cos \beta \pm \sin \alpha \sin \beta$$
$$\sin(\alpha \pm \beta) = \sin \alpha \cos \beta \pm \cos \alpha \sin \beta \qquad \text{(A.75)}$$

From Figure A.14, the following relationship between x_2 and x_1 and y_1 can be derived using the identities of Equation (A.75):

$$\begin{aligned}
x_2 &= -r \cos \beta = -r \cos(180° - \alpha - \theta) \\
&= r \sin 180° \sin(-\theta - \alpha) - r \cos 180° \cos(-\theta - \alpha) \\
&= r \cos(-\theta - \alpha) = r \cos(\theta + \alpha) = r \cos \theta \cos \alpha - r \sin \theta \sin \alpha \\
&= x_1 \cos \theta - y_1 \sin \theta \qquad \text{(A.76)}
\end{aligned}$$

Using the same procedure, the following relationship between y_2 and x_1 and y_1 can be derived:

$$y_2 = x_1 \sin \theta + y_1 \cos \theta \qquad \text{(A.77)}$$

Next, these results are combined to write the following equations relating x_2, y_2, and z_2 to x_1, y_1, and z_1:

$$\begin{aligned}
x_2 &= x_1 \cos \theta - y_1 \sin \theta \\
y_2 &= x_1 \sin \theta + y_1 \cos \theta \\
z_2 &= 0x_1 + 0y_1 + z_1 \qquad \text{(A.78)}
\end{aligned}$$

At this point, the concept of a matrix can be introduced. An $n \times m$ matrix is an array of numbers, functions, or operators that can undergo mathematical operations such as addition and multiplication with one another. The operation of interest to us in considering rotation about the z axis is matrix multiplication. We illustrate how matrices, which are designated in bold script, such as **A**, are multiplied using 2×2 matrices as an example.

$$\mathbf{AB} = \begin{pmatrix} a_{11} & a_{12} \\ a_{21} & a_{22} \end{pmatrix} \begin{pmatrix} b_{11} & b_{12} \\ b_{21} & b_{22} \end{pmatrix} = \begin{pmatrix} a_{11}b_{11} + a_{12}b_{21} & a_{11}b_{12} + a_{12}b_{22} \\ a_{21}b_{11} + a_{22}b_{21} & a_{21}b_{12} + a_{22}b_{22} \end{pmatrix} \quad \text{(A.79)}$$

Using numerical examples,

$$\begin{pmatrix} 2 & 1 \\ -3 & 4 \end{pmatrix} \begin{pmatrix} 1 & 6 \\ 2 & -1 \end{pmatrix} = \begin{pmatrix} 4 & 11 \\ 5 & -22 \end{pmatrix} \quad \text{and} \quad \begin{pmatrix} 1 & 6 \\ 2 & -1 \end{pmatrix} \begin{pmatrix} 1 \\ -1 \end{pmatrix} = \begin{pmatrix} -5 \\ 3 \end{pmatrix}$$

Now consider the initial and final coordinates (x_1, y_1, z_1) and (x_2, y_2, z_1) as 3×1 matrices (x_1, y_1, z_1) and (x_2, y_2, z_1). In that case, the set of simultaneous equations of Equation (A.78) can be written as

$$\begin{pmatrix} x_2 \\ y_2 \\ z_2 \end{pmatrix} = \begin{pmatrix} \cos \theta & -\sin \theta & 0 \\ \sin \theta & \cos \theta & 0 \\ 0 & 0 & 1 \end{pmatrix} \begin{pmatrix} x_1 \\ y_1 \\ z_1 \end{pmatrix} \qquad \text{(A.80)}$$

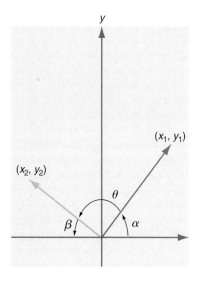

FIGURE A.14

We see that we can represent the operator for rotation about the z axis, R_z, as the following 3×3 matrix:

$$\mathbf{R_z} = \begin{pmatrix} \cos\theta & -\sin\theta & 0 \\ \sin\theta & \cos\theta & 0 \\ 0 & 0 & 1 \end{pmatrix} \tag{A.81}$$

The rotation operator for 180° and 120° rotation can be obtained by evaluating the sine and cosine functions at the appropriate values of θ. These rotation operators have the form

$$\begin{pmatrix} -1 & 0 & 0 \\ 0 & -1 & 0 \\ 0 & 0 & 1 \end{pmatrix} \quad \text{and} \quad \begin{pmatrix} 1/2 & -\sqrt{3}/2 & 0 \\ \sqrt{3}/2 & 1/2 & 0 \\ 0 & 0 & 1 \end{pmatrix}, \quad \text{respectively} \tag{A.82}$$

One special matrix, the identity matrix designated $\mathbf{I}$, deserves additional mention. The identity matrix corresponds to an operation in which nothing is changed. The matrix that corresponds to the transformation $(x_1, y_1, z_1) \rightarrow (x_2, y_2, z_2)$ expressed in equation form as

$$x_2 = x_1 + 0y_1 + 0z_1$$
$$y_2 = 0x_1 + y_1 + 0z_1$$
$$z_2 = 0x_1 + 0y_1 + z_1 \tag{A.83}$$

is the identity matrix

$$\mathbf{I} = \begin{pmatrix} 1 & 0 & 0 \\ 0 & 1 & 0 \\ 0 & 0 & 1 \end{pmatrix}$$

The identity matrix is an example of a diagonal matrix. It has this name because only the diagonal elements are nonzero. In the identity matrix of order $n \times n$ all diagonal elements have the value one.

The operation that results from the sequential operation of two individual operations represented by matrices $\mathbf{A}$ and $\mathbf{B}$ is the products of the matrices: $\mathbf{C} = \mathbf{AB}$. An interesting case illustrating this relationship is counterclockwise rotation through an angle θ followed by clockwise rotation through the same angle, which corresponds to rotation by $-\theta$. Because $\cos(-\theta) = \cos\theta$ and $\sin\theta = -\sin\theta$, the rotation matrix for $-\theta$ must be

$$\mathbf{R_{-z}} = \begin{pmatrix} \cos\theta & \sin\theta & 0 \\ -\sin\theta & \cos\theta & 0 \\ 0 & 0 & 1 \end{pmatrix} \tag{A.84}$$

Because the sequential operations leave the vector unchanged, it must be the case that $\mathbf{R_z R_{-z}} = \mathbf{R_{-z} R_z} = \mathbf{I}$. We verify here that the first of these relations is obeyed:

$$\mathbf{R_z R_{-z}} = \begin{pmatrix} \cos\theta & -\sin\theta & 0 \\ \sin\theta & \cos\theta & 0 \\ 0 & 0 & 1 \end{pmatrix}\begin{pmatrix} \cos\theta & \sin\theta & 0 \\ -\sin\theta & \cos\theta & 0 \\ 0 & 0 & 1 \end{pmatrix}$$

$$= \begin{pmatrix} \cos^2\theta + \sin^2\theta + 0 & \sin\theta\cos\theta - \sin\theta\cos\theta + 0 & 0 \\ \sin\theta\cos\theta - \sin\theta\cos\theta + 0 & \cos^2\theta + \sin^2\theta + 0 & 0 \\ 0 & 0 & 1 \end{pmatrix}$$

$$= \begin{pmatrix} 1 & 0 & 0 \\ 0 & 1 & 0 \\ 0 & 0 & 1 \end{pmatrix} \tag{A.85}$$

Any matrix $\mathbf{B}$ that satisfies the relationship $\mathbf{AB} = \mathbf{BA} = \mathbf{I}$ is called the inverse matrix of $\mathbf{A}$ and is designated $\mathbf{A}^{-1}$. Inverse matrices play an important role in finding the energies of a set of molecular orbitals that is a linear combination of atomic orbitals.

Appendix B   Point Group Character Tables

B.1 THE NONAXIAL GROUPS

C_1	E
A	1

C_s	E	σ_h		
A'	1	1	x, y, R_z	x^2, y^2, z^2, xy
A''	1	-1	z, R_x, R_y	yz, xz

C_i	E	i		
A_g	1	1	R_x, R_y, R_z	$x^2, y^2, z^2, xy, xz, yz$
A_u	1	-1	x, y, z	

B.2 THE C_n GROUPS

C_2	E	C_2		
A	1	1	z, R_z	x^2, y^2, z^2, xy
B	1	-1	x, y, R_x, R_y	yz, xz

C_4	E	C_4	C_2	C_4^3		
A	1	1	1	1	z, R_z	$x^2 + y^2, z^2$
B	1	-1	1	-1		$x^2 - y^2, xy$
E	$\begin{Bmatrix} 1 & i & -1 & -i \\ 1 & -i & -1 & i \end{Bmatrix}$				$(x, y), (R_x, R_y)$	(yz, xz)

461

B.3 THE D_n GROUPS

D_2	E	$C_2(z)$	$C_2(y)$	$C_2(x)$		
A	1	1	1	1		x^2, y^2, z^2
B_1	1	1	-1	-1	z, R_z	xy
B_2	1	-1	1	-1	y, R_y	xz
B_3	1	-1	-1	1	x, R_x	yz

D_3	E	$2C_3$	$3C_2$		
A_1	1	1	1		$x^2 + y^2, z^2$
A_2	1	1	-1	z, R_z	
E	2	-1	0	$(x, y), (R_x, R_y)$	$(x^2 - y^2, xy), (xz, yz)$

D_4	E	$2C_4$	$C_2(= C_4^2)$	$2C_2'$	$2C_2''$		
A_1	1	1	1	1	1		$x^2 + y^2, z^2$
A_2	1	1	1	-1	-1	z, R_z	
B_1	1	-1	1	1	-1		$x^2 - y^2$
B_2	1	-1	1	-1	1		xy
E	2	0	-2	0	0	$(x, y), (R_x, R_y)$	(xz, yz)

D_5	E	$2C_5$	$2C_5^2$	$5C_2$		
A_1	1	1	1	1		$x^2 + y^2, z^2$
A_2	1	1	1	-1	z, R_z	
E_1	2	$2\cos 72°$	$2\cos 144°$	0	$(x, y), (R_x, R_y)$	(xz, yz)
E_2	2	$2\cos 144°$	$2\cos 72°$	0		$(x^2 - y^2, xy)$

D_6	E	$2C_6$	$2C_3$	C_2	$3C_2'$	$3C_2''$		
A_1	1	1	1	1	1	1		$x^2 + y^2, z^2$
A_2	1	1	1	1	-1	-1	z, R_z	
B_1	1	-1	1	-1	1	-1		
B_2	1	-1	1	-1	-1	1		
E_1	2	1	-1	-2	0	0	$(x, y), (R_x, R_y)$	(xz, yz)
E_2	2	-1	-1	2	0	0		$(x^2 - y^2, xy)$

B.4 THE C_{nv} GROUPS

C_{2v}	E	C_2	$\sigma_v(xz)$	$\sigma_v'(yz)$		
A_1	1	1	1	1	z	x^2, y^2, z^2
A_2	1	1	−1	−1	R_z	xy
B_1	1	−1	1	−1	x, R_y	xz
B_2	1	−1	−1	1	y, R_x	yz

C_{3v}	E	$2C_3$	$3\sigma_v$		
A_1	1	1	1	z	$x^2 + y^2, z^2$
A_2	1	1	−1	R_z	
E	2	−1	0	$(x, y), (R_x, R_y)$	$(x^2 - y^2, xy), (xz, yz)$

C_{4v}	E	$2C_4$	C_2	$2\sigma_v$	$2\sigma_d$		
A_1	1	1	1	1	1	z	$x^2 + y^2, z^2$
A_2	1	1	1	−1	−1	R_z	
B_1	1	−1	1	1	−1		$x^2 - y^2$
B_2	1	−1	1	−1	1		xy
E	2	0	−2	0	0	$(x, y), (R_x, R_y)$	(xz, yz)

C_{5v}	E	$2C_5$	$2C_5^2$	$5\sigma_v$		
A_1	1	1	1	1	z	$x^2 + y^2, z^2$
A_2	1	1	1	−1	R_z	
E_1	2	$2\cos 72°$	$2\cos 144°$	0	$(x, y), (R_x, R_y)$	(xz, yz)
E_2	2	$2\cos 144°$	$2\cos 72°$	0		$(x^2 - y^2, xy)$

C_{6v}	E	$2C_6$	$2C_3$	C_2	$3\sigma_v$	$3\sigma_d$		
A_1	1	1	1	1	1	1	z	$x^2 + y^2, z^2$
A_2	1	1	1	1	−1	−1	R_z	
B_1	1	−1	1	−1	1	−1		
B_2	1	−1	1	−1	−1	1		
E_1	2	1	−1	−2	0	0	$(x, y), (R_x, R_y)$	(xz, yz)
E_2	2	−1	−1	2	0	0		$(x^2 - y^2, xy)$

B.5 THE C_{nh} GROUPS

C_{2h}	E	C_2	i	σ_h		
A_g	1	1	1	1	R_z	x^2, y^2, z^2, xy
B_g	1	−1	1	−1	R_x, R_y	xz, yz
A_u	1	1	−1	−1	z	
B_u	1	−1	−1	1	x, y	

C_{4h}	E	C_4	C_2	C_4^3	i	S_4^3	σ_h	S_4		
A_g	1	1	1	1	1	1	1	1	R_z	$x^2 + y^2, z^2$
B_g	1	−1	1	−1	1	−1	1	−1		$x^2 − y^2, xy$
E_g	$\begin{cases} 1 \\ 1 \end{cases}$	$\begin{matrix} i \\ -i \end{matrix}$	$\begin{matrix} -1 \\ -1 \end{matrix}$	$\begin{matrix} -i \\ i \end{matrix}$	$\begin{matrix} 1 \\ 1 \end{matrix}$	$\begin{matrix} i \\ -i \end{matrix}$	$\begin{matrix} -1 \\ -1 \end{matrix}$	$\begin{matrix} -i \\ i \end{matrix}$	(R_x, R_y)	(xz, yz)
A_u	1	1	1	1	−1	−1	−1	−1	z	
B_u	1	−1	1	−1	−1	1	−1	1		
E_u	$\begin{cases} 1 \\ 1 \end{cases}$	$\begin{matrix} i \\ -i \end{matrix}$	$\begin{matrix} -1 \\ -1 \end{matrix}$	$\begin{matrix} -i \\ i \end{matrix}$	$\begin{matrix} -1 \\ -1 \end{matrix}$	$\begin{matrix} -i \\ i \end{matrix}$	$\begin{matrix} 1 \\ 1 \end{matrix}$	$\begin{matrix} -i \\ i \end{matrix}$	(x, y)	

B.6 THE D_{nh} GROUPS

D_{2h}	E	$C_2(z)$	$C_2(y)$	$C_2(x)$	i	$\sigma(xy)$	$\sigma(xz)$	$\sigma(yz)$		
A_g	1	1	1	1	1	1	1	1		x^2, y^2, z^2
B_{1g}	1	1	−1	−1	1	1	−1	−1	R_z	xy
B_{2g}	1	−1	1	−1	1	−1	1	−1	R_y	xz
B_{3g}	1	−1	−1	1	1	−1	−1	1	R_x	yz
A_u	1	1	1	1	−1	−1	−1	−1		
B_{1u}	1	1	−1	−1	−1	−1	1	1	z	
B_{2u}	1	−1	1	−1	−1	1	−1	1	y	
B_{3u}	1	−1	−1	1	−1	1	1	−1	x	

D_{3h}	E	$2C_3$	$3C_2$	σ_h	$2S_3$	$3\sigma_v$		
A_1'	1	1	1	1	1	1		x^2+y^2, z^2
A_2'	1	1	-1	1	1	-1	R_z	
E'	2	-1	0	2	-1	0	(x, y)	(x^2-y^2, xy)
A_1''	1	1	1	-1	-1	-1		
A_2''	1	1	-1	-1	-1	1	z	
E''	2	-1	0	-2	1	0	(R_x, R_y)	(xz, yz)

D_{4h}	E	$2C_4$	C_2	$2C_2'$	$2C_2''$	i	$2S_4$	σ_h	$2\sigma_v$	$2\sigma_d$		
A_{1g}	1	1	1	1	1	1	1	1	1	1		x^2+y^2, z^2
A_{2g}	1	1	1	-1	-1	1	1	1	-1	-1	R_z	
B_{1g}	1	-1	1	1	-1	1	-1	1	1	-1		x^2-y^2
B_{2g}	1	-1	1	-1	1	1	-1	1	-1	1		xy
E_g	2	0	-2	0	0	2	0	-2	0	0	(R_x, R_y)	(xz, yz)
A_{1u}	1	1	1	1	1	-1	-1	-1	-1	-1		
A_{2u}	1	1	1	-1	-1	-1	-1	-1	1	1	z	
B_{1u}	1	-1	1	1	-1	-1	1	-1	-1	1		
B_{2u}	1	-1	1	-1	1	-1	1	-1	1	-1		
E_u	2	0	-2	0	0	-2	0	2	0	0	(x, y)	

D_{6h}	E	$2C_6$	$2C_3$	C_2	$3C_2'$	$3C_2''$	i	$2S_3$	$2S_6$	σ_h	$3\sigma_d$	$3\sigma_v$		
A_{1g}	1	1	1	1	1	1	1	1	1	1	1	1		x^2+y^2, z^2
A_{2g}	1	1	1	1	-1	-1	1	1	1	1	-1	-1	R_z	
B_{1g}	1	-1	1	-1	1	-1	1	-1	1	-1	1	-1		
B_{2g}	1	-1	1	-1	-1	1	1	-1	1	-1	-1	1		
E_{1g}	2	1	-1	-2	0	0	2	1	-1	-2	0	0	(R_x, R_y)	(xz, yz)
E_{2g}	2	-1	-1	2	0	0	2	-1	-1	2	0	0		(x^2-y^2, xy)
A_{1u}	1	1	1	1	1	1	-1	-1	-1	-1	-1	-1		
A_{2u}	1	1	1	1	-1	-1	-1	-1	-1	-1	1	1	z	
B_{1u}	1	-1	1	-1	1	-1	-1	1	-1	1	-1	1		
B_{2u}	1	-1	1	-1	-1	1	-1	1	-1	1	1	-1		
E_{1u}	2	1	-1	-2	0	0	-2	-1	1	2	0	0	(x, y)	
E_{2u}	2	-1	-1	2	0	0	-2	1	1	-2	0	0		

D_{8h}	E	$2C_8^3$	$2C_8$	$2C_4$	C_2	$4C_2'$	$4C_2''$	i	$2S_8^3$	$2S_8$	$2S_4$	σ_h	$4\sigma_d$	$4\sigma_v$		
A_{1g}	1	1	1	1	1	1	1	1	1	1	1	1	1	1		x^2+y^2, z^2
A_{2g}	1	1	1	1	1	−1	−1	1	1	1	1	−1	−1		R_z	
B_{1g}	1	−1	−1	1	1	1	−1	1	−1	−1	1	1	1	−1		
B_{2g}	1	−1	−1	1	1	−1	1	1	−1	−1	1	1	−1	1		
E_{1g}	2	$\sqrt{2}$	$-\sqrt{2}$	0	−2	0	0	2	$\sqrt{2}$	$-\sqrt{2}$	0	−2	0	0	(R_x, R_y)	(xz, yz)
E_{2g}	2	0	0	−2	2	0	0	2	0	0	−2	2	0	0		(x^2-y^2, xy)
E_{3g}	2	$-\sqrt{2}$	$\sqrt{2}$	0	−2	0	0	2	$-\sqrt{2}$	$\sqrt{2}$	0	−2	0	0		
A_{1u}	1	1	1	1	1	1	1	−1	−1	−1	−1	−1	−1	−1		
A_{2u}	1	1	1	1	1	−1	−1	−1	−1	−1	−1	1	1	1	z	
B_{1u}	1	−1	−1	1	1	1	−1	−1	1	1	−1	−1	−1	1		
B_{2u}	1	−1	−1	1	1	−1	1	−1	1	1	−1	−1	1	−1		
E_{1u}	2	$\sqrt{2}$	$-\sqrt{2}$	0	−2	0	0	−2	$-\sqrt{2}$	$\sqrt{2}$	0	2	0	0	(x, y)	
E_{2u}	2	0	0	−2	2	0	0	−2	0	0	2	−2	0	0		
E_{3u}	2	$-\sqrt{2}$	$\sqrt{2}$	0	−2	0	0	−2	$\sqrt{2}$	$-\sqrt{2}$	0	2	0	0		

B.7 THE D_{nd} GROUPS

D_{2d}	E	$2S_4$	C_2	$2C_2'$	$2\sigma_d$		
A_1	1	1	1	1	1		x^2+y^2, z^2
A_2	1	1	1	−1	−1	R_z	
B_1	1	−1	1	1	−1		x^2-y^2
B_2	1	−1	1	−1	1	z	xy
E	2	0	−2	0	0	$(x, y), (R_x, R_y)$	(xz, yz)

D_{3d}	E	$2C_3$	$3C_2$	i	$2S_6$	$3\sigma_d$		
A_{1g}	1	1	1	1	1	1		x^2+y^2, z^2
A_{2g}	1	1	−1	1	1	−1	R_z	
E_g	2	−1	0	2	−1	0	(R_x, R_y)	$(x^2-y^2, xy), (xz, yz)$
A_{1u}	1	1	1	−1	−1	−1		
A_{2u}	1	1	−1	−1	−1	1	z	
E_u	2	−1	0	−2	1	0	(x, y)	

D_{4d}	E	$2S_8$	$2C_4$	$2S_8^3$	C_2	$4C_2'$	$4\sigma_d$		
A_1	1	1	1	1	1	1	1		x^2+y^2, z^2
A_2	1	1	1	1	1	-1	-1	R_z	
B_1	1	-1	1	-1	1	1	-1		
B_2	1	-1	1	-1	1	-1	1	z	
E_1	2	$\sqrt{2}$	0	$-\sqrt{2}$	-2	0	0	(x,y)	
E_2	2	0	-2	0	2	0	0		(x^2-y^2, xy)
E_3	2	$-\sqrt{2}$	0	$\sqrt{2}$	-2	0	0	(R_x, R_y)	(xz, yz)

D_{6d}	E	$2S_{12}$	$2C_6$	$2S_4$	$2C_3$	$2S_{12}^5$	C_2	$6C_2'$	$6\sigma_d$		
A_1	1	1	1	1	1	1	1	1	1		x^2+y^2, z^2
A_2	1	1	1	1	1	1	1	-1	-1	R_z	
B_1	1	-1	1	-1	1	-1	1	1	-1		
B_2	1	-1	1	-1	1	-1	1	-1	1	z	
E_1	2	$\sqrt{3}$	1	0	-1	$-\sqrt{3}$	-2	0	0	(x,y)	
E_2	2	1	-1	-2	-1	1	2	0	0		(x^2-y^2, xy)
E_3	2	0	-2	0	2	0	-2	0	0		
E_4	2	-1	-1	2	-1	-1	2	0	0		
E_5	2	$-\sqrt{3}$	1	0	-1	$\sqrt{3}$	-2	0	0	(R_x, R_y)	(xz, yz)

B.8 THE CUBIC GROUPS

T_d	E	$8C_3$	$3C_2$	$6S_4$	$6\sigma_d$		
A_1	1	1	1	1	1		$x^2+y^2+z^2$
A_2	1	1	1	-1	-1		
E	2	-1	2	0	0		$(2z^2-x^2-y^2, x^2-y^2)$
T_1	3	0	-1	1	-1	(R_x, R_y, R_z)	
T_2	3	0	-1	-1	1	(x,y,z)	(xy, xz, yz)

O	E	$8C_3$	$3C_2(=C_4^2)$	$6C_4$	$6C_2$		
A_1	1	1	1	1	1		$x^2 + y^2 + z^2$
A_2	1	1	1	-1	-1		
E	2	-1	2	0	0		$(2z^2 - x^2 - y^2, x^2 - y^2)$
T_1	3	0	-1	1	-1	$(R_x, R_y, R_z), (x, y, z)$	
T_2	3	0	-1	-1	1		(xy, xz, yz)

O_h	E	$8C_3$	$6C_2$	$6C_4$	$3C_2(=C_4^2)$	i	$6S_4$	$8S_6$	$3\sigma_h$	$6\sigma_d$		
A_{1g}	1	1	1	1	1	1	1	1	1	1		$x^2 + y^2 + z^2$
A_{2g}	1	1	-1	-1	1	1	-1	1	1	-1		
E_g	2	-1	0	0	2	2	0	-1	2	0		$(2z^2 - x^2 - y^2, x^2 - y^2)$
T_{1g}	3	0	-1	1	-1	3	1	0	-1	-1	(R_x, R_y, R_z)	
T_{2g}	3	0	1	-1	-1	3	-1	0	-1	1		(xz, yz, xy)
A_{1u}	1	1	1	1	1	-1	-1	-1	-1	-1		
A_{2u}	1	1	-1	-1	1	-1	1	-1	-1	1		
E_u	2	-1	0	0	2	-2	0	1	-2	0		
T_{1u}	3	0	-1	1	-1	-3	-1	0	1	1	(x, y, z)	
T_{2u}	3	0	1	-1	-1	-3	1	0	1	-1		

B.9 THE GROUPS $C_{\infty v}$ AND $D_{\infty h}$ FOR LINEAR MOLECULES

$C_{\infty v}$	E	$2C_\infty^\Phi$	$\cdots$	$\infty\sigma_v$		
$A_1(\Sigma^+)$	1	1	$\cdots$	1	z	$x^2 + y^2, z^2$
$A_2(\Sigma^-)$	1	1	$\cdots$	-1	R_z	
$E_1(\Pi)$	2	$2\cos\Phi$	$\cdots$	0	$(x, y), (R_x, R_y)$	(xz, yz)
$E_2(\Delta)$	2	$2\cos 2\Phi$	$\cdots$	0		$(x^2 - y^2, xy)$
$E_3(\Phi)$	2	$2\cos 3\Phi$	$\cdots$	0		
$\cdots$	$\cdots$	$\cdots$	$\cdots$	$\cdots$		

$D_{\infty h}$	E	$2C_\infty^\Phi$	...	$\infty\sigma_v$	i	$2S_\infty^\Phi$	...	∞C_2		
Σ_g^+	1	1	...	1	1	1	...	1		$x^2 + y^2, z^2$
Σ_g^-	1	1	...	-1	1	1	...	-1	R_z	
Π_g	2	$2\cos\Phi$	...	0	2	$-2\cos\Phi$	...	0	(R_x, R_y)	(xy, yz)
Δ_g	2	$2\cos 2\Phi$	...	0	2	$2\cos 2\Phi$	...	0		$(x^2 - y^2, xy)$
...	...	...	...	...	...	...	...	...		
Σ_u^+	1	1	...	1	-1	-1	...	-1	z	
Σ_u^-	1	1	...	-1	-1	-1	...	1		
Π_u	2	$2\cos\Phi$	...	0	-2	$2\cos\Phi$	...	0	(x, y)	
Δ_u	2	$2\cos 2\Phi$	...	0	-2	$-2\cos 2\Phi$	...	0		
...	...	...	...	...	...	...	...	...		

Appendix C Answers to Selected End-of-Chapter Problems

Numerical answers to problems are included here. Complete solutions to selected problems can be found in the *Student's Solutions Manual*.

Chapter 1

P1.1 $\Delta v_{H_2} = 0.879 \text{ m s}^{-1}, \dfrac{\Delta v}{v} = 2.89 \times 10^{-4}$

P1.2 $\dfrac{\overline{E}_{osc} - kT}{\overline{E}_{osc}} = -0.0306$ for 800. K. The corresponding values for 500. K and 250. K are -0.0495 and -0.102.

P1.3 $\widetilde{\nu} = 109677 \text{ cm}^{-1}, 27419.3 \text{ cm}^{-1}$, and 12186.3 cm^{-1} and $E_{max} = 2.17871 \times 10^{-18}$ J, 5.44676×10^{-19} J, and 2.42078×10^{-19} J for the Lyman, Balmer, and Paschen series.

P1.4 924 m s^{-1}, 4.13×10^3 m s^{-1}, 9.24×10^3 m s^{-1}, and 2.92×10^5 m s^{-1} for 10^4 nm, 500. nm, 100. nm and 0.1 nm; 959 K, 1.92×10^4 K, 9.59×10^4 K, and 9.59×10^7 K for 10^4 nm, 500. nm, 100. nm, and 0.1 nm.

P1.5 2.178×10^{-18} J

P1.6 1.07×10^{16} electrons, $E = 3.38 \times 10^{-19}$ J $v = 8.61 \times 10^5$ m s^{-1}

P1.7 at 1350. K, 2.51×10^{-3} J m^{-3}; at 5250 K, 0.575 J m^{-3}

P1.8 0.0791 m s^{-1}

P1.9 4.91 m s^{-1}

P1.10 $\dfrac{E - E_{approx}}{E} = -0.0162, -0.0496$, and -0.219 at 6000. K, 2000. K, and 500. K.

P1.11 52 K for H and 0.40 K for Xe

P1.12 $4.16 \times 10^6 \text{ m s}^{-1}$

P1.13 1.59×10^{-10} m for H at 250. K and 9.19×10^{-11} m for H at 750 K. For Ne, 3.56×10^{-11} m and 2.05×10^{-11} m at 250. K and 750. K, respectively.

P1.14 3.38×10^{-6} m, 2.21×10^{-6} m, and 5.23×10^{-7} m for 450. K, 1500. K, and 4500. K

P1.15 5.74 cm

P1.16 29.7 V

P1.17 6.565×10^{-5} m; 3.647×10^{-5} m

P1.18 $\lambda = 0.813$ nm, $n = 1.40 \times 10^{14} \text{ s}^{-1}$

P1.19 $h \approx 7.0 \times 10^{-34}$ J s; $\phi \approx 4.0 \times 10^{-19}$ J or 2.5 eV

P1.20 3.76×10^{26} W

P1.21 $\nu \geq 1.09 \times 10^{15} \text{ s}^{-1}$ v $= 5.96 \times 10^5 \text{ m s}^{-1}$

P1.22 1.84×10^{20}

P1.23 1.21569×10^{-5} m; 9.11768×10^{-6} m

P1.24 $8.59 \times 10^{18} \text{ s}^{-1}$

P1.25 $3.93 \times 10^4 \text{ J s}^{-1}, 0.0222$ m

P1.27 a. $1.0 \times 10^7 \text{ J s}^{-1}$;

   b. 5.0×10^{17}

Chapter 2

P2.2 $N = \sqrt{\dfrac{2}{d}}$

P2.3 a. no

   b. yes, 2

   c. yes, -1

P2.6 a. $r = \sqrt{29}$,
   $\theta = 0.980$ radians, $\phi = 0.464$ radians

   b. $x = -1.03, y = 2.47, z = 6.47$

P2.9 a. $\sqrt{58} \exp(-0.129i\pi)$;

   b. $5 \exp(-0.5i\pi)$

   c. $\dfrac{5}{\sqrt{17}} \exp(-0.217i\pi)$;

   d. $\dfrac{\sqrt{17}}{5} \exp(0.430i\pi)$

P2.12 a. no

   b. no

   c. yes, -1

   d. yes, $-a^2$

   e. no

P2.14 a. yes, -6

   b. yes, -1

   c. yes, -16

P2.16 $x = 0.791$ m; $t_0 = 4.47 \times 10^{-4}$ s

P2.17   a.   yes, -9

b.   yes, 1

c.   yes 1

P2.18 $\dfrac{n_3}{n_1}$ (250. K) = 1.74 $\dfrac{n_3}{n_1}$ (900. K) = 2.58

$\dfrac{n_{15}}{n_1}$ (250. K) = 3.82 × 10^{-6} $\dfrac{n_{15}}{n_1}$ (900. K) = 0.221

P2.20 $a = \pm 1$

P2.21   a.   no

b.   yes, -1

c.   yes, 1

P2.22   6, no

P2.25 $\pm\dfrac{1}{\sqrt{2}}, \pm\sqrt{\dfrac{3}{2}}x, \pm\dfrac{1}{2}\sqrt{\dfrac{5}{2}}, \mp\dfrac{3}{2}\sqrt{\dfrac{5}{2}}x^2$

P2.26   for $n_2/n_1 = 0.225\ T = 145$ K for
$n_2/n_1 = 0.875\ T = 1.63 \times 10^3$ K

P2.27 $b = 1$

P2.28 $N = \dfrac{1}{\sqrt{2\pi}}$

P2.29   a. $N = \dfrac{2}{\sqrt{ab}}$

b. $N = \sqrt{\dfrac{6}{\pi a^3}}$

P2.32 $d_0 = 0$ and $d_1 - d_5 = 0.$ $c_1 = \dfrac{2b}{\pi},$

$c_2 = -\dfrac{b}{\pi}, c_3 = \dfrac{2b}{3\pi}, c_4 = -\dfrac{b}{2\pi},$ and $c_5 = \dfrac{2b}{5\pi}$

P2.35   a. $-2i$

b. $-2\sqrt{5}i$

c. $-1,$

d. $\dfrac{3}{5 + \sqrt{3}}(1 + i)$

P2.36   a.   no

b.   no

c.   yes, $-i$

d.   no

e.   no

Chapter 4

P4.1   a. $\alpha = 6.86 \times 10^{10}$

b. 1.80×10^{-31} J

c. 4.44×10^{-11}

P4.5   c.   1/4, 1/16, and 11/16

d. $\langle E \rangle = 5.25\ E_1$

P4.9 $\sqrt{\dfrac{105}{a}}, \dfrac{5a}{8}, \dfrac{5a^2}{12}$

P4.12   a.   16.6

b.   24.5

c.   0.502

P4.15 3.29×10^{-6} m

P4.16 8.39×10^{35}

P4.20   a. 1.18×10^{-38} J

b. 2.85×10^{-18}

P4.22 4.92×10^{-21} J

P4.25   a.   2

b.   3

P4.29 8.8×10^{-10} m

P4.34   a.   0.028

b.   0.0026

Chapter 5

P5.1   e. $\left|\dfrac{F}{A}\right|^2 = 0.1$ for $E = 1.5 \times 10^{-19}$ J and
$0.02\ E = 1.1 \times 10^{-19}$ J

f.   0.2

P5.2 $T_{Si} = 555$ K; $T_C = 4.6 \times 10^3$ K

P5.3 $\dfrac{\Delta\rho_{total}(x)}{\langle\rho_{total}(x)\rangle} = 0.060; \dfrac{\Delta\rho_{n=11}(x)}{\langle\Delta\rho_{total}(x)\rangle} = 2$

P5.4 $\lambda = 239$ nm

P5.5 $\lambda = 368$ nm

P5.6 2.8×10^6 A/m^2 4.2×10^8 A/m^2

P5.7   b. 4.76×10^{-20} J, 1.86×10^{-19} J, and
3.95×10^{-19} J

Chapter 6

P6.5   c.   for 1.0×10^{-9} s
$\Delta\nu = 8.0 \times 10^7$ s^{-1}, 0.00265 cm^{-1} for
1.0×10^{-11} s, 8.0×10^9 s^{-1} and 0.265 cm^{-1}

P6.10 $p = 5.275 \times 10^{-24}$ kg m s^{-1} $\dfrac{\lambda}{b} = 0.126$

P6.12 $E = 1.6$ eV

P6.18   c. $z = \pm 1.45 \times 10^{-2}$ m

P6.23 $\Delta x = 1.6 \times 10^{-33}$ m

Chapter 7

P7.1 $0, 4.21 \times 10^{-22}$ J

P7.3 1, 2.73, 0.127, 1.40×10^{-7}

P7.4 0.420, 0.752, 0.991, 1.20, 1.39, and 1.57 radians as well as π minus these values

P7.5 $6.92 \times 10^{13}\,\text{s}^{-1}$, $3.91 \times 10^{11}\,\text{s}^{-1}$

P7.8 470. $\text{N}\,\text{m}^{-1}$, 5.21×10^{-2} m

P7.9 0, 3.04, 11.1, 42.5

P7.13 $E_0 = 4.02 \times 10^{-32}$ J, $E_0/kT = 9.76 \times 10^{-12}$ $4.01 \times 10^{-15}\,\text{m s}^{-1}$

P7.14 $5.58 \times 10^{13}\,\text{s}^{-1}$; 5.37×10^{-6} m

P7.16 965 kg s^{-2}; 0.984 kg

P7.17 for $n = 0$, 1, and 2: 0.0565, 0.0979, and 0.126

P7.18 a. 2.01 N m^{-1};

  b. 9.94×10^{-34} J

  c. 2.52×10^{-5} J

  d. 1.27×10^{28}

P7.19 a. 3.57 pm from Cl

  b. 48.35 pm from O

P7.22 3.04×10^{-20} J, $3.04 \times 10^{3}\,\text{m s}^{-1}$ $|v|/|v_{\text{rms}}| = 2.23$

P7.23 2.97×10^{-20} J, $8.97 \times 10^{13}\,\text{s}^{-1}$

P7.24 2.74, 1.02×10^{-9}

P7.25 0, 1.29×10^{-22} J

P7.33 $E_{rot} = 4.58 \times 10^{-20}$ J, 11.1

  $E_{vib} = 4.11 \times 10^{-20}$ J, 9.93

  $T_{rot} = 7.59 \times 10^{-14}$ s

  $T_{vib} = 8.06 \times 10^{-15}$ s, $\dfrac{T_{rot}}{T_{vib}} = 9.41$

P7.34 a. 0, 7.68×10^{-23} J, 2.30×10^{-22} J, 4.61×10^{-22} J, 7.68×10^{-22} J

  b. 0, 3.84×10^{-23} J, 1.54×10^{-22} J, 3.46×10^{-22} J, 6.15×10^{-22} J

Chapter 8

P8.1 $E_0 = 2.61 \times 10^{-20}$ J, $E_1 = 7.69 \times 10^{-20}$ J $E_2 = 1.26 \times 10^{-19}$ J, $E_3 = 1.73 \times 10^{-19}$ J

  $\nu_{0 \to 1} = 7.67 \times 10^{13}\,\text{s}^{-1}$, $\nu_{0 \to 2} = 1.51 \times 10^{14}\,\text{s}^{-1}$

  $\nu_{0 \to 3} = 2.22 \times 10^{14}\,\text{s}^{-1}$

  Error $(\nu_{0 \to 2})$, $(\nu_{0 \to 3}) = -1.8\%, -3.7\%$

P8.2 142 N m^{-1}; 5.19×10^{-14} s

P8.3 1.5 cm, 0.35 cm

P8.5 C—O 116.227 pm, O—S 156.014 pm

P8.7 9.1718×10^{-11} m

P8.9 $E_0^{HCl} = 2.97 \times 10^{-20}$ J

  $E_0^{DCl} = 2.13 \times 10^{-20}$ J

  $\dfrac{E_0^{HCl} - E_0^{DCl}}{kT} = 2.03$

P8.14 $0.437\,\text{cm}^{-1}$; $2.62 \times 10^{10}\,\text{s}^{-1}$

P8.16 For F_2 at 300. and 1000. K, $n_1/n_0 = 0.0123$ and 0.267

  For F_2 at 300. K and 1000. K, $n_2/n_0 = 1.52 \times 10^{-4}$ and 0.0715

  For I_2 at 300. K and 1000. K, $n_1/n_0 = 0.357$ and 0.734

  For I_2 at 300. K and 1000. K, $n_2/n_0 = 0.127$ and 0.539

P8.17 1.62×10^{-10} m, 1.69×10^{-10} m

P8.18 6.1×10^{5} cm, 2.7×10^{2} cm

P8.20 2.32085×10^{-10} m

P8.23 1.133×10^{-20} J

P8.25 0.935

P8.28 $1.33 \times 10^{13}\,\text{s}^{-1}$; 4.42×10^{-21} J

P8.30 $7.94 \times 10^{13}\,\text{s}^{-1}$, 1.26×10^{-14} s 2.63×10^{-20} J, 0.164 eV

P8.31 159.6 pm

P8.33 $4.605 \times 10^{-48}\,\text{kg m}^2$, $1.3177 \times 10^{-33}\,\text{kg m s}^{-2}$, 1.89×10^{-19} J

P8.35 1.738×10^{-18} J, 1.717×10^{-18} J/molecule or $1.034 \times 10^{3}\,\text{kJ mol}^{-1}$

P8.36 11

P8.38 25%

P8.39 $1.938 \times 10^{-46}\,\text{kg m}^2$. $1.2121 \times 10^{-33}\,\text{kg m s}^{-2}$, 3.791×10^{-21} J

Chapter 9

P9.1 most energetic, 109,678; $27419.3\,\text{cm}^{-1}$; $12186.3\,\text{cm}^{-1}$

  Least energetic 82258.5, 15233.1, $5331.57\,\text{cm}^{-1}$

P9.4 -4.358×10^{-18} J

P9.5 0.323

P9.6 0.920, 0.0620, 2.77×10^{-3}

P9.7 $r = 4a_0$

P9.8 $1.5a_0$

P9.10 H: 2.179×10^{-18} J

He$^+$: 8.717×10^{-18} J

Li^{2+}: 19.61×10^{-18} J

Be^{3+}: 34.87×10^{-18} J

P9.11 $\langle r \rangle_H = \left(\dfrac{3}{2}\right) a_0$; $\langle r \rangle_{He^+} = \left(\dfrac{3}{4}\right) a_0$;

$\langle r \rangle_{Li^{2+}} = \left(\dfrac{1}{2}\right) a_0$; $\langle r \rangle_{Be^{3+}} = \left(\dfrac{3}{8}\right) a_0$

P9.12 $26, 0.0201$ eV

P9.13 $1.1 \times 10^{-3}, 0.32, 0.99$

P9.14 $(3/4)(a_0)^2$

P9.15 $0, a_0^2$

P9.16 $\langle F \rangle_{1s} = -\dfrac{e^2}{2\pi \varepsilon_0 a_0^2}$

$\langle F \rangle_{2pz} = -\dfrac{e^2}{48\pi \varepsilon_0 a_0^2}$

9.18 $30 \mu a_0^2$

9.19 $I_H = 13.60$ eV; $I_{He^+} = 54.42$ eV;

$I_{Li^{2+}} = 122.4$ eV; $I_{Be^{3+}} = 217.7$ eV

9.20 $5a_0$

9.26 1.26

9.29 $(3/2)a_0, a_0$

Chapter 10

P10.2 $54.7°$ and $125.3°$

P10.6 $2\hbar^2$

P10.7 $2\hbar^2, 0$

P10.8 $\alpha_{optimal} = \dfrac{m_e e^2}{4\pi \varepsilon_0 \hbar^2}$

Chapter 11

P11.7 $\Delta \nu = 1.70 \times 10^9$ s^{-1}

$\dfrac{\Delta \nu}{\nu} = 3.33 \times 10^{-6}$

P11.8 $-\dfrac{16\sqrt{2}\, e}{81} a_0$

P11.10 b. 3.86×10^{-8}

8.67×10^{-5}

4.65×10^{-5}

P11.11 364

P11.12 Lyman

82257.8 cm^{-1} $\lambda = 121.569$ nm;

97490.7 cm^{-1} $\lambda = 102.574$ nm;

102822 cm^{-1} $\lambda = 972.553$ nm;

109677 cm^{-1} $\lambda = 911.768$ nm

Balmer

15232.9 cm^{-1} $\lambda = 656.473$ nm;

20564.4 cm^{-1} $\lambda = 486.276$ nm;

23032.2 cm^{-1} $\lambda = 434.175$ nm;

27419.3 cm^{-1} $\lambda = 364.707$ nm

Paschen

5331.52 cm^{-1} $\lambda = 1875.64$ nm;

7799.25 cm^{-1} $\lambda = 1282.17$ nm;

9139.75 cm^{-1} $\lambda = 1094.12$ nm;

12186 cm^{-1} $\lambda = 820.591$ nm

P11.13 $E_{max} = 2.17872 \times 10^{-18}$ J;

$\nu_{max} = 3.28813 \times 10^{15}$ s^{-1}

$\tilde{\nu}_{max} = 109677.6$ cm^{-1}

$\lambda_{max} = 9117.63$ nm

$E_{min} = 1.63404 \times 10^{-18}$ J;

$\nu_{min} = 2.4661 \times 10^{15}$ s^{-1}

$\tilde{\nu}_{min} = 82258.2$ cm^{-1}

$\lambda_{min} = 121.568$ nm

P11.18 3.08255×10^{15} s^{-1}; 3.08367×10^{15} s^{-1}

P11.23 $E(3p^2P_{1/2}) = 3.369 \times 10^{-19}$ J $= 2.103$ eV

$E(3p^2P_{3/2}) = 3.373 \times 10^{-19}$ J $= 2.105$ eV

$E(4s^2S_{1/2}) = 5.048 \times 10^{-19}$ J $= 3.150$ eV

$E(5s^2S_{1/2}) = 6.597 \times 10^{-19}$ J $= 4.118$ eV

$E(3d^2D_{3/2}) = 5.797 \times 10^{-19}$ J $= 3.618$ eV

$E(4d^2D_{3/2}) = 6.869 \times 10^{-19}$ J $= 4.287$ eV

P11.28 $4, 3, 2,$ and 1

P11.30 3.4 eV

P11.31 $E(4p^2P) = 6.015 \times 10^{-19}$ J $= 3.754$ eV

$E(5s^2S) = 6.597 \times 10^{-19}$ J $= 4.117$ eV

P11.35 $3.65; 4.70$

Chapter 12

P12.9 For $\varepsilon_2 = 1.94$ eV, $c_{22} = -1.03$, $c_{12} = 1.18$

For $\varepsilon_1 = -20.9$ eV, $c_{21} = 0.71$, $c_{11} = 0.40$

P12.14 8.67×10^{-30} C m $= 2.60$ D

P12.15 $S = 0.1$: -14.5 eV, -12.5 eV

$\qquad$ $S = 0.2$: -15.3 eV, -11.1 eV

$\qquad$ $S = 0.6$: -17.4 eV, $+1.7$ eV

P12.20 0.97, 0.86, 0.67

P12.22 $S = 0.1$: -18.8 eV, -13.2 eV

$\qquad$ $S = 0.2$: -19.1 eV, -12.1 eV

$\qquad$ $S = 0.6$: -20.9 eV, $+1.94$ eV

Chapter 13

P13.14 a. 3.35×10^{-18}

$\qquad$ b. 0.213

Chapter 17

P17.3 5.87 T, 23.3 and 14.5 T

P17.6 -2.04 and 1.12 ppm

Index

Page numbers in italics indicate tables and page numbers in bold indicate figures.